Robotics and Facto
of the Future '87
Proceedings of the Second I

R. Radharamanan
Editor

Robotics and Factories of the Future '87

Proceedings of the
Second International Conference
San Diego, California, USA
July 28–31, 1987

With 260 Illustrations

Springer-Verlag Berlin Heidelberg New York
London Paris Tokyo

R. Radharamanan
Associate Professor of Mechanical Engineering
San Diego State University
San Diego, CA 92182
USA

Library of Congress Cataloging-in-Publication Data
Robotics and factories of the future : proceedings of the second International Conference, San Diego, California, USA, July 28–31, 1987 / editor, R. Radharamanan.
p. cm.
Includes bibliographies and index.
1. Robots, Industrial—Congresses. 2. Automation—Congresses.
I. Radharamanan, R. II. International Conference on Robotics and Factories of the Future (2nd : 1987 : San Diego, Calif.)
TS191.8.R5595 1987 88-16089
670.42′7—dc19

Camera-ready copy provided by the authors.
Printed and bound by Quinn-Woodbine, Woodbine, New Jersey.
Printed in the United States of America.

9 8 7 6 5 4 3 2 1

ISBN 3-540-50025-1 Springer-Verlag Berlin Heidelberg New York
ISBN 0-387-50025-1 Springer-Verlag New York Berlin Heidelberg

The Second International Conference on Robotics and Factories of the Future

OBJECTIVE

In recent years, improvement in manufacturing technology has been accelerated with the advances in CAD/CAM and robotics. The factory of the future is aiming to be fully automated, which demands one unified system integrating design, machining, processing, material handling, assembly, inspection, and packaging to produce and ship a given product. Hence, the objective of the Second International Conference on Robotics and Factories of the Future is to bring together industry leaders, researchers and educators from various sectors including the traditional and modern manufacturing areas to exchange ideas and systematically analyze the problems related to modernization and automation of industries through an integrated approach of education, research, and application of robots and computers in all phases of manufacturing. Since today's technology is highly multidisciplinary and complex, such an exchange is expected to contribute to the evolutionary growth of science and technology and provide an insight into the factories of the future.

CONFERENCE ORGANIZATION

Sponsor
International Society for Productivity Enhancement (ISPE)

Co-Sponsors
College of Engineering, San Diego State University, USA
Department of Mechanical Engineering, San Diego State University, USA
Naval Ocean Systems Center, San Diego, USA
CIPHER Data Products, San Diego, USA

Conference Chairman
R. Radharamanan
San Diego State University, USA

Conference Co-Chairmen
Sanford Baum
University of Utah, USA

Suren N. Dwivedi
University of North Carolina at Charlotte, USA

Conference Coordinator
R. Janardhanam
University of North Carolina at Charlotte, USA

Exhibition Coordinator
George A. Mansfield
San Diego State University, USA

International Organizing Committee

Belgium
M. Becquet (Brussels)
R. Gobin (Leuven)
J. Peters (Leuven)
R. Snoeys (Leuven)
H. Van Brussels (Leuven)
P. Vanherck (Leuven)

Brazil
Maria Emilia Camargo (Santa Maria)
Edgar Pereira (Porto Alegre)

Canada
B. Manas Das (Calgary)
Kalyan Ghosh (Trois-Rivieres)
Mark B. Zaremba (Hull)

Denmark
Finn Fabricius (Lyngby)

France
A. Bourjault (Besancon)
Phillipe Pract (Besancon)
M. Staroswiecki (Villeneuve-D'Ascq)
Claude Vibet (Evry)

India
P. R. Ramalingam (Karaikudi)
C. P. Sarathy (Madras)

Italy
Ario Romiti (Torino)

Japan
T. Yamashita (Torino)

Netherlands
J. A. M Willenborg (Utrecht)

Peoples Republic of China
Qixian Zhang (Beijing)

Poland
A. Morecki (Warsaw)

Republic of China
Shui-Shong Lu (Taipei)

Romania
Voicu N. Chioreanu (Sighetu Marmatiei)
Mircea Ivanescu (Craiova)

Spain
R. Ceres (Madrid)

Thailand
R. Sadananda (Bangkok)

United Kingdom
John Billingsley (Portsmouth)
David Hughes (Plymouth)
David Paul Stoten (Bristol)

USA
Xavier J. R. Avula (MO)
Charles Bernard (DC)
J. Kent Butler (CA)
John B. Cheatham (TX)
Douglas Fishman (MD)
V. J. Hucka (UT)
Gary L. Hurd (CA)
Laveen N. Kanal (MD)
Unny Menon (CA)
Howard Moraff (DC)
Bruce Moxon (MA)
Sachio Nakamura (NJ)
Y. C. Pao (NE)
Robert L. Rosenfeld (VA)
Richard G. Schuler (NY)
Sahjendra N. Singh (NV)
William R. Tanner (MI)
Kamal Youcef-Toumi (MA)
R. Donald Tyler (IN)
Dennis E. Wisnosky (IL)
Andrew Young (CA)
Helmut T. Zwahlen (OH)

Preface

The papers presented at the Second International Conference on Robotics and Factories of the Future held in San Diego, California, USA during July 28-31, 1987 are compiled in this volume. Over two hundred participants attended the conference, made technical presentations and discussed about various aspects of manufacturing, robotics and factories of the future. The number of papers published in this volume and the number of unpublished presentations at the conference indicates the evidance of growing interest in the areas of CAD/CAM, robotics and their role in future factories. The conference consisted of five plenary sessions, twenty three technical sessions, workshops, and exhibits from local industries and educational institutions.

I wish to acknowledge with many thanks the contributions of all the authors who presented their work at the conference and submitted the manuscripts for publication. It is also my pleasure to acknowledge the role of keynote, banquet, and plenary sessions speakers whose contributions added greatly to the success of the conference. My sincere thanks to all session chairmen. I wish that the series of the International Conferences on Robotics and Factories of the Future which was initiated in 1984 in Charlotte, North Carolina will have a major impact on the use of robots and computers in the automated factories of the future.

I acknowledge with deep gratitude the help and the guidance received from the organizing committee. I also wish to extend my gratitude to the sponsoring organizations. Grateful appreciations are due to graduate and undergraduate student volunteers from San Diego State University and University of Utah for their enthusiastic participation and help in organizing this conference. Thanks are also due to all my colleagues, friends, and family members who extended their help in organizing this conference and making it a success.

I would like to appreciate the efficient work done by Springer-Verlag in publishing this proceedings.

R. Radharamanan
Conference Chairman and Editor

Contents

INVITED LECTURES

PLANNING OF AUTOMATION

CAD/CAM

CIM/FMS

KINEMATIC ANALYSIS

DYNAMICS AND CONTROL

TRAJECTORY PLANNING

SENSORS AND VISION SYSTEMS

AI AND EXPERT SYSTEMS

MOBILE ROBOTS/ROBOTIC DEVICES

ROBOT APPLICATIONS

AUTOMATION AND INNOVATION IN MINING

CAD/CAM AND ROBOTICS EDUCATION/TRAINING

SAFETY, HUMAN RESOURCES, SOCIAL ECONOMIC IMPLICATIONS

Invited Lectures

Keynote Speech

"The Factory of the Future : Elements of Effective CIM"
R. Leonard
Head, Total Technology, UMIST, England

Banquet Speech

"Perspective on Manufacturing"
Nam P. Suh
Assistant Director, National Science Foundation
Washington D.C., USA

Plenary Sessions

"Robot of the Future"
Howard Moraff
Program Director, National Science Foundation
Washington D.C., USA

"Telerobotic Services on the Space Station"
S. Ollendorf
Goddard Space Flight Center, NASA, Maryland, USA

"Robotics and Factories of the Future : Too Slow or Too Fast"
D. E. Wisnosky
Wizdom Systems Inc., Naperville, Illinois, USA

"Intelligent Task Automation Manufacturing System"
R. L. Rosenfeld
Project Manager, DARPA, Arlington, Virginia, USA

"Robotic Manufacture of Hazardous Substances"
C. Bernard
Pentagon, Washington D.C., USA

"Pattern Recognition and its Application in Industries"
L. N. Kanal
Department of Computer Science, University of Maryland
College Park, Maryland, USA

"Robotics and Factories of the Future as Applied to Large Mechanical Products"
R. D. Tyler
Detroit Diesel Allison, Indianapolis, Indiana, USA

"Microcomputer Applications for CAD/CAM in Engineering Education"
Y. C. Pao
Department of Mechanics, University of Nebraska
Lincoln, Nebraska, USA

INVITED LECTURES

Elements of Effective CIM

R. Leonard

Total Technology, UMIST, England

(1) SUMMARY

Developments in CAD, CAM, MRP, and FMS, are discussed in terms of system economics, to the point where such technologies now enable true CIM to become an economic reality. From this initial base, the paper proceeds to show that by analysing each sub-set of CIM, with respect to its economic impact on a company, it is possible to determine the appropriate level of technology for a specified industrial situation.

The paper concludes by describing the application, over a wide selection of industries, of a purpose built investment analysis program named IVAN. It is shown that because IVAN enables ALL the factors surrounding an investment to be quantified, including those previously regarded as important but intangible, such as better delivery or improved product quality, most technological investments emerge as being considerably more viable than was believed in the past.

(2) INTRODUCTION

A great deal has been written on the subject of Computer Integrated Manufacture (CIM). Much of this work is speculative, however, and lacks current significance for practical applications. This is reflected in the recent report, 'Integrated Manufacture'(1), which shows that few companies have achieved any worthwhile integration, and even the term CIM fails to find universal acceptance, with at least six conflicting interpretations being discussed. The report reflects the confusion throughout industry, where Flexible Manufacturing Systems (FMS), Automated Assembly (AA) and Robotics are usually included in the definition of CIM. For the purposes of this paper, the above sub-elements will not be termed CIM, as they are only advanced aids to manufacture, rather than overall control systems. However, before true CIM can emerge as a manufacturing concept in its own right, it is necessary to discuss the economic viability of its constituent elements.

(3) ELEMENTS OF CIM

(3.1) Robots

Robots are viewed as 'machines which can replace people', thus Engleberger[2] states, 'the prime issue in justifying a robot is labour displacement'. Ayres and Miller[3] confirm this when they say 'only labour savings are quantifiable and arguments about improving quality or increasing flexibility carry little weight with accountants'. It is even suggested that the method to justify a robot is to calculate an 'equivalent wage' and if this is less than a human wage, the robot is 'hired'. The present research at UMIST has resulted in an extensive lists of costs and benefits for robots[4], where the major gains relate to lower material costs, reduced stock levels and increased flexibility/sales.

(3.2) Flexible Manufacturing

(3.2.1) Flexible Modules

At the single machine level, FMMs consist of units based on CNC technology but which can run independently of an operator. They thus provide an alternative to CNC and, like CNC, do not normally influence the way a factory is operated. Because FMMs are currently based on CNC designs, the benefits of machining efficiency (i.e. improved metal removal rates, tool positioning and combining operations) can be achieved at the lower cost of CNC. A FMM, however, can work independently of the operator, thereby achieving an improvement in utilisation, and when activity diagrams are considered for the machine and the operator, the maximum utilisation is achieved when the diagrams are unconnected. Thus the fact that a FMM costs around 50% more than the equivalent CNC, is offset by the FMM doing the work of 1.5 CNC machines[5].

(3.2.2.) Flexible Systems

A FMS comprises a number of FMMs, integrated with a management control computer and a workpiece/tooling transport system. Because of load balancing difficulties, and losses caused by breakdowns, the output of a FMS will be LESS than that of the same number of FMMs when operated as individual machines, yet the capital cost of a FMS may be 50% more than that of separate FMMs. The main benefit of a FMS derives from the ability to schedule work as if the FMS was a transfer line, thereby gaining the advantages of reliable delivery times, reduced levels of finished stock and the capacity to introduce new products speedily. Thus while FM is justified at the single module level in terms of production savings, full systems must be viewed in a company-wide context.

(3.3) CAD

Companies normally justify CAD in terms of Drawing Office savings. Senker(6), quoting a survey of 34 UK companies, showed that 25 had exclusively projected labour savings to justify the CAD investment, 5 had not carried out any financial analysis, while the remaining 4 companies gave no plausible reason for their investment decision. Estimates of productivity improvements from CAD range from 2:1 to 4:1, thereby giving the implication that at least 50% of the existing draughtsmen will not be needed once a system becomes operational. The chances of CAD getting an enthusiastic welcome in the Drawing Office under these circumstances are low, and therefore the full potential of the investment will not be achieved.

Because of the high initial cost of CAD, and the even higher running costs over the project's life, few companies can actually justify such systems in the context of DO savings. The cost of running the DO, however, is typically less than 1% of the cost of sales. Therefore if CAD can increase sales by only 1%, the extra profit will exceed the savings expected from the DO. Companies which have installed CAD, acknowledge its considerable impact on total company performance, with major benefits coming from faster, better quotations and parts lists, thereby improving the delivery performance while cutting the production costs. Other benefits stem from the standardisation of components and the speed that new products can be introduced. Since the major financial benefits of CAD come from quotations and parts lists, the emphasis in selecting a system must be on its data processing capability, rather than the draughtsman's efficiency. Thus large mini computer based systems, with data processing facilities, become easier to justify than cheaper, micro based alternatives.

(3.4) CAD/CAM

Although it appears attractive to extend CAD for CNC tape preparation, the benefits must be compared with the costs. This includes the cost of any additional terminals and computer power needed to avoid degrading the response time, and the machine tool productivity being reduced if the post-processor lacks features that fully exploit the CNC machine. The cost of extending a CAD system can be considerably higher than purchasing a dedicated system, designed solely for tape preparation.

In the final analysis, the savings from linking CAD/CAM depend upon the application. Thus a company manufacturing complex components that are subject to frequent design changes, will obtain major benefits from reduced throughput times, together with avoiding the need for component geometry to be inputted into multiple computer systems. These savings will be small for most companies, however, and CAD/CAM must show significant improvements before it becomes a better investment than CAD plus a dedicated CAM system.

(3.5) MRP AND MRP2

Material Requirements Planning (MRP) and Manufacturing Resources Planning (MRP2) are different in scale, but they can be similarly treated for financial evaluation, therefore the term MRP will be used for both systems. MRP is not normally perceived as an investment decision, with Boxx[7] suggesting that top management often accept MRP as being 'inevitable'; while Hall and Holden[8] report that MRP is seen as a 'justifiable act of faith'. This attitude often results in a lack of adequate facilities for MRP implementation, or the absence of measurable objectives. The work at UMIST has shown that data systems can be treated like any other capital project, with cost and benefits being identified, and a DCF return calculated. The benefits of MRP fall within four broad categories:-

1. Reduced paperwork and improved control systems.
2. Reduced disruption of production.
3. Controlled stock levels.
4. Shorter and more reliable lead times.

The first two benefits relate to improved productivity and reduced production costs, and these are normally considered when MRP is seen as a production control system. When quantifying the benefits, however, improving the sales performance and reducing the level of finished stocks usually outweigh any direct production savings.

(4) LINKING

Two or more modules are linked when data transfer is possible from a specific module to subsequent modules. This usually involves the transmission of data in some neutral format (ASCII code). For this to occur, software is needed to convert the data to ASCII code in the sending module, and then to reconvert this information in the receiving module. The link may enable the transmission of parts lists to occur from CAD to MRP in a predetermined format when the computers involved are individually incompatible.

(5) INTEGRATION

Integration provides a much broader concept than linking, and genuine two-way communication between modules is required. For example, the designer can access the CAM module tool files. This facilitates better design for manufacture as the CAD user designs for the tooling available, thereby avoiding the need for special tools. To be successful, integration requires a responsive workforce that is free of traditional demarcation lines; such that the designer understands the conceptual and technical needs of the planning department, and they appreciate the disciplines of the design process.

The power of a fully integrated system is readily demonstrated by considering a customer's order. Upon notification to change a specific aspect of an order, the design office generates a new parts list. Next the production and costing modules are interrogated, by a 'what if' analysis, to determine the costs/workload implications of the change, and an accurate cost report is generated for both management and customer.

(6) COMPUTER INTEGRATED MANUFACTURING

The objective of Computer Integrated Manufacturing (CIM) is to combine individual system elements into an integrated whole. This is strongly affected by the needs of sophisticated data transmission networks and advanced control systems, both of which are only just emerging as industry-wide standards for interfacing are established. Because of the costs involved, companies do not pursue integration for its own sake, rather they are forced along this path by increasing competition. Similarly, while the path to integration is different for each enterprise, companies usually follow the evolutionary process of linking modules which already exist.

If individual systems are to behave in the way required for CIM, communication must be facilitated between different computers. To achieve this objective, the International Standards Organisation (ISO) is developing the Open System Interconnection model (OSI)[9]. This is restricted to part of the communications problem, however, and has only recently been supported by computer manufacturers, with IBM announcing its co-operation with the OSI model in 1985.

The OSI model outlines seven layers of protocols to organise transmission and the sharing of resources between computers. The hierarchical structure allows each layer to be served by the layer below and serve the layer above. Local Area Networks, such as Ethernet, is the nearest standard, yet even Ethernet is only for the bottom two layers of the OSI model, and although the seven layers of the model have been defined, most of the standards have not been agreed. Thus recent computers can only incorporate standards for the bottom layers, and any interconnection between company systems will continue to be restricted to the use of compatible machines, purchased from a single vendor.

(7) FINANCIAL EVALUATION

There is a problem of communication in many companies between Engineers and Accountants. Thus Engineers, who may have spent months working on the Technical Specification of a project, see the Financial Evaluation as the final pass/fail hurdle to be overcome. This often leads to the attitude of "how big do I have to make the cash flow figures to get it past the accountant?'

Thus a sound financial evaluation is a critical aid to management decision making, and can answer the following questions:-

1. What area of investment should be considered?
2. What type and technical specification of system is required?
3. Is the investment likely to be profitable?

Evaluation is not a process of merely trying to justify a specific project, but rather should be a means of answering the above questions. Thus the need was identified at UMIST for a new approach for investment appraisal. This resulted in the specification for a system which would enable Engineers and Managers, without detailed financial training, to combine their expertise to produce the correct technical and financial specification for any project. Such an evaluation had to conform to the following:-

1. It must adhere to accepted methods of financial appraisal.
2. The results must be acceptable to Accountants.
3. The evaluation should include ALL the factors affected.

Techniques were subsequently developed and extensively applied which overcame both the technical and conceptual problems of investment appraisal. These have now been converted into a commercially available programme called IVAN(10), which is capable of running on most business microcomputers. The work at UMIST is the first, worldwide, to prove that all "intangible" benefits affecting an investment can be redefined and quantified. IVAN overcomes the traditional problems of investment appraisal, and its use shows that the benefits which had previously been excluded as 'intangible' were much greater than those traditionally included, such as savings in direct labour. IVAN is now used by many major UK and overseas companies as their standard method of investment appraisal(11). This fact has strong implications for the future development of CAD/CAM technology, and is reflected in the statement that several major CAD manufacturers use IVAN extensively, both within their organisation to predict technological developments, and outside their companies to help justify sales.

8. THE FUTURE

Manufacturing technology has developed over the years by a process of trial and error, with the emphasis inevitably being on unit cost reduction. In the future, however, manufacture should not be looked at in isolation, but rather as an integral part of the total company, and investments must reflect a beneficial effect on the interrelationship with design and marketing. Therefore the objective should not be to improve individual departments, nor install the most up-to-date technology, but rather to enhance the company's overall competitive position. This means providing customers with the products they want, at the required specification, quality, cost and delivery.

An example of this new approach to investment can be seen with respect to quality. Companies normally treat quality as a cost, and efforts are aimed at reducing scrap, rework and warrantee payments; thus the aim is a negative, "cost reduction exercise". If product quality is viewed as an asset, however, and the company identifies the effect that improved quality would have on sales, not only are the financial benefits of better quality much greater, but the ways that quality can be improved will change.

A NEDO Report on 'Quality and Value for Money' states that, "product quality has become the key to improved competitiveness". The report emphasised that the customer's perception of quality was strongly influenced by "delivery, packing, installation, product descriptions, instruction manuals and customer support". Therefore changes to improve quality in manufacturing, as perceived by potential customers, are different from changes aimed at the traditional objective of reducing the cost of quality.

Another example of how manufacturing must suit the overall company needs can be seen in organisations which operate in volatile markets. Here the need is to respond to rapid changes in sales volume and product mix, and constantly introduce new products or enhance existing designs. Although improvements can be obtained by installing "islands of automation", such as FMS, there are limits to the benefits obtained. Thus to achieve the overall performance the market requires, it may eventually be necessary to operate the total factory under direct computer control.

It will be a long time before computers can be truly integrated, but it is possible to adapt existing technology to obtain the major financial benefits of an integrated factory. For example, a company using CAD to produce quotations, can automatically generate a parts list. This can be transmitted in ASCII code to an MRP2 system (even if the computers are incompatible), where the list is converted into shop floor documentation. The MRP2 system can transmit instructions to an automated storage and material handling/transport system, controlled by another incompatible computer, and this transports materials to the manufacturing workstations. These may operate under their own control system, with the work sequencing being directed by commands from MRP2. What is really being transmitted by such an overall system is management control information, and the benefits obtained arise from organisational improvements, thus enhancing the overall competitive position of the company.

9. CONCLUSIONS

This paper discriminates between each aspect of CIM, especially the considerable differences that exist between simply linking modules, and the true integration of systems. Major benefits will accrue when integration becomes economically feasible, but in the intervening years, companies are advised to examine the merits of providing working links between existing modules. Significant time is spent in most companies on preparing manufacturing instructions, therefore the ability to quote a reduced delivery period is a major selling advantage, and attempts should be made to reduce the overall order duration. In particular, the time to convert customer orders into manufacturing instructions should be critically appraised.

Research into the economics of CAD systems at UMIST has shown that when the 'intangible' benefits are quantified and included in an investment appraisal, not only is CAD a much more attractive investment but the main financial benefits are external to the Drawing Office. In the same way, when CAM and CIM are viewed in the context of financial viability, the main benefits are in areas not directly related to the technology itself, but to the effect which it has on a company's ability to compete in the market place.

The provision of a CAM module within a company already supporting a CAD system is fashionable because vendors can readily supply the necessary facilities. Yet few benefits result for many companies because the majority of components are geometrically simple. When the attributes of a CAD-MRPII link are identified, however, this is shown to be particularly useful for companies who modify their Bill of Materials to meet customer variations. If a medium sized company can achieve a 1% increase in sales, by linking its CAD-MRPII systems to improve delivery performance, the extra profit could pay for the link in less than a year. In reality, the saving in production costs, from eliminating transcription errors, together with the benefits of increased sales, would make the link into an investment which few companies could not afford to do without. Therefore, companies are advised to appraise financially, in a company-wide sense, the value of providing a CAD-MRPII link.

Acknowledgements

The author wishes to thank the SERC for their financial backing for this work, and to his colleagues, Peter Primrose, David Harrison and David Petty for their most valuable contribution to the overall research programme during recent years.

References

(1) Ingersoll Engineers. 'Integrated Manufacture'. IFS (Publications) Ltd. 1985.

(2) Engleberger, J. R. 'Robotics in practice', 1980 (American Management Association, New York).

(3) Ayres, R. V. and Miller, S. M. 'Robotics applications and social implications', 1983, (Bollinger Publishing Company, Springfield, Illinois, USA).

(4) Primrose, P. L. and Leonard, R. 'Incorporating the "intangible" benefits of robots within a comprehensive financial evaluation'. I.Mech.E. conference, UK Robotics Research 1984, London, Dec. 1984, pp 115-120 (Mechanical Engineering Publications, London).

(5) Primrose, P. L. and Leonard, R. 'The financial evaluation of flexible manufacturing modules'. Proc. 1st Int. Machine Tool Conference, Birmingham, June 1984, pp 61-72 (IFS Conferences).

(6) Senker, P. 'Some problems in justifying CAD/CAM'. Proc. 2nd European Conference on Automated Manufacture, pp 59-66, 1983 (IFS Conferences).

(7) Boxx, D. B. 'How to convince management to invest in MRP'. Production and Inventory Management, 1978, 20(2), 49-62.

(8) Hall, P. E. and Holden, G.K. 'Computerguide 9: production control', 1973, (National Computing Centre, Manchester).

(9) ISO Open System Interconnection Model, ISO/TC97/SC26/DP7498 (January 1981).

(10) IVAN: investment analysis computer program. Sales literature (Organisation Development Limited, Altrincham, Cheshire).

(11) Primrose, P. L. and Leonard, R. 'The Financial Evaluation and Economic Application of Advanced Manufacturing Technology'. Proc. I.Mech.E. Vol.200B, 1986.

Intelligent Task Automation Manufacturing System

Robert L. Rosenfeld

Defense Advanced Research Projects Agency, Arlington, VA, USA

Introduction

The Intelligent Task Automation (ITA) program was initiated in 1982. The program consists of a number of projects related to key enabling technologies of intelligent automation and two major projects to demonstrate the integration of the key technologies. Each of the major projects has completed a research phase to identify and develop enabling technologies. The projects will be completed during 1987 and 1988. The purpose of this paper is to report on the results of the research phase for the two major projects of the ITA program from the perspective of the entire program. More specific and detailed results of the two projects have been and are being made available in other papers.

The philosophy and some of the key objectives of the ITA program were described by Levinthal in 1982 [1]. The ITA program was directed both at the development of more productive manufacturing processes and extending military operational capabilities in the combat environment. The two major ITA projects described here support the ITA goals. The goals of these projects were "to develop and demonstrate generic technologies which

1) form the basis for advanced intelligent task automation systems
2) have near term application to unit processes, component assembly, and many aspects of defense batch manufacturing and
3) establish the technology base which will open up new opportunities to apply intelligent systems to complex military tasks." [2]

The ITA projects have been funded by the Defense Advanced Research Projects Agency and the Materials Laboratory of the Air Force Wright Aeronautical Laboratories. In one project, Honeywell Inc. is developing a programmable system to be demonstrated on the assembly of a microswitch from its 17 parts. In the other project Martin Marietta is developing a programmable system to be demonstrated on the inspection of an F15 bulkhead. The projects have much in common and the differences do not particularly relate to the differences between the demonstration tasks, assembly and inspection. Nonetheless, I will refer to the projects as the "programmable assembly" and "programmable inspection" projects.

Programmable Assembly

The programmable assembly ITA system will be demonstrated on the assembly of a Honeywell 1EN1-6 microswitch [3]. The 17 parts of the microswitch will be dumped from a kit onto a 5" by 5" tray and presented to the robotic system. The system will use vision to identify the parts and assemble the microswitch. The robotic system is shown in Figure 1. It includes vision subsystems, robotic manipulators, and computers. The overhead vision subsystem is used to identify the microswitch parts and consists of a grey-scale video camera and a 3D scanner. The 3D scanner uses light triangulation to measure distance.

One of the robots is a precise and rapid SCARA robot to handle the microswitch parts. It is equipped with a force-sensing wrist and force-sensing fingers. It also holds a light-stripe 3D scanner used for part identification. The second robot is less precise and is used with a light-stripe 3D scanner to inspect parts, to guide the positioning of the first robot for precise alignment of parts before insertion, and for inspection of subassemblies..

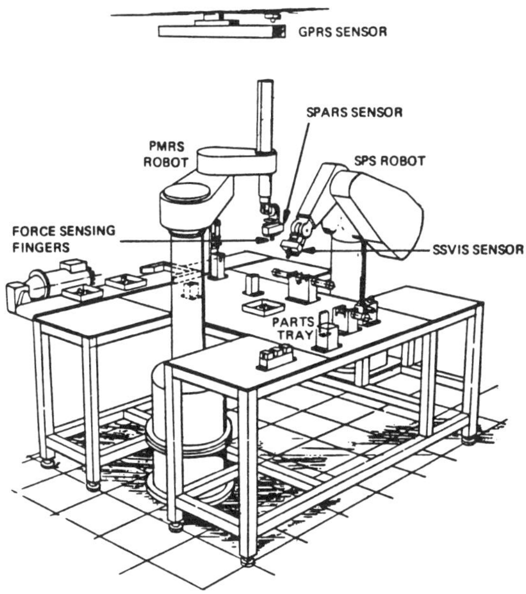

Figure 1 -- Programmable Assembly

Programmable Inspection

Figure 2 shows the programmable inspection ITA system [4]. It includes a single robot with a variety of inspection tools. The major demonstration task is the inspection of an F15 bulkhead. The system utilizes ultrasonic probes and a 3D vision scanner to measure the dimensions of features of the bulkhead. A second robot to cooperate in the inspection process by inspecting the other side of the bulkhead was in the original scope of the project, but will not be utilized in the final demonstration system.

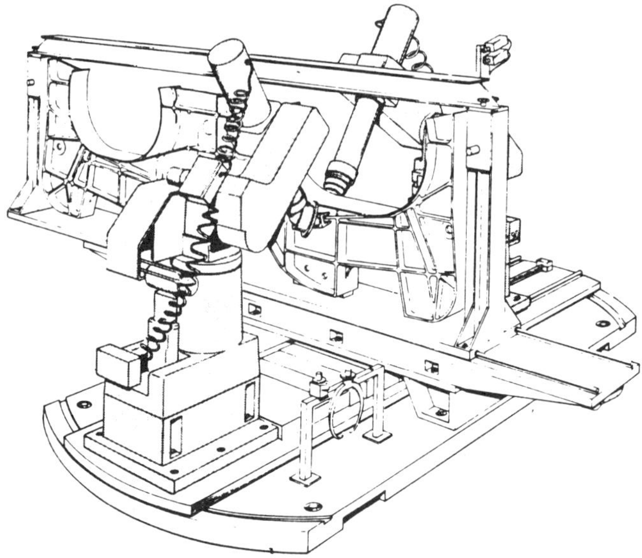

Figure 2 -- Programmable Inspection

Comparison of Programs

Both ITA projects feature advanced computer control and the integration of technologies into a useful and demonstrable system. The programmable assembly system development has emphasized the control of forces during delicate assembly operations and the use of computer vision to identify parts even in a cluttered image. The programmable inspection system development has stressed the planning of robotic motions so that the robot may be directed by high-level commands with the robot planning the detailed steps.

ITA Manufacturing System

We now combine features from the two projects and describe an ITA manufacturing system that does not distinguish between the two ITA projects.

The ITA manufacturing system is sensor-based and intelligently controlled. It includes a variety of robotic machines and sensors, is controlled on-line by a hierarchy of computers, and is programmed off-line. The system is trained off-line in the tasks that it will perform and in the appearance of the parts that it must identify. It receives unfixtured parts in kits, identifies the parts, assembles the parts delicately, and inspects the parts and/or completed assembly. This capability to accept kits simplifies the transfer of parts from the storage area to the assembly area.

The system consists of two (or more) robots. The robots have overlapping work areas so that they can easily cooperate on assembly tasks. The robots are instrumented with force sensors to sense and control applied forces. The robots utilize a variety of quickly interchangeable end-effectors. End effectors include grippers, inspection tools, and vision sensors. At least one of the robots has a vision sensor near its end effector. A quick-change device permits end effectors to be changed quickly by either robot. The ITA assembly project investigated the addition of a micromanipulator for delicate assembly operations, but found that the micromanipulator was not necessary with a high precision robot.

The robots work alongside a table holding flexible fixtures. (Flexible fixtures per se have not been addressed by the ITA projects.) Parts or assemblies are placed on the table and the system is commanded to inspect and/or assemble them. The vision subsystem identifies unobstructed parts, looking particularly for the parts needed first. It moves the larger parts to staging areas so that it can identify the remaining parts. The vision subsystem identifies a grasp location for each part and the manipulator arm moves it to a fixture. Assembly proceeds one part at a time.

Mating of parts involves precise alignment followed by delicate insertion. A precision robot aligns one part with another under guidance of a vision sensor carried by a second robot. The control subsystem limits the forces applied during delicate insertion. Sensors on the robot provide feedback of forces to the control subsystem.

Inspection utilizes a suite of sensors. The quick disconnect device permits the robot to change inspection sensors quickly. A precise three dimensional vision subsystem checks visible dimensions. While versatile, the precision of this vision subsystem limits the precision of measurements. Proper insertion and seating of parts is verified by a three dimensional light stripe vision subsystem.

A computer plans and controls all of the above operations. An operator trains the vision subsystem off line by aiding the computer in finding features in sample images that can be used later for object identification. The operator also describe the prerequisites and results for each kind of task that the system may perform. Off line the system develops sequences of tasks that it can utilize on-line. This planning reduces the amount of on-line planning needed.

On-line the operator simply instructs the system to assemble or inspect a given object or assembly. The system applies the available plan and puts the operations into an optimum order as necessary. Should any unexpected condition occur on-line, the full planning system is available to aid in recovery.

The ITA system was designed with the idea of applying machine intelligence to task automation. Intelligence comes from combining planning and control with versatile sensors and servo-controls under an advanced computer architecture. Together these can perform more kinds of assembly and inspection operations than fixed automation. Planning and control are a direct application of machine intelligence intended to make the application of robotics easier. This easier application and programming make the complex robotic capabilities demonstrated by ITA practical. A major feature of these capabilities is the servo-control of motion by sensed force or vision at the end effector rather than by position sensing at the joints. The computer architecture provides a structure for combining the planning and control functions with the sensor-based operations.

Computer Architecture

The control subsystem for ITA is hierarchical with a system controller responsible for higher-level task control and separate modules for each of the major subsystems. The system is shown in Figure 3. Each major subsystem is either a robot with sensors or a sensor complex enough to require its own separate control subsystem. Communication of information within the control for any of the major subsystems is over a bus for higher bandwidth than communication to and from the system controller. The communication among major subsystems is carried out either over a network (such as an Ethernet) or directly from a multiple port communication controller on the system controller.

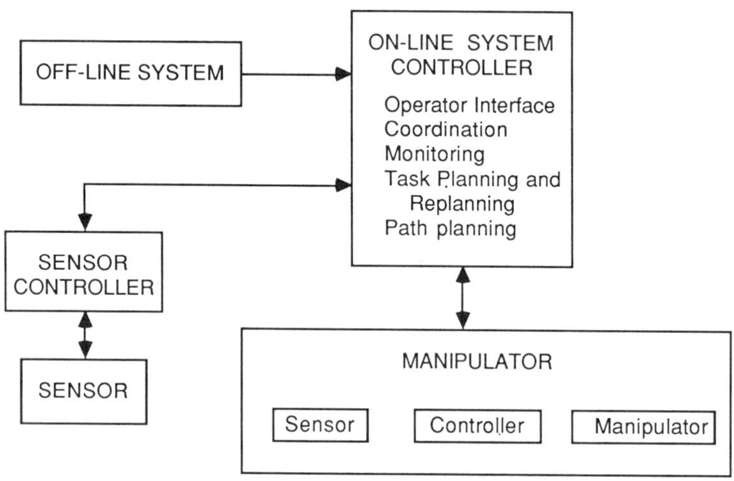

Figure 3 -- Computer Architecture

Each subsystem controller is responsible for the functions of its own robot or sensor subsystem. For example, the vision subsystem controller performs the many object identification functions. Each robot controller controls motion to a prescribed path or force level utilizing sensors of position or force.

Simple sensors and sensors requiring close integration with a particular robot are controlled by the robot's controller. Separate subsystems control more complex sensors, such as for vision, which requires extremely fast computation. The separation introduces a delay into real time control loops including the sensor. However, such complex sensors require relatively slow processing and cannot be used in rapid servo-control loops.

Figure 3 names the basic functions allocated to the the system controller, which has the master role. These function include the operator interface, task planning and replanning, coordination of subsystems, status determination and monitoring, and error recovery. In addition to the tasks shown, the system controller must initialize the subsystems, provide a means for debugging system operation, accept plans prepared by the off-line system, and permit calibration of the robots and their sensors.

The off-line planning subsystem, "off-line planner", is used to program and/or plan the operations of the ITA system for one task while it is performing another. The off-line planner operates independently from the other subsystems and the system controller. Only when plans need to be loaded into the system controller for execution is the interface to the other subsystems exercised. The system controller performs the same kinds of planning on-line for error recovery or in response to unexpected conditions. Thus, the system controller and off-line planning computer share features. The off-line planner also requires a strong 3D graphics output for simulation of robotic motion.

Planning and Task Control

The off-line planner provides both complete plans for on-line use and short plans called "macrops" to enable the rapid creation of new plans on-line. Each macrop is the plan for a common operation. The planner builds macrops from basic operations and stores the macrops off-line. The computer is given the preconditions and results for each macrop. Preconditions are the conditions which must be true in order to carry out an operation. For example, an ultrasonic test can only be performed if the ultrasonic probe is held by the robot. The planner creates full plans off-line to achieve given goals and starting conditions. The planner does this by combining macrops known to result in the desired goals. The planner then considers the preconditions of each selected macrop as goals and plans further macrops until all of the preconditions are satisfied either by further macrops or by the the given starting conditions.

Image understanding

The ITA system includes 2D and 3D vision sensors. These sensors are used to identify parts, determinetheir position and orientation, align parts during assembly, verify assembly, and inspect dimensions.

The off-line system is also used to "train" the vision subsystem to recognize the objects expected for a particular robotic task. The operator trains the vision subsystem interactively using sample images of the parts to be recognized. The operator selects the distinguishing features of each object and tests the selected features on the sample images. The off-line system records the features and clusters of features that will be applied to the matching process. In a more advanced system the distinguishing features would be identified automatically from CAD models.

Objects are recognized in three stages: feature identification, hypothesis generation, and hypothesis verification. *Feature identification* is the searching through the visual image for geometric features, such as line segments, dark or light areas, arcs of circles, etc. During the feature identification stage a catalog is built of the features in the image. During *hypothesis generation* clusters of features in the image are matched to clusters of features for expected objects. This should generate one or more hypotheses. In the *hypothesis verification* stage additional features are matched for the hypothesized objects. The features applied to verification typically take longer to identify than features applied during the first two stages of object recognition.

On-line planning

On-line planning involves exception handling and path planning. Exception handling requires a repetition of planning. However, the exception handler can develop recovery plans quickly. The exception handler starts with the list of preconditions from the original planning process. The exception handler then develops actions to meet the preconditions that have not been met.

The plan, whether prepared on-line or off-line, is not complete. The steps of the plan must be put into an optimum order. The ITA system optimizes by minimizing the time required for a single robot to perform the operations. This problem is the traveling salesman problem of operations research.

Path planning is the determination of the exact path to be followed by a robot to move from one place to another. The macrops created by the off-line planner only identify end points, thus leaving the path planning to be done on-line. The path planning process can call upon complex geometric reasoning, especially for avoidance of interferences. Neither of the ITA demonstration systems requires the kind of complex three dimensional motion that demands the most complex geometric reasoning.

Servo-Control

Both ITA contractors studied state-of-the-art methods for controlling forces applied by robotic manipulators. The ITA system is able to apply a specified force in one direction while maintaining position along another direction. In the ITA demonstrations this will be applied to the control of the force of an ultrasonic probe and to the insertion of one part in another. Neither task is possible merely by control of the position of the robot, no matter how precise the control. Feedback from force sensors enables the necessary control of force. The force control is imposed as an additional servo loop around the position control of the purchased standard robot controller.

Although not being applied to either demonstration, control of dual coordinated arms has been studied. Dual coordinated arms in the ITA system permits two robots to handle large objects cooperatively. To handle a large object, a single robot would need a very strong end effector and/or would need to grasp the object near its center of mass. Dual arms need less strength, but must sense and respond to forces so as not to work against each other and apply excessive force to the object being handled. Separate control over the impedance at each end effector has been investigated.

The insertion of one object into another and the threading of a fastener require not only force control, but also precise alignment of the mating objects. The ITA system aligns the mating parts visually. A structured light sensor held by one robot examines the relative positions of the parts. The second robot moves one of the parts to reduce the amount of observed misalignment. Because vision processing is slow, even with structured light, movement is slow and there are no problems with servo stability.

Conclusion

The ITA projects have developed tools that can be applied beneficially to flexible manufacturing. The development of the ITA systems provides several lessons:

1. Image understanding technology is capable of handling cluttered images, but places great demands on computer resources.
2. Forces can be controlled for delicate assembly, but the required servo-controls are not generally available in commercial robots.
3. A full-scale precision robot can handle the assembly of small objects without the aid of a micromanipulator.
4. Three dimensional images can be used for inspection of dimensions, but the sensors may limit precision.
5. The artificial intelligence approach of goal decomposition enables the robotic system to program itself from relatively few basic operations. The system can improve its self-programming capability by saving and reusing macrops.

References

1. Elliott Levinthal, DARPA Program -- Intelligent Task Automation, in Robotics and Industrial Inspection, San Diego, (August 24-27, 1982) Proceedings of SPIE, Volume 360

2. Internal document

3. Ronald Benton and Donald Waters, Honeywell Inc., Intelligent Task Automation Phase I, Final Report for Period October 1982 - December 1985, Contract F33615-82-C-5092 (July 1986)

4. Martin Marietta Denver Aerospace, Phase I -- Intelligent Task Automation (ITA), Final Report for Period November 1982 -- January 1985, AFWAL-TR-85-4062 (April 1986)

Robotic Manufacture of Hazardous Substances

Charles Bernard

Office of the Deputy Under Secretary of Defense, Tactical Warfare Programs, Washington, D.C., USA

INTRODUCTION

Safe and efficient manufacture of hazardous substances and products is a continuing concern. The development of robotic devices and techniques, and the associated technologies, provides unique opportunities to address these concerns while improving the product and, in the long run, reducing costs.

It is appropriate to note the automation of manufacture of hazardous products has a history. The Smithsonian Museum of American History now has on display the Howe Pin Machine. This ingenious device (figure 1) was developed for the production of straight pins.

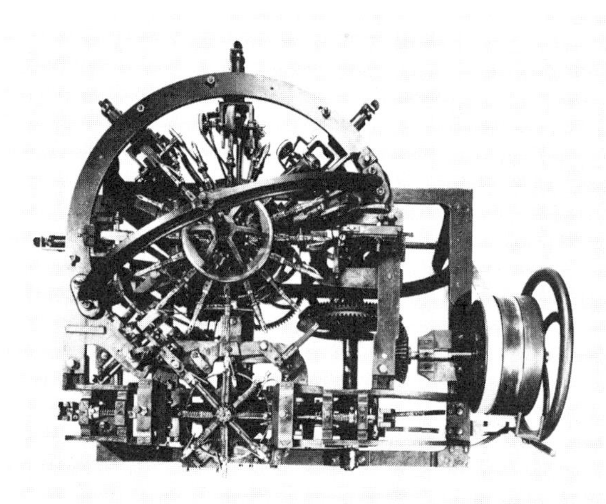

COURTESY OF SMITHSONIAN INSTITUTION 85-7511

Figure 1 HOWE PIN MACHINE

The pin machine was invented by John Howe of New York City in 1842. The machine took in brass wire and cut, headed and pointed it to produce pins. This invention was in competition with manual pin production where up to seventeen people worked on each pin, each performing a single task. The Howe Pin Machine could produce sixty pins a minute, or 24,000 pins in an eleven hour work day, and was in continuous service for about thirty years. It eventually made Howe a wealthy man. One journalist of the time wrote "...It is difficult to believe the machine is not an intelligent being."

Obviously, automated machines that only perform pre-set repetitions are limited to basic operations. We can do much more with today's technology. Today we have intelligent machines that can detect and respond to their environment. Reprogrammable robots with vision systems keep man away from the work site, but in the loop. Man can have a telepresence.

There are five areas of interest where there are special robotic applications to the manufacture of hazardous materials.

CASE OVER PROPELLANT

The large knitting machine, which had been in use for many years to produce composite cases for rocket launcher tubes, inspired the Naval Ordnance Station, Indian Head, Maryland, to robotize it to handle propellant and to build rocket motors inside out.

There are several significant advantages to building a rocket motor by wrapping the motor case on the propellant rather than the conventional way of first building a case, then lining it with a case bonding material, and pouring the propellant into the lined case. In the conventional method, motor case unevenness, case bonding material thickness variations and the resulting total volume uncertainty leads to a variation in propellant weight with a concomitant impulse variation that gives errors in the range of the rocket. In addition, threading the head-end and nozzle of a rocket gives an inescapable thrust malalignment error that detracts from the deflection accuracy of the rocket. The combination of deflection and range dispersion leads to a requirement to fire more than the minimum number of rockets to destroy the target, with logistic and battlefield shortage disadvantages.

Robotizing the knitting machine, so it could be used in an unmanned mode in a hazardous room with the propellant, made the concept of case over propellant possible. Figure 2 shows a propellant grain with end fixtures that are the head end of the motor case and the nozzle fixed to a mandrel that will maximize the straightness of the finished motor. The propellant grain has been end machined to a precise weight for optimum impulse control by a robot end mill operation. The case bonding agent would be robot sprayed in production. Precision in its thickness is also part of the control of the rocket's impulse.

Figure 2 PROPELLANT GRAIN IN KNITTING MACHINE

This entire manufacturing process meets all three of the reasons for going to a robotic process, it is dangerous, dull and dirty. Under normal conditions if this was a manned operation, only one could be built at a time in each room because of the concern that a mishap on any machine would cause potential damage to any of the operators. Roboticized, as many machines can be going on in the same room at the same time as the propellant hazardous quantity-distance standards allow for the work unit.

The rocket that emerges from this process has as precise a control on impulse, thrust malalignment and weight as economics can dictate. This allows a smaller number to be used on the battlefield for the same amount of damage.

INSENSITIVE HIGH EXPLOSIVES, PROPELLANT

Several disastrous fires aboard aircraft carriers spearheaded the efforts to replace the present TNT based explosives and RDX-containing propellants with materials that had less sensitivity to shock and temperature. The existing materials have been around throughout most of the twentieth century with all of the manufacturing facilities matched to their characteristics. For that reason, the nation's propellant and explosives development scientists were tasked to determine the feasibility of inventing new formulations that could be used with the existing manufacturing facilities. After about five years of work they concluded that the best bet for near term results would be to utilize the plastic bonded explosives (PBX's) and castable composite propellants that had been invented during the 1960's.

The decision to use castable composites for insensitive propellants caused no concern for the propulsion industry or the acquisition executives in defense, because the facilities for this work existed in adequate numbers and required improvements in processing could be made on the present plants.

Of considerable concern however, was the problem of the explosives manufacturers. Most of the Defense Department's explosives are manufactured in government owned plants. Though many of these plants are operated by industry, the requirement to modernize, and the cost to be borne in the process, come straight out of the Defense Department, and most of them come out of the Army's budget. Because the Army's budget is a zero-sum game, the effort to re-do the entire manufacturing process from one where explosives are liquefied in a melt-kettle and cast, to one where the ingredients are inserted into a propellant-type mixer and handled like propellants has not been accomplished.

The present method of propellant mixer processing of PBX costs several times the amount per pound of melt-kettle processing. This large increase in unit cost comes after the investment of billions of dollars of facility money that does not appear to be available in the near future. This has driven a search for a process that will enable PBX to be processed without going to the expense of installing the number of large mixers and the associated handling equipment necessary for the changeover of the explosive plants.

A promising technique that appears well suited to robotic operation, while minimizing the investment of the billions necessary for propellant mixers, has been invented by a small business firm and is being avidly studied by several defense agencies. This process uses cryogenic cooling of the materials, including the rubbery binders, grinding them to the proper size (particle sizes of few microns), mixing them as a powder in a simple "Y" blender, putting them into the warhead, and elevating the temperatures to effect the cure. In addition to saving the Defense Department billions of dollars in facilitization and production of PBX, and making the owners of the process extremely rich, this robotic handling of dangerous materials in a super-chilled condition should enable high quality product to be made from materials that are not now able to be processed. Figure 3 shows a propellant mixer.

Figure 3 MIXER

It is conceivable that with the proper use of robotic operation, the entire room could be kept at the liquid nitrogen temperature with a possible saving on energy costs required when mixing men with cold processing.

DE-MILITARIZATION REQUIREMENTS

The damage to the environment caused by open-air burning of propellants (indeed open air burning of anything), burying of explosives or chemical weapons, and ocean dumping of hazardous materials, began to be understood in the 1970's. This led the Defense Department to require the costs of, and plans for, de-mil of ammunitions to be included in the acquisition process. This meant that the person responsible for the design would pay more attention to the ability of somebody to dispose of the ammunition twenty or thirty years hence.

This "un-manufacturing" of dangerous materials then becomes part of the manufacturing process of ordnance and ammunition. De-mil of ordnance needs to be done at as low a cost as possible because it adds nothing to the value of warfare, and is certainly looked on by the Defense Acquisition Executives as a painful but necessary fact of life. During the service life of ammunition, it will have been handled, banged around, dropped, and dunked. The hazardous materials within the item may have become desensitized to elevated temperatures, or other environmental handling. The work of de-mil can usually be considered dull, dangerous and dirty. That points to a robotic process.

De-mil can require a robotic operation of higher level of humanness (figure 4) than that required to manufacture it. The manufacture could have been done on explosion-proofed machines such as presses or screw-thread machines, while the de-mil will require vision systems, sensors for temperature rise detection, devices to detect out-of-tolerance or damage conditions...etc. Taking a page from the cryogenic processing manufacturer, de-mil could be done at sub-freezing temperatures with robots, to decrease the likelihood of any cook-off or process fires.

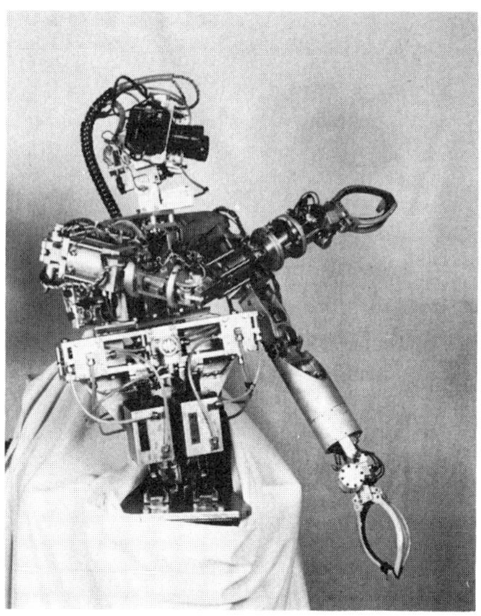

ROBOT CUTTING PALLET BANDS

PROJECTILE HANDLING

Figure 5 ROBOTIC DE-MIL PROCESSING

A fairly standard process for de-mil is to put a closed chamber such as a projectile body into an explosion proofed lathe and part the body with a cutting tool. The buildup of heat at the point where the cutting tool intersects the body must be below propellant cook-off temperature or explosive flash-point. This can be done by a robot with an adequate vision system and the help of temperature sensitive paints or dyes. Pressurized steam wash-out of explosives can also be enhanced by a robot with a specialized vision system that is capable of looking behind obstructions and inside crevices.

FUZES, BUILT UP ASSEMBLIES

Tens of millions of fuzes are manufactured every year in the pursuit of defense in the United States even though we are at peace in the world. New items are built to replenish ammunition used in training, but far larger numbers are built to replace obsolete stocks or to modernize stocks to a later state of the art. In addition to fuzes most ordnance carries a component known as a safe-and-arming device, or an S&A. The purpose of an S&A is to permit a safe separation distance between the firer and the ordnance after launch but before it allowed to explode. This precludes accidental damage to the firer (or aircraft) should the fuze fire and explode the warhead upon first being enabled.

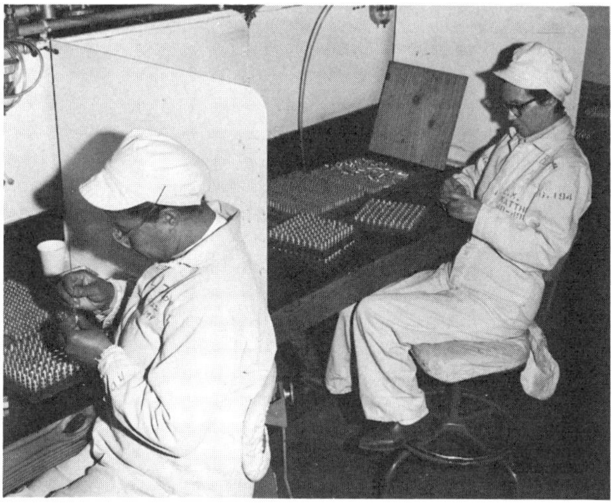

Figure 6 CARTRIDGE ASSEMBLY OPERATIONS

Fuzes and S&A's are usually mechanical and electronic assemblies that contain several electro explosive devices (EEDs). In the past these devices have been built up on special assembly lines where separation partitions kept an explosion at one work station from doing harm to the person at the next station. An example of this kind of work situation is shown in figure 6. The partitions held the damage to the person doing the operation at hand. Soldering in EEDs can be a dangerous operation, and yet with some of the designs, the EEDs needed to be added to the built up assembly during the manufacturing process.

This whole area is ripe for robotization. Clever use of trace radiation elements in components can be used with counters to determine that vital components have been installed, and while the individual dosage is insignificant, a person working in close proximity over a prolonged time can be exposed to radiation overdoses.

A problem to consider is the size of the production run. For mature systems a large assembly line can be set up and can pay off in efficiencies pro-rated against millions of units. However, many operations are developmental, or pilot-line, where no such efficiencies occur. In these situations a limited number of components are needed for testing or promotions, but the large scale production may never come off.

This leads to the need for a robotic capability that will handle the sequential build-up of dangerous materials, but one that is easily programmed for a given operation and easily re-programmed for another in a short time at minimum expenditure of effort. This again points to a human-like robot that has versatility enough to be able to convert from one operation to another easily. Software changes are being used to reprogram robots regularly. Application of reprogramming techniques to robots using dangerous materials should be considered.

DANGEROUS NON-EXPLOSIVE MATERIALS

Manufacture of chemical warheads can kill you quickly if there is a malfunction involving people, and breathing beryllium or dust or particles can kill you slowly if you have not been protected from it during its manufacture. Today's practice is to have the manufacture of chemicals and the loading of chemical warheads done under closed clean room conditions by the use of remote controllers. Beryllium machining as well as the manufacturing of any poisonous heavy metals is done by having the operator work through a glove box. The operators are scrupulously checked for lung vital-area and for general health on a regular basis.

Pyrophoric powders such as fine magnesium, aluminum, zirconium, and red or white phosphorus, though not containing oxidizers of their own, can react violently when exposed to air. These materials are used for spotting charges, tracers for upper atmospheric motions and incendiaries. During the time these components are being handled, the operators have to wear cumbersome protective gear that impedes the work when the material is not ignited, and are probably not adequate to ensure complete protection when the material is ignited.

Energetic materials that could be used for increasing the specific impulse of liquid propellants, such as nitronium perchlorate, have been too hard to handle to make the manufacture of them with standard techniques practical. Nitronium perchlorate is an oxidizer that reacts with most everything including the acid resistant gloves and clothes that a man wears when he is working with most oxidizers.

End users of all of these items are crying out for the roboticizing state of art to catch up with their needs. Versatile robotic manufacturing for dangerous materials in development quantities and pilot quantities is needed first. This will demonstrate to the market and requirement writers that production of the product is possible at manageable cost. When the developers have been able to furnish enough items to show their merits, the roboticized production line can be economical. The ability to work in an inert atmosphere where dangerous materials are quiescent will be one of the robot's greatest contributions to manufacturing dangerous materials.

CONCLUSION

To the people who develop and produce hazardous materials, the dangers and problems are apparent. Robotic techniques have improved safety and solved production problems in many areas. However, there are more areas that need attention. The five areas described above are included.

Perhaps the most significant progress that could be made is replacing the person at the hazardous worksite with a robot capable of projecting the presence of that person back into the production process. In many cases, the robot, with the human in the loop, could do the job better, for less, while removing the risk to human safety.

Robotics and Factories of the Future as Applied to Large Mechanical Products Allison World Transmission Program

R. Donald Tyler

Manager, Advanced Manufacturing Projects, Detroit Diesel Allison Division, Indianapolis, IN 46206-0894, USA

This Division of GM is in the automatic transmission business for large on-highway trucks and buses, off-highway work vehicles, and military tanks and trucks. These products are some of the most sophisticated large mechanical components produced today. Sometimes they are affectionately called 500 lb. watches. As such, there is always a trade-off between the product design requirements versus reliability, durability, and selling price. New programs are continually being evaluated in our Engineering Departments. The program goals are always quite simple and direct -- world class quality, world class competitiveness, and flexibility in manufacturing & assembly.

If we are to consider producing large, complicated mechanical products such as these in a Factory of the Future in which robots will play an important role, there are a number of things that should be considered beyond just the application of robots per se:

- Islands of automation versus an overall plant system.
- Design product and process concurrently.
- Pay particular attention to design for assembly.

Also, experience of those who have started to use advanced manufacturing technology (robots) suggests that the best benefits are to be derived from making complementary changes in organization and management at the same time the robots are introduced.

Several examples are given that show the current problems relating to automation of industries that produce large, complicated mechanical products.

FOR FURTHER INFORMATION CONTACT:
Mr. R. Donald Tyler, Manager
Advanced Manufacturing Projects
P. O. Box 894 (L-26)
Indianapolis, IN 46206-0894
(317) 242-2074

Microcomputer Applications for CAD/CAM in Engineering Education

Y.C. Pao

Professor of Engineering Mechanics, University of Nebraska, Lincoln, Nebraska 68588-0347, USA

INTRODUCTION

For the development of robotics and factories in the future, Computer-Integrated Manufacturing (CIM) plays a central role. Among the broad spectrum of topics covered in CIM, Computer-Aided Design and Computer-Aided Manufacturing (CAD/CAM [1~3]) are addressed in this paper. In particular, views on how should the CAD/CAM courses be taught in engineering colleges are expressed. Expected to have a major impact on advancement of robotics and factories in the future is the microcomputer applications for engineering education of CAD/CAM in response to the rapid increase in availability of microcomputers at homes, in schools, and in the laboratories and factories.

Mainframe computers will continue to be an essential tool for research, for generating software of the expert systems, and for networking of stand-alone computer systems and peripheral devices. But, the CAD/CAM professionals who carry out the CIM tasks, apply the expert systems, and design the robots and manufacturing processes being cultivated in the academe are likely to use microcomputers more than the mainframes. This trend, if not apparent now, will definitely be very clear in the nearest future. Microcomputers are used more in the lower-class courses where the students learn CAD/CAM and other fundamentals than in the upper-class courses where they specialize and hence apt to apply commercial software and the expert systems.

In order to meet the future needs in robotics and factory development, the topics that all engineering students regardless of their majoring fields should learn in the lower-level CAD/CAM courses are discussed in the following sections. It ranges from the addressable dots (pixels) on a display monitor for drafting application to the probabilistic error estimate for path control in robot operation.

DRAFTING

It is inevitable that drafting boards, T squares, compasses, and templates will be completely replaced by the keyboards and display monitors of microcomputer systems. So will be the blueprints which the production stations in the plant must read in order to make the products. Drawing are now stored as data files and viewed on display monitors. Small drawing can be viewed by a single screen while large ones can be handled with multiscreen and through clipping and enlarging [1] of some particular parts of a drawing for closer scrutiny.

There are numerous software packages for engineering drafting needs, such as AutoCAD [4] and VersaCAD [5]. Still being heatedly debated is the question "When should engineering students be introduced to such popular computer drafting tools?" This author's opinion is that teaching the application procedures of commerical drafting software definitely should not precede but follow the teaching of the basic principles of engineering drafting incorporated with modern computer graphics techniques.

Currently, many institutions are teaching one semester of classical engineering drafting, and another semester of computer programming along with AutoCAD or the likes. Engineering drafting, computer programming, and computer-aided drafting should be combined into one semester and the students should be required to write their own computer-aided drafting programs. By developing their own software, though primitive but through which they learn the underlying principles involved in AutoCAD and VersaCAD and so forth, the students will not need a whole semester to learn about the application procedures of such commercial software packages. This one-semester computer-aided-drafting course should be offered during freshman year and for students from all engineering disciplines.

For teaching AutoCAD and the likes, the decision on "In which courses? When? And for how long?" should be made based on the student's chosen field. Design of circuit boards may have to devote more time than mechanical design or vise versa - the upper-level courses's instructors are in better positions to judge what to do! Evidently, the scope of teaching the application procedures of commercial software depends primarily on the extent of intended applications.

COMPUTER GRAPHICS AND INTERACTIVE OPERATIONS

Replacing a drafting board and papers by a graphics display monitor, and replacing the drafting tools such as T-square, compass, and pencils and pens by a keyboard are what make the traditional engineering drafting to become modern computer-aided drafting. But mainframe computers didn't expedite such development as much as the microcomputers! The reason is INTERACTION. Only in recent years, almost every student can have access to a microcomputer and his or her program does not have to be run by batch on a mainframe. This happening really speeds up CAD/CAM applications. Also, BASIC is becoming more popular than FORTRAN, PASCAL, C and other languages because as far as interaction is concerned, BASIC does not require link and edit. BASIC is especially attractive for spontaneous demonstration in the teaching environments.

For effective interaction, CAD/CAM students should have basic understanding on the structures of computer memory and of the display monitor. Only when they have sufficiently comprehended the uses of bytes, bits, and pixels, can they apply the screen graphic resolution and programming commands and statements to the full. They should know how text characters are composed, from which they learn how to create their own characters of rotated or different sizes. Knowing where a picture is on the screen in units of pixels, they can then manipulate to keep it elsewhere on disk, or to change the whole picture or a part of it - clipping! Transfer among computer memory, screen, and flexible/hard disk should also be demonstrated in classroom so the students learn how data and images can be stored, retrieved, and manipulated.

The concepts of spreadsheet and templates in CAD/CAM software development (such as Lotus 1-2-3 [6] and WordPerfect [7], respectively) should also be taught without the peripheral graphics devices such as mouse, lightpen, joystick, tablet, and soft screen being available because they should learn how to create their own cursors by programming. For interactive activities such as digitizing or accessing to a certain modular program, the self-created cursors work as good as the commercial peripheral devices. Such operation may not match the speed of the commercially available interactive graphic hardware, but it help the students grasp the underlying principles involved in the interaction. Figure 1 illustrates that CAD/CAM students should be able to develop a program for interactively placing a self-designed symbol anywhere on the console without the aid of a peripheral graphics device.

Fundamentals of interactive graphics should be taught by a systems-engineering approach [9] as a part of sophomore-, or, junior-level course and for all engineering students. Computer-aided design should also be a part of this course.

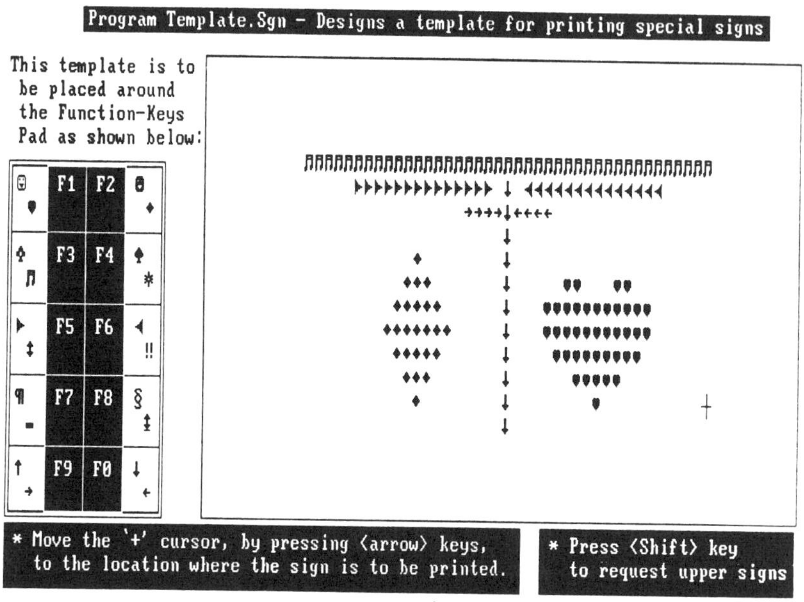

Figure 1

PARAMETRIC AND OPTIMAL STUDIES

Computer-aided drafting applies microcomputers and mainframe computers for physically drawing the design of various products, systems, and processes. Drafting is therefore a means by which a design comes into being, and CAD indicates that it is accomplished with the aid of computer. Hence, whether CAD stands for computer-aided drafting or computer-aided design will be continually debated. As far as teaching CAD/CAM is concerned, it should be the computer-aided <u>design</u> that needs to be emphasized.

Once having learned interactive graphics, engineering students may become overly indulged in merely drawing <u>pictures</u>. That is to say, drafting alone should be therefore de-emphasized and design should be stressed. Using a calculator (computer), user should be more concerned about what is being calculated (design) than which buttons have been pressed (drafting). Of course, pressing wrong buttons gives incorrect result (drawing wrong picture). But the calculation needs a formula to begin with (to design something). So, CAD should be the acronym for computer-aided design. To use CADD for representing Computer-Aided Drafting and Design is a good compromise!

In teaching CAD, the <u>interactive</u> capabilities of microcomputer should be repeatedly stressed and utilized. The students should not merely draw pictures but acutually do the design, and not only design but refine their designs. Parametric and optimal studies [1] must be taught for the students to try for different designs and achieving the best design possible. So long as a program has already been developed and loaded onto a computer, one ought to make most use out of it by experimenting with various inputs which means changing the values of the parameters involved in the design. Through keyboard, interactive input of parameters' values and effective use of the display monitor to exhibit the design result in graphic form enable the trial designs to be compared. Often, the comparision may inspire the proper steps that could be taken for furthering the trial process of design.

Optimization techniques [10] should be taught so the students can try for achieving the best design. Parametric and optimal studies are necessary for them to assess and then improve their design. They ought to keep in mind that while complying with the <u>design specifications</u>, one should strive for the best in this competitive world.

SYSTEMS AND FINITE ELEMENT ANALYSES

For assessing a designed element or system, one has to perform qualitatitive and quantitative evaluations by use of <u>mathematical models</u>. In general, complex geometric shape and dimensions, and material composition are involved in the model analysis, and various parts need to be assembled together following some particular orders. The basic understanding of the behaviors of the constituent components must first be gained also through modelling, and then the <u>block-diagram</u> manipulation can lead to the final analysis of the entire system's behavior [11].

The CAD/CAM students learn the fundamental concepts in selection of various models, about their advantages and limitations, and the consequent results of using them. And they should develop CAD programs for interactive applications of microcomputer systems for modelling analysis. Figure 2 demonstartes how a user-created '+' cursor is employed for interactively selecting constituent blocks in computer-derivation of a system's transfer function [12]. Besides the college training, the students will learn through professional practices to decide whether to adopt a highly sophisticated model that will provide a better prediction of the required analysis but at a higher cost or a less accurate model that is less costly.

Finite Element Analysis [12] is based on the fundamental understanding about the elements which have simple geometric shapes and

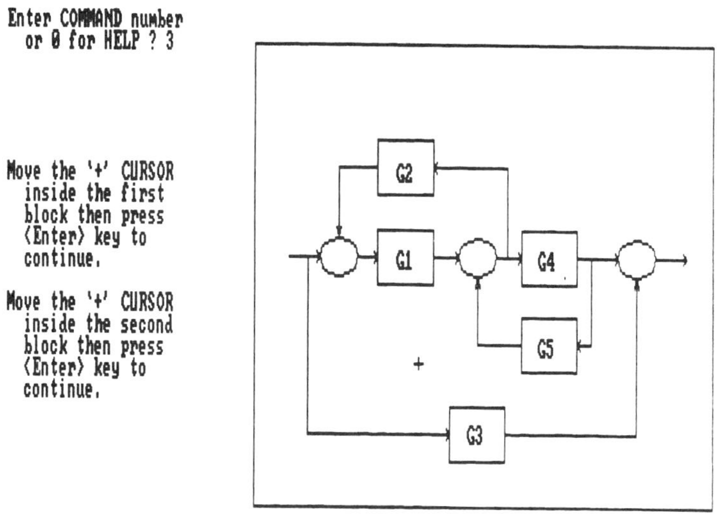

Figure 2

dimensions and each made of a unique material, and uses them as building blocks to reconstruct the investigated object (structure or system). For predicting the behavior of the assembled system, finite element analsyis takes into consideration the interrelationships among the building blocks and also the order in which they are put together.

The market is flooded with software packages for systems and finite element analyses. For CAD/CAM teaching, it is again important to stress the underlying principles involved in such software by clearly explained in classroom their fundamental concepts, and by letting students do software development in the introductory-level courses and apply the commercial software in the upper-level courses. To start with spoon-feeding the students is a wrong practice; they will be completely lost since all they do is to provide inputs and the software prints out the results without any knowledge of which input is used where and why.

Moreover, the analogy among various engineering problems should be stressed so that the CAD/CAM students will not be reluctant to treat problems outside their major fields. CAD/CAM is an interdisciplinary field [9].

NUMERICAL CONTROL AND ROBOTICS

The design of processing control and robots for manufacturing purposes is central to Computer-Aided Manufacturing (CAM). Robots with artificial intelligence (AI) capable of seeing, hearing, and feeling by touch are being developed so that adaptive control systems [14] can be added to correct the errors. Research on probabilistic analyses attempts to predict the effects on the reliability of the product or manufacturing process due to processing and measurement noises [15].

Numerically controlled (NC) machines are being increasingly adopted in the factories for automated manufacturing. Setup of cutting tools and replacement of damaged ones are programmed and handled by computer.

For CAD/CAM teaching, simple classroom robot models and illstrative examples of numerical control program suitable for all engineering students should be included in the lower-level currriculum. On the display monitor of microcomputer system, animations of the robotic motion and manufacturing process should be included as required coverage. Specialized courses at upper level could then be offered to guide the students into their majoring fields. Robot control courses in mechanical and electrical engineering departments and plant process control in industrial engineering department are typical examples.

Figure 3 is a simple model for classroom animation showing on the display monitor of a microcomputer system the painting process of a three-degrees-of-freedom robot. It works out well in conjunction with the teaching of interactive graphics in an introductory-level CADD course [8]. It guides the engineering students into other more sophisticated robot models which should be developed by the individual engineering departments for upper-level, specialized robotics courses. These models should be demonstrable in classrooms using microcomputer systems and by interactively entering commands via keyboard.

CONCLUDING REMARKS

Standarization is another keyword for the development of robotics and factories in the future. For example, suppliers are complying with the General Motors Corporation to create the Manufacturing Automation Protocol (MAP). The MAP specifications when met build multiple data bases, and a network architecture then allow the large-capacity data storage devices along with other resources such as

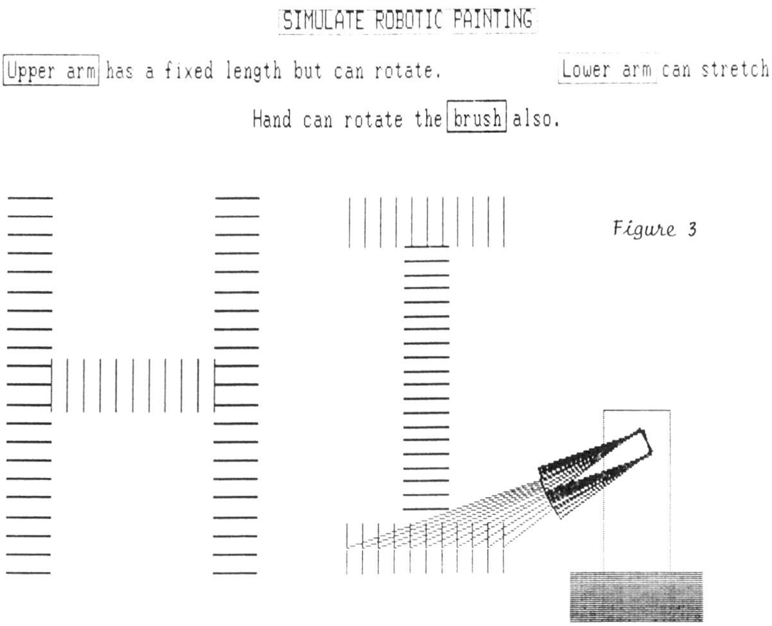

Figure 3

plotters and high-speed printers to be shared by different standalone computers. Though still under development, this protocol will eventually be able to setup different equipment, provide smooth flow of required materials, and then to fabricate and package the products without human intervention.

International efforts for standarization are being made through organizations such as the International Organization for Standardization (ISO) and the International Electrotechnical Commission (IEC). Same as following the design codes in all engineering fields established by the national committees of ASCE, ASME, and IEEE, etc., engineering students must also be taught in the academia about these standarization codes worked out by ISO, or, IEC if they are to work in the fields of CIM, CAD, and CAM.

Hence, in addition to the topics already addressed in the above sections, others such as database operations, networking, material acquisition and inventory, waste and cost reduction should receive equal attention in consideration of CAD/CAM education for the engineering students to prepare them for meeting the future needs in robotics and automated plants.

ACKNOWLEDGEMENT

This work was supported, in part, by the General Motors Corporation through a CAD/CAM Fellowship Program and by the Engineering Research Center, University of Nebraska-Lincoln.

REFERENECES

1. Y. C. Pao, Elements of Computer-Aided Design and Manufacturing, CAD/CAM, John Wiley, New York, 1984.
2. M. P. Groover and E. W. Zimmers, Jr. CAD/CAM Computer-Aided Design and Manufacturing, Prentice Hall, Englewood Cliffs, New Jersey, 1984.
3. M. F. Hordeski, CAD/CAM Techniques, Reston Publishing, Reston, Virginia, 1986.
4. AutoCAD is a trade mark of AutoDesk, Incorporated, Sausalito, California 94965.
5. VersaCAD is a trade mark of VersaCAD Corporation, Huntington Beach, California 92647.
6. Lotus 1-2-3 is a trade mark of Lotus Development Corporation, Cambridge, Massachusetts 02142.
7. WordPerfect is a trade mark of Satellite Software International, Orem, Utah 84057.
8. Y. C. Pao, Introduction to Interactive Microcomputer Applications, a 500+-page manuscript being reviewed by major textbook publishers.
9. Y. C. Pao, "Teaching An Introductory CAD Course with the System-Engineering Approach," CoED Journal, American Society for Engineering Education, Vol. 5, 1985.
10. D. J. Wilde, Globally Optimal Design, Wiley-Interscience, New York, 1978.
11. R. H. Cannon, Jr., Dynamics of Physical Systems, McGraw-Hill, New York, 1967.
12. Y. C. Pao and L. C. Chang, "Manipulating Transfer Functions of Engineering Systems Using Interactive Graphics," Computers in Mechanical Engineering, Vol. 3, 1985.
13. Y. C. Pao, A First Course in Finite Element Analysis, Allyn & Bacon, Boston, Massachusetts, 1986.
14. K. S. Fu, R. C. Gonzalez, and C. S. G. Lee, Robotics: Control, Sensing, Vision, and Intelligence, McGraw-Hill, New York, 1987.

15. L. C. Chang, Probabilistic Error Model of Robot End-Effector's Dynamic State, Ph. D. Dissertation, Department of Engineering Mechanics, University of Nebraska, Lincoln, Nebraska, 1987. (Advisor: Dr. Y. C. Pao)

16. P. Mulvey and R. Sheftic, "Information Flow on the Factory Floor: A Network for Automation," Compuetrs in Mechanical Engineering, Vol. 5, 1986.

PLANNING OF AUTOMATION

Methodology of Assembly Automation: A New Approach

A. Bourjault

Laboratoire d'Automatique de Besançon – (UA CNRS 822), ENSMM, Route de Gray, F – 25030 BESANCON CEDEX

INTRODUCTION

The fundamental problem in automated production appears more and more to be an assembly problem. Whether for "modernizing" existing installations to make them more competitive, or for building others, when a new product appears, we notice that the method services proceed most often by "fits and starts". Empiricism, though not always adapted to the actual contingencies is still widely used. The problem is even more crucial when considering a family of products.

The "Assembly Methodology" team from Laboratoire d'Automatique de Besançon has been working for several years on this problem. We have worked out a methodology for aid the conception of (flexible) assembly system[(1)] and most recently, we have develop ed the software SAGA (Automatical Generation of Assembly Sequenc es)[(2)].

1. PROCEDURE PROPOSED IN VIEW OF THE DESIGN OF ASSEMBLY SYSTEMS

1.1. Analysis of the final product

At this stage must be specified :

- the description of the product, in terms of functional connections and elementary parts (or considered as such) ; certain components may have not been defined to permit an eventual redesign.

- for each functional connection, the assembly processes (if they are chosen) and the operational modes (if they are imposed).

1.2. Elaboration of the set of admissible temporal sequences

At this stage is elaborated a set of assembly sequences which verify a set of constraints established on the basis of the data issued from the previous stage.

At this point, non pertinent temporal sequences can be eliminated by using certain simple criteria relating to the product or requirements imposed by the form ; likewise similar sequences can be merged.

1.3. Determination of operational modes

Each sequence considered express a different strategy whether at the level of the assembly operations or at the structural level of the necessary installations.

Each action of assembly necessitates the implementation of operational modes which differ following the order of realization and therefore depend on the sequence. For each sequence it is advisable to examine if the existing operational mode allows for the realization of corresponding actions.

To be defined particularly are : the relative nominal movements of the concerned components (primary/secondary), the necessary tools, etc... . If the assembly processes have not been chosen they can be at this stage.

This study can be conducted starting with functional diagrams[(3)] representing, not the spatial implementation of the installations but their functionality ; we progressively establish the characteristics which the assembly system must verify.

1.4. Schedule of the assembly system

From the functional diagram we determine the work stations and the transfers around each station. Considering the production imperatives, expressed essentially by cycle time and an output frequency, we can decide at this level what can be automated and how ; if an automatisation with robot is envisaged for example, we determine its characteristics (structure, degrees of freedom, precision, etc...).

This is to be repeated for each sequence elaborated during the preceding stages.

1.5. Redesign of the product

At each stage and considering the results obtained, it is possible to modify the initial product and reiterate the stages already effectuated. In theory we can realize as many repetitions as wanted.

2. ELABORATION OF ASSEMBLY TIME SEQUENCES

2.1. Preliminary definitions

We consider in all that follows, a product with m elementary parts. When two parts are directly linked by one or several mechanical connections, there exists between them a functional connection. A functional connection is established by the implementation of a functional action. The functional actions are corresponding to the realization of assembly processes and are characterized by an operational mode. The concept of functional action can be extended to these operations which do not have a link but do however have a functional character necessary to the final product.

An orderly sequence of functional actions whose realization (in series and/or in parallel) leads to the product represents an assembly sequence.

The realization conditions of a functional action i are represented by the set of initial states authorizing its realization. An initial admissible state is characterized by :

- the compulsory presence or absence of certain connections,
- the indifference towards the other connections.

By identifying each functional connection with binary variables it is possible to code their presence or absence. Any state E of the product can thereby be represented by a logical product of n terms (n = number of connections)

$$E = \tilde{1}.\tilde{2}. \ldots . \tilde{i} . \ldots \tilde{n}$$

with :

$\tilde{i} = i$ if the connection i is established

$\tilde{i} = \bar{i}$ if the connection i is not established.

To the set of initial admissible states of the action i is corresponding the logical function C_i, sum of the functions associated to each corresponding state. C_i is written as a logical sum of logical products of n terms.

From the set of realizability conditions C_1, C_2, ... C_n it is easy to elaborate the corresponding assembly sequences.

The main problem to be solved is therefore the establishment of realizability conditions.

2.2. Strategy for the establishment of realizability conditions

The set of initial admissible states can be determined by the set of forbidden states. A forbidden state is a state with which it is impossible to finish the assembly of the product :

- either because it prevents the realization of a given action
- or because it leads to a state having the same effect.

Given E_p the set of 2^n possible states, E_A the set of initial admissible states, E_F the set of forbidden states ; knowing E_p and E_F we can deduct :

$$E_A = E_P - E_F$$

In practice, we try to determine partial forbidden states ; a partial state being one which implements p connections such that $1 \leqslant p \leqslant n$. For example, if the partial state $(\overline{1}.2)$ is forbidden, all states engendered by $(\overline{1}.2)$ are as well.

2.3. Search for forbidden partial states

We consider a connection i and ask the following questions :

Q : "What suffices to systematically prevent the realization of i ?".

There are four possible solutions corresponding to four different situations :

S1 : the presence of only one connection x

S2 : the presence of several connections x_1, x_2, ..., x_p

S3 : the absence of only one connection y

S4 : the absence of a connection y_1, y_2, ..., y_q

It becomes impossible to realize connection i when one or more of these situations exists : these causes are not exclusive. The presence or absence of connections other than those rendering the establishment of i impossible does not affect this impossibility, this is therefore a systematic prevention.

In question Q, we are interested only in the realization of connection i without worrying about the future impossibilities which the establishment of i could engender. This can be illustrated by the example of the ball point pen represented in fig. 1 with the graph of associated functional connections:

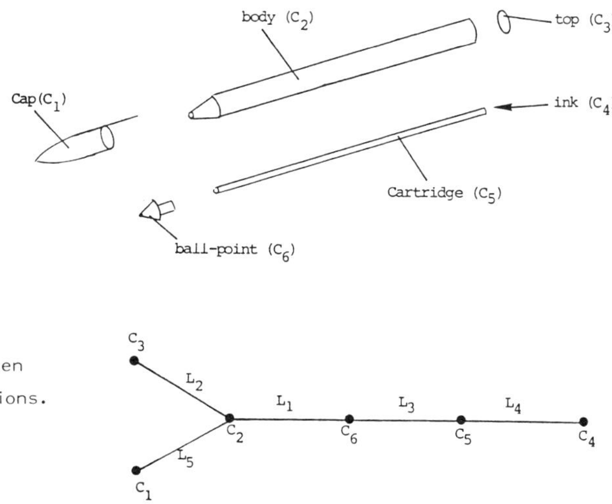

Fig. 1 : Ball point pen and graph of connections.

We consider the connection L5 between the body and the cap ; to the question : "What suffices to prevent the realization of L5 ?", we must answer : "no connection", even if it is evident that later on the assembling of the head on the body will be compromised. In fact, this impossibility will appear naturally when studying the head-body connection. That remark is fundamental ; it shows that it is not necessary to look for a possible sequence to study the realizability of action i.

In practice, it is necessary when asking question Q for each connection i to see which possible answers are verified. To arrange this research, it is preferable to substitute Q with more explicit questions, inspired by potential answers. Four types of questions are thus established, noted Q1, Q2, Q3 and Q4 :

"Is it possible to realize i ...

Q1 : ... when only connection x is established ?"

Q2 : ... when connections x_1, ..., x_p are established ?"

Q3 : ... when only connection y is not established ?"

Q4 : ... when connections y_1, ..., y_q are not established ?"

Comment on Q1 : these questions must be asked by successively replacing x by all the connections other than i. However it is clear that the connections looked at must be adjacent to connection i itself ; in fact, all connections non adjacent to i, if established, produce a sub-assembly apart from the subset which i can realize and therefore cannot prevent the realization of i. In consequence, in all the questions Q1 will be examined only connections adjacent to the connection being studied.

Comment on Q2 : these questions must be asked by successively replacing x_1, x_2, ..., x_p by all the connections other than i. For the same reasons as previously, we are interested only in the possibility of connecting two macro-components, formed by (x_1, x_2, ..., x_3).

2.4. Definition of a specific tool

We have defined a tool using a relational type approach to the problem which permits the easy obtention of corresponding forbidden states.

For Q1 and Q2 : a binary relation R such that :

$R(i,x) \Leftrightarrow$ "it is impossible to realize i if x is established"

$R(i,x_1 + \dots + x_p) \Leftrightarrow$ "it is impossible to realize i if x_1 AND x_2 ... AND x_p are established".

For Q3 and Q4 : a binary relation S such that :

$S(i,y) \Leftrightarrow$ "it is impossible to realize i if y is not established"

$S(i,y_1 + \dots + y_q) \Leftrightarrow$ "it is impossible to realize i if NEITHER y_1, ... NEITHER y_q have been established."

When R(i,x), connection i must be realized <u>before</u> x or <u>at the same time</u> as x ; this can be described as follows :

- the partial state $(\bar{i}.x)$ is forbidden for all actions,
- if i and x are not realizable simultaneously : the partial state $(\bar{i}.\bar{x})$ is forbidden for the action x,
- if i and x are realizable simultaneously : the partial state $(\bar{i}.\bar{x})$ is not forbidden for x, but is reserved uniquely for the simultaneous realization of x and i.

A set of similar properties are likewise available for $R(i, x_1 + x_2 + \dots)$ and for S relations.

Concerning the S relations, we rarely encounter the material impossibility of realizing i if one or more connections have not been established. For our part, we have encountered such situations only in the case of products comporting a liquid (assimilated to a component), such as the ink in a ball-point pen. In consequence the usefulness in keeping this type of relation can be discussed.

2.5. Fundamental properties of relation R

P1 : $R(i,j) \implies R(i, j + k + l + \dots)$

Or, more generally :

P'1 : $R(i,\Omega) \implies R(i,\Omega')$ with $\Omega \subset \Omega'$

In effect, if it is impossible to establish i when a set Ω of connections other than i is realized, this impossibility will all the more remain for all over-set Ω' of Ω. This property leads to the concept of minimal relation : R is said minimal if the right half of its relation is irreductible. It is interesting to keep only the minimal relations as they produce the largest number of forbidden states.

P2 : $R(i,\Omega) \implies R(i,\Omega')$ with $\Omega' \subset \Omega$

In effect, if it is possible to realize i when Ω is established, this possibility exists with all the more reason, for all sub-sets of Ω. This property allows for a great simplification in the study of each connection i, by checking first of all if the relation :

R (i, set of all connections other than i) is true or false.

If false, the study of i is finished ; if true, we must then look for all the minimal relations with i.
In the software we have developed, all these questions are produced by the software itself : according to the users response, the forbidden states are automatically elaborated.

3. THE PROBLEM OF SIMULTANEOUS ACTIONS

This problem is relatively complex to deal with, partly because there exists several types of simultaneities. We consider only the simultaneities said to be compulsory, as opposed to those said to be optional. The latter are easily determined form the temporal sequences, by grouping one or more consecutive actions.

Compulsory simultaneous realizations, when they exist, appear during the elaboration of the sequences ; they do not result from a previous choice and are revealed with the establishment of the realizability conditions. We must distinguish : systematic, conditional and automatic simultaneities.

<u>Systematic simultaneities</u> : two actions i and j are always simultaneously realized, if the following propositions are verified :

$$R(i,j) \quad \text{AND} \quad R(j,i)$$

<u>Conditional simultaneities</u> : two actions i and j must be realized simultaneously when the set of connections Ω is established, if the following propositions are verified :

$$R(i,\ j + \Omega) \quad \text{AND} \quad R(j,\ i + \Omega)$$

In effect, it is clear that these two relations can be expressed by :

- i is to be realized <u>before</u> ($j + \Omega$) or at the sametime as j if Ω is realized.
- j is to be realized <u>before</u> ($i + \Omega$) or at the sametime as i if Ω is realized.

When all the connections of Ω are established, the only possible solution is the simultaneous realization of i and j.

<u>Automatic simultaneities</u> : it happens that in certain cases, the realization of a connection is "automatically" accompanied by the establishment of another connection j (and symetrically). This situation appears notably for all cyclical configurations of the functional connection graph : when (N-2) connections of a cyclical configuration of N connections are established, the last two connections will automatically be realized simultaneously (except if these connections are not rigid). From these remarks, we can also conclude that any state with (N-1) connections realized is impossible, therefore forbidden. For a cycle of 3 connections i, j, k for example, the states $\bar{i}.j.k$, $i.\bar{j}.k$, $i.j.\bar{k}$ are impossible ; this corresponds to the following propositions :

$$R(i,\ j + k) \ ; \ R(j,\ i + k) \ ; \ R(k,\ i + j)$$

In consequence, as soon as a cycle of N connections is detected, it is possible to immediately write N R relations.

4. CONCLUSION

We have presented a methodology which leads to the design of any assembly system ; this methodology must be computer aided. In a first step, we have developed a software for automatically elaborating assembly sequences ; this software has been tested with a lot of examples and is used in some plants. The tool developed to establish the assembly sequences is completely transparent for the software user.

The real utility of the S relations is not proven, given the small number of cases in which they intervene. In the actual version, the production system of S relations is locked but could be reactivated. The automatic treatment of simultaneities, as developed, permits the resolution of most cases.

In conclusion, we must remember that the questions asked by the software are not "brainteasers" ; the answers supplied by the user must result from a choice seriously thought out, which reflects to the sequences obtained.

BIBLIOGRAPHY

(1) A. BOURJAULT, "Contribution à une approche méthodologique de l'assemblage automatisé : élaboration automatique des séquences opératoires", Thèse d'Etat, Université de Franche-Comté, 1984.

(2) A. HOUMMADY, A. BOURJAULT, J.M. HENRIOUD, A. JANEX, "SAGA : un système d'Aide à l'élaboration automatique des Gammes d'Assemblage", Convention Automatique Productique - Paris 1987.

(3) P. OLIVIER, "Analyse et synthèse des systèmes d'assemblage automatisés", Thèse d'Université, Université de Franche-Comté, 1986.

The Application of the Second Law Technique for the Prediction of Trends in the Development of a Factory of the Future

Dr. E.S. Geskin

Mechanical Engineering Department, New Jersey Institute of Technology, 323 King Blvd., Newark, NJ 07102, USA

Abstract

This paper is concerned with the development of a formal technique for the prediction of technology development. The production, shaping and treatment of steel is discussed in order to demonstrate the use of prediction technique.

Exergy function is employed for the estimation of the cost of the manufacture of metal products. The schematic exergy balances for producing, refining and solidifying of liquid steel are constructed. The conditions to minimize exergy consumption during steelmaking are considered and ways to improve energy use are defined. Analysis of novel processes demonstrates that exergy balance enables trend prediction in metallurgical technology development. The paper demonstrates the application of exergy analysis for technology design. As an example, the energy effective process for the conversion of iron ore into a commercial steel product is suggested.

Introduction

A number of alternative routes, available for manufacturing of specific products as well as the cost of experimental examination of each single route, make it necessary to use computational technique as the first step in the evaluation of possible alternatives in the process development.

To make the computed technology assessment practical, the development of a general technique which can be applied with some modifications to different processes is necessary. The first stage of such a procedure can be a qualitative assessment of different available routes of product manufacturing. The effectiveness of such an assessment is demonstrated by the practice of the design of energy systems.

The design of a power plant, regardless of specific details of the plant, is based on the requirements determined by Carnot Theorem. According to this theorem, the principle condition for maximization of the plant efficiency is minimum entropy generation in the course of energy conversion.

The objective of a manufacturing process is material rather than energy conversion. However, the overall energy cost of the process, including the cost of producing and processing of raw materials, can be used for the evaluation and, subsequently, for the design of manufacturing processes [1].

This paper is concerned with the development of a formal procedure for technology design. At this stage a qualitative process assessment is discussed. It is suggested that exergy function [2-6] is an expedient tool for the evaluation of manufacturing processes. Conditions of minimization of exergy consumption enables determining principal features of a process in question.

An analysis of steel production technology is used to demonstrate the utility of the application of exergy function for process design. An analysis of the conditions of minimization of exergy losses enables prediction of trends in the development of steelmaking. This analysis also enables the selection of an effective procedure for conversion of iron ore in steel products. Similar approaches can be used for other manufacturing processes.

First Law Analysis of a Manufacturing Process

It is convenient to characterize a manufacturing process by the energy consumption, E_c, determined by the equation:

$$E_c = \sum_i E_i + Q_i + W_i - Q_e - W_e \qquad (1)$$

where E_i is the energy cost of the production of raw material "i" (i = 1, 2 . . . M); Q_i and Q_e, W_i and W_e correspondingly denote heat and work consumed and produced during of production.

As it follows from Equation 1, first law analysis provides limited information about process. This equation does not identify the quality of energy consumed. There is a distinction between 1kJ of electrical energy and 1kJ of heat rejected at the surrounding temperature. The energy balance does not identify the way of energy degradation in the course of processing. The above information can be obtained if the energy balance will be supplemented by the entropy balance. However, the process is to be characterized by a single objective function. This can be accomplished by the use of exergy function for process characterization.

Exergy Function

Exergy is the function determining the maximum useful work that can be obtained from a system and, consequently, is a measure of the quality as well as the amount of energy. Exergy can be determined as:

$$B = H - T_0 S + \sum_i \mu_i N_i \qquad (2)$$

where B denotes exergy, kJ; H is enthalpy, kJ; S is entropy, kJ/K; μ_i is the chemical potential of material i, kJ/kmole; N_i is the amount of the material i, kmole; and T_0 is the surrounding temperature, K. The exergy of electricity, fuel, air and oxygen consumed in a manufacturing process determines the cost of energy. The exergy of raw materials characterizes the cost of these materials. Consequently, total exergy consumption can be used to assess overall cost of an operation.

The conditions of the energy utilization in an operation are determined by the exergy balance equation:

$$\sum_i B_i + B_s = B_e + B_r + B_w + LW + B_u \qquad (3)$$

where B_i is exergy of a material treated in the unit operation (concentrate, hot metal, molten steel, solid steel, etc.); B_s is exergy supplied by electricity, fuel, air, oxygen, fluxes and other components required for the process accomplishment; B_e is exergy of the product; B_r is exergy of flue gases, slag and other rejected matters; B_w is exergy losses due to heat exchange with surroundings; LW is exergy losses (lost work) due to irreversible processes (chemical reactions, phase transformation, heat exchange, plastic deformation, mixing) occurring in the course of processing; and B_u is exergy output from the energy recovery systems (boilers, vaporizing, cooling systems, gas turbine, etc.). For reversible processes, LW = O and the structure of Equation 3 is similar to that of the energy balance equation.

Energy consumption in the course of a process B_c is expressed by the equation:

$$B_c = \sum_i B_i + B_s - B_u = B_e + B_r + B_w + LW \qquad (4)$$

The initial and final states of the product for each stage of the treatment are assumed to be known. The objective of the process optimization is the reduction of LW, B_r, and B_w.

Exergy Analysis of Steelmaking

In order to demonstrate the use of Equation 4 for the prediction of technology development, let us apply it to an analysis of steelmaking.

The exergy utilization analysis at each stage of steel production can be accomplished by constructing the exergy balances for each stage. The schematics of exergy balances for the making, refining and solidification of steel are given in Figures 1-3.

Let us discuss reducing B_c for each stage during the manufacture of steel products. Lost work is determined by energy dissipation in the course of the operation. There are several methods to reduce LW. Practically, all real processes are irreversible and result in energy dissipation. Consequently, LW can be reduced by terminating processes which do not result in the desired change in a system's thermodynamic state. Among thermodynamically unnecessary processes are coke and pellet production, and dissolving of excessive amounts of carbon, silicon, and manganese in the metal. Different shapes obtained during solidification are almost thermodynamically identical. Because of this, substantial exergy savings can be obtained by net shape casting.

Energy dissipation is also caused by heat and mass transfer in reactors. This dissipation can be reduced by increasing system conductivity. Completing transport processes in intensively mixed liquid baths and reducing the size of solid particles supplied into these baths will increase conductivity. The objectives above can be achieved in steel production by a boiling metal or slag bath supplied by fine coal and concentrate. The use of coke and ore pellets increases exergy losses.

The minimum value of B_r is achieved if flue gases do not contain CO, H_2 or any other combustible components, and flue gases and slag are rejected at temperatures approaching the

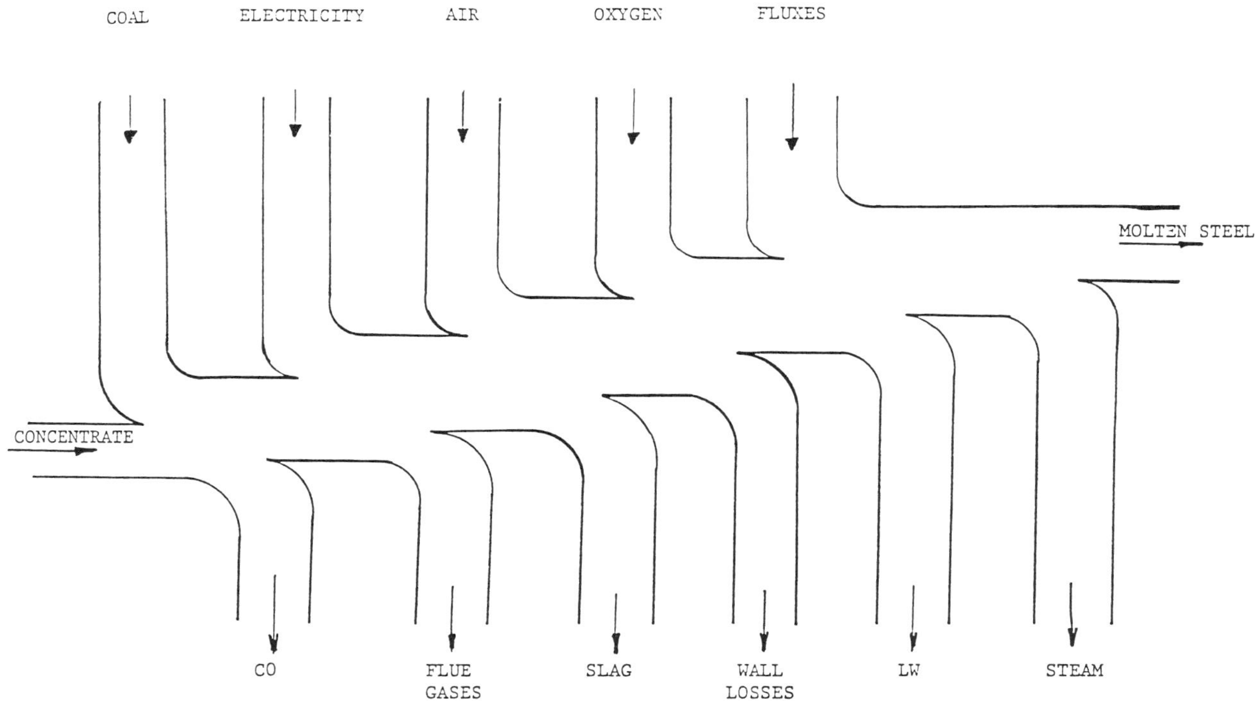

FIG. 1 EXERGY BALANCE OF THE MAKING OF STEEL

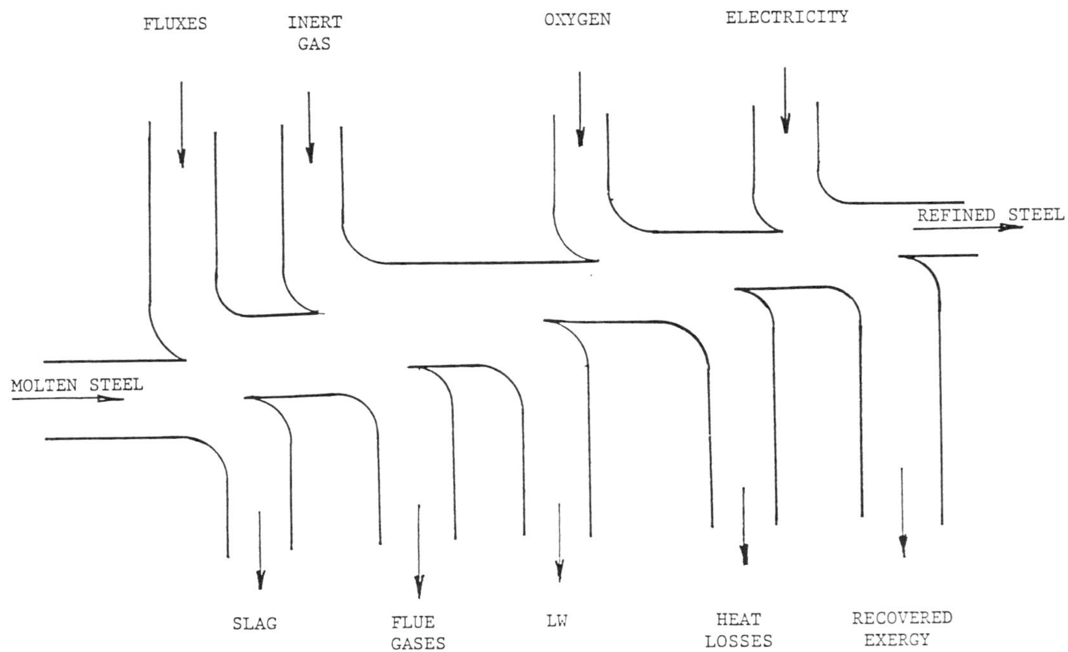

FIG. 2 EXERGY BALANCE OF THE REFINING OF STEEL

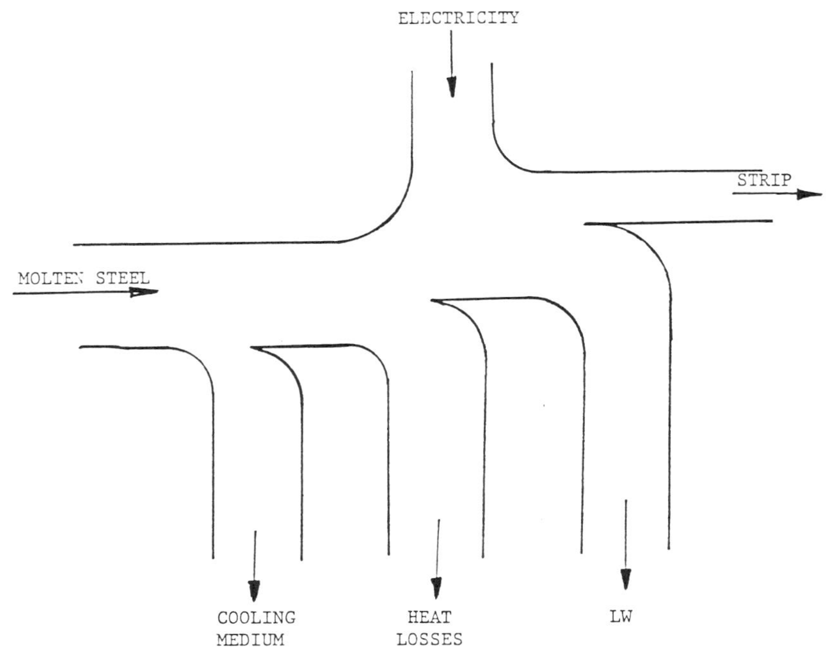

FIG. 3 EXERGY BALANCE OF THE SHAPING OF STEEL

surrounding temperature.

B_w can be reduced by increasing the rate of process, decreasing or terminating the exposure of high temperature metal surfaces to the atmosphere. The rate of process can be increased by the improvement of conductivity, e.g. by carrying out reaction in boiling liquid baths supplied by a fine coal and concentrate.

The making and shaping of steel involves the use of electricity, oxygen, inert gases and other components, production of which requires significant exergy consumption. Because of this, the exergy cost of the process should include the exergy losses in the production of these components.

The practice shows that combustion of carbon-oxygen mixtures enables us to achieve the temperature required for the steelmaking process. The energy balance of combustion shows that oxygen can be replaced by air without a decrease of process thermal efficiency, if air temperature exceeds 1300° C. Consequently, the consumption of oxygen and electricity in steelmaking can be reduced by the use of coal combustion in preheated air as the principle sources of energy. The consumption of inert gases can be reduced by their recycling through cooling and cleaning.

Methods for the reduction of B_c are represented in Table 1. Practical examples, given in Table 2, demonstrate that the development of metallurgical technology is to a great extent, controlled by the conditions of improvement in exergy utilization..

Development of an Effective Steelmaking Process

The conditions needed to minimize B_c presented in Table 1, enable us to outline the following requirements necessary to improve steelmaking and shaping:

* steel production and refining should be accomplished in a sequence of interconnected, sealed liquid baths. Process interruption and metal exposure to the atmosphere are to be restricted.
* baths should be fed by fine concentrate, coal, preheated air, fluxes and inert gases.
* bath temperature and composition, and the conditions of the concentrate, fluxes and coal supplies, determine the speed of absorption by the bath.
* process energy should be generated by coal combustion in air preheated to a temperature of more than 1300° C.
* electricity and oxygen use should be restricted to processes where the use of air and coal are not technologically feasible.
* net shape and near net shape casting are to be used.
* metal heat is to be removed by a fluid, for example, an inert gas, which allows controlled heat extraction and heat recovery.
* gases exhausted from reactors should not contain combustible components.
* heat content of flue gases and slag is to be recovered.

An example of a process implementing the requirements above [11] is shown in Figure 4.

According to this figure, fine concentrate, coal, fluxes and preheated air are injected into a liquid bath, similar to the bath of a steel making converter at the end of the decarburization process. Range of the carbon content is 0.5 - 1.0 percent. Oxygen in the air supply and iron oxides react with

TABLE 1
THE METHODS FOR THE REDUCTION OF THE EXERGY CONSUMPTION IN MAKING AND SHAPING OF STEEL

ITEM OF EXERGY BALANCE	WAY OF IMPROVEMENT	TECHNICAL MEANS
1. Lost Work	1.1 Termination of thermodynamically unnecessary processes.	1.1.1 Replacement of coke by fine coal.
		1.1.2 Replacement of pellets by fine concentrate and fluxes.
		1.1.3 Direct casting of strips.
		1.1.4 Completion of iron oxide reduction at the minimum possible level of impurities content.
		1.1.5 Reduction of the range of metal reheating.
	1.2 Reduction of the entropy production in heat and mass transfer processes.	1.2.1 Use of fine coal, concentrate and fluxes.
		1.2.2 Rapid smelting of solid materials.
		1.2.3 Performing of chemical reactions boiling baths.
2. Exergy Input	2.1 Reduction of the use of electricity and oxygen.	2.1.1 Preheating of the air to temperature above 1300 C.
3. Exergy Output	3.1 Reduction in heat content of outflowing materials.	3.1.1 Complete combustion.
		3.2.1 Recovery of heat of flue gases.
		3.2.2 Recovery of heat of slag.
		3.2.3 Recovery of heat of metal.
4. Exergy Losses	4.1 Increase in reaction rate.	4.1.1 Use of fine coal.
		4.1.2 Rapid melting.
		4.1.3 Use of injection.
	4.2 Recovery of wall losses.	4.2.1 Vaporizing cooling.
	4.3 Reduction of the exposure of high temperature metal surface.	4.3.1 Process performance in sealed interconnected reactors.

TABLE 2
EFFECT OF THE NOVEL TECHNOLOGY ON THE EXERGY CONSUMPTION

THE TYPE OF PROCESS	PROCESS OUTLINE	MEANS TO REDUCE EXERGY CONSUMPTION
1. ELRED Process [6]	1. Direct reduction by mixing fine concentrate with coal dust and preheated air. Reduction completed in an arc furnace	1.1 Use of concentrate and coal. 1.2 No use of oxygen. 1.3 Intensive mixing. 1.4 Recovery of heat losses.
2. INRED Process [7]	2. Fine concentrate is smelted in oxygen and coal streams. The produced sponge iron is refined in an electrical furnace. Heat from waste products used for steam generation.	2.1 Use of fine coal. 2.2 Fast melting. 2.3 Recovery of heat losses.
3. "Ore-Steel" Process [8]	3. Concentrate-coal pellets are melted in boiling slag and reduced. Obtained melt is refined in sequence of reactors, producing steel.	3.1 Use of concentrate and coal. 3.2 Rapid melting. 3.3 High temperature. 3.4 No iron production.
4. Ladle Injection [9]	4. Injection of a powder material in a melt.	4.1 Use of fine fluxes. 4.2 Use of the jet injection.
5. Horizontal and Rotary Casting [10]	5. Casting of strip and billets.	5.1 Reducing the range of metal deformation. 5.2 Reducing heat losses in metal processing.

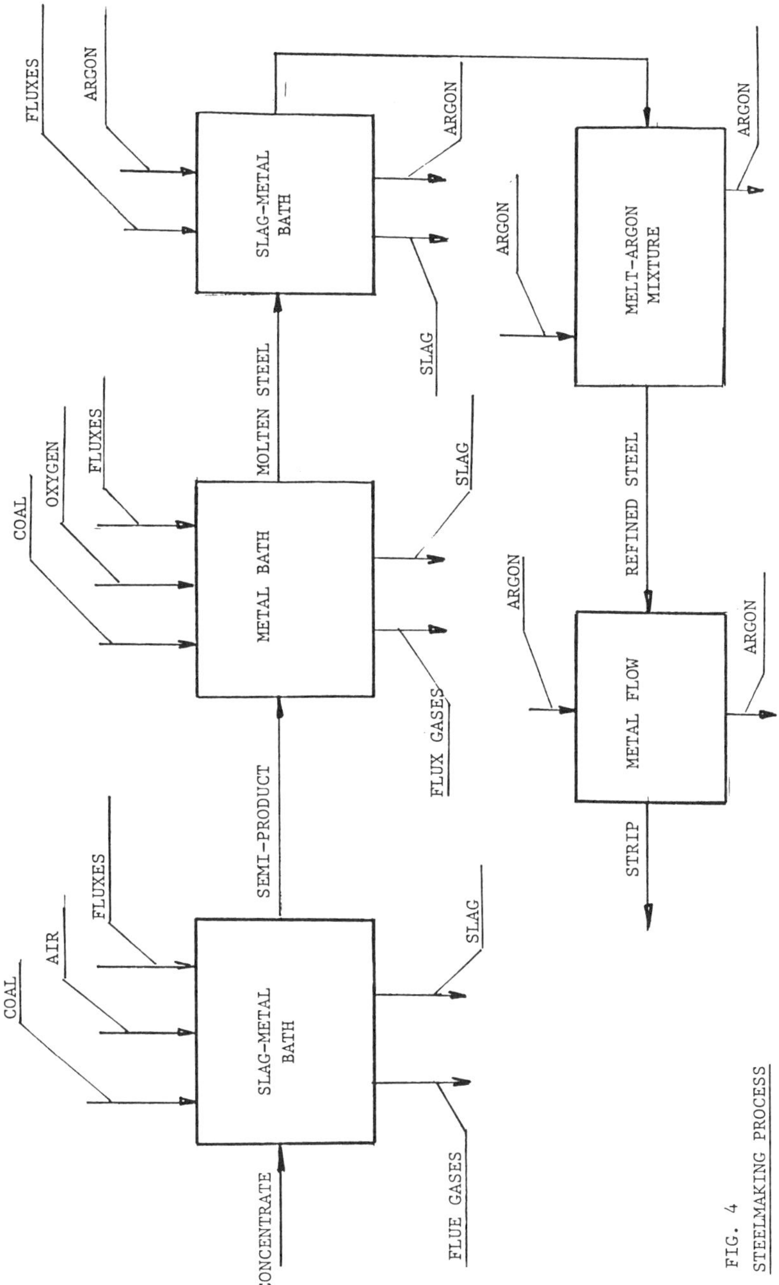

FIG. 4
STEELMAKING PROCESS

carbon dissolved in metal. The evolved CO gas is burned by the injected air during emerging CO bubbles in the bath or above the bath. The carbon content of the metal is sustained by the dissolving of coal. The range of coal, ore and air temperature assure the constant composition of the bath and evolving of the amount of heat, which is required by the energy balance of smelting and reducing the concentrate. Silica and other oxides contained in the concentrate are bonded by lime and other injected fluxes and absorbed by the slag. Sulphur is partially removed from coal prior to the coal injection and partially dissolved into the slag. The evolved metal is directed into the next bath where the carbon, silicon, maganese and sulphur contents are reduced to the required level by the injection of oxygen and fluxes. Further, metal flow is atomized by the expansion of injected inert gas. Metal drops are purified by submersion in a slag bath and in a flow of an inert gas. The slag and gas treatments remove non-metallic inclusion.

Steel solidification is carried out in a thin fluid layer reproducing the shape of a strip. The strip surfaces are blasted by jets of an inert gas. Intensive heat exchange on the impinged solid surfaces ensures controlled solidification. When solidification is completed, the solid metal can be deformed by rolling. Because the metal temperature approaches the temperature of solids, the work of the deformation is negligible. After the completion of solidification, a new portion of liquid metal is supplied to the solid surface and the solidification and deformation processes repeated. Inert gas jets impinge the strip surface to control metal cooling. After metal treatment, the inert gas is cooled in a power plant turbine. Cooled and compressed gas is cleaned by membranes or other cleaning devices.

A slag bath is used for extracting heat from slag and flue gases evolving in the course of steelmaking and refining. Further cooling of the flue gases is completed in a conventional recuperator and boiler. Recovered energy is used for preheating the air supplied into the reactors and for electricity production. The air is preheated first in the recuperator and then heated to 1300° C in the slag bath.

Conclusion

The exergy analysis is used in order to outline requirements to the unit operations involved in the steelmaking and shaping. Some of these requirements, for example, reduction of heat losses or completion of combustion, are obvious and follow straightforward from the energy consideration. Other requirements, however, can be defined only in the framework of second law analysis, which enables us to outline the comprehensive set of conditions for the effective energy utilization in the steelmaking and shaping. For example, only exergy analysis demonstrates the effectiveness of using fine concentrate and coal or the use of a boiling bath.

This procedure is the simplest case for the application of exergy analysis because only tendencies in the exergy change are assessed. However, this simple assessment enables us to predict the technology trends selected by the practice and provides useful information for process design. Development of the suggested procedure consists in the quantitative evaluation of the total exergy consumption for unit operation as well as the further process disintegration into more elemental operations. This improvement will bring about the creation of the formal objective procedure for the design of the manufacturing processes.

References

1. H.H. Kellog, "Energy Efficiency in the Age of Scarcity," Journal of Metals, June 1974.
2. Y.M. El-Saed, and R.B. Evans, "Thermoeconomics and the Design of Heat Systems," Journal of Engineering for Power, Jan. 1970, pp. 27-35.
3. T.J. Kotas, The Exergy Method of Thermal Plant Analysis, Butterworth, 1985, pp. 29-37, 162-171.
4. G.E. Ahern, The Exergy Methods of Energy Systems Analysis, Wiley, 1980, pp. 73-78.
5. R. Gagiolli and S.W. Wepter, "Exergy Economics," Energy, Aug. 1980, pp. 823-839.
6. E. Bengtsson, et. al., "The Chemistry of the ELRED Process," Iron and Steelmaker, Oct. 1981, pp. 30-35.
7. R. Irving, "The INRED Process: An Iron Making Option," Iron Age, July 6, 1981, pp. 21-23.
8. Y. Kozhgevnikov, et. al., "Physical Chemical Fundamentals and Outline of the Technology of Continuous 'Ore-Steel' Process," Physical Chemistry of Direct Iron and Steel Production, Nauka, Moscow, 1977, pp. 11-18.
9. T.L. Mazel, "Ladle Injection Metallurgy, Where It's At, Where It's Going and Why," 33. Metal Producing, April 1981, pp. 53-59.
10. G.I. McManus, "Continuous Casting Continues to Evolve New Techniques," Iron Age, Feb. 2, 1981, pp. 7-11.
11. E.S. Geskin, Steelmaking Method, U.S. Patents: 4, 434, 003 of Feb. 28, 1984 and 4, 480, 373 of Nov. 6, 1984.

A Coherent Model of Production Highlighted from an Epistemological Approach

B. Huet

Université Paris XIII, Faculté Médecine, Département Biostatistiques et Informatique Médicale, 74 Av. M. Cachin, 93012 Bobigny Cedex, France

I - INTRODUCTION

The economical stake of the applications of data processing and knowledge engineering in production systems is very wide: energy, water , communication, transportation, health,...are today's huge budgets in develop ed countries; the best mastership of these expenses is a permanent preoccupation for all responsible persons of these large organizations.

The long term and middle term strategical choices as well as diary management need a permanent evaluation of many parameters by experts working alone or within teams in a cooperative or antagonistic manner.

In this context a real integration of data processing and knowledge engineering seems to be a very important parameter for efficiency.

II - THE METHOD

In fact it is interesting to notice that as early as 1971, Kling proposed a "paradigm for the reasoning by analogy".

The paradigm for reasoning by analogy presented by Kling (1971) is:

Let P be a "Phor", T be a "Theme", the data structures describing two objects (facts, events, states,...) in a knowledge base, A and B two substructures in P; C and D two substructures in T.

The paradigm formalized in this knowledge base is founded on the similarity of substructures of which the more general formulation is:

A is to B as C is to D

It explains a similarity and allows to infer thanks to similarities a new knowledge (D) from already known knowledges (A, B, C).

To increase the universality of the Phor, it was performed an epistemological approach.

Our Phor is founded on the conclusions of an epistemological study done by (Forrester J.W. - 1969) and various communications (Forrester J.W. - 1975) on socio-economic systems: the control loop feed-back is a universal structure of systems.

We applied Forrester's conclusions to biological systems and all features already formulated by Forrester were verified (Huet B. - 1981), this impulsed us to consider structural and behavioral features described by Forrester as the Phor.

At this step the phenomenological study of the control loop feed-back appeared us as necessary:

- way of regulation (positive or negative).
- nature of operators.

Was it possible to explain the "intelligent" behavior of this structure by "intelligent" operators?

Since a long time biochemists distinguish two types of biochemical operators: "Michaellian" operators process information while "Allosteric" operators have a decisional function: they regulate the flow of production of the whole control loop feed-back by stopping or acting the processing of information of Michaellian operators (Lehninger - 1973)

These two types of operators were studied in types of automata (Huet B. - 1980) it was concluded that Michaellian operators function as a Moore machine while Allosteric operators function as a Mealy machine.

This study was pursued on an other criterium: to type biochemical operators (integrated in control loop feed-back of biological systems) in algorithmic operators and non algorithmic operators.
The result (Huet B. - 1986) highlights two operators types:
-algorithmic operators performing data processing, they are the "Michaellian" operators.
-Allosteric operators are mixed operators types, they are constituted with two operators: one is non algorithmic type, the other one (under dependence of the first one) is an algorithmic type.
This allows to present a control loop feed-back structure in types of operators corresponding to a technological view of the feed-back.

Such a loop is constituted with two functional parts:
-a sequence of algorithmic operators processing information;
-a mixed operator (at the head of the loop) composed of a non algorithmic operator regulating the flow of processing and an algorithmic operator processing information.

Afterwards we inferred the structure of the "theme" by applying the reasoning by analogy:
* The substructure A of the Phor and the substructure C of the Theme highlight two identical operators types (algorithmic ones and non algorithmic ones)
* The substructure B of the Phor is a control loop feed-back integrating a functional organization of the two operators types of the substructure A of the Phor; this substructure B can be organized in superimposed cooperative or antagonistic loops of which final equilibrium is regulated by a "balance".
* We infer that production systems (Theme) can be designed on the same substructure D as B.

III - RESULT

The basic structural element is the control loop feed-back constituted with the functional integration of several (five to seven) algorithmic type operators under the dependence of a non algorithmic type operator positionned at the head of the loop (fig. 1)

Several control loops are superimposed and can be cooperative or antagonistic, they allow a dynamic system thanks to a permanent equilibrium between loops of the same level (fig. 2)

* Arguments for this model

Researchers on distributed data processing agree approximately together: they propose to distribute tasks and data, according various strategies.

In fact to consider only distributed data processing is a biased manner to solve the problem because problem-solving-entities are not integrated in this design.

Even in the proceedings of the IJCAI 1985 almost all works relate on the design of cooperative entities in distributed artificial intelligence (D.A.I.) (Durfee E.H. et al.) (Stuart C.), this allows to Rosenschein and Genesereth (1985) to conclude that almost all researchers on D.A.I. work on a simplified hypothesis which restricts severely the consequences of their results: these researchers consider that the agents (Newell A. - 1982) have common or cooperative objectives; that is what Rosenschein and Genesereth call the "paradigm of the benevolent agent".

Even if this paradigm of the benevolent agent is appropriate in large production systems (this has to be demonstrated), we consider that there are ones in which the structure is in opposition with the paradigm, especially because of the conclusions extracted from our researches in this domain (Huet B. - 1981) (Huet B. - 1985) because in the two systems which were analysed, there are two kinds of relations between agents: cooperative ones and antagonistic ones.

In conclusion our results and ways of research agree with Rosenschein and Genesereth ideas and are in opposition with "the paradigm of the benevolent agent".

In such a model, the major problem is the coherence of this system, this confirms the idea of Durfee, lesser, Corkill (1985) who consider that "globally coherent behavior is the holy grail of distributed problem solving network".

IV - COHERENCE OF THIS MODEL

This model is coherent in certain conditions.

* Study of the behavior of a single control loop feed-back.

** Goodwin (1965) proposed a set of equations describing the dynamics of a negative control loop feed-back, these ones define a non-linear oscillator; he simulated oscillations with damping and oscillations without damping.

** Hsu and Meyer (1968) analyse this control loop feed-back into two functions: a transfer function representing the linear part of the system, a non-linear function. From this basic work several mathematical analysis were performed: 1)- by Popov criterion (Othmer - 1976); 2)- by Lyapunov method (Walter - 1969); 3)- by Hopf bifurcation theorem (Mees Rapp -1976).

** Sparrow (1980) studied precisely the conditions of a chaotic behavior; he defined a transfer function and presents steady points and unsteady points with their values, finally he presents a classical scheme of the Hopf bifurcation.

** For positive control loop Bartha (1980) studied a set of n differential equations and supply a complete study on the instability.

** The aim of Berding and Haubs (1985) is to find a stability criterium at dynamical equilibrium, they built the stability matrix and applied the Gershgorin theorem (which informs on the range of stability).

** Palsson and Lighfoot (1985) studied a set of kinematic parameters, it exists several steady states corresponding to several flows and several values of parameters.

** From the differential equations An der Heyden and Mac Kay (1982) exhibit the existence of steady and unsteady limit cycles and chaos.

* Study of the behavior of several control loops feed-back.

** Allwright (1977) proposes a method to study control loops which exhibit the global stability of Goodwin equations (1965) when the Hill coefficient is equal to one.

** Walter (1970) highlighted that limit cycles are observed in several negative loops even if the reaction speed is a kinetic of the first order.

** Hyver (1984) proposes a method of prediction of certain types of global limit cycles.

** Thomas (1978) performed a Boolean analysis of the dynamical aspects of control loops.

** Mees and Rapp (1976) exhibit that for a system with several negative loops; they present afterwards several studies on a Lyapunov function and conclude on the existence of periodic solutions.

** Glass and Pasternak (1978) modelled the non-linear part of the loops in terms of linear Piece Wise equations.

V - CONCLUSION

This coherent model presents a function ally integrated view of data processing entities and problem-solving-entities based on feed-back. A higher structure is proposed between antagonistic loops, regulated by a balance.

- REFERENCES -

Allwright, D.J. (1977). A global stability criterion for simple control loops. J. Math. Biology . 4 , 363-373

An der Heyden, U., and M. C. Mackey (1982). The dynamics of production and destruction: analytic insight into complex behavior. J. Math. Biology . 16 , 75-101

Bartha,F. (1980). On a criterion of instability for positive feed-back control systems. J. Theor. Biol. 82 , 141-147

Berding, C. and G. Haubs (1985). On the stability of equilibria in metabolic feed-back systems. J. Math. Biology . 22 , 349-352

Chouraqui, E. (1986). Le raisonnement par analogie: sa problématique, ses applications. Actes I Jour. nat. PRC IA , Nov. 1986, Aix les Bains, France, Cepadues: Toulouse, 107-117

Desoer, C.A., and M. Vidysagcar (1975). Feed-back systems: input-output properties Academic Press: New York

Durfee, E.H., V.R. Lesser, and D.D. Corkill (1985). Increasing coherence in a distributed problem solving network. Proc. IXth Int. Joint Conf. Artif. Intel. IJCAI 85 , Aug. 1985, Los Angeles, USA, 1025-1030

Forrester, J.W. (1969). Urban Dynamics . M.I.T. Press, Cambridge, Mass.

Forrester, J.W. (1975). Dynamics of socio-economics. Proc. Int. IFAC Conf. . Boston, USA, 1-8

Glass, L., and J.S. Pasternak (1978). Stable oscillations in mathematical models of biological control systems. J. Math. Biology . 6 , 207-223

Goodwin, B.C. (1965). Oscillatory behavior in enzymatic control processes. Advances in enzyme regulation 3 , 425-438

Hsu, J.C., and A.V. Meyer (1968). Modern control principles and applications . Mac Graw Hill: New York

Huet, B. (1980). Modélisation des processeurs du vivant par la machine de MOORE et la machine de MEALY et leur correspondance métabolique. Proc. IXth Intern. congress for Cybernetics . Sept. 1980, Namur, Belgique, Cybernetica: Namur, 477-487

Huet, B., (1981). Contribution à l'étude de l'isomorphisme dans le systèmes complexes, à partir des systèmes biologiques et des caractères extraits par Forrester. Kybernetes , 10 , 43-47

Huet, B., (1985). Semantical modelling of biological subsystems by means of multi level control structure concept. Kybernetes , 12 , 93-99

Huet, B. (1986). Basic propositions for an intelligent system design. Proc. Ist IFIP IFAC Int. Conf. Economics and A.I. . Sept. 1986, Aix en Provence, France, 93-98

Hyver, C. (1984). Local and global limit cycles in biochemical systems. J. theor. Biol. 107 , 203-209

Kling, R.E. (1971). A paradigm for reasoning by analogy. Artificial Intelligence , 2 , 147-178

Lehninger, A.L. (1973). Biochemistry . Wirth, New York.

Mees, A.I., and P.E. Rapp (1976). Periodic metabolic systems: Oscillations in multiple-loop negative feed-back biochemical control networks. J. Math. Biology . 5 , 99-114

Newell, A. (1982). The knowledge level. Artificial intelligence . 18 , 87-127

Othmer, H.G. (1976). The qualitative dynamics of a class of biochemical control circuits. J. Math. Biology . 3 , 53-78

Palsson, B.O., and E.N. Lighfoot (1985). Mathematical modelling of dynamics and control in metabolic networks. V. Static bifurcations in single biochemical control loops. J. Theor. Biol. 113 , 279-298

Rosenschein, J.S., and M.R. Genesereth (1985). Deals among rational agents. Proc. IXth Int. Joint Conf. Artif. Intel. IJCAI 85 , Aug. 1985, Los Angeles, USA, 91-99

Sparrow, C. (1980). Bifurcation and chaotic behavior in simple feed-back systems. J. Theor. Biol. 83 , 93-105

Stuart, C. (1985). An implementation of a multi agent plan synchronizer. Proc. IXth Int. Joint Conf. Artif. Intel. IJCAI 85 , Aug. 1985, Los Angeles, USA,

1031-1033

Thomas, R. (1978). Logical analysis of systems comprising feed-back loops. J. Theor. Biol. 73 , 631-656

Walter F. (1969). The absolute stability of certain types of controlled biological systems. J. Theor. Biol. 23 , 39-52

Walter F. (1970). The occurence and the significance of limit cycle behavior in controlled biochemical systems. J. Theor. Biol. 27 , 259-272

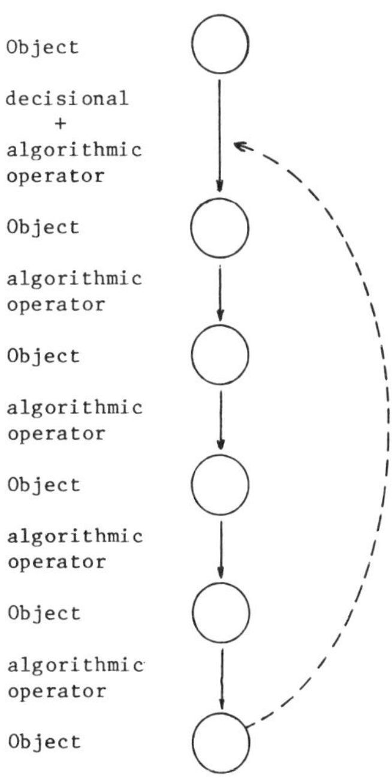

figure 1

Control loop feed-back

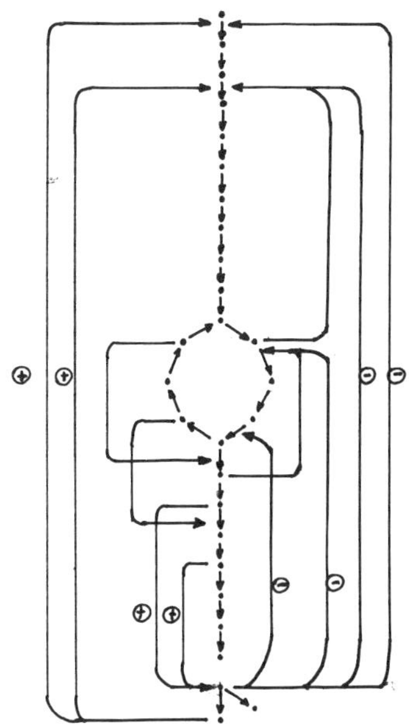

figure 2

Higher structure of feed-backs

Probabilistic Calculation of Critical Clearing Time in a Large Power System

Edgar Pereira

Universidade Federal do Rio Grande do Sul, Departamento de Engenharia Elétrica, Av. Osvaldo Aranha, esq. Sarmento leite, 90.210–Porto Alegre–RS, Brazil

ABSTRACT

In the present work a methodology of calculation was developed that uses statistical behavior of load magnitudes in each bus and makes statistical association with transient stability of the system in a sample population of critical clearing time to a given perturbation. Using this method, the designer of protection and control systems will have the advantage of obtaining a solution more reliable than deterministic one.

1 - INTRODUCTION[1,5,6]

An electric power system is subjected to continuous changes in load magnitude. Fortunately, these changes of load magnitude can be described statistically. There has been continous development of better techniques for assessing transient stability in an electric power system. It has also been recognized that there is a need for additional development of a technique to determine transient stability which includes important factors that influence the system stability. Normally, stability calculations have been deterministic rather than statistical. The purpose of this work is to analyze the statistical changes of load magnitude in the electric power system, and determine a reasonable clearing time to be used in the design of protection and control systems.

A digital computer program was developed to determine a statistical distribution of critical clearing time.

In this program, load magnitudes were treated as random variables with an assumed statistical distribution with a standard deviation of 10%.

Load flows were computed to provide initial conditions for transient stability calculation. Critical clearing time for each random load magnitude variation was determined to form a statistical sample population. The probability of critical clearing time was compared to the POISSON and NORMAL distributions. The best statistical distribution was verified using the GOODNESS-OF-FIT hypothesis test. A confidence interval estimate of critical clearing time was used to determine the required speed of the protection system.

2 - THE STABILITY PROBLEM

Two points must be clear when they treat transient stability problems in an electric power system.

a - Deterministic point of view for determination of critical clearing time in part of the system under consideration;

b - Probabilistic point of view for determination of critical clearing time in the same part of the system under consideration.

The first situation,[4] can be solved with the help of equal-area criterion of stability that uses the swing equation (2.1):

$$M \frac{d^2\delta}{dt^2} = P_u - P_e \qquad (2.1)$$

Where:

M = Inertia Constant

δ = Angular displacement of the rotor from the synchronous rotating reference axis

t = Time

P_u = Shaft power input

P_e = Electric power output of generator

Therefore, if the critical clearing time (CCT) and time operation of protection system (T) are known the system can work safety if the inequality (2.2) is observed:

$$CCT > T \qquad (2.2)$$

Where:

$T = T_1 + T_2 + T_3 + T_4$; and

T_1= Fault detection time for primary relay,

T_2= Breaker operate time for primary protection scheme,

T_3= Back-up operate time for relay,

T_4= Breaker operate time for back-up protection scheme.

In the second situation[1,2,3], many points must be observed, in the following sequence:

a - Variation of load magnitude $S_i = P_i \pm jQ_i$, to i=1,... ...,NB, where NB represents the number of busses of the electric power system,

b - Statistical distribution of load with its average value and standard deviation, or assume a normal statistic distribution, also with its average value and correspondent standard deviation;

c - If the normal statistics distribution of load (S_i) is assumed, then the random numbers can be used to determine the variables of the loads;

d - A complete statistic table, containing the critical clearing times population and its frequency of occurence, to a given fault on the system, is elaborated;

e - Through the table indicated in item "d" a cumulative probability distribution p(t) is calculated by equation (2.3),

$$p(t) = \int_{t \leq CCT_i} g(t)\,dt \tag{2.3}$$

where:

g(t) = function that described the cumulative probability

f - At this point a "Probability Level Distribution function (PLD)", is defined by expression (2.4),

$$PLD(t) \triangleq 1 - p(t) \tag{2.4}$$

g - The calculation of the cumulative probability item "e" can be substituted for a POISSON or NORMAL distribution, because if the loads S_i were assumed with NORMAL statistics distribution, this does not imply that the universe of critical clearing time population may also be normal;

h - A doubt is found at this point, so one of the distribution used will have to be selected. The best selection of distribution, through statistics tests with confidence interval is achieved by the GOODNESS-OF-FITT hypothesis test.

3 - DIGITAL COMPUTER PROGRAM[1]

To perform all the activities described previously a computer program was written. This program has two models,one electric and the other statistic and operate in an integrated form.

The electric model generates the random numbers, calculates the load flows, assemble transitory nets, estimate transient stability to each (S_i) and gives as results a statistic population of critical clearing times for each point in fault in the electric power system.

The statistic model, through many subroutines, assemble tables and estimate the probability levels distribution PLD(t), by POISSON and NORMAL distribution and makes the hypothesis tests by GOODNESS-OF-FIT.

A simplified flow chart of general program is shown in figure 3.1.

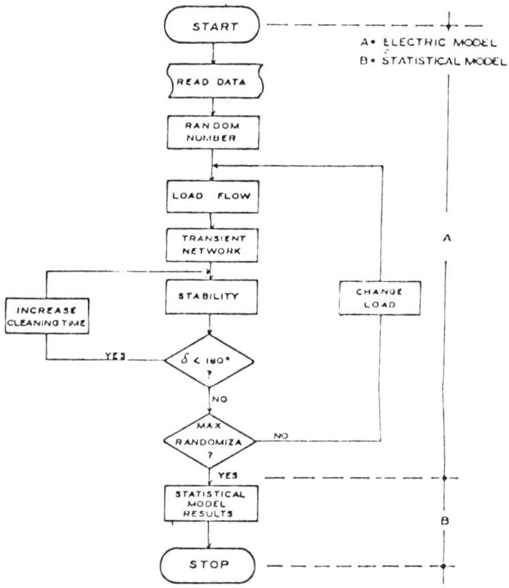

Figure 3.1 - Simplified Flow Chart of General Program

4 - COMPARISONS[1]

Comparisons for the test programs were made for several parts, such as:

4.1 - LOAD FLOW

The program described in the present work, in the load flow segment, was compared with the PHILADELPHIA program, reaching the results that are shown in table 4.1.

4.2 - COMPARISON OF REDUCED Y MATRICES

In Power System Control and Stability, Volume I, by P.M. Anderson and A.A. Fouad, first edition, 1977 on page 44, Table 2.6 shows reduced Y matrices. These matrices were calculated for an electric power system of nine busses (Figure 2.19) with three generators.

The program developed for this investigation made the same calculations and the results are illustrated in Table 4.2.

4.3 - COMPARISON BETWEEN CALCULATION OF CRITICAL CLEARING TIME WITH PHILADELPHIA ELECTRIC PROGRAM AND THE PRESENT INVESTIGATION

The BPA System was used to make a comparison between the calculation of critical clearing time with Philadelphia Electric and investigation

Table 4.1 - Comparison Between Solution of Load Flow, By Philadephia and This Present work

Bus #	Philadelphia (Voltages)		Present Work (Voltages)	
	Magnitude (0/1)	Angle (Degrees)	Magnitude (0/1)	Angle (Degrees)
1	1.000	0.0	1.0000	0.0
2	1.070	- 0.9	1.0700	- 0.40003
3	0.992	- 0.7	0.9920	- 0.22136
4	0.971	- 0.4	0.9710	- 0.17801
5	0.969	8.8	0.9690	8.86838
6	0.971	-14.6	0.9710	-14.07092
7	1.037	-54.1	1.0370	-53.59187
8	0.996	1.0	0.9960	1.53793
9	0.998	- 5.0	0.9980	- 4.55325
10	0.990	-18.5	0.9901	-18.08481
11	0.978	-19.5	0.9791	-19.00453
12	0.981	-20.5	0.9822	-19.99804
13	0.983	-22.1	0.9837	-21.64075
14	0.971	-24.8	0.9719	-24.28799
15	1.010	-11.0	1.0108	-10.73726
16	1.005	- 4.3	1.0051	- 4.15977
17	0.997	- 1.6	0.9972	- 1.48358
18	0.944	-12.5	0.9448	-12.15572
19	0.948	- 9.2	0.9489	- 8.94410
20	0.949	-18.2	0.0497	-17.78091
21	0.968	-18.7	0.9685	-18.26216
22	0.970	-22.7	0.9711	-22.28529
23	1.040	- 4.7	1.0405	- 4.27495
24	0.981	4.3	0.9807	4.43001
25	0.981	- 4.6	0.9812	- 4.54887
26	0.951	- 6.5	0.9537	- 6.32183
27	0.966	- 2.7	0.9664	- 2.43826
28	0.977	- 3.8	0.9967	- 3.38531
29	0.966	- 7.5	0.9959	- 7.13262
30	0.966	-45.7	0.9967	-45.21704

Table 4.2 - Reduced Y Matrix (this study)

Type of Network	Node	1	2	3
Prefault	1	.84334 - j2.98597	.28566 + j1.51444	.20550 + j1.22673
	2	.28566 + j1.51444	.41869 - j2.72260	.21233 + j1.08885
	3	.20850 + j1.22673	.21233 + j1.08885	.27628 - j2.36742
Faulted	1	.65581 - j3.81499	.0 + j.0	.06984 + j.63082
	2	.0 + j.0	.0 - j5.48546	.0 + j.0
	3	.06984 + j.63082	.0 + j.0	.17383 - j2.79576
Fault Cleared	1	1.23190 + j2.14411	.14479 + j.74306	.19981 + j1.09659
	2	.14479 + j.74306	.39471 - j1.91948	.20180 + j1.24248
	3	.19981 + j1.09659	.20180 + j1.24248	.27476 - j2.33551

programs. Results of this comparison are shown in Table 4.3.

Table 4.3 - Comparison of Critical Clearing Time Calculations

Cases	Critical Clearing Time (CCT) and CPU Time in Seconds			
	Philadelphia		Present work	
	CCT	CPU	CCT	CPU
Fault on bus number 17 and open transformer 12	.24	10.95	.24	3.11
Fault on bus number 10 and open line 30	.16	11.17	.16	2.64
Fault on bus number 11 and open transformer 29	.12	7.43	.12	2.66

Three fault cases were run to verify the program developed. As can be seen in Table 4.3 the critical clearing times compare well for each of the three fault cases.

5 - ELECTRIC SYSTEM STUDIED

The electric power system used in this investigation is a modified real system located in the Northwest of the United States of America, and is shown in figure 5.1.

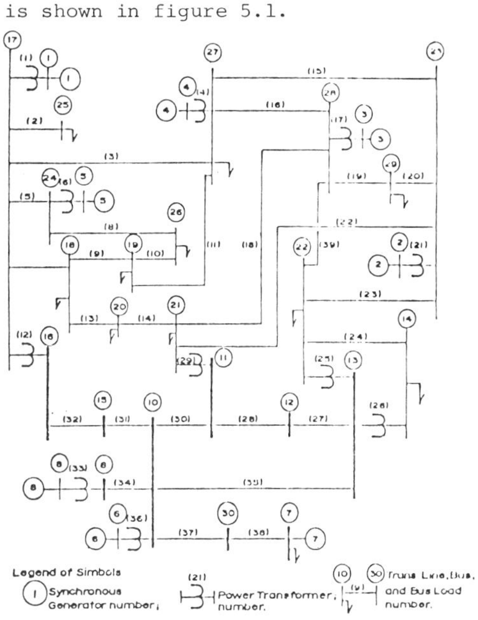

Figure 5.1 - One Line Diagram of Electric Power System

6 - STUDIED CASES AND RESULTS[1]

Table 6.1 shows the cases studied and results.

Table 6.1 - Summary of Results

Contingency	Cases studied	H.P.[1] A*	H.P.[1] R**	H.N.[2] A	H.N.[2] R	CCT cycle	CONFI (%)	PL (%)	Initial Values CCT cycle	Initial Values PL (%)
A	6.6.1.1 Three-Phase Fault on Bus Number 17	X				1	89.99	100.00	17	11.06
	6.6.1.2 Three-Phase Fault on Bus Number 10		X	X		1	89.99	98.19	9	56.83
	6.6.1.3 Three-Phase Fault on Bus Number 11		X	X		1	89.99	97.82	7	57.48
B	6.6.2.1 Three-Phase Fault on Bus Number 17 & Clearing by Opening Transformer 12		X	X		1	89.99	98.55	14	28.87
	6.6.2.2 Three-Phase Fault on Bus Number 10 and Clearing by Opening Line 30		X	X		1	89.99	98.18	9	56.30
	6.6.2.3 Three-Phase Fault on Bus Number 11 and Clearing by Opening Transformer 29		X	X		1	89.99	97.81	7	56.82
C (65% of Inertia Constant Generator at Bus 8)	6.6.3.1 Three-Phase fault on Bus Number 17 and Clearing by Opening Transformer 12		X	X		1	89.99	98.55	14	28.87
	6.6.3.2 Three-Phase Fault on Bus Number 10 and Clearing by Opening Line 30		X	X		1	89.99	97.92	9	45.89
	6.6.3.3 Three-Phase Fault on Bus Number 11 and Clearing by Opening Transformer 29		X	X		1	89.99	97.55	6	63.44
D (70% of Generation at Bus 8)	6.6.4.1 Three-Phase Fault on Bus Number 17 and Clearing by Opening Transformer 12	X				1	89.99	100.00	12	41.68
	6.6.4.2 Three-Phase Fault on Bus Number 10 and Clearing by Opening Line 30	X				1	89.99	100.00	11	84.64
	6.6.4.3 Three-Phase Fault on Bus Number 11 and Clearing by Opening Transformer 29	X				1	89.99	100.00	8	87.89

1-Hypothesis Poisson; 2-Hypothesis Normal; * Accept; ** Reject.

7 - CONCLUSIONS[1]

7.1 - Contingency A cases are more stable than the contingency B cases studied.

7.2 - Contingency C with variation inertia constant in 35% at generator number 8 requires more protection than contingency B cases studied.

7.3 - Contingency D with decreasing of generation at generator number 8 requires less protection than contingency B cases studied.

7.4 - The past behavior of load magnitude and its statistical distribution is important to determine the sample population of CCT.

7.5 - Design of the protection and control system can not be done only by deterministic calculation.

8 - ADVANTAGE OF THE METHOD[1]

The advantage of the present method, is the possible use of past behavior of load magnitudes, statistically associated with sample population of critical clearing time. Using this method the designer of the protection or control system will have the advantage of obtaining a solution that is more reliable than the deterministic solutions.

9 - POINTS OF FURTHER STUDIES[1]

9.1 - Load flow using GAUSS-SEIDEL method could be changed to another faster and more reliable method, such as, Newton-Raphson;

9.2 - Transient stability using Range-Kutta 4th order method could be changed to "Fast transient stability solutions", by Dommel and Sato;

9.3 - The statistical distribution of loads needs to be determined from past load data, and more investigation is advised in terms of the use of the correct distribution;

9.4 - Poisson and Normal were assumed for CCT distributions in the statistical model. Other types of distributions are recommended to be analyzed;

9.5 - Studies of PLD for unsymmetrical faults, such as single line to ground, line to line, and double line to ground, are recommended;

9.6 - Load type variations should be included.

10 - CONTRIBUTIONS[1]

10.1- Development of a digital computer program which can use FORTRAN or WATFIV language in an IBM 360/370, to perform calculations of random numbers, load flows, network transformation, transient stability, critical clearing time sample population, statistic, probability, and hypothesis test;

10.2- Determination of the most reliable protection and control system by a statistical method, when past behavior of load magnitude busses in know;

10.3- Providing the electric power system engineer with information to determine which areas of the system need added control and protection.

REFERENCES

1 - PEREIRA, E., Probabilistic Calculation of Critical Clearing Time in a Large Electric Power System, Doctoral dissertation,University of Missouri, Columbia, 1981.

2 - BILLINTON, R., and KURUGANTY, P.R.S., Probabilistic Evaluation of Transient Stability in a Multimachine Power System, Proc. IEEE, vol. 126, nº 4, April 1979, pp. 321-6.

3 - BILLINTON, R., and KURUGANTY, P.R.S., Index for Transient Stability, IEEE Trans. PA&S, vol. PAS-99, nº 1, Jan./Feb. 1980.

4 - ANDERSON, P.M. and FOUAD, A.A., Power System Control and Stability, Iowa State University Press, Ames, Iowa, 1977.

5 - WALKER, L.N., Transient Security Monitoring ou Stochastic Basis, University of Missouri, Columbia, June, 1977.

6 - RAHIMI, A., STANTON, K.N., and SALMON, D.M., Dynamic Agregation and the Calculation of Transient Stability Indices, IEEE, Winter Meeting 1971, New York, NY.

Critical Observations on "The Japanese Manufacturing Competitive Advantage"

Jay M. Pandya[*], Robert L. Nolan[†], S. N. Dwivedi[‡]

[*]Manager, Product Engineering, Westinghouse Electric Corp., PO Box 7002, Charlotte, NC 28217, USA
[†]Manager, Quality Assurance, Westinghouse Electric Corp., PO Box 7002, Charlotte, NC 28217, USA
[‡]Asso. Professor, Mechanical Engineering, University of North Carolina at Charlotte, Charlotte, NC 28223, USA

ABSTRACT

This paper provides insight into the subject of understanding the major factors behind the Japanese competitive edge. This edge enables them to maintain the selling price of their products delivered to the U.S. lower than our production cost in many cases.

Various areas which are important considerations in the quality and productivity improvement have been identified. Present Japanese trends in managing these issues are described and are frequently compared with U.S. methodology. The main areas which have been determined to be of significance are discussed in details. These areas are the management system, technology and techniques and productivity through people. Other observations which are of some significance are appended to the main observations. These are followed by the conclusions of the authors' summary.

INTRODUCTION

The authors visited a number of Japanese manufacturers in October of 1986. The main objective was to determine what gives the Japanese manufacturers the competitive edge that enables them to set the selling price of their products delivered to U.S. lower than our production costs in many instances.

During this trip several observations were made while conducting hours of formal and informal discussions with the cross-section of Japanese employees, watching employee work habits, studying numerous charts posted on shop floors and analyzing many project reports. Efforts are made here to relate these observations to the main issue - What are the major factors behind the Japanese competitive edge? The authors' conclusions are stated at the end of this paper.

MANAGEMENT SYSTEM

Experience has shown that the management system has a profound effect on quality and productivity improvement process.

Employment Practice: All the Japanese companies we visited were major corporations and claimed to follow **"Life Time Employment"** for their employees. Assured of permanent employment, the employee is more likely to be loyal and very productive. The management assumes labor cost to be a "fixed" cost and tries to lower this with the help of subcontract workers.

The subcontract workers are more widely employed than in the U.S. At one turbine manufacturer we saw many subcontract draftsmen and programmers for numerically controlled machine tools. The subcontractors do more "routine" production work, thereby freeing the permanent employees to concentrate more on new work. The subcontractors get substantially lower wages and have no guaranteed life time employment.

Early retirement is the most common means of reducing employment levels when there is a steadily reducing work load. One electrical manufacturer reduced its workforce from 1876 to 1250 employees mainly by encouraging early retirement and employee transfers. There was no layoff.

Employee Compensation: The total compensation consists of salary, bonuses and liberal benefits. The salaries are based upon the years of service, and are generally comparable to those in the U.S.. For instance, at an integrated steel plant, an average production worker earned ¥ 3.5 million ($23,300) in annual salary, while, a graduate engineer with ten years service at a turbine plant earned ¥ 4.8 million ($32,000).

In addition to salary, bonuses are generally awarded twice a year. At an integrated steel plant an average employee earned a total of ¥ 1.0 million ($6700) in bonuses. Managers decide how much bonus individual employees in their groups will receive. The main considerations for bonuses are years in service, position, performance, and employee initiatives such as cost improvement suggestions. The bonuses for individual employees varied between 80 to 100 percent of the maximum.

There is no incentive pay linked to productivity primarily because the management does not know how this can be used in an equitable manner.

Labor Union: One union represents the entire workforce at a given plant. The employees elect union leaders in a direct election. These leaders meet with the plant management a couple of times a year. The union leaders make suggestions to the management on holidays, vacation, benefits, etc. The management only considers these suggestions and is in no way obligated to agree to them. The unions are not a dominant force and generally work with management as a team. Work stoppages are unheard of.

Performance Management System: There was no evidence of a formal Performance Management System (PMS) comparable to that used in the U.S. where the manager evaluates employee performance against specific criteria and provides feedback. This may have something to do with the Japanese culture where personal confrontation and criticism are generally avoided.

Communication: Japanese managers take great care to establish good communication with their employees. A consensus is vital to their decision making process.

Front End Planning: Attention to minute details and thorough planning are remarkable strengths one notices at every manufacturing facility. For example, at an electrical equipment manufacturing plant a check list identified the requirements which must be met before starting the manufacturing process. The check list included:

- Is the raw material available?
- Are the tools and fixtures available?
- Is the tape for NC machine available?
- Is the manufacturing line up available?
- Is the manufacturing process qualified?

At one plant, we noticed the NC programmers operating from the shop floor, providing excellent interface between the production worker and NC machines. This kind of meticulous planning greatly reduces out of sequence manufacturing operations, thereby providing improved utilization of man and machine.

Strategic Planning: With an eye set on achieving a long term worldwide strategic advantage, Japanese management considers strategic planning to be very important. To formulate these plans, it was observed that a number of employees worked overseas for long periods to assess the market and their competitive positions.

The strategic plans aim at several years into the future. The plans are in modular form so that progress is achieved each year along with the resulting benefits. At a turbine manufacturer, the master plan for factory automation spanned from 1981 to 1988, and included:

- Flexible Manufacturing Systems (FMS)
- Robot Cell
- PC Based System to control Jigs and Fixtures
- PC Based System for Production Control
- CAD/CAM/CAE

One manufacturer was able to increase its Machine Tool Operating Time, defined below, from 62% to 75% in five months. A further increase to 82% is planned in the next phase. Machine Tool Operating Time is defined as:

$$\frac{B}{A} = \frac{\text{Hours during which the tool is actually producing chip}}{\text{Hours producing chip + set up time + idle time + downtime}}$$

To provide increased visibility, this manufacturer prominently displayed on the shop floor the critical machines that were inoperative and the elapsed time.

TECHNOLOGY AND TECHNIQUES

Japanese management places significant emphasis on "high tech" for increasing machine tool output, improving quality, and reducing labor costs.

Advanced Technology: Japanese manufacturers moved fast in applying laser technology and factory automation. At a turbine manufacturing plant, laser technology was used to attach stellite on steel blades for hard surfacing. Laser technology was also used to measure blade profile.

At another turbine plant, blade tenon pitch is recorded electronically, and the data is transmitted directly to a Flexible Manufacturing System (FMS) that drills holes into the shroud piece depending upon the measured pitch. The same FMS, deburrs holes, rounds the edges, and rolls the shroud to a predetermined radius.

Machine Tool Upgrade and Automation: It is impressive to see how the Japanese have reduced labor cost by employing "high yield" machine tools, One Man Multiple Machines (OMMM), Unmanned Machines, FMS and Robotic Tools. The number of such automated systems are a testimonial to their commitment to automation and modernization.

- Existing machine tools are regularly upgraded to increase their productivity. At a plant in Nagasaki, the three axis lathe for machining large forgings was modified to have two carriages. Each contained a cutting tool which doubled the productivity. Likewise, a manufacturer upgraded the milling machine to contain eight milling cutters operating simultaneously.

- OMMM's are common on the factory floor. At one factory three production workers we observed were operating eight machine tools. One manufacturer of electrical machines claimed to have increased labor productivity between 20 to 70% through the use of OMMM.

- At a turbine blade manufacturing plant, a robot was observed measuring and recording final blade dimensions, measuring and recording material hardness, and stamping a serial number on the finished blade.

- Flexible Manufacturing Systems (FMS) integrate several manufacturing operations at one location. This reduces material handling costs and manufacturing cycle time due to continuous operation. The automated material transporters, special jigs, and fixtures reduce labor costs and product variations are virtually eliminated because of the rigid process control. One manufacturer planned to increase the number of FMS from 23 to 36 between 1983 and 1988.

Work In Process (WIP) Inventory: A lot is said about Just-In-Time (JIT) and KANBAN systems to reduce the inventory carrying costs. They are, however, more beneficial to mass production items. The manufacturers of heavy machines are equally successful in reducing the WIP inventory costs by using other techniques.

One manufacturer of electrical generators was able to reduce inventory costs by 20% in two years. The method was simple but effective.

- Good Housekeeping - scrap unnecessary parts, reduce storage space and stockers.
- Stickers - Managers put special stickers on inactive parts on the shop floor. The area manager is then obligated to come up with an action plan within three days to keep the part moving through the manufacturing cycle.
- Production Planning - Improved material delivery schedule so that the material is delivered exactly when it is needed.
- Communication - Any changes in schedules are promptly communicated to everybody so that the production planning can be altered accordingly.

Error Prevention: The plants we visited took extraordinary precautions to avoid catastrophic machining errors. For example, at one plant the NC lathe for machining large forgings was equipped with a redundant system. In this system, the

NC tape for turning was prepared in a conventional manner using CAD data base. Independently, another data base was created from engineering drawings representing the machined surface. This data base was loaded into its own dedicated computer. This dedicated computer constantly monitored the actual tool path and compared it against the required outline of machined surface. The lathe stopped when the difference exceeded a predetermined value.

PRODUCTIVITY THROUGH PEOPLE

While automation, planning and effective implementation of programs have contributed greatly to the Japanese competitive advantage, the people factor still holds the key. It soon becomes apparent that the employee's work habits, disciplined approach to meeting requirements, adaptability to high technology, loyalty, attention to details, and their active involvement in the management initiated programs are critical to their success.

Training: At every location, training was considered very important. At an electrical equipment manufacturing plant, each newly hired engineer underwent six months of practical training before being assigned to a specific job responsibilities. The same practice was followed at an integrated steel plant.

Safety: The Japanese industries have an outstanding safety record. Excellent housekeeping and physical exercise before the start of work may be the contributing factors. Their safety record is measured, not by manhours without disabling injury, but by manyears. At one large manufacturing plant there was no reported injury during the last five years.

Accountability: Lifetime employment increases the employee's accountability for his actions. Career progress is slow. Frequent "Job Hopping" is out of the question. Publicizing objectives and routinely displaying progress against objectives adds to the peer pressure to meet objectives.

Quality Awareness: Japanese management is a great believer in "Building Quality Into the Product". This also means that the primary responsibility for meeting requirements lies with the production worker and not the quality assurance personnel. Slogans such as **"Check and Refresh"** encourage employees to recheck the job requirements rather than relying on memory.

At another manufacturing plant, the management stressed **STEP** program which stood for:

Service
Technology
Export
Profit

Employee Involvement: At every facility, there were formalized and active programs to encourage employee participation in quality and productivity improvement. For instance, at a steel plant there were 330 quality circles working on the problems or improvement suggestions coming from the workers. Each circle generally had six to seven employees. The quality circles received an average of one suggestion per month from every worker. The employee was awarded ¥ 200 ($1.33) for submitting a suggestion. The reward went up to ¥ 500 ($3.33) if the suggestion was implemented. If the suggestion produced a substantial benefit, the worker could benefit by as much as ¥ 100,000 ($667). Thirty percent of all suggestions received were implemented.

Productive Hours: In general, the Japanese worker spends significantly more hours on the job compared to his counterpart in the U.S.. For example, at the integrated steel plant, each employee worked an average of 2100 hours/year. Although the workers were entitled to 20 days of paid vacation, it was reported that only 50 percent of the workers and an even smaller number of professionals and managers took full vacation. When asked to explain why the Japanese worker does not take the full vacation to which he is entitled to, one engineer replied **"That is our tradition"**. It was obvious that peer pressure, and the fact that each hour of overtime is a paid overtime, are important considerations. In view of the high cost of living, additional compensation through paid overtime is an economic necessity for some.

Employee Productivity: The key factor of high productivity still is the industrious worker. He is disciplined, well trained, loyal and hardworking. At a large turbine manufacturing facility it was reported that between 1975 and 1986, the production of equipment went up by 150 percent, while the plant population decreased by 47 percent. It is recognized, however, that the use of Robots and FMS also contributed to the productivity increase.

SUMMARY AND CONCLUSIONS

The discussions with the Japanese managers and employees, as well as the plant tours, leave no doubt about the manufacturing superiority the Japanese have attained. The rapid and continued growth of Robotics, FMS, OMMM, and particularly the productivity of the Japanese worker, are the key factors to their success.

The salaries of the average productive worker in Japan appear comparable to that in the U.S. The same can be said about engineers and managers.

A comprehensive, long range strategy to upgrade manufacturing technology and its effective execution, in a step by step manner, has improved their competitive position. The workforce is disciplined, industrious, and is well adapted to high technology.

Their capital investment for modernization is high. The results are improved productivity, reduced production costs, and shortened manufacturing cycle time. This in turn has helped increase the business volume and provided adequate return on investments.

Preliminary Study on Design Methods for Large Scale Systems Applications: Integrating Distributed–Problem–Solving

B. Huet

Université Paris XIII, Faculté de Médecine, Département Biostatistiques et Informatique Médicale, 74 Av. M. Cachin, 93012 Bobigny Cedex, France

I - INTRODUCTION

The integration of data processing and knowledge engineering needs a model and a method for large scale systems (LSS) applications.
The model is based on control loop feed-back constituted with a functionnally integrated non algorithmic operator type activating a sequence of algorithmic operators types. A higher structure is built by a hieraracy of cooperative and antagonistic loops regulated by a balance (Huet B. - 1987).

It is necessary to drive in coordination on one hand non algorithmic operator type design and on an other hand the algorithmic operator type design of this model, i.e. a total design method is constituted with two functionnally integrated modules:

- a design method for the algorithmic operators subsystem.
- a design method for the non algorithmic operators subsystem.

The aim of this paper is to present the premisses of this total design method.

II - METHOD

The necessary integration of these two design methods imply, according to us is to build them on a conceptual common basis to result to coherent doublets.

This conceptual common basis means that both methods project the same "view of the world", i.e. design tools have to be based on the same level of reductionnism opposite to perceived reality.

For example at a semantical level a coherent total design method is constituted with a semantics based design method for the algorithmic subsystem and a semantics based design method for the non algorithmic operator subsystem.

It is surely necessary to dispose of a family of design methods to allow various cases for LSS applications.

Several good operational tools exist already for the design of the algorithmic subsystem (Information System design methods = ISDM); the main ones were evaluated by several I.F.I.P. workshops (1982) (1983), we consider it is necessary to type them to assign each one within coherent doublets.

There is today no design method for Expert-systems (ES), however it is surely possible to type E.S. to assign each one to doublets.

This common typing on the same criterium result to coherent design methods for the whole system.

* Arguments.

** The information systems and expert-systems relationships.

In the objective of a technological implementation, a subsystem constituted with algorithmic operators induce our thinking towards "classical" computers while a subsystem constituted with non algorithmic operators induce our thinking towards expert-systems.

Is it correct?

Moreover:

Do information systems design methods cover exactly the subsystem of algorithmic operators?

Do E.S. cover exactly the subsystem of non algorithmic operators?

We highlighted that according the tools of the design method and the definition of E.S., information systems relate with expert-systems by complementation or by intersection (Huet B. - 1986)

In the following paragraphs, we consider only ISDM and ES in strict complementation.

** Typing of the information systems design methods (ISDM).

We consider that the selected ISDM by IFIP TC8 (1982) (1983) are good operational tools which allow to implement algorithmic operators in organizations.

The chronological analysis of these ISDM exhibit that the older ones are the functional methods, followed by the process-based-methods, followed by the semantics-based-methods.

All these tools differ by a more or less reductionnist view of perceived reality.

However this variety is necessary because, the world is complex and organizations are more or less well known, with various degrees of details, in a manner that it is necessary to dispose of a family of design methods allowing to integrate more or less of the perceived reality of the organization to design.

These design methods can be classified on the following manner:

semantics-based-methods	"less reductionnist view of the world" ↑
process-based-methods	↕
functional methods	↓ "more reductionnist view of the world"

III - RESULTS

* Information Systems design Methods selected by IFIP (1982) (1983) were typed on this criterium (Huet B. - 1986):

- three methods are based on functional analysis: D2s2, Isac, Sysdoc

- seven methods are based on processes or Petri nets: Dades, Draft/prop., Edm, Iml, Sdla, Use, Merise

- four methods are based on semantical modelling: Remora, Cim, Acm/pcm, Niam

* The typing of expert-systems is more difficult.

According Gallaire (1987), "knowledge representation is a major subject of research far from being solved".
An E.S. must dispose of the necessary knowledges for problem-solving, this implies the description of its universe of working, i.e. a representation of the objects of the real world involved in the problem solving and of the relationships between them.

This representation is possible under various types,"No universal solution to the problem of knowledge representation is outlined on the horizon, it is then one of the key problems in A.I." (Gallaire H. - 1987)

What are in these conditions the Criteria for a good typing?

Gallaire (1987) thinks that there are several possible criteria: power of expression, efficiency, nature of representation,...
By nature of formalism, he means "the more or less large possibilities that it supplies to represent what we want explain and,..., the easiness it gives to guide the choice of what it is useful to represent".

In these conditions a first typing based upon a more or less reductionnism view is interesting.

We have demonstrated a correspondance between knowledge representation of ES and basis of design of ISDM (Huet B. - 1986)

* At the semantical level there is an identity of content of representation between semantical modelling of ISDM (Remora, Cim, Acm/pcm,...) and cause effect relationship networks as Casnet

* At the process level, there is an identity of content of representation between process-based-ISDM (Sdla) and rule-based-ES (Mycin).

* At the functional level, there is a correspondance between function-based-ISDM (Sysdoc) and the recent propositions of Levesque for a functional view of ES (1985).

It is then possible to highlight three sets of coherent tools:

type of modelling	Information Systems design method type	Expert-systems type
Semantics	Remora	Casnet
Process or Petri nets	Sdla	Mycin
Functional	Sysdoc	Levesque

Moreover we looked for an other typing oriented towards tasks.

In 1982, Stefik and coll. published a tutorial on Expert-Systems organization; after the observance of a lack in ES design methods, they propose six types of tasks for ES: interpretation, diagnosis, monitoring, planning, prediction,design.
We highlighted (Huet B. - 1986) such a second typing

IV - UNSOLVED PROBLEMS

They concern almost exclusively the non algorithmic operators subsystem:

* Dedicated conceptual tools.

We plan to go further in three steps:

- definition of a descriptive model
- choice of parameters for normalization
- definition of a conceptual model

* Analysis of a hierarchy of levels.

* Parallel architecture and learning

* Level of antagonistic subsystems

V - CONCLUSION

This preliminary study has highlighted the following elements:

- necessity to build coherent doublets constituted with one information systems design method (ISDM) plus one non algorithmic operator subsystem design method;
- many good operational ISDM are available;
- necessity to define conceptual tools for non algorithmic operator subsystem design method.

REFERENCES

Gallaire H. (1987). La représentation des connaissances. La recherche en intelligence artificielle , coll. Point Sciences, 265-283

Huet B. (1986). Basic propositions for an intelligent system design. Proc. Ist IFIP, IFAC Int. Conf. Economics and A.I. Sept. 1986, Aix en Provence, France, 93-98

Huet B. (1987). Integration of problem-solving-entities and processing entities in a coherent model; basis of a method of application. Proc. 1987 A.C.M. XVth annual Comp. Sc. Conf. Feb. 1987, St Louis, USA, 376

I.F.I.P. (1982). Information Systems design methodologies: a comparative review. Proc. IFIP TC8 Conf. May 1982, Noordwijkerhout, ND, North-Holland: Amsterdam.

I.F.I.P. (1983). Information Systems design methodologies: a feature analysis. Proc. IFIP TC8 Conf. Jul 1983, York, BG, North-Holland: Amsterdam.

Levesque, H.J. (1984). Foundations of a functional approach to knowledge representation. Artificial Intelligence . 23 , 155-213.

Stefik, M., J. Aikins, R. Balzer, J. Benoit, L. Birnbaum, F. Hayes-Roth, and E. Sacerdoti (1982). The organization of expert-systems, a tutorial. Artificial Intelligence . 18 , 135-173.

Producibility Awareness

Gary L. Hurd

Manager, Advanced Manufacturing Engineering, Ciper Data Products, Inc., San Diego, CA 92131, USA

DESIGN FOR ASSEMBLY -- This sounds like the subject of a seminar talk, doesn't it?--But it's not! Design For Assembly is a topic of national defense.

A while back a fictional TV series titled "Amerika" was broadcast portraying the great United States of America conquered by apathy and lack of action, then totally taken over by a foreign power. I submit that we're deep into an economic war. It's not fictional like TV; it doesn't have bullets, but it does have casualties (watches, cameras, televisions, VCR's, radios, power tools, steel, automobiles, electronic components, IC's, disk drives, etc.).

My task is to REACH you. If you've already been reached and believe in producibility, then my task is to MOTIVATE you--to send you back to your workplace to motivate others.

Well, how do I do that? How do I reach you? --

How many of you like you boss, or your boss' boss, etc., sending in some outside expert to tell you how to do your jobs. After all, the expert doesn't know your product, history, or your company's internal political constraints. How dare he? This scenario happens all too often and it builds an instant wall between us and the outsider. Even if he did have good information for us, under these conditions it seldom makes the transition and even less often gets implemented with the wholehearted effort generally required for success. I would like us not to build that wall today--we don't have time to tear it down!

I would like for you to answer a couple of questions for me, but for now answer them only in your minds.

Q: How much training did you get to be a good American?
A: I would call that your primary education, so it's about 12 years.

Q: How much training did you get to become a competent engineer or manager? (After all, this occupation is a big chunk of your life.)

A: For most engineers it is probably 4-6 years of college.

Q: Now, with a show of hands, how many people are or have been married?

A: Most of us.

Q: How many got training to be thoughtful, caring, loveable, competent spouses before you got married?

A: Almost nobody--It's generally shocking to realize what a significant portion of our lives we go through by trial and error, with little or no training.

Q: How many of you have children?

A: Almost everyone again.

Q: How many received formal training on how to be a good parent before that first baby arrived?

A: Again, almost nobody--And yet our every action in that young child's mind is stored away forever and has far reaching ramifications.

Answering these questions negatively for myself struck an emotional nerve, but it didn't trigger any of my normally over-active defense systems. I hope you can tell yourself the samething I did--"It's not my fault, nobody taught me, I didn't even know there was training."

The same condition was true for my entire engineering training. All I was ever taught or learned to do was to create parts. Never during my education did anyone teach me how to eliminate parts.

Generally the ultimate part or product depends--as do most things--on your point of view. If we make our examination from the production department's point of view, the situation normally looks like this: "Who does engineering think they are--don't they know we're their customer? We have to live with their product as long as it's being produced." Occasionally they accuse engineering of sarcastically quoting the designers' motto:

It can't be formed
It can't be milled
I'm sure it can't be cast
Eureka!
Success at last.

Sometimes manufacturing feels like the only good part is the one that's not there. After all, that's the only part that never gives them any problems.

There needs to be some middle ground to bring development and production together. We must unify in the common goal of getting product from the drafting board to the customer as quickly as possible, at the highest quality and the lowest cost possible. Marketing is development engineerings' customer, and after all, marketing has product specifications that must be met or there's no product. Some companies have design manuals for development engineering, but rarely do production and manufacturing input to the specifications. These production departments need producibility guidelines that can aid engineering in designing a product that both meets the marketing specifications and that can be built. It is imperative that management support such guidelines. Guideline compliance often needs to compete on equal terms along side product specifications when conflicts with the project schedule arise. Guidelines need to be clear, concise, non-restrictive, user-friendly, and to minimize individual interpretation. If you can't do this, if your guidelines resemble the Los Angeles yellow pages, they will wind up on a shelf gathering dust, or in the back of a filing cabinet bottom drawer--or, in the trash. My best example of this is the package of three manuals that must be digested before you can use a new personal computer. You have to wade through a 2" installation manual, a user's guide, and a DOS manual. The first time I had to go through that I ended up frustrated, and hated the company that produced the PC. That's not the emotion which should exist between engineering and production.

Producibility is a very old concept that employs nothing but common sense and common-place techniques. Designing for Producibility is really just optimizing the design for the production process that will bring the design to its final state. Processes such as parts fabrication, parts inspection and gauging, assembly tooling, assembly, product testing, and service should all be considered.

It takes a little more effort to consider those aspects of a product if you're not used to it, but like anything that's worth going the extra mile for, there are many significant reasons or benefits for doing so.

A primary benefit of producibility is a significant increase in product quality. One time in a production issues meeting a development engineering vice president said, "If manufacturing just knew their jobs, they could build our design at the quality and cost standards it was designed for." This brings up a significant issue that must be understood if total quality on a product is ever to be achieved. Actually it's not an issue, it's a fact! QUALITY cannot be BUILT into a product--it must be DESIGNED IN. Manufacturing can destroy all the quality that engineering designs into a product, but they can't build in 1/10th of 1% more quality than the design allows. If you think seriously about that for a minute, you'll know it's true. If any of you saw the academy award-winning best picture of the year "Chariots of Fire," you may recall Mr. Abrahms' coach, Sam Masibini. Sam was talking over his coaching terms with the hopeful track trainee, Mr. Abrahms, when he said, "You can't put in what God's left out." The same is just as surely true of quality with respect to the design.

It's in this "designed-in quality" concept where the serious problems occur for the design engineer. Traditional mechanical engineering education has had no way to define or quantify the quality or producibility of part or product design. It can't be done alone with Roark's stress and strain equations, or a finite element analysis. Neither La place transformation, nor a fourier series, nor a pretty CAD drawing will resolve the answer to this aspect of the design status. The most significant reason for the dilemma is that quality is more than the functional characteristics we were taught to calculate; it encompasses HOW we do our job, or make our part, or assemble our product. All those HOW's are really "the process" by which we produce. If there were a quality design criteria that engineering could use, it would be a list of process considerations--it might even be called Producibility Guidelines!

Another benefit of designing for producibility is a significant decrease in the direct and indirect costs. These cost reductions are generally attributable to a dramatic reduction in the number of mechanical parts in the product. When properly using the producibility guidelines on products or subassemblies that have been designed by traditional engineering principles, we have been able to reduce the mechanical parts count by 30%-40%.

The significance of reducing the number of parts in a product while maintaining all the functional characteristics is literally awesome. You don't have to tool up or maintain molds for parts that don't exist. It doesn't take any labor to assemble the part that is not there, and you don't have any problems with it--no vendor, workmanship or field failures. The part that's not there is the only part that does its job perfectly every time. It never fails or causes something else to fail; the deliveries are never late, it costs nothing to produce, and very little to design. When we design a part into a product we always think it is necessary, but rarely do we consider the business consequences of each part. Think about these--designers design the part and the inevitable changes; drafters draw it; checkers review the drawing; documentation must maintain the drawing; material planners must plan the production requirements; purchasing agents buy it; receiving checks it in against the purchase order; receiving inspection must check it to the print; stockers must put it up in the stockroom; kitters must pick it for production; kit auditors must confirm that it's in the kit; trackers in production control monitor the kit status;

expediters are required, miraculously, to get the parts that don't show up when they should; assemblers put the part on the product; and inspectors confirm the assembler put it on the product without damaging it. This list is just a highlight of sixteen functions (mostly overhead) triggered by <u>every</u> part. It's no wonder our factories have tremendous overhead rates!

My last point, in why we should design for producibility, is short--<u>it's our responsibility</u>. It's our professional responsibility to design the best parts and products that our companies can get. After all, they're paying for it and think they're getting it!

The natural reaction to all this is often, "We always design our products with all those things in mind--he must be talking about somebody else's development engineering group." You may not see the actual situation in your company as being exactly this way, but how does this real scenario sound? A product is designed for a 12 hour labor standard and a given material cost. When pilot production started, it took <u>THOSE PEOPLE</u> 100 hours to build it. Somewhere in the second year <u>THOSE PEOPLE</u> were still taking 60+ hours to build our design. Well, it's the fourth year and <u>THOSE PEOPLE</u> finally started a labor cost reduction effort and got the labor standard down from 39 to 19 hours. What do you know--it's the sixth production year and <u>THOSE PEOPLE</u> are finally building it in 8 hours--4 hours less than we designed it for. I always knew our design standard was a little fat and they could build it for what we said, and I don't think those 2000 engineering changes they made us process were all really necessary.

How about this product case study: Marketing places orders and a production schedule is laid out on a product that really looks great, except there is only one prototype and the testing program has just started. Full rate production starts a year late and losses are running $1M a month in unfilled orders. The "out of box audit" yields are only 60%. The second production year there is a 100% tool redesign, injection molds, die casting molds, stamping dies, etc.

This happens, in varying degrees, to small companies, large companies; one product after another, over and over and over again. It drives the small companies out of business; the big companies just absorb the losses within the profits of older, established products. Next they try to pass those losses along to all their customers with price increases across other product lines. This is great as long as you're the only game in town--and we used to be--but we are no longer the only game in town! We're no longer the worlds' supplier of choice; our own people are tired of opening a brand new box on game day and finding the football soft and low on air.

So what can we do, what does it take to get our producibility act in gear? Not much, really. We need only to identify the yardstick of design measurement (from the production process point of view) and have the sincere desire, the ownership-abandoning willingness, the MOTIVATION to implement as much of the yardstick into each part, each subassembly, each product, as is technically possible.

There are generally two sources of this motivation. The first is the best, and that is an individual personal desire to design for producibility. The main problem with this source of motivation is that it only randomly affects the entire body of design people in an organization. Because employees come and go, there can be no long term, consistent results. The second source is from management--top down. It generally takes a lot of pressure to cause this to happen because the status quo design methods carry a lot of inertia. If a company's products are high volume (500-1000/hour), that volume itself generally drives the issue. It's exciting for a designer to shave 10 cents off a part and save his company $200,000-$300,000/year. In this arena, engineers even go after pennies.

The other hot button that motivates management is a competitor underselling your marketing group. When upper management sits through a sales status meeting and everyone realizes the competition is selling his version at about the price you pay for materials alone, they get real motivated! The road back from this point is littered with casualties--some, like Black and Decker, recover--but it's a long, slow, expensive recuperation. The good aspect of management motivation is that it lasts forever. Everyone gets on board (or jumps ship) and the company is truly marching forward toward being a world class producer of high quality, low cost, profitable products.

After all of that, the actual Producibility Guidelines are anti-climactic. When we started I said they were all common sense, and they are. The key is implementing the concepts instead of debating them and not using the latitude they allow as an excuse for non-compliance.

There are only eight main areas of guidelines to focus in on and they are:

a. MINIMIZE PARTS by combining mating parts, eliminating fastening hardware and harnesses.
b. DESIGN PARTS TO BE SYMMETRICAL without internal orientations.
c. DESIGN SELF-LOCATING FEATURES to align and locate mating parts.
d. ELIMINATE ADJUSTMENTS and decisions from the assembly process.
e. DESIGN FOR MATERIAL HANDLING by avoiding features that tangle or misfeed.
f. PROVIDE ACCESS FOR TOOLS during assembly, test, and service.
g. MODULARIZE SUBASSEMBLIES so they can be tested separately and moved without coming apart.

h. UTILIZE A MOUNTING BASEPLATE to provide mounting stability and facilitate top down assembly.

In conclusion--and back to the war--it is my hope that we can cast off the national and governmental apathy for our world competitors; cast off our slightly over-inflated feelings of superiority; open wide our eyes and make the decision to survive (economically). Seek out your design engineers who have a basic hatred of parts, who have experience and know how parts are fabricated, assembled and tested; who can apply Design for Producibility principles. Listen to them. Let them teach your development staffs; and if you need help to get it kicked off or with training, there are a lot of us available to help you.

Finally, I would like to leave you with this:

"KNOW AND UNDERSTAND THE ASSEMBLY SEQUENCE, PROCESS AND EQUIPMENT.

ACCEPT RESPONSIBILITY FOR THE ASSEMBLABILITY OF YOUR PRODUCT."

Application of a Stochastic Inventory Model

M. Ali Montazer, Ph.D.[*] and A.J.G. Babu, Ph.D., P.E.[†]

[*]Department of Industrial Engineering and Computer Science, University of New Haven, West Haven, CT 06516, USA
[†]Department of Industrial and Management Systems Engineering, University of South Florida, Tampa, FL 33620, USA

Abstract

The inventory system at the maintenace office of the Amherst Physical Plant of SUNY at Buffalo was investigated. The electrical items were selected for modeling purposes. Analysis of the records showed that demand sizes and leadtimes for those items were probabilistic in nature. A stochastic inventory model was constructed from which the optimal operating policy for those items can be determined.

I. Description of The System Under Investigation

The inventory system under consideration is a stock room located in the maintenance office of the Amherst Physical Plant, State University of New York at Buffalo. Over 9,000 different types of items are stocked. The items are broadly categorized according to different trade categorizations for maintenance of the buildings and facilities of the University's Amherst Campus. They can be best ranked by decreasing stocked quantities as follows: Electrical goods; Plumbing goods; Mechanical goods; Refrigeration goods; Carpentry goods; Masonry goods; Cleaning goods; and Miscellaneous.

There are a number of vendors supplying the orders. The vendors are broken down into two broad categories for the purpose of model development. This was done by recognizing the differences in the contractual obligations to the State of New York. The two categories are as follows:

(i) Contracting vendors who are contracted yearly through the state of New York with considerable price discounts.

(ii) Local and non-contracting vendors who have no obligations to the State of New York, and who operate independently.

Stockroom personnel are to deal with the contracting vendors at all times. However, the non-contracting vendors are occasionally utilized under special circumstances such as emergencies, special projects, and for commodities not included in the contract of the vendors in Category (i) above.

An Advantage of employing the contracting vendors is the price break gained through contracts; however, because many of these vendors are not in Western New York, leadtimes are generally greater than that of the non-contracting vendors.

To reduce the complexity of the problem and to gain a better understanding, the highest usage items (i.e., Electrical items) were chosen for further analysis and eventual model development. Each item is identified by its name and a multi-digit number on a formated inventory card. The card is formated so that it can hold the historical record of demand, leadtime, quantities ordered, prices, tentative reorder point, order quantity, and addresses of prospective vendors who could fill the orders. The records of all Electrical items were studied and it was observed that they fall into two broad categories of high-usage and low-usage. The records of the items belonging to the former category were further studied for demand and leadtime characteristics. It was determined that monthly demand sizes and leadtimes were both probabilistic in nature for almost all the items investigated. It was decided that the operating rule for such items could be most appropriately determined by a stochastic model.

II. Model Formulation

Given that demand and leadtime are uncertain, let their probability density functions (pdf) be denoted by $f(x)$ and $g(t)$, respectively. The probability density function for demand during the lead time (here on referred to as 'leadtime demand') is denoted by $h(v) = f(x)g(t)$, by using the assumption that demand and leadtime are independent. The only necessary information required regarding the leadtime demand distribution are the first four moments, which are assumed to exist and be known. It is also understood that the system under consideration consists of a single installation, i.e., it consists of just one stocking point with a single source for resupply. Customer demands arrive at the single stocking point, and at appropriate times orders are placed with the source to replenish the inventory. The basic assumptions of the stochastic inventory model to be developed are as follows:

1. The unit cost of an item is a constant and independent of the order quantity, Q.
2. Back orders are permitted.
3. The ordering cost, unit inventory holding cost, and unit stockout cost are all constant irrespective of quantity.
4. Lot sizes are purchased rather than produced internally.
5. Demand is uncertain with a pdf $f(x)$.
6. Leadtime is uncertain with a pdf $g(t)$.
7. There is never more than a single order outstanding.

The assumptions stated above were to a greater or lesser extent found to be the characteristics of the situation investigated. The cost parameters involved in this model are only the ordering costs, inventory holding costs, and stockout costs denoted by C_p ($ per order), C_h and C_s ($ per unit per year) respectively. Other cost parameters are independent of Q and R, and will not be included in the model. In a stochastic cost expression one is dealing with expected values of components like demand, leadtime, leadtime demand, and expected number of stockouts rather than simple deterministic values. Due to the unpredictable nature of demand it is usually necessary to carry safety stock (S) in order to reduce the chances of stockouts. However, in model formulation when dealing with probabilistic models, it is convenient to write the safety stock (S) in terms of the reorder point R, by an appropriate transformation. The maximum inventory level during the leadtime period is the reorder point R, which consists of the safety stock (S) and expected leadtime demand. A stockout will occur if leadtime demand is greater than the reorder point R. If leadtime demand is less than the reoder point R, not only will there not be any stockout, but there may even be a surplus. Hence the safety stock can be written as:

$$S = \int_0^{\infty} (R-v)\ h(v)dv = R - \int_0^{\infty} v\ h(v)dv = R - E(LTD) \qquad (1)$$

An important parameter of the cost model is the ordering cost. The cost of placing an order was denoted by C_p. Due to the probabilistic nature of demand all one can say is that the expected number of demands per year is E(D). Since an order is placed after every Q demands (i.e., after Q units are withdrawn), the Expected Annual Ordering Cost, EAOC can be written as:

$$EAOC = C_p \ E(D)/Q \qquad (2)$$

Note, if the demand history is daily, weekly or monthly; the average yearly demand is obtained by multiplying those values by 250, 50, or 12, respectively, to determine the expected annual demand. The second component of the cost expression is the expected inventory holding cost. The inventory holding cost was taken to be C_h. The simplest way of setting up this cost component is to argue that this cost is equal to the cost of carrying one unit in inventory per unit of time (e.g., one year) multiplied by the expected number of units in inventory. However, the expected number of units in inventory is no longer a constant Q/2 as in the case of deterministic models. The expected number of units (per cycle) in inventory for this model is (Q/2 + S) where S is the safety stock as defined in expression (1).

Consequently, the expected number of units in inventory is a random variable; and more specifically it is a function of reorder point R and the expected leadtime demand denoted by E(LTD) as follows:

$$\begin{aligned} E(LTD) &= \int_0^\infty v\, h(v)\,dv \\ &= \int_0^\infty \int_0^\infty x \;\; t \;\; f(x)\; g(t)\; dx \;\; dt \\ &= \int_0^\infty t\; g(t)\; dt \int_0^\infty x\; f(x)\; dx = E(LT)\; E(D) \end{aligned} \qquad (3)$$

where E(LT) and E(D) are expected leadtime and demand, respectively (using the independence of demand and lead time). The Expected Annual Holding Cost, EAHC can then be written as:

$$EAHC = C_h \;\; (Q/2 + \text{Safety Stock})$$

Substituting the mathematical definition of safety stock one obtains:

$$\begin{aligned} EAHC &= C_h \; [Q/2 + R - \int_0^\infty v\, h(v)\,dv] \\ &= C_h \; [Q/2 + R - E(LTD)] \end{aligned} \qquad (4)$$

The next step is to find the final cost component of the model, which is the expected annual stockout costs. The average number of stockouts incurred per year is nothing but the expected number of stockouts in each cycle multiplied by expected number of cycles per year. Specifically, this cost is a function of random variable, v representing leadtime demand where v itself is the convolution of two independent random variables x and t, denoting demand and leadtime, respectively. Thus the stockouts, S can be stated mathematically as:

$$S = \begin{cases} v - R & \text{if} \quad v - R \geq 0 \\ 0 & \text{if} \quad v - R < 0 \end{cases}$$

The expected stockouts per period (year) is therefore:

$$E(S) = \int_0^\infty S\; h(v)\; dv = \int_R^\infty \;(v-R)\; h(v)\,dv \qquad (5)$$

From this, the Expected Annual Stockouts Cost, EASC is written as:

$$\begin{aligned} EASC &= C_s \;\; E(D)/Q \int_R^\infty \;(v - R)\; h(v)\,dv \\ &= Cs \;\; E(D)/Q \;\; E(S) \end{aligned} \qquad (6)$$

At this stage all the components of the Annual Total Expected Cost have been determined and one can write TEC as follows:

$$
\begin{aligned}
TEC &= \text{Expected Ordering Cost} + \text{Expected Holding Cost} + \text{Expected} \\
&\quad \text{Stockout Cost} \\
&= C_p \; E(D)/Q + C_h \; [Q/2 + R - \textstyle\int_0^\infty v \; h(v) \; dv] \\
&\quad + C_s \; E(D)/Q \; \textstyle\int_R^\infty (v-R) \; h(v) \; dv \qquad (7) \\
&= C_p \; E(D)/Q + C_h \; [Q/2 + R - E(LTD)] + C_s \; E(D)/Q \; E(S)
\end{aligned}
$$

To compute E(S), first four moments of leadtime demand are obtained from the first four moments of leadtime and demand using the formulas presented below, see Kottas and Lau [1,5]. Let the leadtime demand, v be defined by:

$$v = \sum_{i=1}^{t} X_i$$

where t is a random variable denoting leadtime in days (or months, etc.) and X_i, a random variable, represents the demand size in the ith day of leadtime, t. Then the first four moments for v are written in terms of the first four moments of demand and leadtime as follows:

$$
\begin{aligned}
\mu_v &= \mu_x \; \mu_t \\
\mu_2(v) &= \mu_x^2 \mu_2(t) + \mu_2(x) \; \mu_t \\
\mu_3(v) &= \mu_x^3 \; \mu_3(t) + \mu_3(x) \; \mu_t + 3\mu_2(x) \; \mu_x \; \mu_2(t) \\
\mu_4(v) &= \mu_x^4 \; \mu_4(t) + 6\mu_x^2 \; \mu_2(x)(\mu_t \mu_2(t) + \mu_3(t)) + 4\mu_x \mu_3(x)\mu_2(t) \\
&\quad + 3(\mu_2(x))^2 \; (\; \mu_t^2 - \mu_t + \mu_2(t)) + \mu_4(x) \; \mu_t
\end{aligned}
$$

These four moments can then be used to approximate the density function, h(v), by the Schmeiser-Deustch family of distributions [4] which has four parameters; a, b, c, and d. To apply the S-D distributions to a data set, $(x_1,\ldots,x_n)$ the four parameters should be estimated. The process of estimating the four parameters requires that c and d be estimated first. To estimate c and d, measures of skewness and kurtosis of the data are needed. Having these two measures, a nonlinear programming problem can be solved to find accurate estimates of c and d. However, a somewhat less accurate but faster and more convenient approach is to obtain two new measures, one being in terms of skewness and the other in terms of kurtosis. Having obtained the two measures, parameters c and d can be obtained from a graph. The first (a) and second (b) parameters are then easily estimated from formulas which are written in terms of c and d (see Schmeiser and Deustch [4]).

Expected stockouts, E(S) can now be computed as in equations (9) and (10) suggested by Kottas and Lau [2]. Note that a,b,c, and d are the four parameters of the Schmeiser-Deutsch family of distributions while R is the reorder level.

If $R \geq a$, then

$$E(S) = (a-R)(1-H) + (b/(c+1)) [(1-D)^{(c+1)} - (H-d)^{(c+1)}]$$

$$\text{where} \quad H = d + [(R-a)/b]^{(1/c)} \tag{9}$$

If $R < a$, then

$$E(S) = (a-R)(1-H) + (b/(c+1)) [(1-d)^{(c+1)} - (d-H)^{(c+1)}]$$

$$\text{where} \quad H = d - [(a-R)/b]^{(1/c)} \tag{10}$$

Having obtained E(S), total expected cost, TEC is nothing but a nonlinear function of two variables (Q and R), after all other parameters are provided. To obtain the mimimum cost, the partial derivatives of TEC with respect to Q and R are equated to zero and solved to find the optimal values of Q and R which are denoted by Q^* and R^*, respectively. Due to the limited space the optimization steps are not given here. A sensitivity analysis was also performed to see how the optimal solution changes with respect to small changes in the cost parameters. For the details of the optimization steps and the sensitivity analysis, one should refer to Montazer [3].

III. Summary

Inventory systems of the Amherst Physical Plant (at SUNYAB) were studied. A stochastic inventory control model was developed based on the characteristics of the system. In the model demand and leadtime were treated as uncertains, each having a probability distribution independent of one another. Hence, the first four moments of leadtime demand distriution were computed using the method proposed by Kottas and Lau [1,5], from which the leadtime demand distribution was approximated using the system of four parameters of Schmeiser and Deustch [4]. Finally an annual variable cost function in terms of the order quantity, Q and reorder level, R was developed, taking only the relevant cost parameters into consideration, namely ordering costs, holding costs and stockout costs. The cost function was then minimized by taking the partial derivatives with respect to Q and R, setting the results to zero, and then solving for the optimal order quantity, Q^*, and reorder point, R^*. A computer program (coded in Fortran IV, and tested on CDC CYBER 174) was developed (see Montazer [3]) which can be used interactively to determine Q^* and R^* for an item. The input requirements of the program is minimal and one needs only to provide the relevant cost parameters and the frequency distributions of demand and leadtime corresponding to an item. The frequency distributions of demand and leadtime could be in terms of either days, weeks or months. Finally optimal policies for a number of items were determined and compared with the operating policies presently used by the system.

References

1. Kottas, J.F. and Lau, H.S. Inventory Control with General Demand and Leadtime Distributions. Int. J. Sys. Sci., 1979, Vol. 10, No. 5, pp. 85-492.

2. Kottas, J.F. and Lau, H.S. A realistic Approach for Modeling Stochastic Leadtime Distribution. AIIE Trans.,1979, Vol. 11, No. 1, pp. 54-60.

3. Montazer, M.A. A Probabilistic Inventory Control Model. Unpublished MS Thesis, SUNY at Buffalo, 1982.

4. Schmeiser, B.W. and Deutsch, S.J. A Versatile Four parameter Family of Distributions Suitable for Simulation. AIIE Trans., 1977, Vol. 10, No. 5, pp. 176-182.

5. Wan, W. and Lau, H.S. Formulas for Computing Moments of Stochastic Leadtime Demand. AIIE Trans., 1981, Vol. 13, No. 3, pp. 281-282.

A Just-In-Time/Group Technology Based Scheduling Model

Behnam Pourbabai

Department of Statistics and Operations Research, New York University, New York, NY 10006, USA

Abstract: A mathematical programming model is developed for short term planning and scheduling of a batch manufacturing system. This model is based on the Just-in-Time manufacturing philosophy and the Group Technology. Furthermore, job splitting is allowed and the processing order of jobs is prespecified based on a dispatching rule. In this model, the makespan during the short term planning time horizon is minimized such that production of each job will be equal to its demand.

Key Words: Planning, Scheduling, Lot-sizing, and Group Technolgy.

Introduction

In this paper a mathematical programming model for planning and scheduling of a batch manufacturing system during a short term planning time horizon will be developed. This model will be based on the Just-in-Time manufacturing philosophy and the Group Technology. Furthermore, job splitting will be allowed and processing order of jobs will be specified based on a dispatching rule. In this model, the makespan during the short term planning time horizon will be minimized such that production of each job will be equal to its demand. For a review of some related models, see [1 to 4].

The above model can then be embedded as a component of a decision support system for on-line control of the operational activities of a compatible manufacturing system. The advantage of such a decision support system is that if the input parameters of the system (e.g., the processing priorities, the demand order sizes, processing capabilities, etc.) change, the decision makers will be able to respond to those changes at the beginning of each planning period by using the decision support system as often as necessary.

In this paper, it is assumed that the batch manufacturing system consists of a set of heterogeneous work stations linked by a compatible material handling system. Furthermore, the following assumptions regarding the operational characteristics of the system are made; 1) The planning and the scheduling time horizon should be selected to be sufficiently short, to enable the decision makers to explicitly or implicitly consider all the changes due to the internal and the external factors at the beginning of each planning period. Hence, new operational plans can be generated as frequently as they are needed; 2) Based on the bill of materials, the following parameters can be identified; i) the number of units of each part. (Notice that each part type could be used in different products. In this paper, the term "job" will be referred to a batch of identical parts); ii) the due date associated with each job; and iii) the availability time of each job. (That is the earliest time that each job can be considered for processing. This parameter can be influenced by the availability of the raw material and the process plan of the corresponding part); 3) The marketing strategy is based on the Just-in-Time manufacturing philosophy. That is, no job will be considered for manufacturing unless its potential customer is identified. It is noted that this perspective has many implications in the planning and organiztional stucture of the system. For a review of the basic idea of this concept and its applications, see [5]; 4) Each work station is designed based on a Group Technology approach. That is, at each work station, the number, the type, and the layout of the machines for processing a compatible group of parts belonging to the same family of parts are specified. Furthermore, each job can be processed at least at one of the work stations. This assumption enables us to treat a job, which may require multiple stages of operation at different machines of each work station, as a job which only requires a single stage of operation at that work station. That is, the Group Technology assumption enables us to require only an estimate (e.g., preferably the expected value) of the total required processing time of each job at each compatible work station. Hence, there is no need to keep track of the processing times for all the necessary stages of operation of each job at each compatible work station. For a review of the related literature, see [6]; 5) If there are jobs which can be processed at compatible (but not necessarily identical) work stations, job splitting will be allowed. That is, each batch of identical parts can be split among all the compatible work stations which are individually capable of processing all of its necessary stages of operation; 6) The processing order of jobs are prespecified according to a dispatching rule during the short term planning time horizon. The reason for specifying the processing order of jobs is that, because of the dynamic nature of the manufacturing process and some of the internal and the external factors which were described before, the recommended operational plan by the operation managers may not be implemented by the shop floor supervisors at the floor of the factory. That is, even if a realistic (e.g., an optimal) scheduling and loading plan could have been developed by the operations managers, because commonly by the time the plan is going to be implemented, the parameters used for generating it may change, the operational schedule may become nonimplementable. For this reason, it is recommended that during the short term planning time horizon, the processing sequence of jobs be specified (e.g., preferably based on a dispatching rule which is mutually acceptable by the operations managers and the shop floor supervisors).

Based on the above technological and operational characteristics, a mathematical programming model will be suggested to minimize the makespan such that production of each job will be equal to its demand.

The organization of this paper is as follows; in section 2, the components of the model will be delineated; and in section 3, the concluding remarks will be discussed.

2. The Model

Throughout the remainder of this paper the following notations will be used.

2.a. Notation

Let, i (1 to N) be the job index; j (1 to M) be the work station index; J_{ij} be the job i which has to be processed at the work station j; L_i be the number of units in the job i; a_i be the availability time of the job i; t_{ij} be the required time to perform all of the necessary stages of operation of each unit of the job i at the work station j; D_i be the due date of the job i; s_{ij} be the required time for setting up the work station j for processing job i; X_{ij} be the number of units of the job i to be produced at the work station j; Z_{ij} be the completion time of the job i to be produced at the work station j; $Y_{ij} = 1$, if $X_{ij} > 0$, otherwise $Y_{ij} = 0$; α_{ij} be a sufficiently large constant (e.g., $\alpha_{ij} = L_i$); β_{ij} be a sufficiently small constant (e.g., $\beta_{ij} = 0$); and W be the makespan during the planning time horizon.

2.b. The Model Components

Herein the basic idea of allocating the indices of the parameters and the decision variables of the model according to a general dispatching rule is described. Then, the components of the model are described.

In general, if any job can not be processed at any work station, delete its corresponding parameters and decision variables from the model. Then, assign the indices to the parameters and the variables according to a desirable dispatching rule. More specifically, if a job has the highest [lowest] processing priority among the jobs, assign the index 1 [N] to that job, and in the same way, for the job with the next highest [lowest] processing priority, assign the index 2 [N - 1] to that job, etc. It is recommended that the priorities of the jobs during the short term planning time horizon should be assigned based on an acceptable agreement between the operations managers and the shop floor supervisors. After appropriately allocating the indices among the jobs, the processing order of jobs at each work station will be as follows; $J_{1.}, J_{2.}, \ldots, J_{N.}$.

Now, to obtain the optimal lot-sizes such that the makespan during the short term planning time horizon will be minimized while the production of each job will be equal to its demand, the following model is developed.

$$\text{Min } W \qquad (1)$$

Subject to

$$W - Z_{Nj} \geq 0 \qquad j = 1 \text{ to } M \qquad (2)$$

$$\sum_{j=1}^{M} X_{ij} = L_i \qquad i = 1 \text{ to } N \qquad (3)$$

$$Z_{ij} - (t_{ij} X_{ij} + s_{ij} Y_{ij}) \geq a_i \qquad i = 1 \text{ to } N,\ j = 1 \text{ to } M \qquad (4)$$

$$Z_{ij} - (t_{ij} X_{ij} + s_{ij} Y_{ij}) \geq Z_{i-1,j} \qquad i = 1 \text{ to } N,\ j = 1 \text{ to } M \qquad (5)$$

$$\beta_{ij} Y_{ij} \leq X_{ij} \leq \alpha_{ij} Y_{ij} \qquad i = 1 \text{ to } N,\ j = 1 \text{ to } M \qquad (6)$$

$$X_{ij},\ Z_{ij}, \text{ and } W \geq 0 \qquad i = 1 \text{ to } N,\ j = 1 \text{ to } M \qquad (7)$$

where, $a_0 = 0$ and Y_{ij} is a binary variable (8)

In the above model, Z_{Nj} in constraint (2) identifies the completion time of the last job (e.g., N) to be processed at the work station j during the planning time horizon, and W indicates the makespan (e.g., the completion time of the last job) during the short term planning time horizon: then, the makespan is minimized in expression (1). Constraint (3) guarantees that production of each job at all work stations will be equal to its demand. Constraint (4) indicates that the completion time of each job must be greater than or equal to the summation of its availability time, its total required processing time, and its required setup time. Constraint (5) shows that the completion time of each job also must be greater than or equal to the summation of the completion time of its preceding job, its total required processing time, and its required setup time. Constraint (6) is given to serve three objectives; Firstly, to prevent consideration of the setup times in the completion time of the jobs which will not be manufactured at the corresponding work stations. That is, if $X_{ij} = 0$, then $Y_{ij} = 0$; Secondly, to identify the work stations which will be used for processing of each job; Thirdly, to specify the preprocessed inventory capacity of each work station by appropriately selecting α_{ij}. Furthermore, the value of β_{ij} can appropriately be specified such that the allocation of unwanted small batches to each work station is prevented. In addition, α_{ij} and β_{ij} could also implicitly indicate the limitations of that segment of the material handling system which transport parts to the work station j.

It is noted that in the above mixed binary linear programming model, X_{ij} should be an integral variable. However, based on our numerical experimentation, it has been concluded that to improve the computational performance of the model, X_{ij} should be treated as a continuous variable. Our numerical results indicate that the above relaxation significantly reduced the required computation time. For a review of the mixed binary linear programming algorithms, see [7].

2.c Simplification

In this section, to simplify the model, given that the availability time of each job is equal to zero, Z_{ij} variables are deleted from the model and the constraints 2, 4 and 5 can are replaced by the following constraint,

$$W - \sum_{i=1}^{N} (t_{ij} X_{ij} + s_{ij} Y_{ij}) \geq 0 \qquad j = 1 \text{ to } M \qquad (9)$$

It is noted the above simplification reduces number of continuous variables from (2MN+1) to (MN+1), and also reduces number of constraints from (M + N + 2MN - 1) to (N + M).

3. Concluding Remarks

In this paper a decision making model for planning and scheduling of a batch manufacturing system is suggested such that the makespan is minimized during the short term planning time horizon while production of each job will be equal to its demand. The unique features of this model are; Firstly, the model is based on the Just-in-Time manufacturing philosophy which in the recent years has become a popular concept for low to medium volume manufacturing; Secondly, the model is based on the Group Technology concept, which has also become widely accepted in various modern manufacturing environments; Thirdly, the processing order of jobs can be changed as frequently as necessary. That is, various dispatching rules can be considered; Fourthly, the limitations of the material handling system and the local inventory of each work station can be considered; Fifthly, setup times are explicitly considered and job splitting is allowed; Sixthly, aggregate scheduling for combining all of the necessary stages of operation of each job at each work station is permitted; Finally, a relaxation method and a simplification perspective are suggested to improve the computational performance of the model.

The model which is presented in this paper or its enhanced version in conjunction with its corresponding mixed binary linear programming subroutine (which is commercially available world wide) should be embedded in a compatible computer support system for on-line control of the operational activities of a compatible batch manufacturing system.

REFERENCES

1. Pourbabai, B., Scheduling Models for Combining and Splitting Lots, Studies in Management Sciences and Systems Series: Production Management: Methods and Studies, (Ed. B. Lev), Elsevier Science Publishing Co., Inc., 1986-a.

2. Pourbabai, B., An Operation Model For Automated Batch Manufacturing Systems, Proc. of V IFAC/IFIP/IMAX/IFORS Symposium on Information Control Problems in Manufacturing Technology Robotics, and Flexible Manufacturing Systems, (Ed. Kuznetsov), Pergamon Press, 1986-b.

3. Pourbabai, B., A Production Planning and Scheduling Model for a Flexible Manufacturing System, Proc. of the 2nd ORSA/TIMS Conference on Flexible Manufacturing Systems: Operations Research Models and Applications, (Eds. K. Stecke and R. Suri), Elsevier Science Publishing Co., Inc., 1986-c.

4. Kabak, I., and B. Pourbabai, A Just-in-Time Aggregate Lot-Size Scheduling Model, Proc. IXth ICPR Conference, Cincinatti, OH, (1987).

5. Schonberger, R. J., Japanese Manufacturing Techniques, The Free Press, New York, NY, 1982.

6. Waghodekar P.H., and S. Sahu, Group Technology: A Research Bibliography, Opsearch, 20 (4), (1983) 225-249.

7. Salkin, H.M., Integer Programming, Addison Wesley Publishing Co., Reading, Massachusetts, 1975.

Single Facility Location Problem With Non–Traditional Cost Function

Wilfred V. Huang*, Rajan Batta†, and A.J.G. Babu‡

*Division of Industrial Engineering, Alfred University, Alfred, NY, USA
†Department of Industrial Engineering, State University of New York at Buffalo, Buffalo, NY, USA
‡Department of Industrial Engineering/Management System, University of Central Florida, FL, USA

INTRODUCTION

The traditional problem of location theory is to seek the location which yields the minimum total transportation costs with fixed demands and known market locations; see Louveaux et al. [1]. For the case of a network Hakimi [2] shows that it is possible to reduce the infinite number of alternatives (any point on the network) to a finite number (nodes of the network). We refer the reader to a recent survey paper by Tansel et al. [3] and the location theory text by Handler and Mirchandani [4] for a review of location problems on networks. For the location problem on the plane we are unaware of a recent survey paper. We refer the reader to the location theory text by Francis and White [5], Hansen et al. [6] and a recent paper by Larson and Sadiq [7].

More recently several authors have emphasized profit maximization instead of transportation cost minimization as an objective. In this context, Thisse and Perreur [8] demonstrated that Hakimi's nodal optimality principle is quite adequate.

The Weber problem is perhaps the best studied of all location problems. Now we shall discuss transportation cost assumptions made in the Weber problem, and motivate the research effort in this paper.

Transportation costs typically depend on:

(i) Distance.

(ii) Transportation mode (e.g., truck, plane, rail or water).

In the Weber problem, typically one of the assumptions is that transportation cost is linearly proportional to distance. Alternatively, this assumption can be stated as the existence of a constant transportation rate. We feel that this is a restrictive assumption in some important real-world applications, which we elaborate below.

The transportation of goods is often a complex operation involving shipment by one or more kinds of goods carrier. For example, trucks, planes, railroads or a combination thereof could be used for transporting goods. On any given type of transportation medium, transportation cost (usually) increases less rapidly than in direct proportion to distance.

Hoover [9] points out that this is primarily because terminal costs (and some other expenses) are independent of the length of the haul. In other words, given a particular transportation medium there is a fixed cost plus a variable cost proportional to distance. This leads quite naturally to a piecewise linear cost function, where transportation modes with heavy investment (or high fixed cost) have long-haul economies (i.e. cheaper for longer hauls), and similarly those with low investment (or low fixed cost) have short-haul economies (i.e. better for shorter hauls).

Weber problem can be formulated as follows:

$$\min_{X \in \mathbf{R}^2} \sum_{i=1}^{n} V_i C_i(d(P_i, X)),$$

where n is the number of demand points, $C_i(.)$ is the cost function of demand point i, P_i is the location of demand point i, X is the location of the facility, d(,.,) is the distance function, and V_i is the weight of the demand point i.

WEBER PROBLEM WITH PIECEWISE LINEAR CONCAVE COST FUNCTION

Formulation

Let the function C be a r-segmented piecewise linear concave function. Then:

$k_1=g_1$,

$k_2=(s_1-s_2)g_2$,

$k_3=(s_1-s_2)g_2+(s_2-s_3)g_3$,

$\vdots$

$k_r=(s_1-s_2)g_2+\ldots+(s_{r-1}-s_r)g_r$.

Then,

$$C(d(P_i,X))=\begin{cases} k_1+s_1d(P_i,X) & \text{if } g_1<d(P_i,X)\leq g_2,\\ k_2+s_2d(P_i,X) & \text{if } g_2<d(P_i,X)\leq g_3,\\ \vdots & \vdots \\ k_r+s_rd(P_i,X) & \text{if } g_r<d(P_i,X). \end{cases} \qquad (E1)$$

where s_i is the slope of line i and g_i is the point when slope changes; $g_1=0$.

So the Weber problem with a piecewise linear concave cost function (P1), can be expressed as a problem with equation E1 as constraints.

Analysis

Draw r-1 concentric circles for each demand point, with the demand point as the center, and radii $g_2,\ldots,g_r$ respectively. Figure 1 shows an example with 3 demand points, 3 transportation rates, and 16 regions. Each region represents a unique set of transportation rates for each demand point. For example, if the facility lies within region #7, the transportation rates for demand points 1,2 and 3 are s_2, s_2 and s_3 respectively.

Proposition 1

If the set of transportation rates $(s_{[1]},\ldots,s_{[n]}$ where [i]=1 or 2 ... or r for i=1,...,n) of all demand points is known, then P1 is a metric constrained Weber problem.

The number of combinations of transportation rates in P1 is finite, and its maximum value is equal to r^n, where r is the number of linear segments of the piecewise linear concave cost function, and n is the number of demand points.

From Proposition 1, we know that there are at most r^n constrained Weber problems to be solved to solve problem P1. Algorithms for solving the constrained Weber problem has been developed by Schaefer and Hurter [10], Hurter et al. [11] and Hansen et al. [12,13].

We would like to point out that a piecewise linear function is monotonically increasing function, and a positive linear combinations of strongly quasi-convex functions need not be strongly quasi-convex. This leads us to the following proposition.

<u>Proposition 2</u>

There may exist many local solutions to problem P1.

Proposition 2 points out that local solutions exist. To get the global optimum all local solutions need to be exhausted.

For a given combination set of transportation rates, we have a constrained Weber problem. A constrained Weber problem is a Weber problem with a constrained feasible region. In three dimension it can be seen that Weber function is a convex cone.

<u>Proposition 3</u>

Objective function of problem P1 consists of convex cones.

<u>Algorithm 1</u>

1. Find the convex hull as formed by $[P_1, P_2, \ldots, P_n]$.
2. Draw r-1 circles for each demand point with $g_1, \ldots, g_r$ (grid points) as the values of the radii respectively. Find intersection points of these circles. Also find intersection points of each circle with the convex hull.
3. Eliminate any intersection points outside the convex hull.

4. Determine the regions adjacent to each intersection points inside the convex hull. Eliminate any duplicate region.
5. Solve a constrained Weber problem for every region obtained.
6. The point with the minimum objective function value is the optimal solution if all regions obtained in 4 are convex.

In step 5 since most of the constraints are nonconvex and non-polygonal, We propose a penalty function method to solve step 5, which need not guarantee optimal, and examples were found in computational experience.

<u>Algorithm 2</u>

Let f(.) be the objective function of Pl.

1. Choose a random starting point X_1.
2. Compute the corresponding set of transportation rates S.
3. Solve an unconstrained Weber problem based on S, to obtain a new point X_2.
4. If X_2 is within the feasible region, that is, the set of transportation rates is still S, Stop. Else go to 5.
5. Find the intersection points, $I_1, I_2, \ldots, I_k$, that separate regions between X_1 and X_2.
6. Line search separately between X_1 and I_1, between I_1 and $I_2, \ldots,$ between I_k and X_2. Find the minimum on each segment: using appropriate transportation rates in each segment. We note that since the objective function is strictly convex on each such segment an exact line search is possible.
7. Let X_3 be the minimum of the minima amongst the segments. If the set of transporation rates associated with X_3 is S, Stop. Else set $X_1=X_3$, and go to 2.

We note that Algorithm 2 possesses a descent property.

Computational Experience

Since there are only two decision variables (coordinates of the facility), exact solutions can be obtained by grid search though it is time consuming and inefficient. Results of grid search have been used to compare with those of Algorithm 1. We tested 15 problems, 5 each for n=10,r=3; n=3,r=2; and n=10,r=2. The computation was done on a DEC VAX 785/RMS. The coding was in Pascal. Average CPU time for the classes of problems tested was as follows:

(n-r)	CPU seconds
(3-2)	1.348
(10-2)	39.362
(10-3)	122.462

Only 3 of the tested problems yielded sub-optimal solutions; the maximum difference in objective function value was less than 5%. Two of these were of the class of (n=3,r=2). We feel that it was due to the nonconvexity of the regions that the penalty function method could not reach the optimal.

Algorithm 2 does not guarantee a global optimum solution. To enhance it, we follow this strategy in its implementation:

1. Start the heuristic method with each demand point.
2. Start the heuristic method with the point obtained by solving an unconstrained Weber problem which uses the average of the transportation rates.
3. Find the minimum amongst solutions obtained in 1 and 2.

Out of 120 randomly generated problems, implementation of the heuristic by the proposed strategy yields the same result as that obtained by Algorithm 1. The generated problems consisted of 30 problems with 3 demand points and 2 transportation rates (n=3,r=2), 30 problems with 10 demand points and 2 transportation rates (n=10,r=2), 30 problems with 10 demand points and 3 transportation rates (n=10,r=3) and finally, 30 problems with 10 demand points and 4 transportation rates (n=10,r=4).

The coding was in Pascal and was run on a DEC VAX 780/RMS. Average CPU time for the classes of problems mentioned was as follows:

(n-r)	CPU seconds
(3-2)	0.4943
(10-2)	4.9936
(10-3)	4.2290
(10-4)	5.6696

Only 2 out of 120 problems gave an optimal solution by step 1 of the heuristic strategy. The average CPU time was as follows:

(n-r)	CPU seconds
(3-2)	0.0076
(10-2)	0.0136
(10-3)	0.0123
(10-4)	0.0120

Only 97 out of 120 problems gave an optimal solution by step 2 of the implementation of the heuristic strategy. The chief drawback of the proposed heuristic strategy is that we have to start the heuristic method with each demand point. Thus its efficiency decreases as the number of demand points increases.

CONCLUSION

We have showed that solving the Weber problem with a piecewise linear concave cost function is equivalent to solving a finite, albeit large, number of constrained Weber problems. Unfortunately, algorithms for solving the constrained Weber problem with nonconvex and nonpolygonal regions cannot guarantee an optimal solution. A heuristic algorithm with a descent property was also proposed, along with suggestions to enhance the quality of solutions obtained by the heuristic. Finally, computational experience was reported, and results obtained did well in finding a "good" quality solution. .LS1

REFERENCES

1. Louveaux, Francois, J-F Thisse and Hubert Beguin,"Location Theory and Transportation Costs," Regional Science and Urban Economics, 12, pp. 529-545, 1982.
2. Hakimi, S. L., "Optimum Location of Switching Centers and the Absolute Centers and Medians of a Graph," Operations Research, 12,pp. 450-459, 1964.

3. Tansel, B. C., R. L. Francis and T. L. Lowe, "Location on Networks : A Survey, part I and II," Management Science, 29, pp. 482-510, 1983.
4. Handler, G. Y. and P. B. Mirchandani, Location On Network, MIT Press, 1979.
5. Francis, Richard L. and John A. White, Facility Layout and Location: An Analytical Approach, Prentice-Hall, 1974.
6. Hansen, P., D. Peeters and J-F Thisse, "Public Facility Location Models: A Selective Survey," Locational Analysis of Public Facilities,eds. J-F Thisse and H. G. Zoller, North-Holland, 1983.
7. Larson, Richard C. and Ghazala Sadiq, "Facility Locations with the Manhattan Metric in the Presence of Barriers to Travel," Operations Reserach, 31, 4, pp. 652-669, 1983.
8. Thisse, J-F and J. Perreur, "Relations between the Point of maximum Profit and the Point of Minimum Total Transportation Cost : A Restatement," Regional Science, 17, 2, pp. 227-234, 1977.
9. Hoover, Edgar Malone, The Location of Economic Activity, McGraw-Hill, 1948.
10. Schaeffer, Margaret K. and Arthur P. Hurter Jr., "An Algorithm for the Solution of a Location Problem with Metric Constraints," Naval Research Logistic Quarterly, 4, pp. 625-636,1974.
11. Hurter, A. P. Jr., M. K. Scharffer and R. E. Wendell, "Solutions of Constrained Location Problems," Management Science, 22, 1, pp. 51-56, 1975.
12. Hansen, P. and J-F Thisse, "Some Localization Theorems for A Constrained Weber Problem," Regional Science, 21, 1, pp. 103-115, 1981.
13. Hansen, P., D. Peeters and J-F Thisse, "An Algorithm for a Constrained Weber Problem," Managaement Science, 28, 11, pp. 1285-1295, 1982.

Coordinates of localities are (2,4), (6,8) and (7,2)

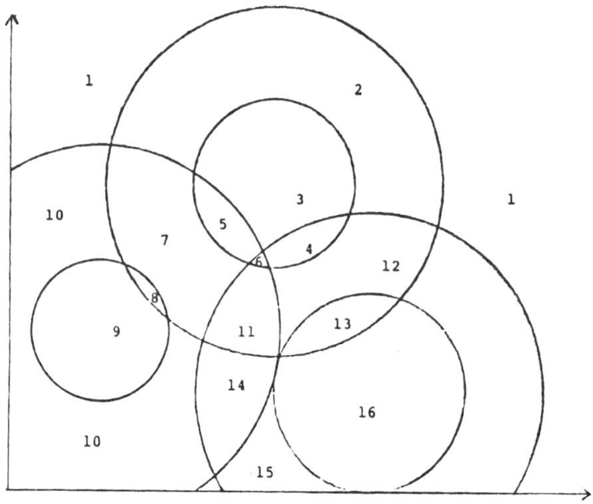

Fig. 1 An Example with Three Localities and Three Transportation Rates (i.e. n=3, r=3)

Cellular Manufacturing—Some Design Considerations

R. Meenakshi Sundaram

Department of Industrial Engineering, Tennessee Technological University, Cookeville, TN 38505, USA

Abstract

Cellular manufacturing is making rapid inroads as a viable production system for improving productivity in discrete lot manufacturing. To make the system operate efficiently, several important factors have to be closely examined at the design phase. The paper outlines some of the important factors one cannot afford to overlook in the design of such a system.

Introduction

Cellular manufacturing may be defined as a production system consisting of several manufacturing equipment of different types to manufacture a family or group of parts requiring similar processing. A cell may have anywhere from two to fifteen machines. Cellular manufacturing is very much prevalent in metal cutting/machining. The concept is also used in certain assembly operations. Cells are the norm in a flexible manufacturing system. Cells for fabrication of families of parts are being developed for sheet metal parts.

According to Huang and Houck [1985], the concept of cellular manufacturing has been around since 1925. The European countries, Russia, and Japan have adopted the concept very well and the American companies have been less enthusiastic in the past. Recent interest in Cellular manufacturing among companies in the U.S. can very well be inferred from the paper by Palframan [1987]. Several companies seem to prefer manufacturing cells over a flexible manufacturing system because of the cost and manageability. Therefore, the author shares the opinion of the marketing manager of a major CNC machine tool maker in the U.S. that there will be lot of growth in the manufacturing cell area. It is forecast by a consulting company that the total U.S. market for cells, flexible manuacturing systems and flexible assembly systems will be about $2 billion in 1990.

With the anticipated growth and interest in the cellular manufacturing system, it is appropriate to develop some guidelines useful for effective designing of cells. The basis for any cell design is the concept of group technology (GT). This concept is based the on sameness principles to help identify parts having the same manufacturing requirements. A coding technique my be used to code the parts to be manufactured and the codes in turn can be used to form part families. Other methods based on process routing are also used for designing cells. Whatever be the method, the cell designed must be able to operate efficiently for the intended objective.

Advantages and Disadvantages of Cellular Manufacturing

One of the primary advantage of cellular manufacturing as claimed by several proponents is the reduction in production time up to 50%. This is possible by the reduction in setup time. Other advantages that are claimed are:

- Reduced lead time
- Reduced in-process inventory
- Smaller lot sizes
- Reduction in number of production equipment
- Reduced labor cost
- Reduced material handling
- Reduced tooling requirements
- Simplified production control procedures
- Better human relations and improved organizational behavior
- Improved productivity
- Better overall control of the operations

Of course, there are some disadvantages one has to be cognizant of when considering implementing cellular manufacturing. They are:

- Increased capital investment
- Lower machine utilization for some production equipment
- Impossibility of having 100% cellular manufacturing

In any case, the advantages of cellular manufacturing far outweigh the disadvantages.

Review of Literature

Several cell design techniques have surfaced for effectively configuring cells. Black and Schroer [1986] have discuss the axiomatic approach for designing manufacturing systems. Included are several corollaries to cover additional aspects in cell design. These axioms pretty much cover all the different aspects of a cell design. Greene and Cleary [1985] discuss about axioms that are especially appliable to designing, controlling and operating of a cellular manufacturing system. These axioms are based on the principle of maximizing the benefits and minimizing the drawbacks. One has to decide whether these axioms are applicable to a particular facility under consideration.

From the literature search on the design of cells, several considerations can be gleaned. One of the documents often referred in the GT literature is the process or route sheet. A process or route sheet is essential in tracing the flow of a part in the shop. By grouping items with similar routing, it will be possible to form part families. Also, the production equipment required for a family can be grouped together to form a cell. It should be emphasized that it is impossible to convert the entire manufacturing operations into cellular manufacturing. Also, some specialized production or inspection equipment have to be shared with several cells in spite of the need for having a smooth flow. Oftentimes production machines such as a heat treatment facility, paint booth or degreaser must be isolated from the rest of the machines in the cell because of their special requirements. This isolation may affect the nature of the flow.

Another major consideration is the maximum number of parts and number of machines to group in a cell. What is the optimum number of machines to have in a cell? The answer depends upon the number of parts the cell is expected to process. The number of parts depend upon the number of machines and the number of machines in turn depend upon the number of parts. The decision is strictly based on the economics involved. A few studies related to this aspect may be found in the literature. The current practice seems to be to select the necessary equipment with the exception of specialized production machines to form a manufacturing cell. A technique such as cluster or rank order cluster analysis may be used to form cells. In order to apply such a technique, the process sheet is mandatory. Application of this technique usually results in exceptional job or bottleneck machines. To handle this situation, several techniques have surfaced in the recent literature on GT. In any case, it is possible to decide on the number of machines and parts usually based on strict economics. In practice, this will make more sense than any other criterion.

Some Overlooked Design Considerations

None of the design techniques currently available handles the alternate routing in considering process routing for the design of cells. Alternate routing, if considered, may result in an efficent production cell for the least cost. Wemmerlov and Hyer [1987] in a recent paper laments the absense of a step-by-procedure to cell design based on process routing. In this attempt, consideration should be given to alternate routing. This approach will also tend to minimize the idle time of production equipment and improve the machine utilization. In using the existing route sheets for cell designs, one has to make sure that the current production methods are in agreement with the route sheets. When selecting alternate routings, availability of the chosen machines for processing must be checked. If not, the possibilites and cost of acquiring them must be assessed. Alternate routing offers a very great potential in improving the cell design.

Another improvement factor often overlooked in the design of cells is the capacity factor. At present cells are formed using any one of the available techniques based on the routing data. Cells so designed are not examined for possible balance of load and capacity. Setup and processing times can be obtained from the route sheet. The number of hours equipment are available must be known to compute the load and check to see if the load is balanced. With these, it is hard to find the load and capacity with a view to balance them.

Tools and fixtures required are another aspect in cell design being overlooked. Wemmerlov and Hyer [1987] in their recent study also stress the importance of selecting tools, fixtures, and pallets in the design of cells. Of course, the selection of tools depend upon the processes that are to be carried in the cell. The selection of fixtures very much depend on the size of the components that are to be machined in the cell. The selection of pallets also depend upon the size of the components to be machined in the group. The number of pallets required depend upon the lot sizes to be produced in each of the parts in the family.

Loading and scheduling policies are often overlooked operational problem rather than the design problem. If these policies are considered at the time of designing cells, it will be easier to operate a cell with maximum efficiency Greene and Cleary [1985] discuss the problem of cell loading. According to them, cell loading involves the determination of the right cell among the feasible cell a part should be assigned for processing. This is especially important from the point of balancing the load among the cells and improving the machine utilization. If a part is flexible, such that it can be loaded into more than one cell, loading of cell becomes more complex. But at the same time, the importance of obtaining maximum cell and machine utilization with minimum work in process increase significantly.

Scheduling is concerned with the determination of the order of in which parts are to be processed in the cell. It also helps to determine the start and stop times of each job in the cell. According to Wemmerlov and Hyer [1987], the potential use of finite sceduling procedures such as OPT for scheduling in cellular manufacturing needs to be investigated. It is probable that more sophisticated scheduling techniques can be applied in cellular manufacturing since the size of the scheduling problem is considerably reduced compared to a larger job shop. In any case, the loading and sceduling policies should address the issue of balancing the workload and improving the machine utilization in the cell.

Lot sizing is another operational issue that has to be considered in the design of cells. Sundaram and Sundrarajan [1987] discuss an approach that can be used in determining the lot sizes of parts to be made in a cell. The approach presented considers both the work-in-process and finished inventory costs in addition to the setup costs. If the lot sizes are determined at the time of designing a cell, it will lead to opportunities for reducing the setup time and thus will pave way for manufacturing lot sizes of closer to one. The cell configuration can be looked at carefully for the possibilty of reducing the work in process inventory so that economical lot size of one can become a reality.

Practical Considerations

Majority of the cell design techniques, according to Wemmerlov and Hyer [1987], need further refinements as some of them are incompletely reported and others are ambiguous. They are of little help to either practitioners or researchers. It would definitely be of interest to both the groups, if interactive procedures are developed for cell design to meet the desired design objectives. In a practical sense, practitioners will be interested in obtaining the best possible solution under a given circumstance. Thus rational methods are to be developed rather than trial and error methods. The rational methods will provide consistent results. The factors that are commonly overlooked as referred in the previous section should be carefully considered. For example, desirable design goal could be to create cells with a smooth flow, balanced work loads and flexibility. Flexibility is an important practial consideration in the design of cells. For industrial appliations, step-by-step procedures and guidelines must be developed considering several complicated and 'messy' situations occuring in industry. A computerized method to interactively design cells will be a boon to practitioners. Of course, the method used must check to see if data used is reliable. However sophisticated the method might be, if the data is incorrect, there is no guarantee the design will be sound. Therefore, in the design of cells, considerations to several practical aspects must be given.

Conclusions

In conclusion, it is emphasized that the design of cells in cellular manufacturing system is a very serious problem and requires several important considerations. If some of the important factors are overlooked, the design will turn out to be poor and will not help to meet the intended objective. A cell may be manned or automated. The degree of automation must be known at the time of designing a cell. A manned cell will not lend itself to automating it at some point in the future. Also, space requirement and layout will be completely different for an automated cell compared to a manned cell. The procedures used for designing a manned cell may prove to be a farce for an unmanned cell and vice versa. A designer of cells in cellular manufacturing has to be aware of the common pitfalls in designing cells. Objectives must be stated clearly. The objectives stated should be met in the final design. Design alternatives can be evaluated using simulation approaches. With the availabilty of several graphical simulation packages, it should not be difficult to evaluate the alternatives. Finally,the user should be happy with the design as he is the one that will be operating the system.

References

1. Askin, R.G., and Subramanian, P.S, 1987, A Cost-Based Heuristic for Group Technology Configuration, International Journal of Production Research, Vol. 25, No. 1, pp 101-113.

2. Black, J.T., and Schroer, B.J., 1986, Decouplers Improve Flexibilty in Cellular Manufacturing Systems, Proceedings of the Workshop on Automation and Robotics for Military Applicatons, U.S. Army Missile Command, Huntsville, Alabama.

3. Green, T.J., and Cleary, C.M., 1985, Is Cellular Manufacturing Right for You?, Preceedings of the 1985 Annual International Industrial Engineering Conference, Institution of Industrial Engineers, Atlanta, Georgia, pp 181-190.

4. Huang, P.Y. and Houch, B.L.W., 1985, Cellular Manufacturing: An Overview and Bibligraphy, Production and Inventory Management, Fourth Quarter, 1985, pp 83-93.

5 Palframan, D., 1987, FMS: Too Much, Too Soon, Manufacturing Engineering, Vol. 98, No. 3, pp 34-38.

6. Sundaram, R.M., Sundrarajam, R.M, 1987, Lot Sizing in Cellular Manufacturing Systems, Proceedings of the 9th Annual Conference on Computers and Industrial Engineering, Atlanta, Georgia, March 18-20.

7. Wemmerlov, U., 1987, Research Issues in Cellular Manufacturing, International Journal of Production Research, Vol. 25, No. 3, pp 413-431.

Single–Source, Single–Truck Models for the Location—Production—Distribution Problem

Jai W. Kang*, Turgut Aykin†, A.J.G. Babu‡, and J.W. Chrissis‡

*Ecology and Environment, Inc., Buffalo, NY, USA
†Alfred University, Alfred, NY, USA
‡College of Engineering, University of South Florida, Tampa, FL, USA

Abstract

This paper addresses the joint problem of location, production, and distribution. These problems occur in a variety of contexts and can be characterized by considering the production, inventory and distribution costs. More specifically, the problem is one of locating a source and then sequencing delivery vehicles so as to minimize the effects of holding and transportation costs. Thus, the problem is one of cost minimization where the cost components are interrelated. Each of the elemental costs, i.e., production, inventory, and distribution, is considered and its relationship to the others discussed.

Introduction

A determinisic location problem is the problem of finding the locations of sources in a specified area to serve certain destinations whose locations and requirements are known, such that an objective is optimized. A common objective is the minimization of transportation costs. When a source, such as a plant where production takes place and finished products are inventoried, is considered, there is another cost factor involved in addition to transportation; namely, inventory holding costs.

Suppose a truck delivers a truckload of products to a destination. The inventory level at the supply location is reduced by that truckload. The remaining inventory bears holding costs at least until the truck returns to the source and reduces the inventory level further. Since it is anticipated that total inventory holding costs would differ depending on when and which destination is serviced by the truck, this becomes a distribution problem. Further, the inventory level depends on the initial stock stored when the system is ready to operate. That is, the initial inventory level depends on how much of the required products have been produced. Thus levels of production add another feature to the location problem.

This paper considers location-production-distribution problems for the single-source, single-truck situation and addresses exact and heuristic algorithms to solve such problems.

Production, Location, and Distribution Aspects

A deterministic location problem can generally be defined as the problem of situating sources in a specified space so that they may serve certain destinations whose locations and requirements are known, such that some objective is optimized. Sources are supply points and destinations are demand points. A more complete set of references can be found in Kang[1984].

The location model was stated assuming that the costs of operating the sources do not vary as its location is changed in the solution space. When the source considered is a manufacturing plant, the nature of its operating costs can be related to holding inventories. Dudick and Cornell [1979] states that inventory can represent as much as 50 percent of the total investment, and the floor space required for receiving, storing, and staging inventories can amount to more than 45 percent of total floor area at the factory location. The investment and rent costs alone indicate that holding inventory is costly.

The inventory holding cost is considered a major cost factor in inventory system analysis. Tersin [1976] and describes the aspect of holding cost in inventory system analysis and Dudick and Cornell [1979] break down the inventory holding cost into various categories.

If an entire lot of products is produced in one batch, the lot is held in inventory until dispatched for delivery. Maintaining inventories for a considerable length of time will increase holding costs. Such inventory holding costs become a major factor in the process of finding the source's location as they constitute a considerable portion of the total cost. One way of reducing holding costs can be accomplished by determining an optimal sequence of visisting destinations; that is, by solving an appropriate distribution problem.

Single-batch production creates large inventory-holding costs throughout the system. Inventory holding costs may, however, be reduced if the source is allowed to produce the required products in more than a single batch. This will reduce total inventory holding costs significantly, since the inventory level is less at any given time. Also, multiple-batch production would be more realistic than single-batch in real world problems since the input resources needed to produce all the products in one batch may make the process either impossible or financially prohibitive.

Incorporating these aspects of production and distribution into the deterministic location problem yields the location-production-distribution (LPD) problem. LPD problems can be classified by the number of sources, trucks, and batches using the nomenclature (number of sources / number of trucks / number of batches). This paper primarily deals with the (1/1/1) and (1/1/m) situations and makes references to the others.

The (1/1/1) Problem

The Weber problem is the problem of finding the single-source location which minimizes the sum of positively weighted Euclidean distances from the source to the given number of destinations located at known, distinct points in the plane. That is, transportation costs are the only decision criterion considered.

In this single-source location-distribution (1/1/1) problem, the inventory-holding costs incurred by the source are also considered. The objective of the problem is to minimize the sum of transportation and holding costs. As it was mentioned, different holding costs are incurred by the source depending on the sequence of distribution to the destinations. Therefore, the problem is not only to find the optimal source location, but also to find the optimal sequencing system so as to serve a given number of destinations whose locations and product requirements are known.

The following assumptions are made in formulating the (1/1/1) problem:

1) Distance Function: The Euclidean distance function is used.

2) Holding Costs: Holding costs will be incurred for all products remaining at the source for the entire period under consideration while the truck(s) are delivering products to destinations. The time necessary for loading and unloading is not counted as part of inventory holding costs. Each destination is assumed to required one type of product line, which implies that there is a holding cost rate for each destination.

3) Fixed Cost Rate: Any cost involved in the problem is assumed both time and location invariant, which means that such costs as holding and transportation are independent of the source location in the solution space and are fixed throughout the period under consideration.

4) Availability of Products: It is assumed that sufficient product to satisfy all requirements are available when the system is operational. This might be accomplished by purchasing or producing all product requirements at one time.

Problem Formulation

The single-source location-distribution problem is formulated as follows. Let m destinations be located at known distinct points P_1, P_2, ..., P_m and let the source be located at point X in the plane. Let $d(X,P_i)$ represent the distance between the location of the source and destination i. Let the transportation cost per truckload per unit distance traveled between the source and destination i be denoted by C_i with W_i being the number of trips required between them for the total time period of consideration. Then the total transportation cost (TC) associated with the source located at X is given by

$$TC(X) = \sum_{i=1}^{m} c_i w_i d(X, P_i) \qquad (1)$$

Let $h_{ij}(\cdot)$ be the inventory holding cost incurred when destination i is reached on the j-th trip, which is a complicated function of $d(X,P_i)$ and the inventory level. Applying a zero-one variable, Z_{ij}, as unity if destination i is reached from the source in the j-th trip, and zero otherwise, the total holding cost, $HC(\cdot)$, is:

$$HC(\cdot) = \sum_{i=1}^{m} \sum_{j=1}^{m} Z_{ij} \; h_{ij}(\cdot) \qquad (2)$$

subject to:

$$\sum_{i=1}^{m} Z_{ij} = 1, \quad j = 1, 2, \cdots, m$$

$$\sum_{j=1}^{m} Z_{ij} = 1, \quad i = 1, 2, \cdots, m$$

$$Z_{ij} = 0 \text{ or } 1, \text{ for all } i \text{ and } j$$

Summing TC and HC from equations (1) and (2) the single-source location-distribution problem can be written as:

$$\text{Minimize } \psi = \sum_{i=1}^{m} c_i \cdot w_i \cdot d_i + \sum_{i=1}^{m} \sum_{j=1}^{m} Z_{ij} \cdot h_{ij}(\cdot) \qquad (3)$$

subject to:

$$\sum_{i=1}^{m} Z_{ij} = 1, \quad j = 1, 2, \cdots, m$$

$$\sum_{j=1}^{m} Z_{ij} = 1, \quad i = 1, 2, \ldots, m$$

$$Z_{ij} = 0 \text{ or } 1 \text{ for all } i \text{ and } j$$

where:

ψ = the total cost for the total time horizon in which all demands are satisfied by a given sequence of visition destinations

c_i = the transportation cost per truckload per unit distrance traveled between the source and distination i, = 1 (assumed without loss of generality)

w_i = number of trips required during the total time horizon between the source and destination i

d_i = $d(X, P_i)$ = the distance traveled per trip between two points X (source) and P_i (destination i) = $[(x-a_i)^2 + (y-b_i)^2]^{1/2}$ for Euclidean distance

Z_{ij} = 1, if destination i is reached from the source on the jth trip, 0 otherwise

$h_{ij}(\cdot)$ = the inventory holding cost incurred while destination i is being reached on the jth trip, a function of $d(X, P_i)$ and the inventory level while destination i is visited in the jth trip of the sequence. It is also a function of which trips preceded it.

The decision variables of Problem (3) are the coordinates (x and y) of the source and the sequencing assignment (Z_{ij}). Each constraint ensures that each destination is positioned only once in a given sequence of $([1], [2], \cdots, [m])$ where [i] denotes i-th destination to be delivered for a required number, $w_{[i]}$, of times by the truck.

A Heuristic Method

The location-distribution problem (3) is difficult to solve because it is non-linear in the continuous variables x and y and has the 0-1 variables Z_{ij}. Suppose the sequencing is known, but the location of the source is to be determined, then the solution procedure would be quite simplified. Because of the values of the 0-1 variables, if Z_{ij} for all i and j are specified, that is , if we know the sequence of destinations to be reached by the source, then, holding costs are calculated as follows.

For the basic single-source/single-truck/single-batch (1/1/1) model the behavior of inventory-holding cost h(t) is easy to visualize. At time zero there are $w_{[1]} + w_{[2]} + \cdots + w_{[m]}$ truckloads of products available in the source. Defining h(t) by the inventory holding cost rate at time t, $h(o) = \sum_{k=1}^{m} w_{[k]} h_{[k]} \tilde{=} wh(1)$ where $h_{[k]}$ is the inventory holding cost per truckload per unit time for the kth destination. But h(t) is reduced by $h_{[1]}$ as soon as the truck leaves the source for the first destination; $h(t) = wh(1) - h_{[1]}$. There is no change in h(t) until the completion of the first trip, which occurs at time $2d_{[1]}$. Then h(t) drops to $wh(1) - 2h_{[1]}$ and remains there until the completion of the second trip which occurs at $4d_{[1]}$. Continued in this manner, it is easy to see that h(t) is a decreasing step function over the interval $[\,0, \sum_{k=1}^{m} 2\, w_{[k]}\, d_{[k]}]$.

Suppose the truck is now finished delivering products to the first j-1 destinations in a given sequence of $([1], [2], \cdots, [m])$, and that destination i is reached from the source on the truck's jth trip. The current inventory level is $\sum_{k=j}^{m} w_{[k]}$. The holding cost rate wh(j) becomes $\sum_{k=j}^{m} w_{[k]}h_{[k]}$. But h(t) is reduced by $h_i (=h_{[j]})$ as soon as the truck departs for destination i (=jth destination) for the time period $2d_i$. The inventory holding cost is therefore given by

$$2d(X, P_i) \left(\sum_{k=j}^{m} w_{[k]} h_{[k]} - h_i \right) = 2d_i(wh(j) - h_i) \qquad (4)$$

which assumes a unit time taken for a unit distance. During the next trip to the same destination, h(t) is reduced by another h_i, that is, $wh(j) - h_i - h_i$, and so on until the truck finishes all w_i truckloads. Therefore, total inventory holding cost incurred while destination i is being reached in the jth trip of the source is formulated mathematically as:

$$h_{ij}(t) = 2d_i \,[wh(j+1) + (w_i-1)w_ih_i/2],\ i = 1, \cdots, m;\ j+1 = 2, \cdots, m \qquad (5)$$

where:

$h_{ij}(t)$ = inventory holding cost incurred while destination i is being visited in the jth trip from the source,

d_i = $d(X, P_i)$,

$w_{[k]}$ = number of trips required for the total time horizon between the source and the kth destination,

$h_{[k]}$ = inventory holding costs per truckload per unit time for products to be delivered to the kth destination,

h_j = holding cost of products for destination j per truckload per unit time,

$wh(l) = \sum_{k=1}^{m} w_{[k]} h_{[k]}$, $l = 1, \cdots, m$.

Then the total holding cost (HC) would be:

$$HC = 2 \sum_{j=1}^{m} d_{[j]} \,[wh(j+1) + (w_{[j]}-1)\, w_{[j]}h_{[j]}/2 \qquad (6)$$

where

$$wh(j+1) = \begin{cases} \sum_{k=j+1}^{m} w_{[k]}h_{[k]}, & \text{if } j = 1, 2, \cdots, m-1, \text{ and} \\ 0 & \text{if } j = m. \end{cases}$$

Summing (6) and (2), problem (3) in the case of known sequencing becomes:

$$\text{Minimize } \psi = \sum_{j=1}^{m} w_{[j]}d_{[j]} + 2\sum_{j=1}^{m} d_{[j]}\left[\sum_{\substack{k=j+1\\ j\neq m}}^{m} w_{[k]}h_{[k]} + (w_{[j]}-1)w_{[j]}h_{[j]}/2\right] \quad (7)$$

Algorithm for Solving Problem(3)

Step 0. Let X^0 be the starting location of the source chosen arbitrarily, and ε be a small positive number. Set k=0, and go to step 1.

Step 1. Perform sequencing of m destinations with the source located at X^k by non-increasing order of h_i/d_i, i=1, 2, ⋯, m. Let 0^k = ([1], [2], ⋯, [m]) be the resulting sequence. If k=0, go to step 2. If $0^k=0^{k-1}$, then stop: the solution is X^k and 0^k. Otherwise, go to step 2.

Step 2. Calculate weights for the single-source Euclidean location problem:

$$V_i = w_{[i]}\{1 + 2[(\sum_{\substack{k=i+1\\ i\neq m}}^{m} w_{[k]}h_{[k]}) + (w_{[i]}-1)h_{[i]}/2]\}$$

Step 3. Solve the location problem using the interative Hyperboloid Approximation Procedure (HAP) (Francis and White 1974). Let the optimal solution be X^k with its functional value of f^k. If $|(f^k - f^{k-1}|<\varepsilon$, stop: the solution is X^k and 0^k. Otherwise, increase k by 1 and go back to step 1.

The Single-Source Problem on a Straight Line

The L-P-D problem is simplified when all destinations are located on a straight line and the source is also to be located somewhere on the same straight line. Such a case would be where all destinations are located along a highway or a seashore line. An exact algorithm is developed using some properties of the rectilinear-distance problem illustrated by Francis and White [1974].

The objective of this straight line problem is again to minimize the transportation and holding costs. Since different holding costs are realized at the source depending on the distribution requirements for destinations(as discussed), the problem is not only to find the optimal source location, but also to find the optimal sequencing scheme so as to best serve a given number of destinations whose locations and requirements are known.

Since the solution space lies on the straight line, the problem formulation from (3) differs in source and destination location coordinated and distance function.

$$\text{Minimize} = \sum_{i=1}^{m} c_i w_i d_i + \sum_{i=1}^{m}\sum_{j=1}^{m} Z_{ij}h_{ij}(\cdot)$$

$$\text{subject to } \sum_{i=1}^{m} Z_{ij} = 1, \qquad j = 1,2, \cdots, m$$

$$\sum_{j=1}^{m} Z_{ij} = 1, \qquad i = 1,2, \cdots, m$$

$$Z_{ij} = 0 \text{ or } 1 \qquad \text{for all } i \text{ and } j$$

where $d_i = |x - a_i|$ and all other variables are the same as in (3).

Due to the way this problem simplifies, an exact algorithm for solving the straight-line m-destination problem is available (Kang[1984]).

The Single-Truck/Two-Batch Location-Production (1/1/2) Problem

If the entire lot of requirements is produced in one batch, inventories are held until they are dispatched for delivery to destinations. Holding inventories for a considerable time will generate holding costs originatng from such sources as capital cost, taxes, insurance, handling, storage, shrinkage, obsolescence, and deterioration. Such inventory costs become a major factor in the process of determining the source location.

One way of reducing the inventory holding cost is to determine an optimal sequence for visiting destinations. That is, when the source location is known with respect to a given number of destinations, the holding cost varies depending on sequence. Hence, an optimal sequencing rule for the single-truck/single-batch problem is: visit the destinations in non-increasing order of h_i/d_i, $i = 1, \ldots, m$.

Even though the inventory holding cost can be reduced using the optimal sequencing rule, there is an upper limit on such savings. Further savings may be obtained by not requiring that all the requirements be produced in a single batch. In other words, single-batch production generates large holding costs throughout the system. This holding cost may be reduced if it is possible to produce the entire requirements in more than a single batch, i.e., multiple-batch production. Doing so will reduce total inventory holding costs significantly since the inventory level is less at any given time. Also, multiple-batch production would be more realistic than single-batch in the real world as obtaining rèsources needed to produce all the requirements in one batch may make it either impossible or financially prohibitive.

The Single-Truck, Single-Source, Multi-Batch Problem (1/1/m)

Extending the concepts of the (1/1/2) model gives the (1/1/m) situation. In addressing this problem the following assumptions are made:

1. The single source is to be located at a specific point and operates a single truck.

2. All input commodities are obtained only immediately prior to production with zero holding costs assumed.

3. The first batch production should be completed immediately prior to time 0 (i.e., $T_{B1} = 0$) and the k-=th batch production time should be completed just prior to to time T_{Bk}. The time required for batch production is not considered and there is no in-process inventory involved until all production in a batch is complete.

4. Loading/unloading time in negligible and assumed zero.

5. Batch set-up cost is less than the savings in inventory holding cost.

6. Demand at each destination is satisfied by dispatching trucks continuously to that destination. If a specific destination i requires w_i truckloads, then trucks will be dispatched to destination i only after a first trip has been made to i.

In this multi-batch case, the optimal sequencing rule for the (1/1/2) model does not necessarily yield an optimal solution. A branch-and-bound methodology approach for solving this problem is presented in Kang[1984].

Some Results and Conclusions

The basic LPD problem and a number of variants have been presented. Some of the results for these models are now summarized.

(1/1/1) Problem. As a basic model of the LPD problem, the single-source operates a single truck to deliver products to destinations. Since different holding costs (HCs) are incurred by the source depending on the sequence of distribution to the destinations, the problem is not only to find the optimal source location, but also to find the optimal sequencing to serve a given number of destinations with known locations and product requirements.

(1/1/1) on the Straight Line. A special case of the (1/1/1) problem where the solution is located on a straight line instead of the plane. The solution is the destination having the minimum function value. The sequencing is by non-increasing order of h_i/d_i.

The (1/1/2) Problem. A basic model of multiple batch production, this is the two-batch production problem with a fixed sequence of visiting destinations in a single-source/single-truck case. Given a number of destinations, the location of the single source is to be set so as to minimize the costs of transportation and holding inventories. The difference between this problem and the single batch case is to decide how much and when each batch should be produced.

An exact algorithm was developed to determine the optimal time point (T_{B2}) for producing the second batch. When T_{B2} is determined, the location problem becomes the Weber problem.

(1/1/m) Problem. An extension of the (1/1/2) problem which relaxes such assumptions as fixing the sequence of visiting destinations and two-batch production. The main focus was on how to assign destinations to batches so as to save the most HC. By giving a counter-example, it was refuted that the single-batch sequencing rule might apply to (1/1/m) problems. Thus, another method was developed to determine both the sequencing and batch production requirements using a branch-and-bound approach. After solving a number of problems using the branch and bound method, a heuristic method has been devised. It has been found that the heuristic algorithm provides an extremely close approximation to the solutions using the exact method in dramatically less computation time.

Selected References

Dudick, T. S. and R. Cornell[1979] Inventory Control for the Financial Executive. John Wiley & Sons, New York, N.Y.

Eyster, J. W., J. A. White and W. W. Wierwille[1973] On Solving Multi-facility Location Porblems Using a Hyperbolic Approximation Procedure. AIIE Transaction, 5, 1-6.

Fisher, M. L., A. J. Greenfield, R. Jaikumar, and J. T. Lester[1982] A Computerized Vehicle Routing Application. Interfaces, 12, 4, 42-51.

Francis, R. L. and J. A. White[1974] Facility Layout and Location: An Analytical Approach. Prentice-hall, Englewood Cliffs, New Jersey.

Kang, J. W. [1984], Location-Production-Distribution Problems, Ph.D. Dissertation, State University of New York - Buffalo.

Tersine, R. J.[1976], Material Management and Inventory Systems, North Holland, New York.

CAD/CAM

An Interface Between CAD/CAM Softwares

John F. Stephens, III, R.I. Noorani, and L.Y. Chen

Department of Mechanical Engineering, The University of Southwestern Louisiana, Lafayette, LA 70504-4170, USA

ABSTRACT

Ever since the introduction of commercially available computer-aided design (CAD) and computer-aided manufacturing (CAM) systems, the need for an interface between the two systems has become apparent. Despite efforts by several software companies and the American National Standards Institute (ANSI), most vendors of CAD/CAM have not made their softwares compatible with graphics standards. It is extremely essential to have a single recognized structure for communication of data between CAD/CAM systems.

The Initial Graphics Exchange Specification (IGES), managed by the National Bureau of Standards (NBS), established the initial base for direct digital exchange of graphics data. The IGES is a communication file structure for data produced on and used by CAD/CAM systems.

The purpose of this paper is to define and develop a data interface between CAD and CAM software packages using IGES file structure. The interface will be achieved by (1) the establishment of the standard procedure for processing the IGES file, (2) the definition of the standard data base for the storage of the data in the IGES file, and (3) the generation of the communication file for the CAM software.

INTRODUCTION

Although CAD and CAM have developed tremendously over the last several years, these two systems have been developed as separate islands of development with little attention given to system integration and interdisciplinary capability and data communication. For total system integration, it is essential that data be shared and exchanged among all disciplines, that data redundancy be minimized, and that a universally acceptable standard be established.

The important element that provides the bridge between CAD and CAM is the interface--the database and communication linkage. Although the needs and problems of CAD/CAM integration have been suggested by many researchers {1 - 9}, many challenges must be faced before a complete integration will be practical. A recent report {10} of the committee on the CAD/CAM interface provides an excellent summary of the current issues in the interface challenge.

Recent efforts to integrate CAD/CAM generally generated from the CAD data, a process plan, NC tool path or set of manufacturing information about the part. One of these efforts is a software for automated drafting and manufacturing program called AD 2000 {11}. Using AD 2000, one can create a 3-D part by defining all surfaces. Some NC tool paths can be calculated by storing a limited amount of topology, but the resulting paths are always subject to visual verification by the user. The limitations of AD 2000 are that the program does not check for tool interference nor does it guarantee that a designed part is geometrically and topologically realizable. The work of Grayer {12}, Chan {13}, Chang {14} have used a solid modeling data base. Their research efforts have attempted either to determine the machine sequence for some previously known feature for NC path determination or to decompose the part into primitive volumes of arbitrary shapes. The applications of these methods are limited because they have been developed to only certain specific feature types. The research efforts of Woo {15}, Armstrong {16}, and Kyprianou {17} made significant contribution toward the simplification and improvement of a link between CAD/CAM systems.

Although a complete integration of CAD/CAM systems is yet to be achieved, some U.S. and overseas companies have made substantial moves toward such integration {18, 19}. Three major government programs (NASA's Integrated Program for Aerospace Vehicle Design, the Air Force's Integrated Computer-Aided Manufacturing, and the National Bureau of Standard's Automated Manufacturing Research Facility) target the types of integration problems that much of industry faces. The Product Definition Data Interface (PDDI) project of ICAM was developed to bridge the barrier between CAD and CAM. So far none of the commercial software systems has completely implemented the PDDI interface to process the IGES file.

This paper presents the interface program that was developed as the communication medium between the two CAD/CAM software systems that are currently being used as teaching aids on CAD/CAM courses. These software packages are the GEODRAW (a computer-aided design and drafting program from Structural Dynamics Research Corporation) and the NICAM (a computer-aided manufacturing program from Numeridex).

THE INTERFACE PROBLEM AND IGES

The problem of linking CAD and CAM systems is that electronic and mechanical designers know so little about each others work. Traditionally, CAD/CAM system manufacturers have tended to be electronic designers who are able to build unified CAD data bases, but are unfamiliar with manufacturing operations. Moreover, manufacturing information is much more complex than design information and more difficult to codify in a data base.

To stem the proliferation of software translators, IGES was developed by a joint government/industry committee that includes NBS. IGES is a neutral data format that describes product design and manufacturing information for different CAD/CAM systems. At present, IGES handles more design than manufacturing information.

The interface program is designed as the communication medium between GEODRAW and NICAM software. Each IGES file from GEODRAW software contains a view entity which is defined as a two-dimensional projection of a selected subset of the model, together with non-geometric information such as text. The IGES file output from the GEODRAW contains only geometric elements of the model. These geometric models fall into two categories. One is the geometric definition of the physical shape of the model and the other is the dimensioning or annotation geometry in the drawing without the text. Usually, these dimensioning geometries are of no use to the CAM software. The user should only select the geometric definition model to be output to the IGES file.

The NICAM software is composed of a number of modules. Each of these modules is a special purpose program. The SHAPE module of NICAM is a program that can accept the two-dimensional unbounded geometry definition from an ASCII file. This unbounded geometry definition is then used to construct a bounded geometry that represents the desired part profile. Due to the fact that most NC machines can only process linear and circular interpolation, the geometry that can be defined in the SHAPE module are point, line and circle. Also, there are a maximum of 250 symbols or discrete elements (combinations of points, lines, circles and scalers) allowed within any one running session.

THE INTERFACE PROGRAM

The purpose of this program is to read the IGES file generated from the GEODRAW software, convert the data in the IGES file to a format that is acceptable to the SHAPE module of the NICAM software package. To do a successful data transformation, a data tolerance must be selected according to the gap size that exists in the data. The value of the default tolerance is 0.0625 for INCH system and 0.5 for mm system. If users change the default tolerance to a new value, the program would automatically check all the data again and report the error data. The major functions of the interface program are listed below.

1. Read the IGES file.
2. Display data in the IGES file.
3. Change data tolerance and check data errors.
4. Write the geometry definition file.

The interface program includes several input/output modules. A general structure was selected to construct the body of the program including all of its major subroutines. The main program (called BRIDGE) contains several routines in it. These routines are divided into three groups. The first group contains the several service routines. The second group contains the data handling and display routines. The third group contains the screen display initialization routines. Figure 1 shows the BRIDGE program which shows the major routines and modules.

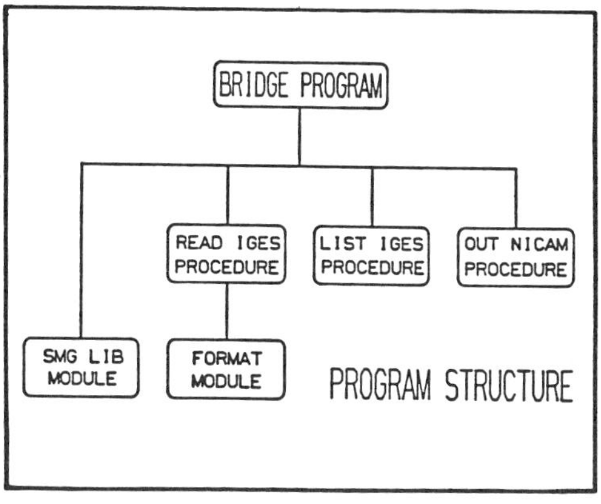

Figure 1. The structure of interface program

RESULTS

To study the performance of the interface program, a test part is created using GEODRAW software as shown in Figure 2. The IGES file that was generated from this test part and the corresponding geometry definition file that was generated by the interface program can be seen in {20}. Figure 3 shows the resulting geometry that was transferred to NICAM SHAPE module. It is important to note that L_1 and L_2 have not been transferred because of their duplication with L_9 and L_{10}.

Although the results of the program are not extremely exciting at this point, it does represent a step in the right direction. The interface program does help us handle arc and spline which normally could not have been handled by NICAM software. The program also saves duplicating time because the object did not have to be created both by CAD and CAM softwares.

CONCLUSIONS

The very purpose of CAD and CAM software make them handle data in two different ways. While Computer-Aided Design/Drafting (CADD) supplies many powerful tools in creating man-readable drawings, it does not provide anything to check the data integrity. At present, there exists no reliable program to make a low quality data into a high quality data. The solution to this data transfer probelm may be traced to the development of an interface program that not only takes the CAD data, but also generates CAM compatible data of high quality.

In this interface program, a method of limited power is used to verify the quality of data. If the data are simple such as a closed profile, user can verify the data quality very quickly. If the data are composed of many independent elements, the verification of data integrity is very difficult.

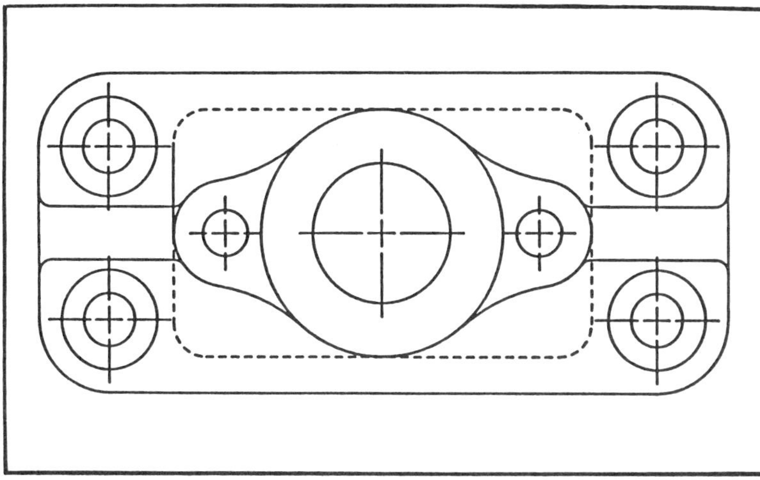

Figure 2. Drawing of a Test Part From GEODRAW

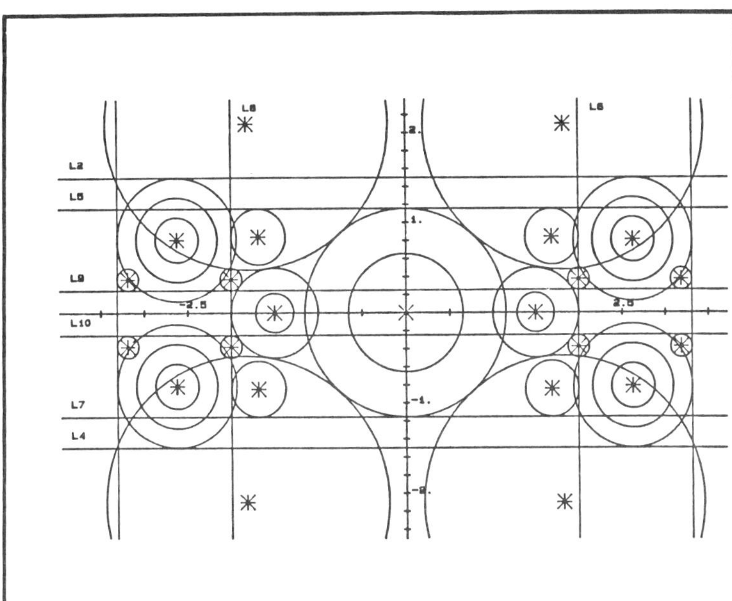

Figure 3. Geometry Transferred From Interface Program

The interface program integrates data from different files so that inconsistencies and inefficiencies from duplicated data do not arise.

REFERENCES

(1) Nilson, E. M., "CAD/CAM Interface: Problems and Solutions," Proceedings of Numerical Constrol Society, Chicago, IL., April 9-12, 1978, pp. 6 - 28.

(2) Wilson, P. R., Faux, I. D., Pasquill, K. G., "Interfaces for Data Transfer Between Solid Modeling Systems," IEEE CG&A, January, 1985, pp. 41 - 51.

(3) Zecher, J. E., Lehman, N. W., Graves, M. T., Brown, C. W., "Developing a Desktop Computer-Based Three-Dimensional Modeling System," Mechanical Engineering, Vol. 105, No. 11, Nov., 1983, pp. 50 - 61.

(4) Kretch, S. J., "CAD/CAM for Robots," SME Technical Paper, Series, MS. 82-135, 8p, 1982.

(5) Sakurai, H., Gossard, D. C., "Solid Modeling through Orthographic Views," Computer Graphics (ACM), Vol. 17, No. 3, 1983, pp. 243 - 252.

(6) Foley, J. D., Wallace, V. L., Chan, P., "Human Factors of Computer Graphics Interaction Techniques," IEEE Computer Graphics and Applications, Vol. 4, No. 11, Nov., 1984, pp. 13 - 48.

(7) Wilson, P. R., Faux, I. D., Ustrowski, M. C., Pasquill, K. G., "Interfaces for Data Transfer Between Solid Modeling Systems," IEEE Computer Graphics and Applications, Vol. 5, No. 1, Jan., 1985, pp. 41 - 51.

(8) Liewald, M. H., "Initial Graphics Exchange Specification: Success and Evolution," Computers & Graphics (Pergamon), Vol. 9, No. 1, 1985, pp. 47 - 50.

(9) Morley, B. D., "Mechanical Product Design with Computer-Aided Engineering," Manufacturing Engineering, Vol. 93, No. 3, Sept., 1984, pp. 75 - 78.

(10) Thomson, A. R., Buffman, H. E. Devlin, S. S., Frech, W. P., Christop, W., McHale, M. J., Melankoff, M. A., Merchant, M. E., Van Sant, R. W., "Interface Challenge," NRC Committee on CAD/CAM Interface, U.S.A., American Machinist, Vol. 129, No. 1, Jan., 1985, pp. 95 - 102.

(11) AD 2000 User's Manual, Control Data Corporation, 1980.

(12) Grayer, A. R., "The Automatic Production of Machined Components Starting From a Stored Geometric Description," PROLAMAT Proceedings, 1976, North Holland Publishing Company.

(13) Chan, B. T. F., "ROMAPT: A New Link Between CAD and CAM," Computer-Aided Design, Vol. 14, No. 5, Sept., 1982, pp. 261 - 266.

(14) Chang, T. C., TIPPS: A Totally Integrated Process Planning System, Ph.D. Thesis, Virginia Polytechnic Institute, Blacksburg, VA., November, 1982.

(15) Woo, T. C. H., Computer Understanding of Designs, Ph.D. Thesis, University of Illinois, Urbana-Champaign, IL., 1975.

(16) Armstrong, G. T., Carey, G. C., and de Pennington, A., "Numerical Code Generation From a Geometric Modeling System," in Solid Modeling by Computers From Theory to Applications, M. S. Pickett and J. W. Boyse, Editors, Plenun Press, 1984

(17) Kyprianou, L. K., Shape Classification in Computer-Aided Design, Ph.D. Thesis, University of Cambridge, Cambridge, ENGLAND, July, 1980.

(18) Goldstein, B. C., Heller, A. R., Moss, F. H., and Wladawsky-Berger, I. (1984): "Directories in Cooperative Processing Between Workstations and Hosts," IBM Systems Journal, 23 (3), pp. 236 - 244.

(19) Manufacturing Automation Protocol (MAP) Specifications V_x2.0, Warren, MI., GM Technical Center, Feb. 29, 1985.

(20) Chen, L-Y, The Development of an Interface Between Commercial Computer-Aided Design and Manufacturing Software Packages, M.S. Thesis, The University of Southwestern Louisiana, Lafayette, LA., May, 1986.

Isoparametric Finite Elements for Micropolar (Cosserat) Elasticity

Sachio Nakamura, Ph.D.* and Funag-Yuan Huang, E.Sc.D.†

*AT&T Bell Laboratories, Holmdel, NJ 07733, USA
†Mechanical Engineering Department, New Jersey Institute of Technology, Newark, NJ 07102, USA

1. Introduction

As the need for new man-made light-weight materials increases in industries, the development of CAE/CAD tools that can handle such materials is becoming crucial, since the conventional FEM fails to efficiently analyze composite materials with microstructures. In this study, isoparametric finite elements for micropolar (Cosserat) elasticity have been developed.

A micropolar elasticity differs from a classical elasticity in that each material point has extra degrees of freedom independent of translation, and can transmit couple stress as well as the usual force stress. The theory of micropolar elasticity has been extensively studied by Eringen and his coworkers [1] and is known to be a valuable theory to explain the behavior of bone type materials with microstructures [2,3].

2. Variational Formulation for Micropolar Elasticity

Previsouly, total potential energy $\prod$ for micropolar elasticity was derived [4]:

$$\prod = \frac{1}{2} \int_V (\tau_{ji}\epsilon_{ji} + m_{ji}\phi_{i,j})dV - \int_V (G_j u_j + C_j \phi_j)dV - \int_S (T_j^{(\nu)} u_j + M_j^{(\nu)} \phi_j)dS \tag{2.1}$$

here τ_{ji} and m_{ji} in the first term are the force stress and couple stress, respectively, and $\epsilon_{ij} = u_{j,i} - e_{ijk}\phi_k$ is the *micropolar strain* of Eringen [1]. G_j and C_j in the second term are the applied body force and body couple, respectively, and $T_j^{(\nu)}$ and $M_j^{(\nu)}$ in the third term are the surface force and couple traction, respectively.

Extremizing the above total potential energy expression with respect to displacement field u_j and micro-rotation field ϕ_j gives the following set of equations for micropolar elasticity derived by Eringen [1]:

Force equilibrium equation:

$$\tau_{ji,j} + G_i = 0 \tag{2.2}$$

Moment equilibrium equation:

$$e_{ijk}\tau_{jk} + m_{ji,j} + C_i = 0 \tag{2.3}$$

Cauchy's formula for force stress:

$$T_i^{(\nu)} = \tau_{ji}\nu_j \tag{2.4}$$

Cauchy's formula for couple stress:

$$M_i^{(\nu)} = m_{ji}\nu_j \tag{2.5}$$

here ν_j in equation (2.4) and (2.5) is a j-th component of unit normal vector.

3. *Finite Element Formulation for Micorpolar Elasticity*

Displacement type finite element methods for linear micropolar elasticity can be formulated using equation (2.1). One uses appropriate shape functions N_u and N_ϕ to interpolate the displacement and microrotation fields in element V_e in terms of nodal field variables,

$$u(x) = N_u(x)u^e, \qquad \phi(x) = N_\phi(x)\phi^e \tag{3.1}$$

where $u(x)$ and $\phi(x)$ are field variable vectors corresponding to displacement and microrotation in element V_e and u^e and ϕ^e are nodal field variable vector corresponding to them. Micropolar strain of Eringen can be expressed symbolically

$$\epsilon = Lu(x) + M\phi(x) \tag{3.2}$$

where L is a differential operator and M is the permutation matrix. Substituting equation (3.2) into equation (3.1), one obtains

$$\epsilon = \left[LN_u, MN_\phi\right] \begin{Bmatrix} u^e \\ \phi^e \end{Bmatrix} = B_0 U^e \tag{3.3}$$

where $U^e = [u^{e^T}, \phi^{e^T}]^T$ is a nodal value vector. Similarly for $\phi_{i,j}$, one can express

$$\nabla\phi = \nabla\,(N_\phi\phi^e\,) = \left[0, \nabla N_\phi\right] \begin{Bmatrix} u^e \\ \phi^e \end{Bmatrix} = B_1 U^e. \tag{3.4}$$

Expressing the constitutive equations as

$$\tau = D_0\,\epsilon, \qquad m = D_1\nabla\phi, \tag{3.5}$$

and substituting eautions (3.3) and (3.4), one obtains

$$\tau = D_0B_0U^e, \qquad m = D_1B_1U^e. \tag{3.6}$$

Thus the discretized version of total potential energy now becomes

$$\Pi_k = \frac{1}{2}\, U^{e^T} \int\int_{V_k} (B_0^T D_0 B_0 + B_1^T D_1 B_1)\, dV\, U^e - (F_V^e + F_S^e)^T\, U^e \tag{3.7}$$

where

$$F_V^{e^T} = \int\int_{V_k} \left[G_i N_u, C_i N_\phi\right] dV, \qquad F_S^{e^T} = \int_{S_k} \left[\tau_i^{(\nu)} N_u, M_i^{(\nu)} N_\phi\right] dS \tag{3.8}$$

Taking the first variation of the above equation with respect to the nodal field variable vector, U^e, and equating it to zero yields the discretized equilibrium equation

$$k^e U^e = F_V^e + F_S^e \tag{3.9}$$

in which

$$k^e = \int\int_V (B_0^T D_0 B_0 + B_1^T D_1 B_1)\, dV \tag{3.10}$$

is the element stiffness matrix for micropolar elasticity. The first and second terms in the above equation correspond to strain energy due to classical displacement-strain and micro-rotation gradient, respectively.

4. *Isoparametric Elements in Three Dimensional Cartesian Coordinate System*

In this section, general finite element formulation described in the previous section is used to develop sepcific isoparametric elements for micropolar elasticity in Cartesian coordinate system. Three diemsnional location in Cartesian coordinate system can be interpolated as

$$x = \sum_{i=1}^{q} N_i x_i, \qquad y = \sum_{i=1}^{q} N_i y_i, \qquad z = \sum_{i=1}^{q} N_i z_i, \tag{4.1}$$

where x, y, z are the coordinates at any point in the element and x_i, y_i, and z_i, $i = 1,\ldots,q$, are the coordinates of q element nodes. The interpolation function N_i are defined in the natural coordinate system (r,s,t) of the element, whose ranges are between -1 to +1. The interpolation functions for isoparametric elements in Cartesian coordinate system are given in Ref. [6].

In the isoparametric formulation, the same interpolation functions can be used to express displacement field:

$$u = \sum_{i=1}^{q} N_i u_i, \qquad v = \sum_{i=1}^{q} N_i v_i, \qquad w = \sum_{i=1}^{q} N_i w_i, \tag{4.2}$$

and microrotation field:

$$\phi_x = \sum_{i=1}^{q} N_i \phi_{x_i}, \qquad \phi_y = \sum_{i=1}^{q} N_i \phi_{y_i}, \qquad \phi_z = \sum_{i=1}^{q} N_i \phi_{z_i} \tag{4.3}$$

here u, v, and w are displacement and ϕ_x, ϕ_y, and ϕ_z are micro-rotation along x, y and z directions, respectively, and u_i, v_i, and w_i are nodal displacement and ϕ_{x_i}, ϕ_{y_i}, and ϕ_{z_i} are nodal micro-rotation at node i along x, y and z directions, respectively.

To evaluate element strain, one needs derivatives of element displacement with respect to Cartesian coordinate system. The differential operator in Cartesian coordiante system is

$$L = [\frac{\partial}{\partial x}, \frac{\partial}{\partial y}, \frac{\partial}{\partial z}]^T. \tag{4.4}$$

In isoparametric formulation, displacement field and micro-rotation field in equation (4.2) and (4.3) are expressed in terms of natural coordinate system. To obtain derivatives, $\partial/\partial x$, $\partial/\partial y$, and $\partial/\partial z$, one first uses chain rule:

$$\begin{bmatrix} \frac{\partial}{\partial r} \\ \frac{\partial}{\partial s} \\ \frac{\partial}{\partial t} \end{bmatrix} = \begin{bmatrix} \frac{\partial x}{\partial r} & \frac{\partial y}{\partial r} & \frac{\partial z}{\partial r} \\ \frac{\partial x}{\partial s} & \frac{\partial y}{\partial s} & \frac{\partial z}{\partial s} \\ \frac{\partial x}{\partial t} & \frac{\partial y}{\partial t} & \frac{\partial z}{\partial t} \end{bmatrix} \begin{bmatrix} \frac{\partial}{\partial x} \\ \frac{\partial}{\partial y} \\ \frac{\partial}{\partial z} \end{bmatrix} \tag{4.5a}$$

or in matrix form:

$$\frac{\partial}{\partial R} = J \frac{\partial}{\partial X} \tag{4.5b}$$

where $R = [r,s,t]^T$ and $X = [x,y,z]^T$ and coefficient matrix J, is called Jacobian matrix. Taking the inverse of equation (4.5), one can evaluate derivatives of each displacement and micro-rotation:

$$\frac{\partial u}{\partial X} = J^{-1}\frac{\partial u}{\partial R}, \qquad \frac{\partial v}{\partial X} = J^{-1}\frac{\partial v}{\partial R}, \qquad \frac{\partial w}{\partial X} = J^{-1}\frac{\partial w}{\partial R} \tag{4.6}$$

$$\frac{\partial \phi_x}{\partial X} = J^{-1}\frac{\partial \phi_x}{\partial R}, \qquad \frac{\partial \phi_y}{\partial X} = J^{-1}\frac{\partial \phi_y}{\partial R}, \qquad \frac{\partial \phi_z}{\partial X} = J^{-1}\frac{\partial \phi_z}{\partial R} \tag{4.7}$$

Substituting equations (4.2) and (4.3) into equations (4.6) and (4.7) and using equations (3.3) and (3.4), one obtains strain-displacement matrix, B_0 and microrotation-gradient matrix, B_1. The rest of the procedures are the same as classical FEM [5].

5. *Isoparametrtic Elements in Cylindrical Coordinate System*

Applying differential operator in Cylindrical coordinate system (r,θ,z)

$$L = [\frac{\partial}{\partial r}, \frac{1}{r}\frac{\partial}{\partial \theta}, \frac{\partial}{\partial z}]^T \tag{5.1}$$

to equation (3.2), onbe obtains micropolar strain tensor in cylindrical coordinate system,

$$\epsilon = \begin{bmatrix} \frac{\partial u_r}{\partial r} & \frac{\partial u_\theta}{\partial r} - \phi_z & \frac{\partial u_z}{\partial r} + \phi_\theta \\ \frac{1}{r}\frac{\partial u_r}{\partial \theta} - \frac{u_\theta}{r} + \phi_z & \frac{1}{r}\frac{\partial u_\theta}{\partial \theta} + \frac{u}{r} & \frac{1}{r}\frac{\partial u_z}{\partial \theta} - \phi_r \\ \frac{\partial u_r}{\partial z} - \phi_\theta & - \frac{\partial u_\theta}{\partial z} + \phi_r & - \frac{\partial u_z}{\partial z} \end{bmatrix} \tag{5.2}$$

The coordinate transformation from (x,y,z) to (r,θ,z) is done using relationship

$$x = r\cos\theta, \qquad y = r\sin\theta, \qquad z = z \tag{5.3}$$

Corresponding to equations (4.2) and (4.3) in the previous section, displacement field are interpolated:

$$u_r = \sum_{i=1}^{q} N_i u_{r_i}, \qquad u_\theta = \sum_{i=1}^{q} N_i u_{\theta_i}, \qquad u_z = \sum_{i=1}^{q} N_i u_{z_i} \tag{5.4}$$

$$\phi_r = \sum_{i=1}^{q} N_i \phi_{r_i}, \qquad \phi_\theta = \sum_{i=1}^{q} N_i \phi_{\theta_i}, \qquad \phi_z = \sum_{i=1}^{q} N_i \phi_{z_i} \tag{5.5}$$

where u_r, u_θ, u_z, ϕ_r, ϕ_θ, and ϕ_z are displacement and microrotation along (r,θ,z), respectively. u_{r_i}, u_{θ_i}, u_{z_i}, ϕ_{r_i}, ϕ_{θ_i}, and ϕ_{z_i} are nodal displacement and microrotation.
Defining the nodal value vector, U^e,

$$U^e = [u_{r_1}, u_{\theta_1}, u_{z_1}, \phi_{r_1}, \phi_{\theta_1}, \phi_{z_1}, \ \ldots\ldots, \ u_{r_q}, u_{\theta_q}, u_{z_q}, \phi_{r_q}, \phi_{\theta_q}, \phi_{z_q}]^T \tag{5.6}$$

one can express equation (3,4) in compact form:

$$\begin{bmatrix} u_r \\ u_\theta \\ u_z \\ \phi_r \\ \phi_\theta \\ \phi_z \end{bmatrix} = \begin{bmatrix} N_1, & & & & & & \ldots\ldots & N_q & & & & & \\ & N_1, & & & & & \ldots\ldots\ldots, & & N_q & & & & \\ & & N_1, & & & & \ldots\ldots\ldots & & & N_q & & & \\ & & & N_1, & & & \ldots\ldots\ldots & & & & N_q & & \\ & & & & N_1, & & \ldots\ldots\ldots & & & & & N_q & \\ & & & & & N_1 & \ldots\ldots\ldots & & & & & & N_q \end{bmatrix} U^e \tag{5.7}$$

Substituting above equation into equation (5.2) and using similar formula to equation (4.6) and (4.7), one obtains strain-displacement matrix, B_0.

$$[\epsilon_{rr}, \epsilon_{\theta\theta}, \epsilon_{zz}, \epsilon_{r\theta}, \epsilon_{\theta r}, \epsilon_{rz}, \epsilon_{zr}, \epsilon_{\theta z}, \epsilon_{z\theta}]^T = B_0 U^e \tag{5.8}$$

Similarly, for microrotation gradient, one obtains the following B_1 matrix:

$$[\phi_{r,r}, \frac{\phi_{\theta,\theta}}{r} + \frac{\phi_r}{r}, \phi_{z,z}, \phi_{\theta,r}, \frac{\phi_{r,\theta}}{r} - \frac{\phi_\theta}{r}, \phi_{z,r}, \phi_{r,z}, \frac{\phi_{z,\theta}}{r}, \phi_{\theta,z}]^T = B_1 U^e \tag{5.9}$$

Substituting above B_0 and B_1 matrices into equation (3.10), one obtains stiffness matrix for micropolar elasitcity in cylindrical coordinate system. Actual expression of B_0 and B_1 matrices are given in Ref. [7].

6. *Isoparametric Elements for Axi-Symmetric Micropolar Elasticity*

For the axisymmetric case, dependency on θ vanishes and therefore all the derivative terms with respect to θ in (5.2) disappear. Furthermore, for axisymmetric load, microrotation about r and z direction vanish and strain tensor can be further reduced. Furthermore, for axisymmetric load, microrotations about r- and z- axis vanish and i strain tensors can be further reduced :

$$(\epsilon_{rr}, \epsilon_{\theta\theta}, \epsilon_{zz}, \epsilon_{rz}, \epsilon_{zr}) = (\frac{\partial u_r}{\partial r}, -\frac{u_r}{r}, \frac{\partial u_z}{\partial z}, \frac{\partial u_z}{\partial r} + \phi_\theta, \frac{\partial u_r}{\partial z} - \phi_\theta) \tag{6.1}$$

That is, in axi-symmetric problems for micropolar elasticity, each material point has only three degrees of freedom (u, v, ϕ), where:

$$u = u_r, \qquad v = v_z, \qquad \phi = \phi_\theta \tag{6.2}$$

which represent displacement in the r and z directions and microrotation about θ direction, respectively.

The same shape functions that have been used in equations (5.4) and (5.5) are used to interpolate above three field variables. Defining the nodal value vector, U^e;

$$U^e = [u_1, v_1, \phi_1, \ldots, u_q, v_q, \phi_q]^T, \tag{6.3}$$

one can obtain reduced form of equation (5.7). Substituting the expression into (6.1), one obtains B_0 matrix. Similarly, for microrotation gradient, one obtains the following B_1 matrix:

$$[\frac{\partial\phi}{\partial r}, -\frac{\phi}{r}, \frac{\partial\phi}{\partial z}]^T = B_1 U^e \tag{6.4}$$

These B_0 and B_1 matrices can be substituted into equation (3.10) to evaluate stiffness matrix. Actual expression for these matrices can be found in Ref. [7].

7. *Numerical Examples*

To verify the formulation, algorithm and computer program, numerical experiments are conducted.

(1) Stress Concentration around a Hole

The same example problems previously used in Ref. [4] to verify the FEM for plane micropolar elasticity using triangular constant strain element were solved. Numerical values were close to the one in Ref. [4] although less element were used. This was expected, since isoparametric element were supposed to be more efficient than triangular constant strain element.

(2) Simple Tension of Cylindrical Micropolar Solid

Using 8-node and 20-node cubic elements, simple tension problems of cylindrical micropolar solid were solved and compared with analytical results. Material constants were chosen for Coupling factors were varied : N = 0.0, 0.25, 0.50, 0.75, and 0.90. Characteristic length was fixed to l = 8.333x10^{-3} (inch). Other material parameters were: μ = 1x10^3 (psi), κ = 0.0 (psi), β = 666.6666 (psi), and γ = 0.1851852 (pound).

Analytical solutions are given by Gauthier [8],

Force Stress: r-displacement: z-displacement:

$$\tau_{zz} = -\frac{P_z}{A} \qquad u = -\frac{\nu_m P_z r}{E_m A} \qquad v = \frac{P_z z}{E_m A}$$

where A is a cross-sectional area and Young's modulus and Poisson's ratio are defined by

$$E_m = \frac{(2\mu+\kappa)(3\lambda+2\mu+\kappa)}{(2\mu+2\lambda+\kappa)}, \qquad \nu_m = \frac{\lambda}{(2\mu+2\lambda+\kappa)}$$

Both FEM and analytical solution gave the exactly same numbers:

Displacement: u = - 0.34615386x10^{-3} (inch), v = 0.11538462x10^{-2} (inch),

Stress: τ_{zz} = 1.0 (psi), τ_{rr} = 1.0x10^{-15} (psi)

The same problem was solved using 4-node and 8-node axisymmetric elements and exactly the same values were obtained.

(3) Micropolar Easltic Cylinder with a Semi-Circular Groove

An isotropic micropolar elastic cylinder with a semi-circular groove is shown in Fig. 7.1. The diameter of a cylinder is fixed to 0.1 inch. The radius ratio k is defined by $k = \frac{r}{d}$, where r is the radius of groove, and $d = D - 2r$. k is varied in the range of 0.05 to 0.5. Finite element mesh used is shown in Fig. 7.2 for radius ratio k = 0.5. The material properties used for isotropic case are shown in Table 7.1, with characteristic length: l = 8.3333x10^{-3} and Poisson's ratio = 0.3. Numerical results of stress concentration factor, K_c, is plotted in Fig. 7.3.

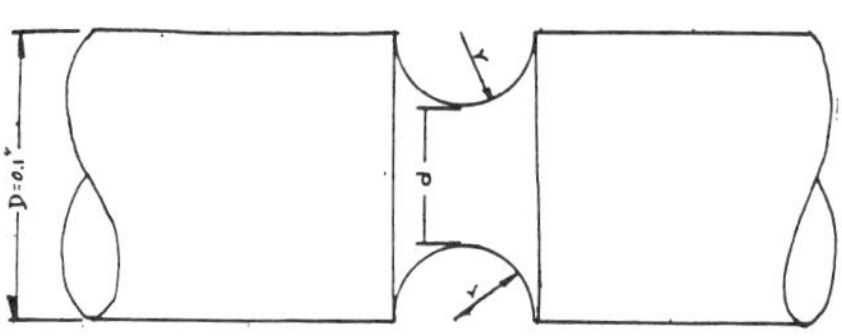

Fig. 7.1 Micropolar Elastic Cylinder with Semi-Circular Groove

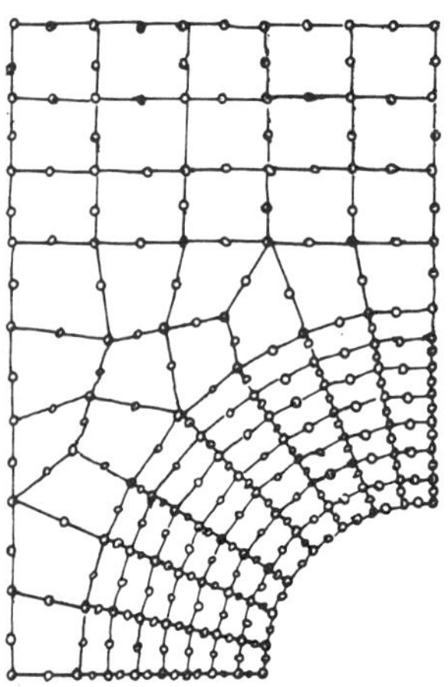

Fig. 7.2 Finite Element Mesh

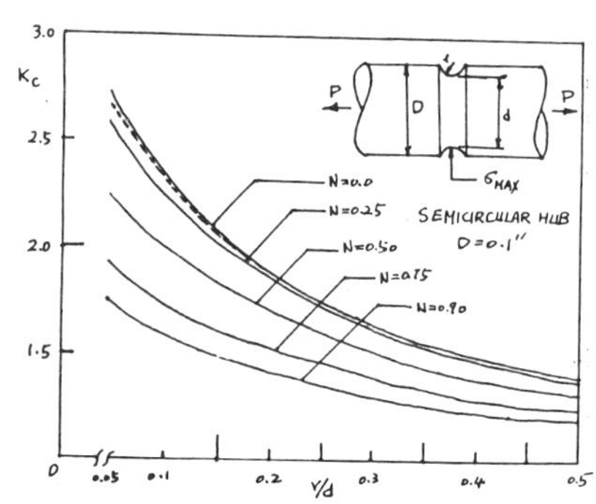

Fig. 7.3 Stress Concentration Facator, K_c for Isotropic Micropolar Elastic Cylinder with Semi-Circular Groove

The same model was used to solve for orthotropic material cases. However, because of the limitation of the space, those results are undable to show here.

N	λ	μ	κ
0.0	1.000×10^3 (psi) 6.895×10^6 (N/m^2)	6.666×10^2 (psi) 4.596×10^6 (N/m^2)	0.0 (psi) 0.0 (N/m^2)
0.25	1.125×10^4 (psi) 7.757×10^7 (N/m^2)	7.000×10^3 (psi) 4.826×10^7 (N/m^2)	1.000×10^3 (psi) 6.895×10^6 (N/m^2)
0.50	2.250×10^3 (psi) 5.063×10^6 (N/m^2)	1.000×10^3 (psi) 6.895×10^6 (N/m^2)	1.000×10^3 (psi) 6.895×10^6 (N/m^2)
0.75	5.833×10^2 (psi) 4.022×10^6 (N/m^2)	1.111×10^2 (psi) 7.660×10^6 (N/m^2)	1.000×10^3 (psi) 6.895×10^6 (N/m^2)
0.90	1.795×10^2 (psi) 1.213×10^6 (N/m^2)	3.827×10^2 (psi) 2.639×10^5 (N/m^2)	1.000×10^3 (psi) 6.895×10^6 (N/m^2)

N	γ	α	β
0.00	0.185185185 (pound)	0.0 (pound)	0.0 (pound)
0.25	2.083333333 (pound)	0.0 (pound)	0.0 (pound)
0.50	0.416666666 (pound)	0.0 (pound)	0.0 (pound)
0.75	0.108024691 (pound)	0.0 (pound)	0.0 (pound)
0.90	0.032578875 (pound)	0.0 (pound)	0.0 (pound)

Table 7.1 Material Properties for Isotropic Micropolar Elasticity

8. *Concluding Remarks*

Based on the variational formulation for micropolar (Cosserat) theory, following two and three dimensional isoparametric finite elements were developed:

[1] 4- and 8-node quadratic elements for two-dimensional analysis in Cartesian coordinate system, [2] 8- and 20-node cubic elements for three dimensional analysis in Cartesian coordinate system, [3] 4- and 8-node quadratic elements for axi-symmetric analysis, and [4] 8- and 20-node cubic elements for three-dimensional analysis in cylindrical coordinate system.

Finite element formulations and corresponding computer programs developed have been verified through numerical experiments. Agreements between known analytical/experimental results and numerical results obtained by the present FEM are excellent for isotropic micropolar elastic solids. Several numerical experiments are also conducted for orthotropic micropolar elasticity whose analytical solutions are unknown.

This research has been conducted while the first author was a faculty of New Jersey Institute of Technology and has been sponsored by New Jersey State Budget Research. Computer programs developed are copy-protected and the copyright for the current version belongs to New Jersey Institute of Technology.

9. *References*

[1] Eringen, A.C., Theory of micropolar elasticity, In Fracture, vol. 1, pp. 621-729, (edited by H. Liebowitz), Academic Press, 1968.

[2] Lakes, R.S, and Young, J.F.C., "Micropolar elasticity in bone: rotational modulus κ, Developments in Mechanics, vol. 12, pp. 233-237, 1983.

[3] Behiri, J., Bonfield, W., Nakamura, S., and Lakes, R.S., "Fracture mechanics of bone with short cracks", (preparing).

[4] Nakamura, S., Benedict, R., and Lakes, R.S., "Finite element method for orthotropic micropolar elasticity", Int. J. Engng Sci., vol. 22, pp. 319-330, 1984.

[5] Nakamura, S. and Zeng, Y.Z., "Two- and three-dimensional isoparametric finite elements for micropolar elasticity in Cartesian coordinate system", (unpublished paper), New Jersey Institute of Technology, 1985.

[6] Bathe, K.J., Finite Element Procedures in Engineering Analysis, Prentice-Hall, 1982.

[7] Huang, F.Y., Two- and Three-Diemsnional Isoparametric Finite Elements for Axisymmetric Micropolar Elasticity, Doctoral Thesis of Mechanical Engineering Department, New Jersey Institute of Technology, 1986.

[8] Gauthier, R.D., Analytical and Experimental Investigations in Linear Isotropic Micropolar Elasticity, Doctoral Thesis, Univ. of Colorado, 1974.

[9] Nakamura, S. and Lakes, R.S., "Finite element analysis of stress concentration around a blunt crack in a Cosserat solid", Comp. Meth. Appl. Mech. Engng., (submitted), 1985.

An Interface Standard for Communicating Between Dimensional Measuring Equipment and CAD Data Bases

J. Kent Butler, PhD* and Robert Neshta†

*Cal Poly—Industrial Engineering Department, San Luis Obipo, CA 93407, USA
†DMIS—Technical Director, P.O. Box 1034, Fort Richey, FL 34288-1034, USA

The specification described in this report is the result of a joint effort by users and vendors of dimensional measuring equipment to develop a standard for the communication of inspection data among automated systems. The specification was first developed by IIT Research Institute (ITTRI) with funding provided by the Quality Assurance Program of Computer Aided Manufacturing - International, Inc. (CAM-I). Version 1.0 was released to the public April 1986. Pratt&Whitney is currently under contract with CAM-I to expand and refine the specification. The result of the Pratt&Whitney effort will be DMIS Version 2.0 . It is anticipated that DMIS will become an American National Standard Institute (ANSI) standard during the last quarter of 1987.

The objective of the Dimensional Measuring Interface Specification (DMIS) is to provide a standard for the bidirectional communication of inspection data between computer systems and inspection equipment. The specification is a vocabulary of terms which establishes a neutral format for inspection programs and inspection results data. DMIS then provides a common vocabulary through which inspection programs may be passed to dimensional measuring equipment (DME) and through which results may be passed back to a receiving system. A representative configuration of the DMIS environment is shown in Figure1. An inspection program can be written on any of three dissimilar CAD systems (A,B or C). Each would have a preprocessor which converts the inspection program into DMIS format. The inspection program can then be run on any of four dissimilar DME's. DME 1 and DME 2 have DMIS postprocessors which converts the DMIS program and data into their own unique formats. Also, a host computer is being used to drive DME 2 and DME 3. The host has a postprocessor wich decodes the DMIS program and drives the two DME's, either through DMIS formats, or through some user-defined data exchange format.

Results data is converted to DMIS format by a preprocessor, on each DME, and then fed back to the CAD system for analysis, process control or archiving. The manual interface indicates that DMIS programs can be written, and results analyzed, without the use of computers.

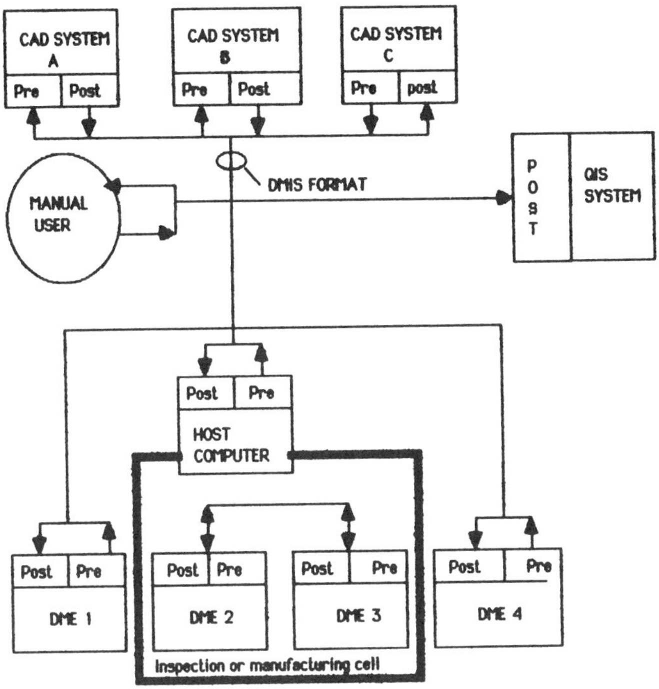

FIGURE 1. DMIS ENVIRONMENT

The DMIS vocabulary is similar in syntax to the APT NC programming language, with major and minor words seperated by slashes. Some DMIS words are the same as APT words, however the may have different meanings. DMIS is a structured programming language designed to be processed sequentially by a DME. Dimensions and tolerences are defined to the ANSI Y14.5m-1982 standard. DMIS vocabulary has been designed to be compatible with the Initial Graphics Exchange Specification (IGES), ANSI Y14.26M-1978. Characters used in the DMIS vocabulary are from the 128 character ASCII set.

DMIS STATEMENTS

There are two types of DMIS statements; process-oriented commands and geometry-oriented definitions.

Process-oriented commands are commands which are unique to the inspection process, such as, motion and machine parameter commands. Geometry-oriented definitions are used to describe geometry, tolerences, coordinate systems, and other quantifiable data required by the CAD database.

Presently, CAD part models do not include all the data needed in the DMIS interface (tolerences in particular), so supplementary data must be added manually. The evolution of CAD systems, though, is in the direction of complete part models, and DMIS has been designed to be compatible with complete part models.

The DMIS vocabulary consists of ASCII characters which are combined to form words, labels, parameters and variables. These combine to form definitions and commands, which in turn are combined to form program subunits. Program subunits , along with additional commands and definitions are combined to form complete DMIS programs.

WORDS - Major 2 to 6 alpha characters serving as a DMIS command or to indicate a class of commands.
Minor 1 to 6 alpha characters used to modify a major word or describe parameters in a command.

LABELS - 1 to10 alphanumeric characters used to name features, tolerences, coordinate systems, sensors, etc. All labels are enclosed in parenthesis. Datum labels are permitted one or two characters.

PARAMETERS - Positive or negative numeric values

REAL VARIABLES - Single alpha character or single alpha character followed by an integer between 0 and 9. These are used to compare measured results against some value.

COMMANDS - direct the DME or receiving system to perform some function. A command may contain a pointer to a definition. The definition is then used in the execution of the command. The basic structure of a command is:

MAJOR_WORD/MINOR_WORD(S),PARAMETER(S)

Reresentative Major Words used as **commands** include:

GOHOME Used to position the sensor at the coordinates defined in the last FROM statement.

GOTARG	Initiates execution of a sensor safe move and defines the end point to which the sensor will travel.
ARMEAS	Signifies that an automatic arc measurement is to be performed.
LNMEAS	Signifies that an automatic line segment measurement is to be performed.
PTMEAS	Signifies that an automatic point measurement is to be performed.
CONST	Causes the DME to construct a feature from previously defined features.
DISPLY	Specifies the device to which output data will be sent and the format in which it will be sent.
OUTPUT	Causes results of a DME measurement or evaluation to be output.
CALL	Calls and invokes execution of a macro.
ELSE	Controls a branch in a conditional.
ENDIF	Signifies the end of a conditional
ROTAB	Explicitly controls the motion of a rotary table on the DME.

FEATURES - Geometry - oriented definitions which may or may not be on a part. DMIS supports a set of feature definitions which are simple geometric elements. Combinations of features may be used to define complex two dimensional curves and complex three dimensional surfaces. Representative MAJOR/MINOR words used in **geometry - oriented definitions** include:

FEAT/ARC	Defines the actual (measured) arc of a previously defined nominal arc
FEAT/CIRCLE	Defines the actual (measured) circle of a previously defined nominal circle

Other features (FEAT/) currently supported include:

CONE	CUBE	CYLNDR	ELLIPS
GCURVE (complex curve)		GSURF (complex surface)	
LINE	PLANE	POINT	SPHERE

TOL/ANGL	Specifies an evaluated angular tolerance for a feature
TOL/DIAM	Specifies an evaluated diameter tolerence for a feature

Other tolerences (TOL/) currently supported include:

Radial	Circularity	Cylindricity	Flatness

Straightness Profile(curve) Profile(surface)
Runout(circular) Runout(total)
Angularity Distance Parallelism
Perpendicularity Concentricity
Position(coordinate system axis)
Position(feature)

MEASUREMENT SUBPROGRAMS

Consider the measurement of a single feature; The format of a measurement sequence begins with the major word MEAS, indicating the start of a measurement sequence, followed by a series of motion commands and PTMEAS commands and ending with the major word ENDMEAS. This indicates the completion of the measurement sequence. A series of valid DMIS statements programmed to measure the inner diameter of a circle is listed below;

```
1   FINPOS/ON
2   F(CIRCLE1)=FEAT/CIRCLE,INNER,CART,10,10,5,0,0,1,4
3   MEAS/CIRCLE,F(CIRCLE1),3
4   GOTO/10,10,5
5   PTMEAS/CART,12,10,5,-1,0,0
6   PTMEAS/CART,8,10,5,1,0,0
7   PTMEAS/CART,10,12,5,0,-1,0
8   ENDMEAS
```

The first line is a command consisting of only a major word and a minor word, with no parameters. This directs the DME fine positioning feature to be turned on. The second line is a definition containing a major word, a label, minor words and parameters. It specifies a unique label (CIRCLE1) to the generic feature (FEAT/CIRCLE/INNER) to be measured in cartesian coordinates (CART), with centerpoint located at x=10, y=10, and z=5. The next three parameters indicate the direction vector of the plane that the circle lies in. Therefore the circle is in the x-y plane since the direction vector of the plane is z=1. Finally the nominal diameter of the circle is 4.

Line 3 directs the DME to take 3 measurements of the circle, specified in line 2 and labeled CIRCLE1. Line 4 is a motion command directing the DME probe to a specific point from which measurements will be made. In this example the reference point is the center of the circle (x=10, y=10, z=5).

The next three lines direct the DME to take point measurements at given locations [(x=12, y=10, z=5), (x=8, y=10, z=5) and (x=10, y=12, z=5)].

The trailing three parameters of the PTMEAS command designate the approach direction vector to be used in making the measurement. For example, line 5 directs the probe to come from the negative x direction, line 6 directs the probe to come from the positive x direction and line 7 directs the probe to come from the negative y direction. Line 8 is a major word indicating that the measurement sequence is complete.

OTHER SUBPROGRAMS

A second type of subprogram is the motion sequence. Motion sequences direct a series of non-measurement moves. They begin with a GOTARG command and are terminated with an ENDGO. The motion sequences may be designate as under DME control, DMIS program control or manual control.

A third type of subprogram is the conditional. Conditionals are used to compare a measurement result to some value and then branch execution of the program based on the result. Conditionals begin with an IF command and are terminated with an ENDIF. Branching must always be forward in a program. A branch cannot cause execution of the program to be passed back to a command which occurs earlier in a program. Branching cannot branch into the middle of another subprogram.

Macro routines comprise the fourth type of subprogram and can be executed with different values in the place of dummy parameters. Macros begin with a O definition and are terminated with an ENDMAC.

PROGRAMS AND OUTPUT FILES

A DMIS inspection program consits of commands, definitions, and subprograms. Programs have a FILNAM command as the first line in the program and an ENDFIL as the last line in the program. The program is not limited in size. Measurement results are passed in the form of actual feature, FA(label), and tolerence, TA(label),definitions, that are similar to the nominal input feature, F(label), and tolerence, T(label). The output file always has a FILNAM as the first line and an ENDFIL as the last line. The output file name is the same as the file name of the program generating measurements and making tolerence comparisons.

CURRENT DEVELOPMENTS

Pratt & Whitney, a division of United Technologies has been under contract to since August 1986 to expand and enhance the software interface specifications of DMIS Version 1.0 to include additional dimensional measuring equipment and to provide expanded capabilities for equipment currently supported. This development effort will result in DMIS Version 2.0 which in turn will be offered as an ANSI standard in late 1987.

REFERENCES

1. Aubin, Richard F., Dimensional Measuring Interface Specification, SME Proceedings.
2. IIT Research Institute, DMIS I: CAM-I Dimensional Measuring Interface Specification Project Final Report, R-86-DMIS-04, March 1986.
3. Neshta, Robert, DMIS Version 1.1 Update, CAM-I Internal Communication,January, 1987.

Integrating the Management of Product Design

John E. Nast

Technical Documentation Consulting Services, Los Gatos, CA 95030, USA

ABSTRACT

The factory of the future is not complete without program and design management methods that result in a cost effective and stable design. As the complexity of design and production technology increases, both the rate of design changes and their impact throughout the company also increases. We must fully coordinate design management and design control with the objective of optimization and change avoidance, without imposing the delays of complex administrative systems. This paper examines integrating design management and coordinating design decisions as part of the factory system; it also looks at the use of Project Management Programs and Artificial Intelligence to aid this process.

INTRODUCTION

A QUESTION OF SCOPE -

In spite of the claims of tooling, computer and software salesmen, successfully operating a factory is not just a matter of integrating CAE/CAD/CAM/MRP/CAPP/NC/CNC/Bar Coding/and Robotics. Anyone who has implemented (or tried to implement) MRP knows the importance of:

- A well defined and controlled product design that is acceptably stable and cost effective to produce,
- A market for the product.
- Effectively integrated program and design management. (that coordinates interrelated activities among all affected organizations throughout the company.)

Without these prerequisites, successful factory operation is an exercise in futility, and some new technology makes it even more difficult. Many of today's computer based systems seem to be more concerned with change than control, or their approach to control is too simplistic. Using CAD system data bases in production without controlling changes to it is frightening. Establishing these data files, and changing them, have too broad an effect on the company, to allow free access for changes.

THE NEED FOR COORDINATION -
In the past, a first class designer knew all there was to know about his product: it's design, how it was built, how it would be used, maintained, etc. The designer created an initial design and then determined when changes were needed and what they would be. However, today's rapidly changing technology and its resulting specialization and increasingly complex products can no longer be effectively managed with unilateral design control by Engineering. Not only does product design influence every part of the organization, but many of these functions have requirements that must be incorporated into the design in order to be effective and efficient, or sometimes even to do their job.

The design of today's complex products requires decisions that are based on an evaluation of information from throughout the corporation. When you consider the number of specialists and different systems that have requirements, information, and guidance that can influence design decisions, no designer can be expected to develop an optimum design without the benefit of their consultation, advice, and information. In the highly competitive environment that exists today, many industries cannot afford not to have optimized designs. To achieve this we must use an integrated approach to coordinating the management of products and product design through their design and production life.

This approach must include control and coordination among interrelated activities throughout the corporation, and at times, even its suppliers and customers. Accomplishing this requires a broader perspective than CAD/CAM or even CIM. We must consider the entire industrial process.

GENERALIZED ORIGINS OF DESIGN PROBLEMS
A severe problem facing industry today is the increasing rate at which design changes are required. Changes will always be with us; however, we must minimize them and be able to process them efficiently. To do this we must first understand that they have three basic causes:

CHANGING CONDITIONS - Problems result from changes to the conditions on which the design was based, such as:

- New requirements from the marketplace or customer.
- Discontinued, revised, or price changes for purchased parts.
- Changes to production, inspection, test, or customer support.

These changes are a fact of life in today's industry. There is little that can be done to avoid them except to design enough flexibility in the product to make them easier to implement.

ACCELERATED SCHEDULES - Meeting a product introduction date often requires a development schedule that does not allow adequate time for analysis and testing. The problems that result from tight schedules and the changes to correct them are the indirect result of scheduling decisions; hopefully, based on an informed tradeoff of risk and benefits.

DESIGN INADEQUACIES - Problems that result from inadequate design practices, generally include:

- o Design errors, interference fits, missing or unclear information, conflicting requirements, etc..
- o Functional inadequacies, unnecessary complexity.
- o A need to improve producibility, maintainability, testability, inspectability, availability, etc.
- o Documents not meeting the needs of the users.

Much can, and should, be done to minimize these problems with better testing, review, checking, and by having other functions communicate their needs to design during development.

INTEGRATING PRODUCT DESIGN MANAGEMENT

My use of the term Product Design Management refers to a commercial adaptation of the principles of Configuration Management. [1] This term more clearly communicates its application and scope to the management of commercial companies. It also helps overcome the stigma of bureaucratic requirements associated with Government systems. Product Design Management applies to four areas of activity:

- PRE-DEVELOPMENT PLANNING,
- PRODUCT DEVELOPMENT,
- PRODUCTION START-UP, and
- PRODUCTION CHANGE CONTROL

It applies a corporate perspective to the coordination and review of design development and design changes by:

- o Establishing design objectives and requirements for development.
- o Providing information and guidance needed by the design.
- o Managing release of the design to production with a plan based on the users need for the information.
- o Coordinating design decisions and activities among those that have something to contribute, and by having specialists available to provide information and consult with design as needed.

---

[1] Configuration Management is the name applied to requirements for design management by Government contracts. The subject is divided into CONFIGURATION IDENTIFICATION (Baseline Management), CONFIGURATION CHANGE CONTROL, and CONFIGURATION STATUS ACCOUNTING.

COORDINATION DURING DEVELOPMENT FOR AN OPTIMUM DESIGN

When should design be given producibility inputs? After it's in production, during start-up, during Release to Production? - or - while it is still in its formative stage? - The key to effective product design, and avoiding changes, lies in coordination during product development. (Lets do it right the first time.) We must provide input to designers before the design is finished. Long ago many companies accomplished this by having Process Engineering coordinate closely with Design Engineering while the design was still on the board. A designer would be visited by a process specialist; together they would work out any problems and end up with a design that was cost effectively producible.

Although some companies still do this many others have little or no communication between Design and Manufacturing or Process Engineering until the design is submitted to release. This can leave the designer feeling that the design represents his best guess, and then having to submit it for judgement and criticism. This can and does create ill will, and does little to promote a spirit of teamwork. To be effective a designer needs coordination, advice, and information (whether he knows it or not), in a cooperative way during the design process. Some companies broaden this approach to include a Product Introduction Team of representatives from materiel, manufacturing and quality engineering, customer support, and other contributing functions.

Design Review Meetings are also a part of coordinating the development process, but they are not a substitute for "one on one" consultation mentioned above. Design Reviews are not the time or place to discuss details related to manufacturing processes or selection of datums. Design Review Meetings are for management level consideration and coordination to resolve problems, evaluate technical risks, determine schedule status, evaluate test results, reevaluate the program plan, and decide on corrective action (if needed).

PROJECT MANAGEMENT PROGRAMS FOR DEVELOPMENT AND PRODUCTION START-UP

Project Management Programs like PERT (Program Evaluation and Review Technique) provide effective planning, scheduling, coordinating, and cost control tools for managing projects with a large number of interrelated activities. They involve dividing the project into small, manageable tasks and defining each task and its starting and concluding events. Each task requires information that includes: inputs and constraints to starting and completing it, a descriptive name, the time it will take, the manhours required, etc..

PERT/Cost adds material and labor costs for each task and integrates financial information into its reports. PERT Networks illustrate the interrelationships among the activities and identify which tasks are critical to meeting the schedule. Gannt charts provide schedule information in a simplified form for day-to-day management. A series of other reports are prepared as needed to manage the project. As a program progresses, actual dates and costs are added, and the system identifies areas where the plan is not being met and corrective action is needed. These programs also provide the ability to do "what if" analysis.

PERT/Cost is a valuable tool for managing Development, Design Release, and Production Start-up; especially when linked with other systems to capture data about the product structure, document status, procurement status, material status, labor charges, and financial transactions. Even in very small projects, where cost tracking is not needed, PERT networks and Gannt Charts are useful management tools. Much of the benefit of using PERT is in the discipline that it applies during planning. Many times an initial PERT network has illustrated relationships that caused major changes in the approach to the project.

IMPROVING CONTROL OF DESIGN CHANGES

There are two major problem areas associated with the control of design changes during production. One is the review of proposed changes and planning their implementation. The other is maintaining control of the simultaneous implementation of 50, 100, or more changes.

COORDINATED CHANGE REVIEW AND APPROVAL

Review of proposed changes should include technical, cost, and schedule analysis prior to their authorization and incorporation implementation.
TECHNICAL ANALYSIS should include the following, as applicable:

- MODELING and TEST - Most changes should be modeled. If the change involves any functionally sensitive areas it should be tested and the report submitted for review.
- USERS REVIEW - specialists from each affected area should be given a chance to determine if their needs are met.
- CUSTOMERS REVIEW - Sometimes a customer requires an opportunity to review changes, and perhaps approve them.
- MARKETING REVIEW of changes to program design requirements.

COST ANALYSIS based on part disposition and change effectivity should include all one time costs of implementing the change, rework or scrap costs, and its net effect on the unit cost of the product.

SCHEDULE ANALYSIS of changes and a proposed implementation schedule may also be required for change authorization. But the need for it varies widely. Many changes require no scheduling, other simple changes can be quickly implemented with only one or two events. However, some changes require elaborate scheduling.

Not all of this analysis is required for every change. Many are clear cut and obvious without any question that they should be implemented. Many have no significant cost or schedule impact. Providing a simple process for simple changes allows more time for analysis of complex changes; the problem is be able to expedite the simple ones and assure that adequate analysis of the complex ones cannot be bypassed.

A policy for Change Authorization should be similar to the financial policy that limits the authority to commit funds. This policy would identify the cost limit for change approval by the chairman of the change board, and establish progressively higher approval levels for more costly changes.

MANAGING CHANGE IMPLEMENTATION

Many companies have difficulty implementing changes, although some of them don't even know it. They may recognize problems meeting production schedules, unacceptably high inventory levels, constant shortages, etc. and not realize that a major contributor to these problems is a lack of control of change implementation.

Many factories process 100 or more changes per month. Each change includes between 1 and 50 or more documents, and may require from one day up to more than 9 months to implement. Implementation may only require an update to the MRP or involve as many as 200 or more inter-related tasks or events by 10 or more different departments. A part in a design change that takes 6 months or more to implement may suddenly have a requirement for an emergency change that must be implemented immediately. But even this is manageable if change implementation was well planned and everything was done on schedule (which almost never happens). Many MRP systems claim to handle change implementation but I know of none that manage the environment and detail mentioned above.

PERT scheduling and reporting can do a great deal to ease this problem, especially if it is integrated with the MRP system, payroll account reporting; status records for design, quality, and manufacturing documents; procurement, receiving, stores, and other related activities. It would provide the ability to identify problems early enough to allow effective corrective action to be taken before the problem becomes severe, or impacts product delivery or the production schedule. Projecting the problems that changes give us today into the future, with even more complex higher technology products, it is hard to imagine a Factory of the Future operating without such a system.

WHAT DO WE NEED IN THE FUTURE

If you look at our needs from the perspective I have presented, it is apparent that we need a coordinated team approach to the development of product design and control of its changes; but it must include a clear delegation of responsibilities.

We could benefit from a computer program using Artificial Intelligence to support of the review, analysis, scheduling, and tracking of design changes. Such a system would have to be integrated with MRP, design status records, status records for Manufacturing, Quality and Test Engineering documents; and other Engineering, and Manufacturing and Customer Support systems. It would first consider the change objective, priority, urgency, parts & processes affected, disposition of parts, etc. With that it could assemble data on stock status, lead time, production setbacks, tools and fixtures affected, instructions and procedures affected, cost, design sensitivity, process sensitivity, related changes in-process, impact on customer support and spares, etc. and conduct a preliminary analysis. This analysis would provide a list of questions, additional data requirements, etc. and identify who can best respond to each item. Once the new data is entered the system could provide a cost analysis and implementation schedule. Management would use these and any technical comments, results of testing and modeling, etc. to decide whether or not a change should be approved. After approval, it would then track implementation. Other benefits could include historical data that would provide information to improve new designs and their production. This is a long way off but I believe it can, and will be developed. Since each company and product is unique and constantly changing, this program would use some level of Artificial Intelligence to learn and keep up-to-date with its changing environment.

To illustrate how the necessary rules and logic for such a system could be developed, consider just their classification by primary objective, using the following categories:

- o ERROR CORRECTION - Correct design error, missing data, etc.
- o DESIGN REQUIREMENT - Make the product perform to spec.
- o MANUFACTURING SERVICE - Facilitate or improve Manufacturing, Quality Assurance, or Test costs or practices.
- o CUSTOMER SUPPORT SERVICE - Facilitate or Improve Customer Support costs, performance, or practices.
- o PRODUCT ENHANCEMENT - Improve performance above required levels, add features or capabilities, etc.
- o CUSTOMER REQUESTED - Changes to requirements that were previously requested. (Paid for by the customer.)

Each category has its inherent characteristics and requirements for analysis. Cost is not usually a factor for an ERROR CORRECTION or PRODUCT REQUIREMENT change. Except for availability and other problems that could stop production, the decision to approve a MANUFACTURING SERVICE or CUSTOMER SUPPORT SERVICE change is based on cost savings. PRODUCT ENHANCEMENTS must be justified by Marketing to assure that their benefits offset the cost. And it is the customer that must accept the cost of a CUSTOMER REQUESTED change. It becomes clear that each class could have its individual rules for cost evaluation and justification.

This same analysis can be applied to other change criteria and decisions and be used to derive a set of practices, recommendations, and rules for change analysis. Consider the consequence of not implementing a change, urgency for a change, planning the most cost effective implementation, the fastest implementation, etc.

IN CONCLUSION

Today we face the problems of design deficiencies and a constant barrage of design changes. These problems will not go away, in fact, they will get progressively worse as our technology becomes more complex. Some companies have found that design changes represent about 25% of the gross production labor and material cost, other companies loose up to 50% of their capacity from problems caused by poor control of change implementation. If we are to be successful in the future we must minimize the need for change and learn to handle them better. I hope this paper will give you some ideas on how this can be accomplished.

Use of CAD Systems in Design of Space Station and Space Robots

Suren N. Dwivedi[*], P. Yadav[*], Gary Jones[†] and Elmer W. Travis[‡]

[*]College of Engineering, University of North Carolina at Charlotte, Charlotte, NC, USA
[†]Mechanical Engineering Branch, Goddard Space Flight Center, Greenbelt, MD, USA
[‡]Head, Mechanical Engineering Branch, Goddard Space Flight Center, Greenbelt, MD, USA

ABSTRACT

In the last twenty five years, enormous advances have been made in the area of computer aided design (CAD) resulting from the improvements in the related hardware and software over this period - from the first expensive mainframes of the 1960's to the present inexpensive microcomputer level hardware system to run its program, the emergence of various geometrical modeling techniques like the wire frame, 2-D, 3-D, and solid modeling etc., and the various user friendly Finite Element Analysis software packages. The important features offered by CAD are computer graphics and its ability to store, retrieve, manipulate and display graphical information with speed and accuracy.

CAD has really revolutionized the concept of design. The downward trend of prices and the upward trend of performance of computers of all levels and the phenomenal growth of the CAD industry during the last few years portend a much wider implementation of CAD facilities in industries. NASA, as a whole, and Goddard Space Flight Center in particular, a pioneer in the field, has highlighted the work areas for concentration, and, having the most advanced CAD systems and facilities available, will continue to use and develop advanced software for special and general purpose use.

In the first part of this paper, a review of the state-of-the-art is given, whereas the second part deals with the various advanced CAD facilities and supplementing systems being used at the Goddard Flight Center of NASA, one of the leading Government agencies in developing and applying numerous CAD softwares and protocols, especially for design and development of space station and robots to be used in space station.

INTRODUCTION

In the last twenty five years, enormous advances have been made in the area of computer aided design (CAD) resulting from the improvements in the related hardware and software over this period - from the first expensive mainframes of the 1960's to the present inexpensive microcomputer level hardware system to run its program, the emergence of various geometrical modeling techniques like the wire frame, 2-D, 3-D, and solid modeling etc., and the various user friendly Finite Element Analysis software packages. The important features offered by CAD are computer graphics and its ability to store, retrieve, manipulate and display graphical information with speed and accuracy. In the first part of this paper, a review of the state-of-the-art is given, whereas the second part deals with the various advanced CAD facilities and supplementing systems being used at the Goddard Space Flight Center of NASA, one of the leading Government agencies in developing and applying numerous CAD softwares and protocols.

1.0 STATE OF THE ART

CAD has its origin in the government supported research and development programs for flight simulation, radar detection and other related purposes, during the 1960's. As of today, CAD encompasses a wide range of activities, a sequential listing of which may look like the following:

- Product structure definition.
- Concept sketching.
- Design parameter definition.
- Design logic rules definition.
- Dynamic dimensioning.
- 3-D model building.
- Analysis and optimization.
- Viewing and verification.
- Layout and production drawings and lists.
- Process planning.
- Numerical control programming.
- Quality control.

This substantial increase in capabilities has been made possible due to the improvements in hardware and software utilized in CAD.

1.1 Hardware

1.1.1 Computers

The first computers to run programs for CAD were the mainframes of the early 1960's. The high costs involved limited their use mainly to Universities and large corporations. Among the first non governmental entities to adopt CAD were the aerospace and automotive industries. However, in the 1970's, CAD was made accessible to a broader range of users with the advent of less expensive and more compact minicomputers based on integrated circuit technology. Today, machines running CAD software span an even broader range, from low cost personal computers that make the technology affordable to even the smallest of firms to the most powerful supercomputers that routinely perform tasks formerly considered too computation-intensive to be handled at a reasonable time and cost. The characteristics that once clearly differentiated one type of computer from another have recently become blurred as low-end equipment gets more powerful and high-end machines become less expensive and more compact. But broadly classifying, the four main categories of computers generally used for CAD are; microcomputers, minicomputers, mainframes and super computers. Selecting one kind over the other is almost always a trade-off between the cost of the machine and the computation load it can handle. However, most CAD systems continue to rely on minicomputers as the processing machine. These are generally 32 bit computers that access up to 16-M bytes of main memory. To enhance the capabilities, considerable third-party software is available, spanning the range of CAD/CAM programs including geometric modeling drafting, finite element analysis, kinematics and NC programming as well as electronic circuit design, analysis and testing programs.

On the other hand, mainframes are still used in those applications requiring substantial data processing and large memory capacities. They typically support numerous peripheral devices such as printers, plotters, terminals, tape drives and disc drives.

1.1.2 Peripherals

The peripheral devices used for obtaining a hard copy are sophisticated too, of which a number of choices are available. As with the selection of a computer, here too selection is a trade off between cost and capabilities of the device.

1.1.2.1 Pen Plotters

Many vendors have electro mechanical pen plotters available for CAD systems, of which there are two categories:

(I) Flat bed type, and

(II) Drum type

In the flat bed x-y plotters, the servo controlled pens are moved along both the axes while the paper remains stationary, whereas in the drum type, the pens or ink jets move along one axis and the paper mounted on a revolving drum moves along the other axis.

1.1.2.2 Impact Printers

This type of device is an outgrowth of the established monochrome matrix line printer in which tiny hammers in a print head strike a ribbon in a manner similar to that of a typewriter, producing a dot matrix image on the paper. Various colors are produced with multiple ribbons or with a single ribbon have bands of different colors. Some colors are produced with a single strike on the ribbon, while other shades are created with multiple strikes that combine primary color in the dot matrix.

1.1.2.3 Ink-Jet Printers

This type of device creates images with a matrix of colored dots made by ink propelled toward the drawing medium in fine droplets. Ink-on-demand systems generate individual droplets of ink with a discrete pulse only when required to generate the image. Alternatively, continuous-jet printers constantly accelerate ink in a continuous stream, which is broken into droplets and then deflected away from the paper (usually electrostatic repulsion) and recirculated. When an individual dot is called for, the charge is momentarily removed and the ink proceeds to the paper.

1.1.2.4 Camera Systems

Camera systems are typically terminal driven and normally tap directly into the video signal of the color terminal. This system has originated from one of the oldest methods used to produce color CRT hardcopy. The resolution of camera system copies is superior to all other approaches. However, the high cost of the system and cost per copy restrict its use to high level organizations.

1.1.2.5 Electrostatic Printers

An array of densely spaced writing nibs place point charges on a dielectric paper, which is then passed through an applicator containing a toner (liquid or dry depending on the printer type). The toner is attracted to the charged areas and fused onto the paper to produce the permanent color copy. The paper is charged and exposed to each toner separately, with one final pass required for each primary color and black.

1.1.2.6 Thermal Transfer Systems

In this technique, tiny nibs are heated to melt pinpoint spots of color pigment from a wax-impregnated sheet onto the underlying paper. The paper makes three passes under the thermal head, one for each primary color. The ink-sheet-roll is segmented into consecutive, page size bands of color.

1.2 Software

With the availability of today's software, a design engineer can develop a product design on an interactive graphic display terminal, monitor the work continuously and on completion, store the data in a tape or disc. These data can subsequently be used by a control medium to manufacture the desired product. But before the product is created actually, the whole process can be simulated and the results of the model analysis be obtained, which portend the true characteristics of the finished product. But in CAD, the vital objective is to create the models quickly, which is understandable to the user. Various kinds of geometric modeling techniques are available now for interactive computer graphics.

1.2.1. CAD/CAM Drafting

The most basic CAD system provides drawings created much like manual drawings, i.e., orthographic projections of an object can be created with the support of interactive computer graphics software. Points, lines and circular arcs are described. To draw a circular arc, the circle options and the input parameters, as prompted by the software, is specified. The most basic CAD systems provide drawing construction capabilities with little or no intelligence concerning the geometric characteristics of the drawings.

1.2.2. Wire Frames

Wire frames give a skeletal outline of an object, much like the frame of a model airplane or a bird cage. With the substantial data, as is required by this technique, being available in the computer, it is easy to provide software that facilitates the visualization of the wire frame from different viewing positions. The modeled object can also be viewed in perspective and in scaled formats. These functions can be accommodated without making multiple drawings in the conventional sense. The primary facility, however, is that of visualizing as we look through the plane of an object. Automatic removal of hidden line and other 3-D enhancement may or may not exist. Although wire-frame modeling is generally not suitable for analysis, there are many applications where all that is needed is the wire frame.

Wire frame representation is an aid to complete visualization, to NC tool paths generation, to inspection of finite element models prior to analysis and as an added advantage computational requirements are minimal.

1.2.3. 2-D Mathematical Modeling

This is a more design oriented modeling system, and the part is mathematically modeled in 2-D. Here the system provides a large number of mathematical forms from which the designer makes a selection - perhaps by trial and error. In some design problems, this might be desirable. But usually, the designer is not a mathematician and has no idea which mathematical functions should be tried. However, such 2-D mathematical modeling may serve as boundaries for 3-D surfaces.

1.2.4. 3-D Modeling

This modeling scheme creates both a comprehensive pictorial presentation and a comprehensive geometric model of what is presented. By incorporating surface and/or solid presentation techniques, the object appears on the display much like the actual physical object would appear. Thus, surface texturing, shading, coloring, hidden line removal, and other visual enhancements may be provided.

2.0 Softwares and Protocols

CAD softwares and protocols have been developed to a great extent and applied in diverse areas by NASA. This part of the paper describes some of the most recently developed and used CAD tools at the Goddard Space Flight Center of NASA.

CAD at NASA/GSFC is being used to support all phases of spacecraft design from early conceptual to the final design. In the conceptual phase the CAD work product normally consists of 3-D solid models of the various alternative concepts together with the associated plots illustrating such things as instrument fields of view and clearance studies. CAD is also being used at GSFC to produce the final dimensioned drawings for spacecraft part fabrication.

CAD is currently being used at GSFC to support the space station robotics tasks. This includes the development of solid models representing the Man Equivalent Robot (MER), development of drawings of the proposed GSFC Robotics development facility, and the redesign of existing spacecraft modules to be robot serviceable.

GSFC is making extensive use of the Initial Graphics Exchange Specifications (IGES) to support Space Station tasks. Engineering drawings in the IGES format have been exchanged with the NASA/Johnson Space Flight Center and General Electric Corporation. GSFC working under the auspices of the NASA Office of Chief Engineer has developed IGES test files that have been extensively used within NASA and in industries to test the various implementation of IGES translators.

The available facilities in the form of software and hardware at the Goddard Flight Center of NASA to work on the above mentioned areas have been categorized as follows:

(1). 2-D Drawing systems.

(2). 3-D Wire frame drawing systems.

(3). Solid modeling systems.

(4). Finite element analysis systems.

(5). Hardcopy output/reproduction services.

(6). Transfer/translation services.

(7). Display devices.

(8). Annotation/modification.

The details of the various systems and software packages and their features follows.

2.1 2-D Drawings

The various tools used to obtain 2-D drawing are described below:

2.1.1 APPLICON BRAVO & NASCAD

This system provides the capability to generate fully dimensioned 2-D drawings for production use. In general, use of these systems for the generation of 2-D drawings is only appropriate on one or both of the systems.

2.1.2 AUTOCAD 2.1

A single user version of the AUTOCAD 2.1 software package with drawings extension I, II, III is available for use by arrangement with the TMIS Space Station Technical Operation Center (SSTOC). This package runs on standard configuration TMIS workstations and provides a 2-D computer aided drawing capability. Output can be directed to workstation printers and/or to a myriad of other peripherals if required. Drawing files produced using AUTOCAD can be exported to other system through the use of third party vendor produced conversion.

This version features a more refined user interface and a range of new utilities including an integral IGES file import/export capability.

2.2 3-D Wire Frame Models

2.2.1 NASCAD & APPLICON BRAVO Systems

This system provides the capability to produce fully dimensioned 3-D models for conceptual and preliminary design work.

The ability to create 3-D wire frame models using the two systems are essentially equivalent. However, BRAVO has a more advanced user interface and supports raster graphics (i.e. screen repaint is not required to see each change). Accordingly, BRAVO can be slightly faster for a CAD designer to use.

Figures 1A-1B illustrate the Space Station concept configuration. The model was developed at the Johnson Space Flight Center and sent to the Goddard Space Flight Center in an IGES data file.

2.2.2 AUTOCAD 2.5

Construction of limited 3-D wire frame models using this system is possible. All 3-D models must be constructed as a set of extrusions along the z-axis of a 2-D model.

2.3 Solid Modeling

Solid modeling at NASA is achieved by employing the following facilities:

2.3.1 APPLICON BRAVO System

This system is a design iteration tool useful for conceptual and preliminary design work. Mass modeling and moment analysis is supported (results are +/- some delta due to offsets required to improve the output image). Two analysis models are available: Precise analysis, which is an off line batch mode capability; and faceted analysis, which is an online quick look capability.
Figures 2, 3, and 4 illustrate concepts of a robot arm and a robot developed on the Applicon Bravo System using solid modeling.

Illustrated in Figures 5, 6A, 6B, 6C and 6D are conceptual drawings of a proposed EUVE spacecraft. These conceptual layouts were developed on the Applicon Bravo CAD system using solid modeling.

The CAD support section has the ability to produce finite element models in has a variety of formats and has a variety of finite element modeling tools available for use. However, the CAD support section does not take responsibility for completing actual analysis work on a production basis.

2.3.2 APPLICON BRAVO/MSC NASTRAN

This capability has been tested one. Finite element model construction (mesh/element structure development). Using the Application Bravo utility is a semi-automated iterative process. Appropriate subsets of element can be used in COSMIC NASTRAN. The CAD/CAE support section facility supports the MSC NASTRAN software system.

2.3.3 APPLICON BRAVO/ANSYS

The capability of producing an ANSYS format Finite Element Model file exists but has not been tested. The CAD support section facility does not support the ANSYS software system.

2.3.4 APPLICON BRAVO/SUPERB

The capability of producing a SUPERB format Finite Element Model file exists but has not been tested. The CAD support section facility does not support the SUPERB software system.

2.3.5 NASCAD/COSMIC NASTRAN

The capability to display a cosmic NASTRAN model in NASCAD has been implemented. The capability to generate a model for use in COSMIC NASTRAN from NASCAD exists but is not currently in use. The COSMIC NASTRAN system is supported by the CAD support section facility.

Figure 7 illustrates a concept layout of a Robot Development Facility being constructed at the GSFC; the drawing was generated via NASCAD.

2.4 Hardcopy Output/Reproduction Services

Hardcopy reproduction at NASA is done by employing the below mentioned because of their high resolution and clarity.

2.4.1 TEKTRONIX 4129

This stand alone workstation/plotter supports the online near real-time production of full-color, shaded, and hidden line plots on
8 1/2"X11" paper size.

2.4.2 VERSATEC 9242 Color Plotter

This plotter supports the offline production of full color, shaded, and hidden line plots. Plots are available in standard sizes A through E as well as arbitrary formats up to 40" wide. The plotter uses 42" wide roll paper. This plotter is capable of generating plots from either the NASCAD or the Applicon Bravo systems.

2.4.3 Large Format Xerox

A large format Xerox machine is available for Black & White drawing reproduction (standard paper & vellum).

2.4.4 35MM Slide Production

The capability to easily produce color 35MM slides from color terminal display is available.

2.5 Transfer/Translation Services

These services in various capabilities are obtained by one of the following facilities available at NASA.

2.5.1 IGES 2.0 TO APPLICON BRAVO 3-D Wireframe Translation

This utility provides a partial translation capability. Surface entities are not supported. A maximum file size limit may exist.

2.5.2 NASCAD TO IGES 3.0 Translation

This is a tested capability currently in use by the CAD support service facility. Software is iterated as required to insure compatibility.

2.5.3 APPLICON BRAVO SOLID MODEL TO APPLICON BRAVO 3-D Wireframe Translation

This is a semi-automated process. The system pulls the boundary curves from all surfaces of the solid mode to establish a 3-D wire frame model. Inaccuracies due to inconsistencies/assumptions made during the development of the solid may cause problems.

2.5.4 APPLICON BRAVO SOLID MODEL TO APPLICON BRAVO 2-D Wireframe Translation

This capability exists and can be effectively employed on appropriately constructed solid models. The limitations of the solid models may prevent the direct generation of 2-D wireframe engineering drawings for production use. If the generation of production quality drawings are going to be required from a solid model the CAD designer should be notified prior to the start of the solid model construction process.

2.5.5 APPLICON BRAVO 3-D Wireframe To APPLICON BRAVO Solid Model Translation

This is predominantly a manual process. The basic geometry and part placement information from the 3-D wireframe is used as an on-line reference by the CAD designer.

2.5.6 APPLICON BRAVO TO IGES 2.0 Translation

This capability is also available. A partial translation is possible; solids and surface entities are not supported. A maximum file size limit may exist.

2.5.7 NASCAD IGES File Editor

This is a functioning NASCAD utility which will operate on any IGES file. The utility provides diagnostics on all non-conforming entities in the IGES file being processed.

2.5.8 NASCAD Analysis Program

The IGES Analysis program insures that an IGES file is fully conforming. Three aspects of conformity are evaluated:

(A) Produce of list of all entities. Unrecognized entities are flagged.

(B) Ascertain if the physical structure of the file is correct.

(C) Insure that the geometric information is consistent.

2.5.9 AUTOCAD 3.0 TO IGES Translation (Future)

This capability is under test by the TMIS office as a means to allow drawings produced, annotated and/or modified on TMIS work stations to be uploaded into the GSFC TMIS CAD catalog system.

2.5.10 IGES TO AUTOCAD 3.0 Translation (Future)

This capability is under test by the TMIS office as a means to allow drawings to be downloaded from the GSFC CAD catalog system to TMIS work stations for display and annotation/modification.

2.5.11 Magnetic Tape Transfer Files

Here, VMS tape file structure is employed. Unlabeled IGES files consisting of pure character data with a blocking factor can be produced. It uses 80 character records and 10 records per block. All characters are in ASCII code. It does not use headers.

2.5.12 VAX 11/785 TO IBM 4341 Electronic Transfer

This is an 9-6 KB link via the Applitek/CCTV system and the Code 520 VAX 8600 Gateway. Work to establish this communication path is in progress. The major work item remaining is the development of the VAX 8600 Gateway software to manage the DECNET to XNS bi-directional protocol conversion.

2.5.13 VAX 11/785 TO JSC CYBER Electronic Transfer

This is a 9.6 KB link to JSC. The communication line is in place. Various testings are being performed.

2.6 Display

The different kind of displays and workstations employed at NASA are described below:

2.6.1 4129 TEKTRONIX Workstation

Stand alone workstation used as terminal for accessing the NASCAD system on the VAX 11/785. (Future) A second fully configured 4129 workstation is on order.

2.6.2 4016 TEKTRONIX Terminals

There are currently two of these storage tube type terminals still operating. They are used on an infrequent basis.

2.6.3 4111 TEKTRONIX Workstation (Future)

A new 4111 color raster, graphics/text editor terminal with a VT-100 emulation capability is on order.

2.6.4 APPLICON Terminals

There are currently two Applicon color terminals, used for accessing the Applicon Bravo system on the VAX 11/785. (Future) A third Applicon color terminal is on order.

2.6.5 TMIS Workstations

An IBM/PC AT is being used to support AutoCad tasks. It is configured with a data tablet, plotter support and a high resolution display.

2.7 Annotation/Modification

The facilities available at NASA are:

2.7.1 NASCAD, APPLICON BRAVO & AUTOCAD 2.1

Currently, all annotation/modification is handles via mark-up of hardcopy output and/or online discussions with the CAD designers. Efforts are underway to develop an electronic interface for exchanging markups of all products (AUTOCAD, NASCAD or APPLICON BRAVO).

The ability of AUTOCAD to import and export IGES files will be tested. If successful, the annotation and/or modification of drawings contained in the GSFC TMIS CAD catalog system at TMIS workstations will be possible. Furthermore, the changed drawings can then be uploaded into the unapproved CAD catalog.

CONCLUSION: CAD has revolutionized the concept of design. The downward trend of prices and the upward trend of performance of computers of all levels and the phenomenal growth of the CAD industry during the last few years portend a much wider implementation of CAD facilities in industries. NASA, a pioneer in the field, has highlighted the work areas for concentration, and, having the most advanced CAD systems and facilities available, will continue to develop advanced software for special and general purpose use.

REFERENCES

1. Advancing Automation and Robotics Technology for Space Station and for US Economy. NASA Technical Memorandum 87566. Submitted to US Congress, April, 1985.

2. Clarke M., Bronez, Mark A., "Telerobotics for the Space Station", Journal of Mechanical Engineering, February, 1986.

3. Robotics Servicer Workshop, Goddard Space Flight Center, Greenbelt, May, 1986.

4. Space Station Automation and Robotics Studies - Operator Systems Interface D483-10027-1, Boeing Co., 1984.

5. Space Station Automation Studies - Automation Requirements derived from Space Manufacturing Concept, Contract NAS 5-25182, 1984.

6. Automation Studies for Space Station Subsystems and Mission Ground Support, Final Report submitted by Hughes Aircraft Co., Reference No. F5713, Contract 82-14F, 1984.

7. NASA Space Station AI Based Technology Review Project 7268, SRI International, Menlo Park, CA, 1985.

8. Space Station Automation Study - Autonomous Systems and Assembly Contract NAS 8-35042, Martin Marietta, 1984.

9. Space Station Automation Study - Satellite Servicing Contract NAS 8-35081. TRW Inc. 1984.

10. Nevins, J.L. and Whitney, D.E., Information and Control Issues of Adaptable Programmable Assembly System for Manufacturing and Teleoperator Applications? Journal of Theory of Mechanisms and Machine Theory, Vol. 12, pp. 27-43, 1977.

11. Man Space Flight Study No. 981-10-30-04 Manipulator Systems for Space Application, Vol. I and II, Argan National Laboratory, April 1967.

12. NASA GFSC Plan for Space Station Robotics Program, Private Communication.

13. Dwivedi, S.N., "Guidelines for Design for Manufacturability and Automation", Proc. of International Congress in Technology and Technology Exchange, Pittsburgh, Oct. 6-8, 1986.

14. Dwivedi, S.N. and Klein, B.R., "Design for Manufacturability Makes Dollar and Sense", Journal of CIM Review - A Journal of Manufacturing and Management, Vol. 2, No. 3, pp. 53-59, 1986.

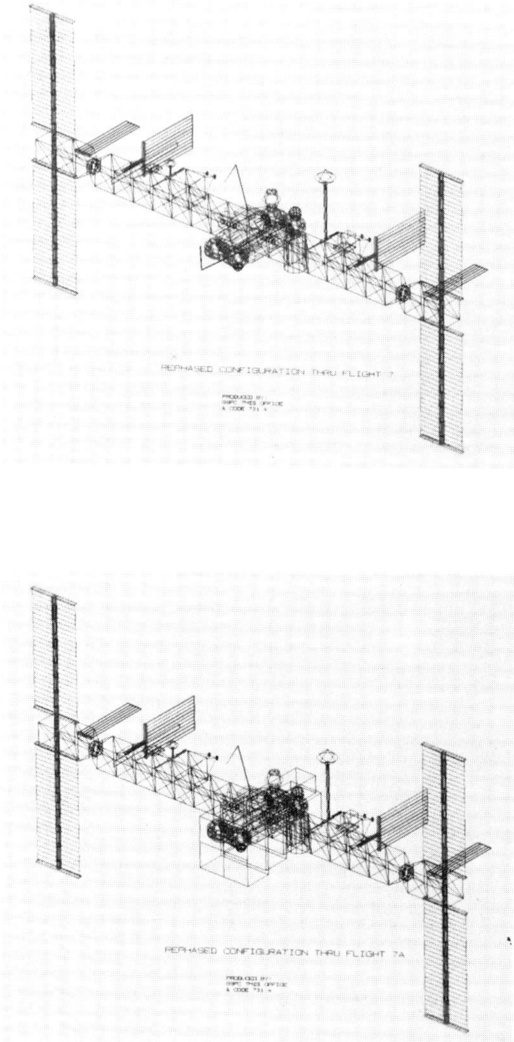

Fig 1. (A) & (B) SPACE STATION CONCEPT

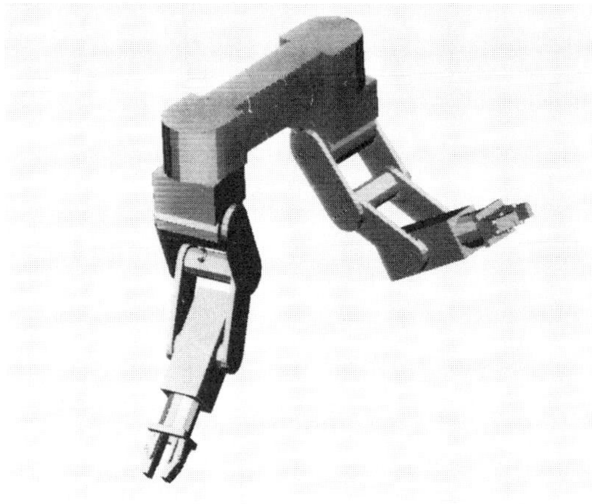

Fig 2. DUAL ARM ROBOT CONCEPT

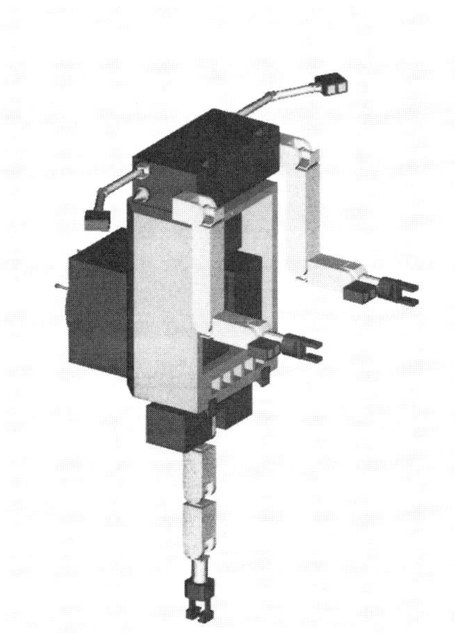

Fig 3. DESIGN OF TWO ARM ROBOT
FOR SPACE APPLICATION

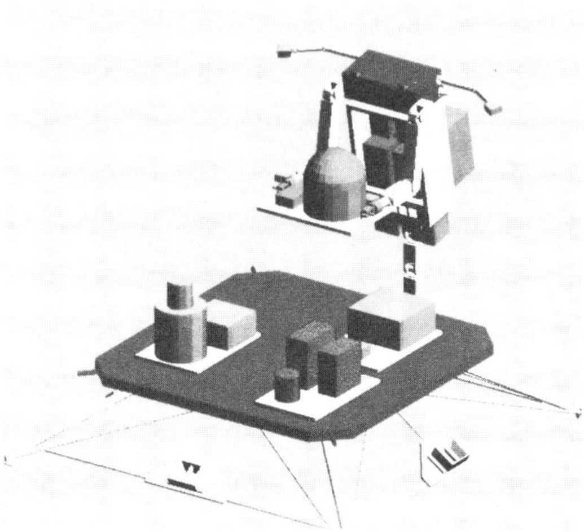

Fig 4. ASSEMBLY BY ROBOT IN SPACE STATION

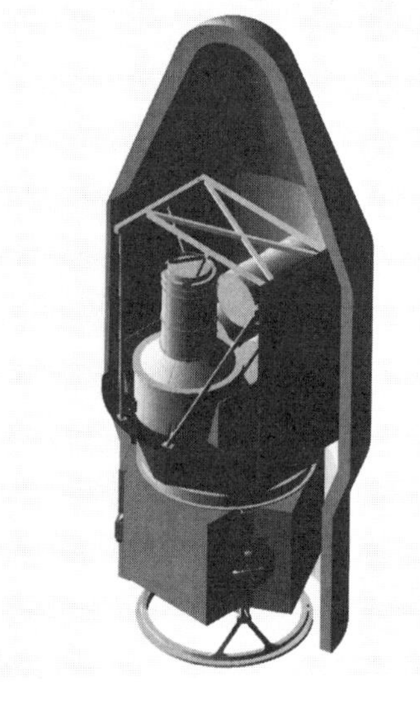

Fig 5. EUVE SPACECRAFT IN DELTA FIRING

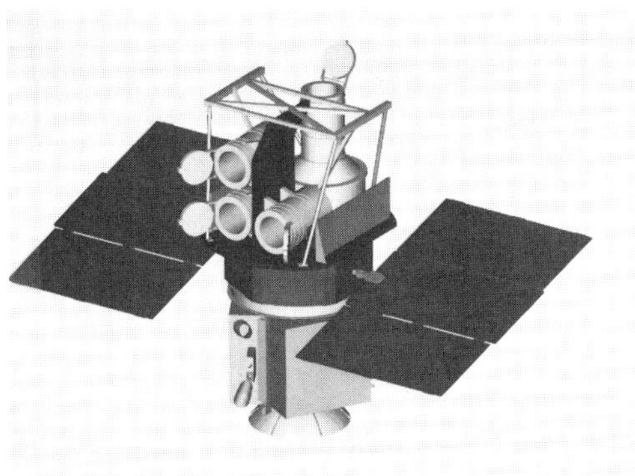

Fig 6 (A). EUVE SPACECRAFT CONCEPT A

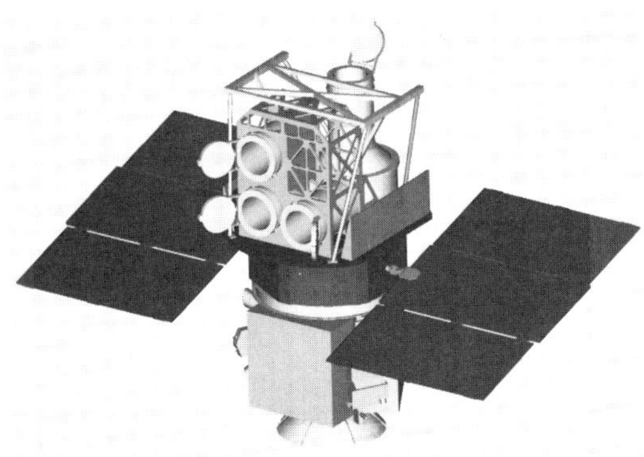

Fig 6 (B). EUVE SPACECRAFT CONCEPT B

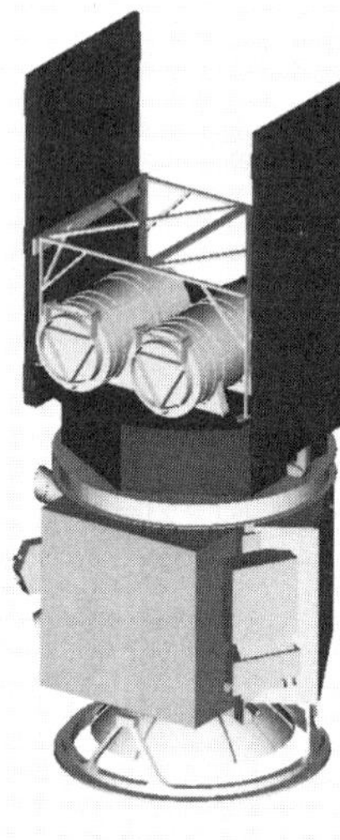

Fig 6 (C). EUVE SPACECRAFT CONCEPT C
(SOLAR ARRAYS STOWED)

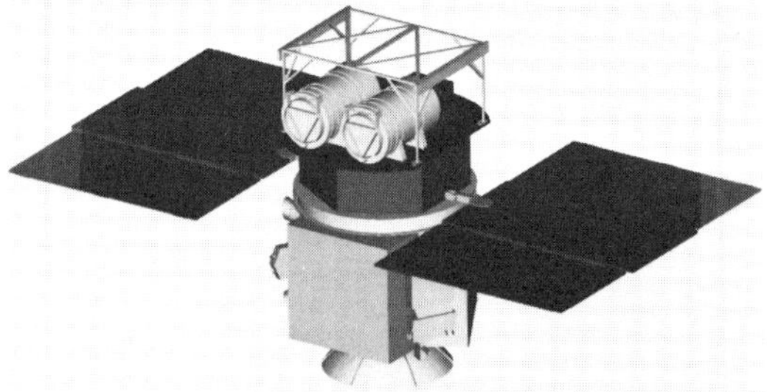

Fig 6 (D). EUVE SPACECRAFT CONCEPT D

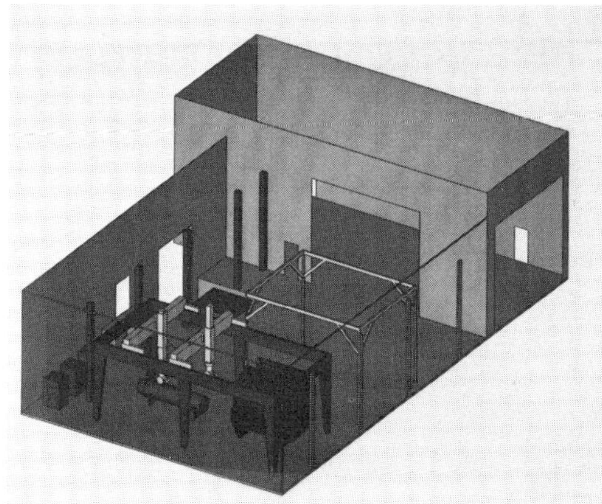

Fig 7. PROPOSED ROBOT FACILITY AT GSFC

Relevance of Numerical Controlled and Computer Controlled Machine Tools in India

Rakesh Sagar and K.V. Ramakrishna

Mechanical Engg. Dept., IIT, Delhi, India

ABSTRACT

In this paper a techno-economic evaluation of the Indian Industries using numerical controlled and computer controlled machine tools is made. For this purpose questionnaries are prepared for the organization, for the engineering personnel connected with the N.C. and C.N.C. machines and for the operators working on these machines. The comparison of productivity of N.C. and C.N.C. machines with conventional machine tools is also made. The survey has shown that the number of indigeneously produced CNC and N.C. machine tools being used by Indian Industry as compared to the imported ones is very small. The reasons for this are lack of variety, high prices and lack of advanced technology. For improving this situation it is suggested that there should be a simple licensing policy, generous offers for collaboration be made, taxes to be reduced for a few years, higher rate be allowed for depreciation and allowing foreign equity capital. There is a demand for quality products and introduction of this technology will help in certain critical manufacturing activity. Initial cost of the machines is an inhibiting factor for small and medium scale industry. At present very few training facilities exist for the operator. So the number of trained operators is less and there is a feeling of superiority among the trained personnel which is resented by their other colleagues working on conventional machines. The trained people demand higher wages. The production levels of C.N.C and N.C machines are very high as compared with conventional machine tools. In some cases the ratio is as high as 23.5. The high productivity is a result of combining of the operations, reduction in transportation and handling times. In some cases there is poor utilization of machine tools because of improper scheduling, and more time is taken for prove-outs. Improper handling and breakdown account for some of the lost time. Power cuts and operator absentism is another cause for lost time. The operators are happy with the working conditions as they work in air conditioned chambers. One can observe the pride associated with working on N.C. machines.

INTRODUCTION

Indian N.C. scene starts from early sixties. The current rate of import of N.C. and C.N.C. machine tools is more than 100 per year. This is from various countries such as Japan, U.S.A., U.K., West Germany, Switzerland etc. Japanese machines are more in India as they are inexpensive compared to the machines from other countries.

By and large Indian machine tool industry started with machine tools designed by foreign collaborators during 1950's or earlier. During the last 30 years indigenous development has taken place in some areas leading to the replacement of the old designs. During 1960's industry added to its range highly productive machines such as copying lathes, automats, gear cutting machinery etc. again through collaborators. By 1970 indigeneous industry had sufficiently mastered machine building technology and 70% of the domestic demand was met. Highest being in 1978 (86%). The application of numerical control in machine tools brought a revolution in machine tool technology in advanced countries as early as 1960. But in India the application of this Technology started on a very limited scale after ten years. Even today maximum percentage of our requirement is met through imports. This technological gap has its effect on increased imports as well as reduced exports. Imports have increased from Rs. 400 million in 1978 to Rs. 1100 millions in 1982. The increasing imports have obviously brought down the percentage share of domestic production in the country's consumption of machine tools. Indian exports of machine tools were Rs. 200 millions in 1978 and Rs. 240 millions in 1982. The increase was due to higher unit value realization rather than on account of increase in the volume of exports. Indian C.N.C. machine tool production started in 1978 by H.M.T. Ltd.. Three more companies are on the way for the production of C.N.C. machine tools with foreign collaborations. A committee investigated various factors affecting the Indian machine tool industry (1).

METHODOLOGY

Engineering industries desirous of switching from a conventional machining technology to highly advanced N.C. Technology face the problem of high capital investment, lack of trained personnel and availability of spare parts besides numerous other difficulties. It is essential to realise that N.C. technology is not just another metal cutting technique: it is a management technique and a discipline, which if applied correctly can result in improving the overall operation. Alfred Tack (2) states that N.C. is a dictator; benevolent if its organizational needs are met; tyrant if they are not. When the installation of N.C. is being considered it is essential to be prepared to revise conventional ideas of manufacture and to rethink about the whole factory configuration. The procedure adopted here is to look at each organization separately. The common part of the investigation being the questions asked. Three questionnaires are prepared-one for the organization, one for the engineers concerned with N.C. and one for the operators.

QUESTIONNAIRE FOR THE ORGANIZATION:- This questionnaire is prepared to study the following factors.

1. Facilities existing which support the functioning of the C.N.C. machines- such as computers for programming etc.

2. Machine tool utilization and break-up of the idle time.
3. Comparison of the production times taken for some sample parts on conventional as well as on N.C. machines.
4. Savings that are obtained by going for N.C. machines.
5. Reasons for purchasing N.C. machines and also going for imported N.C. machines.
6. Reasons mentioned in the preinvestment purchase analysis if at all done and deviation from the expectations.

ENGINEERS QUESTIONNAIRE:- The main basis for the engineer's questionnaire was the survey conducted during 1971 (3). Training given to the engineers is asked for in the questionnaire. This gives a chance to understand the interest of the company in establishing a new technology in their organization. Team work is understood to have a major impact in the success of managing the N.C. machines. The engineers are asked to choose on a five point scale, their replies. This questionnaire incorporates the response of engineers to all the aspects which are important for NC operations.

OPERATOR ATTITUDE ANALYSIS:- C.N.C. machines are considered as high technology machines. These machines are given a special status on the shop floor. The reasons for this being it is a new and expensive technology which demands higher utilization, requires greater team effort to make it economically viable. The questionnaire for operators is mainly designed to study and assess their level of understanding of the N.C. machines, changes in expectations like salary, changes in relations with co-workers and to find out whether they are satisfied with existing facilties.

STUDY FINDINGS

With an objective to investigate how is N.C. in India, seven Public Sector industries and one Private Sector industry have been selected. The findings from the organizational questionnaires are given separately for each industry visited. Because of the difference in organizational norms such as recording the information and organizational structure, some deviation can be observed in adopting the model chosen in these industries. The data obtained from the engineer's questionnaires and the operator's questionnaires are combined.

BHARAT ELECTRONICS LTD. BANGALORE:- B.E.L. is an industry which essentially makes defence communication equipment has a good experience in the field of C.N.C. machining with the latest machines available in their Bangalore unit. The following reasons were obtained after going through the records and the recommendations made by the Technology Development Department. i) N.C. will replace 3-4 conventional machines ii) better accuracy and consistancy iii) reduced operator dependency etc. But no economical evaluation was made before going for N.C. except the idea that it will be definetly economical. Reasons for purchasing N.C. machines were reduction in throughput time, consistant accuracy, the parts are more in number and of small and medium batch sizes, the parts require more number of operations, closer tolerances, the machining time is low compared to set up time and the parts are subjected to frequent design changes. The reasons for going for imported machines are reliability, non availability in India, dimensional requirement, advanced technology, more economical in operation.

A comparitive statement for conventional and N.C. turret punch press for 10 parts and for a batch size of 100 indicates that average productivity increases by 8.5 i.e. the time taken by conventional machine tool is 8.5 times the time taken for the same operations by N.C. machine tool.

TABLE 1

(I) Conventional Machine Tool			(II) N.C. Machine Tool		Productivity
Part No	Setup time per piece	Machining Time per piece	Setup time per piece	Machining time per piece	Ratio of the cycle times I/II
1.	.034	0.84	0.02	0.1	7.28
2.	.025	0.95	0.01	0.15	6.093
3.	.042	0.212	0.015	0.03	5.6
4.	.02	0.135	0.01	0.03	3.88
5.	.032	4.24	0.025	0.4	10.0
6.	.04	1.6	0.02	0.08	16.4
7.	.029	0.205	0.01	0.05	3.9
8.	.025	0.07	0.01	0.005	6.33
9.	.33	2.79	0.02	0.444	6.72
10.	.03	0.675	0.01	0.02	23.5

A similar comparison of times taken by 11 parts for a batch of 100 on conventional machines and C.N.C. machine (Maho milling machine) shows an increase in productivity of 4.76.

The cost analysis was done and it was found that the pay back in the case of the N.C. turret press was two years although the actual life of the machine is expected to be 10 years. In the case of the Maho milling machine the pay back time was 3 years.

HINDUSTAN AERONAUTIC LTD, BANGALORE:- H.A.L. is an industry under the ministry of defence which makes the aircrafts and the supporting system. H.A.L. is using N.C., machines in the production of complicated parts which require high precision and are of complex profiles. These are difficult to be made by conventional machines. H.A.L. being under the ministry of defence most of the information asked was not provided as it was termed classified. However, it can be said that the reasons for purchasing N.C. machines are, the parts require more number of operations, the operations are complex, consistant accurancy is required, the parts are more in number and of small and medium batch sizes, they require closer tolerances, there are frequent changes in design and the machining time is low compared to setup times. Information regarding the preinvestment analysis which checks the economic viability of the machine is not provided as it is classified. As it was done in the other industries comparison of timing between conventional and N.C. machines are done in H.A.L. also, which shows 3 to 4 times increase in productivity which is a healthy sign.

A Comparative Statement for Conventional and C.N.C. lathe was done, the batch size was variable, the results are given below:-

TABLE 2

	(I) Conventional Machine Tool Timings (Minutes)			(II) N.C. Machine Tool Timings (Minutes)			Productivity
Part No	Initial setup time	Machining time	Cycle time per piece	Initial setup time	Machining time	Cycle time per piece	Ratio of the cycle times I/II
1.	1.25	2.35	2.41	1.30	1.21	1.27	2.89
2.	2.15	2.45	2.62	1.30	1.20	1.26	2.07
3.	1.15	1.25	1.30	1.00	0.57	0.58	2.24
4.	1.00	0.48	0.53	1.30	0.30	0.32	1.65

A Similar study was done for C.N.C. milling machine, the results are given below:-

TABLE 3

	(I) Conventional Machine Tool			(II) N.C. Machine Tool			Productivity
Part No	Initial setup time	Machining time	Cycle time per piece	Initial setup time	Machining time	Cycle time per piece	Ratio of I/II
1.	10.15	15.00	16.69	4.00	3.15	3.35	4.98
2.	9.45	11.25	12.80	7.00	4.00	4.46	2.87
3.	5.15	39.00	39.80	6.30	13.45	13.85	2.87
4.	6.00	22.00	23.00	3.15	4.45	4.66	4.93

In the case of N.C. lathe there is a two fold increase in productivity. This is inspite of the fact that the in process waiting time on conventional line is not taken into account. When that is considered the ratio will be still higher. The time per piece is reduced by 3 to 5 times in case of the milling machine.

HINDUSTAN MACHINES TOOLS LTD., BANGALORE:- H.M.T. Bangalore is using Horizontal Machining Centre (OERLIKON), Vertical Machining Centre (OERLIKON), Horizontal Machining Centre (KTM), and N.C. Plano Milling Machine (WALDRICH COBOURG) for making machine tool parts. On plano milling machine, tool parts such as beds, columns, slides etc. are machined. On machining centres, parts such as main and feed drive, gear boxes etc. are machined. These machines are utilised on a two shifts basis and two operators are associated with each machine. H.M.T. Bangalore made the first N.C. Machine tool in India which was introduced in the market in 1975.

The cycle time savings of some parts made on N.C. machines are given in the following table.

Table 4

Part	Conventional Machine Tool cycle time (days)	K.T.M. 1500 cycle time (days)
1. Housing	7	5.5
2. Feed Box housing	6	4.5
3. Apron	21	17.5
4. Headstock	21	16.5
5. Gear Box	6	4.5
6. Feed Box	5	3.5
7. Speed Box	5	4.0

The reduced time is because of combining of the operations.

HINDUSTAN MACHINE TOOLS LTD. HYDERABAD:- Hindustan Machine tools Limited, a pioneer in the Indian machine tool industry has started the production of N.C. machines from 1976, though it conceived the idea of making N.C. machines in India at an earlier date. It was a welcome start by H.M.T. Limited in India going for the production of N.C. machine tools. In every division of H.M.T. Limited, N.C. machines are being used. H.M.T. Limited (Hyderabad) is having N.C. lathes (HMT) two numbers, V.M.C. (KTM), Horizontal Machining Centre (KTM), Horizontal Machining Centre (OERLIKON) and Prece Ram (PEGARD). The parts made on the above machines are housings, gear boxes, bevel gears, shafts and other small parts. Columns and beds are machined on Preceram. The utilization of this machine was poor as there was too much of absentism on the part of the operators. The other factor was that there were only two operators associated with each machine. If one operator was absent, the machine could not be used in second shift. The comparison of production times for conventional machines and KTM 1500 is given in the following table.

Table 5

Part Name	C.N.C. Machine Tool Timings (Hours)		Conv. Machine Tool Timings (Hours)		No. of operations
	Set up time	Machining time	Set up time	Machining time	
1. Head Stock	15	18.8	21.7	90.1	8
2. Head Stock	14	37.5	14.9	83.75	7
3. Housing	13.33	17.33	11.43	56.47	5
4. Feed Housing	4.33	10.33	10.33	62.66	5
5. Face Plate	6.0	34.0	9.46	66.83	5
Total:	52.66	117.96	67.82	359.81	30

Batch size for C.N.C. is 5 and for conventional line only single part is machinined.

C.N.C. Machine Tool: Total time = Setup + Prove out + Machining

= 52.66 + 4 x 117.96 + 5 x 117.96 = 1114.3

Conventional Machine Tool: Total time = Setup + Machining + Waiting

= (67.82 + 359.81) 5 + 4 x 25 x 25 = 4638.15.

There is four hours of waiting time between two operations. Prove out is assumed as four hours for every one hour of machining. The significant feature is the reduction in transportation and waiting time. The total time for producing five parts using conventional machine tools is 4.16 times the time taken by using KTM 1500.

N.C. Machines in H.M.T. are not fully utilised. This can be seen from the high percentage of operator absentism. The backup facility such as a computer for programming is not available.

HINDUSTAN AERONAUTICS LIMITED, HYDERABAD:- This organization is concentrating on the manufacture of electronic components, it is having 4 CNC machine tools. These are Horizontal Machining Centre (MITUSI SEIKI), C.N.C., PCB drilling machine (EXCELLON) and two Milling Machines (MAHO). The increase in productivity using these machines was of the order of 2-3 times over conventional machines.

BHARAT HEAVY ELECTRICALS LIMITED, HYDERABAD:- This organization is using C.N.C. machines for machining turbine rotors, parts of pump sets. Oil rig equipment, switch gear components, drilling of plates and pipes. The C.N.C. machines are C.N.C. lathe (MAX MULLER), C.N.C. lathe (WOHLENBERG), C.N.C. Horizontal Machining Centre (K.T.M.), C.N.C. Vertical Multi Spindle Drilling machine (WADIKAN), and C.N.C. deep hole drilling machine (KOLB). The productivity analysis was done for C.N.C. lathe (MAX MULLER) and it was found that its productivity was 2 to 2.5 times that of the conventional machine tools used for performing same tasks.

BHARAT ELECTRONICS., GHAZIABAD:- Bharat Electronics Limited a pioneer in communication equipment is having 4 C.N.C. machines in their Ghaziabad unit. These are Vertical Machining Centre (MITUISI SEIKI), Horizontal Machining Centre (WASINO), N.C. Turret punch press (RASKIN) and PCB drilling machine (EXCELLON). A six fold increase in productivity is observed in the case of VMC over the conventional machine tools.

ESCORTS LIMITED., FARIDABAD:- This is a company engaged in making tractors and it has two N.C. machines, these are N.C. lathe (HMT) and Vertical Machining Centre (YASDA). They make engine cylinder block of the tractor on the VMC. They are using it for only 4 jobs as they have 4 types of cylinders. On the H.M.T. lathe rotating parts are machined. Performance of the V.M.C. is studied by taking the process planning of one type of cylinder block which they were making on conventional as well as N.C. machine. The productivity using this machine is three times that of the conventional machine tools.

ENGINEER'S AND OPERATOR'S QUESTIONNAIRE:- The findings from the questions put to the engineers and the operators are as follows:- The team work in the industries studied can be stated as good. Reorganization of the various departments concerned is needed to suit the N.C. machines. A good amount of confidence is reflected in the answers to the questions like understanding the working of the N.C. machines, preparing a N.C. programme and finding the fault in case of breakdowns. N.C. machine tool operators are feeling superior to the conventional machine tool operators. They expect a higher salary as they are working on more advanced machines. They are very satisfied about the chance to learn new things and fairly well satisified about the variety of jobs they do.

CONCLUSIONS

From the survey conducted in India it can be seen that there is a heavy demand for C.N.C. machine tools. Inspite of the low cost of machining that an N.C. machine offers, the initial cost of the machine is high. The present provision for depreciation of C.N.C. machine is that it can be depreciated in 24,000 hours of working which comes to 6 years on a two shift operation. The actual pay back can be obtained within 3 years. But the demand for higher depreciation provision is high because procuring a new machine requires considerable effort in generating funds. The reduction of various costs such as cost of the machine tools, running costs, programming costs etc., is desirable. This can be achieved to some extent by reducing or eliminating the import duty on the imported machines. The productivity increase observed in B.E.L. (Bangalore) is 8.5 in case of C.N.C. Turret punch press, highest being 23.5 (Table 1). This productivity increase is not only because of combining the operations but also because of the elimination of transportation time. This trend can also be seen in the study conducted at H.M.T. Hyderabad where the observed average waiting time between two operations is 4 hours (Table 5). The reduction in the cycle time is also observed for the parts machined on C.N.C. machining centres installed at H.M.T. Bangalore (Table 4). On the whole there is an increase in productivity as a result of using C.N.C. machine tools. Introduction of N.C. Technology is going to change the manufacturing environment. It is inevitable. So every one concerned with it i.e. Government of India, machine tool industry and users of the machine tools should be prepared to welcome N.C. Technology.

REFERENCES

1. Indian Machine Tool Industry, 1983-1993: A Perspective. C.M.T.I. Publication (1983).
2. Nee. A.Y.C., and Venkatesh, V.C. The Present and Future Trends of Production Engineering in Singapore. Annals of CIRP, Vol. 30/1, (1981) 519.
3. Davies B., Bruce W. and DeBar A.E. NEL/MTIRA Technological Forecast for the Computer Control of Machine tools. Fifteenth M.T.D.R. Conference Proceedings, (1974), 97.

CAD/CAM Data Representation and Interchange: An Assessment of Current Practice and Future Trends

Melvin R. Corley, Dale O. Anderson, R. Michael Harnett, and Robert O. Warrington

College of Engineering, Louisiana Tech University, Ruston, LA 71272-0046, USA

This paper presents the findings of research conducted in the College of Engineering at Louisiana Tech University. The objective of the project was to study the effects of emerging technological developments within the CAD/CAM industry and university environments. The research project consisted of four work tasks: three which involved extensive surveys of published literature in areas associated with CAD/CAM data representation, translation and interchange, and the fourth which included surveys of CAD/CAM system vendors.

CAD/CAM Data Translators

The exchange of CAD/CAM data between systems is becoming increasingly important as the number of CAD systems and users continues to grow. A recent study by the Building Services Research and Information Association in the construction industry indicated that 84% of the users wanted to exchange data, however, only 27% had [1]. Most of these data exchanges had been accomplished with what has been referred to as level 1 translators, that is, translators made to special order such as the GDS/Intergraph version. The next level of translator is the neutral file format, such as the Initial Graphics Exchange Specification (IGES). The IGES concept was originated at a Department of Defense Manufacturing Technology Advisory Group meeting in 1979 and is based on Boeing's CAD/CAM Integrated Information Network and General Electric's work on a neutral data base. Further coordination of the efforts to develop IGES has been the responsibility of the US

National Bureau of Standards (NBS) [2,3,4].

The chart shown in Figure 1 illustrates the progress since 1979 on efforts for data exchange in CAD/CAM. The level of activity and interest in the country and abroad is definitely increasing. Most of the research and development is currently being done by industry, particularly by large companies with a need and smaller companies looking to market their software. Relatively few research publications on data translators from universities were uncovered in this search. Several user groups have been formed (Federation of Computer-Aided Design User Groups, Information Technology Users Standards Association and are pushing harder for standards and for vendors to adhere to these standards. As can be seen in Figure 1, the International Organization for Standardization is working toward combining the best features of several translator standards to create one standard for the exchange of product definition data. The Standard for the Exchange of Product Data (STEP), version 1 is due for release soon. The major contributors appear to be IGES (version 3, with version 4 to be released in April, 1987), the Product Definition Exchange Standard, the Product Definition Data Interface, Standard d'exchange et Transfert, and VDA/VDAMA-FS from the Federal Republic of Germany.

Considerable work is being done in the area of data base translators. Warman from K. Four Ltd., describes work on a network data base system called the Engineering Records Organisation System from Applied Research Cambridge (UK). According to Warman, this system can handle both graphical and alphanumeric data and can be interfaced with any turnkey CAD system. British Aerospace is interested in the development of a master database for all design information with the purpose of avoiding data exchange translation and their inherent errors [5,6]. Researchers at the State University of New York, Buffalo and Carnegie Mellon are working on relational DBMS schemes to adapt to CAD/CAM integration. In a report by Emond and Marechal [7], their work on an Integrated Computer Aided Design System also uses a relational DBMS. They point out that almost all of the research to date in data base management systems has been for business applications and that applications to CAD/CAM are just beginning.

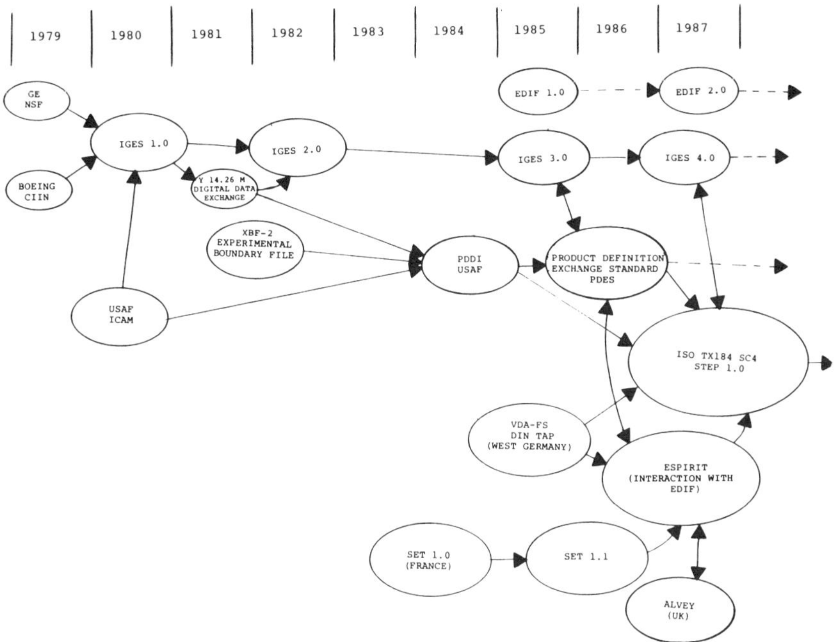

Figure 1
Summary of Graphical Data Exchange Standards

Several organizations have developed or are developing neutral file format translator systems separate from IGES or are expanding IGES because the IGES specification does not meet their particular needs. This presently appears to be a major drawback to the acceptance of IGES as an international standard. Voeller and Keller, working for the AE firm of Black and Veatch, describe the Neutral File System and associated translators. Their system provides for the complete translation of graphical and associated database information and retains the object intelligence in the graphics files. Their neutral file is an ASCII flat file but is supposedly very man-readable. Their system is commercially available. Hampel, Garner and Matthews, of Lawrence Livermore Laboratories, reported on the work by Jones with the DOE Neutral Data Exchange Format and their work on the Intelligent Gateway Processor. Part of the function of this processor involves the translation of dissimilar data formats. This group is also developing mechanisms for the verification of translated CAD data files, a topic which, as pointed out later in this paper, has not received very much attention.

Affuso and Sevak [8], have developed translators which they say are similar in function to IGES and Weissflog [9,10] and others are doing research with IBM IGES translators. The Interactive Graphics Research Group [11] was enlisted by the German automobile industry to develop a common interface for the exchange of product definition data for their industry. The automaker's view is that the curve and surface definition of IGES is not right for the auto industry. This group has worked on sculpted surfaces and on representing design information in a common data base.

Many CAD vendors are developing IGES translators for their products and companies, such as Graphic Information Exchange Services, Inc., are dedicated to developing products for the exchange of CAD/CAM data. Bloor and Owen [12] have been working with the implementation and extension of IGES for several years. Their work with the Data Exchange Center on the NEDO project originally considered the transfer of 2-D drawings with text. They are now looking at a much broader picture, that of "design intent". CAM-I, an international company, has been considering the transferring of solids modeling data and will have significant input to IGES version 4. In addition, this company has recently started a project called IMPACT for the automatic generation of NC control data.

Relatively recently (1983) the electronics industry [13-18] formed the Electronic Design Interchange Format Steering (EDIF) and Technical Committee. The EDIF committee was made up of representatives of Motorola, National Semiconductor, Texas Instruments, Daisy Systems, Mentor Graphics, Tektronix, and Professor Richard Newton from the University of California, Berkeley. EDIF started out as a consolidation of several systems and was a response to the deficiency of IGES that it did not define an exchange language or a hardware definition language. The electronics industry felt that the IC designer needed to give meaning to the symbols put on the screen. Version 1.0 of EDIF was released in 1985 and subsequent version are forthcoming. It is not clear at this time if enhancements to IGES to include printed circuit board design will result in a future consolidation of EDIF and IGES.

Several papers hint at an expert systems approach to the entire problems of CAD/CAM data transfer and integration. Japan and Europe appeared to be particularly interested in this approach [19]. The

ALVEY project in the UK and ESPIRIT program in Europe (nine collaborative partners) both appear interested in the expert system approach in CAD data transfer [20-23].

CAD/CAM Data Verification

Of 177 abstracts identified through a computer index search of publications since 1980, 80% qualified for the key word set with "verification" as opposed to "validation". "Conformity" was not observed in any of the abstracts.

Usages of the terms verification and validation were wide ranging and reflected some imprecision in distinguishing between them. Commonly observed usages included:

- logic verification
- design verification
- mask/artwork verification
- design validation
- model validation
- prototype verification
- system verification
- layout verification
- timing verification
- control algorithm verification
- computer code verification

Obviously, several of these usages relate to the specialized process of designing and manufacturing microelectronic products.

The disappointing finding in reviewing these 177 abstracts is that none of them contained information relating to the verification or validation of CAD/CAM data. Therefore, it is concluded that this is an area of activity that has not been discussed in the open literature.

Solids/Surface Modeling

Current constructive solid geometry (CSG) modelers are [24-29] limited to a few primitive solids with generalized mathematical definitions: cone, cylinder, plane, sphere, torus, etc. Arbitrary shapes are then "constructed" [30,31] using the operations of union, intersection and difference on scaled primitive solids. The construction technique of creating an arbitrary shape out of simple primitive shapes provides a relatively uncomplicated, easily understood user interface for solids modeling. Surfaces of intersection can be readily computed. Structural/inertial properties of the primitive solids are usually defined in the code so the volumetric properties of the constructed solid can be readily computed. The memory requirements are minimal, in that only the instance transformation for a primitive solid need be stored, since the "generic" description of the primitive solid is coded into the program. One of the disadvantages is that complex curved surfaces sometimes must be approximated by a patchwork of planar polygons (facets).

Current boundary representation (B-rep) modelers utilize [25-28,32-33] planar surfaces, parametric cubic surfaces and swept curve surfaces to form the faces of a solid model. Parametric cubic [34-37] and swept curve surfaces provide a very general basis of representation for arbitrary curved surfaces because they can be interactively adjusted to obtain the desired shape. Typically, a surface will be modeled as a patchwork of parametric cubic surfaces where the slope (tangent) of each surface at a common edge is constrained to be the same, thus the patchwork surface will have continuous first derivatives. The memory requirements for the boundary representation of an arbitrary solid are large, each bicubic patch requires 16 control points or vectors as well as calculated coefficients.

Commercial software appears to be taking a hybrid approach [25-27] to solids modeling that supports both constructive solid geometry and boundary representation solids. The complete merging of CSG and B-rep would necessitate the extension of CSG construction operators (union, intersection and difference) to operate on the boundary representation solids, which involves the calculation of intersections of parametric and nonparametric curves/surfaces [38-41]. Some success has been achieved in the translation between different representations [27,42].

The publicly reported research and development in solid modeling is primarily being done in universities. Private enterprise is, for the most part, involved in transforming the morass of published solids modeling ideas into marketable software, and is apparently guarding most new internal developments as trade secrets. The number of new concepts in CSG, B-rep construction, swept surface construction and data structures appears to be decreasing while the number of new concepts in parametric surface patches, surface intersection calculation, surface blending, translation between models and model display techniques is increasing. The trend indicates an increasing diversity of curve/surface models, a gradual merging of CSG and B-rep modeling and an increased focus on translation between models. References to IGES or any other standard are rare, indicating that standards are lagging far behind current research.

CAD/CAM Vendor Survey

A telephone survey of CAD/CAM system vendors was done to determine: (1) the level of commitment each company has to the published IGES standard, (2) technical incompatibilities between a CAD/CAM system's internal data structure and representation using the IGES format and (3) other mechanisms the company provides for transferring CAD/CAM data with systems supplied by other vendors. For each interview, a person directly associated with the technical support of data translators was sought. However, in most cases, the interviewee was a person in sales or in technical support of sales.

The survey questions used are the following:

1. What level of the IGES Standard for CAD data interchange does your company support?
   COMMENT: The current standard is at revision level 3.0.
2. Are your IGES translators maintained in-house or by outside contractors?
3. How may full-time equivalent technical personnel are dedicated to IGES translator support?
4. What features of your system's internal data base are not adequately supported by the current IGES standard?
5. What other translator standards or direct bridges to other CAD systems does your company support?
6. Do you have any representatives on the IGES standards committee?

Summary of Vendor Survey

All the CAD system vendors contacted expressed some degree of support for generic data format standards such as IGES. Motivations for the support are apparently many. Some vendors regard it simply as a necessary tool that has to be provided to customers who demand it. Others heavily advertise IGES capabilities as major features of their CAD systems. Almost all the system vendors supplement IGES translators with direct bridge translators for one or more competing systems. These seem to have been developed in response to (potential) customer demands for compatibility. While it is by no means unanimous, there exists a general sense of optimism regarding the role of IGES and its successors as a successful vehicle for CAD/CAM data interchange. There is some significant pressure to freeze IGES and proceed with the development of a more encompassing standard such as PDES.

The impetus for IGES translators seems not to be competition among vendors so much as demands of major customers. General Motors, for instance, apparently has mandated specific levels of IGES compatibility for CAD systems it uses. Target dates for implementing this compatibility have also been established by GM. None of the vendors mentioned the aerospace industry in general nor any aerospace company in particular as being a driving force in the implementation of CAD system data interchange standards.

CAD system vendors are making responses to the rapid rise to prominence of PC-based personal CAD systems such as AutoCad, Versacad, and Cadkey. Most of the vendors contacted now offer bundled PC-based hardware/software systems which offer most, if not all, of the functionality of their dedicated systems. Generally priced at under $10,000 per unit, personal CAD systems are expected to have broad appeal to small and mid-size firms.

Conclusions

At the present time, IGES is not an adequate data exchange standard for surface/solids modeling. It does not provide a sufficiently general vehicle for defining 3-dimensional parametric surfaces. However, the NBS has stated that a new, expanded version of IGES is forthcoming, probably in April, 1987, which will hopefully remedy that situation. Enthusiasm for the standard is not universal, but enough exists to predict that its successor (PDES) will be widely supported by both CAD/CAM system vendors and many manufacturing industries. In the interim, IGES can be used to integrate several modeling and drafting systems by developing user-supplied extensions to accommodate entities not covered by IGES.

It is highly unlikely that a solids modeling system based solely on CSG would serve the needs of the modern manufacturer. Even though it is estimated that 90% of all production solids models for mechanical engineering can be represented by CSG, modern manufacturing routinely requires that complex surfaces be modeled and analyzed.

The problem of CAD/CAM data format proliferation will be exacerbated in the near future if PC-based CAD systems become widely used in industry. IGES translators for these systems may not be well refined since they are primarily targeted for smaller firms where the data interchange requirements are more modest.

Rererences

1. McCarthy, J., "Industry Searches for a Standard to Draw (Computer-Aided Design)," Building (GB), V 2, N 12, 21 March 1986.

2. Smith, B. M., "IGES: A Key to CAD/CAM Systems Integration," IEEE Computer Graphics and Applications, V 3, N 8, PP 78-83, Nov 1983.

3. Smith, B. M. and J. Wellington, "IGES: A Key to CAD/CAM Systems Integration," CAD/CAM Technology, V 3, N 1, PP 33-35, Spring 1984.

4. Smith, B. M. and J. Wellington, "IGES, A Key Interface Specification for CAD/CAM Systems Integration," SME Report MS84-188, 1984.

5. Mason, H., "Searching For Standards," Engineering (London), V 225, N 3, PP 154-156, Mar 1985.

6. Mason, H., et al., "Introducing IGES (CAD/CAM Data Transfer)," Engineering (GB), V 225, N 3, PP 154-8, March 1985.

7. Emond, Jean-Claude and G. Marechal, "Experience in Building Arcade, A Computer Aided Design System Based on a Relational DBMS," "Database Week: Engineering Design Applications," Proceedings of Annual Meeting, 1983 May 23-26.

8. Affuso, T. and N. Sevak, "Integration through Translation: The CAD/CAM Pilt Probe at Xerox," Computers in Mechanical Engineering, V 2, N 2, PP 14-24, Sept 1983.

9. Weissflog, U., "Product Data Exchange: Design and Implementation of IGES Processors," Proceedings of the SEAS Spring Meeting 1985, 15-19 April 1985.

10. Weissflog, U., "Experience in Design and Implementation of an IGES Translator," Computers & Graphics, V 8, N 3, PP 269-273, 1984.

11. Pasemann, K., "Interfaces for CAD/CAM-Integration," CAMP '85 - Computer Graphics: Applications for Management and Productivity, 24-27 Sept 1985.

12. Bloor, M. S. and J. Owen, "IGES: Its Impact on Computer-Aided Engineering," International Conference on Computer-Aided Engineering, PP 87-91, 10-12 Dec 1984.

13. Marx, E. R., "EDIF: The Standard For Workstation Intercommunication," IEEE Micro, V 5, N 5, PP 68-75, Oct 1985.

14. Crawford, J. D., "EDIF: A Mechanism for the Exchange of Design Information," IEEE Design & Test of Computers, V 2, N 1, PP 63-69, Feb 1985.

15. Burson, D. C., "Use of EDIF for Exchanging PCB Design Data," Automated Design and Engineering for Electronics: Proceedings of the Technical Sessions, 15-17 Oct 1985.

16. Kahn, H. J., "EDIF at the University of Manchester," IEEE Colloquium on Electronic Design Interchange Format (EDIF), 27 Feb 1986.

17. Elliot, S., "EDIF and Daisy Systems Corporation," IEEE Colloquium on Electronic Design Interchange Format (EDIF), 27 Feb 1986.

18. Hillawi, J. I. and K. R. Bennett, "EDIF - An Overview," Computer Aided Engineering, V 3, N 3, PP 102-7, June 1986.

19. Sato, K., et al., "Integrated Custom VLSI Design System," IEEE International Conference on Circuits and Computers, 28 Sept-1 Oct 1982.

20. Hope, A., "Room for Improvement," Engineering (London), V 225, N 3, P 158, Mar 1985.

21. Courtieux, G. and D. Guibert, "Using a Conservational Translator Writing System for Generating Computer-Aided Design Systems in Architecture," Inf Process '77 (Proc of IFIP), PP 383-387, 8-12 Aug 1977.

22. Encarnacao, J., et al., "A Reference Model for Components and Interfaces of a CAD System," Proceedings of the IFIP WG 5.2/WG 5.3 Working Conference, PP 135-159, 8-10 Nov 1984.

23. Sun, Y. S., "Discussion of Standard Interfaces of Geometrical Modelers," Proceedings of Trends and Applications 1984, PP 226-229, 23-24 May 1984.

24. van Wijk, J. J., "SML: A Solid Modelling Language," Computer-Aided Design, V 18, N 8, Oct 1986.

25. Casale, M. S., and E. L. Stanton, "An Overview of Analytic Solid Modeling", IEEE Computer Graphics and Applications, Feb 1985.

26. Kimura, Fumihiko, "GEOMAP-III: Designing Solids with Free-Form Surfaces", IEEE Computer Graphics and Applications, June 1984.

27. Requicha, A. A. G., H. B. Voelcker, "Solid Modeling: Current Status and Research Directions", IEEE Computer Graphics and Applications, Oct 1983.

28. Pratt, M. J.,"Solid Modeling and the Interface Between Design and Manufacture", IEEE Computer Graphics and Applications, July 1984.

29. Voelcker, H., et al., "The PADL-1.0/2 System for Defining and Displaying Solid Objects", Computer Graphics, V 12 N 3, 1978

30. Laidlaw, D. H., et al., "Constructive Solid Geometry for Polyhedral Objects", Computer Graphics (ACM), V 20 N 4, Nov 1986.

31. Requicha, A. A. G., H. B. Voelcker, "Boolean Operations in Solid Modeling: Boundary Evaluation and Merging Algorithms", Production Automation Project, Univ of Rochester, Technical Memorandum N 26, Jan 1984.

32. Kimura, F., "GEOMAP-III: Designing Solids with Free-Form Surfaces", IEEE Computer Graphics and Applications, June 1984.

33. Kalay, Y. E., "Modelling Polyhedral Solids Bounded by Multi-curved Parametric Surfaces", Computer-Aided Design, V 15, N 3, May 1983.

34. Boehm, W., "Multivariate Spline Methods in CAGD", Computer-Aided Design, V 18, No 2, Mar 1986.

35. Boehm, W., "Curvature Continuous Curves and Surfaces", Computer-Aided Design, V 18, No 2, Mar 1986.

36. Tan, S. T., K. C. Chan, "Generation of High Order Surfaces over Arbitrary Polyhedral Meshes", Computer-Aided Design, V 18 No 8, Oct 1986.

37. Barskay, B. A., "A Description and Evaluation of Various 3-D Models," IEEE Computer Graphics and Applications, Jan 1984.

38. Koparkar, P. A., S. P. Mudur, "Generation of Continuous Smooth Curves Resulting from Operations on Parametric Patches", Computer-Aided Design, V 18, N 4, May 1986.

39. Lasser, D., "Intersection of Parametric Surfaces in the Bernstein-Bezier Representation", Computer-Aided Design, V 18 No 4, May 1986.

40. Putnam, L. K., P. A. Subrahmanyam, "Boolean Operations on N-Dimensional Objects", Computer Graphics (ACM), V 20 No 4, Nov 1986.

41. Varady, T., M. J. Pratt, "Design Techniques for the Definition of Solid Objects with Free-Form Geometry", Computer-Aided Geometric Design, V 1, N 3, 1984.

42. Peterson, D. P., "Boundary to Constructive Solid Geometry Mappings: A Focus on 2D Issues", Computer-Aided Design, V 18, No 1, Jan 1986.

A Program in Basic to Calculate Electronic Box Moment of Inertia

H.A. Saleh, Ph.D.

TRW Electronic Systems Group, Redondo Beach, CA, USA

ABSTRACT

For a preliminary dynamic analysis of a complex electronic chassis it is often assumed as a beam with relatively constant cross-section. With this approach, many weak points can be revealed in the initial design. Design changes can then be incorporated early in the program. The moment of inertia of the cross-section along the X and Y axis will be needed for determining the resonant frequency for vibration along the Y-axis and the X-axis. Also dynamic loads, deflections stresses can be determined. The chassis cross-section is broken into elements. The neutral axis of the whole section is first found. Each element has a moment of inertia about its own centroid or center of gravity. In addition, there is a much greater moment of inertia for each element because of the distance of its center of gravity to the neutral axis of the whole section. This moment of inertia is equal to the area of the element multiplied by the distance of its C.G. to the neutral axis squared.

Thus, the moment of inertia of the entire section about its neutral axis equals the summation of the two moments of inertia of the individual elements. This along the Y-axis and the X-axis. If there are many elements in the chassis cross-section, this process becomes very tedious. A program written in Basic, using Apple IIe simplifies this process. This is an interactive program, which asks the user questions for input data about the chassis cross-section, then it tabulates the input data followed by the results.

INTRODUCTION

The program calculates the center of gravity and moment of inertia of a rectangular chassis. It allows the engineer to make a quick estimate of mode of vibrations and dynamic loads to see whether panels will buckle rivets shear, or weld crack. This preliminary analysis can be done in a very short time before an expensive computerized analysis is made to determine the frequency response characteristics of the chassis.

CENTER OF GRAVITY AND MOMENT OF INERTIA FORMULAE

Considering an electronic box as shown in Figure 1., the whole section is broken into rectangular elements. The neutral axis of the whole section is defined by the location of the centroid of the area, which can be expressed by equations 1 and 2.

$$\bar{X} = \frac{\sum_{i=1}^{N} \left(A_i X_i \right)}{\sum_{i=1} \left(A_i \right)} \qquad (1)$$

$$\bar{Y} = \frac{\sum_{i=1}^{N} \left(A_i Y_i \right)}{\sum_{i=1} \left(A_i \right)} \qquad (2)$$

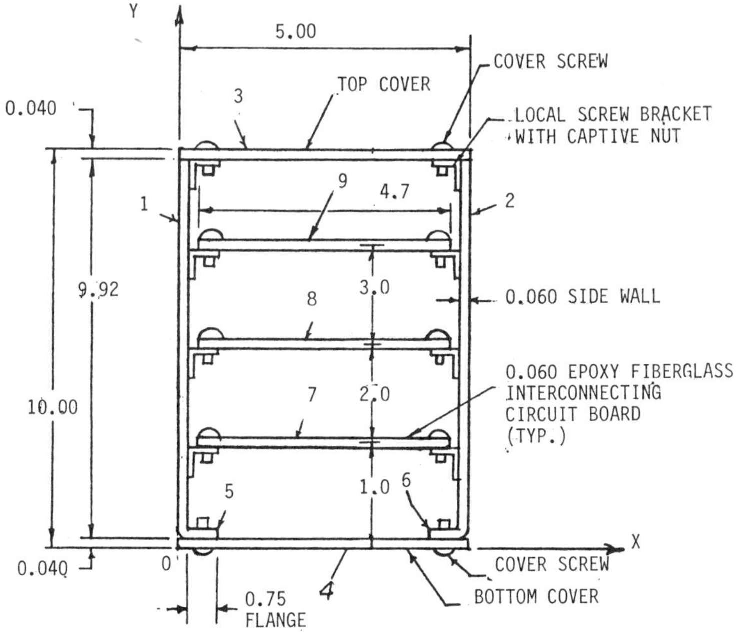

FIG. 1 CROSS SECTION OF ELECTRONIC CHASSIS (ALUMINUM)

The moment of inertia of each element about its own centroid or center of gravity (C.G.) is given by the formulae for rectangular sections (3), and (4) and figure 2. Equations (5) and (6) used for the moment of inertia of the entire section.

$$I_{xi} = \frac{b\,d^3}{12} \tag{3}$$

$$I_{yi} = \frac{b^3\,d}{12} \tag{4}$$

$$I_x = \sum_{i=1}^{N} \left(I_{xi} + A_i C_{yi}^2 \right) \tag{5}$$

$$I_Y = \sum_{i=1}^{N} \left(I_{yi} + A_i C_{xi}^2 \right) \tag{6}$$

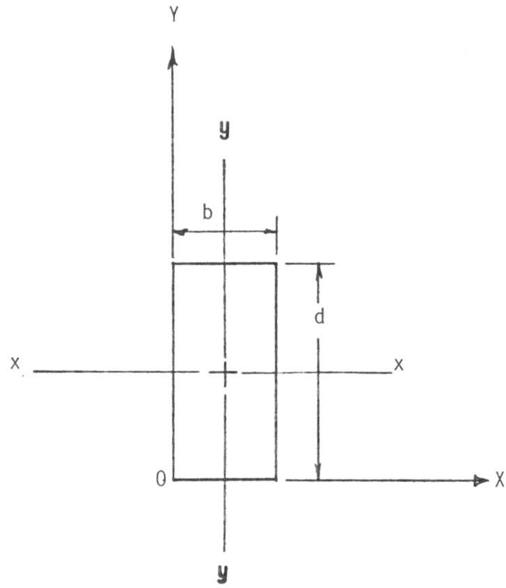

FIG. 2 RECTANGULAR CROSS-SECTION

Where I_x and I_y are the total moment of inertia about the X-axis and the Y-axis respectively. Ai equal area of element, Cxi, Cyi are the distances from the neutral axes of the whole section to the centroid of each element, and N equals number of elements.

D. Steinberg (1) in his book explained that bolted covers are used to provide access to the top and bottom sections of the chassis. The general tendency in the electronics industry is to provide convenient access to components. This usually means quarter-turn screws, so that the covers can be removed and replaced quickly. Tests on this type of fastener indicate it is only about 10% efficient in its ability to hold two members together during vibration. This efficiency factor ranges from 0% for a joint that has no connection to 100% for a joint that is welded. He suggested that bolt efficiency factor of 25% for epoxy fiberglass circuit-interconnecting board, and 50% for heavy flanged cover, with many high strength cap screws.

Since there are two different materials included in the cross-section, the epoxy fiber glass board and the aluminum chassis, consideration must be made for the difference in modulus of elasticity. This can be done by adjusting the epoxy plate area and moment of inertia with the ratio of the modulus of elasticity of the two materials. The following two equations from reference (1).

$$A_{al} = A_{epox} \times \frac{E_{epox}}{E_{al}} = A_{epox} \times \frac{2 \times 10^6}{10 \times 10^6} \qquad (7)$$

$$I_{al} = I_{epox} \times \frac{E_{epox}}{E_{al}} = I_{epox} \times \frac{2 \times 10^6}{10 \times 10^6} \qquad (8)$$

(For members with the same height)

The program was made to be interactive, such that it will ask the user what is the number of elements in the cross-section, the bolt efficiency factor, the element X-ordinate of its centroid from a reference axis, element Y-ordinate, and it will continue to repeat those questions until the total number of elements is consumed. After the last element the computer start printing the results as shown below. The program listing is also shown below.

BOX MOMENT OF INERTIA

BOLT EFF	A	B	X	Y
1	.06	9.92	.03	5
1	.06	9.92	4.97	5
.1	5	.04	2.5	9.98
.1	5	.04	2.5	.02
1	.75	.06	.435	.07
1	.75	.06	4.565	.07
.05	4.7	.06	2.5	1
.05	4.7	.06	2.5	3
.05	4.7	.06	2.5	6

```
BOX TOTAL AREA OF SECTION        = 1.3627  IN2
PRODUCT OF AREA*X-ORD           =3.40675 IN3
BOX CENTRIOD X-ORD              =2.5 IN
SUM OF MOMENT OF INRTIA OF CENTRIODS ALONG X-AXIS = .165776453 IN4
SUM PRODUCT OF AREA*(X-XBAR)^2   = 7.64629162  IN4
MOMENT OF INERIA ALONG X-AXIS    = 7.81206808 IN4

PRODUCT OF AREA*(Y-ORD)          = 6.2993 IN3
SUM OF MOMENT OF INERTIA OF CENTRIOD Y-AXIS      = 9.76195992 IN4
BOX CENTRIOD Y-ORD              =4.6226609IN
SUM PRODUCT OF AREA*(Y-YBAR)^2   = 3.28152924 IN4
MOMENT OF INERTIA ALONG Y-AXIS   = 13.0434892 IN4
```

```
LIST

10  PRINT "BOX MOMENT OF INERTIA"
20  PRINT " A PROGRAM IN BASIC TO CALCULATE ELECTRONIC BOX MOMENT OF INER
    TIA IN TWO AXES."
30  PRINT "THIS PROGRAM CALCULATES THE
40  PRINT "MOMENT OF INERTIA OF THE "
50  PRINT "CROSS-SECTION ALONG THE "
60  PRINT "Y-AXIS AND THE MOMENT OF"
70  PRINT "INERTIA ALONG THE X-AXIS"
80  DIM EF(100),A(100),B(100),X(100),Y(100)
81  DIM AR(100),AX(100),IXO(100),AY(100),IYO(100)
82  DIM CX(100),CY(100)
83  DIM XCSQ(100),YCSQ(100),SQACX(100),QSACY(100)
90  PRINT "WHAT IS THE NUMBER OF"
100  PRINT "ELEMENTS IN THE CROSS-SECTION"
110  INPUT N
120  REM  ALL DIMENTIONS IN INCHES
130  FOR I = 1 TO N
140  PRINT "BOLT EFFICIENCY FACTOR,EF(";I;")"
141  INPUT EF(I)
150  PRINT "ELEMENT X-ORDINATE,A(";I;")"
151  INPUT A(I)
160  PRINT "ELEMENT Y-ORDINATE,B(";I;")"
161  INPUT B(I)
170  PRINT "MEMBER CENTROID X-ORDINATE,X(";I;")"
171  INPUT X(I)
180  PRINT "MEMBER CENTROID Y-ORDINATE,Y(";I;")"
181  INPUT Y(I)
190  NEXT I
191  PRINT
200  FOR J = 1 TO N
210 AR(J) = EF(J) * A(J) * B(J)
220 AX(J) = AR(J) * X(J)
221  REM  AX(J) IS THE PRODUCT OF ELEMENT  AREA * X-ORDINATE OF ELEMENT C
     ENTRIOD.
230 IXO(J) = (EF(J) * B(J) * (A(J) ^ 3)) / 12
231  REM    IXO(J) IS THE MOMENT OFINERTIA OF ELEMENT ALONG X-AXIS, THROU
     GH ITS CENTRIOD
240 AY(J) = AR(J) * Y(J)
241  REM   AY(J) IS THE PRODUCT OFELEMENT  AREA * Y-ORDINATEOF ELEMENT CE
     NTRIOD.
250 IYO(J) = (EF(J) * A(J) * (B(J) ^ 3)) / 12
251  REM   IYO(J) IS THE MOMENT OF INERTIA OF ELEMENT ALONG Y-AXIS THROUG
     H ITS CENTRIOD.
260 ASUM = ASUM + AR(J)
270 XASUM = XASUM + AX(J)
280 YASUM = YASUM + AY(J)
290 XIOSUM = XIOSUM + IXO(J)
300 YIOSUM = YIOSUM + IYO(J)
400  NEXT J
500 XBAR = XASUM / ASUM
600 YBAR = YASUM / ASUM
601  PRINT
700  FOR K = 1 TO N
710 CX(K) = X(K) - XBAR
720 CY(K) = Y(K) - YBAR
730 XCSQ(K) = CX(K) ^ 2
740 YCSQ(K) = CY(K) ^ 2
750 SQACX(K) = AR(K) * XCSQ(K)
760 QSACY(K) = AR(K) * YCSQ(K)
770 NUMACX2 = NUMACX2 + SQACX(K)
780 TUMACY2 = TUMACY2 + QSACY2(K)
```

```
790  NEXT K
800 XXX = XIOSUM + NUMACX2
810 YYY = YIOSUM + TUMACY2
811 D$ =  CHR$ (4)
812  PRINT D$;"PR# 1"
813  PRINT  CHR$ (27); CHR$ (33)
814  PRINT  CHR$ (14)"BOX MOMENT OF INERTIA"
815  PRINT  CHR$ (27); CHR$ (34)
816  PRINT  CHR$ (27); CHR$ (89)
817  PRINT  CHR$ (15)
820  PRINT "BOLT EFF","A","B","X","Y"
830  FOR K = 1 TO N
840  PRINT EF(K),A(K),B(K),X(K),Y(K)
850  NEXT K
851  PRINT
852  PRINT
860  PRINT "BOX TOTAL AREA OF SECTION"," = ";ASUM;"  IN2 "
870  PRINT "PRODUCT OF AREA*X-ORD","=";XASUM;" IN3"
880  PRINT "BOX CENTRIOD X-ORD"," =";XBAR;" IN"
890  PRINT "SUM OF MOMENT OF INRTIA OF CENTRIODS ALONG X-AXIS";" = ";XIOS
     UM;" IN4"
900  PRINT "SUM PRODUCT OF AREA*(X-XBAR)^2"," = ";NUMACX2;"  IN4"
910  PRINT "MOMENT OF INERIA ALONG X-AXIS"," = ";XXX;" IN4"
920  PRINT
930  PRINT
940  PRINT "PRODUCT OF AREA*(Y-ORD)","= ";YASUM;" IN3"
950  PRINT "SUM OF MOMENT OF INERTIA OF CENTRIOD Y-AXIS"," = ";YIOSUM;" I
     N4"
955  PRINT "BOX CENTRIOD Y-ORD","=";YBAR;"IN"
960  PRINT "SUM PRODUCT OF AREA*(Y-YBAR)^2"," = ";TUMAY2;" IN4"
970  PRINT "MOMENT OF INERTIA ALONG Y-AXIS"," = ";YYY;" IN4"
980  PRINT
981 D$ =  CHR$ (4)
982  PRINT D$;"PR# 0"
990  PRINT "TYPE IN A '1' IF YOU WANT ANOTHER SOLUTION OTHERWISE TYPE '2'
     "
1000  INPUT S
1010  IF S = 1 THEN  GOTO 90
1015 D$ =  CHR$ (4)
1020  END
```

CONCLUSIONS

A program in basic to calculate the centroid and moment of inertia of a rectangular electronic chassis is presented in this paper. This program provides the design engineer with the necessary data for the preliminary dynamic and stress analysis of electronic chassis.

REFERENCES

1. Steinberg, Dave S., "vibration analysis for electronic equipment", John Wiley and sons, Inc. 1973, P. 323-325.

A Computer Aided Procedure for the Selection of Manufacturing Materials

Raj S. Sodhi

Associate Professor, Department of Mechanical Engineering, New Jersey Institute of Technology, Newark, NJ 07102, USA

ABSTRACT

Material selection is a key step in the engineering design process. Engineering systems contain many components and design engineers are constantly faced with decisions regarding the selection of materials for these components. The selection of the best material from two or more competing alternatives depends upon the decisions made by the design engineer. The selection process becomes enormous because of a large number of alternatives and the factors affecting the selection. This paper presents a technique which employs computer as an a important link between the decision process and the necessary calculations to determine the best selection of the manufacturing material for designing a component.

INTRODUCTION

At present, most engineering development is taking place in the direction of design automation. Currently, "Expert Systems" such as Structural Analysis Systems or Design Simulation Systems are being introduced that computerize specific tasks. Computer-Aided-Design Systems are automatically designing parts and supplying Nemrical Control (NC) programs to automated milling machines with instructions to manufacture the parts. The selection of engineering materials for the manufacture of these parts is a key step in the design process. But there is no well-developed methodology for material selection which can be used in an overall computer-aided-design system. Hence it is the purpose of this paper to present a computer based procedure for the selection of manufacturing materials. It is expected that such a procedure can be integrated with an overall computer-based design scheme.

The importance of the use of computer in the material selection process can be appreciated considering the fact that there are over 40,000 useful metallic-alloys and thousands of non-metallic engineering materials available from which the selection can be made.

Moreover, tha data on the material performance characteristics based on the past experience of an industrial organization can also be stored in the computer.

FACTORS AFFECTING THE DESIGN

Selecting a material for a component involves much more than just matching the properties of the materials with the functional requirements in the service. For example the designer needs to know if the material selected can be machined into the finished part with the machinery available. Is the selected material available ? What would be the manufacturing cost ? Is design data available on the material ? Would the material for the component be acceptable to the consumer ? These are some of the numerous questions which must be taken into account while selecting manufacturing materials. Some of the factors which affect the material selection are included in Table 1.

COMPONENT SIZE	MATERIAL AVAILABILITY	ERGONOMICS
COMPONENT COST	MANUFACTURING PROCESSES	POLITICS
PRODUCT WEIGHT	CONSUMER ACCEPTANCE	SERVICE LIFE
COMPONENT LIFE	STANDARD SPECIFICATIONS	MATERIAL COST
PERFORMANCE	VARIABILITY IN PROPERTIES	AESTHETICS
ENVIRONMENT	MANUFACTURING FACILITY	PATENTS
SHIPPING	DESIGN DATA AVAILABILITY	RELIABILITY
QUANTITY	MATERIAL PROPERTIES	SAFETY
TOLERANCES	MATERIAL FORM NEEDED	SHELF LIFE

Table 1 Factors Affecting the Material Selection

The functional requirements and performance of the materials based on some of these factors are not always expressable in quantifyable numbers. For example, the reliability, consumer acceptance and aesthetics are difficult to quntify. Hence some kind of a subjective rating is required here. One scheme can be:

	Rating			
Material Property	A	B	C	D
Machinability	Excellent	Good	Fair	Poor

In addition the performance of a material can be expressed in terms of physical, mechanical, thermal, chemical or electricl properties. The failure mode of a material in a practical situation can be related to one or more of these material properties. For example, the failure by buckling is dependent on the compressive yield strength and the modulus of elasticity of the material.

The above factors and the material properties are called attributes of the design alternatives in our program.

MATHEMATICAL MODELLING FOR THE MATERIAL SELECTION PROBLEM

The mathematical model used in the computer program is based on the work by several researchers (1,2,3). The key words used in the model are described below.

Alternatives: Various manufacturing materials which are being considered from which the selection is to be made. There are n materials.

Attributes: These are the factors based on the functional requirements which need to be considered while making the decisions. There are j attributes.

Attribute Rating: A numerical value associated with a particular material.

Attribute Importance: The relative importance of j attributes.

Merit value: This is the calculated value of the relative performance of each material based on the attribute rankings decided by the designer. The material with highest merit value is the best choice.

Let A_{im} represent the mth attribute of the ith alternative. Also $AMAX_m$ and $AMIN_m$ are determined to be the maximum and minimum values of the attribute m. If R_{im} represents the normalized rating of attribute m for alternative i, then these normalized ratings can be determined by using,

$$R_{im} = (A_{im} - AMIN_m / (AMAX_m - AMIN_m) \qquad (1)$$

Equation (1) is used in design situations where a larger value of an attribute rating represents preference. For example, to maximize the strength in a design, we will be preferring a material with higher yield strength. On the other hand, if a lower value of an attribute rating represents preference, as the machining cost of a material, then the normalized rating is determined by:

$$R_{im} = 1 - (A_{im} - AMIN_m) / (AMAX_m - AMIN_m) \qquad (2)$$

Two different methods which can be used for determining the relative importance of each attribute to the decision are described below:

Comparison by Pairs:

The j attributes are listed and compared in every combination taken two at a time. The attribute which is considered to be more important of the two is given a 1 and the less important is given a 0. The total number of possible comparisons are $N = j(j-1)/2$. If the total number of positive decisions for mth attribute are p_m, then the normalized importance for the atrribute m is

$$I_m = p_m / N \qquad (3)$$

The method is illustrated below for j = 4, where N = 6 decisions are made.

Attribute	Decision						Positive Decisions	Im
	(1)(2)	(1)(3)	(1)(4)	(2)(3)	(2)(4)	(3)(4)		
1) Strength	1	1	1				3	.5
2) Cost	0			0	1		1	.166
3) Life		0		1		0	1	.166
4) Stiffness			0			1	1	.166

Attribute Ranking Method:

The method used in our program is based on the rankings of the attributes according to the preference of the designer. For example, high material strength may be more important to the designer than its cost and low cost may be preferred over stiffness. We assign 1 to the most important attribute and 2 to the next important and so on. If two attributes are equally important, then the same rank is assigned to both of them. Let R_m denote these rankings. Then the normalized importance I_m for each attribute is determined as follows:

$$I_m = (j - R_m) / \sum_j R_m \qquad (4)$$

The maximum performance or merit value of a design material choice is calculted as:

$$M_i = \sum_i (R_{im} \times I_m) \quad i = 1 \text{ to } n, \; m = 1 \text{ to } j \qquad (5)$$

The candidate materials are rated according to the calculated merit value and the material with the highest merit value is considered to be the best choice.

THE MATERIAL SELECTION PROCESS

The problem of selecting a manufacturing material for a component deals with the selection of the best choice from a group of materials. It is assumed that the data on various attributes for this group of materials is available. This data will depend on the objectives of an organization, their design philosophy and the nature of their manufacturing methods. For example, in highly automated industries, the material cost is critical and the savings in material cost are preferred over overhead costs. On the otherhand, in advanced aerospace applications, the components are going to be subjected to the limits of their properties and material cost is not an overriding factor. The attributes which can be expressed in quantifyable numbers will need a subjective rating as explained earlier.

Attributes / Alternatives	Tensile Strength (KSI)	Elastic Modulus (10^6 PSI)	Density (lb/in^3)	Corrosion Resistance	Machinability	Formeability	Availability of Design Data	Material Availability	Fracture Toughness	Cost ($/lb)
Al 2014 T6	70	10	0.101	D	B	B	A	B	D	0.6
Al 2024 T3	68	10	0.100	D	B	B	A	A	D	0.6
Al 2024 T4	68	10	0.100	D	B	B	A	B	D	0.6
Al 5052	38	10	0.096	A	D	A	C	A	A	0.6
Al 7075 T6	83	10	0.101	C	B	B	A	A	D	0.45
St 4340 HRA	101	30	0.28	D	C	C	B	B	D	0.45
St 4340 CDA	110	30	0.28	D	C	C	B	C	C	0.45
St 4340 N	185	30	0.28	D	C	C	B	D	B	0.45
St INC 718	193	31	0.296	A	D	D	C	C	A	0.9
Ti 6Al 4V	130	30	0.160	B	D	D	B	B	D	1.6
Attribute Rankings	6	7	1	9	5	4	8	10	2	3

NOTE A,B,C,D are relative ratings

Table 2 Data for the Aircraft Material Selection Problem

COMPUTER PROGRAM FOR THE MATERIAL SELECTION PROBLEM

The mathematical model described earlier for the material selection problem has been programmed for use on a microcomputer. The BASIC program was used on a Philips P3100 personal computer with 256K memory. The user can change the relative preferences of the attributes for a material selection problem without changing the data on the attributes.

DESIGN EXAMPLE

Suppose it is desired to select material for the skin of a small aircraft. The choice for the materials is limited to 10 materials listed in Table 2. The ten attributes and their ratings are also listed in the table. Six of the attributes do not have numerical values associated with them. Larger values indicate preference for 8 of the attributes whereas lower values indicate preference for cost and density. The relative rankings for the attributes are also shown in Table 2. The computer printout of the solution is given below:

MATERIAL SELECTION FOR AN AIRCRAFT SKIN

DATA MATRIX

DESIGN	ATTRIBUTE NAMES									
	T.Str	E.Mod	CorrR	Machi	Forme	DesDa	MatAv	Fract	Densi	Cost
AL14T	70.00	10.00	1.00	3.00	3.00	5.00	3.00	1.00	0.10	0.60
AL24T	68.00	10.00	1.00	3.00	3.00	5.00	5.00	1.00	0.10	0.60
AL24T	68.00	10.00	1.00	3.00	3.00	5.00	3.00	1.00	0.10	0.60
AL505	38.00	10.00	5.00	1.00	5.00	2.00	5.00	5.00	0.10	0.60
AL7JT	83.00	10.00	2.00	3.00	3.00	5.00	5.00	1.00	0.10	0.60
ST40H	101.00	30.00	1.00	2.00	2.00	3.00	3.00	1.00	0.28	0.45
ST40C	110.00	30.00	1.00	2.00	2.00	3.00	2.00	2.00	0.28	0.45
ST434	185.00	30.00	1.00	2.00	2.00	3.00	1.00	3.00	0.28	0.45
STINC	193.00	31.00	5.00	1.00	1.00	2.00	2.00	5.00	0.30	0.90
TI6AL	130.00	30.00	3.00	1.00	1.00	3.00	3.00	1.00	0.16	1.60

ATTRIBUTE RANKINGS ARE

T.Streng	6
E.Modulu	7
CorrResi	9
Machinab	5
Formeabi	4
DesDataA	8
MatAvail	10
FractTou	2
Densit9	1
Cost	3

ALTERNATIVE	MERIT	RANK
AL14T6	.759796	4
AL24T3	.7768048	3
AL24T4	.758623	5
AL5052	1.053755	1
AL7JT6	.7937842	2
ST40HR	.4662143	8
ST40CD	.503311	7
ST434N	.5791175	6
STINC7	.4612648	9
TI6ALV	.3377685	10

SOLUTION ANALYSIS

The results clearly indicate that the best choice for the material for the aircraft skin is Al 5052. This material has the highest merit value which is about 25 percent greater than the next choice. The user can change the attribute ratings to consider their affect on the best choice.

CONCLUSIONS

This paper has presented a computer-based method for the selection of manufacturing materials. Engineers are constantly faced with decisions regarding the selection of materials while designing components for engineering systems. The number of material choices and the performance requirements tend to be very large. This paper employs the computer as an important link between the decision process and the necessary calculations to determine the best selection of the manufacturing materials. With the increased emphasis on design automation, it is expected that the proposed technique will generate more interest in integrating the material selection in the overall computer-aided design methods.

References

1. Johnson, R. C.,"Mechanical Design Synthesis With Optimization Applications", Van Nostrand Reinhold Company, New York, 1980.

2. Mistree, F., Lyon, T. D., Shupe, J. A., "Computer Based Design Synthesis, An Approach to Problem Solving," Proceedings of Frontiers in Education Conference, Columbia, SC, October 8-20, 1982.

3. Dieter, G. E., "Engineering Design: A Materials and processing Approach", McGraw-Hill, 1983.

4. Sodhi, R. S.,"Selection Process in Engineering Design Using Personal Computer", Proceedings of the International CAD/CAM, Robotics and Automation Conference, Tucson, AZ, Feb 11-15, 1985.

Theory and Practice of Minimum Weight Design

J.-M. Lagache

PSA Etudes & Recherches, 10 av J.-B. Fortin, F-92220 Bagneux

INTRODUCTION

The present paper relates experiments that have taken place in Ecole Centrale de Paris, within the frame of annual projects, that are freely discussed between teachers and students. The project began about five years ago, with the initial,- and rather naive,- aim of illustrating Michell Theory [1,2,3,4] on simple mechanical examples. Preliminary discussions with students revealed that the design of a windmill was an appealing subject. Integration of a Michell torque transmission cantilever in a low speed vertical axis Savonius engine [5] was thought to be a gentle exercise, because of the low degree of mechanical sophistication of the machine.

There are relatively few valuable textbooks on real world design,- at the notable exceptions, by example, of refs [6,7] ,- and we were quite unaware of the very special kind of difficulties that are encountered in practical problems [8] , and, especially, of the strong feed back that physical reality exerts on theoretical concepts.

Part 1 of the paper is a short summary of the basic principles of minimum weight design in Michell Theory. Results concerning the application of Fenchel's duality |9, 10| to the design problem, have already been published, and useful complements to the present text can be found in refs [11,12,13,14,15,16] .

Part 2 of the paper is devoted to the simplest frames that can be generated by Michell Theory : torque transmission cantilevers. We hope this paragraph will be a little more than one of the sempiternal variations on Michell's second example, that can ordinarily be found in literature. The stress, here, is put on practical realizations. A shape generator, running on the Apple II e , is described, and many examples of generated frames and composites, that were built by students of Ecole Centrale, are given. By comparison with earlier presentations [17,18] , three new technical points are essentially developed : use of polyvinylchloride, plastic moulding of cores, and fiber reinforced polyurethan Michell composites.

Part 3 of the paper, at last, tries to determine the place that could be inoccupied by Michell Theory in modern design : that of an (intelligent) artificial architect, in dialectical interaction with practical realizations and classical methods of structural analysis and design.

1. MINIMUM WEIGHT DESIGN

Suppose one has to carry a load to a rigid core, with the minimum material consumption. Considering that almost every state of stress in a material, can be described by a set of elementary tractions and compressions, a possible way of finding the minimum weight design, consists in decomposing the load in elementary self-equilibrated pairs of forces, in evaluating the material cost of each pair,- namely, the weight of a small material bar sustaining the considered pair,- in making the summation of all elementary costs, and, finally, in taking to the infimum with respect to all possible decompositions.

Mathematically, let X be the (infinite dimensional) vector space of all finite collections of concentrated loads throughout the design region, and consider that X is equipped with the following cost function :

$$f(x) = \begin{cases} |L|\,|a-b| \,/\, s_o & \text{when } x = (a,L;b,-L) \text{ is a self-equilibrated 2-force loading at the ends, } a \text{, } b \text{, of a feasible segment ,} \\ & \text{or, alternatively, when } x = (a,L) \text{ is a 1-force loading along a boundary segment resting at b on the support} \\ +\infty & \text{in any other circumstance ,} \end{cases} \quad (1)$$

where a and b denote spatial locations ; L , a vector force ; and s_o , the Tresca yield stress. It should be clear that, given an arbitrary system of loads, z , the design problem equivaults to find

$$F(z) = \operatorname*{Inf}_{p} \operatorname*{Inf}_{z = x_1 + \dots x_p} f(x_1) + \dots f(x_p) \quad , \quad (2)$$

where the infimum is first taken with respect to all possible decompositions in p terms, and, then, with respect to all possible integers p's.

Consideration of Fenchel's duality [11,14] straightforwardly leads to the saddle-point equality

$$F(z) = \operatorname*{Sup}_{w \in C} \; z \cdot w \quad , \quad (3)$$

where C is the canonical subset of the dual X' of X , that is defined by

$$C = \{ w \in X' \mid w \cdot x \leqslant f(x) \text{ , for each } x \in X \} \quad . \quad (4)$$

Substitution of definition (1) in the last inequalities, shows that C , in the present design problem, is the set of kinematically admissible virtual displacements whose principal strains nowhere exceed in magnitude $1/s_o$. The right hand side in (3) obviously corresponds to the synthesis of a state of maximum virtual work, under the preceding kinematical restriction. More precisely, a detailed study of the saddle-point condition reveals that in the state of maximum virtual work, optimum material lines are arranged along principal strain lines.

Formula (4) is the very definition of *Maxwell-Michell virtual strain fields*. In most original papers, the saddle-point relation is only recognized as an inequality [1,2,4]. Hemp [3] considers the maximization of the virtual work as an heuristics. The combinatorial formulation, and the canonical transformation to the maximization of the virtual work are of our own[11,12,13,14] . This straightly leads to a non-linear finite element approach of optimum shapes [18] . Another advantage lies in the possibilities of generalization. Instead of considering assemblies of bars, we can decide, by instance, to incorporate in the design those uniformly stretched ribbed membranes, that were first imagined by Cox [11] , and found to be lighter than trusses. This only requires the addition in (1) of new clauses, stipulating that f is the volume of the Cox membrane for any consistent system of loads. One derives in this way a new saddle-point equality, with the main theoretical result that 2D Michell frames are indeed generalized solutions to the *elastic continuum* design problem, provided principal virtual strains satisfy the strict inequality

$$e_1 + e_2 \qquad 1/s_o \tag{5}$$

2. TORQUE TRANSMISSION CANTILEVERS

a) Introduction. 2D frames that carry a torque, or tangential loads to a rigid core inoccupy a quite special place in Michell Theory [1,4] : they are of the highest practical interest in the design of various turbines ; analytical methods can be substituted to the maximum work principle, owing to the fact that the Maxwell-Michell strain field has everywhere opposite maximum principal strains. In other words, the pure kinematical synthesis of a virtual strain-field satisfying $e_1 = 1/s_o$, $e_2 = -1/s_o$, yields a state of maximum virtual work whose principal strain lines correspond to the optimum distribution of material. Condition (5) being obviously satisfied, the considered truss-like continua are absolute elastic designs.

The basic model,- known as Michell's second example,- is a network of equiangular spirals around a circular core. Arbitrary convex cores lead to more general *Hencky-Prandtl nets* [2,3,4] ,- comparable to slip-lines fields [19],- at angles of 45° with the cores (dense diagrams on fig 1). The distinguishing mark of such networks,- known as Hencky's first theorem,- is that lines form bundles of constant angular magnitude that deviate one another in proportion of their relative magnitudes.

Complete analytical solutions, based on the underlying telegraphy equation, have been obtained in the past, for geometrical, kinematical, and static analysis of networks. Those contributions seem to have reinforced the idea, among practical engineers, that Michell Theory was an esoterical matter of few practical interest. Indeed, practicians will never believe that the abstract discussion of elementary tractions and compressions can deliver an optimum realistic design, with a detailed description of the state of stress ! By contrast, we will now see how some relaxations of the initial models can bring life to reasonable Artificial Architects, that are quite able to work in an engineering environment.

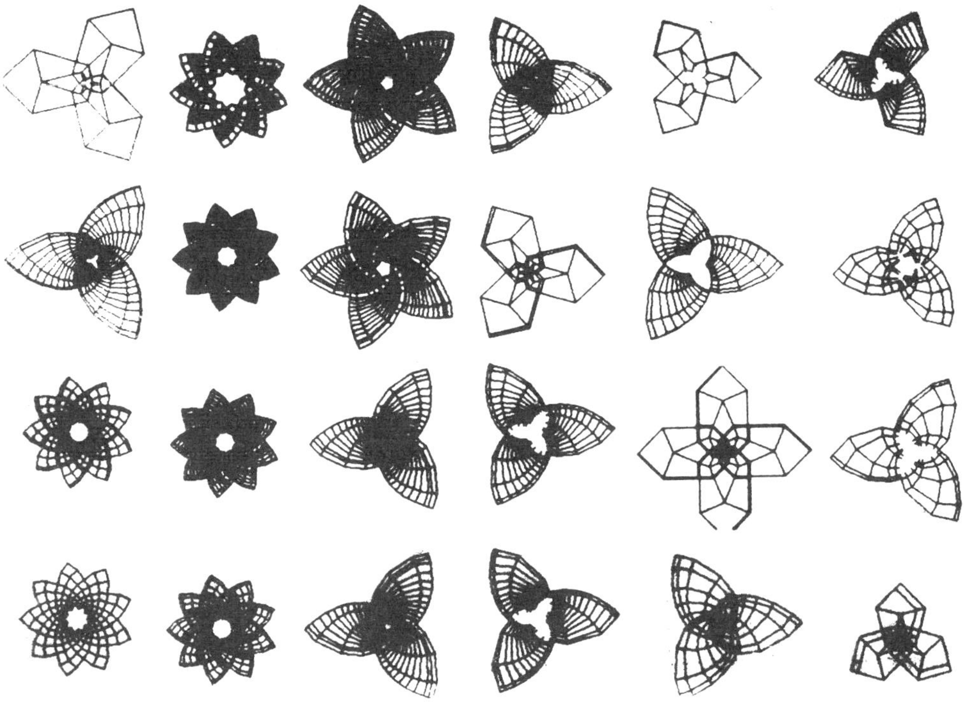

Fig 1

Fig 2

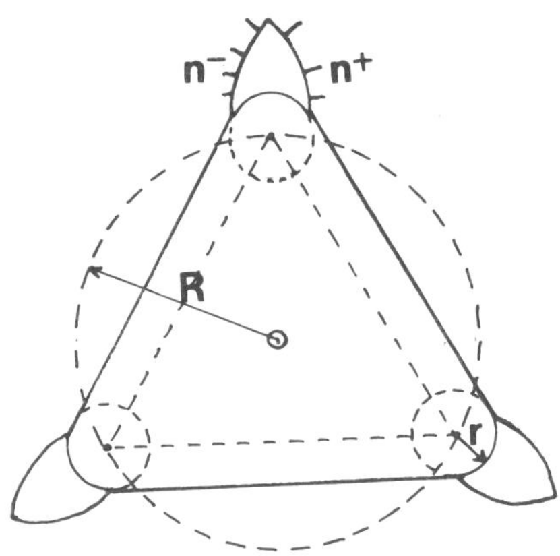

Fig 3

b) MICHELL : a small shape-generator on the Apple II e

Fig 1 shows some specimen of a 6 parameter family of Hencky-Prandtl nets that can be used in the design of wind turbines (fig 2). The shapes were generated by MICHELL, a 1 000 line UCSD Pascal program running on the Apple II e, 128 K. The program delivers graphic outputs, together with complete tables of nodal coordinates, lengths, and bending angles along positive and negative fibers. The diskette and source listing can be freely obtained from the author.

The main difficulty in describing the work that is achieved by the program lies in that practical considerations now begin to interfer with pure Michell design. We are placed in the rather paradoxal situation, where rules receive their full justification from practical results, while practical results,- the frame on Fig 4, by example,- take their origin and justification in the application of rules. This is typically a *dialectical* situation,- in the sense proposed by the Swiss psychologist Jean PIAGET [20],- which in fact conceals the considerable amount of time, money, and work we have spent in the mutual equilibration of theoretical models and practical realizations.

The program has been written to synthetize torque transmission frames, able to sustain, in a fully symmetrical manner, an arbitrary number of identical cylindrical blades. The first input parameter is the number, N , of blades.

For the sake of simplicity, MICHELL assumes the core has the same order of rotationnal symmetry as the overall design : 3 blades correspond to a triangular core, and 4, by example, to a square one. The convex hull of N "secondary" circles uniformly distributed along a "primary" circle centered at the spindle, constitutes the theoretical rigid core. The radii of primary and secondary circles,- respectively denoted R and r ,- are the second and third input parameters. The theoretical Hencky-Prandtl net presents dense networks of logarithmic spirals in the vicinity of secondary circles, which have been very soon recognized as perfectly uncompatible with practical noding. MICHELL incorporates the networks in the continuous rigid core, which, in its final form, is a convex polygon equipped with natural assembly "ears" (Fig 3).

Positive and negative fibers are placed by the program in the middle points of regular angular intervals, along the right and left spirals that delineate each assembly ear. Two parameters characterize the distribution : the number, n^+ , of positive fibers per ear ; the number n^- , of negative ones. Each fiber is considered to represent the mean line in the bundle of lines that inoccupiesthe same angular interval in the theoretical Hencky-Prandtl net, and receives a weight whose magnitude is that of the interval, and whose sign corresponds to the type of the considered fiber. The key of the network synthesis is a discrete counterpart of Hencky's first theorem : each time a fiber of the weight a meets with a fiber of the weight b , the former is deviated by an amount of b , and the latter, by an amount of a .Writing this central part of the program in such a way that rotationnal symmetries are preserva-

Fig 4

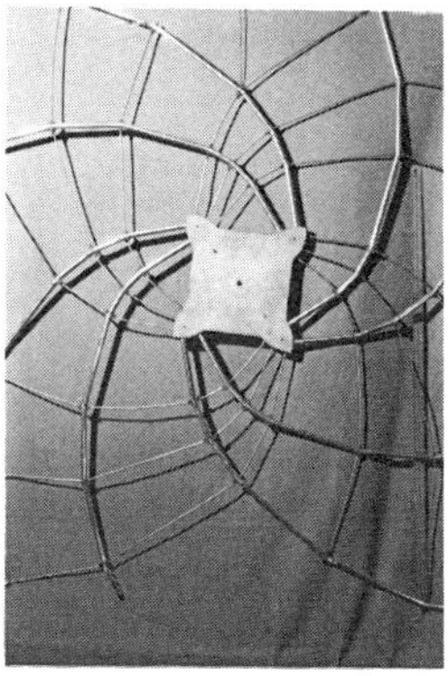

Fig 5

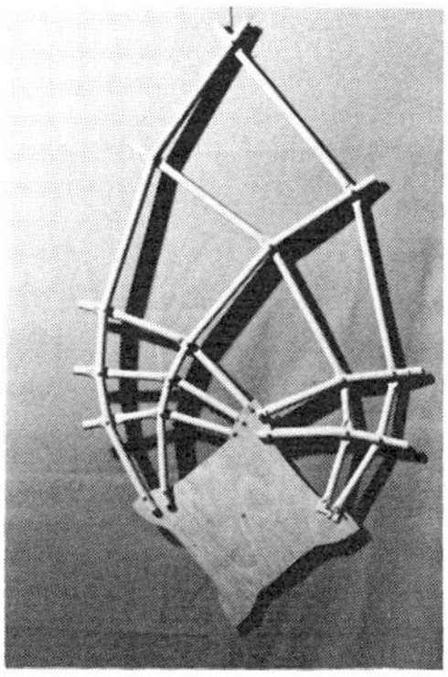

Fig 6

ted, and fibers are continuously described from their origins to their extremities, is an interesting programming exercise. Just say that the program computes the intersection grid of n^+ current positive fibers with n^- negative ones,- which in fact is the discretization of the theoretical boundary problem relative to 2 principal arcs in slip-lines Theory [19],- performs the rotationnal symmetries, and takes to the next grid, once boundary conditions have been rearranged.

The number of elementary grid computations fully characterizes the extension of the frame. This is the 6th, and last, input parameter.

Degenerate lines exist in the theoretical Hencky-Prandtl net that take their origins on the rectilinear boundary segments between the assembly ears. They present the amazing particularity of keeping parallel in the whole field, which can be of a great help in the assembly of blades(fig 2). Degenerate lines do not perturbate the angular geometry of the network and can be either present, as on fig 2, or absent, as on fig 4. Their behaviour is simulated in the discretized model by additional fibers of zero weight, that are deviated by other fibers, but do not deviate them. MICHELL systematically places such zero weight fibers at the extremities of each boundary spiral, which materializes the boundary of the area inoccupied by parallel curves. Finally, the program carefully closes the design, in such a way that ribs called to receive blades are constitued by pairs of parallelzero weight fibers.

c) Practical realisations

Figs 2, 4, 5, 6 show 4 examples of practical frames based upon the preceding geometry.

The frames on fig 2 have a diameter of 1 m, and are made of balsa wood, coated with polyester resin. Positive fibers are rectangular rods of 1 cm height. Negative fibers are constituted by pairs of small square rods of 2mm. Each ear receives 3 positive fibers and 4 negative ones. Rectangular rods are stuck to small rods inside small notches. Dissymmetry in the numbers, n^+ and n^-, of positive and negative fibers is just a practical trick to prevent an excessive weakening of notched bars.

The frame on fig 4, is a 3m diameter triangular frame whose fibers are polyvinylchloride tubes ordinarily used by plumbers and electricians. Thick fibers have a diameter of 3 cm, and thin ones, a diameter of 1 cm. The 3 cm tubes are pierced, at intervals computed by the program, to provide outlets to the thin tubes. The general assembly is more difficult than for the notched balsa rods, but results in a relatively stronger structure : tubes are first heated and bent at each nodal point, thick tubes are assembled to the core, and thin ones are then pushed through the assembly holes from the outside toward the core. The very reason why fibers are placed by the program at regular angular intervals along boundary spirals is that all fibers in one family have the same weight, and, thus, induce a uniform bending of the other family. Only two bending templates are thus required to build our polyvinylchloride truss : a template of 10° for thick tubes, and one of 15° for thin ones.

The frame on fig 5 is a 2,40 m square frame built, on the same principle, in aluminium tubes and wires. A more simple realization is shown on fig 6 where thick fibers are constituted by pairs of PVC tubes enclosingthin fibers constituted by single tubes, the assembly being then obtained by bolts and screws.

Fig 7 and 8 show a piece of empirical chemistry, which has little to do with Hencky-Prandtl nets, although it is of the highest practical importance in the realization of torque transmission frames. The problem on hand was to substitute to the prototypical wooden cores that equip the preceding frames, more convenient plastic cores including standard coupling sleeves, for connection to iron spindles. Apossible solution appears on fig 8. The basic material is a 2 component thermosetting polymer that is poured, at ambiant temperature, in lubricated silicon rubber moulds.The core is constituted of 2 parts : a triangle, that receives the coupling sleeve, and is independent of the geometry of the frame to be connected ; and a crown wheel in which fibers of any type can be inserted by means of small mouldedplots. Fig 7 shows the series of preliminary wood and rubber mouldsthat are necessary to obtain the final moldsfor the triangle and the crown wheel. Note that the time of production of a core, which is already reduced to 10 or 15 minutes with a single mould,can be considerably reduced by successive duplications of the initial mould.This method was used for the production of plots (fig 8) : it only consists in building a 2 place mould with 2 initial casts, and then a 4 place mouldwith the 2 casts and 2 new ones obtained from the 2 place mould,and so on, in a geometrically accelerated process.

8

Fig 7

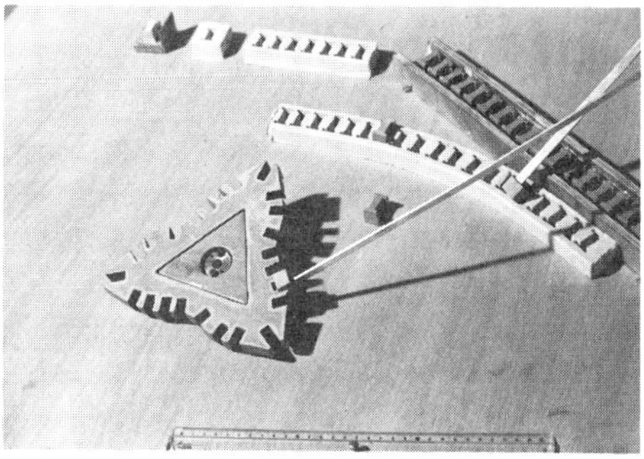

Fig 8

Fig 9

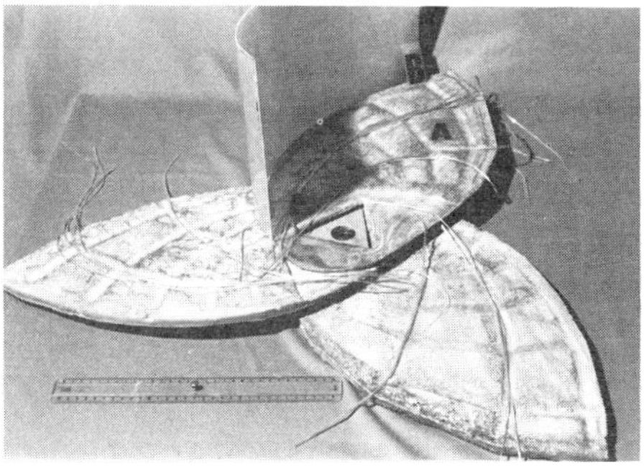

Fig 10

Figs 9 and 10show an attempt to build either stronger fiber-reinforced cores for large structures or easily reproducible small scale composite structures. The method consists in moulding rigid polyurethan loafs in a wooden flat mould whose 2 opposite faces reproduce an approximate Hencky-Prandtl net. Glass fibers can then be placed along the grooves, as indicated on fig 10.A final coating with polyester or epoxy resin, together with a central reinforcement with several layers of chopped strand mat and roving, will then achieve the realization of efficient torque transmission composites. The work is presently under achievement and we cannot, unfortunately, show any picture of the final result...

3. THE OVERALL DESIGN PROCESS

Up to now, only problems concerning shape generation, practical junctions, and materials, have been taken into account. Note, by example, that no precise stress analysis has been proposed, and that the considered structures have thus lost their initial meaning of "optimal structures". Nevertheless, they have inheritated of the underlying minimum weight layout theory geometrical rules of good construction that impose themselves with some kind of self evidence (assembly ears, parallel ribs,...). In other words, Michell Theory has made the job that is ordinarily performed by architects, and cannot be here considered, in any way, as a branch of Optimization.

In parallel with material experiments, complementary tasks of Structural Engineering have been proposed to students of Ecole Centrale. Various programs have been developed in this occasion on PC compatible microcomputers. We can just mention in this short paper the three main realizations :

- a UCSD Pascal program for the interactive computation of efforts in fibers ;
- Fortran program, based on Brebbia's boundary element programs [21] for the computation of stress concentrations in the cores ;
- a Turbo-Pascal program, at last, to compute simplified 2D steady flows around Savonius rotors (1986).

Although perfectly valuable,- the latter program, by example, is able to deal with 600 degrees of freedom on the IBM-XT,- those works are still uneffective and, truely speaking, have little influence on the current developement of Michell frames and windmills. We have already mentioned in ref [18] that stress analysis of a windmill aluminium component had revealed that the maximum tolerable windspeed was about 100 mph ! In the same way as preliminary design could not be viewed as a simple application of Michell rules, it is clear that obtention of effective computational models can only result from the careful, and lengthy, mutual equilibration of material prototypes and methods of computation. Even in the computer ages, the only reasonable method still remains direct observation of physical objects (not simulation!) and detection of striking features and failures mechanisms.

Real life design should finally look like a dynamical harmonization process between 3 antagonistic partners : architectural concepts,- and especially Michell Theory,- on the left ; practice, in the middle ; Engineering Analysis and Design, on the right. As was noted by a Chinese author in a celebrated treatise on Chinese gardens, there are basically two kinds of gardens : small gardens that can be observed from a single viewpoint, and large ones that require multiple points of view.

REFERENCES

1 Michell A.G.M. (1904). On the Limits of Economy of Material in Framed-Structures. *Phil Mag* 8 : 589-597
2 Cox H.L. (1965). *The Design of Structures of Least Weight*. Pergamon, Oxford.
3 Hemp W.S. (1973). *Optimum Structures*. Clarendon Press, Oxford.
4 Prager W. & Rozvany G.I.N. (1977). *Optimization of Structural Geometry*. University of Florida.
5 Botrini M. (1982). *Etude Aérodynamique d'une Eolienne Savonius*. Thesis, University of Aix-Marseille.
6 Higgins G. (1926). *Centrifugal Pumps, their Design, Operation and Testing*. Crosby Lockwood & Sons. Stationers' Hall Court, Ludgate Hill.
7 Benjamin B.S. (1969). *Structural Design with Plastics*. Van Nostrand, New York.
8 Chasen S.H. (1985). Productivity improvement : the Quest for the Strategy. Proc of CAD/CAM Int Conf. Tucson, Arizona. pp 3-7 .
9 Laurent P.-J. (1972). *Approximation & Optimisation*. Hermann, Paris.
10 Rockafellar R. (1972). *Convex Analysis*. Princeton.
11 Lagache J.-M. (1982). Abstract Convolutíon & Optímum Layout. Proc Euromech 164 , University of Siegen. Wissenschaftsverlag, Mannheim.pp 340-345 .
12 Lagache J.-M. (1983). Convolution & Géométríe Optímale. Proc of 3rd Int Conf on Num Methods in Engineering. Pluralis, Paris. pp 415-424.
13 Lagache J.-M. (1985). Minimum weight design of trusses and some other bodies. Proc of Int Conf on Nonlinear Mechanics, Shanghaï. Science Press, Shanghaï. pp 1239-1244 .
14 Lagache J.-M. (1986). Theory & Practice of Michell Frames. Proc of IASS Symposium on Membrane Structures & Space Frames, Osaka. Elsevier. pp 221-228 .
15 Lagache J.-M. (1981). Treillis de Volume Minimal dans une Région Donnée. *J de Méc* 20 (3) : 415-448 .
16 Lagache J.-M. (1980). A Geometrical Procedure To Design Trusses in a Given Area. *Eng Opt* 5 : 1-12 .
17 Lagache J.-M. & Menet P. (1984). Michell To-Day. Proc Int Conf on Space Structures, Guildford. Elsevier, London. pp 700-705 .
18 Lagache J.-M., Menet P. & Kasharherou J. (1985). Michell Windmills. Proc of CAD/CAM Int Conf. Tucson, Arizona. pp 301-306 .
19 Hill (1950). *The Mathematical Theory of Plasticity*. Oxford.
20 Piaget J. (1980). *Les formes élémentaires de la dialectique*. Gallimard, Paris.
21 Brebbia C.A. (1972). *The Boundary Element Method for Engineers*. Pentech Press, London.

CIM/FMS

A Work-Orientation Device for Castings in an Automated Foundry

R. Givens* and J. Jeswiet†

*Engineer, GSW, London, ON Canada
†Associate Professor, Queen's University, Kingston, ON Canada

This paper describes a work orientation device which eliminates the need for vibratory equipment in a small foundry and which presents a workpiece in a known orientation to a robot at a grinding work station. The mechanism described is part of an automation study which was made for a local high volume foundry.

In any robot application or flexible manufacturing system, key parts are the handling tools which present the object in a known orientation in a cell. In many cases, this requires special designs and innovations, a feature which usually occurs in situations where one considers installing a robot (1). Foundrys always have a workstation where the flash on the casting is removed, commonly called fettling. The study (2) indicated the fettling area to be one where automation of the system with a robot would provide the most benefit to the manufacturer. However, whether the robot fettles by picking the casting up, or manipulating a tool against a fixed casting, a human element is always needed to orient the part. A proposed method to keep the casting oriented from the mold station through the shakeout process to the fettling operation and also maintaining flexibility for the different product lines is outlined.

Foundry Overview

The foundry casts copper faucet bodies which are then assembled with other faucet parts and packaged and shipped to North American sales locations. There are six basic types of faucet bodies which must be handled of which two are illustrated in figure 1.

The foundry produces 1260 castings per eight hour shift based upon the following figures: 7 melts per shift, 2 ladles per melt, +30 molds per ladle, +3 faucets per mold.

The area that was found to be "ripe" for automation, if a fettling robot was introduced, was a conveyor belt system which came from the casting area and led to a shakeout table with subsequent normal labour. The processes which were carried out were: cleaning mold material from castings, cutting off the sprue, shot peening and grinding off the flash. A schematic of the existing foundry layout is shown in figure 2. The proposed alternative, the schematic work station shown in figure 3, requires space which is approximately one-tenth the existing area and it contains: a press to eliminate flashing, a cutoff position and shot blasting.

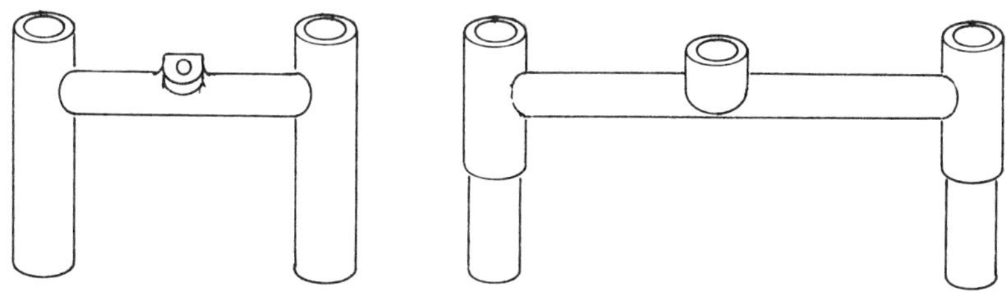

Figure 1. Two examples of cast copper faucet bodies being considered.

Work Orientation Equipment

A specially designed device is proposed for part cleaning and orientation. The device, which is specially designed for this station, has been dubbed a "mold abrader". It eliminates the need for a shakeout/cleaning station, however its primary advantage is that it leaves the casting in a fixed, oriented position after mold sand has been removed.

The mold abrader consists of a clamp (figure 4) that rides on a track, and of the mold abrader itself, which removes sand mold. The clamp rides on a track made of two sections of angle iron, as in fig. 5.

A carriage, incorporating spring loaded rollers allows the clamp to roll up and down the track. The carriage is driven along the track by a cable which is driven by a pneumatic cylinder, in indexing steps.

The operation of the abrader system is illustrated in figure 6. The clamp comes to a fixed position directly above the sprue of the mold sitting on the mold raiser. The mold raiser is simply a table mounted on pneumatic cylinders that forces the mold into the jaws of the clamp. The jaws actually pierce the mold sand in the vicinity of the sprue, and then close on the sprue.

The clamp drags the casting, with attached mold sand, into the abrader compartment where stiff wire brushes, turning at a high rate, wear away mold material. Because of the messy type of operation the abraders must be in an enclosed compartment. Note that the mold raiser lifts the mold only to a level where the bottom board is prevented from continuing on with the mold.

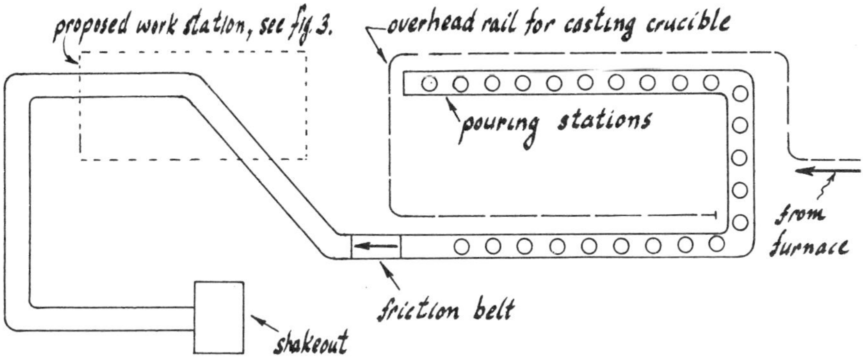

Figure 2. Existing foundry layout.

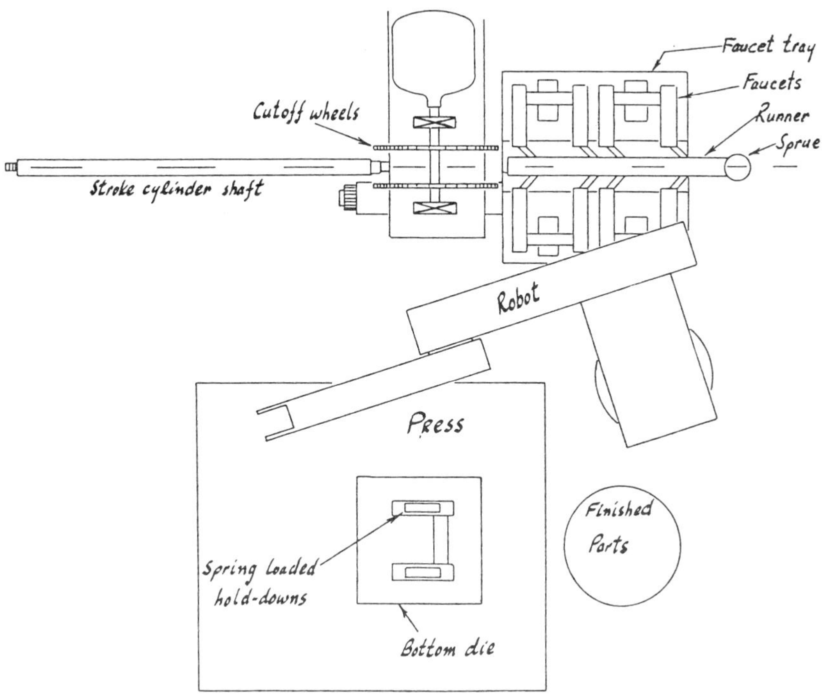

Figure 3. Proposed alternative work station.

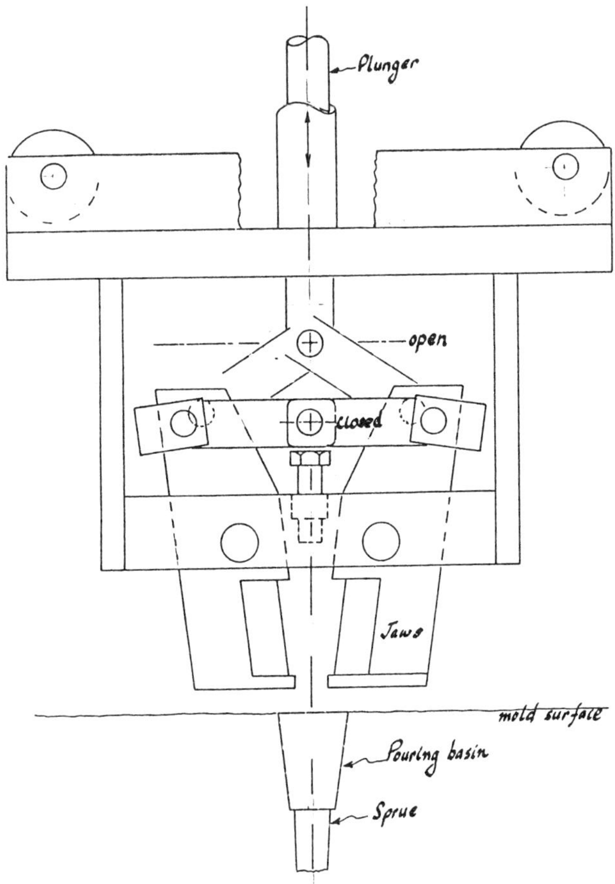

Figure 4. Special sprue clamp for grabbing the casting sprue.

The Clamp and Clamp Carriage:

As can be seen in Figure 4, the clamp locks onto the pouring basin of the casting. The jaws would come to rest in rounded grooves in the pouring basin. The liquid metal could not be allowed to overflow the pouring basin, which is allowed to happen occasionally, as this would interfere with the operation of the clamp. This would also force employee's to produce less waste. The basin need not be filled to the top of the mold, but it should be at least 50% full. A pouring basin of the shape shown in Figure 4 can be obtained fairly easily on an automatic

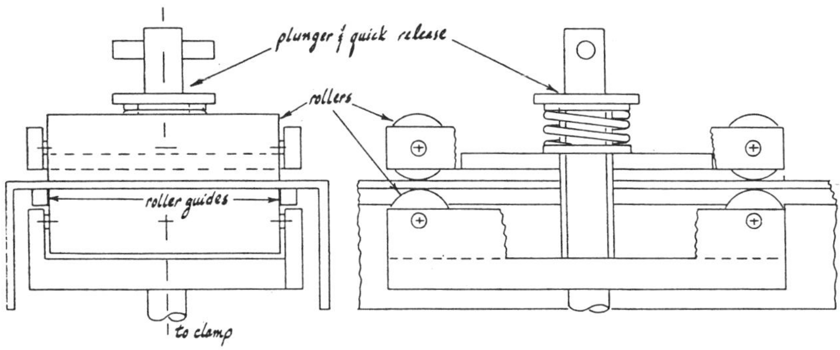

Figure 5. Clamping carriage details.

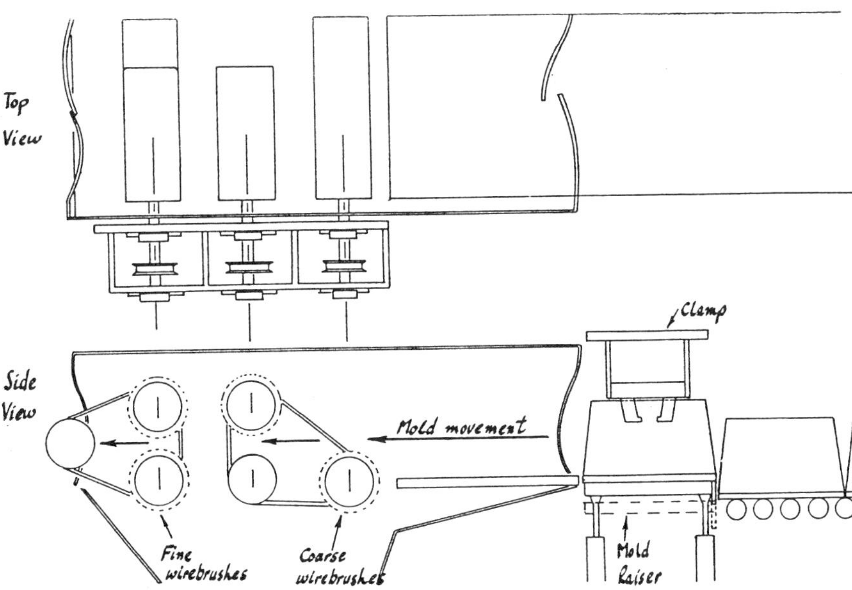

Figure 6. Proposed mold abrader.

moldmaker, but on a squeeze-jolt moldmaker some other configuration, possibly a wide sprue with flat, vertical lugs, might be more workable.

The clamp construction allows for a "Vise-Grip" lock on the pouring basin. As the wear on the jaws would be very high, sheetmetal jaw covers, held in place by small screws, or by friction alone, would protect the jaws as they moved through the sand. To fine-tune the locked position of the jaws, different gages of sheetmetal can be used, and a bolt, located directly under the plunger, adjusts the distance the plunger travels past the "point of greatest locking force."

Figure 5 shows the simple rail track that the clamp rides on. Rollers exert a constant pressure on two angles that form the track. The pressure comes from a heavy spring, so that no vertical movement, or twisting about an axis along the track is allowed. Guides on either side of the bottom roller prevent any lateral movement. The carriage-track construction is designed to provide a very rigid support for the clamp, but at the same time, be simple and inexpensive.

The track would run about a foot above the 3 work stations: the mold abrader station, the shotblast station and the press station. It would be about 15 feet long to accommodate all three stations, plus a "sprue-release" station. Most of the time, there would be 3 clamps holding castings being worked on, one at each of the stations. They would begin to move once all work was complete. the clamp above the mold raiser would "work" only when it carried the mold through the mold abrader, however.

The track would return the clamps by doubling back over itself. That is, an individual clamp, coming out of the last station would be turned upside-down by the track for the return to the starting point. Thus the return track would be above the "working" track. In all, seven equally spaced clamps would probably be needed, three in the work stations and four empty clamps in transit around the top of the track.

To make sure that the clamps came to stops in exact, repeatable locations, the best way to drive the carriages along the track would be by means of pneumatic cylinders indexing a cable. At intervals the cable would have lugs attached so that the cylinders have something to apply force to. An alternative is to use a chain and motor, but making sure the clamps came to exact stopping locations would be difficult.

The clamp is closed by depressing the plunger with a pneumatic cylinder. It is sprung open after the press station so that the sprue and runner fall into a bin. To spring the clamp open, the carriage simply drives against runners that gradually lift the plunger (see Figure 5) out of the locked position. These runners are two inclined rigid surfaces in a fixed position.

Summary

This paper illustrates a design for a small parts foundry which makes the casting system more automatic and potentially more flexible, requires much less space, eliminates some cleaning equipment, fully automates the fettling operation and provides proper part orientation for the workstation robot.

The need for this type of design innovation is important in any conversion to automation or in setting up flexible manufacturing stations.

References

(1) K. Lau and R.J. Hocken. " A Survey of Current Robot Metrology Methods ". National Bureau of Standards, Gaitersburg, MD, USA, 1984.

(2) R. Givens and J.Jeswiet." A Study of Small Foundry Automation". Abstract in the proceedings of the 8th International Conference on Production Research, Stuttgart, 1985, p 935, ©Springer-Verlag.

CIM Framework for Electronics Manufacturing Systems

Unny Menon, Ph.D. and Andrew Young[†]

[*]Professor and Director of CIM Center, Industrial Engineering Department, California Polytechnic State University, San Luis Obispo, CA 93407, USA
[†]Northern Telecom, Santa Clara, CA 95054, USA

INTRODUCTION

Computer Integrated Manufacturing (CIM) provides a structured basis for establishing connectivity between sub-systems to facilitate product data delivery and control from the conceptual design stage through the many intermediate stages in producing the finished product. In this paper we focus on a CIM architecture applicable to electronics manufacturing for printed circuit boards based on studies completed by the authors under the sponsorship of the Northern Telecom University Interaction Program. Our presentation commences with the conceptual considerations of Computer Integrated Manufacturing and the special process characteristics of printed circuit board manufacture which must be considered in formulating the appropriate CIM architecture. We then address the system design issues and problems involved in implementing a viable CIM framework for integrating the many islands of automation in all areas of the manufacturing enterprise.

CIM STRUCTURAL CONSIDERATIONS

The term CIM originates from Harrington (1973) who proposed the following elements of discrete parts manufacture for integrated computer aided control:

* Designing the product
* Engineering the manufacturing process
* Deciding how many and when to manufacture
* Scheduling the steps in the process
* Controlling the tools and energy used
* Monitoring the execution
* Collecting and processing data on accomplishment

At a conceptual level the principal elements of contemporary CIM architectures are included in the above list, which was formulated long before the adoption of CIM by industry. A contemporary perception of CIM can be found in the "CIM wheel" published by CASA/SME (see example in CIM Technology, Vol 4, No 3, p3), which expands the organizational scope of CIM to include Marketing, Finance, Human Resource Management and Strategic Planning which were specifically excluded in the Harrington (1973) perception of the scope for CIM.

The key characteristic element of CIM is the middle letter "I" in the acronym, requiring an "Integrated" approach, with a boundary which far exceeds the technological elements of manufacturing and seeks to incorporate a total system framework for the entire manufacturing enterprise. The opportunity for organizational cohesiveness which CIM represents, is attractive and represents the ideal. As with any "ideal" the attainment of such total integration presents an immense challenge for complete implementation in practice. Organizations which have made the necessary commitment to CIM and have the infrastructure conducive to such integration, will make progress towards attaining the ideal as a long term development activity.

The key observation that can be made with respect to the CIM structure, is that we are not attempting to design a completely new single entity, but rather we seek to establish a framework for interconnecting many of the well established "islands of automation" (as reviewed by Appleton (1986), Ranky (1986) and Saxe (1986)), where each sub-system represents a substantial accomplishment of integration within that functional domain. However, the CIM task involves far more than merely establishing the connections between functional sub-systems, because we now need effective control systems which manage the inter-system relationships from the corporate perspective of the complete manufacturing enterprise and the dynamics of situation dependent operations strategies.

THE ISLANDS OF AUTOMATION

In general each "island of automation" represents one or more computer aided system which enable some functional activity to be completed efficiently. Groover and Zimmers (1984), Hordeski (1986), Ranky (1986) and Yeomans *et al.*(1985) provide details of these sub-systems and stress the need for an integrating framework. The functional areas outlined below have been the focus of developments to provide computer aided systems during the last two decades of progress in engineering and manufacturing automation.

CAD: Computer Aided Design is increasingly the norm for structuring many aspects of the product definition process, providing opportunities for productivity gains and connectivity with other elements of the overall organization. From the early beginnings of CAD in computer aided drafting, we have now moved on to encompass analytical tools to validate the design, construct solid models and incorporated some intelligent aids for the design process, with scope for including manufacturability considerations.

CAM: Computer Aided Manufacturing with its early origins in Numerical Control has progressed along with the emergence of increasing levels of intelligence and connectivity available in CNC controllers and local area networks. This provides a viable framework for integration on the shop floor, with respect to the manufacturing process. FMS (Flexible Manufacturing Systems) provides a working example of the current capability to accomplish a high level of integration and flexible automation for CAM.

CAPP: The transition from design to manufacturing requires a process plan which provides the necessary interface between CAD and CAM. This plan provides the logistics for the ordered sequence of operations to be executed in producing a part. Although some Computer Aided Process Planning systems with limited capabilities are available, the search for adequate systems continues with an increasing trend towards generative approaches using artificial intelligence, *versus* the alternative variant approach using group technology.

CAI/CAT: The development of systems for Computer Aided Inspection and Computer Aided Testing provides the basis for some automation for quality assurance functions. The process of automatically inspecting geometric dimensions using either contact probes, vision or laser beams provides CAI. Similarly CAT, with its emphasis on functional testing of the finished product or interim sub-assemblies using computer based systems enables automated testing for conformance to product specifications. Both CAI and CAT, provide improved connectivity with other processes.

Manufacturing Planning and Control: An integrated framework for manufacturing system planning and control is described by O'Grady and Menon (1986). We require control systems for demand forecasting, release of orders, purchasing, inventory control, material control, resource allocation, scheduling, shop floor control and process monitoring. Computer based data processing systems which automate (to varying degrees) the operational planning and control functions for the manufacturing system are available. However, in most instances (see Saxe (1986)) they are not inter-connected and could be viewed as islands of sub-functional automation.

If we review the currently available islands of automation outlined above, it would appear that the scope for "computer aided control" in each of the areas listed by Harrington (1973) and cited earlier in our introduction, has been addressed within each functional area. The missing element is of course the integrative "I" in CIM; which remains to be accomplished (see Farnum (1985)).

CHARACTERISTIC DIFFERENCES OF ELECTRONICS MANUFACTURING

The term electronics manufacturing can refer to both the micro-electronic integrated circuit (IC) chip manufacture or the printed circuit board (PCB) level of manufacture. Our focus in this paper is at the PCB level which is reviewed here with respect to some differences in the islands of automation that have emerged in this industry.

CAD systems for PCB design have evolved in parallel with similar systems for mechanical design. The CAD systems for PCB design focus on the functional circuit design as well as the board layout for different types of components (resistor, capacitor, diode, transistor, transformer, DIP IC chips, surface mount devices, etc.). In general there is greater commonalty of component specification, description and physical characteristics in electronics, *versus* the generally dissimilar components in mechanical design. However, the PCB designer faces a complex design task, in accommodating all the components on densely packed boards with several inter-connections between components, which have to be routed via copper track and soldered connections. The current generation of CAD systems for PCB provide some intelligent aids for the designer in various aspects of this design process. CBDS (Circuit Board Design System) is one example of such a system developed by BNR (Bell Northern Research) and currently marketed by IBM, with a large PCB design user population.

CAM for PCB manufacture uses numerical control concepts for directing auto-insertion machines, but does not have the counterpart to APT as used in computer aided NC programming for machining and there is less standardization of machine controller level NC code for automated PCB manufacture. This necessitates greater user effort to develop custom post-processors, to establish connectivity for each PCB CAD system.

CAPP for PCB manufacture is simpler in some respects and more complex in other ways, when compared to CAPP for machining. It is simpler because the bill of material for a PCB has a two tier hierarchy, with level zero being the board and level one contains all of the components which have to be inserted. The complexity of the CAPP problem for PCB arises from the difficulty of determining the component insertion sequence which will be feasible with respect to proximity, machine/process dependent considerations and the placement efficiency with respect to minimum distance traversed on the board layout. In addition, the distribution of components at specific machines for multi-product environments presents logistical problems in the CAPP decision sequence. Attempts by CAPP vendors to adapt their systems (which were designed initially for machining) for electronics appear to be unsatisfactory at the present time.

Although, the framework for CIM will have features that are common to many types of industries, the characteristic differences which arise in particular cases as with PCB manufacture cited above, require careful consideration, in the design and implementation of the specific CIM architecture for each organization. The need for CIM in electronics manufacturing has been recognized and adopted in general. This has been reported by Shapiro (1986), based on the responses from several industry leaders, who anticipate reductions in manufacturing cost and improved product quality as the principal benefits of a CIM strategy.

THE NORTHERN TELECOM CIM ARCHITECTURE

The overall framework of the Northern Telecom CIM architecture is depicted in the system diagram, indicating the integrative linkages between design (CAD), design engineering (CAE), manufacturing engineering (CAMENG), production planning (MRP and MRP II), production control, and the data control and administration subsystem (which acts as the centralized data storage facility for the entire environment). This system architecture provides a basis for the overall integration of large sub-systems, by establishing linkages between functional entities and processes in this manufacturing enterprise. It is recognized that the actual implementation of this overall CIM framework is a long term process and requires not only technological adaptations but also human resource/organizational culture adaptations and strong management commitment to facilitate the transition to CIM (see Slem, Levi and Young (1986).

The integration of design engineering functions includes the co-ordinated use of independent CAD tools for circuit board design (CBDS), for documentation of mechanical design (CADAM, CATIA) and other internally developed CAD systems for VLSI and associated software. To ensure high quality designs for VLSI and PCB's, Northern Telecom uses internally developed CAE simulators for both analog and digital circuitry. CAEDS is used for finite element mesh analysis and

CIM FUNCTIONAL ARCHITECTURE

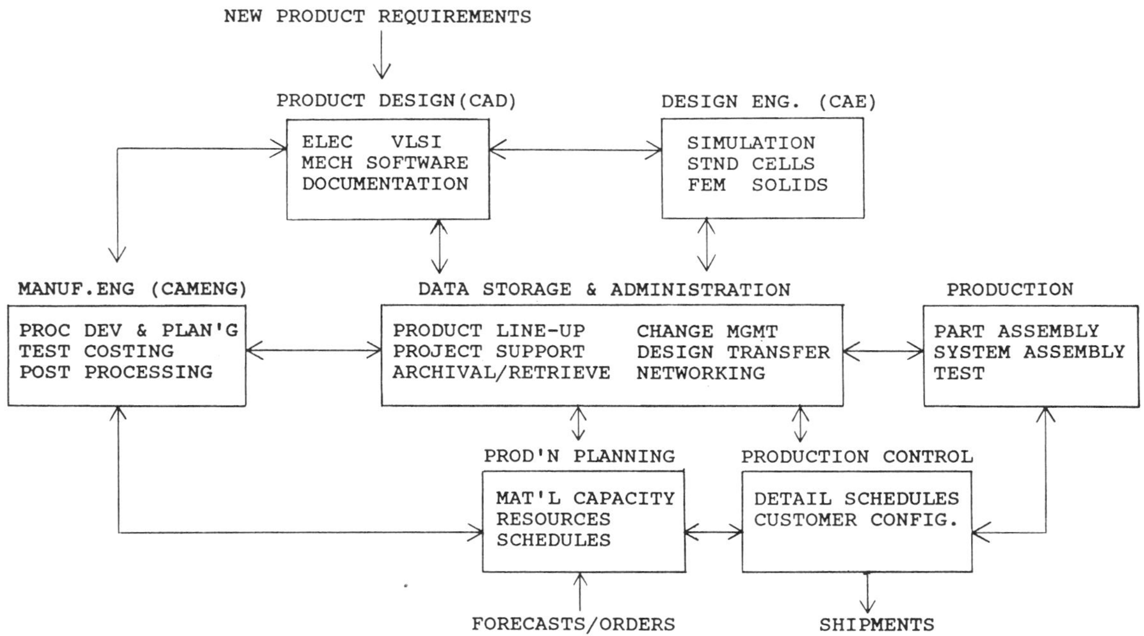

LASAR for functional test simulation. Product design is subject to innovative change and revision on a continuing basis which cannot be avoided. The control and management of this change process will be accomplished by an internally developed module which is part of the current structure for data storage and the administration subsystem. Some of the principal elements of CAMENG such as post processors to convert design data for use on the shop floor and CAPP for process planning are available, while facilities for test, costing, and other manufacturing engineering activities are in place using internally developed systems.

OTHER INTEGRATION ISSUES

The shop floor environment in electronics manufacture is a mixture of automated and manual processes with varying levels of interfacing capability. The problems associated with the establishment of satisfactory interfaces between shop floor systems and higher level systems within the CIM structure has been examined in one of our recent studies. Our findings indicate that the development of viable interfaces will continue to present difficulties. Some of these problems may be reduced if, and when, the concepts in MAP become an operational reality. In the interim, CIM implementation strategies should include contingency provisions of adequate time and software engineering resources to address interfacing problems during commissioning. Our initial studies on CAPP for PCB production has now broadened in scope to include the framework for integrating manufacturability and quality assurance within CIM. We anticipate that knowledge based development systems will facilitate the process of integration which cuts across the functional boundaries of design, manufacturing and quality assurance.

SUMMARY

We have reviewed the conceptual framework of CIM and the islands of automation that have evolved within functional areas of the manufacturing enterprise. The characteristic differences of printed circuit board manufacture as it affects the system design of a CIM architecture have been discussed and provides some indication of the need for some modifications to the CIM architecture for electronics manufacturing industry. The Northern Telecom CIM architecture provides a good example of the kind of framework that is applicable to electronics manufacture. In addition to overcoming the technological difficulties, a successful CIM strategy requires an organizational culture conducive to such integration including strong management support and a CIM oriented corporate strategy to sustain such endeavor.

REFERENCES

D. Appleton, "Integration Technology", CIM Technology, Vol 5, No 1, Spring 1986.

G. Farnum, "AUTOFACT'85: Putting the "I" in CIM", p21, (also CASA/SME CIM wheel page 3), CIM Technology, Vol 4, No 3, Fall 1985.

G. Goos and J. Hartmanis, "Methods and Tools for Computer Integrated Manufacturing", Springer-Verlag, 1984.

M.P. Groover and E.W. Zimmers, "CAD/CAM: Computer Aided Design and Manufacturing", Prentice-Hall, 1984.

J. Harrington, "Computer Integrated Manufacturing", Industrial Press, 1973.

J. Harrington, "Understanding the Manufacturing Process: Key to successful CAD/CAM Implementation", Marcel Dekker, 1984.

J. Harrington, "Why Computer Integrated Manufacturing", Commline Jnl. Comp. Mfg., Vol XV, No IV, P.O.Box 11000, Wheeling Illinois 60090, July/August, 1986.

W. Herroelen, Z. Degraeve and M. Lambrecht, "Justifying CIM: Quantitative Analysis Tool", CIM Review, Vol 3, No 1, Fall 1986.

M.F. Hordeski, CAD/CAM Techniques, Reston Publishing, 1986.

R.H.F. Jackson and A.W.T. Jones (Eds.), "Real-time Optimization in Automated Manufacturing Facilities", NBS Special Publication 724, U.S. Govt. Printing Office, 1986.

D.L. Judson, "Integrated Information Support System IISS", CIM Technology, Vol 5, No 2, Summer 1986.

J.W. Nazemetz, W.H. Hammer and R.P. Sadowski (Eds.), "Computer Integrated Manufacturing Systems: Selected Readings", Industrial Engineering & Management Press, Georgia 30092, 1985.

P.J. O'Grady and U. Menon, "A Hierarchy of Intelligent Scheduling and Control for Automated Manufacturing Systems", p 15-30, in Proceedings of Real-time Optimization in Automated Manufacturing Facilities, NBS Special Publication 724, U.S. Govt. Printing Office, 1986.

P.G. Ranky, "Computer Integrated Manufacturing", Prentice-Hall Intnl. 1986.

N.E. Ryan, "Implementing CIM: AUTOFACT Leads the Way", CIM Technology, Vol 5, No 4, Winter 1986.

K. Saxe, "MRP-II into CIM: The Interface phase", CIM Technology, Vol 5, No 3, Fall 1986.

M. H. Seigel, "CIM Strategy: Integrated Systems Play Leading Role in Chrysler's Comeback", CIM Review, Vol 3, No 1, 1986.

S. F. Shapiro (Ed), "CIM: The Right Medicine for Electronic Manufacturing? ", Computer Design, p 73-88, November 1, 1986.

C. Slem, D. Levi and A. Young, "Evaluating an Organization's Potential for Adopting Computer Integrated Manufacturing", in O. Brown and H. Hendrick (Eds.), Human Factors in Organizational Design and Management II, North-Holland, 1986.

R.W. Yeomans, A. Choudry, & P.J.W. ten Hagen (Eds.), "Design rules for a CIM system", North-Holland, 1985.

FMS and Organization

Jos A.M. Willenborg

KPMG Klynveld Bosboom Hegener, Management Consultants, Kaap Hoorndreef 54, 3563 AV Utrecht, The Netherlands

ABSTRACT

The effective use of new technologies has become a key factor in international competition. Introduction of a Flexible Manufacturing System (FMS) should make it possible to combine the traditionally conflicting objectives productivity and flexibility. But experience shows that it is hard to realize the anticipated benefits. Especially organizational factors impede the successful implementation of FMS.

The most important practical implication of computerized manufacturing is the need of a perfectly controlled production process. Improvizations on the shopfloor can no longer compensate for imperfections in the supply of materials and information. Given the wide product range and low production volumes of Dutch metalworking companies, compared to the United States and Japan, this need for standard conditions presents specific problems for the implementation of FMS.

1. INTRODUCTION

The effective use of new technologies in products and processes has become a key factor in international competition. Western companies have rediscovered the importance of a productive yet flexible production system. Much attention is paid to computerized manufacturing. A Flexible Manufacturing System (FMS) is a group of machine tools with automatic exchange of tools and parts under control of a computer. An FMS should make it possible to combine the traditionally conflicting objectives productivity and flexibility (3, 7, 9, 16).
But in spite of the potential benefits, only a limited number of systems has actually been installed world-wide. Experience shows that not only financial and technical factors, but also organizational aspects impede the successful implementation of FMS. This study concentrates on the question:

What is the influence of FMS on the organization ?

The empirical research originally contains eight FMS projects in the Dutch and British metal industry. For analytical reasons we concentrate here on four case studies. The firms range from large batch to one-off production (see figure 1.). Especially experiences of firms with a relative high part variety and low production volumes are interesting, since most of the literature on FMS deals with systems that produce only a small number of different parts.

	MINE	PRINT	COMPRESS	TRUCK
TOTAL COMPANY				
end product	coal mining equipment	textile printing machinery	compressors	trucks
number of employees	1900	700	475	7000
production	one-off	small batch	small batch	large batch
NC experience	high	high	low	medium
machine layout	process	group	process	product
machines before FMS	CNC	CNC	universal	special
FMS				
number of machines	6	2	2	6
major parts	gear box	cubic parts	crank house	rear axle
average volume (cm)	160*90*50	60*30*20	25*25*12	200*50*50
weight (kg)	2000	5-200	1-200	100-130
operators/shift	10	3	2	2-5
part variants	10	315	100	6
production volume	low	medium	medium	high

Figure 1: Major characteristics of the firms.

The introduction of FMS is considered as a change in the technical system, which implicates a change in the work that has to be done. This consequently requires adjustments in the way the work is divided and coordinated: the organization structure. In this process of change, management decisions play a crucial role.

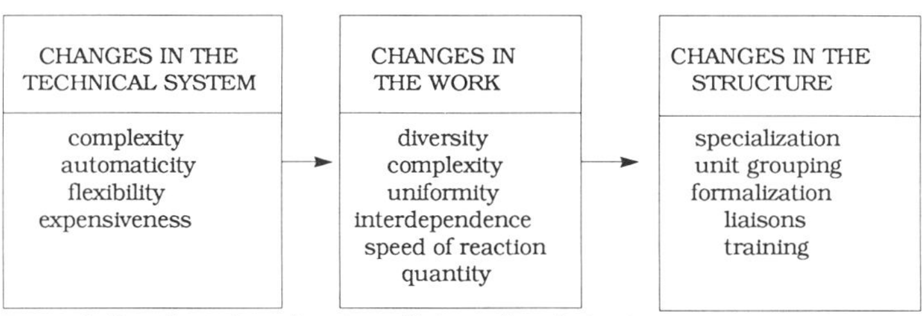

Figure 2: How the technical system effects work and structure.

2. THE TECHNICAL SYSTEM

We concentrate on the transition from conventional machine tools (universal and special-purpose) or stand-alone CNC to the integrated FMS.
The FMS is more complex than the other machine tools as for electronics and software. It is also more automated, because of the computerized operations control and automatic exchange and transport of tools and parts.
From financial analyses the FMS appears to be highly expensive: software and peripheral equipment, like transport device and tooling system, are so expensive that the machines in the system often account for only half the total investment.

Less clear is the comparison of flexibility. We distinguish:

- *product flexibility:* ease of changes in product design;
- *process flexibility:* ease of changes in process plan (tool path, feeds and speeds etcetera);
- *production flexibility:* ease of changes in production (set-up time).

The product flexibility of FMS is often overestimated. Ross (18, p. 32) for example mistakenly asserts:
"An FMS combines the consistency and accuracy of numerical control with the flexibility usually associated with highly skilled operators". This is not true. A skilled machinist and a universal machine tool form an extremely flexible combination. With FMS changes in product design are only easier within the critical limits of the system capabilities. A system is more than the aggregate of its components: every element in the system, like machines, transport device, tooling system and fixtures, must be able to handle a new product design.
The process flexibility refers to the possibility to adapt cutting conditions to unanticipated variations in raw materials. This is very easy with an operator on a universal machine. With FMS this is less convenient than with stand-alone CNC since changes in the NC program for one machine may have consequences for other elements in the system.
One of the most important advantages of FMS is the drastic reduction of set-up times: the production flexibility is extremely high. When the necessary parts, tools, programs and fixtures are immediately available, products can be processed in chaotic sequence.

FEATURES OF THE TECHNICAL SYSTEM	UNIVER-SAL	SPECIAL-PURPOSE	CNC	FMS
complexity	1	1	3	4
automaticity	1	2	3	4
expensiveness	1	2	3	4
product flexibility	4	1	3	2
process flexibility	4	1	3	2
production flexibility	3	1	2	4

Figure 1: Distinctive features of metal cutting equipment
1 = very low 3 = very high
2 = low 4 = high

By changing these features of the technical system, introduction of FMS has several organizational implications.

3. PROCESS PLANNING

The most radical implication of FMS is that the high expensiveness and automaticity, together with the low process flexibility, require a perfectly controlled production process. All potential causes of errors have to be eliminated: defects in NC programs, cutting tools, raw materials, fixtures, machines and so on. Process planners must carefully prepare the production process, because improvizations on the shopfloor can no longer compensate for imperfections in the supply of materials and information: it is too expensive, takes too much time or human control is even absent. Process planning becomes more complex, not so much because of the NC programming, but because of the need of a profound knowledge of metal cutting technology. This is required for the:

- efficient use of machines, transport and tooling system;
- standardization of cutting tools, jigs and fixtures;
- realization of unmanned cutting processes (cutting rates, tool life, swarf).

The empirical data suggest that the transition from universal machines to FMS increases the time needed for the process planning of a particular product (see figure 4.). There appears to be only a slight difference between CNC and FMS. The firm TRUCK, where the FMS replaces special purpose machines, reports a reduction in process planning time: less machines and standardization of tools.

	MINE			PRINT		COMPRESS		TRUCK	
	CONV	CNC	FMS	CNC	FMS	CONV	FMS	CONV	FMS
TIMES									
set-up (h)	180	80	40	0,3	0	4	0,5	6	0,2
cutting (h)	.	.	.	0,8	0,3	6	4	.	.
througput (d)	120	70	35	20	10	15	10	.	.
process-planning (d)	5	60	65	4	4	8	10	30	5
NUMBERS									
batch size	1	1	1	50	35	16	2	450	8
operators	42	2	10	4	2	.	2	6	2

Figure 4: Important figures in the old situation and with FMS
(h = hours, d = days, . = data are missing)

4. COMPONENT MANUFACTURING

FMS means a drastic reduction in direct labour costs. In the firms we studied the number of operators decreased with 40 through 80 percent.
As FMS requires a perfectly prepared production process, decisions on the process plan (operation sequence, tools, feeds, speeds etc) inevitably shift from operators to experts on a separate department. But this specialization does certainly not imply deskilling of operators. On the contrary, in all firms the FMS operators were highly qualified, often more than operators on conventional machines.
The main reason for this is the need of organization and coordination: all parts, programs, cutting tools, pallets and fixtures must be available in the prescribed time and quality. With introduction of FMS the character of the work on the shopfloor shifts from manually operating one machine to organizing a complete system. This requires knowledge of planning and logistics, and is a rather complex task.
But there remain also simple tasks, like (un)loading parts and exchanging cutting tools. The work on the FMS can be characterized as very diverse: both complex and simple tasks have to be performed. This means that managers have to make a doubtful choice as regards to the labour organization: job specialization or job rotation ?
Brödner (2, p. 303) asserts that most installed systems have job specialization. Contray to this, all publications we have studied advocate job rotation in order to increase motivation and problem solving capacity (1, 2, 4, 6, 12, 13, 19).
Our empirical data however showed neither extreme forms of job specialization, nor of rotation. All firms had FMS operators and a system manager. The operators mainly were responsible for monitoring, (un)loading parts and setting up fixtures, pallets and tools. They also have to do some planning to coordinate parts, fixtures and tools. In some firms the operators do the tool presetting and minor program adjustments. The function of the system manager contains production scheduling, supervision of operators and reacting to unanticipated deviations from normal operations.

Two conclusions are possible now:

1) Our empirical data are not representative: most installed systems do indeed have the traditional form of specialization. If this is true, there apparently exists a remarkable gap between theory and practice. Why are the theoretical recommendations not applied in practice ?
This situation can be compared with the debate around CNC programming. Most theoretical publications advocate programming by the operator, while in practice most machines are programmed by a separate programmer (17).

2) Our data are representative. If this is the case, the proposition that most installed systems have specialization is based on false data, or has become obsolete. It is possible that recently an evolution has occured, or is occurring: The disadvantages of high specialization may have urged managers to seek an alternative, more flexible work organization in order to raise motivation and problem solving capacity.

Any way, more practical experience and objective research seems to be needed for reliable and normative conclusions on effective organizational forms for FMS.

5. PRODUCTION MANAGEMENT

As objectives for purchasing the FMS the firms especially reported decreasing delivery time and stock, besides reducing direct labour costs. An outstanding advantage of FMS is the high production flexibility: the drastic reduction of set-up time allows for the reduction of batch size, stock, throughput time and planning horizon. In most cases, management saw FMS as a means to apply the `*Just In Time*' principle (JIT), which requires a flexible and reliable technical system. FMS implies then a shift from an MRP approach (Materials Requirements Planning/ Manufacturing Resources Planning) to JIT. The method of production management shifts from `push' to `pull' and component manufacturing shifts from batch production to flow production.
One of the problems for production managers was that the FMS meant high flexibility in only a limited part of the total logistic chain. They stressed the importance of an integral investment plan to raise flexibility of the whole manufacturing process. FMS and JIT require short and reliable delivery time of raw materials and a hundred percent quality assurance. So purchasers have to shift the emphasis from price to quality and delivery time. Relations with suppliers become more long-continued, so that purchasers lose their trump in negotiations, namely threatening to go to another supplier.

6. DESIGN

During interviews production engineers stressed that product designers must not think that with FMS every design is possible. The designers have more freedom than with conventional machines, as long as their designs do not cross the critical limits of the system capabilities. When for example a part cannot be processed by the transportation device of the FMS, the whole system is useless. Design standardization is important for reducing the number of cutting tools, jigs and fixtures. The traditional sequential interdependence between the departments of design and production engineering turns into a reciprocal interdependence: designers need feedback information on the extent to which their designs can be manufactured and production engineers need information on future product lines for the sake of capacity planning. This growth in reciprocal interdependence results in multi-disciplinary project teams, formal procedures for design changes and contracts between designers and production engineers.

7. MAINTENANCE

Because the FMS is so complex and expensive, maintenance becomes more complex and critical. The emphasis shifts from mechanical to electrical maintenance and programming. Preventive maintenance becomes more important. Technicians must, more than with conventional machines, rely on formal protocols and documents for the prevention and diagnosis of failures. They have to attend to not only the machine tools, but also to other devices like robots, guided vehicles, computers and so on. Especially for firms with little NC experience, like COMPRESS, skill requirements of electrical technicians rise sharply.

8. DIFFERENCES IN PRODUCTION SITUATION

Most of the firms were early adopters: they bought a system that was quite new for the supplier as well. They run into technical problems with software, tolerances, swarf, tool variety, tool life and so on. But, as the plant manager of COMPRESS puts it:
"We can solve the technical problems; more difficult are the organizational problems". His goal is to raise the machine utilization from the actual 60 % to 75 %. He regards the 85 % that the supplier promised not realistic. In his company the FMS was initially operated by men from the mechanical department. Later on it appeared necessary to appoint an older, experienced production engineer as system manager in order to assure the supply of materials and information. This was necessary because: *"the required standard conditions were not yet fulfilled".*
With conventional machines, improvizations on the shopfloor compensated for failures: incorrect work instructions; lack of jigs and fixtures; bad quality or even absence of raw materials and so on.
The other firms appeared to have less problems with the standard conditions on the shopfloor. MINE and PRINT already had much NC experience. Above that, PRINT had a group technology layout, so that the FMS took over production of only one, distinctive production group. TRUCK can be characterized as high volume production, so that the manufacturing process was already more planned, prepared and standardized before the introduction of FMS. The conclusion from our analysis is that FMS requires more drastic organizational adjustments when:

- the firm does not have much experience with numerical control;
- the machine shop has a process type layout instead of a group technology layout;
- the firm produces many product variants in low volumes.

It is to be expected that technological developments gradually expand the FMS application area to production processes with higher part variety than is usual nowadays (see 7, p. 443). FMS will become more appropriate then for small batch/one-off production. Our empirical research shows that for these firms standardizing the production process will be a major task in order to realize the anticipated benefits.

REFERENCES

1. BESSANT, J., 1986, *'Flexible skills for flexible manufacturing'*, In: Brödner, P. (ed.), Skill based automated manufacturing, IFAC workshop (preprints), Karlsruhe, september 3-5.
2. BRÖDNER, P., 1984, *'Improved working conditions and productivity in batch production'*, In: T. Lupton (ed.), Human factors in manufacturing, (Kempston: IFS Conferences), 303-311.
3. BULLINGER, H.J., H.J. WARNECKE (ed.), 1985, *Toward the factory of the future* , (Berlin: Springer-Verlag).
4. CHAKRAVARTY, A.K., D. GERWIN, *Management of programmable automation*, University of Wisconsin, Milwaukee, (personal communication).
5. FERDOWS, K., 1983, *'Technology-push strategies for manufacturing'*, Tijdschrift voor Economie en Management, 2,125.
6. GRAHAM, M.B.W., 1986,*'A tale of two FMSs'*, In: Voss, C.A. (ed.), Managing advanced manufacturing technology, (New York: IFS Publications).
7. GROOVER, M.P., E.W. ZIMMERS jr., 1984, *CAD/CAM: computer-aided design and manufacturing*, (Englewood Cliffs: Prentice Hall).
8. HARTLEY, J., 1984, *FMS at work*, (North Holland: IFS Publications).
9. HERROELEN, W., LAMBRECHT, M.R., 1983, *'Management aspects of computerized manufacturing'*, Economisch en Sociaal Tijschrift, 6, 723.
10. IFS, 1982, *Flexible manufacturing systems*, Proceedings of the 1st international conference, (Kempston: IFS Conferences).
11. INGERSOLL ENGINEERS, 1985, *Flexible Fertigungssysteme*, (Berlin: Springer-Verlag).
12. D'IRIBARNE, A. a.o., 1983, *'Work organization in FMS, First findings from international comparisons'*, In: T. Martin (ed.), Design of work in automated manufacturing systems, IFAC workshop, FDR, november 7-9.
13. JONES, B., P. SCOTT, 1986, *'Working the system: A comparison of the management of work roles in American and British flexible manufacturing systems'*, In: Voss, C.A. (ed.), Managing advanced manufacturing technology (New York: IFS Publications).
14. OTA, 1984, *Computerized manufacturing automation*, Office of Technology Assessment, OTA-CIT-235, U.S. Congress, Washington D.C.
15. QUINTRELL, J. (ed.), 1983, *The strategic benefits of computer integrated manufacturing* (London: Arthur D. Little).
16. RANKY, P.G., 1985, *Computer integrated manufacturing* (Englewood Cliffs: Prentice Hall International).
17. REMPP, H., a.o., 1982, *Wirtschaftliche und soziale Auswirkungen des CNC-Werkzeugmaschinen - einsatzes* (Eschborn: Rationalisierungs Kuratorium der Deutschen Wirtschaft).
18. ROSS, M.H., 1981, *'Automated manufacturing, Why is it taking so long?'*, Long Range Planning, 3, 28.
19. TOIKKA, K., 1986, *'Development of work in FMS - Case study on new manpower strategy'*, In: P. Brödner (ed.), Skill based automated manufacturing, IFAC Workshop (preprints), Karlsruhe.
20. WILLENBORG, J.A.M., 1987, *Flexibele automatisering en organisatie: Veranderingen in de organisatie bij invoering van een flexibele fabricage systemen*, Twente University of Technology .

Universal Machine Tools Make Flexible Manufacturing Systems More Efficient

Vikram Cariapa

Department of Mechanical Engineering, Marquette University, Milwaukee, WI 53233, USA

The free market principle in the United States has affected it's economy tremendously. High quality, low cost imported goods are available in abundance and, hence, consumers have become extremely discerning in their choices. Local manufacturers have had to face this competition along with the added handicaps of higher labor costs and more importantly, lower productivity. These two factors have caused many firms to move abroad. This movement has two serious drawbacks, since jobs were not only being exported but the burden of generating real wealth for the country fell on fewer and fewer sectors of the economy. It is a well known fact that manufacturing is one of the real-wealth generating sectors of the economy. Thus, the United States was gradually losing it's means of creating wealth.

Manufacturing in the United States consists of three major areas -- job shop, batch production, and mass production. Batch production is an important sector of the economy contributing more than $1,400 billion to the economy and is the most susceptible to foreign competition. Most modern manufacturing strategies have focused on improving batch production. The earliest strategy was the use of numerically controlled machine tools. Analysis of plant operations showed that superior machine tools did not offer a solution (6). Quality was definitely upgraded but only 15% of a product's manufacturing cycle time was spent on the machine. The bulk was spent in transit or storage. Thus, strategies shifted to planning and scheduling in addition to building sophisticated manufacturing hardware.

The trend was clear -- a combination of integrated computerized machine tools and computerized planning systems would be the most efficient method of meeting competition head on. These systems were expected to be highly capital intensive, and to be expensive to set up (15), but would be independent of direct human intervention. A three-pronged strategy of integrating manufacturing facilities with supporting facilities like Process Planning, Scheduling, and core facilities like Design and Engineering became a goal called Computer Integrated Manufacturing. The integration of computerized machine tools begins by the creation of cells (6). A cell consists of a group of machine tools whose sole function is to produce a complete component. Raw material is fed in at one end and a finished component is produced by the machines in the cell. By combining many cells together, a complete manufacturing system could be developed where a whole series of components could be produced. These systems are called Flexible Manufacturing Systems (FMS) and represent the premier organization of production machines and material handling equipment for meeting both schedule demands and product quality.

The FMS which currently exists (7) have been mostly constructed with Machining centers and Automated guided vehicles for product movement. Some features on these machines include automated inspection, random tool selection, random loading of parts and finally, the ability to run twenty-four hours a day. Problems with these FMS include integration of different equipment (3), under-utilization of various machining centers (8), backtracking of Automated Guided Vehicles and most important of all, restriction on the types of products that can be produced because of size and geometry. It is to these problems that the Universal Machine Tools (UMT's) addresses itself as a viable alternative.

In order to display the advantages of the Universal Machine Tool, it is proposed to machine one product from the family of parts shown below in Figure 1. The parts which are shown require turning, milling, drilling, and tapping. As such, they do not constitute a "common process" family, but represent typical geometries which can be produced on this machine.

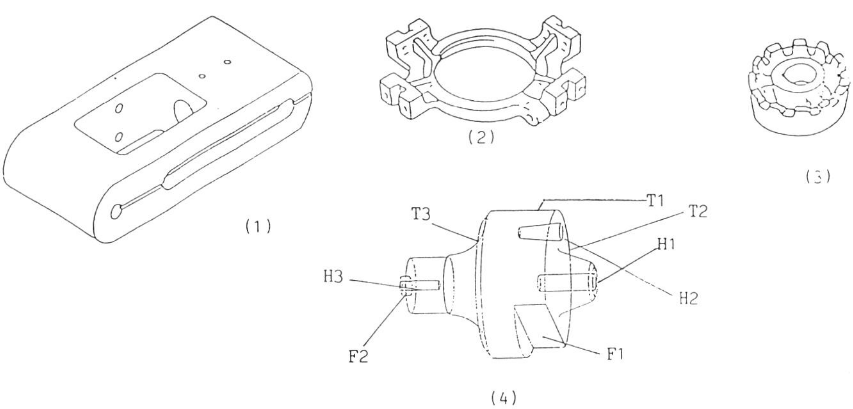

Fig. 1. Jobs Under Analysis

Figure 2 depicts a conventional FMS which has a vertical machining center and turning center. These machines are required as the jobs shown in Figure 1 require turning, milling, drilling and tapping. The FMS operations includes a pallet which would take the job to the turning center first and then to the vertical machining center to have the sides faced and drilled and tapped. The pallet would then have the job indexed by a robot for machining the fourth face. After completion, the robot would unload the part from the pallet on to the conveyor. This two-machine operation appears straightforward. The two workstations are served by one robot (not shown) and a pallet conveyor system (if twenty-four hour system operation is required). The loading of the two machines would be based on the machine cycle which would be based on the machine cycle which would be running at that time. If reference surfaces are interchangeable, it is possible that one can load either machine center independently and build up a stock of semifinished parts. This would enable either of the machines to be loaded at will, depending on which one is running. On the other hand, this would mean a compromise on minimum inventory storage and probable loss of component delivery time. Planning and scheduling becomes more difficult as the product variety and machine quantities increase.

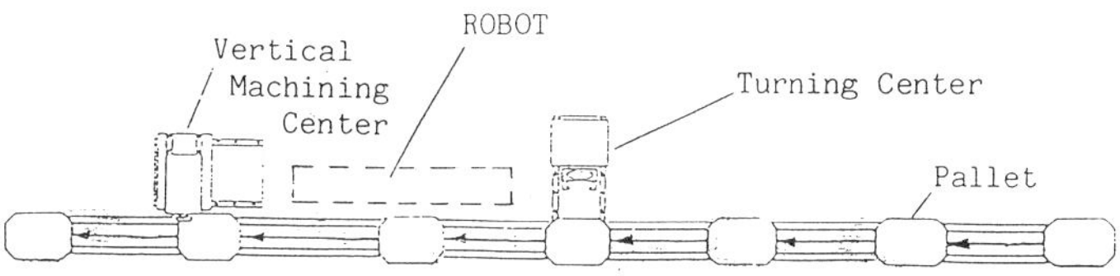

Fig 2. Conventional Flexible Manufacturing System

One Universal Machine Tool would replace both these machine tools. The Material Handling robot becomes redundant and the pallet system becomes simpler. No random programming of pallet moves are required as only one machine need be loaded and unloaded. Product cycle times from the Universal Machine Tool and the conventional system are analyzed taking product 4 from Figure 1 as a common reference. The concept of a universal machine is based primarily on the current capabilities of modern machine controllers. Modern controllers are capable of monitoring the movements of 5 axes simultaneously (6). These movements could generate complex three dimensional contours in addition to the normally required mutually perpendicular 3 axis moves. It is the structure of the universal machine that differentiates them from current machine tools. Current Flexible Manufacturing Systems consist of numerous combinations of machining and turning centers (7). Machining centers may be either horizontal or vertical and have capacities for infinite tool access or limited tool access depending on the tool magazine capability (6,7). Turning centers are usually horizontal axis machines with limited quantities of tooling. The structure of a Universal Machine Tool is different in the sense that both turning and milling type operations can be done in one station. At present it is a research concept and the developed working form is expected to vary with each product family. It's purpose in the current example will be to alternate between turning and milling as and when the programmer issues the necessary commands. Hence, a three-column machine is proposed with the columns in the North, East and West positions. Each column will have a spindle and three mutually perpendicular motions. Each spindle in turn will have a standardized bore for tool location and driving. All spindles access a common "random" position tool magazine as and when required by the program. The South side of the machine will be the pallet entry side with the "smart" fixture on it.

The two tables, Tables 1 and 2, show the operations and times associated with the conventional FMS shown in Figure 2. Table 3 shows the operations and times associated with a universal machine tool system.

Table 1
Turning Center Operations

Operation	Operation Type			Time
	Load/Clamp	Index	Machining	
1. Load Product	*	-	-	T1
2. Clamp	*	-	-	T1
3. Rough Turn T1	-	-	*	Tm
4. Rough Turn T2	-	-	*	Tm
5. Finish Turn T1	-	-	*	Tm
6. Finish Turn T2	-	-	*	Tm
7. Drill H1	-	-	*	Tm
8. Unclamp	*	-	-	T1
9. Robot Unload	*	-	-	T1
10. Robot Index	-	*	-	Ti
11. Robot Load	*	-	-	T1
12. Clamp	*	-	-	T1
13. Rough Turn T3	-	-	*	Tm
14. Finish Turn T3	-	-	*	Tm
15. Drill H3	-	-	*	Tm
16. Robot Unload	*	-	-	T1

The robot unloads the turning center and places the job on to a pallet for transfer to the vertical machining center, where the remaining operations will be performed.

Table 2
Vertical Machining Center Operations

Operation	Operation Type			Time
	Load/Clamp	Index	Machining	
1. Load Product	*	-	-	T1
2. Clamp Product	*	-	-	T1
3. Face Mill F2 End	-	-	*	4 x Tm
4. Index	-	*	-	4 x Ti
5. Unload	*	-	-	T1
6. Reload Vertically	*	-	-	T1
7. Clamp	*	-	-	T1
8. Mill F1	-	-	*	Tm
9. Drill H2	-	-	*	Tm
10. Unclamp	*	-	-	T1
11. Unload	*	-	-	T1

On the Universal Machine Tool System, the part is loaded on to the "smart" fixture and from then onwards, there is no manual interaction.

Table 3
Universal Machine Tool

Operation				Time
	Load/Clamp	Index	Machining	
1. Pallet moves in	*	-	-	T1
2. Fixture unclamps	*	-	-	T1
3. Machine chucks	*	-	-	T1
4. Rough Turn T1, T2	-	-	*	Tm
5. Finish Turn T1, T2	-	-	*	Tm
6. Drill H1	-	-	*	Tm
7. Live center moves in	*	-	-	T1
8. Face mill flat (F1)	-	-	*	Tm
9. Turn T3	-	-	*	Tm
10. Mill flats F2	-	-	*	4 x Tm
11. Index	-	*	-	4 x Ti
12. Fixture moves in	*	-	-	T1
13. Unclamp	*	-	-	T1
14. Fixture clamps	*	-	-	T1
15. Drill H3	-	-	*	Tm
16. Moves out to unload part	*	-	-	T1

An analysis of the three tables shows that the Universal Machine Tool has a greater percentage of machining time as compared to loading time or clamping or transport time. This itself means that the product cycle time gets reduced. In addition the fixture does away with the need for the robot. It is this "smart" fixturing along with the other characteristics which are listed below, which make universal machine tools such a benefit to modern manufacturing. The other characteristics of UMT's which contribute to the efficiency of FMS are explained below. They include:

1) Programming
2) Process conversion
3) Control and communication

Programming Characteristics

Programming is expected to be the most difficult aspect of the UMT's. The UMT's differ from conventional machine tools because their form is programmable. This mandates that the programmer fix up a suitable hard home for the lathe or milling mode and reference moves from there. As the configuration may change for different products, this skill becomes even more necessary in order to avoid crashes. The language programming is not critical as any standard language could be used. It is pertinent to note, however, that the language must have a facility for multi-tool usage, and communication with auxiliary equipment. The greatest advantage here is that once familiarity has been achieved, a wide range of products can easily be programmed.

Process Conversion

Conversion from milling to turning would be by loading chucks from a tool holder on to the spindles. These chucks would have auto clamping and spindles would have indexing provisions. Thus, in the milling mode the job could be faced and center drilled on either end. These would form axial references for the auto clamping during the lathe mode. Turning would be done using a chuck and a center, as in the conventional sense. Indexing of the job would be done by bringing in the "smart fixture" reclamping the job, indexing and a continuation of the turning. Turning tools would be changeable in the same manner as milling tools, i.e., they would be called up from the lathe program using requisite tool numbers. Flange drilling would be accomplished in two ways. If the job is short, and hence, held as a cantilever, the drilling tools would be on the opposite machine center post and drilling would be done by x, y movement for positioning and z movement for drilling. In case the job is a simply supported beam, the problem becomes more complex, and a 90 degrees-bend-indexable toolholder becomes necessary. Tool changing is still possible though the machine moves become more complex. The chuck will have to index along with x, y head positioning to get to the correct hole. Threading of the components will be accomplished by moving the job against the tool at the correct rate of motion. Tools for different thread profiles would be stored in toolholders in separate magazines.

Components which do not need end supports, can be turned even more easily as the two heads can approach the product simultaneously. This would lead to faster production rates. In addition, this cantilever type mounting allows milling to occur from three sides if the job demands it. In case both ends of the job demand a machining operation, each side will be machined in turn. The fixture would do the indexing in a horizontal plane between operations.

Control and Communications

Control of these machines would be dependent on the configuration of the UMT. Unlike conventional FMS, these UMT's have the additional complexities of "smart" figures, indexable spindles, and oriented tooling. Unlike milling cutters, lathe tools will have to be loaded in a fixed orientation to the job. As these tools could be loaded at random on any of the heads, two methods could be used for ensuring orientation. The first and more foolproof method is to have a key or lug to orient the tool, with a spindle locking system. A spindle indexing system is the second method and this offers more flexibility as milling operations could be executed on several faces of a job.

Programming of the UMT is expected to be more complex than conventional programming. The command code is expected to be standardized, but the number of units being synchronized is different. In the case cited in this paper, the pallet with the "smart" fixture will need programmed control for movement, indexing, release of job, and monitoring of different sensory inputs. The machining heads would need programmed head positioning for proper toll usage. Tool choice would be based on operation sequence, and the desired geometry. A dry simulation of the program on a plotter is expected to be more difficult as the tool motions are different. In a conventional dry run, the tool is shown moving against the job. In the UMT the heads and jobs move relative to each other for different operatons at different periods of time. Thus, dry runs would be misleading. The most efficient method of displaying the program would be a computer graphic image. This would mandate some investment in software development.

Conclusion

In conclusion, it would appear that UMT's do offer tremendous gains in efficiencies of FMS. The two machine example forms a basis for the argument that UMT's offer better utilization than conventional machine tools. There are expected problems in configuring a UMT -- as the heads must be configured for a product family. This has the advantage of suiting current product lines and the disadvantage of not being totally "flexible". This latter point is also true for conventional FMS -- all machines will have some limitation on physical movement. "Smart" fixturing is proposed as a necessary component of UMT's Proper design of these fixtures is an important prerequisite for the success of these UMT's. Careful planning of product families and processes will be necessary in order to make workable fixtures and also reduce the variety of fixtures. Programming of these machines is definitely going to be more complicated than conventional programming. Integration is anticipated to be less of a problem than is currently available as all the machines in a system are essentially duplicates of one model. It is because of all these points that it is proposed that Universal Machine Tools would definitely make Flexible Manufacturing Systems more efficient.

References

1. Bryce-Graham, A., "Flexible Manufacturing Sytems in the USA", Flexible Manufacturing Systems. Proceedings of the 1st International Conference, Brighton, UK, 1982.
2. Buzacott, J.A., "The Fundamental Principles of Flexibility in Manufacturing Systems", Flexible Manufacturing Systems. Proceedings of the 1st International Conference, 20-22, October 1982.
3. Carrie, A.S., Adhami, E., Stephens, A., Murdoch, I.C., "Introducing a Flexible Manufacturing System", Flexible Manufacturing Systems. Recent Developments in FMS, Robotics, CAD/CAM, CIM, New York, Elsevier Science Publishers B.V., 1985.
4. Cutkosky, M., Fussel, P., Milligan, R. Jr., "The Design of Flexible Machining Cell for Small Batch Production", Journal of Manufacturing Systems, Vol. 3, No. 1, pp. 39-59, 1984.
5. Gustavsson, Sten-Olaf, "Flexibility and Productivity in Complex Production Processes", Flexible Manufacturing Systems. Recent Developments in FMS, Robotics, CAD/CAM, CIM, New York, Elsevier Science Publishers B.V., 1985.
6. Haas, Paul, "Flexible Manufacturing Systems A solution for the Mid-Volume, Mid-Variety Parts Manufacturer", Understanding Manufacturing Sytems, Milwaukee, WI, USA, Kearney & Trecker Corp.
7. Hatvany, J., Merchant, M.E., Rathmill, K., and Yoslukawa, "World Survey of CAM", Kent, UK, Butterworth and Co., 1983.
8. Kusiak, A., "Loading Models in Flexible Manufacturing Systems", Flexible Manufacturing Systems. Recent Developments in FMS, Robotics, CAD/CAM, CIM, New York, Elsevier Science Publishers B.V., 1985.
9. Linder, Jan., "Design of Production Systems for High Flexiblity and High Capital Utilization". Flexible Manufacturing Systems. Recent Developments in FMS, Robotics, CAD/CAM, CIM, New York, Elsevier Science Publishers, B.V., 1985.
10. Ohmi, T., Yoshida, Y., "Flexible Manufacturing Systems in Japan - Present Status". Flexible Manufacturing Systems, Proceedings of the 1st International Conference, Brighton, UK, 1982.
11. Stecke, K.E., and Talbot, F.B., "Heuristic for Loading Flexible Manfacturing Systems". Flexible Manufacturing Systems. Recent Developments in FMS, Robotics, CAD/CAM, CIM, New York, Elsevier Science Publishers B.V., 1985.
12. Suresh, N., Meredith, J., "Justifying Multimachine Systems: An Integrated Strategic Approach". Journal of Manufacturing Systems, Vol. 4, No. 2, 1985, pp. 117-134.
13. Wood, L., "Flexible Automation or Liquidation", Flexible Manufacturing Systems, Proceedings of the 1st International Conference, Brighton, UK, 1982.

Universal Assembly Systems—Making Low Volume Assembly Pay

Finn Fabricius

Institute for Product Development (IPU), Building 423, DTH, DK-2800 Lyngby, Denmark

ABSTRACT:

The decreasing market life of products combined with the increasing number of product variants, have made the concept of flexibility essential in assembly automation.

Many companies, however, have found themselves unable to justify flexible assembly systems due to low annual production volumes. The development of "Universal Assembly Systems" makes it possible to assemble a variety of entirely different products and sub-assemblies on the same system, thus increasing the total volume artificially and reducing the investment per assembly.

Three Danish companies in cooperation with the IPU are working on establishing guidelines for the design of such modular assembly systems. The project is to provide answers to the questions:

* Which products are suitable?
* What modules does a Universal Assembly System consist of?
* How should Universal Assembly Systems be implemented?

In connection with this project, a robot-based, modular, experimental system is being built up in IPU's laboratory. This system is used for full scale assembly trials and as a basis for the development of assembly systems for the three participating companies. The design of the products to be assembled is examined at the same time as the assembly process, thus reducing the uncertainty connected with subsequent investment decisions.

Since Universal Assembly Systems bring a variety of new possibilities, many of which relate to dynamic aspects of assembly, the payback period will be relatively short. The classic methods for cost evaluation have, however, to be revised.

1. THE CONCEPT OF UNIVERSAL ASSEMBLY SYSTEMS.

The constantly increasing need to make changes in the assembly process during the production of industrial products has caused many companies and research groups to take an interest in flexible automatic assembly systems. Up till now, efforts to "build flexibility into" assembly systems have as a rule lead to the construction of flexible assembly cells or assembly lines intended for a particular family of products (covering a number of product variations), or perhaps just to the depressing observation that flexibility was too expensive in comparison with its often rather tenuous advantages. Many companies have been obliged to discover that their products, even if all variations are taken into account, are not produced in sufficiently large numbers for automatic assembly to be a paying proposition.

In order to analyses alternative principles for automatisation within such companies, IPU is carrying out a project together with DAMIXA A/S, KOSAN TEKNOVA A/S and HARTVIG JENSEN & Co. A/S, with a view to investigating how universal assembly systems can be used to automatize the collected assembly of a group of quite different products, each of which is produced in too small numbers to justify automatisation.

The aim of the project is, amongst other things by using a full-scale experimental system, to develop a series of system modules which can be combined with as few product-specific parts as possible, in order to build up universal assembly cells for each of the participating companies.

It would be meaningless to try to develop assembly cells with unlimited universality. The assembly cells for each company must contain a degree of universality which is necessary and sufficient for the production task in hand.

2. THE PRODUCTION TASKS.

Whereas the participating companies' requirements for versatility in the assembly process are very much the same, their production tasks differ considerably from one another.

DAMIXA A/S produce water taps, using a variety of working principles, and in a huge number of variations. The variant properties are the various colours and surface coatings, various washer materials and numbers, various types of connection and so on. The taps are assembled to order, with batch sizes in principle down to single items. The annual production varies between 30,000 and 150,000, depending on the type of tap.

HARTVIG JENSEN A/S produce equipment for spraying fields and gardens. The sprays are built up from a number of different sub-assemblies, none of which are produced in numbers large enough to justify automatic assembly; typical production volumes vary between 8000 and 100,000 per year.

KOSAN TEKNOVA A/S produce a large variety of valves and regulators for bottled gas. Assembly involves a number of critical leak tests. Their products are produced in numbers from 12,000 per year and up.

3. LAYOUT CONSIDERATIONS.

The choice of products for the pilot universal assembly system for each of the three companies involved considering the following factors:

* The number of components involved
* The number of different components
* The number of directions of assembly
* The complexity of the assembly
* The auxiliary equipment required
* The quality and form of the components
* The production volumes
* Costs for manual assembly

None of the products examined at the three companies has a volume of production large enough to justify independent automatisation. Automatisation can here only be made profitable by allowing the different sub-tasks to share the investment-intensive portions of the assembly system. The investment in those parts of the assembly system which can be used for the majority of the sub-assemblies is called the general investment, while the remainder of the investment is the task-specific part. If we look at the functional modules required in the assembly system, we find that these have different proportions of general and task-specific parts.

3.1 Sub-systems

SCARA-robot: A large 4-axis robot is used. The robot is considered as a general investment. (If a 6-axis robot were used, the internal return on investment would be reduced by 14 percentage points.)

Sub-systems: A module chassis with a steel top-plate is used. Roughly 2/3 of the chassis system's parts are general; the remainder are used as a basis for task-specific sub-systems.

Feeder equipment is in many ways the bottle neck from a flexibility point of view. Traditional feeder systems, such as bowl feeders, are costly and usually have to be made to measure for each individual component. In this analysis we have worked with three basic ideas for feeder systems:

1. Separate feeder systems for all components.
2. Feeder bases partly in common.
3. Manual feeding of components.

Re point 1: Separate feeder systems, such as circular feeders for each component, are an expensive solution, and should be avoided.

From the point of view of running a project, this way of doing things is a form of emergency solution, since it is known that one can "get through" by using it.

Re point 2: This type of system is based on the assumption that about half of the investments in feeder equipment can be made general. The idea is to do this by dividing the feeder lines up into fixed and replaceable parts, by the use of replaceable liners for the feeders, and so on. The use of feeder bases which are partly common gives a much better financial result in projects than the use of separate feeder systems.

Re point 3: Manual filling of feeder tracks, magazines and so on is for two reasons a real alternative to automatic feeding. Firstly, because it only requires small investments per component, and secondly because the efficiency of the assembly system will rise due to the elimination of the primary source of faults: the automatic feeder equipment. It is assumed that the efficiency will rise around 10 percentage points, which in certain cases can make room in the production plan for yet another sub-task. In the following calculations we assume that manual feeding of a component takes on average 3 seconds.

3.2 Layout Possibilities

The changeover between sub-tasks can basically take place in three ways:

* Fixed task and moveable general equipment. In the current project this solution is unrealistic, since not only the robot but in fact most of the equipment is general. So either the robot, general chassis modules, local cell control system, air systems, control panels and so on should all be multiplied (which would ruin the economics of the project), or else all this equipment should be arranged on a moveable platform, while the tasks remained where they were.

* Fixed general equipment and moveable tasks. This is to be recommended in this case for HARTVIG JENSEN A/S and KOSAN TEKNOVA A/S, since transport of the tasks is easier and cheaper to implement.

* No change (where everything is available to the robot). This is the ideal way of doing things. This is a possibility for DAMIXA A/S, where the family relationship between the products is very close, and where universal assembly cells can be coupled to a central transport system.

3.3 Proposed Layouts

For HARTVIG JENSEN A/S and KOSAN TEKNOVA A/S, it is advisable to use a layout where the general equipment is fixed and the task-specific equipment is moved to and from the cell.

For these two companies changeover starts with the filling up of the moveable part with components, after which it is moved on a pallet truck or similar to the cell, and is made fast. Gripper fingers, fixtures, feeder tracks and so on are replaced (this takes less than an hour), the sensors and cylinders for the sub-task are coupled onto the cell with snap-action plugs, and the robot and PLC programs are read in. If necessary, the robot carries out a search procedure (one of these has been developed by IPU for Bosch robots), which registers the exact position of the task and corrects its movement program accordingly. After this, the desired batch size is coded in via a control panel, and the task is started.

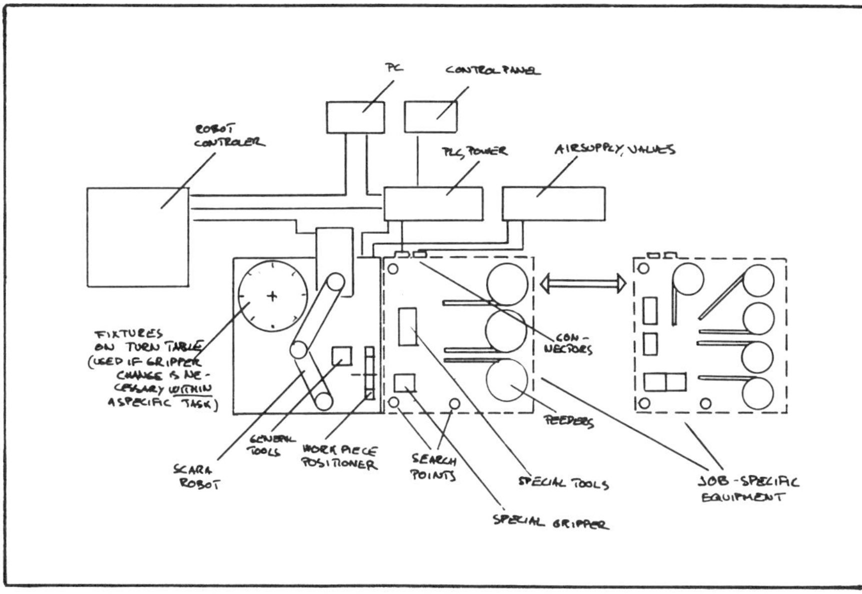

Fig.1 Proposed layout of Universal Assembly Cell

In the case of DAMIXA A/S mechanical changeovers of the cells are unnecessary, since the products more or less can be built by combining a limited number of components in different ways. The cells could here be linked to a central transport system, that brings along a product mix according to the orders of the day.

4. ECONOMIC CALCULATIONS.

4.1 Assumptions

A long series of assumptions have been made with relation to the calculation of the economic aspects of such projects.

* Assembly costs: Costs for manual assembly have been set to between 12$/hour and 27$/hour, depending on the company.
* Number of shifts: It is assumed that the systems is run for two shifts, each amounting to 1760 hours over a year.
* Efficiency: It is assumed that the system performs useful production during 85% of its running time. The remaining 15% is used for fault correction, changeovers and so on. If manual

	component feeding is used, the efficiency is assumed to be 95%, as feeding problems are then reduced.
* Interest rates:	The rate of interest on investment loans is put at 9%.
* Inflation:	The annual rate of inflation is put at 3%.
* Lifetime:	Useful production life is taken to be 8 years. (The equipment is expected to have a physical lifetime of at least 10 years.)
* Scrap value:	The scrap value is taken to be zero.
* Other savings:	These are in the first instance taken to be zero. If possible, savings due to improved product quality, reduction of the amount of goods under production, and similar factors should be taken into account here.

4.2 Procedure

For each individual sub-task, budgets are made for general and task-specific equipment. The general equipment is that equipment which is used in the majority of tasks, such as the robot, control system, basic chassis and so on. The task-specific equipment, such as gripper fingers and fixtures, on the other hand, has no general use.

In the first instance, internal return on investment and cost reduction are calculated for each sub-task individually, using the assumptions given above.

On the basis of these calculations, the sub-tasks can be given a priority - an ordering on a "scale of attractiveness".

This ordering is not fixed, but depends on the pre-suppositions with respect to equipment.

After this, an overall calculation is made, in which the economics of the sub-tasks are evaluated by considering marginal investments. The first sub-task "orders" a complete assembly cell, and therefore has to pay for it. For the second task, we typically only need a marginal extra investment, corresponding to this task's specific investment contribution. This marginal calculation of consequences is performed for the whole sequence of sub-tasks; the first task that overruns the first cell's capacity must carry the costs of the basic investment in cell number two.

If we now work out the total internal return on investment for sub-task 1, sub tasks 1+2, sub-tasks 1+2+3 and so on, then we obtain a measure of the total economic consequences of these groupings of tasks.

The total internal return on investment will rise as we move through the series of tasks, until we reach a point where the "doubtful tasks" begin to make the total result worse again.

This methods of calculation puts us in a position simultaneously to get rid of the unattractive sub-tasks and to obtain a measure of the total economic result. This method will lead to a choice of equipment with a certain unused capacity. Since this extra capacity is both bought and paid for, it can be freely used for new initiatives.

4.3 Results.

We can now work out the real total economic dividend from automatisation. Since the various companies have different ways of calculating the value of the engineering effort in such tasks, the profitability is worked out without taking engineering planning costs into account. If these costs should be included, they must be incorporated in the first task.

Let us now look at some typical results from the calculations which we have carried out.

When partly common feeders are used:

Product types	Total assembly time (h)	Total invest-ment (USD)	Total cost reduct. (USD)	Number of cells	Total pay-back (year)	Total internal rate of return (%)
Product A	572	69600	20900	1	3.8	28
A and B	817	77000	32100	1	2.7	41
A to C	1278	83300	39900	1	2.3	48
A to D	2200	96500	55300	1	1.9	58
A to E	2987	116000	68500	1	1.8	61
A to F	3144	125100	71200	1	1.9	58
A to G	3377	139900	74900	1	2.0	54
A to H	3552	207800	77200	2	3.0	36
A to I	3623	208200	77500	2	3.0	36

Fig.2 Typical results of the economical calculation concerning a Universal Assembly Cell, capable of producing a number of entirely different assemblies

As can be seen in the table, the maximum total internal return on investment is obtained by automatisation of tasks A to E. It will not in this case pay to include tasks F and G, even though the first cell in this case has some unused capacity. Similarly, the calculation shown here gives no grounds for investment in 2 cells.

The maximum total internal return on investment of 61% will typically fall by 10-12 percentage points if separate feeders are used, while manual feeding will give a drop of only 4-6 percentage points.

At present, manual feeding of components is only profitable for large, voluminous components. If, on the other hand, the components have been put into magazines as they were produced (at a time when handling is often necessary in any case), then the necessary manual effort at the assembly cell will be reduced considerably. If this principle were to be followed, then the marginal investment for a new sub-task would likewise be able to be reduced considerably. Investigations of these possibilities should form a part of a company's long-term development strategy.

5. CONCLUSION

As continually increasing demands are being made of companies' ability to compete and to adapt to the market, flexible automatisation has become a central concept. In the areas of administration and the production of components, computers, CNC machines and the like have contributed markedly to the realisation of "the flexible company". In the area of assembly, on the other hand, people have been more hesitant in relation to flexible automatisation, which to a considerable degree is due to product designs being poor from an assembly point of view, together with uncertainty with respect to how flexible automatic assembly should be justified

The analyses which have been carried out for HARTVIG JENSEN A/S, KOSAN TEKNOVA A/S and DAMIXA A/S indicate that the introduction of universal assembly cells is one of the most attractive ways in which to realise flexible automatisation.

The calculations which we have carried out show repayment times for the investments themselves of 0.8 to 3.6 years, and total internal returns on investment of 29 to 130%.

Thus it appears to be justifiable to use universal assembly systems for assembly of a number of different products, none of which on its own is produced in numbers large enough for automatic assembly.

Compared with ordinary flexible systems for the assembly of a family of products, universal assembly systems can be considered as having a particularly high content of the type of flexibility which we denote "versatility" (3). The value of this versatility can here be evaluated in dollars and cents, whereas it is normally not quantifiable.

If we compare universal assembly systems with conventional inflexible assembly systems, the universal systems' greatest strength is that they, in addition to being able to assemble different products, can take into account changes in matters related to production technique.

- Will the working week be shortened?
- Will wages rise faster than productivity?
- Will the product be changed?
- Will more variations be introduced?
- Will the rate of changeover rise? (for example due to changing to a system of production to order)

If you find it difficult to answer one or more of these questions, then you have at the same time recognised the value of the ability which flexible/universal automatisation has to reduce dependency on these matters.

REFERENCES

(1) Ahm, T., Andreasen, M.M.: "Design for Flexibility", Proceedings to 8`th ICAA, IFS (Publication), Kempston, 1987

(2) Andreasen, M.M., Kahler, S., Lund, T.: "Design for Assembly", IFS (Publication), Kempston, 1983

(3) Fabricius, F.: "An Overview on Flexible Assembly Automation", Proceedings to 6`th ICAA, IFS (Publication), Kempston, 1985

A Simulation System for Flexible Manufacturing Cells

Jyh-Cherng Shieh and Shui-Shong Lu

Department of Mechanical Engineering, National Taiwan University, Taipei, Taiwan 10764, ROC

ABSTRACT

Based on the Local Area Network concept, four sets of IBM-PC are used to form a simulation system for a flexible manufacturing cell, which consists of a milling machine, a turning machine, an industrial robot and two pallet stands.

In this simulation system, one of the four PC´s is used as the cell controller, which is responsible for monitoring and controlling the flow of workpieces within the cell, and for responding the machining status or results to upper-level controller. Two PC´s are used as the simulators of the milling machine and the turning machine, real-time graphic animations of the tool-path are produced according to the scheduled part programs. The fourth PC simulates the action of the robot, which loads and unloads workpieces among pallet stands and machine stations within the cell.

All events occurred in the simulation are viewed in graphic images. Upon completion of the simulation, a statistical performance report is produced.

INTRODUCTION

For the automation of small batch production of small parts, the development of flexible manufacturing systems is very important.

From the point of view of the static structure, a "FLEXIBLE MANUFACTURING CELL" may be defined as a fixed entity which provides the necessary manufacturing information to its upper-level controller to assist the control of the material flow. As for the setup and content of the FMC, e.g., the kinds of adopted machines or the capacities of local buffers are decided by the long-term considerations of the production type and the manufacturing methodology. A FMC within the factories may act as a workstation or an orderly manufacturing cell which can process the products thoroughly. From the above descriptions, the FMC can be represented by a machine or by a group of machines.

Simulation has been employed increasingly to aid the design and implementation of new manufacturing systems[1]. Also, if adjusted to the current state of planning or operational strategies, it will allow a short-term analysis of system alternatives and provide a powerful decision aid for the design engineer[2].

In manufacturing, the simulation methodology can be catagorized as: analytic model[3,4], physical model[5,6] and simulation model[7,8]. Wang and Davies[9] simulated a flexible manufacturing cell including a cell controller, a robot, a turning machine and a turning machine by use of four terminals of VAX-11/750 minicomputer.

Existing simulations are far from the real system. Much time has to be expent in order to realize them. In this paper, the distributed processing and hierachical control concepts are introduced into simulation system. It divides the real FMC into two levels: supervisor and slaves. Personal computers are then used as simulators to emulate the responsibilities of the real devices. Between them, the commands and status feedback are transmitted through the local area network system. Under these concepts, the simulation system can easily be realized by installing an interface between the simulator(the personal computer) and the real device.

In this paper, the FMC which consists of a cell controller, one CNC turning machine, one CNC milling machine, an industrial robot and pallet stands with variable capacities as the local buffer, acted as a workstation [10] in FMS is taken as the objective of simulation.

SYSTEM STRUCTURE

The simulation system is implemented based on a local area network system, in which fours PC´s are included. One of the four PC´s is used as the cell controller, which is responsible for monitoring and controlling the flow of workpieces within the cell, issuing commands to machine tools and the robot, and responding the machining status to its upper-level controller. Two PC´s are used as the simulators of the milling machine and the turning machine. The fourth PC simulates the action of the robot, which loads and unloads the workpieces among pallet stands and machine stations within the cell. These PC´s are connected to the Trans-Net[11] trunk at network taps (Fig. 1). A netwok tap consists of a tap box on the network trunk, and a tap cable which connects the PC to the Trans-Net trunk line. The Trans-Net is a local area network of common bus topology. The transmission line is shared, as on a telephone party line, by the simulators.

All the events including the status of stations, and the flow of workpieces can be viewed graphically. The pallet stands/fixtures/workpieces are represented by boxes. An empty box indicates that the station is idle, or that the part is removed. A filled box indicates that the station is busy, or occupied by parts. The path of gripper of the robot and the tool-path of the machine tools are traced and displyed on the screen of simulator dynamically.

Upon completion of the simulation, the statistical performance reports are produced to show the occurred time of events, the machining time of the CNC machine tools and the flow of workpieces.

SIMULATION OF THE CELL CONTROLLER

The tasks of the cell controller are:

1. Monitoring the input/output of workpieces at pallet stand,
2. Monitoring and controlling the flow of workpieces within the cell,
3. Issuing commands to machine tools and the robot.
4. Responding the machining status to its upper-level controller.

When the cell controller is notified by its upper-level controller that the cart has arrived at pallet stands, it will decide which pallet stand to use and commands the robot to unload the workpieces to the appropriate pallet stand. After unloading, when the machine tools and the robot are ready, the cell controller will then command the robot to load one workpiece from pallet stand to the first machine tool. The machining action will then be initialized. When the first machining step is completed, the cell controller will request the robot to move the workpiece to the next machine tool, and initiate the second machining. Then, the other workpiece will be loaded onto the first machine tool. When the last machining step is completed, the finished part will be unloaded and moved to pallet stand. After parts on pallet stand are processed, the cell controller will set a finished status flag as an indication to its upper-level controller in order to request for a cart to move away the finished parts.

According to the above descriptions, the activity flowchart of the cell controller are drawn up as shown in Fig.2 by tracing the actions of moving entities - workpieces. 17 kinds of events (Table 1) may be extracted from this activity chart, and the corresponding event relationship is shown in Fig. 3.

The responsibilities of the cell controller is to monitor and adjust the activities of the stations within the FMC. The occurrence of an event is initiated randomly. The occurrence of an event means the end of certain event in certain station. Therefore, the control goal (next event) will be decided by the status of other stations within the FMC, and commands are issued to the appropriate station for initiating the next event. The control goal diagram is shown in Fig. 4.

In order to handle the flow of workpieces within the FMC and to clarify the control goals, three respective data tables are designed for pallet stands, workpieces, and machine tools. Fig. 5 shows the file formats, in which the indeces of pallet stand, workpiece and machine tool are used as keys, the other items are constantly updated to reflect the status and are used to assist the decision of control goals.

SIMULATION OF THE ROBOT

The robot within the cell is responsible for handling the workpieces among pallet stand and machine tools. Therefore, the configuration of pallet stand and the relative location coordinates of work-stations must be known beforehand. These consist of the position of the pallet stand and the clamping position of machine tools.

According to the information described above, the visual layout is displayed on the screen of the robot simulator for the purpose of graphic animation. The path of the gripper is traced by a series of dots which are displayed dynamically, and the elapsed time is measured by dividing the distance of two stations by the scaled velocity. The source and destination stations, between which the workpieces are moved, are decided by the received command issued from the cell controller.

For the realistic purpose, an originally filled box which represents a part being carried away is firstly blanked when the gripper approaches. After the gripper arrives at the destination station, an empty box located at that station is filled and a job-finished message is issued to the cell controller to indicate that the part has been picked up, moved or released.

The robot is a slave device within the FMC. The start of service event is initialized by the cell controller. At the end of service, the next event (control goal) is scheduled, as shown in Fig. 6. In order to handle the status of the robot and to clarify for the control goal, an event data table is designed and shown in Fig. 7. In the data table, event item is used as the key.

SIMULATION OF THE CNC MACHINE TOOLS

Thc CNC machine tools within the cell are responsible for machining workpieces. Corresponding real systems, Two PC´s are used to act as the machine interface terminals and the CNC simulators, one representing the turning machine, and the other the milling machine. For the realistic purpose, the simulation proceeds according to the scheduled part program associated with the workpieces, and the tool-paths are traced and displayed on the screen of the simulator dynamically.

All displays consist of the scaled elapsed machining time, part program, coordinates of tool-path and cutting length, etc. The action of clamping and releasing of workpiece are merely simulated by setting a delayed time.

The CNC machine tools are also slave devices within the FMC, and the control goals are shown in Fig. 8.

ILLUSTRATION

As an illustration, simulation of six workpieces to be machined in the FMC will be shown here. The workpieces on a pallet are transported by a cart to the FMC. Designate the turning machine as the first machining station, and the workpiece flow proceeds from the pallet stand to the turning machine, milling machine, and back to the pallet stand.

Fig. 9 to 12 show the status of FMC at two minute mark. After completion of simulation, an statistical records of the operation are obtained such as the utilization of the machine tools (See Table 2,3) and the flow of the first workpiece (See Table 4).

CONCLUSIONS

This simulation system can be converted to part of real operational FMC. For example, one can install an interface between the simulator and the real devices (e.g., the robot and CNC controllers), so that the simulator can communicate with and control the devices.

In this simulation system, the start of activity is viewed as an occurred event. From the operational rule of each device and its relationship with others, the control goals(next occurred events) are scheduled and programmed. According to the relationship between the events and the entities within the FMC, the data tables for the events and entities are then designed. This simulation system is expandable and can be easily modified for a different configurations or material flow policies.

REFERENCES

1. H. Rahnejat , "Simulation Aids Design for Flexible Automation," The International Journal of Advanced Manufacturing Technology, Vol.1, No.2, pp.91-108(1986).
2. G. Spur, W. Hirn and G. Seliger, "The Role of Simulation in Design of Manufacturing Systems," Advances in CAD/CAM, North-Holland Publising Company, pp.349-373(1983)
3. J. A. Buzacott and J. G. Shanthikumar, "Models for Understanding Flexible Manufacturing Systems," AIIE Transactions, Vol.12, No.4, pp.339-349(1980).
4. K. E. Steke ,"Formulation and Solution of Nonlinear Integer Production Planning Problems for Flexible Manufacturing Systems," Management Science, Vol. 29, No.3, pp.273-288(1983).
5. S. Y. Nof, M. P. Deisenroth and W. L. Meier, "Using Physical Simulator to Study Manufacturing System Design and Control," AIIE Proceedings, Spring Annual Conference, pp.219-227(1979).
6. M. P. Deisenroth and S. Y. Nof,"A Robot Machine Loader Physical Simulator," Technical Report, School of Industrial Engineering, Purdue University (1980).
7. L. E. Lin and S. L. Chung,"A Systematic FMS Model for Real-time On-line Control and Question-Answer Simulation using Artifical Intelligence," Proceedings of the Second ORSA/TIMS Conference on Flexible Manufacturing Systems Operations Research Models and Applications, Ann Arbor, MI, USA, Aug. 12-15, pp.567-579(1986)
8. H. Wakai, S. Sato, C. Sakamoto, Komatsu Ltd. and T. Sata,"KOSMO - a Simulator for Flexible Manufacturing Systems," Annal of the CIRP, Vol. 35/1, pp.347-350(1986).
9. N. Wang and B. J. Davies, "Simulation of a Flexible Manufacturing System," Proceedings of 26th Internation MTDR Conference, pp.81-87(1985).
10. J. J. Shieh and S. S. Lu,"Simulation of Computer Integrated

Manufacturing System Based on Personal Computers," 3rd Nat. Conf. on Mech. Eng. Chinese Society of Mechanical Engineering, Chung-Li, Dec. 13, pp.1169-1180(1986)
11. TRANS-Net User Guide, R.P.T. Intergroups International Ltd., Taipei, R.O.C.(1986)

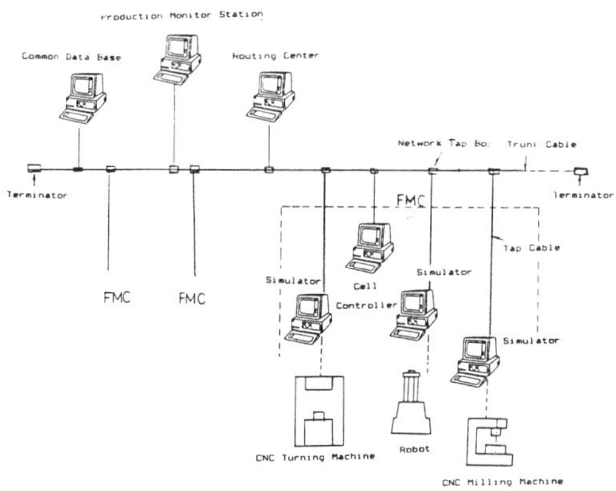

Fig. 1 Configuration of the Simulation System

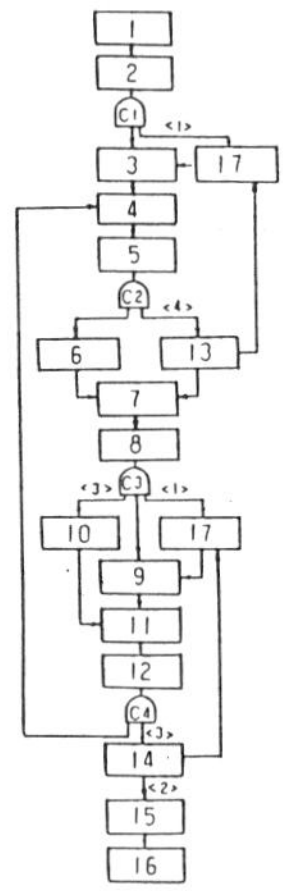

Fig. 2 Activity Flowchart of the Cell Controller

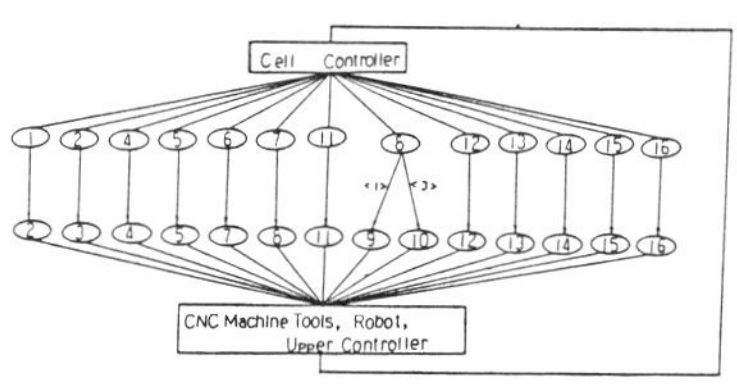

Fig. 4 Control Goals of the Cell Controller

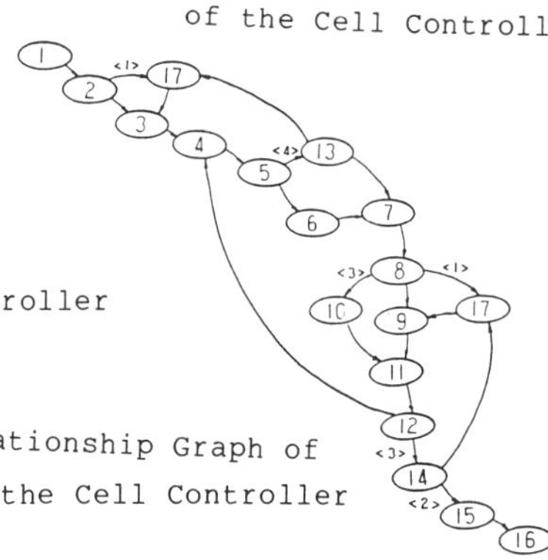

Fig. 3 Event Relationship Graph of the Cell Controller

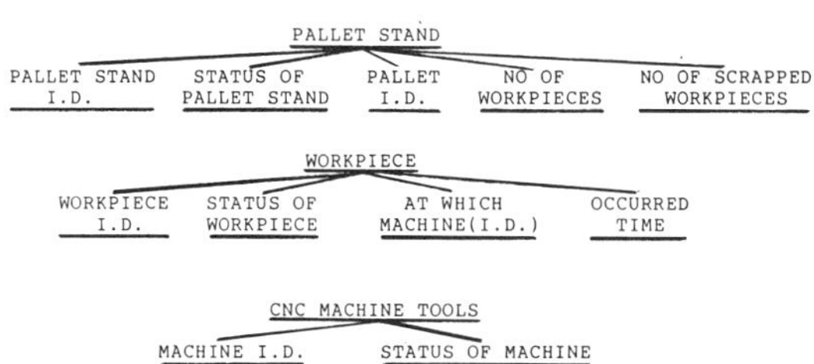

Fig. 5 File Formats of the Cell Controller

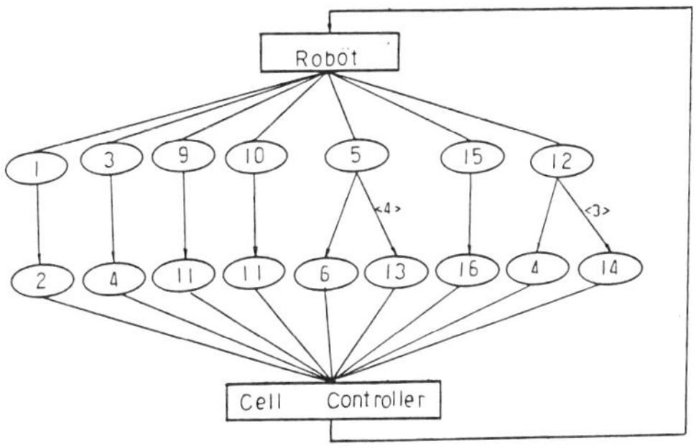

Fig. 6 Control Goals of the Robot

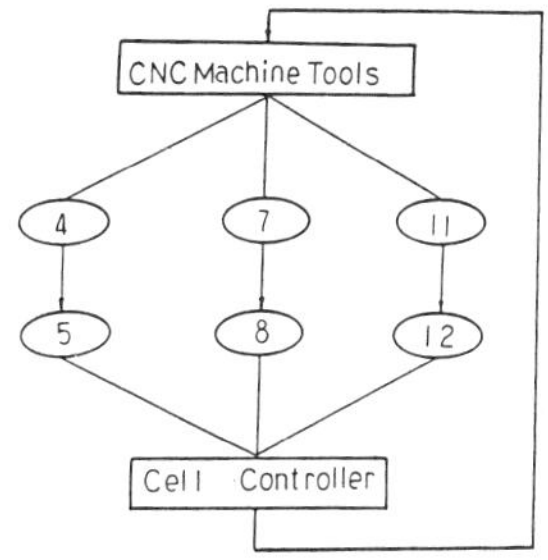

Fig. 8 Control Goals of the CNC Machine Tools

Fig. 7 File Formats of the Robot

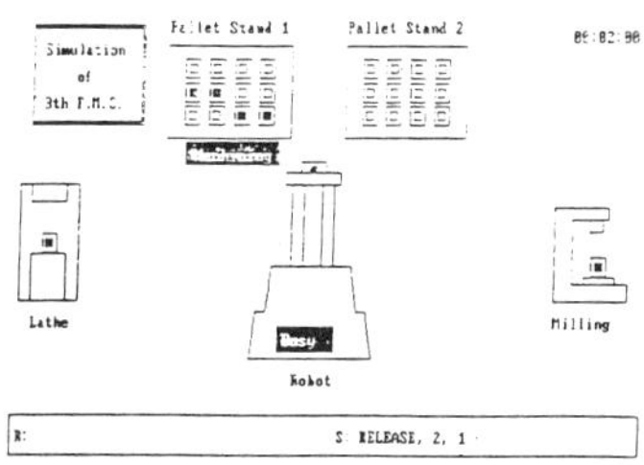

Fig. 9 The Status of the Cell Controller at Two Minute Mark

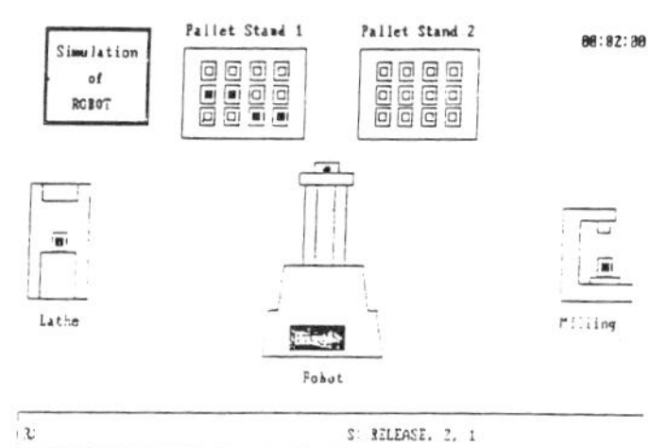

Fig.10 The Status of the Robot at Two Minute Mark

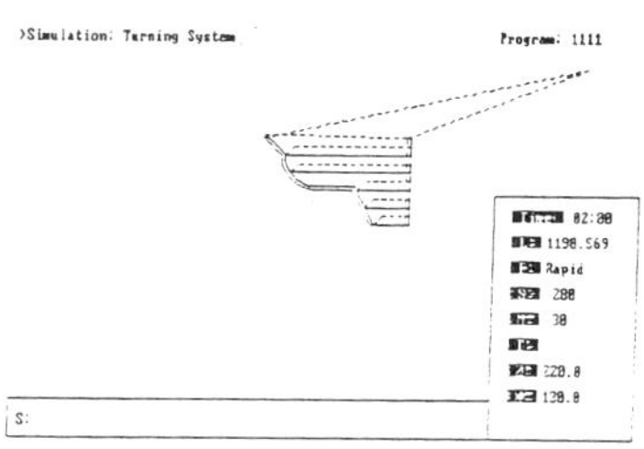

Fig.11 The Status of the CNC Turning Machine Tools at Two Minute Mark

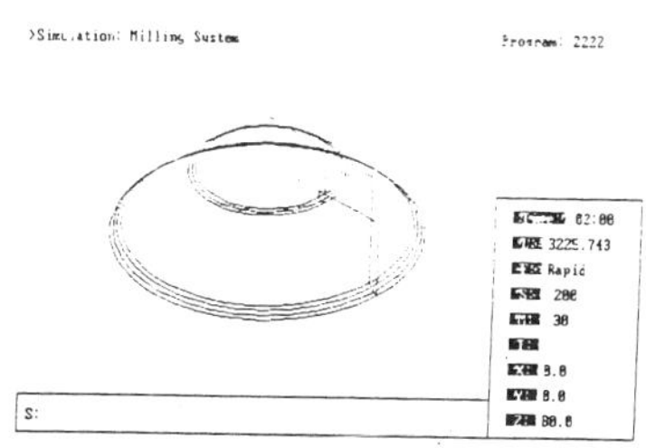

Fig.12 The Status of the CNC Milling Machine Tools at Two Minute Mark

Events

1. Begin unloading pallet
2. Complete unloading pallet
3. Loading workpiece to machine
4. Clamping workpiece
5. Complete clamping
6. Complete loading workpiece
7. Begin machining
8. Complete machining
9. Moving workpiece to another machine
10. unloading workpiece from machine
11. Releasing workpiece
12. Complete releasing
13. Complete moving
14. Complete unloading workpiece
15. Begin loading pallet
16. Complete loading pallet
17. Waiting for ready of machine

<Condition>

1. The machine is busy
2. All parts on pallet stand are finished
3. The workpiece is finished
4. The workpiece will be moved to another machine

Control Point

1. The pallet has been unloaded at the pallet stand
2. The workpiece has been clamped at the machine
3. The workpiece has been finished at the machine
4. The workpiece has been released at the machine

Table 1 Events, Conditions and Control Point of the Cell Controller

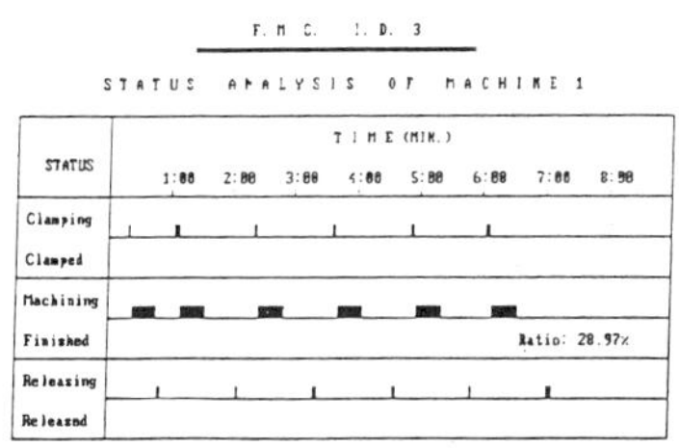

Tabel 2 Utilities of the CNC Turning Machine

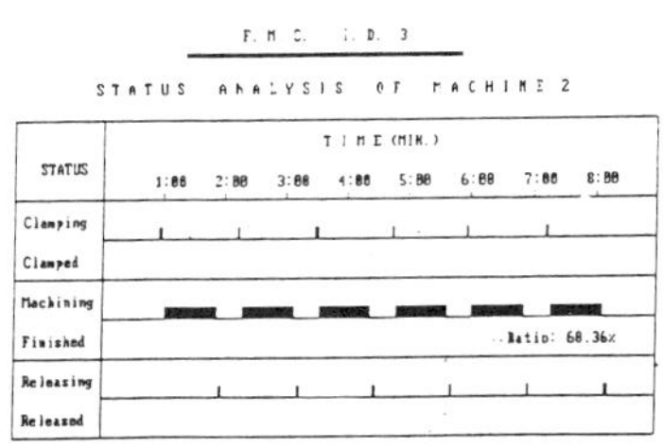

Table 3 Utilities of the CNC Milling Machine

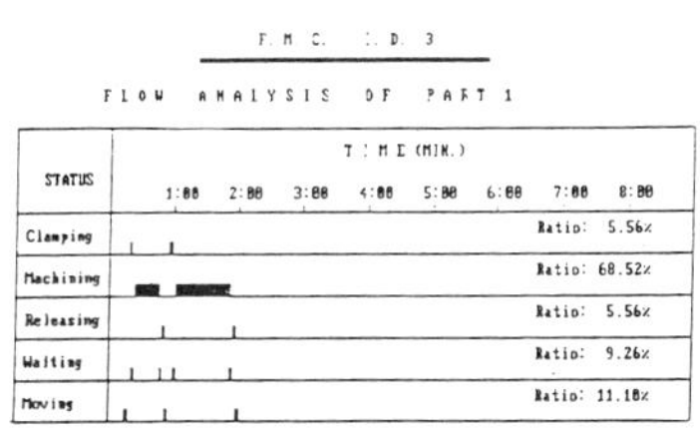

Table 4 The Flow of the First Workpiece

Defining Requirements for Computer Integrated Manufacturing Systems

Roger Maull and David Hughes

The CIM Institute, Plymouth Polytechnic, England

INTRODUCTION

The work described in this paper stems from collaboration between a number of U.K. academic institutions and companies pursuing the integration of discrete manufacturing systems.

Traditional definitions of Computer Integrated Manufacture (CIM) which relate to the transfer of Computer Aided Design (CAD) data to Computer Numerically Controlled (CNC) machine tools are unsatisfactory. CIM will have an impact on the organization as a whole. In addition to CAD and Computer Aided Manufacturing (CAM), CIM encompasses; process planning and control, material selection, materials handling, manufacture, quality assurance, inventory management and production planning and control. Perhaps most importantly, it also interfaces with marketing, finance and the general business function. Thus CIM will affect all aspects of the business and must, therefore, be designed to suit idiosyncratic business needs. Furthermore, since no one supplier has yet mastered the skills of developing the vast array of software required to satisfy all individual requirements, it follows that CIM systems are likely to consist of a number of discrete applications supplied by competing vendors.

In order to achieve a system tailored to individual company requirements each organization needs to identify an appropriate subset of applications and combine them in a way which will satisfy its own specific needs. To ensure that an effective system is developed ways must be found of:

Improving current requirements definition practice

Facilitating the transition from an existing manufacturing system to an appropriate integrated system

The issues associated with the latter problem have been addressed in a number of papers e.g. Hughes (1) and Hughes and Maull (2). This paper focuses on the issues associated with the practice of requirements definition.

The authors argue that the approach taken to requirements definition will depend on the view of the organization held by the individual analyst. This view will, itself, arise from a particular philosophical perspective, rarely articulated, yet underlying and influencing each phase of the requirements definition process. To illustrate this point we shall examine two conflicting philosophies and explore their limitations before presenting an alternative approach, based within a third philosophy, which the authors believe considerably improves upon current practice.

REQUIREMENTS DEFINITION

One dichotomy central to the theoretical basis for requirements definition is the rational/natural dichotomy. This was summarised by Scott (3) as;

"The Rational system model assumes that organizations are purposefully designed for the pursuit of explicit objectives. The Natural systems model emphasizes the importance of unplanned and spontaneous processes. Organizations are viewed as coalitions of participants who lack concensus on goals, sharing only a common interest in the survival of the organization"

Underpinning the naturalistic and rationalistic orientations are two very different philosophical bases. Whilst rational views are based in positivism natural views can be seen to be based within phenomenology.

Organizations as Rational Systems and Positivism

In contemplating organizations as rational systems the analysis begins with "a positive problem". Simplifying assumptions are made about the nature of the situation being investigated and a limited number of alternative solutions are examined. Understanding is tested through the use of quantitative methods (simulation, statistical analysis etc.) and similar techniques are used to evaluate alternative solutions utilising some defined quantitative criteria. Such an approach is the traditional method of the systems analyst and usually consists of four stages.

- The specification of an objective or objectives.
- The specification of alternatives by which these objectives may be accomplished.
- The costing of these alternatives.
- The establishment of criteria for evaluating alternatives.

A number of criticisms can be made of this approach. Firstly, it assumes that it is always possible to specify future objectives. Implicit in the decision to commit resources to automate an existing process is an assumption that the process will still be required in the future or that the payback period on the project is very short. This in turn assumes a knowledge of future product trends. The prediction of future product demand in many businesses is notoriously difficult. Determining the quantity and mix of products, over a period of 5-8 years (a conservative estimate of the life of a machine tool) with any degree of precision has been found by most practitioners to be more of a black art than a science.

Secondly, the approach ignores an examination of the current situation. Without agreement on "where we are now" it is extremely difficult to specify future requirements. Experience indicates that a multitude of perceptions of the current situation may exist. Unless these differences are accommodated systems will be developed which satisfy only the dominant coalition's view of requirements and disenfranchise other valid contributions to the decision.

Thirdly, it assumes that a "cost" can be applied to future systems. Many studies have shown the highly complex nature of CIM investment decisions. Attempts have been made to cope with this complexity by structuring the decision process; and there have even been attempts to quantify the "intangible" benefits of CIM (4). In practice such attempts amount to no more than a feeble representation of the complex decision processes involved. Furthermore, they rarely consider the opportunity lost in not investing. Although many studies (5) have pointed out the considerable benefits to be obtained from investing in CIM, few if any, focus on this aspect.

Finally, it ignores the "political" aspects of decision making. In presupposing that decision making is carried out rationally, this analysis ignores the practical realities of organizational life. Recent organizational theorists have concluded that decisions are events that occur when problems and solutions come together in a "fortuitous confluence" rather than on the basis of a logical application of reasoning.

The building of theories on the basis of positivism ignores the historical, cultural and personal dimensions of organizational life and as a result is a poor philosophical base from which to attempt the definition of CIM requirements.

Organizations as Natural Systems and Phenomenology

Positivism is based on the assumption that the world exists independent of those who live in it, whilst phenomenology concentrates on the nature and content of the individual's (actor's) beliefs. The central focus of phenomenology is on the interpretation imposed by actors in giving meaning to the state in which they live. Phenomenology requires that analysts place the real world in parentheses and do not take for granted assumptions regarding the nature of social reality.

Methodologically the focus is on the construction of the actors' social reality rather than the investigation of structural relations. This is done by gathering accounts and interpretations of actors through the process of what Checkland (6) describes as "action research" or more colloquially "living with the natives".

Because the phenomenological approach lays such stress on actor's interpretations whilst ignoring the structural context, the results of such investigations are a narrative of events. To turn such a narrative into a precise definition of requirements would not only be extremely difficult but to miss the purpose of a phenomenological investigation.

In comparing the two dimensions there appears to be considerable differences. Rationalistic views are concerned with structural relations whilst limiting models to exclude complex and dynamic variables in the search for universal truths. The phenomenological/naturalistic views are more concerned with non-deterministic, non-programmable variables as if there were no structural context to the meanings attributed by actors. It is clear that for an area such as CIM both phenomenological and positivistic approaches are unsuitable.

We believe that the philosophical base required to define the requirements for CIM must relate to both the real world and the meanings ascribed by actors, both analysts and users. In this sense a philosophical base is required that is mid-way between positivism and phenomenology.

HERMENEUTICS

According to Gadamer "hermeneutics" is the confluence of three important traditions.....namely "the phenomenology of Husserl, the historicism of Dilthey and the Hermeneutic existentialism of Heidegger", Wolf (7). Hermeneutics seeks to add to phenomenology by identifying the structural and cultural contexts which surround meaningful action. Hermeneutics develops phenomenology further by incorporating the past in the structure and also by acknowledging that the analyst brings his/her own historicity to bear in interpreting the problem situation. In reality the analyst will fall back on his values, prejudices, and beliefs when attempting requirements definition and the client/user will do the same. By each outlining his own position the analyst and client/user engage in a mutual learning process, as Gadamer points out "hermeneutics is not a mysterious communion of souls, but the taking part in a common meaning", the generation of common meaning being dependent upon the historicity of each of the individuals involved.

This is not to suggest that there is, or should be, an obligation on the analyst to collude in the suicide of the client/user by endorsing a solution that merely compounds existing difficulties. Sharing a common understanding of the problems does not imply that a solution can be found that will have the support of all those involved, nor that the solution itself will be correct. The analyst's role is to provide a forum in which problems and solutions can be explored in a mutual learning process.

The approach to requirements definition presented here is based within the philosophical tradition of Hermeneutics.

THREE STAGE PLANNED EVOLUTIONARY APPROACH

Four premises underlie the approach developed. First, that software will provide the backbone of CIM; a premise well supported in the literature (8). Secondly, that the majority of companies will implement CIM by an incremental development of their existing systems. Thirdly, that an essential pre-requisite to the definition of requirements is a clear understanding of the existing system. Finally, that considerable effort must be expended in generating and evaluating realistic alternative design strategies to derive the best system. (Recent research (9) has shown that 80% of manufacturing costs are inherited from the systems design specification).

Based on these premises the following approach has been developed.

- Construction of an IDEF model of the existing system.
- Construction of an IDEF model of the proposed system.
- Comparison of the models of the existing and proposed systems to identify the planning space and the selection of a strategy to move from the existing to the proposed system.

The ICAM Definition Method

IDEFO was developed by SOF'TECH for the U>S. Air Force's ICAM (Integrated Computer Aided Manufacturing) programme. ICAM was established to improve the productivity of aerospace contractors through the systematic application of computer technology. The programme, which began in 1977, was initially aimed at sheet metal fabrication and has received funds in excess of $100 million.

IDEFO is a derivative of the Structured Analysis and Design

Technique, SADT, developed by SOF'TECH. SOF'TECH sold SADT to the Air Force Materials Laboratory who released a slightly amended version, IDEFO, to Computer Aided Manufacturing - International (CAM-i). The Air Force's intention was to see IDEF accepted as a standard for the communication of complex systems definitions. CAM-i was expected to expand the standard into new areas thereby improving the ability to understand manufacturing concepts being developed throughout the world.

IDEFO encompasses three subsets;

- A set of methods that assist in understanding a complex subject.
- A graphical language for communicating that understanding.
- A set of management and human-factor considerations for guiding and controlling the use of the methodology.

The methods of IDEFO are based on several concepts. Top-down decomposition is used to break-up complex topics into small pieces which can be more readily understood. The model building process serves two purposes - it facilitates understanding and provides a means of communicating that understanding. Establishing and using explicit viewpoints and purposes for each model helps to control and limit the information in the model.

An IDEFO model is an ordered collection of diagrams. The diagrams are related in a precise manner to form a coherent model of the subject. The number of activities in a diagram is determined by the breadth and depth of analysis required for the purpose of that particular model, although three is felt to be a reasonable minimum (for below a diagram of two can usually be incorporated into a higher level diagram) and six as a maximum because of individual cognitive limitations. At all times the system and the relationship of any part of the whole remains graphically visible.

The graphical language of IDEFO uses boxes and arrows coupled together in a simple syntax. Boxes on a diagram represent activities. The arrows that connect to a box represent real objects or information needed or produced by the activity. The side of the box at which an arrow enters or leaves shows the arrows role as an input, a control or an output. Incoming arrows (which are shown on the left and top of the box) show the data needed to perform the activity. Outgoing arrows (right of the box) show the data created when the activity is performed. An Input is converted by the activity into the output. A control describes the conditions/circumstances that govern the transformation. The bottom of a box is reserved to indicate the mechanisms or means (person, device, computer model etc.) used to carry out the activity.

In summary the input and output show what is done by the activity, the control shows why it is done and the mechanism shows how it is done.

THE CASE

The company involved is a major European manufacturer of mainframe computers, with a turnover in excess of £1 billion. In recent years the company has been engaged in a major redevelopment of its manufacturing facilities. Current developments are aimed at establishing a complete CIM to include systems for; Production Control, Quality Control, Computer Aided Test, Computer Aided Manufacture and Planning, Automated Materials Handling and Computer Aided Design all of which are to be integrated with the Company's existing information systems.

One of the authors (Dr Maull) joined the Company's manufacturing systems team for a period of 18 months in order to apply, test and refine the approach developed.

First Stage

In the first stage IDEFo was used to construct a model of the existing system. Particular attention was given to the flow of information between activities and the interfaces.

IDEF diagrams were used to reconcile different viewpoints about the existing system. This was achieved by "walkthroughs" of the diagrams with staff operating or affected by the system. Staff commented on the accuracy of the diagrams with respect to their own knowledge of the system. Individual differences were usually minor points of detail. However, occasional major conflicts of opinion did occur. By diagrammatically representing each view, the authors were able to encourage a process of debate. These discussions were successful on each occassion and agreement was reached on how the existing systems operates. The iterative review procedure ensured that the model accurately represented the Company's existing systems.

Having established "where we are now", stage two was used to specify future CIM requirements.

Second Stage

This stage of the methodology was considerably influenced by the adoption of the Hermeneutic approach. In this stage IDEF was used to model the proposed alternative CIM system. The analyst's contribution lay in generating individual dialogues with managers and staff associated with the system (users).

The authors began by outlining their perceptions of requirements and encouraged the user to do the same. These views were represented utilising IDEF diagrams and debated until the user was satisfied that the diagrams clearly represented his requirements. This generation of a "common meaning" took place with each user. The authors then compared the resultant diagrams from "round one" and conflicts of opinion were identified. In round two those individuals whose requirements clashed were brought together. By each user outlining his/her own position and engaging in a debate, a mutual learning process ensued. Further IDEF diagrams resulted. In turn these were compared and conflicts identified. These individual engaged in round three and the iterative process continued until accommodation was reached on the final requirements specification.

This specification was then taken to CIM designers who in turn debated the proposals. Modifications were made and taken back to the users for comments. Eventually through this process, termed the "hermeneutic circle", requirements were defined which realistically address the company's needs and which would secure the support of company staff.

Third Stage

In the final stage, the extent of the change required was identified by comparing the model of the existing system with the model of the proposed CIM system. Having established the nature and extent of the changes required, a strategy of planned evolution was adopted to ensure effective progress towards integration. The strategy, like the design itself took into account policy and managerial considerations as well as social, technical, and economic factors.

CONCLUSIONS

The intention of this paper was to discuss two broadly opposite philosophical standpoints from which requirements definition of aims may emerge. We believe that current requirements definition practice tends to be broadly based within rationalistic organizational assumptions and as such is open to a number of important criticisms. The phenomenological standpoint does not facilitate the precise specification of requirements essential in CIM and, therefore, does not offer an improvement over current practice. Such an improvement can be obtained by adopting an intermediate position based upon a different philosophical tradition - Hermeneutics. Such an approach encourages the key issues of debate and concensus, essential in an area where realistic alternatives must be explored and the support of clients and users is required if an effective system is to be developed.

Hermeneutics cannot provide the richness of reality provided by phenomenology nor the precision of positivistic approaches. It provides an intermediary position between the formal mathematical tools of software engineering which are alien to most clients/users and a narrative description which is difficult, if not impossible, to encapsulate in a requirements definition.

The operationalization of the hermeneutic philosophy is through the use of a three stage planned evolution approach. In taking this approach, analysts are encouraged to partake in a process which reveals further prejudices/beliefs through participation and iteration. IDEF's limitation on the amount of information that should be included in a diagram and its ease of comprehension is of particular value. The result of the procedure is the production of a requirements definition which reflects the idiosyncratic needs of a particular company.

References

1 Hughes, D.R., Computer Integrated Manufacture by Design, Computer Applications for Management and Productivity, Berlin, Sept., 1985.

2 Hughes, D.R. and Maull, R.S. A Framework for the Design of CIM System Architecture, Computers in Mechanical Engineering, Vol 4, No 2, 1985.

3 Scott, W.R., Developments in Organizational Theory 1960-1980, American Behavioural Scientist, 24, 3, 1981.

4 Primrose, P and Leonard, R. The Price to Pay for AMT, Industrial Computing February, 1986.

5 Baxter, R., Companies Must Integrate or Stagnate, The Production Engineer, April, 1984.

6 Checkland, P., Systems Thinking Systems Practice, London, Wiley, 1983.

7 Wolff, J., Hermeneutic Philosophy and the Sociology of Art, Routledge and Kegan Paul, 1975.

8 Milacic, V.R., Computer Based Informatisation of Manufacturing Engineering Activities, International Journal of Research, 20, 3, 1982.

9 Parnaby, J., Manufacturing Systems Design for Competitiveness. 4th International Conference on Systems Engineering. Coventry, September 1985.

KINEMATIC ANALYSIS

Differential Relationships with Applications to Robot Inverse Joint Solution and Motion Correction

P.P. Lin

Assistant Professor, Mechanical Engineering, Cleveland State University, Cleveland, OH 44115, USA

ABSTRACT

The general concept of differential relationships is introduced. PUMA 560/600 robot was chosen for study from which the inverse joint solution is presented and the joint variables due to small change in pose (position and orientation) are solved by directly differentiating the inverse joint solution in order to avoid the complexity and difficulty of inverting a Jacobian matrix. In this paper, it is to investigate what change can be considered as differential change in terms of desired accuracy. By specifying a small positional change at a robot end-effector, the overall positional error and the error of each joint movement resulting from the use of differential concept are presented. An algorithm for motion correction during insertion is developed and the differential concept is employed to this algorithm.

INTRODUCTION

Given the joint variable of each robot joint, the final pose (position and orientation), of a robot end-effector can be calculated, which is called the forward kinematic solution. On the other hand, if a robot is to move to a specified pose in a three dimensional space, all the joint variables can be calculated by inverting the Jacobian matrix, which is called the <u>inverse joint solution</u>.

Once the robot has reached the approximate pose, it is still required to solve the inverse joint solution even if the robot motion is quite small. In this case, each joint movement may be small enough to be considered as differential change so that the joint rotational or translational change can be obtained via differential relationships.

INVERSE JOINT SOLUTION

In the inverse problems the pose of the last link (sixth link for PUMA 560/600) is specified, six joint variables are to be solved. In this paper, the method for solving the inverse problems will not be described which can be found in [1] [2] [3]. Only the inverse joint solution is presented (Figure 1 & Table 1)

APPLY DIFFERENTIAL RELATIONSHIPS TO INVERSE JOINT SOLUTION

The end of a six link manipulator with respect to the base frame can be expressed as $T_6 = A_1 A_2 A_3 A_4 A_5 A_6$ The differential change in positon and orientation of T_6 as a function of all six joint coordinates is written as a six-by-six matrix consisting of differential rotation and translation vector elements and is known as Jacobian. Each column of the Jacobian consists of the differential translation and rotation vector corresponding to differential changes of each of the joint coordinates [4]. The required changes in joint coordinates dq_i can be calculated if a desired differential change in position and orientation is given.

This usually requires many arithmetic operations and the

procedure is complicated by singularities which occur whenever the manipulator becomes degenerate. In [4] it is suggested that a better approach is to differentiate the inverse joint solution. The differential joint solution for PUMA 560/600 is as follows:

$$d\theta_1 = (C_1 dp_y - S_1 dp_x)/(C_1 p_x + S_1 p_y)$$

$$d\theta_3 = (A dp_x + B dp_y + P_z dp_z + C d\theta_1)/D \quad \text{where } A = p_x C_1^2 + p_y C_1 S_1 \quad B = p_y S_1^2 + p_x C_1 S_1$$
$$C = S_1 C_1 (p_y^2 - p_x^2) + p_x p_y (C_1^2 - S_1^2) \qquad D = a_2 d_4 C_3 - a_2 a_3 S_3$$

$$d\theta_2 = (E d\theta_3 + dp_z)/F \quad \text{where } E = a_3 C_{23} + d_4 S_{23} \qquad F = -a_3 C_{23} - d_4 S_{23} - a_2 C_2$$

$$d\theta_5 = -[GH + S_{23}K - La_z + C_{23} da_z]/S_5 \quad \text{where } G = C_{23}(d\theta_2 + d\theta_3) \quad H = C_1 a_x + S_1 a_y$$
$$L = S_{23}(d\theta_2 + d\theta_3) \qquad K = (-S_1 a_x + C_1 a_y) d\theta_1 + C_1 da_x + S_1 da_y \quad C_{23} = \cos(\theta_2 + \theta_3)$$

$$d\theta_4 = [-S_4 C_5 d\theta_5 + C_1 (da_y - a_x d\theta_1) - S_1 (da_x + a_y d\theta_1)] / C_4 S_5$$

$$d\theta_6 = -[C_4 (U - R d\theta_4) + S_4 (LQ - T + O_Z G + S_{23} dO_Z - S d\theta_4)]/S_6 \quad U = -Q d\theta_1 - S_1 do_x + C_1 do_y$$
$$R = C_{23} Q - S_{23} o_z \quad Q = C_1 o_x + S_1 o_y \quad T = C_{23}(S d\theta_1 + C_1 do_x + S_1 do_y) \quad S = C_1 o_y - S_1 o_x$$

ERRORS DUE TO THE EMPLOYMENT OF DIFFERENTIAL RELATIONSHIPS

Differential change is an abstract term which is not useful unless it is well defined in terms of accuracy. In order to understand what change can be considered as differential change, PUMA 560/600 robot was employed. It is assumed that the acceptable error in each joint variable due to the employment of differential concepts is approximately 5 percent. It was found from computer simulation results that the errors in joint variables 1,2, and 3 are less than 5 percent. For joint variables 4 and 5, the errors are less than 5 percent if the change in position is approximately 2 percent. For joint variable 6, the error is relatively large compared to the previous five joint variables. In this case. the differential change in joint variable 6 may be set to zero. In spite of the error in each joint variable, the overall positional error is quite small (see Table 2). The overall orientational error is also small. However, this error is somewhat difficult to express in terms of percentage.

ROBOT MOTION CORRECTION VIA DIFFERENTIAL RELATIONSHIPS

The pose of the gripped object in base frame is given by $T = T_6\ E\ O$ where E is for end effector and O is for the object. The robot hand motion will be corrected after the first two-point contact occurs. To correct the hand motion, the inserting object will be rotated about some axis passing through the contact point p in a plane parallel to the bottom surface of the hole.(Fig.2) [5]

Define G as the coordinate transformation from the reference frame of robot base to the frame of target (or hole, in this case) Thus the tip of peg can be expressed by $G^{-1}\ T$ in frame H which is attached to the bottom center of the hole. Let $Z = G^{-1}T$
The center position of the tip of the inserting peg is the first three elements of the fourth column of matrix Z. A vector V can then be formed, which connects point $C(p_{x1}, p_{y1}, p_{z1})$ and point $(0, 0, p_{z1})$ $V = p_{x1}\ \underline{i} + p_{y1}\ \underline{j}$
A frame T' is attached to the contact point P, then T' is expressed by $T' = T_6\ E\ O$ Translation(0,r,0) where r is the radius of peg. There exists a unique vector which is perpendicular to vector V and passing through the contact point P. This unique vector U is

$U = -p_{y1}\ \underline{i} + p_{x1}\ \underline{j} + P_z'\ \underline{k}$

To correct hand motion, the object is rotated about a vector U. A transformation matrix representing a rotation around a vector k at the origin of a specific coordinate had been developed by [4]. Note that vector K has to be a unit vector.

$K = K_x\ \underline{i} + K_y\ \underline{j} + K_z\ \underline{k}$

$K_x = -p_{y1} / K, \quad K_y = P_{x1} / K, \quad K_z = P_z' / K$

Define $Rot(k, \theta)$ as a rotation matrix representing a rotation around vector K by an angle θ, then $Rot(k,\theta)$ is expressed by

$$Rot(k,\theta) = \begin{pmatrix} k_x k_x \operatorname{vers}\theta + \cos\theta & k_y k_x \operatorname{vers}\theta - k_z \sin\theta & k_z k_x \operatorname{vers}\theta + k_y \sin\theta & 0 \\ k_x k_y \operatorname{vers}\theta + k_z \sin\theta & k_y k_y \operatorname{vers}\theta + \cos\theta & k_z k_y \operatorname{vers}\theta - k_x \sin\theta & 0 \\ k_x k_z \operatorname{vers}\theta - k_y \sin\theta & k_y k_z \operatorname{vers}\theta + k_x \sin\theta & k_z k_z \operatorname{vers}\theta + \cos\theta & 0 \\ 0 & 0 & 0 & 1 \end{pmatrix}$$

where $\operatorname{vers}\theta = 1 - \cos\theta$

Rotating frame T' about vector K by an angle θ, with respect to base coordinate, can be expressed by $T_p = Rot(K,\theta)\ T'$

Frame T_p is then transformed back to frame T_6 from which the inverse joint solution is obtained. $T_6' = T_p\ Trans^{-1}(0,r,0)\ O^{-1}\ E^{-1}$

Given a transformation whose elements are functions of some variable, the differential transformation with respect to that variable is the transformation whose elements are the derivatives of the original transformation elements. Given a frame T', the differential change in terms of the base coordinate can be expressed as $dT_p = [Rot(k,d\theta) - I]\ T'$

CONCLUSIONS

The error due to the employment of differential relationships depends on the robot geometry, the current pose and the change in pose. It is recommended that the errors be estimated beforehand for a certain frequently used workspace. The employment of differential relationships is useful not only in inverse joint solution but also in many aspects such as robot vision, and robot motion correction during insertion where changes in position and orientation are generally known and small enough to be considered as differential changes.

REFERENCES

[1] Paul, R., Shimano, B., and Mayer, G., "Kinematic Control Equations for Simple Manipulators", IEEE Trans. on Systems, Man, and Cybenetics, Vol. SMC - 11, No. 6, June 1981, pp. 449-460.

[2] Lee, C. "Robot Arm Kinematics, Dynamics, and Control," Computer, Vol. 15, No. 12 Dec., 1982, pp. 62-80.

[3] Rathnaswami, V.R., and Palm, W. J., "An Inverse Joint Solution for PUMA Manipulators", Report 26, Robotics Research Center, University of Rhode Island, September, 1983

[4] Paul, R. P., Robot Manipulators: Mathematics, Programming and Control, MIT Press, Cambridge, MA, 1981, PP. 103-109

[5] Lin, P., "Gripper-Based Position/Force Sensor Design and Robot Hand Motion Control During Assembly," Doctoral Dissertation, Mechanical Engineering and Applied Mechanics, University of Rhode Island, 1985

TABLE 1 PUMA 560/600 Robot Inverse Joint Solution

$$\theta_1 = \tan^{-1}(p_y/p_x) - \tan^{-1}\left(d_2/\pm\sqrt{r_1^2-d_2^2}\right) \qquad \text{where } r_1 = +\sqrt{p_x^2+p_y^2}$$

$$\theta_3 = \tan^{-1}\frac{a_3}{-d_4} - \tan^{-1}\frac{d}{\pm\sqrt{r_3^2-d^2}} \qquad \text{where } r_3^2 = 4a_2^2\,(a_3^2 + d_4^2)$$

$$d = -a_2^2 - a_3^2 - d_4^2 + p_x^2C_1^2 + p_y^2S_1^2 + p_z^2 + 2p_xp_yC_1S_1$$

$$\theta_2 = \theta_{23} - \theta_3 \qquad \theta_{23} = \tan^{-1}\frac{W_2F_{11} - W_1p_z}{W_1F_{11} + W_2\,p_z}$$

Let

$$C_1p_x + S_1p_y = F_{11}$$
$$a_2C_3 + a_3 = W_1$$
$$a_2S_3 + d_4 = W_2$$

$$\theta_4 = \tan^{-1}((-S_1a_x + C_1a_y)/[C_{23}(C_1a_x + S_1a_y) - S_{23}a_z])$$

$$\theta_5 = \tan^{-1}\frac{C_4\left[C_{23}(C_1a_x + S_1a_y) - S_{23}a_z\right] + S_4(-S_1a_x + C_1a_y)}{S_{23}(C_1a_x + S_1a_y) + C_{23}a_z}$$

$$\theta_6 = \tan^{-1}\frac{-C_5\left[C_4\left\{C_{23}(C_1o_x+S_1o_y)-S_{23}o_z\right\} + S_4(-S_1o_x+C_1o_y)\right] + S_5\left[S_{23}(C_1o_x+S_1o_y)+C_{23}o_z\right]}{-S_4\left[C_{23}(C_1o_x + S_1o_y) - S_{23}o_z\right] + C_4(-S_1o_x + C_1o_y)}$$

TABLE 2

Overall Positional Error Due to the Use of Differential Change

Px, Py, Pz are the components of initial position vector P

Px = 250 Py = 250 Pz = 300

$P = SQRT(Px^2 + Py^2 + Pz^2) = 463.681$

New P: New position due to the exact change in position

New PP: New position resulting from the differential changes in joint variables

% PC: Percentage of position change, [(New P)-P] x100/P

% ERROR: (New PP-New P)x 100/New P

The orientation of the PUMA robot is intended to remain constant. The initial approach vector A has the following components.

Ax (1) = -0.125 Ay(1) = 0.400 Az(1) = 0.900

Ax (2), Ay(2), Az(2) are the new components of the approach vector A due to the use of differential change.

%PC	New P	New PP	%ERROR	Ax(2)	Ay(2)	Az(2)
0.990	468.318	468.310	0.002	-0.126	0.403	0.907
1.961	472.955	472.925	0.006	-0.126	0.403	0.907
2.913	477.591	477.524	0.014	-0.126	0.403	0.907
3.846	482.228	482.108	0.025	-0.126	0.403	0.907
4.762	486.865	486.676	0.039	-0.126	0.403	0.907
5.660	491.502	491.228	0.056	-0.126	0.403	0.907
6.542	496.139	495.763	0.076	-0.126	0.402	0.907
7.407	500.775	500.281	0.099	-0.126	0.402	0.907
8.257	505.412	504.782	0.125	-0.126	0.402	0.907
9.091	510.049	509.266	0.154	-0.126	0.402	0.907

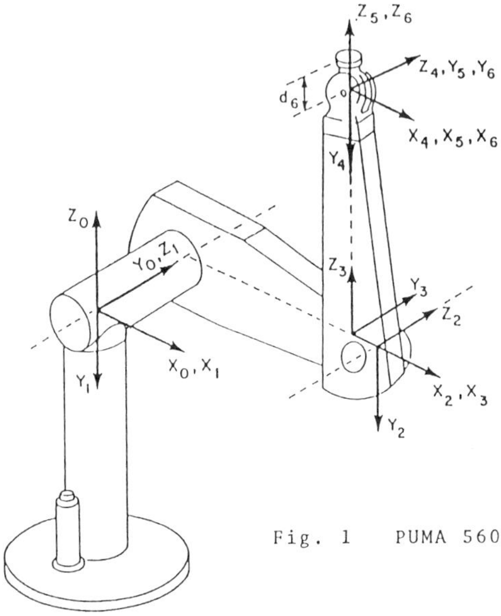

PUMA 560/600 Link Parameters

Joint	α	θ	d(mm)	a(mm)	Range(deg)
1	-90	θ_1	0	0	-160/160
2	0	θ_2	149.5	432.0	-225/45
3	90	θ_3	0	-20.3	-45/225
4	-90	θ_4	432.0	0	-110/170
5	90	θ_5	0	0	-100/100
6	0	θ_6	0	0	-266/266

Fig. 1 PUMA 560/600 Joint Coordinate Systems and Link parameters

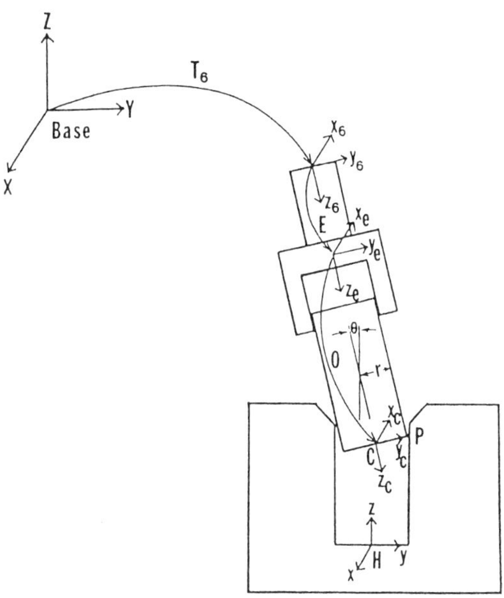

Fig. 2 Coordinate Transformation Used for Motion Correction

A Kinematic Model Of Flexible Robot Arms

Behnam Pourbabai

Department of Statistics and Operations Research, New York University, New York, NY 10006, USA

Abstract: A robot's posture can be identified by identifying the angles of each of its joints, and the movement of the robot arm (e.g., Stanford robot arm) can be caused by changing its appropriate joint angles and by expanding the lenght of its appropriate links.

In this paper, a kinematic model of the movement of a flexible robot arm with multiple joints and expandable (contractable) links, in a cartesian coordinates is developed, to analyze the movement of the robot arm to move along a collision free path from one point in space to another point inside a convex polyhedral. In this model the only decision variables are the coordinates of each joint of the robot arm with respect to the base of the robot arm, and the only necessary information besides the geometric parameters of the robot arm, are the coordinates of the origin and the destination points in space.

Key Words: Inverse kinematic problem, Obstacle Avoidance, and Flexible Robots.

Introduction

Our primary objective in this paper is to present a kinematic model of the movement of a flexible robot arm with multiple joints and expandable (contractable) links, in a cartesian coordinates is developed, to analyze the movement of the robot arm along a collision free path from one point in space to another point inside a convex polyhedral. In this model the only decision variables are the coordinates of each joint of the robot arm with respect to the base of the robot arm, and the only necessary information besides the geometric parameters of the robot arm, are the coordinates of the origin and the destination points in space. In this model, the analytical description of the spatial displacement of the robot arm is not modeled as a function of time, and the the forces/moments that cause the motion are ignored.

Based on our model, the following questions can be answered. Given a desired position and orientation of the end-effector of the manipulator and the geometric link parameters, can the manipulator reach the desired prescribed manipulator hand position? And if it can, what is a feasible configuration (e.g., posture) for the robot arm? It is noted that, these questions are directly related to the inverse kinematic problem.

The inverse kinematic problem can be solved either by an algebraic, iterative, or geometric methodology. Several scholars have suggested algorithms for the PUMA and Stanford robot arms using the algebraic methodology, see [9], [7], and also [8]. As discussed in [4], the algebraic approach suffers from the fact that, the solution does not give a clear indication of how to select the correct solution from the several possible solutions for a particular arm configuration. That is, the user often needs to rely on his intuition to select the correct answer. On the other hand, in the iterative methodology, more computations are needed and the convergence to the correct solution can not be guaranteed, see [12] and [6]. In the geometric methodology, a more efficient algorithm can be suggested, given that the manipulator under consideration is simple and certain geometrical specifications are satisfied, for further details, see [4]. For additional reading, see [2], [7], and [3]. The last authors have selected and provided an excellent collection of the related papers and references.

This paper is basically an extention of Pourbabai's [10] reformulation of the classical inverse kinematic problem. The main difference between this paper and Pourbabai's [10] work is that, here the length of each link of the robot arm could also change. Hence, this new formulation enables us to analyze the inverse kinematic problem for the robot arms which behave similar to the Stanford robot arm.

The organization of this paper is as follows. In section 2, the components of the model are described, and an extension of the model to avoid obstacles and to prevent collision is presented. Finally, in section 3, a summary is provided and the concluding remarks are discussed.

2. The Model

In this section the notation and the components of the model are presented.

2.a. Notation	Description
n	joint or link index.
m	constraint index.
N	number of links.
N + 1	number of joints.
M	number of linear constraints (e.g., half spaces).

L_n	length of link n connecting the adjacent joints n and n+1, for n=1 to N.
x_n', y_n', and z_n' (x_n, y_n, and z_n)	the initial (final) coordinates of joint n+1 along x,y, and z axes with respect to the coordinates of joint n, respectively for n=1 to N+1.
X_n', Y_n', and Z_n' (X_n, Y_n, and Z_n)	the initial (final) coordinates of joint n along x,y, and z axes with respect to the coordinates of joint 1, respectively for n=1 to N+1.
$\hat{X}_{n,n+k}$, $\hat{Y}_{n,n+k}$ and $\hat{Z}_{n,n+k}$ ($\tilde{X}_{n,n+k}$, $\tilde{Y}_{n,n+k}$ and $\tilde{Z}_{n,n+k}$)	the minimum (maximum) allowable projected distance between the joints n and n+k along x,y and z axes, respectively for n=1 to N-1 and k=2 to N-n+1.
$\hat{L}_{n,n+k}$ ($\tilde{L}_{n,n+k}$)	the minimum (maximum) allowable distance between the joints n and n+k, for n=1 to N-1 and k=2 to N-n+1.
a_m, b_m, c_m, and d_m	the coefficients of the linear constraint (e.g., half space) m, which characterizes at least one obstacle, for m=1 to M.
α_n (α'_n)	the initial (final) proprtionality factor of link n.

Throughout this paper, it is assumed that, the reference point is at joint 1 which is positioned at the base of the robot arm (i.e., $X_1 = x_1 = 0$, $Y_1 = y_1 = 0$, and $Z_1 = z_1 = 0$) and joint N+1 is positioned next to the end-effector of the robot arm (i.e., $X_{N+1} = \sum_{n=1}^{N+1} x_n$, $Y_{N+1} = \sum_{n=1}^{N+1} y_n$, and $Z_{N+1} = \sum_{n=1}^{N+1} z_n$). Furthermore, each two adjacent joints of the robot arm is connected by a link. Hence, there are a total of N links interconnecting N+1 joints.

2.b. Components of the model

Before describing the componenets of the model it is noted that; Firstly, the proportionality constants are greater than or equal to zero and less than or equal to one; Secondly, the role of the initial (final) proportionality factors are to identify the initial (final) length of link n. For example, in a Stanford robot arm, if $\alpha'_n=0$ (1) and $\alpha_n=1$ (0), it implies that the initial length of link n has been equal to 0 (L_n) and then its final value has been extended (contracted) to be equal to L_n (0). Notice that for the robot arms whose their links' length do not change, $\alpha_n=1$ for n=1 to N.

In this model, the values of the parameters of N, L_n, X'_p, Y'_p, Z'_p, X_p, Y_p, and Z_p must be specified for n=1 to N, p=1 and N+1, and the only decision variables are x_n, y_n, and z_n, for n=1 to N+1. The model in a cartesian coordinates can be formulated as follows,

$$\sum_{n=1}^{N+1} x_n = X_{N+1} \qquad (1)$$

$$\sum_{n=1}^{N+1} y_n = Y_{N+1} \qquad (2)$$

$$\sum_{n=1}^{N+1} z_n = Z_{N+1} \qquad (3)$$

$$\sqrt{x_n^2 + y_n^2 + z_n^2} = \alpha_{n-1} L_{n-1}, \qquad n=2 \text{ to } N+1 \qquad (4)$$

$$0 \leq \alpha_n \leq 1 \qquad n=1 \text{ to } N \qquad (5)$$

In the above model, the left hand side of the first three equations presents the projected distance between joint 1 (e.g., the base of the robot arm) and joint N+1 (e.g., the end-effector of the robot arm) along x, y, and z axes respectively. Expression 4 is given to guarantee that the identified coordinates of joint n+1 along x, y, and z axes with respect to the coordinates of joint n result in the magnitude which is equal to the length of link n, for n=1 to N. Expession 5 is provided to gurantee that the proportionality factor is not less than zero and not greater than one. It is noted that, in the above set of linear and nonlinear equations x_n, y_n, and z_n can assume values between $-L_n$ and $+L_n$, for n=1 to N. Hence, movement of the robot arm in the space is feasible and allowed.

The above set of linear and nonlinear equations can efficiently be solved by one of the well known algorithms, such as the Newton's method, for further details, see [1]. After selecting an appropriate algorithm, if a feasible solution is resulted, the coordinates of each joint along x, y, and z axes with respect to the base of the robot arm can be identified as follows,

$$X_n = \sum_{q=1}^{n} x_q , \qquad n=1 \text{ to } N+1 \tag{6}$$

$$Y_n = \sum_{q=1}^{n} y_q , \qquad n=1 \text{ to } N+1 \tag{7}$$

$$Z_n = \sum_{q=1}^{n} z_q , \qquad n=1 \text{ to } N+1 \tag{8}$$

Furthermore, if necessary, the values of the joint angles at joint n along x, y, and z axes with respect to joint n-1, and the values of the angles between any two adjacent joints can be identified, based on the values of x_n, y_n, and z_n, for n=2 to N+1. On the other hand, if a feasible solution is not resulted, it is then concluded that the final coordinates of joint N+1 (e.g., X_{N+1}, Y_{N+1}, and Z_{N+1}) can not be reached from the initial coordinates of joint N+1.

It is noted that, the above model can be improved by including additional constraints to consider the movement limitations of each joint with respect to another joint. For this purpose the following constraints are suggested for n=1 to N-1 and k=2 to N-n+1.

$$\hat{X}_{n,n+k} \leq \left|\sum_{q=n+1}^{n+k} x_q - x_n\right| \leq \tilde{X}_{n,n+k} \tag{9}$$

$$\hat{Y}_{n,n+k} \leq \left|\sum_{q=n+1}^{n+k} y_q - y_n\right| \leq \tilde{Y}_{n,n+k} \tag{10}$$

$$\hat{Z}_{n,n+k} \leq \left|\sum_{q=n+1}^{n+k} z_q - z_n\right| \leq \tilde{Z}_{n,n+k} \tag{11}$$

$$\hat{L}_{n,n+k} \leq \sqrt{\left(\sum_{q=n+1}^{n+k} x_q - x_n\right)^2 + \left(\sum_{q=n+1}^{n+k} y_q - y_n\right)^2 + \left(\sum_{q=n+1}^{n+k} z_q - z_n\right)^2} \leq \tilde{L}_{n,n+k} \tag{12}$$

Constraint sets 9, 10 and 11 are provided to limit the maximum and the minimum allowable projected distance between joints n+k and n along x, y, and z axes, respectively. Furthermore, constraint set 12 is presented to limit the maximum and the minimum allowable distance between joints n+k and n. It is noted that, in expressions 9 to 12, the following substitutions have been made, for n=1 to N-1 and k=2 to N-n+1.

$$X_{n+k} - X_n = \sum_{q=n+1}^{n+k} x_q - x_n \tag{13}$$

$$Y_{n+k} - Y_n = \sum_{q=n+1}^{n+k} y_q - y_n \tag{14}$$

$$Z_{n+k} - Z_n = \sum_{q=n+1}^{n+k} z_q - z_n \tag{15}$$

The above expressions are resulted by cancelling the unnecessary terms at joints 1 to n-1. Notice that, eventhough expressions 9 to 12 form a set of linear and nonlinear inequality constraints, the model can still be solved by the Newton's alogrithm or its variations.

2.c. Avoiding obstacles

In this section a set of additional constraints are introduced to avoid certain obstacles and to prevent collision, for a review of the related literature, see [5]. For this purpose, it is assumed that the feasible region in which the robot arm is allowed to freely move, is in the interior of a convex polyhedral. It is noted that, a convex polyhedral is the intersection of a finite number of half-spaces (e.g., $a_m X + b_m Y + c_m Z \leq d_m$, m=1 to M), for further details, see [11]. Besides, it is assumed that, each half-space can be used to represent the infeasible region which is created due to the presence of at least one obstacle.

The model can then be modified by adding the following constraint set.

$$a_m \sum_{q=1}^{n} x_q + b_m \sum_{q=1}^{n} y_q + c_m \sum_{q=1}^{n} z_q \leq d_m \qquad \begin{matrix} n=1 \text{ to } N+1 \\ m=1 \text{ to } M \end{matrix} \tag{16}$$

Notice that, the above constraint set or its alternative representation in (16)

$$a_m X_n + b_m Y_n + c_m Z_n \leq d_m \qquad \begin{matrix} n=1 \text{ to } N+1 \\ m=1 \text{ to } M \end{matrix} \tag{17}$$

represent the feasible region in which the robot arm is allowed to freely move within it, which is formed of M half-spaces. Hence, the coordinates of any joint of the robot arm along x, y, and z axes with respect to the base must satisfy the above constraint set, to avoid the obstacles and to prevent collision.

3. Concluding Remarks

In this paper a kinematic model of the movement of a flexible robot arm with multiple joints and expandable (contractable) links, in a cartesian coordinates is developed, to enable the robot arm to move along a collision free path from one point in space to another point inside a convex polyhedral. Furthermore, two extensions of the model are also presented based on Pourbabai [10] paper to consider the movement limitations of the robot arm at each joint, and to also consider the movement limitations of the robot arm in a confined (e.g., convex polyhedral) space. An algorithmic solution was then suggested to solve the problem (i.e., identify a feasible solution). It is noted that, this model can further be improved to also consider certain dynamic effects of the forces/moments, on the movement of the robot arm.

REFERENCES

1. Demidovich, B.P. and I.A. Maron, Computational Mathematics English Translation, Mir publishers Moscow, USSR, (1981).

2. Duffy, J., Analysis of Mechanisms and Robot Manipulators, John Wiley & Sons New York, NY, (1980).

3. Lee, C.S.G., R.C., Gonzalez, and K.S., Fu, Tutorial on Robotics, Second Edition, IEEE Computer Society Press, (1985).

4. Lee, C.S.G. and M. Ziegler, Geometric Approach in Solving Inverse Kinematics of PUMA Robots, IEEE Transactions on Aerospace and Electronic Systems, AES-20, No. 6, pp. 87-97, (1984).

5. Lozano-Perez, T., Spatial Planning: A Configuration Space Approach, IEEE Transactions on Computers, C-32, No. 2, pp. 26-38, (1983).

6. Milenkovic, V. and B. Huang, Kinematics of Major Robot Linkages, Proceedings of the 13th Int. Symp. on Industrial Robots, April 17-19, Chicago, Illinois, pp. 16-31 to 16-47, (1983).

7. Paul, R.P., Robot Manipulators: Mathematics, Programming and Control, M.I.T. Press, (1981).

8. Paul, R.P., B.E., Shimano, and G., Mayer, Kinematic Control equations for Simple Manipulators, IEEE Transaction on Systems, Man and Cybernetics, SMC-11, 6, pp. 449-455, (1981).

9. Pieper, D.L. and B. Roth, The Kinematics of Manipulators Under Computer Control, Proceedings of II Int. Cong. on the Theory of Machines and Mechanisms, 2, pp. 159-168, (1969).

10. Pourbabai, B., Nonlinear Modeling of the Movement of The Robot Arm, Recent Trends in Robotics: Modeling Control and Education, Jamshidi, Shahinpoor, and Luh (eds.), Elsevier Science Publishing Co.

11. Stoer, J. and C., Witzgall, (1970), Convexity and Optimization in Finite Dimensions I, Springer-Verlag New York, (1970).

12. Uicker, J.J., Jr., J., Denavit, and R.S., Hartenberg, An Iterative Method for the Displacement Analysis of Spatial Mechanism, 31, Series E, pp. 309-314, (1964).

Closed-Form Solution of Inverse Kinematics for the Puma Manipulator

Marko I. Vuskovic and Socorro C. Hunt

Department of Mathematical Sciences, San Diego State University, San Diego, California 92182-0314, USA

Abstract

A closed-form solution of inverse kinematics for the Unimation PUMA manipulator, which is an improved version of standard algorithms initially proposed by R. Paul, is discussed. Standard algorithms use a polar transformation to solve an equation of the form $a\ cos(x) - b\ sin(x) = p$. Here, the equation is considered as a part of two-dimensional orthogonal transformation. This yields more straightforward and more efficient solution for inverse kinematics. The improvement of the time-efficiency for single solutions is more than 20% and for multiple solutions more than 40%.

1. Introduction

The most commonly used algorithm in manipulator control is that of inverse kinematics, which determines the values of joint displacements for a given Cartesian position and orientation of the manipulator. Since the process of the Cartesian-to-joint mapping is a part of the control loop with relatively high sampling rate (60 Hz or more), the time-efficiency is extremely important. Therefore the closed form solution for the inverse kinematics is preferred. It is shown [1] that for kinematically simple manipulators, that is, the manipulators with the last three joint axes intersecting at the point, the closed form solution is always possible. A comprehensive study of kinematically simple manipulators with six degrees of freedom is given in [2]. There are also other similar approaches as in [3], [4], [5] and others. In order to minimize the computation time new multiprocessor architectures are also considered. For example, in [6] a maximum pipelined CORDIC architecture based on the closed form solution given in [2] has been proposed.

In this paper a modification of the algorithm from [2] is considered. The major differenece is in treatment of the trigonometric equation $a\ cos(x) - b\ sin(x) = p$, which is considered as a rotation in two-dimensional space, instead of solving it by trigonometric substitutions as used in [2] and later in [3-5]. This would result in fewer floating-point operations and fewer transcendental function calls. The algorithm has been presented for the PUMA manipulator, but it can also be applied to a general class of kinematically simple manipulators.

2. Solution For Arm Joints

The inverse kinematics of a nonredundant 6 DOF rotary manipulator is formulated as follows: Find joint angles $\theta_i\,(i=1..6)$ such that the manipulator wrist has position $\mathbf{p}$ and orientation $\mathbf{R}=[\mathbf{n}\ \mathbf{s}\ \mathbf{a}]$, where $\mathbf{p}=(p_x, p_y, p_z)$, $\mathbf{n}=(n_x, n_y, n_z)$, $\mathbf{s}=(s_x, s_y, s_z)$, and $\mathbf{a}=(a_x, a_y, a_z)$ are given position, normal, orientation and approach vectors defined relative to the base frame of the manipulator. This reduces to the solution of the transcendental equation

$$\begin{bmatrix} \mathbf{R} & \mathbf{p} \\ \mathbf{o}^T & 1 \end{bmatrix} = \prod_{i=1}^{6} \begin{bmatrix} \mathbf{R}_i(\theta_i) & \mathbf{p}_i(\theta_i) \\ \mathbf{o}^T & 1 \end{bmatrix} \tag{1}$$

where $\mathbf{R}_i(\theta_i)$ and $\mathbf{p}_i(\theta_i)$ are relative orientations and positions of manipulator links, while $\mathbf{o}$ is zero vector. For the class of manipulators with spherical wrist $\mathbf{p}_5 = \mathbf{p}_6 = \mathbf{o}$, thus (1) reduces to

$$\mathbf{p} = \mathbf{R}_1\mathbf{R}_2\mathbf{R}_3\mathbf{p}_4+\mathbf{R}_1\mathbf{R}_2\mathbf{p}_3+\mathbf{R}_1\mathbf{p}_2+\mathbf{p}_1 \tag{2}$$

$$\mathbf{R} = \mathbf{R}_1\mathbf{R}_2\mathbf{R}_3\mathbf{R}_4\mathbf{R}_5\mathbf{R}_6. \tag{3}$$

In order to find arm joints θ_1, θ_2 and θ_3 we rewrite (2) as

$$\mathbf{R}_2^T(\mathbf{R}_1^T(\mathbf{p}-\mathbf{p}_1)-\mathbf{p}_2) = \mathbf{R}_3\mathbf{p}_4+\mathbf{p}_3. \tag{4}$$

Using actual expressions for $\mathbf{R}_i$ and $\mathbf{p}_i$ for the PUMA manipulator as given in [2], (4) yields

$$c_2(c_1p_x+s_1p_y - c_2a_2) - s_2(p_z+s_2a_2) = s_3d_4+c_3a_3 \tag{5}$$

$$s_2(c_1p_x+s_1p_y-c_2a_2) + c_2(p_z+s_2a_2) = c_3d_4-s_3a_3 \tag{6}$$

$$c_1p_y-s_1p_x = d_3 \tag{7}$$

where $c_i = cos(\theta_i)$, $s_i = sin(\theta_i)$, while a_i and d_i are link parameters.

Equation (7) provides the solutions for the first joint. As shown in the Appendix, the companion equation for (7) is

$$s_1p_y+c_1p_x = \xi \tag{8}$$

where

$$\xi = \pm(p_x^2+p_y^2-d_3^2)^{\frac{1}{2}},$$

which gives us

$$\theta_1 = atan2(\xi, d_3) - atan2(p_x, p_y). \tag{9}$$

The sign of ξ can be chosen according to the desired global arm configuration (right-hand or left-hand shoulder).

Substituting (8) into (5) and (6) we obtain

$$\begin{aligned} c_2\xi-s_2p_z &= s_3d_4+c_3a_3+a_2 \\ s_2\xi+c_2p_z &= c_3d_4-s_3a_3. \end{aligned} \tag{10}$$

The left side of (10) represents an orthogonal transformation in two-dimensional space. Consequently

$$\xi^2+p_z^2 = (s_3 d_4+c_3 a_3+a_2)^2 + (c_3 d_4-s_3 a_3)^2,$$

which reduces to

$$s_3 d_4+c_3 a_3 = \frac{d}{2a_2}, \tag{11}$$

where

$$d = \xi^2+p_z^2-[a_2^2+a_3^2+d_4^2].$$

(Square brackets indicate constant expressions which can be computed in advance.) The companion equation for (11) is

$$c_3 d_4-s_3 a_3 = \frac{\eta}{2a_2}, \tag{12}$$

where

$$\eta = \pm([4a_2^2(a_3^2+d_4^2)]-d^2)^{\frac{1}{2}}.$$

Since a_2 is always positive, (11) and (12) yield

$$\theta_3 = atan2(d, \eta)-atan2(a_3, d_4). \tag{13}$$

The sign of η can be chosen according to the desired global arm configuration (elbow-up or elbow-down).

Substituting (11) and (12) into (10) we get

$$\begin{aligned} c_2\xi-s_2p_z &= \frac{d}{2a_2}+a_2, \\ s_2\xi+c_2p_z &= \frac{\eta}{2a_2}, \end{aligned} \tag{14}$$

thus

$$\theta_2 = atan2(\eta, d+[2a_2^2])-atan2(p_z,\xi). \tag{15}$$

Angles θ_1 and θ_3 have two solutions, while angle θ_2 has four solutions according to the chosen signs of ξ and η.

3. Solution For Wrist Joints

Having arm joints θ_1 , θ_2 and θ_3 solved, we can find solutions for wrist joints θ_4, θ_5 and θ_6. Therefore, we rewrite (3) as

$$\mathbf{R}_4^T\mathbf{G} = \mathbf{R}_5\mathbf{R}_6, \tag{16}$$

where

$$\mathbf{G} = [g_{ij}]_3 = (\mathbf{R}_1\mathbf{R}_2\mathbf{R}_3)^T\mathbf{R}. \tag{17}$$

Using actual expressions for R_i [2] , we obtain

$$\begin{aligned} g_{1j} &= c_1 c_{23} r_{1j} + s_1 c_{23} r_{2j} - s_{23} r_{3j} \\ g_{2j} &= -s_1 r_{1j} + c_1 r_{2j} \\ g_{3j} &= c_1 s_{23} r_{1j} + s_1 s_{23} r_{2j} + c_{23} r_{3j} \end{aligned} \tag{18}$$

where $r_{11} = n_x, r_{21} = n_y, r_{31} = n_z, r_{12} = s_x, r_{22} = s_y, r_{32} = s_z, r_{13} = a_x, r_{23} = a_y, r_{33} = a_z$ and $s_{23} = sin(\theta_2+\theta_3)$, $c_{23} = cos(\theta_2+\theta_3)$.

Equation (16) evaluates to

$$\begin{bmatrix} c_4 g_{11}+s_4 g_{21} & c_4 g_{12}+s_4 g_{22} & c_4 g_{13}+s_4 g_{23} \\ -g_{31} & -g_{32} & -g_{33} \\ -s_4 g_{11}+c_4 g_{21} & -s_4 g_{12}+c_4 g_{22} & -s_4 g_{13}+c_4 g_{23} \end{bmatrix} = \begin{bmatrix} c_5 c_6 & -c_5 s_6 & s_5 \\ s_5 c_6 & -s_5 s_6 & -c_5 \\ s_6 & c_6 & 0 \end{bmatrix} \tag{19}$$

Equating elements (3,3) of (19) we get

$$c_4 g_{23} - s_4 g_{13} = 0. \tag{20}$$

The corresponding companion equation is

$$s_4 g_{23} + c_4 g_{13} = \zeta \tag{21}$$

where

$$\zeta = \pm (g_{13}^2 + g_{23}^2)^{\frac{1}{2}}.$$

The signs of ζ correspond to two wrist configurations (flipped and nonflipped). Provided ζ is not equal to zero, (20) and (21) yield

$$\theta_4 = \begin{cases} atan2(g_{23}, g_{13}) & \text{for } \zeta > 0 \\ \pi + atan2(g_{23}, g_{13}) & \text{for } \zeta < 0 \end{cases} \tag{22}$$

If ζ is close to zero, i.e. if $g_{13} = g_{23} = 0$, θ_4 will be undefined. This is called the degenerate case, which is considered in the following section.

Substituting (21) into (19) and comparing elements (1,3) and (2,3) we obtain

$$s_5 = \zeta \qquad c_5 = g_{33}$$

thus

$$\theta_5 = atan2(\zeta, g_{33}). \tag{23}$$

The last joint, θ_6 , can be solved by equating elements (2,1) and (2,2) of (19)

$$s_5 c_6 = -g_{31} \qquad s_5 s_6 = g_{32}$$

giving

$$\theta_6 = \begin{cases} atan2(g_{32}, -g_{31}) & \text{for } \zeta > 0 \\ \pi + atan2(g_{32}, -g_{31}) & \text{for } \zeta < 0. \end{cases} \tag{24}$$

4. The Degenerate Case

For the degenerate case we return to (16) which can be written as

$$\mathbf{G} = \mathbf{R}_4\mathbf{R}_5\mathbf{R}_6. \tag{25}$$

Substituting $s_5 = 0$ and $c_5 = \pm 1$ into (25) we get

$$G = \begin{bmatrix} c_4 & 0 & -s_4 \\ s_4 & 0 & c_4 \\ 0 & -1 & 0 \end{bmatrix} \begin{bmatrix} \pm 1 & 0 & 0 \\ 0 & 0 & \mp 1 \\ 0 & 1 & 0 \end{bmatrix} \begin{bmatrix} c_6 & -s_6 & 0 \\ s_6 & c_6 & 0 \\ 0 & 0 & 1 \end{bmatrix} = \begin{bmatrix} \pm cos(\theta_4 \pm \theta_6) & -sin(\theta_4 \pm \theta_6) & 0 \\ \pm sin(\theta_4 \pm \theta_6) & cos(\theta_4 \pm \theta_6) & 0 \\ 0 & 0 & \pm 1 \end{bmatrix}. \tag{26}$$

There are several possibilities to solve (26). We chose the one which requires a minimum of floating-point operations

$$\theta_5 = \begin{cases} 0 & \text{if } g_{33} > 0 \\ \pi & \text{if } g_{33} < 0 \end{cases}$$

$$\theta_6 = \begin{cases} atan2(g_{21}, g_{22}) - \theta_4 & \text{if } g_{33} > 0 \\ \theta_4 + atan2(g_{21}, g_{22}) & \text{if } g_{33} < 0. \end{cases}$$

Angle θ_4 can be chosen arbitrarily.

5. Efficiency Analysis

As shown, the solution for the arm joints θ_1, θ_2 and θ_3 does not require evaluation of trigonometric functions sine and cosine. This enables fast arm positioning. The solution for wrist joints θ_4, θ_5 and θ_6 requires evaluation of sine and cosine of θ_1 and $\theta_2+\theta_3$. The minimal number of required floating-point operations for the proposed algorithm and for the original algorithm [2] are given in Table 1. Note, that constant expressions which depend on the link parameters only, as for example $atan2(a_3,d_4)$ or those indicated by squared brackets, are excluded from the operation count.

Table 1. Number of Floating-Point Operations for Inverse Kinematics

		add sub	mpy div	sqrt	sin cos	atan2	#of equivalent additions VAX780	Motorola
Ref.[2]	single	25	39	2	8	7	145	194
	closest	43	51	3	12	10	223	284
This paper	single	20	24	3	4	8	113	153
	closest	26	24	3	4	8	120	159

For comparison purposes the number of various floating-point operations is reduced to the number of equivalent floating-point additions. This has been done for two systems: Vax 11/780 with Floating Point Accelerator and for Motorola's 68020-based System 1131. The equivalent characteristics for the two machines are given in Table 2.

Table 2. Number of Equivalent Floating-Point Additions

	add sub	mpy div	sqrt	sin cos	atan2
VAX 780	1	1.1	3	4.5	5
Motorola	1	1.1	1.6	7.5	9

The results in Table 1 are given for two cases denoted by "single " and by "closest ". First case is if the global manipulator configuration, i.e. the signs of ξ, η and ζ, are known in advance. Therefore, only a single set of solutions has to be determined. If this is not the case, we have to choose one out of 8 possible solutions using some additional criteria. The closest solution refers to the criteria of "minimal joint travel" which minimizes the difference between actual joint displacements and the solution that corresponds to the given position and orientation. Since the arm joints can be considered as "large joints" they can be assigned higher weights in the optimization process, this practically reduces to two separate optimizations, one for the arm and one for the wrist. Thus we find the closest solution for the arm first. The optimal solutions for θ_1, θ_2 and θ_3 can then be used to determine the rest of the manipulator joints. This would require computation of 4 solutions for arm joints and only 2 solutions for wrist joints.

The results in Table 1. are obtained for such an approach. As seen, the proposed algorithm is particularly convenient if multiple solutions are required. This is due to the fact that the algorithm does not require the evaluation of sine and cosine for θ_1, θ_2 and θ_3 while determining their optimal values.

Finally, we show that the evaluation of sine and cosine of θ_1 and $\theta_2+\theta_3$ can be avoided completely. Applying the method from Appendix it follows directly:

$$s_1 = \xi w_1 - d_3 w_2 \qquad c_1 = \xi w_2 - d_3 w_1$$

$$s_2 = \xi\eta w_3 - p_z w_4 \qquad c_2 = \xi w_4 + \eta p_z w_3 \tag{27}$$

$$s_3 = d\, w_5 - \eta w_6 \qquad c_3 = d\, w_6 + \eta w_5$$

$$s_{23} = s_2 c_3 + s_3 c_2 \qquad c_{23} = c_2 c_3 - s_2 s_3 \tag{28}$$

where

$$w_1 = \frac{p_y}{p_x^2 + p_y^2} \qquad w_2 = \frac{p_x}{p_x^2 + p_y^2}$$

$$w_3 = \frac{[1/2a_2]}{\xi^2 + p_z^2} \qquad w_4 = w_3(\xi_2 + p_z^2 + [a_2^2 - a_3^2 - d_4^2])$$

$$w_5 = [\frac{d_4}{2a_2(a_3^2 + d_4^2)}] \quad w_6 = [\frac{a_3}{2a_2(a_3^2 + d_4^2)}].$$

If we assume that multiplications and divisions are equivalent to 1.1 additions and that sine/cosine are equivalent to σ additions, the number of equivalent additions required to compute s_1, c_1, s_{23} and c_{23} are as shown in Table 3.

Table 3. Number of Equivalent Additions for sin/cos of θ_1 and $\theta_2+\theta_3$

	Using sin/cos	Using exp (27) and (28)
Single solution	$1+4\sigma$	33.2
All solutions	$4+16\sigma$	58.4

Note that subexpressions $p_x^2+p_y^2$ and $\xi^2+p_z^2$, which have been evaluated previously in Section 2, are not counted here.

Table 3 has two entries, one for single solution and one which includes four solutions for θ_1 and $\theta_2+\theta_3$. Since different solutions are defined by different signs of ξ and η, the terms in expressions (27) are evaluated only once. The decision whether to use trigonometric functions or expressions (27) and (28) should be based on the available floating-point hardware. Table 3 shows that the trigonometric functions are superior over expressions for single solution if $\sigma < 8$ and for multiple solution if $\sigma < 3.4$.

It should be noted, that the use of expressions (27) and (28) can save two atan2 functions, since having s_i and c_i we have $\theta_i = \mathrm{atan2}(s_i, c_i)$ instead of (9), (13) and (15).

6. Conclusion

The closed form solution of inverse kinematics for the PUMA manipulator has been discussed. The algorithm uses a particular approach to the solution of the basic trigonometric equation which is treated as an orthogonal transformation in two-dimensional space. This has resulted in a faster algorithm with fewer floating-point operations and fewer transcendental function calls. It has been shown that the time efficiency is improved more than 20% for single solution and more than 40% for multiple solution, compared to the standard algorithms. The approach presented can also be applied to other types of kinematically simple manipulators.

Appendix: Solving Equation a cos(x) - b sin(x) = p

In inverse kinematics we often deal with transcendental equations of the form

$$a\ \cos(x) - b\ \sin(x) = p. \tag{A-1}$$

Standard solution methods use trigonometric substitutions $a = r\ \sin(\psi)$ and $b = r\ \cos(\psi)$, where $r^2 = a^2+b^2$ and $\tan(\psi) = a/b$.

We consider (A-1) as a part of rotation in two-dimensional space. Therefore, we accompany (A-1) by a similar equation

$$a\ sin(x)+b\ cos(x) = q \tag{A-2}$$

such that

$$q = \pm(a^2+b^2-p^2)^{\frac{1}{2}}.$$

We shall call (A-2) the companion equation of (A-1).

Clearly, (A-1) and (A-2) define a positive rotation of the vector (a,b) by the angle x giving vector (p,q). Magnitudes of both vectors are equal, i.e.

$$a^2+b^2 = p^2+q^2,$$

and the angle between them is

$$x = atan(q, p) - atan(b, a) \tag{A-3}$$

where atan(η, ξ) is the quadrant-sensitive version of atan(η/ξ). Since atan2(η, ξ) = $\pi/2$ - atan2(ξ, η), (A-3) can also be written as

$$x = atan2(a, b) - atan2(p, q).$$

Sometimes we need sine and cosine of x. They can be found directly by using sine and cosine functions or by soving (A-1) and (A-2) for cos(x) and sin(x)

$$cos(x) = \frac{ap+bq}{a^2+b^2}$$

$$sin(x) = \frac{aq-bp}{a^2+b^2}. \tag{A-4}$$

From (A-4) we can also write:

$$x = atan2(sin(x), cos(x)) = atan2(aq-bp, ap+bq).$$

Acknowledgement

The authors wish to thank Ms. Ana Marjanski who has implemented and checked the presented algorithm on the VAX 11/780 and the Motorola System 1131.

References

[1] D. Pieper, "The Kinematics of Manipulators Under Computer Control", Ph.D. Thesis, Stanford University, 1968.

[2] R.P. Paul, B Shimano and G.E. Mayer, "Kinematic Control Equations for Simple Manipulators", *IEEE Trans. on Systems, Man, and Cybernetics*, vol. SMC-11, No. 6, June 1981.

[3] C.S.G. Lee and M. Ziegler, "A Geometric Approach in Solving the Inverse Kinematics of PUMA Robots", *Proc. of the 13th International Symposium on Industrial Robots*, Chicago, Illinois, 1983.

[5] J.J Craig, "*Introduction to Robotics: Mechanics and Control*", Addison-Wesley Publishing Co., Reading, Massachusetts, 1986.

[6] M. Shahinpoor, M Jamshidi and Y.T. Kim, "Exact Solution to the Inverse Kinematics Problem of a Standard 6-Axis Robot Manipulator", *Journal of Robotic Systems*, Vol. 3, No. 3, 1986.

[7] C.S.G. Lee and P.R. Chang, "Computational Structures for Robotic Computations", *Workshop on Space Telerobotics*, Pasadena, California, January 20-22, 1987.

Inverse Kinematic Analysis for a Five Axis Robot Using Monte Carlo Method

Hong Shen and R. Radharamanan

Department of Mechanical Engineering, San Diego State University, San Diego, CA 92182, USA

ABSTRACT:
In this paper, an inverse kinematic model has been used for the computation of joint coordinates, given the Cartesian coordinates of five-axis manipulators of arbitary architecture. The model leads to a formally overdetermined nonlinear algebraic system of six equations in five unknowns. Their least-squares approximations produce the desired joint coordinates. Monte Carlo method is developed to find the initial guess for these nonlinear equations. This procedure is illustrated on a five axis robot. The algorithm developed is found to be more efficient when compared with that of continuation techniques. The method developed can also be used to analyze high nonlinear equations in kinematics. Then using an optimization algorithm the inverse kinematic solutions have been obtained.

KEY WORDS: Joint coordinates, kinematic analysis, Monte Carlo method, optimization algorithm, least-squares approximations, robot.

1. INTRODUCTION

In general terms, an inverse kinematic problem for a serial-link manipulator is to find joint variables according to the given motion to the end-effector. Closed-form solutions to the problem are known for only simple nonredundent manipulator geometries, but for the linkages with complex geometries, the inverse position/orientation problem can be solved iteratively using inverse solutions. Various methods have been developed for these problems, such as inverse transform (Paul et al [8]), screw algebra (Kohli and Soni [6]), dual matrices (Uicker et al [12]), and geometric approaches (Lee and Ziegler [7]). Frequently in industry, manipulators with five axis are so commonly being used for pick and place operations, and the computer programming of their motions requires an efficient inverse kinematic algorithm. Jorge Angeles [1] proposed a kinematic model which is applicable to manipulators of arbitrary architecture containing five joints. The numerical programming of this model involves application of optimization method. The series of points along the desired trajectory of the end effector are considered as the optimum points in each respective iteration. Once the joint coordinates have been computed at any point of this series in the Cartesian-coordinate space, the computed joint coordinates can be used as an initial guess of the iterative procedure for the next point on the trajectory. The desired solution can be determined by certain converge criteria. Thus, the only point of the Cartesian trajectory for which no close enough initial guess available is the first point on the trajectory. Jorge Angeles [2] used condition-number minimization and continuation methods to find the initial guess. His algorithm yields two other optimization procedures, which need considerable iterative calculations and also on-line parameter's adjustments. In this paper, Monte Carlo method [11] is used to detect the initial guess as the minimum of multi-variable nonlinear algebraic equation. This procedure requires less computational time than continuation methods and the results are accurate enough for engineering usage.

2. DISPLACEMENT EQUATIONS

The general notations used in this section are those of Hartenberg and Denavit [12] as shown in Fig. 1. Let $\mathbf{Q_i}$ be the pairwise rotating matrix from link i+1 to link i. Further, the vector $\mathbf{A_i}$ which connects the origin O_i and O_{i+1} is directed from the former to the latter (Fig. 1). Then $\mathbf{Q_i}$ and $\mathbf{A_i}$ are given by:

$$\mathbf{Q_i} = \begin{bmatrix} \cos\theta_{i+1} & -\sin\theta_{i+1} & 0 \\ \sin\theta_{i+1}\cos\alpha_i & \cos\theta_{i+1}\cos\alpha_i & -\sin\alpha_i \\ \sin\theta_{i+1}\sin\alpha_i & \cos\theta_{i+1}\sin\alpha_i & \cos\alpha_i \end{bmatrix} \tag{1}$$

$$\mathbf{A_i} = (a_i,\ -\sin\alpha_i d_{i+1},\ \cos\alpha_i d_{i+1})^T \tag{2}$$

According to Jorge Angeles [1], since the five-axis robot arm cannot both arbitrarily locate one point of a body and orient the body, one has to find a constraint which could be either the position or orientation of the body.

Thus, let **AB** be the line of the end effector to be positioned and oriented in space. If the distance between A and B, denoted by d, is constrained, then the problem is fully defined and thus, they must meet the rigid body condition:

$$(\mathbf{R_A} - \mathbf{R_B})^T (\mathbf{R_A} - \mathbf{R_B}) = d^2 \tag{3}$$

Now, let $\mathbf{R_i}$ and $\mathbf{S_i}$ be the values of the position vectors of points A and B, expressed in the ith joint coordinates, henceforth they are functions of vectors $\theta_1, \theta_2, ..., \theta_5$. Furthermore, let $\mathbf{R_A}$ and $\mathbf{R_B}$ be the Cartesian coordinates of the end effector, hence they are given functions of time. Vectors $\mathbf{R_A}$ and $\mathbf{R_B}$ are most efficiently computed by a recursive scheme.

$$\mathbf{R_4} = \mathbf{A_4} \tag{4}$$

$$\mathbf{R_k} = \mathbf{A_k} + \mathbf{Q_k R_{k+1}} \qquad k=3, 2, 1, 0 \tag{5}$$

$$\mathbf{S_5} = \mathbf{A_5} \tag{6}$$

$$\mathbf{S_k} = \mathbf{A_k} + \mathbf{Q_k S_{k+1}} \qquad k=4, 3, 2, 1, 0 \tag{7}$$

At the base coordinate i=0, $\mathbf{R_0}$ and $\mathbf{R_A}$ represent the same physical value where $\mathbf{R_0}$ is the function of θ and $\mathbf{R_A}$ is the function of time; in the same manner $\mathbf{S_0}$ and $\mathbf{R_B}$ are defined. Then:

$$\mathbf{R_0} = \mathbf{R_A}\ ; \qquad \mathbf{S_0} = \mathbf{R_B} \tag{8}$$

Now, the six-dimensional vector **f** is defined as

$$\mathbf{f} = [\ \mathbf{f_1}^T,\ \mathbf{f_2}^T\ ]^T \tag{9}$$

where $\mathbf{f_1}$ and $\mathbf{f_2}$ are the three-dimensional vectors

$$\mathbf{f_1} = \mathbf{R_0} - \mathbf{R_A}, \qquad \mathbf{f_2} = \mathbf{S_0} - \mathbf{R_B} \tag{10}$$

Thus, the equations (8) lead to the following nonlinear algebraic system of six equations and five unknowns:

$$\mathbf{f}(\theta) = 0 \tag{11}$$

The inverse kinematic problem reduces to finding the roots of equations (11). These represent an overdetermined nonlinear algebraic system. However, if the data $\mathbf{R_A}$ and $\mathbf{R_B}$ are given consistently, the least-squares approximation of equations will be, in fact, their solution.

The least-squares approximation to equations (11) can be found via an optimization technique aimed at finding the values of θ that minimize the objective function Z, defined as:

$$Z = \mathbf{f}^T \mathbf{f} \tag{12}$$

3. ALGORITHM FOR OPTIMIZATION

(1) If no inequality constraints are considered, one straight forward way of minimizing would be to solve the normal equations $\nabla Z = 0$, a nonlinear algebraic system of five equations and five unknowns. This is not attempted here because this method requires computing $\nabla^2 Z$, which increases the computational costs of the procedure.

(2) The Newton-Gauss method, which was used by Jorge Angeles, requires calculation of both objective function and its Jacobi matrix iterations. It is known that calculation of Jacobi matrix is time consumming. On the other hand, Jacobi matrix will not exist or cause ill-condition at singularity area within the workspace of end-effector. Since this method is classified as first-order method, it is necessary to find another more efficient first-order method, in order to reduce computation time and simplify the calculation.

(3) Variable Metric Methods are efficient and reliable methods among first-order methods[13]. The basic concept here is to create an array which approximates the inverse of the Hessian **H** matrix as the optimization progresses. Let S^q be the search direction vector for iterating from $\mathbf{X}^q$ to $\mathbf{X}^{q+1}$, then:

$$S^q = - \mathbf{H}\, \nabla \mathbf{Z}\,(\mathbf{X}^q) \tag{13}$$

and

$$\mathbf{X}^{q+1} = \mathbf{X}^q + \alpha\, \mathbf{S}^q \tag{14}$$

here α is the parameter to be adjusted to make $Z(\mathbf{X}^{q+1})$ minimum.

At the initial design point, the **H** matrix is taken as the identity matrix $\mathbf{H} = \mathbf{I}$ and at the iteration q, a new **H** matrix is defined as

$$\mathbf{H}^{q+1} = \mathbf{H}^q + \mathbf{D}^q \tag{15}$$

where **D** is a symmetric update matrix and defined as

$$\mathbf{D}^q = (1/\sigma)\, \mathbf{P}\, \mathbf{P}^T - (1/\tau)\, \mathbf{H}^q\mathbf{Y}\, (\mathbf{H}^q\mathbf{Y})^T \tag{16}$$

The change vectors **P** and **Y** are defined as:

$$\mathbf{P} = \mathbf{X}^q - \mathbf{X}^{q-1} \tag{17}$$

$$\mathbf{Y} = \nabla \mathbf{F}(\mathbf{X}^q) - \nabla \mathbf{F}(\mathbf{X}^{q-1}) \tag{18}$$

The scalars σ and τ are defined as:

$$\sigma = \mathbf{P}\,\mathbf{Y} \tag{19}$$

$$\tau = \mathbf{Y}^T\, \mathbf{H}^q\, \mathbf{Y} \tag{20}$$

4. ALGORITHM FOR INITIAL POINT SEARCH

All the optimization algorithms are based on the theorem that the iteration is at the near region so that Taylor series can be only processed in certain small regions. If initial point is chosen arbitrarily, the optimum point will not be approached. So the important thing left here is how to find an initial point which is very near to the optimum point. As Monte Carlo method is considered as a powerful method to solve multi-variable, higher order nonlinear equations, it is introduced here for the above purpose.

Given any function $f(\mathbf{X})$, let the distribution function be

$$P(\mathbf{X}) = (e^{-\beta f(\mathbf{X})}) / \left(\int_{-\infty}^{\infty} e^{-\beta f(\mathbf{X})}\, d\mathbf{X} \right) \qquad \beta \geq 0 \tag{21}$$

here β is a parameter to be adjusted while calculating. It is proven that $P(\mathbf{X})$ is an equilibrium distribution function. In the following algorithm, trial numbers are used to find the minimum of $f(\mathbf{X})$.

For given **X**, let $\Delta\mathbf{X} = a[\,2\,\mathrm{ran}(i) - 1\,]$, where $\mathrm{ran}(i)$ is a random number $0 < \mathrm{ran}(i) < 1$ which can be generated at computer as an internal function. So, by choosing $\Delta\mathbf{X}$ as it is defined above, $-a < \Delta\mathbf{X} < a$, where a is a parameter which will be ajusted to make statistics more efficiently.

Next, calculate $f(\mathbf{X})$ and $f(\mathbf{X} + \Delta\mathbf{X})$:

if $f(\mathbf{X} + \Delta\mathbf{X}) < f(\mathbf{X})$, take the new $\mathbf{X}$ as a sample point.

if $f(\mathbf{X} + \Delta\mathbf{X}) > f(\mathbf{X})$, take $\mathbf{X} + \Delta\mathbf{X}$ with probability $e^{-[f(\mathbf{X}+\Delta\mathbf{X}) - f(\mathbf{X})]}$

reject $\mathbf{X} + \Delta\mathbf{X}$ otherwise. (22)

With this algorithm, a distribution function has been established which takes the events; among these events the minimum of the function $f(\mathbf{X})$ happens at the great possibility. The following will prove this statement and also show how to use statistical values to evaluate $\mathbf{X}_{min}$ vs f_{min}.

Suppose $f(X)|_{\mathbf{x}=\mathbf{xo}} = f_{min} \leq f(\mathbf{X})$ (23)

By definition, the mean of random sample is

$$<\mathbf{X}> = \sum_N \mathbf{X_i} P(\mathbf{X_i}) = (1/M) \sum_M \mathbf{X_i} \qquad (24)$$

let $Z = \sum_N e^{-\beta f(\mathbf{X})}$

$$= e^{-\beta f_{min}} + \sum_{(\mathbf{x} \neq \mathbf{x_0})} e^{-\beta f(\mathbf{X})}$$

$$= e^{-\beta f_{min}} [1 + \sum_{(\mathbf{x} \neq \mathbf{x_0})} e^{-\beta [f(\mathbf{X}) - f_{min}]}]$$

as $\beta \to \infty$, the above expression tends to $e^{-\beta f_{min}}$ (25)

Also $<\mathbf{X}> = (1/Z) \sum_N \mathbf{X} e^{-\beta f(\mathbf{X})}$

As $\beta \to \infty$, the above expression tends to

$$[1/(e^{-\beta f_{min}})] [X_o e^{-\beta f_{min}} + \sum_{(\mathbf{x}=\mathbf{x_0})} \mathbf{X} e^{-\beta f(\mathbf{X})}]$$

$$\to (\mathbf{X_o} / e^{-\beta f_{min}}) e^{-\beta f_{min}} = \mathbf{X_o} \qquad (26)$$

In the same manner,

$< f(\mathbf{X}) > \to f_{min}$ (27)

5. EXAMPLE ON THE 5-AXIS ROBOT

A 5-axis table robot is taken as an example (Fig.2). According to the notations of Hartenberg and Denavit, the link coordinate system is in Fig. 3. Table 1 lists the link parameters of the 5-axis robot.

To make the illustration easy, the desired trajectory has been chosen in the Cartesian-coordinate space as a quarter of a circle, as show in Fig.4. The methematical expression of $\mathbf{R_A}$ and $\mathbf{R_B}$ are:

$$\mathbf{R_A} = \begin{bmatrix} a\ \cos(t) \\ a\ \sin(t) \\ h \end{bmatrix} \tag{28}$$

$$\mathbf{R_B} = \begin{bmatrix} (a+d)\ \cos(t) \\ (a+d)\ \sin(t) \\ h \end{bmatrix} \tag{29}$$

Here, $0 \leq t \leq \pi/2$, $a = 250$ mm, $h = 200$ mm, $d = 100$ mm, where t is a reference parameter. As expressed before, let $\mathbf{R_i}$ and $\mathbf{S_i}$ be the values of the position vectors of point A and B, produced by the current values of the ith joint coordinates.

$$\mathbf{R_5} = \begin{bmatrix} 0 \\ 0 \\ 0 \end{bmatrix} \tag{30}$$

$$\mathbf{S_5} = \begin{bmatrix} d\ \sin\beta \\ 0 \\ d\ \cos\beta \end{bmatrix} \tag{31}$$

Using the recursive equations (4-7), parameters from table 1, and applying the equation (8), it is obtained:

$$\mathbf{S_o} = \begin{bmatrix} L_2\cos\theta_2\cos\theta_1 + L_3\cos(\theta_2+\theta_3)\cos\theta_1 + a_5\cos\theta_5\cos(\theta_2+\theta_3+\theta_4)\cos\theta_1 + b_5\cos\theta_5\sin\theta_1 \\ L_2\cos\theta_2\sin\theta_1 + L_3\cos(\theta_2+\theta_3)\sin\theta_1 + a_5\cos\theta_5\cos(\theta_2+\theta_3+\theta_4)\sin\theta_1 - b_5\cos\theta_5\cos\theta_1 \\ -L_2\sin\theta_2 - L_3\sin(\theta_2+\theta_3) - a_5\cos\theta_5\sin(\theta_2+\theta_3+\theta_4) + b_5\sin\theta_5\sin(\theta_2+\theta_3+\theta_4) \end{bmatrix}$$

$$\begin{bmatrix} -b_5\sin\theta_5\cos(\theta_2+\theta_3+\theta_4)\cos\theta_1 - d_5\sin(\theta_2+\theta_3+\theta_4)\cos\theta_1 + a_5\sin\theta_5\sin\theta_1 \\ -b_5\sin\theta_5\cos(\theta_2+\theta_3+\theta_4)\sin\theta_1 - d_5\sin(\theta_2+\theta_3+\theta_4)\sin\theta_1 - a_5\sin\theta_5\cos\theta_1 \\ -d_5\cos(\theta_2+\theta_3+\theta_4) - L_4\cos(\theta_2+\theta_3+\theta_4) \end{bmatrix} \quad (32)$$

$$\mathbf{R_o} = \begin{bmatrix} \cos\theta_1(L_2\cos\theta_2 + L_3\cos(\theta_2+\theta_3)) - L_4\sin(\theta_2+\theta_3+\theta_4) \\ \sin\theta_1(L_2\cos\theta_2 + L_3\cos(\theta_2+\theta_3)) - L_4\sin(\theta_2+\theta_3+\theta_4) \\ -L_2\sin\theta_2 - L_3\sin(\theta_2+\theta_3) - L_4\cos(\theta_2+\theta_3+\theta_4) \end{bmatrix} \quad (33)$$

$$Z = (\mathbf{S_o} - \mathbf{S_B})^T (\mathbf{S_o} - \mathbf{S_B}) + (\mathbf{R_o} - \mathbf{R_A})^T (\mathbf{R_o} - \mathbf{R_A}) \quad (34)$$

This is the objective function with five unknowns $\theta_1, \theta_2, \ldots, \theta_5$.

6. RESULTS AND DISCUSSIONS

The robot's home position has been chosen as the initial point. The quarter circle is divided into eleven points, which represent the objective position of the end-effector during each movement. The home position is

$$\theta_1 = 0.0,\ \theta_2 = -90^o,\ \theta_3 = 90^o,\ \theta_4 = -90^o,\ \theta_5 = 0.0 \quad (35)$$

Table.2 shows Cartesian coordinates of these 11 objective point positions. By Monte Carlo method the first point reached is:

$$\theta_1 = 3.2276\ E-3,\ \theta_2 = -101.0240,\ \theta_3 = 106.7335,\ \theta_4 = -78.5447,\ \theta_5 = 2.2545\ E-3$$

By the continuation technique, to reach the first point it is necessary to develop ten intermediate points which are listed in Table 3. After reaching the first objective point, the next step is to use the optimization method. For this, an optimization algorithm has been developed and used for obtaining the inverse kinematic solutions. Table 4 shows the objective points obtained by solving the joint variables compared with those calculated in Cartesian coordinates.

7. CONCLUSIONS

From the example above, using continuation technique at least ten iterations are required to obtain the initial point and the accuracy is 1.7%, but the calculation time could be hours or longer, depending on experience and intuitive judgement of the operator.

Using Monte Carlo method, with $\beta = 1.0$, number of samples = 1,000, the accuracy is 7%, yet calculation time is just 5 seconds using VAX 11/780. The accuracy could be improved by using more samples. As a rule, if one wants to improve the results by a factor p, then the number of trials must be increased by a factor p^2. However, the results obtained are good enough for engineering purposes. Also one can improve the accuracy by a few iterations after Monte Carlo method. This is the suggestion for further work to accomplish real-time analysis when interfacing a computer with the robot.

8. REFERENCES

1. Angeles, J. "Iterative Kinematic Inversion of General Five-Axis Robot Manipulators", Int. J., Robotic Research, Vol. 4, No.4, pp59-70, Winter 1986.

2. Angeles, J., Rojas, A. A. "Manipulator Inverse Kinematics via Condition-number Minimization and Continuation", Int. J., Robotics & Automation, Nov. 1986.

3. Denavit, J. "Description and Displacement Analysis of Mechanisms Based on 2x2 Dual Matrics", Ph.D Thesis, Mechanical Engineering, Northwestern U., Evanston, Ill., 1956.

4. Frobery, C. E. "Numerical Mathematics", The Benjamin/Cummings Publishing Company, Inc., 1985.

5. Fu, K. S., Gonzalez, R. C. and Lee, C. S. G., "Robotics: Control, sensing, vision, and intelligence", McGraw-Hill Book Company, 1987.

6. Kohli, D., Soni, A. H. "Kinematic Analysis of Spatial Mechanisms via Successive Screw Displacements", J. Engr. for Industry, Trans. ASME, vol.2, series B, pp.739-747, 1975.

7. Lee, C. S. G., Ziegler, M. "A Geometric Approach in Solving the Inverse Kinematics of PUMA Robots", IEEE Ans. Aerospace and Electronic Systems, vol. AES-20, no.6, pp.695-706, 1986.

8. Paul, R. P. "Robot Manipulator: Mathematics, Programming and Control", MIT Press, Cambridge; Mass. 1984.

9. Pieper, D. L. "The Kinematics of Manipulators under Computer Control", Artificial Intelligence Project Memo No.72, Computer Science Department, Stanford University, Pelo Alto, Calif. 1968.

10. Ronald, E. W., Raymond H. M. "Probability and Statistics for Engineers and Scientists", Macmillan Pub. Co. 1978.

11. Rubinstein, R. Y. "Simulation and the Monte Carlo Method", John Wiley & Sons, 1981.

12. Uicker, J. J., Denavit, J., Hartenberg, R. S. "An Iterative Method for the Displacement Analysis of Spatial Mechanisms", Trans. ASME, J. Appl. Mech., vol.31, series E, pp.309-314, 1964.

13. Vanderpleats, G. N. "Numerical Optimization Techniques for Engineering Design", McGraw-Hill Book Company, 1984.

14. Yang, A. T., Freudensteim, R. "Application of Dual Number Quaterniam Algebra to the Analysis of Spatial Mechanisms", Trans. ASME, J. Appl. Mech., vol.31, series E, pp.152-157, 1964.

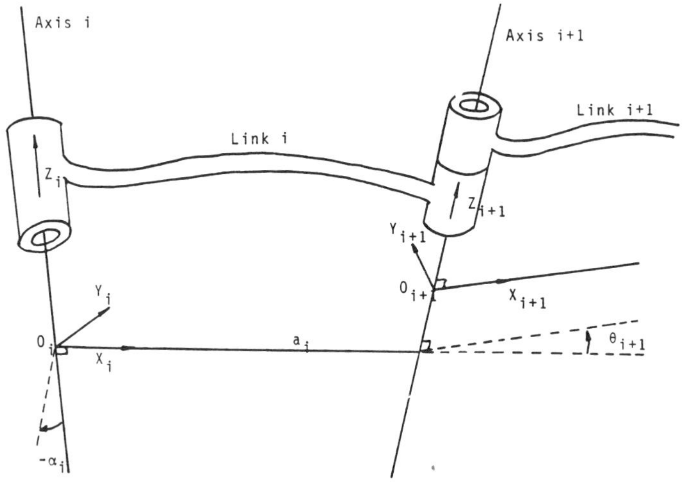

Fig. 1 Coordinates and parameters of linkages i and i+1.

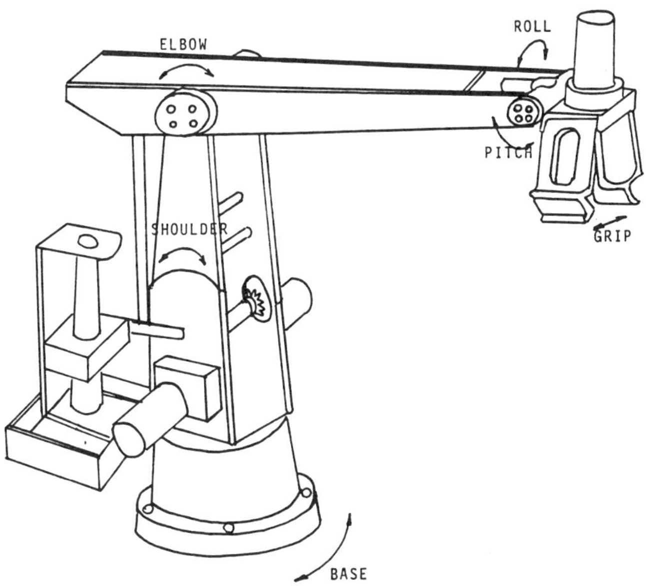

Fig. 2 A 5-axis robot.

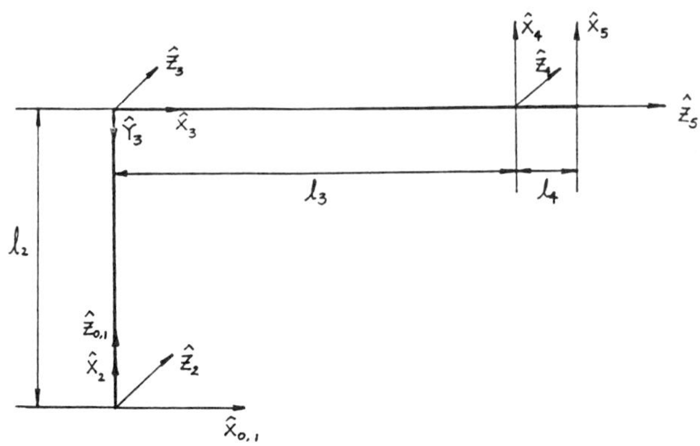

Fig. 3 Notations of 5-axis robot.

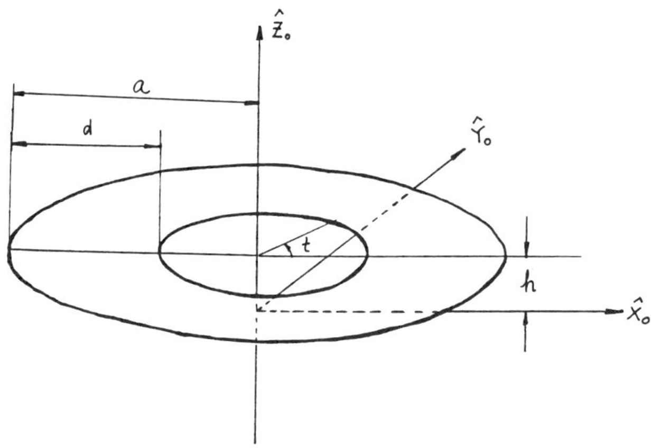

Fig. 4 Trajectory function on X_0-Y_0-Z_0 coordinate.

Table 1. Link Parameters of the 5-axis Robot

Axis	α_i	a_i	d_i	θ_i
1	0	0	0	θ_1
2	-90°	0	0	$-\theta_2$
3	0	L2	0	θ_3
4	0	L3	0	$-\theta_4$
5	-90°	0	L4	$-\theta_5$

Table 2. Coordinates of the Objective Points

	i	0	1	2	3	4	5	6	7	8	9	10
	x	250.00	246.92	237.76	222.75	202.25	176.78	146.95	113.50	77.25	39.11	0.00
R_A	y	0.00	39.11	77.25	113.50	146.95	176.78	202.25	222.75	237.76	246.92	250.00
	z	200.00	200.00	200.00	200.00	200.00	200.00	200.00	200.00	200.00	200.00	200.00
	x	350.00	345.69	332.87	311.85	283.16	247.49	205.73	158.90	108.16	54.75	0.00
R_B	y	0.00	54.75	108.16	158.90	205.73	247.49	283.16	311.85	332.87	345.69	350.00
	z	200.00	200.00	200.00	200.00	200.00	200.00	200.00	200.00	200.00	200.00	200.00

Table 3. Initial Point through Iterations

Point		Angle in °	Vector*	Desired Vector*	Gradient
	1	00.00	281.49	282.40	
	2	-91.24	000.00	000.00	
1	3	92.04	198.49	198.60	
	4	-87.62	370.73	370.34	197.71
	5	00.00	000.00	000.00	
			243.61	243.60	
	1	00.00	277.38	278.80	
	2	-92.33	000.00	000.00	
2	3	93.95	192.66	193.20	
	4	85.06	369.12	368.08	1,291.31
	5	00.00	000.00	000.00	
			232.43	233.20	
	1	00.00	274.42	275.20	
	2	-93.08	000.00	000.00	
3	3	95.70	187.32	187.80	
	4	84.68	367.10	365.82	1,557.06
	5	00.00	000.00	000.00	
			224.87	222.80	
	1	00.00	270.86	271.60	
	2	-93.95	000.00	000.00	
4	3	97.80	180.67	182.40	
	4	-84.10	364.67	363.56	1,535.66
	5	00.00	000.00	000.00	
			215.29	212.40	
	1	00.00	267.28	268.00	
	2	-94.78	000.00	000.00	
5	3	99.83	174.02	177.00	
	4	-83.41	362.19	361.30	1,366.71
	5	00.00	000.00	000.00	
			205.52	202.00	
	1	00.00	263.72	264.40	

	2	95.57	000.00	000.00	
6	3	101.79	167.48	171.60	
	4	-82.64	359.65	359.04	1,251.52
	5	00.00	000.00	000.00	
			195.69	191.60	
	1	00.00	260.16	260.80	
	2	-96.32	000.00	000.00	
7	3	103.67	161.05	166.20	
	4	-81.69	357.05	356.78	1,178.39
	5	00.00	000.00	000.00	
			185.80	181.20	
	1	00.00	256.62	257.20	
	2	-97.02	000.00	000.00	
8	3	105.48	154.71	160.80	
	4	-80.65	354.37	354.52	1,137.42
	5	00.00	000.00	000.00	
			175.83	170.80	
	1	00.00	253.14	253.60	
	2	-97.68	000.00	000.00	
9	3	107.19	148.53	155.40	
	4	-79.50	351.62	352.26	1,201.91
	5	00.00	000.00	000.00	
			165.87	160.40	
	1	00.00	248.16	250.00	
	2	-98.54	000.00	000.00	
10	3	108.73	141.79	150.00	963.73
	4	-74.33	347.90	350.00	
	5	00.00	000.00	000.00	
			149.00	150.00	

* In columns 3 and 4, the first three values correspond to vector $\mathbf{R_A}$ and the last three $\mathbf{R_B}$.

Table 4. Objective Points In Each Iteration

Point		Angle in °	Vector*	Desired Vector*	Gradient
	1	09.21	246.13	246.92	
	2	-98.38	039.89	039.11	
1	3	107.73	145.53	150.00	
	4	-73.53	344.71	345.69	926.75
	5	00.94	055.03	054.75	
			152.81	150.00	
	1	18.28	237.07	237.76	
	2	-98.30	078.29	077.25	
2	3	107.27	146.68	150.00	672.34
	4	-72.27	332.37	332.87	
	5	01.88	108.03	108.16	
			152.41	150.00	
	1	27.28	222.36	222.75	
	2	-98.13	114.67	113.50	
3	3	106.46	148.60	150.00	
	4	-69.91	312.36	311.85	265.87
	5	02.96	158.18	158.90	
			151.30	150.00	

	1	36.38	201.04	202.25	
	2	-98.24	148.12	146.95	
4	3	106.42	148.91	150.00	
	4	-69.23	283.33	283.16	318.10
	5	03.54	204.91	205.72	
			150.67	150.00	
	1	45.37	175.52	176.78	
	2	-98.20	177.80	176.78	
5	3	106.30	149.17	150.00	
	4	-68.92	248.01	247.49	166.66
	5	03.66	246.68	247.49	
			150.52	150.00	
	1	54.36	145.53	146.95	
	2	-98.22	202.99	202.25	
6	3	106.25	149.34	150.00	
	4	-68.67	206.18	205.72	69.28
	5	03.40	282.50	283.16	
			150.37	150.00	
	1	63.29	112.27	113.50	
	2	-98.21	223.08	222.75	
7	3	105.89	150.36	150.00	
	4	-67.45	159.36	158.90	255.99
	5	02.75	311.30	311.85	
			149.90	150.00	
	1	72.22	076.21	077.25	
	2	-98.22	237.74	237.76	
8	3	105.95	150.18	150.00	
	4	-67.52	108.38	108.16	302.39
	5	01.98	332.43	332.87	
			149.79	150.00	
	1	81.10	038.72	039.11	
	2	-98.09	247.13	246.92	
9	3	105.87	150.06	150.00	
	4	-67.68	055.07	054.75	157.57
	5	01.02	345.78	345.69	
			149.87	150.00	
	1	90.06	-00.24	000.00	
	2	-98.21	249.73	250.00	
10	3	105.98	150.04	150.00	
	4	-67.67	-00.21	000.00	306.70
	5	00.15	349.73	350.00	
			149.86	150.00	

* In columns 3 and 4, the first three values correspond to vector $\mathbf{R_A}$ and the last three $\mathbf{R_B}$.

Part Relative Robot Inaccuracy Compensation Using Kinematic and Stochastic Modeling

Brady R. Davies and W. Edward Red

System Automation Laboratories, Brigham Young University, Provo, UT 84602, USA

ABSTRACT

In spite of its programming limitations, most robot tasks are planned using teach pendent programming. Off-line simulation enhances robot programmability, but it has one residual flaw: most industrial robots are too inaccurate to precisely follow paths generated off-line. Robot inaccuracy results in position deviation between the simulated path and the implemented path. An efficient and practical methodology to significantly improve robot accuracy of paths planned off-line has yet to be devised.

The purpose of this paper is to describe a methodology for practical robot calibration intended to reduce robot inaccuracy. Calibrated models of robot position are determined by measuring robot poses in relation to a calibrated table or plate. Pose measurements are obtained with a vision-system camera and touch probe mounted adjacent to the robot end-effector. Kinematic and stochastic models of robot position are derived from the pose measurements. These models are incorporated into simulation software to provide inaccuracy compensation for paths planned off-line.

INTRODUCTION

Teach Pendent Programming

Current robot programming or task planning is primarily accomplished using the teach pendent method. The teach pendent method requires the programmer to manually jog the robot joints until the tool point is located and oriented (posed) at the task position desired by the programmer. Once the desired position is reached, the programmer saves the pose joint values. A series of desired positions are accumulated by jogging the robot joints to consecutive poses until the desired task has been fully described. The robot is thus "taught" a task or path by storing a series of intermediate poses.

The teach pendent approach to robot task programming has several serious limitations including:

1. Programs taught on one robot are not transportable to other robots of the same model, due to structural and manufacturing deviations between robots.
2. Complex paths requiring many taught points can become very tedious to program and inefficient in operation.
3. Paths taught in a dynamic or congested workspace can be difficult to teach and the probability of an inadvertant collision with the workspace hardware is greatly increased in such environments.

Robot Repeatability vs. Accuracy

The primary reason for the use of teach pendent path planning is related to a fundamental characteristic of most industrial robots: robots are inaccurate but repeatable. The inaccuracy of an industrial robot is demonstrated in Figure 1. An assumed (kinematic) model of the robot manipulator provides joint values for specified target positions. The parameters used in the assumed model are based on the nominal structural dimensions of the robot. However, the robot mechanism will deviate structurally from nominal due to manufacturing tolerances (other causes of structural deviation are also probable). The deviations between the structural parameters of the actual robot and the assumed robot contribute significantly to the position inaccuracy shown in Figure 1.

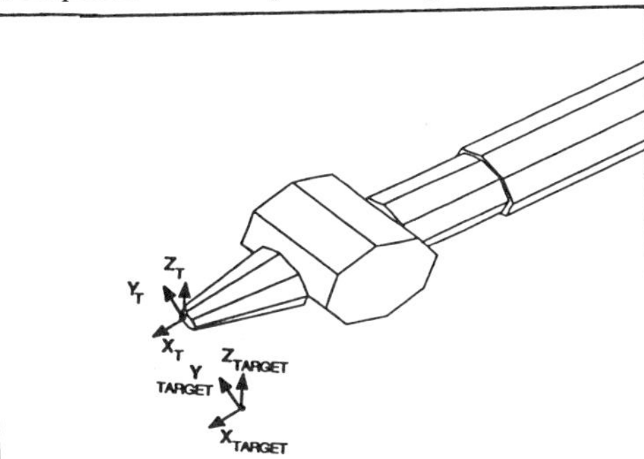

Figure 1: Robot Position Error

Although industrial robots are inaccurate, most robots are very repeatable. As demonstrated in Figure 2, the inaccuracy is not completely random but exhibits definite bias. A robot commanded to position itself at the target position of Figure 2 will pose in one of the the positions designated as achieved positions. When the robot is again commanded to position itself at the target position, it will attain a different achieved position. However, the deviation between achieved positions is significantly less than the deviation between the target position and the mean achieved position. A vector between the target position and an achieved position denotes position inaccuracy. A vector between any two of the achieved positions denotes repeatability. The repeatability of a typical industrial robot might be 1 or 2 orders of magnitude less than its inaccuracy.

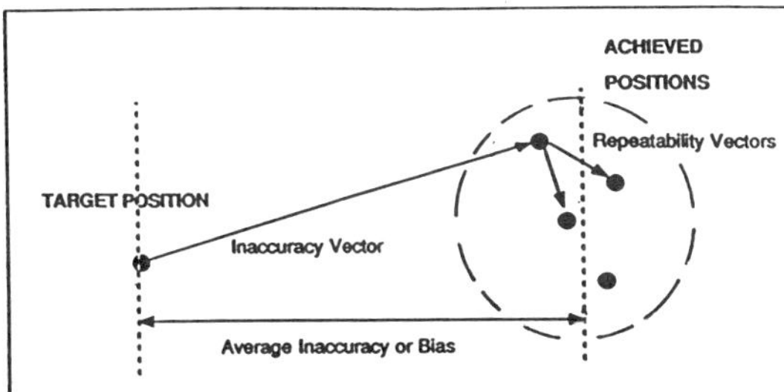

Figure 2: Robot Repeatability vs. Inaccuracy

Off-Line Programming (Simulation)

Many of the limitations of teach pendent programming are overcome if tasks are planned off-line using robot simulators such as the **SMART.BYU** simulation program developed at Brigham Young University. Using mathematical models of robot manipulators and workcell elements, simulators enable programmers to define tasks and paths precisely and efficiently. Workcell congestion and dynamics can also be simulated and incorporated into the path definition.

However, structural parameters of actual robots and workspace hardware will deviate from the mathematical models incorporated into a robot simulation package. These parameter deviations will cause the robot to deviate from the planned task or path. For many robot tasks, the path deviation will be unacceptable and require the robot task to be programmed with a teach pendent to take advantage of robot repeatability.

Part Relative Task Planning

Most robot systems are designed to relate robot position to a global reference frame, but operations performed on manufactured parts are specified relative to a part datum or part reference feature. This implies that robot poses should be referenced and robot inaccuracy characterized relative to part datums. In this research, calibration of the robot manipulator, kinematic modeling, statistical modeling, path planning, and path compensation will be defined with respect to part datums and reference features.

RESEARCH PURPOSE

The purpose of this paper is to describe a methodology for practical robot calibration intended to compensate for robot inaccuracy relative to a part datum or reference. The compensated or corrected paths will result in improved robot position accuracy and reduce path deviation between simulation planned tasks and hardware implemented tasks.

MEASUREMENT HARDWARE

Part relative robot position inaccuracy is measured with a vision camera and touch probe mounted adjacent to the robot end-effector as shown in Figure 3. In conjuction with the vision and touch sensors, a calibration plate or table containing reference features (Figure 4) serves as a template to measure robot position.

Robot position is measured by aiming the vision camera and touch probe at a target reference feature of the calibration table as shown in Figure 4. The vision camera and touch probe then measure the position and orientation of the robot relative to the target reference feature. Since the position of the reference feature is referenced to the calibration table datum, robot poses are measured with respect to the calibration table datum.

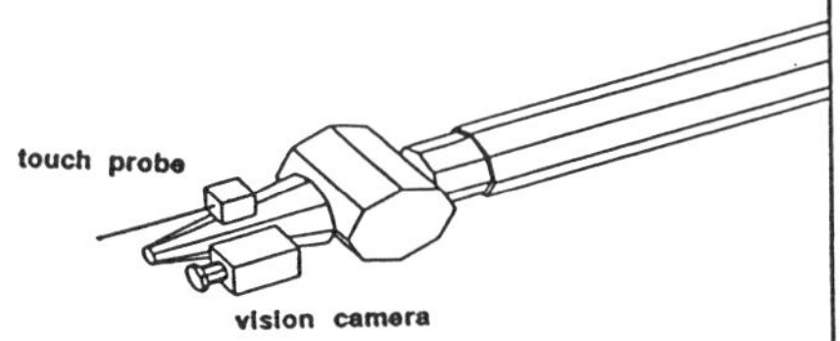

Figure 3: Vision and Touch Sensors

Figure 4: Calibration Table

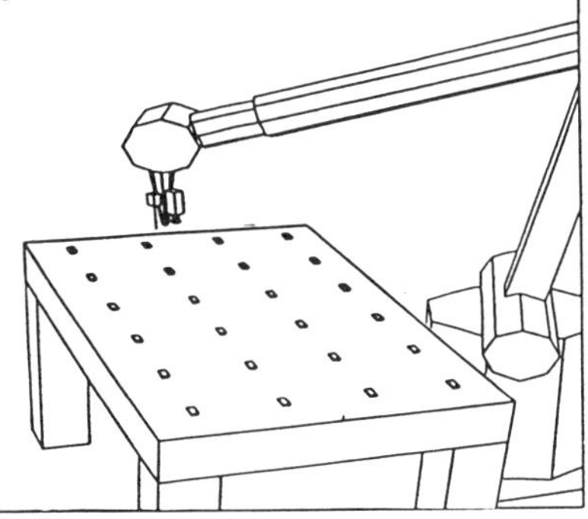

KINEMATIC MODELING

The methodology for measuring the kinematic model of the robot is a slight modification of the procedures described by Mooring and Tang [1,2]. Using these procedures , the kinematic model of a robot manipulator can be calculated from measured pose data. The calculated kinematic model is formulated by determining the location and orientation of each joint axis comprising the robot. This kinematic formulation is based on the generalized screw displacement matrix formulation described by Suh and Radcliffe [3] where a location vector **p** and an orientation vector **u** defines each joint. The methodologies for determining the actual robot kinematic model were tested in simulation [1,2].

The methodogy of Tang was readily adaptable to the method of pose measurement used in this research (described in "Measurement Hardware" section). The mathematical basis for his formulation was derived by approximating the robot forward kinematic equations as a first order Taylor series expansion. The robot forward kinematic equations were differentiated with respect to each of the generalized screw displacement parameters defining the robot joint model. The actual joint model parameters can be iteratively obtained by solving the following matrix equation for $\Delta\mathbf{x}$;

$$\mathbf{J}\,\Delta\mathbf{x} = \Delta\mathbf{T} \qquad (1)$$

where the **J** matrix contains the partial derivatives of the robot pose with respect to each robot model parameter, $\Delta\mathbf{x}$ denotes a column vector containing the differential robot model parameters, and $\Delta\mathbf{T}$ is a column vector containing the differences between the robot pose predicted by the estimated values of the robot model parameters and the measured robot pose. Once the actual joint model parameters are determined, a calibrated kinematic model is developed. This model is defined with respect to the part datum referenced during pose measurement as shown in Figure 5. The calibration plate shown in Figure 4 and described in the "Measurement Hardware" section is the reference part.

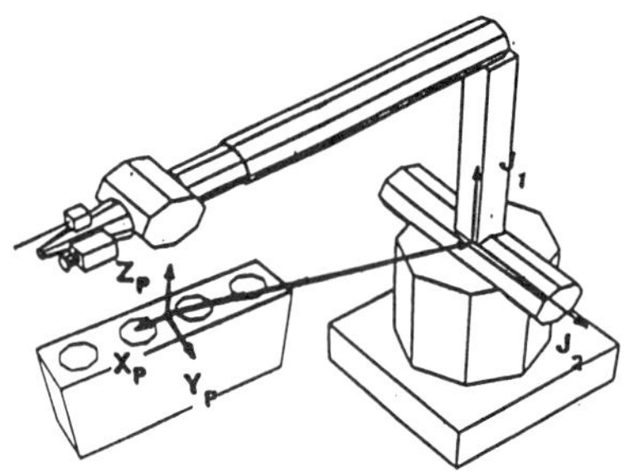

Figure 5: Part Relative Joint Frames

Kinematic Compensation

When the robot kinematic model is accurately known (calibrated model) and each robot joint value is specified, the robot pose can be accurately predicted. Inverse-kinematic models can be developed which calculate joint values necessary to accurately position the robot at specified poses. Inverse-kinematic formulations can be based on Denavit-Hartenburg [4] parameters or any valid formulation of robot kinematics. Most robot controllers utilize such an inverse-kinematic transformation and many robot models are published in the literature.

Several kinematic compensation algorithms were developed by the author. These algorithms make use of inverse-kinematic models (based on nominal robot parameters) and improved pose prediction accuracy of calibrated kinematic models. These algorithms are based on predictor-corrector and inverse-Jacobian formulations.

The simple predictor-corrector method utilizes the calibrated kinematic model of the robot to predict robot position. Robot pose joint values are estimated using a nominal inverse-kinematics model. The resulting pose is then predicted using the calibrated model. The difference between the target pose and the predicted pose is pose inaccuracy. The target position is then corrected to account for pose inaccuracy. The process is repeated iteratively until the predicted robot inaccuracy is sufficiently reduced.

The inverse-Jacobian method incorporates the inverse-Jacobian into a predictor-corrector iteration scheme by again using the calibrated kinematic model to predict robot position and the inverse-kinematic transformation to correct. The mathematical formulation is similar to the calibrated kinematic model formulation. The partial derivatives of the robot position with respect to each joint value are determined and differential joint correction is estimated by solving the following linear matrix equation for $\Delta\mathbf{q}$:

$$(\partial\mathbf{T}/\partial qi)\ \Delta\mathbf{q} = \Delta\mathbf{T} \tag{2}$$

where $\partial\mathbf{T}/\partial q_i$ denotes the partial derivative of the robot pose with respect to the ith joint value q_i, $\Delta\mathbf{q}$ denotes the differential joint value, and $\Delta\mathbf{T}$ is the difference between the robot position predicted by the calibrated joint model parameters and the target robot pose. Thus, the kinematically corrected joint values are obtained by successive iteration of equation (2).

A slightly different inverse-Jacobian formulation was also developed for kinematic compensation. This formulation is contained in Paul [5]. Both inverse-Jacobian formulations were based on the approximation:

$$(\partial\mathbf{T}/\partial q_i)_c \approx (\partial\mathbf{T}/\partial q_i)_n \tag{3}$$

where $(\partial\mathbf{T}/\partial q_i)_c$ denotes the change in robot position with respect to joint value q_i from the calibrated robot model, and $(\partial\mathbf{T}/\partial q_i)_n$ denotes the change in robot position with respect to joint value q_i from the ideal or nominal robot model.

The convergence of the predictor-corrector and inverse-Jacobian methods for a typical robot configuration is shown in Figure 6. The rms positioning error shown on the vertical axis of Figure 6 represents the root-mean-square location error of the simulated robot with successive iterations.

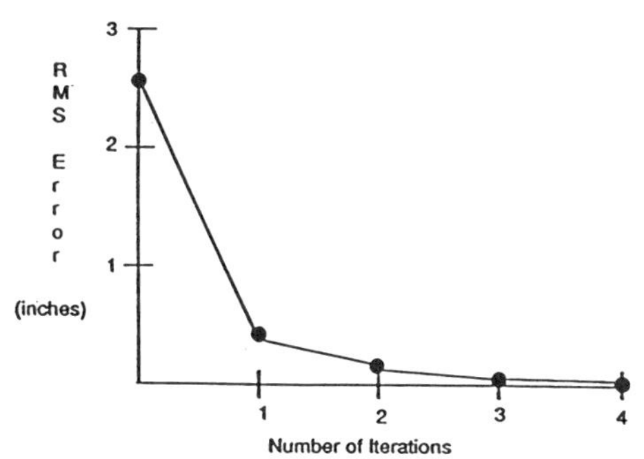

Figure 6: Position Error Convergence

STOCHASTIC MODELING

Stochastic modeling provides additional inaccuracy compensation capability to further reduce robot inaccuracy. The method of stochastic modeling used in this research is based on the Response Surface Methodology (RSM). RSM is used to plan the robot poses necessary to develop a comprehensive statistical model of robot inaccuracy (Box-Behnken or Central Composite) [6,7].

The design space applicable to a 6 degree-of-freedom robot manipulator is a 2 or 3 level, 6 factor space. The number of levels in the design space determines the order of the statistical model (2 level for a linear model and 3 level for a quadratic model). The number of factors reflects the number of robot joints. The factors represent the independent variables of the linear or quadratic model of robot inaccuracy.

Stochastic Compensation Models

Developing the statistical compensation models requires another set of part relative measurements as described in the "Measurement Hardware" section. Before the statistical calibration poses are sent to the robot controller, they are corrected using kinematic compensation (described in Kinematic Compensation subsection). The robot then moves to the corrected poses and the residual inaccuracy is measured.

The residual inaccuracy is transformed to differential joint values by again applying the kinematic compensation algorithms (predictor-corrector or inverse-Jacobian). The residual differential joint perturbations are determined from the statistical compensation equations (4):

$$\Delta q_i = a + b_{i1}q_1 + c_{i1}q_1^2 + b_{i2}q_2 + c_{i2}q_2^2 + \ldots + b_{i6}q_6 + c_{i6}q_6^2 \qquad (4)$$

where a, b, and c represent quadratic model coefficients obtained from regression analysis, q represents the joint values, and Δq_i denotes a differential joint perturbation. The first subscript of the model coefficients designates the joint compensated (i^{th}), and the second coefficient subscript designates the joint. Likewise, the joint value subscript designates the joint.

COMPENSATION PROCESS

The compensation process occurs as follows:

1. The calibration poses are planned off-line on a simulator.
2. The robot positions itself in the calibration poses required by the calibrated kinematic model, and the inaccuracy is measured.
3. The robot positions itself in the calibration poses required by the statistical model (as determined from RSM), and the inaccuracy is measured.
4. The desired task is planned off-line relative to the part datum.
5. The off-line planned task is corrected using the models developed in 2. and 3. and the compensated task information is sent to the robot controller.

(Steps 1. and 2. are repeated infrequently. Step 3. is repeated periodically. Steps 4. and 5. are repeated for each robot position planned on an off-line simulator.)

SUMMARY

This research deals with calibration of an industrial robot to reduce position inaccuracy relative to a part reference frame. A calibrated kinematic model of the robot is determined and used to accurately predict robot position. A statistical model is also developed to further reduce the residual inaccuracy. Subsequent robot positions are corrected using the kinematic and statistical compensation models.

REFERENCES

[1] Mooring, B. W., "The Effect of Joint Axis Misalignment on Robot Positioning Accuracy", Proceedings of the ASME Conference on Computers in Engineering, Chicago, Illinois, 1983.

[2] Tang, G., "Calibration and Accuracy Analysis of a Six-Axis Industrial Robot", Doctoral Dissertation, Dept. of Mechanical Engineering, Texas A & M University, December, 1986.

[3] Suh, C. and Radcliffe, C. W., Kinematics and Mechanism Design, New York: John Wiley & Sons, 1978.

[4] Denavit, J. and Hartenberg, R.S., "A Kinematic Notation for Lower-Pair Mechanisms Based on Matrices", ASME Journal of Applied Mechanics, vol. 77, pgs. 215-221, June, 1955.

[5] Paul, R., Robot Manipulators: Mathematics, Programming, and Control, Boston, Massachusetts, M.I.T. Press, 1982.

[6] Lawson, J. and J. Erjavec, "FMC Experimental Design Stategies", Process and Statistical Analysis Center Technical Department, Princeton, 1978.

[7] Box, G.E.P., W. G. Hunter, J.S. Hunter, Statistics for Experimenters, John Wiley and Sons, Inc., 1978.

Optimization of Angular Position and Location Accuracy in Robots

A. Harnoy and S.S. Sood

Department of Mechanical Engineering, New Jersey Institute of Technology, Newark, NJ 07102, USA

Abstract

The use of robots in manufacturing requires a high degree of accuracy in holding a workpiece at a certain position relative to the machine tool. A robot with an extra degree of freedom can bring a workpiece to a predetermined position in infinite ways, as measured by the rotation of its different linkages. This study shows that it is possible to optimize the movement of these linkages, so as to maximize accuracy of angular position and location of the workpiece.

The optimization method is described and examples of practical cases are analysed. It is shown that this method reduces significantly the manufacturing errors, without upgrading the accuracy of the robotic rotation mechanism.

Introduction

The object of this study is to introduce an extra degree of freedom in a robot, in order to optimize accuracy, cycle time, inertia forces, or any other criteria. To keep the capital expenditure low, designers will have to introduce flexible robots, capable of doing diversified operations; that is fewer robots doing more operations, accurately.

In this study it is shown how, by providing an extra degree of freedom with appropriate software, a robot can be made to choose its movements in such a manner, that the criteria of paramount importance for a particular operation is satisfied. It can even be made to optimize its movement, thereby satisfying to a certain extent more than one criteria. This can be compared to our body motions being controlled by the brain.

Optimization, for minimal time, using redundant degrees of freedom of a manipulator was solved for two dimensional motion lately by Kaplanov (1987). But, the ideas of optimization of time, and decreasing angular positioning errors, by providing an extra degree of freedom, have been presented by Friedman and Harnoy(1972),(1974). We shall analyze the three dimensional motion of a practical robot.

Location Analysis

Our first example is to compare a robot having three members and three axes of rotation with a robot having four members and four axes of rotation, which is an example of an extra degree of freedom.

It is required to move a workpiece held by a robot to a point x,y,z in the work space and to analyze the effect of angular errors in the rotation of different members of the arm, on the resulting location of the work piece.

Consider the robot arm shown in figs.1 and 2 having members 1,2,3,4 each of length L_1, L_2, L_3, L_4 respectively. A cartesian orthogonal system (x_0, y_0, z_0) is fixed to the sta-

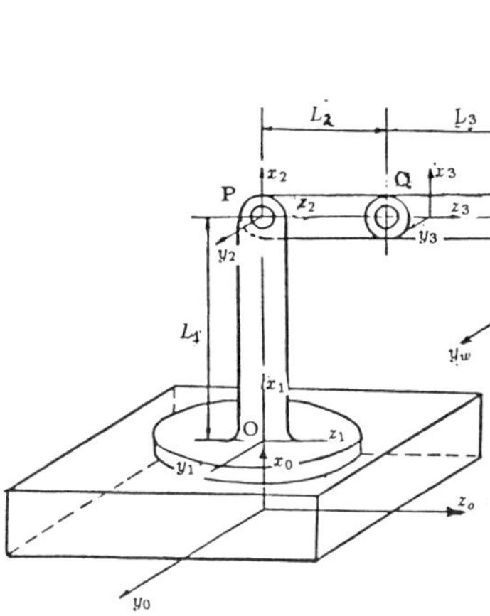

Figure 1

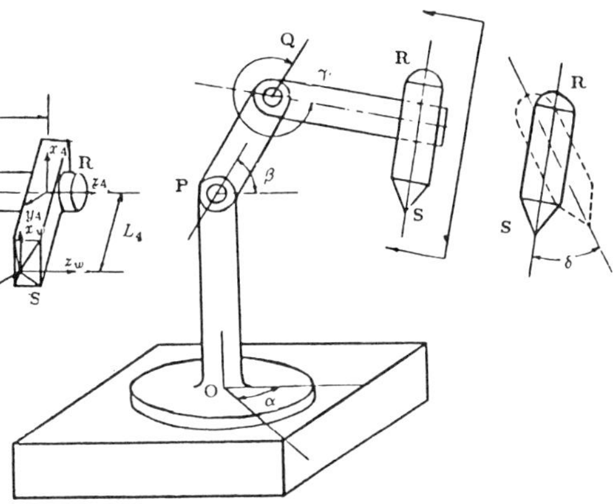

Figure 2

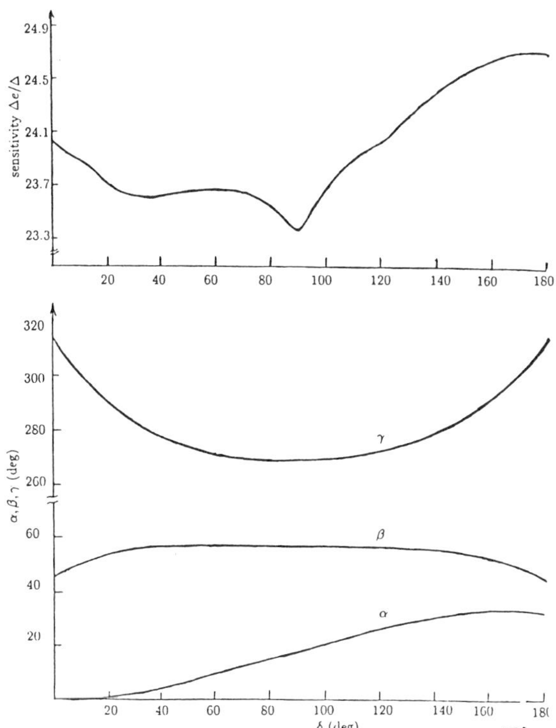

Figure 3

tionary base, while (x_1, y_1, z_1), (x_2, y_2, z_2), (x_3, y_3, z_3) and (x_4, y_4, z_4) are fixed to members 1,2,3,4 respectively. Member 1 rotates around x_0 by an angle α, member 2 around y_1 by an angle β, member 3 around y_2 by an angle γ and member 4 around z_3 by an angle δ. The initial position where $\alpha = \beta = \gamma = \delta = 0$ is shown in figure 1. The relation between the angles of rotation $(\alpha, \beta, \gamma, \delta)$ and the final location of point S (origin of (x_w, y_w, z_w)), measured in the reference system (x_0, y_0, z_0), is derived using the following rotation matrices,

$$[M_\alpha] = \begin{pmatrix} 1 & 0 & 0 \\ 0 & cos\alpha & sin\alpha \\ 0 & -sin\alpha & cos\alpha \end{pmatrix} \qquad [M_\beta] = \begin{pmatrix} cos\beta & 0 & sin\beta \\ 0 & 1 & 0 \\ -sin\beta & 0 & cos\beta \end{pmatrix}$$

$$[M_\gamma] = \begin{pmatrix} cos\gamma & 0 & sin\gamma \\ 0 & 1 & 0 \\ -sin\gamma & 0 & cos\gamma \end{pmatrix} \qquad [M_\delta] = \begin{pmatrix} cos\delta & sin\delta & 0 \\ -sin\delta & cos\delta & 0 \\ 0 & 0 & 1 \end{pmatrix}$$

$$\begin{pmatrix} x \\ y \\ z \end{pmatrix} = [M_\alpha] \begin{pmatrix} L_1 \\ 0 \\ 0 \end{pmatrix} + [M_\alpha][M_\beta] \begin{pmatrix} 0 \\ 0 \\ L_2 \end{pmatrix} + [M_\alpha][M_\beta][M_\gamma] \begin{pmatrix} 0 \\ 0 \\ L_3 \end{pmatrix} + [M_\alpha][M_\beta][M_\gamma][M_\delta] \begin{pmatrix} 0 \\ L_4 \\ 0 \end{pmatrix} \tag{1}$$

Expanding equation (1) we get,

$$x = L_1 + L_2 sin\beta + L_3 sin(\beta + \gamma) + L_4 cos(\beta + \gamma) sin\delta, \tag{2}$$
$$y = L_2 sin\alpha cos\beta + L_3 sin\alpha cos(\beta + \gamma) + L_4 (cos\alpha cos\delta - sin\alpha sin(\beta + \gamma) sin\delta, \tag{3}$$
$$z = L_2 cos\alpha cos\beta + L_3 cos\alpha cos(\beta + \gamma) + L_4 (cos\alpha cos(\beta + \gamma) sin\delta - sin\alpha cos\delta). \tag{4}$$

For the first case consider member 4 fixed rigidly to 3 and δ equal to zero. Substituting $\delta = 0$ in equations (2),(3) and (4) we have three equations and three unknowns α, β and γ. There is only one (in some cases two) solution which satisfies these equations. In other words, the robot can reach a point in its work space only by rotating its members 1,2,3 through a particular set of angles α, β, γ respectively.

Now consider that the robot is able to rotate arm 4, making an angle δ. From equations (2),(3) and (4), we have three equations and four unknowns $\alpha, \beta, \gamma, \delta$. Varying any one of the angles from 0^0 to 360^0 will result in an infinite number of solutions. Of these the robot can be made to follow one, which best satisfies the most important criteria for the operation being performed.

Optimization of Location Errors

Let $\Delta x, \Delta y, \Delta z$ be the location errors in the x_0, y_0, z_0 directions. The location errors are affected by $\Delta\alpha, \Delta\beta, \Delta\gamma$ and $\Delta\delta$ which are the errors in the angles α, β, γ and δ. The equations for small errors are ,

$$\Delta x = |\frac{\partial x}{\partial \alpha}|\Delta\alpha + |\frac{\partial x}{\partial \beta}|\Delta\beta + |\frac{\partial x}{\partial \gamma}|\Delta\gamma + |\frac{\partial x}{\partial \delta}|\Delta\delta, \tag{5}$$

$$\Delta y = |\frac{\partial y}{\partial \alpha}|\Delta\alpha + |\frac{\partial y}{\partial \beta}|\Delta\beta + |\frac{\partial y}{\partial \gamma}|\Delta\gamma + |\frac{\partial y}{\partial \delta}|\Delta\delta, \tag{6}$$

$$\Delta z = |\frac{\partial z}{\partial \alpha}|\Delta\alpha + |\frac{\partial z}{\partial \beta}|\Delta\beta + |\frac{\partial z}{\partial \gamma}|\Delta\gamma + |\frac{\partial z}{\partial \delta}|\Delta\delta. \tag{7}$$

For a common driving mechanism for the different members of the robot arm, we assume that the angular errors are equal. $\Delta\alpha = \Delta\beta = \Delta\gamma = \Delta$. Equations (5),(6) and (7) are reduced to,

$$\Delta x = \left(|\frac{\partial x}{\partial \alpha}| + |\frac{\partial x}{\partial \beta}| + |\frac{\partial x}{\partial \gamma}| + |\frac{\partial x}{\partial \delta}|\right)\Delta, \tag{8}$$

$$\Delta y = \left(|\frac{\partial y}{\partial \alpha}| + |\frac{\partial y}{\partial \beta}| + |\frac{\partial y}{\partial \gamma}| + |\frac{\partial y}{\partial \delta}|\right)\Delta, \tag{9}$$

$$\Delta z = \left(|\frac{\partial z}{\partial \alpha}| + |\frac{\partial z}{\partial \beta}| + |\frac{\partial z}{\partial \gamma}| + |\frac{\partial z}{\partial \delta}|\right)\Delta. \tag{10}$$

The following analysis can easily be extended for a robot with members having different angular errors. Let Δe be the total linear error, where

$$\Delta e = (\Delta x^2 + \Delta y^2 + \Delta z^2)^{\frac{1}{2}}. \tag{11}$$

The ratio $\Delta e/\Delta$ is the sensitivity of the error of location. The angular error Δ is determined by the driving mechanism, therefore our object is to minimize the sensitivity of the error. This can be achieved by selecting the combination α, β, γ and δ which results in the desired location and at the same time will minimize the error Δe. This optimization program should be included in the software which guides the robot.

Example 1

We demonstrate this principle by the following example. A robot as shown in fig.1 having its linkages L_1, L_2, L_3, L_4 equal to 12,6,6,3 length units respectively, was given an initial position $\alpha = 0^0, \beta = 45^0, \gamma = 315^0$ and$\delta = 0^0$. Equations (2),(3) and (4) gave the location of the work piece $x = 16.24, y = 3, z = 10.24$. For optimization, δ was varied from 0^0 to 180^0 and corresponding angles α, β, γ calculated to give the same location. The results are shown in figure 3.

For each set of values of α, β, γ and δ, the error sensitivity $\Delta e/\Delta$ was calculated, using equations (8),(9),(10) and (11). The results are shown in figure 3 and indicate that the error is reduced and has its minimum value at $\delta = 90^0$. Of course any desired location would

Figure 4

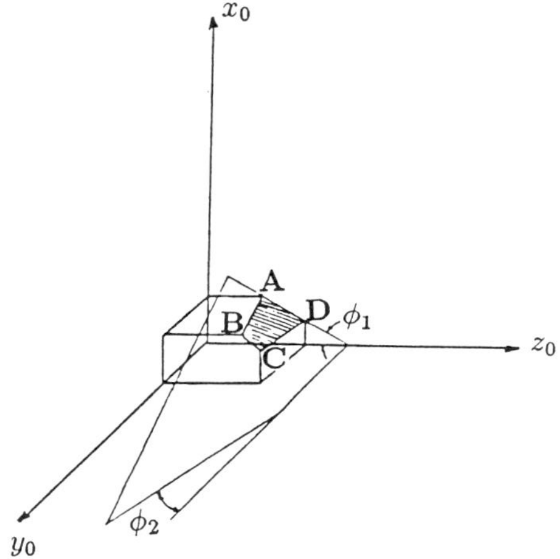

Figure 5

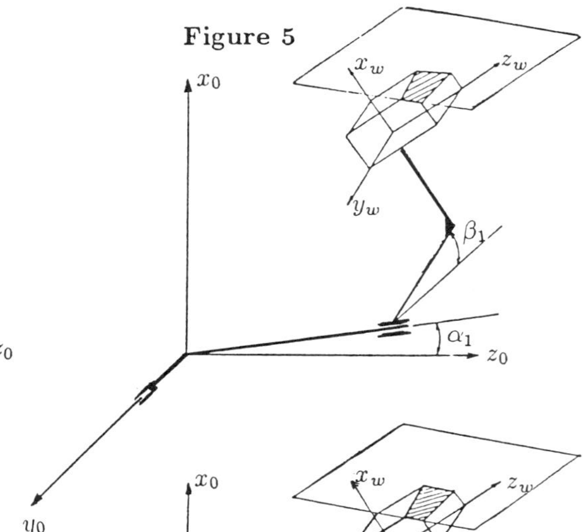

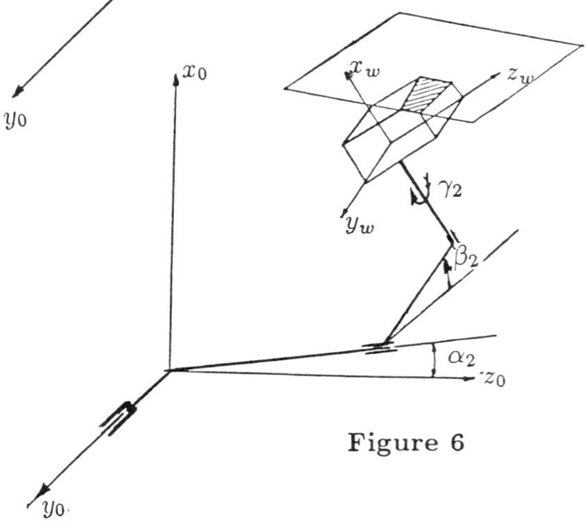

Figure 6

require different value of δ to optimize for minimum error, which is the offset tolerance from the required location of the work piece.

Angular Positioning

In the second example we show how by providing an extra degree of freedom, the robot can be made to choose the best possible positioning of its linkages, so as to give maximum angular positioning accuracy. $x_0y_0z_0$ and $x_wy_wz_w$ are cartesian orthogonal systems fixed to the machine-tool and work piece respectively. It is required to machine a plane ABCD inclined by angles ϕ_1 and ϕ_2 to the axes z_0 and y_0, respectively (refer fig.4). To achieve this, the plane ABCD must be made parallel to the plane $y_0z_0$of the machine-tool.

The machining operation can be done effectively by using a robot arm having only two degrees of freedom as shown in fig. 5. The relation between the angles of rotation α_1, β_1 of the robot and ϕ_1, ϕ_2 are,

$$tan\phi_1 = \frac{tan\alpha_1}{\cos\beta_1} \tag{12}$$

$$\phi_2 = \beta_1 \tag{13}$$

This angular positioning can be achieved in an infinite number of ways using a robot arm with an additional degree of freedom (refer fig.6). Here the relations between the angles of rotation $\alpha_2, \beta_2, \gamma_2$ and ϕ_1, ϕ_2 are,

$$tan\phi_1 = tan\beta_2 sin\gamma_2 + \frac{tan\alpha_2 cos\gamma_2}{cos\beta_2} \tag{14}$$

$$cos\phi_2 = \frac{cos\alpha_2 cos\beta_2}{\sqrt{1 - cos^2\alpha_2 sin^2\beta_2 sin^2\gamma_2 - sin^2\alpha_2 cos^2\gamma_2 - (sin2\alpha_2 sin2\gamma_2 sin\beta_2)/2}} \tag{15}$$

From equations (14) and (15) we get an infinite set of angles $\alpha_2, \beta_2, \gamma_2$ to achieve the desired location.

Let $\Delta\phi_1, \Delta\phi_2$ be the difference between the actual angular positions and ϕ_1, ϕ_2. These errors occur because of angular errors $\Delta\alpha_2, \Delta\beta_2$ and $\Delta\gamma_2$ made by the robot in angles $\alpha_2, \beta_2, \gamma_2$ respectively. Our object is to make the robot select that particular set of $\alpha_2, \beta_2, \gamma_2$, to which ϕ_1 and ϕ_2 are the least sensitive. For small angular errors we have ,

$$\Delta e_1 = \Delta\phi_1 = |\frac{\partial\phi_1}{\partial\alpha_2}|\Delta\alpha_2 + |\frac{\partial\phi_1}{\partial\beta_2}|\Delta\beta_2 + |\frac{\partial\phi_1}{\partial\gamma_2}|\Delta\gamma_2, \tag{16}$$

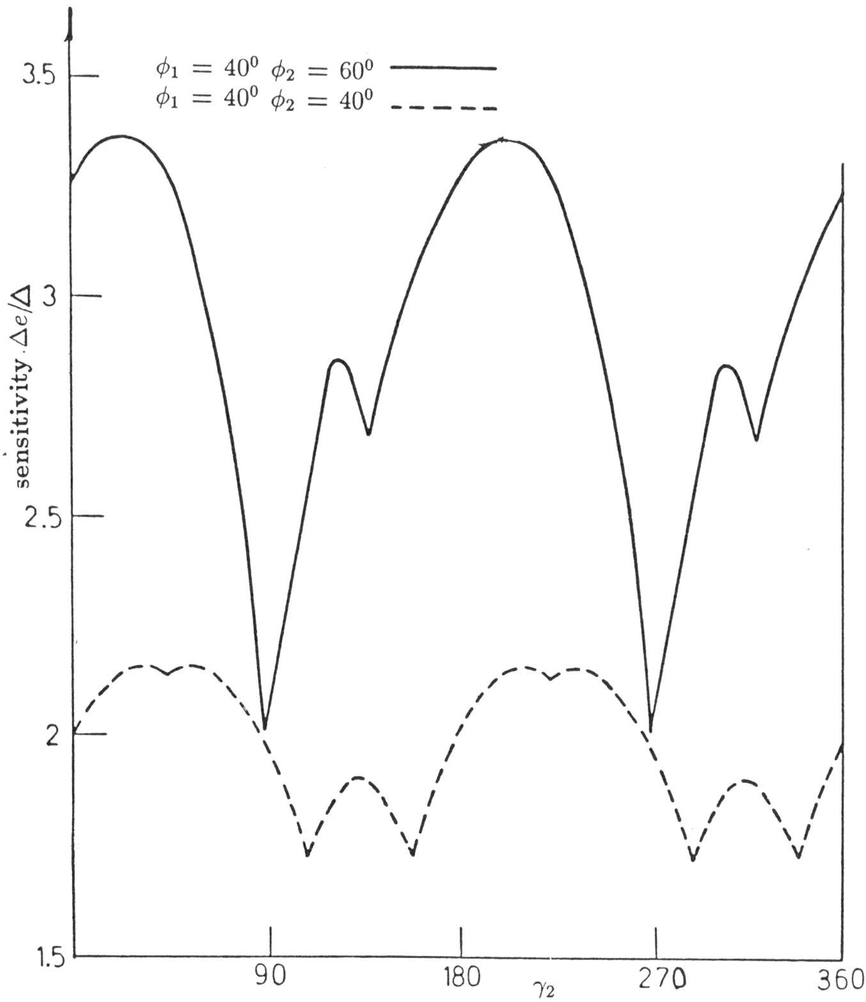

$\phi_1 = 40^0$ $\phi_2 = 60^0$
$\phi_1 = 40^0$ $\phi_2 = 40^0$
3.5
3
2.5
2
1.5
sensitivity $\Delta e/\Delta$
90
180
270
360
γ_2

Figure 7

$$\Delta e_2 = \Delta\phi_2 = |\frac{\partial\phi_2}{\partial\alpha_2}|\Delta\alpha_2 + |\frac{\partial\phi_2}{\partial\beta_2}|\Delta\beta_2 + |\frac{\partial\phi_2}{\partial\gamma_2}|\Delta\gamma_2. \tag{17}$$

Assuming equal angular errors $\Delta\alpha_2 = \Delta\beta_2 = \Delta\gamma_2 = \Delta$ equations (16) and (17) are reduced to

$$\Delta e_1 = \Delta\phi_1 = \left(|\frac{\partial\phi_1}{\partial\alpha_2}| + |\frac{\partial\phi_1}{\partial\beta_2}| + |\frac{\partial\phi_1}{\partial\gamma_2}|\right)\Delta, \tag{18}$$

$$\Delta e_2 = \Delta\phi_2 = \left(|\frac{\partial\phi_2}{\partial\alpha_2}| + |\frac{\partial\phi_2}{\partial\beta_2}| + |\frac{\partial\phi_2}{\partial\gamma_2}|\right)\Delta. \tag{19}$$

Example 2

Varying γ_2 from 0^0 to 360^0 the values for the ratio $\Delta e_1/\Delta$ and $\Delta e_2/\Delta$ are obtained. Of the two, the higher one is plotted against γ_2 for two sets of ϕ_1 and ϕ_2 (refer fig. 7).

Conclusions

The two examples demonstrate that a redundant degree of freedom can be used to reduce manufacturing errors. The same procedure can be used to optimize any other criterion like cycle time or inertia forces. If a robot is required to move along a well defined path, the extra degree of freedom can be used to minimize any deviation from the prescribed path. Any such optimization program can be included in the software which controls the robot. The best of the infinite solutions available to the robot, will be selected by the computer.

Most robots are provided with redundant degrees of freedom. If however, more than one redundant degree of freedom are provided, they can be used to optimize a few parameters simultaneously. As an example, for a robot with two redundant degrees of freedom, a program can be prepared which minimizes the positioning errors and the cycle time. Such optimization can lead to a significant improvement in the versatility and accuracy of the robots.

References

1. Friedman, M.Y., and Harnoy, A., Isr. Jou. of Tech., vol. 10, no. 4, 1972, pp 243-246.
2. Harnoy, A., and Friedman, M.Y., Isr. Jou. of Tech., vol. 13, 1975, pp. 287-291.
3. Kuplunov, A. A., Eng. Cybern. vol. 22, no.2, March, Apr. 1987, pp. 101-108.

DYNAMICS AND CONTROL

Controller Design and Dynamic Simulation of Elastic Robot Arm Mounted in Spacecraft in Presence of Uncertainty

Sahjendra N. Singh

Dept. of Computer Science and Electrical Engineering, University of Nevada, Las Vegas, Las Vegas, NV 89154, USA

ABSTRACT

We present in this paper an approach to the control of an uncertain nonlinear flexible robotic system. A robot arm (PUMA-type) with three rotational joints is considered. The third link is assumed to be elastic. A torquer control law is derived for controlling the joint angles. For damping of the elastic vibration, a force control law using modal velocity feedback is synthesized. Simulation results are presented to show that combination of the torque and force control law accomplishes reference joint angle trajectory tracking and elastic mode stabilization in spite of the uncertainty in the system.

I. INTRODUCTION

We consider the control of a flexible robot arm mounted on a space vehicle with three rotational joints (see Fig. 1). It is assumed that the first two links of the arm are rigid, but the third link is highly flexible. The elastic link is allowed to undergo small, elastic transverse bending in $x_3 - z_3$ and $y_3 - z_3$ planes. For the purpose of control, it is assumed that two forces are applied along two orthogonal directions at the tip of the beam besides the three joint torques. We are interested in designing a control system such that in the closed-loop system joint angle trajectory control can be accomplished and elastic vibrations can be stabilized.

Recently, the control of elastic robotic systems has been considered [1-9]. However, these published works on control of elastic robotic systems are based on a rather restrictive assumption that the dynamics of the system is completely known. Since model uncertainty exists, it is desirable to design controllers which do not require the complete knowledge of the dynamics of the system. Of course, considerable effort has been made in designing controllers for uncertain rigid robotic systems [10-18].

We consider in this paper the question of control system design for the robotic arm in the presence of model uncertainty using joint torquers and force controllers. A nonlinear torquer control law for the control of joint angles is derived. The derivation of the torquer control law is based on a recent result of [19] for the adaptive control of uncertain nonlinear systems. In the closed-loop system, the joint angle trajectory asymptotically converges to the reference trajectory. Such a control law, asymptotically decomposes the elastic dynamics of the robotic arm from the rigid one and this allows force controller design separately using the feedback of elastic modes for vibration stabilization. The design of the stabilizer using linear control theory is obtained. For the case of collocated actuators and sensors at the end effector, the vibration stabilizer design does not require any information on the system dynamics [20-21]. Extensive digital simulation results are presented. These results show that in the closed-loop system the joint angle error asymptotically converges to zero in spite of the model uncertainty and the spacecraft motion and the elastic vibration asymptotically decays to zero.

II. MATHEMATICAL MODEL OF THE ARM AND CONTROL PROBLEM

In this section, the mathematical modelling of the arm is briefly described (see [8,9] for details). We assume that the motion of the arm has no effect on the motion of the space vehicle. A solution of the flexible motion is assumed to be a linear combination of admissible functions $\phi_i(z_3)$ and $\psi_i(z_3)$ multiplied by time-dependent generalized coordinates $p_i(t)$ and $q_i(t)$. The flexible motion of the third link at any point $(0, 0, z_3)$ of the undeformed arm is given by

$$u_e = \left(\sum_{i=1}^{n} p_i(t)\ \phi_i(z_3),\ \sum_{i=1}^{n} q_i(t)\ \psi_i(z_3),\ z_3 \right). \tag{1}$$

The $\phi_i(z_3)$, $\psi_i(z_3)$, $i = 1, \ldots, n$ are assumed to be the eigenfunctions of a clamped-free beam.

Equations of motion are derived using the Lagrangian approach. Define $\theta = (\theta_1, \theta_2, \theta_3)^T$, $p = (p_1, \ldots, p_n)^T$, $q = (q_1, \ldots, q_n)^T$, $z = (\theta^T, p^T, q^T)^T \in R^m$, $m = 3 + 2n$, and $v = (V_{0x}, V_{0y}, V_{0z}, \omega_{0x}, \omega_{0y}, \omega_{0z})$ where V_{0i} and ω_{0i} represent the space vehicle velocity and angular velocity in the x_0, y_0, z_0 frame.

The equations of motion of the elastic arm are given by

$$D(z)\ddot{z} + (\dot{D}(z,\dot{z}))\dot{z} + \dot{R}(z,v,\dot{z},\dot{v}) - \frac{\partial K}{\partial z} + \frac{\partial P}{\partial z} = B_1\ u + B_2\ F + B_1\ T_d \tag{2}$$

where $D(z)$ is the inertia matrix; K and P are kinetic and potential energy of the system; R depends on the vehicle motion; $u = (u_1, u_2, u_3)^T$ represents the generalized torque vector; $T_d = (T_{d1}, T_{d2}, T_{d3})^T$ is joint disturbance torque vector acting at the joints; $B_1 = (I\ 0)^T$; $F = (F_1, F_2)^T$; $\phi_{ie} = \phi_i(\ell)$, $\psi_{ie} = \psi_i(\ell)$; and $B_2 \in R^{m\times 2}$ is

$$B_2 - \begin{bmatrix} 0 & \phi_{1e}\cdots\phi_{ne} & 0\cdots\cdots 0 \\ 0 & 0 \quad\quad 0 & \psi_{1e}\cdots\psi_{ne} \end{bmatrix}^T . \tag{3}$$

The matrix D is positive definite symmetric and, therefore, one obtains from (2)

$$\ddot{z} = D^{-1}\,[-\dot{D}\dot{z} - \dot{R} + \frac{\partial K}{\partial z} - \frac{\partial P}{\partial z} + B_1(u + T_d) + B_2\,F]. \tag{4}$$

We consider a reference model given by

$$\ddot{\hat{\theta}} = - S\hat{\theta} - C\dot{\hat{\theta}} + Sr, \tag{5}$$

where S and C are diagonal matrices with positive diagonal elements and $r\varepsilon\, R^3$ is an external input. We are interested in deriving a control law such that in the closed-loop system, the joint angle asymptotically tracks the reference trajectory $\hat{\theta}$ and the elastic oscillation damps out in spite of the model uncertainty. By a proper choice of $\hat{\theta}(t)$, desirable joint-angle control can be attained.

III. TORQUER CONTROL LAW AND ELASTIC MODE STABILIZATION

First, we consider the question of joint angle control using a result of [19]. Using (6) gives

$$\ddot{\theta} = h_0(\theta,\dot{\theta},p,q,\dot{p},\dot{q},F,v,\dot{v},T_d) + M_{11}\,u, \tag{6}$$

where M_1 is a submatrix of D^{-1} and

$$h_0 = M_1\,[-\dot{D}\dot{z} - \dot{R} + \frac{\partial K}{\partial z} - \frac{\partial P}{\partial z} + B_1\,T_d + B_2\,F]. \tag{7}$$

We treat M_{11} and h_0 as unknown functions.

Define $e = \theta - \hat{\theta}$, $y = (e^T, \dot{e}^T)^T$. Then one obtains

$$\dot{y} = \begin{bmatrix} 0 & I \\ -S & -C \end{bmatrix} y + \begin{bmatrix} 0 \\ I \end{bmatrix} [h_1 + M_{11}\,u] \tag{8}$$

$$\triangleq A_M y + B_M\,h,$$

where $h_1 = h_0 + S\theta + C\dot{\theta} - Sr$, and $h = h_1 + M_{11}\,u$.

Following [22], it can be shown that for some $b_i>0$, $i = 1,\ldots, 4$,

$$u^T\,(h_1 + M_{11}\,u) > \beta_0||u||^2 - ||u||\,(b_1||e|| + b_2||\dot{e}|| + b_3||\dot{e}||^2 + b_4)$$
$$= \beta_0||u||\,(||u|| - \Pi(e,\dot{e},\beta)), \tag{9}$$

where

$$\beta_0 = \inf_{z\varepsilon R^m} \lambda_{min}[M_{11}(z)],\quad \beta_i = b_i/\beta_0,$$

$$\Pi(e,\dot{e},\beta) = \beta_1||e|| + \beta_2||\dot{e}|| + \beta_3||\dot{e}||^2 + \beta_4$$

Here $\lambda_{min}(M_{11}(z))$ denotes the minimum eigenvalue of the matrix $M_{11}(z)$. Since A_M is a stable matrix, for a given Q>0, there exists a unique P>0 satisfying the Lyapunov equation

$$PA_M + A_M^T P = -Q. \tag{11}$$

Following [18-19], we choose the control law (C1) for joint angle control [22]

$$\begin{aligned}
&\dot{\hat{\beta}} = N[||e||,\ ||\dot{e}||,\ ||\dot{e}||^2,\ 1]^T\ ||\alpha(y)|| \\
&\dot{\varepsilon} = -n_5\ \varepsilon(t) \qquad\qquad (C1) \\
&\hat{\beta}(t_0)\ \varepsilon\ (0,\infty)^4,\ \varepsilon(t_0)\ \varepsilon\ (0,\infty),\ n_5>0 \\
&u(t) = -\Pi(y,\hat{\beta})\ \gamma(y,\hat{\beta},\varepsilon),
\end{aligned} \qquad (12)$$

where N is a diagonal matrix $N = \text{diag}(n_{ii})$ with $n_{ii}>0$, $i=1,\ldots,4$, $\alpha(y) = 2B_M^T\ Py$, and the function $\gamma : R^6 \times (0,\infty)^5 \to R^3$ is given by $\gamma(y,\hat{\beta},\varepsilon) = \text{sat}\ 2\Pi(y,\hat{\beta})B_M^T\ Py/\varepsilon$ where for any $\eta\ \varepsilon\ R^3$,

$$\text{sat}\ \eta = \begin{cases} \eta, & ||\eta||\leqq 1 \\ \eta/||\eta||, & ||\eta||>1 \end{cases}$$

A different control law (C2) for the joint angle control is obtained if we express Π in the following way

$$\Pi(e,e,\beta) = \beta_1||y|| + \beta_2||\dot{e}||^2 + \beta_3 \qquad (13)$$

The control law in this case is

$$\begin{aligned}
&\dot{\hat{\beta}} = N[||y||,\ ||\dot{e}||^2,\ 1]^T\ ||\alpha(y)||. \\
&\dot{\varepsilon} = -n_5\varepsilon(t) \qquad\qquad (C2) \\
&\hat{\beta}(t_0)\varepsilon(0,\infty)^3,\ \varepsilon(t_0)\ \varepsilon\ (0,\infty),\ n_5>0 \\
&u(t) = -\Pi(y,\hat{\beta})\ \ \gamma(y,\hat{\beta},\varepsilon).
\end{aligned} \qquad (14)$$

where γ, N, n_5 are appropriately defined as in (C1). Notice that the order of the dynamic compensator is four in (C2).

Using the control law (C1) or (C2) of this section, the joint angles can be controlled to track the reference trajectories $\hat{\theta}(t)$. However, as the arm is controlled, the elastic modes are excited. Thus, it is necessary to rapidly damp out the elastic vibration. To this end, the design of the force control law F will be considered for the stabilization of the elastic modes. The torquer control law asymptotically decomposes the elastic dynamics in two subsystems, representing the elastic vibration in two orthogonal planes as $\theta(t) \to \theta^*$, the desired terminal joint angle vector and $\dot{\theta}(t)$, $\ddot{\theta}(t) \to 0$. This allows the design of the force control law F_1 and F_2 for the stabilization of the elastic modes p, and q independently using low order models. In the following it will be assumed that $\dot{v} = 0$.

For simplicity, the vibration stabilization using only modal velocity feedback is considered here. For the stabilization of the elastic motion let us choose a control F_1 and F_2 of the form

$$\begin{aligned}
F_1 &= -\lambda_p(\phi_{1e},\ldots,\ \phi_{ne})\ \dot{p} \\
F_2 &= -\lambda_q(\psi_{1e},\ldots,\ \psi_{ne})\dot{q}
\end{aligned} \qquad (15)$$

Now consider the system (2) with control law (C1) or (C2) with

$\dot{v} = 0$. Then the closed-loop system is stable and $\theta(t) \to \theta^*$, $\dot{\theta}(t) \to 0$, $(p,q) \to (p^*,q^*)$, and $(\dot{p},\dot{q}) \to 0$ as $t \to \infty$, where p^*,q^* are the steady state values of p, q.

IV. SIMULATION RESULTS

In this section, the results of digital simulation using control (C1) are presented. The link parameters and the expressions for the Cartesian coordinates of the robot arm are given in [8]. A spherical payload of mass $m_4 = 80$ kg and moment of inertia of 25 kg m^2 is assumed to be attached to the end effector.

For the purpose of simulation, we set $s = \omega_n^2 I$, $C = 2\zeta\omega_n I$ in the reference model with $\zeta = 0.707$, and we take $n = 3$ in (1). The initial conditions and the parameters of the control (C1) are selected as $\hat{\beta}(0) = (10^5,\ 5x10^4,\ 4x10^4,\ 4x10^5)$, $\varepsilon(0) = 4.5$, $N = \text{diag}(500)$, and $n_5 = 0.005$. The gains of the vibration stabilizer were selected as $\lambda_p = \lambda_q$ in (17) for simplicity.

To show the capability of joint angle control and elastic mode stabilization, simulation was done using the torquer and force control law. A command trajectory was generated by setting $\hat{\theta}(0) = \hat{\dot{\theta}}(0) = 0$ $\omega_n = 0.5$, and $\theta^* = (80°,\ 60°,\ 50°)$. The modal velocity feedback gains were set to $\lambda_p = \lambda_q = 700$. The selected responses are shown in Fig. 3. Smooth θ-responses were obtained and smaller elastic deflections of the end effector were obtained in spite of a large joint angle command. This shows the stabilization capability of the simple force controller using linear modal velocity feedback.

V. CONCLUSION

A torquer control law for large angle rotational maneuver of an elastic robotic arm and an elastic stabilizer for vibration damping were derived. The derivation of the torquer controller does not require any information on the dynamics of the system and model uncertainty. The torquer controller includes a dynamic system in the feedback path and a reference model. Asymptotically decomposed subsystems representing the vibrations were used to derive linear force control laws using modal velocity feedback independently to stabilize the elastic modes relatively easily.

REFERENCES

[1] Book, W.J., Maizza-Neto, O, and Whitney, D.E., "Feedback control of two-beam, two-joint systems with distributed flexibility," J. of Dynamic Syst., Meas., and Contr., 97, 4, Dec. 1975, pp.424-431.

[2] Maizza-Neto, O., Modal analysis and control of flexible manipulator arms., Ph.D. Thesis, Department of Mechanical Eng, M.I.T., Cambridge 1973.

[3] Judd, R.P. and Falkenberg, D.R., "Control strategy for flexible robot mechanism," Proc. 15th Southeastern Symp. on Syst. Theory, Univ. of Alabama, Huntsville, March 1983.

[4] Book, W.J. and Majette, M., "Controller design for flexible, distributed parameter mechanical arm via combined state space and frequency domain techniques," J. of Dynamic Syst., Meas., and Contr., 105, pp. 245-254, Dec. 1983.

[5] Usora, P.B., Nadira, R., and Mahil, S.S., "Control of lightweight flexible manipulators: a feasibility study," Proc. 1984 American Cont. Conf., pp.1209-1216, San Diego, CA, June 1984.

[6] Chalhoub, N.G. and Ulsoy, A.G., "Dynamic simulation of a flexible robot arm and controller," Proc. 1984 American Contr. Conf., 2, pp.631-638, San Diego, CA, June 1984.

[7] Usoro, P.B., Nadira, R., Mahil, S.S., and Mehra, R.K., "Advanced control of flexible manipulators," Final Report ECS-8260419, Scientific Systems, Inc., MA, April 1983.

[8] Singh, S.N. and Schy, A.A., "Elastic robot control: nonlinear inversion and linear stabilization," IEEE Trans. on Aerospace and Electr. Syst., 22, July 1986.

[9] Singh, S.N. and Schy, A.A., "Control of elastic robotic systems by nonlinear inversion and modal damping," J. of Dynamic Syst., Meas., and Contr., Sept. 1986.

[10] Dubowsky, S. and DesForges, D.T., "The application of model referenced adaptive control of robotic manipulators," J. of Dynamic Syst., Meas., and Contr., 101, pp.193-200, Sept. 1979.

[11] Koivo, A.J. and Guo, T.H., "Adaptive linear controller for robotic manipulators," IEEE Trans. on Automat. Contr., AC-28, pp162-171, Feb. 1983.

[12] Lee, C.S.G. and Chung, M.J., "An adaptive control strategy for mechanical manipulators," IEEE Trans. on Automat. Contr., AC-29, pp.837-841, Sept. 1984.

[13] Kim, G.K. and Shin, K.G., "An adaptive model following control of industrial manipulators, IEEE Trans. on Aerospace and Electr. Syst. AES-19, pp.805-814, Nov. 1983.

[14] Samson, C., "Robust nonlinear control of robotic manipulators," Proc. IEEE on Decision and Control, pp.1211-1216, 1983.

[15] Spong, M.W., Thorp, J.S., and Kleinwaks, J.W., "The control of robot manipulators with bounded control, part II: robustness and disturbance rejection," Proc. International Conf. on Robotics and Factories of the Future, Charlotte, NC, Dec 1984.

[16] Nicosia, S. and Tomei, P., "Model reference adaptive control algorithms for industrial robots," Automatica, pp.635-644, 1984.

[17] Sastry, S.S. and Slotine, J.J., "Tracking control of nonlinear systems using sliding surfaces and application to robot manipulators," International J. on Contr., 38,2, pp465-492, 1983.

[18] Singh, S.N., "Adaptive model following control of nonlinear robotic systems," IEEE Trans. on Automa. Contr., 30, pp.1099-1100, Nov 1985.

[19] Corless, M. and Leitmann, G., "Adaptive control of systems containing uncertain functions and unknown functions with uncertain bounds," J. of Optimization and Theory Applications, 41, pp155-168, Sept 1983.

[20] Benhabib, R.J., Iwens, R.P., and Jackson, R.L., "Stability of large space structure control system using positivity concepts," J. of Guid. Contr., 4, pp487-494., 1981.

[21] Arbel, A. and Gupta, N.K., "Robust collocated control of large flexible space structure," J. of Guidance Contr., 4, pp480-486, 1981.

[22] Singh, S.N., "Nonlinear adaptive control of an elastic robotic arm," AIAA Guid., Nav. and Control Conf., Williamsburg, VA, Aug 1987.

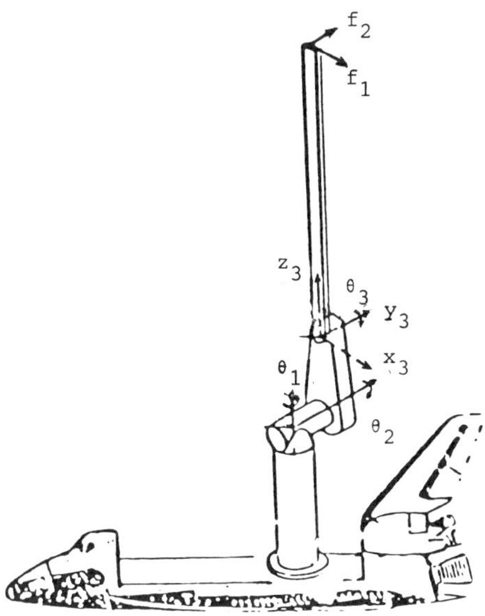

Fig. 1 Elastic Robotic Arm Model

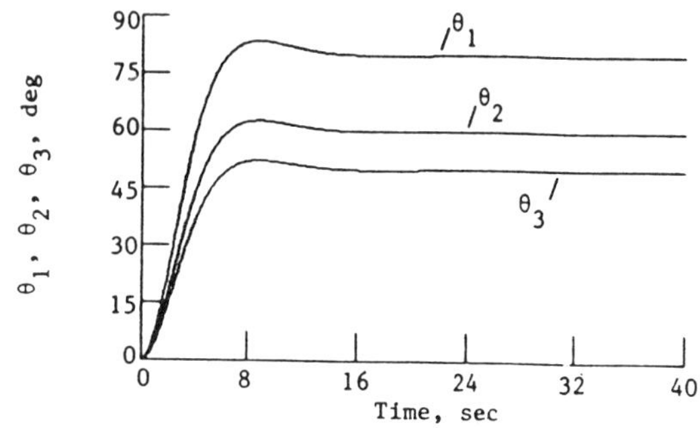

Fig. 2(a) Joint Angles

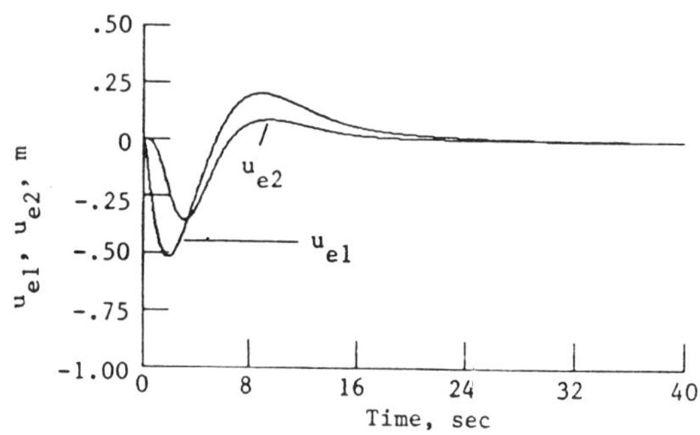

Fig 2(b) Elastic Deflection of the End Effector

Fig. 2 Joint Angle Control and Elastic Mode Stabilization

An Adaptive Control Algorithm for Robot Machine and Process Control

David Paul Stoten

Department of Mechanical Engineering, University of Bristol, Queen's Building, University Walk, Bristol, England

ABSTRACT

This paper describes a Model Reference Adaptive Control (MRAC) algorithm that has been successfully applied to robot, machine and process control problems.

The general dynamical principles of the algorithm are presented first of all, both in continuous and discrete-time. In particular, it is emphasized that the computational and memory requirements are very small compared with self-tuning controllers (for example).

Secondly, it is shown how the algorithm can be applied to the control of robots with prismatic, revolute and a combination of such joints. Simulation studies show that the complex problems of non-linear dynamics, link coupling, parameter variations and external disturbances are overcome by the MRAC algorithm in a relatively straightforward manner.

Finally, it is shown by the algorithm can be applied to more general machine and process control problems. Implementation studies are included to demonstrate the ease of application, and the excellent control that can be achieved.

1. INTRODUCTION

Conventional linear control policies assume that the plant to be controlled is modelled 'reasonably' well, that plant parameter changes are 'reasonably' small and that external disturbances to the plant are also 'reasonably' small. Quite often, some, or all, of these assumptions are violated in practice and the performance of a linear controller is thus degraded from the design specifications.

In such circumstances, a traditional remedy in process industries, for example, has been to retune the controller manually. Alternatively, one can make use of look-up tables (or gain schedules) for controller retuning under varying environmental conditions. The former method suffers from the subjective element of the process operator, whilst the latter only offers open-loop scheduling of gains.

What is really required is an automatic gain scheduler, or retuner, that synthesises the 'optimal' controller in the face of such environmental changes, based on feedback information from the plant response. Work has been conducted in this area for many years now[1,2], but only recently have worthwhile, potentially implementable, schemes emerged. The *self-tuning regulator* approach [3] to the problem is one favoured by some workers. The basic idea is to automatically *identify* the dynamics of the plant and use this changing model to synthesise optimal control strategies. The method can suffer from convergence problems and certainly requires large computational power to implement it in real-time. On the other hand, *Model Reference Adaptive Control* (MRAC)[4,5] requires little in the way of storage and computational power and is the method advocated in this paper.

MRAC strategies, especially in discrete-time, have been the issue of current debate in control engineering circles[6,7]. However, it is emerging that the *continuous-time* MRAC system described in the sequel is well-founded from a theoretical point-of-view. Hence, this continuous-time strategy is 'discretised' to obtain an algorithm that can be implemented with confidence on a digital machine. Simulation and implementation studies of the MRAC algorithm, contained here and elsewhere, also support the view that the strategy is an extremely powerful method of control.

2. THE MRAC ALGORITHM [4,8-10]

The fundamental strategy behind MRAC is to implement a computer-based *reference model* which defines the ideal closed-loop response of the plant. Should an error develop between the reference model and plant responses, an *adaptive mechanism* changes the controller gains in order to drive this error to zero.

We start by taking the standard state-space representation of a multivariable plant:

$$\Sigma_p : \dot{x}_p(t) = A_p x_p(t) + B_p u_p(t) + d_p(t)$$

where x_p, u_p and d_p are the state, control and unknown disturbance (*i.e.* external disturbances, plant non-linearities and plant parameter changes) vectors, respectively. In addition, the reference model is the *exact* linear representation:

$$\Sigma_m : \dot{x}_m(t) = A_m x_m(t) + B_m r_m(t)$$

where x_m ($\dim x_m = \dim x_p$) and r_m are the model state and reference vectors, respectively. The purpose of the model is to define an ideal plant state response (x_m), following a change in the reference signal (r_m). Then, the control problem to be solved is to drive the *actual* plant state (x_p) as close to x_m as possible: this situation is illustrated in Fig.1. for a simple first-order case.

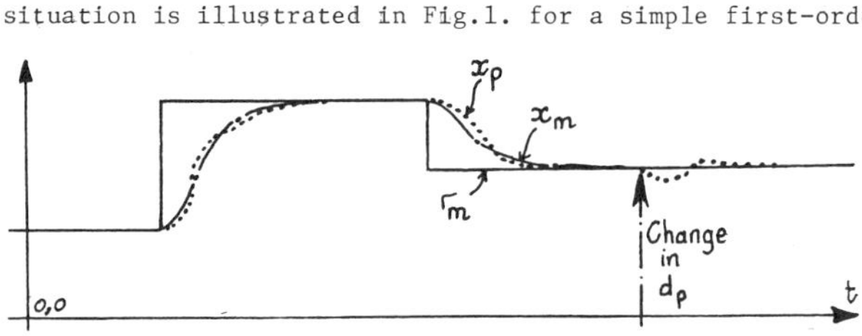

Fig 1.
Ideal Plant Response

2.1. *The Continuous-Time MRAC Algorithm*

The synthesis of the MRAC algorithm is most conveniently split into two parts: the Linear Model Reference Control (LMRC) problem, and the adaptive loop. Thus, the LMRC problem assumes that $d_p = 0$, and on introducing the tracking control equation

$$u(t) = -K_p x_p(t) + K_{pm} r_m(t)$$

one can show that the error dynamics are

$$\Sigma_e \;:\; \dot{x}_e(t) = A_m x_e(t); \quad x_e(t) \triangleq x_m(t) - x_p(t)$$

if $\quad K_p = B_p^{\dagger}(A_p - A_m) \quad$ and $\quad K_{pm} = B_p^{\dagger} B_m$

Here, $B_p^{\dagger}$ is the *left-Penrose inverse* of B_p, and the closed-form solutions for the controller gains K_p and K_{pm} thus solve the LMRC problem, since Σ_e describes a stable homogeneous system. On substituting the solutions for K_p and K_{pm} back into the control and plant equations, one can show that two subsidiary conditions (Erzberger's Conditions[11]) must also be satisfied, *viz:*

$$(I - B_p B_p^{\dagger})(A_p - A_m) = 0 \quad \text{and} \quad (I - B_p B_p^{\dagger}) B_m = 0$$

The LMRC block diagram is shown in Fig.2.

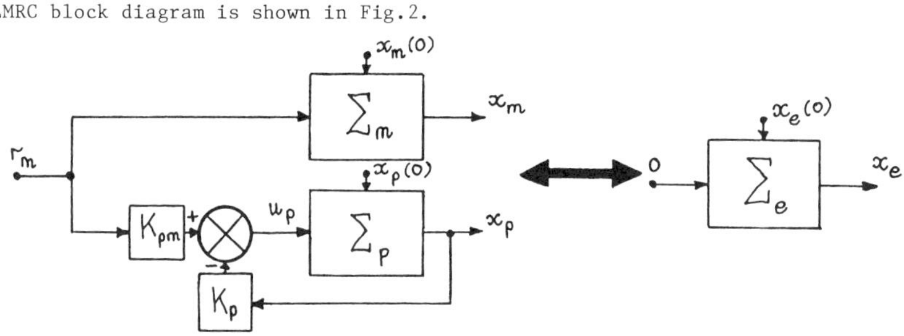

Fig.2. Alternative Visualisation of the LMRC Strategy

However, the disturbance vector $d_p \neq 0$ in general, and the second component of the MRAC strategy - the adaptive loop - must be introduced on top of the LMRC policy, in order to negate the effects of these disturbances. The tracking control equation is modified to

$$u(t) = -[K_p - \delta K_p(t)]\, x_p(t) + [K_{pm} + \delta K_{pm}(t)]\, r_m(t)$$

where δK_p and δK_{pm} are adaptive changes to the LMRC gains (*note:* δ does not imply 'small'). In these circumstances, one can obtain the desirable error dynamics Σ_e if the adaptive changes can be made to cancel the unknown (and unmeasurable) vector d_p. The solution (which is not unique) is based on the concept of *hyperstability* of non-linear feedback systems[12], and the corresponding solution of the *Popov inequality*[4]. The solution, in a simplified (and implementable) form is summarised below:

$$\delta K_p(t) = \int_o^t \phi_1(\tau)\, d\tau + \phi_2(t); \quad \delta K_{pm}(t) = \int_o^t \psi_1(\tau)d\tau + \psi_2(t)$$

where $\phi_1(\tau) = \alpha y_e(\tau)x_p^T(\tau); \quad \phi_2(t) = \beta y_e(t)\, x_p^T(t)$

$$\psi_1(\tau) = \alpha y_e(\tau)r_m^T(\tau); \quad \psi_2(t) = \beta y_e(t)r_m^T(t)$$

$$y_e(t) = C_{pm}x_e(t); \quad C_{pm} = B_p^T P;$$

$$PA_m + A_m^T P = -Q; \quad Q > 0.$$

The latter Lyapunov equation can be solved, for example, by the method of Jameson[13]. The terms α and β are scalar constants which influence the speed of adaption: typically $\beta = 10\alpha$, $\alpha > 0$. Thus, the complete MRAC strategy is as shown in Fig.3.

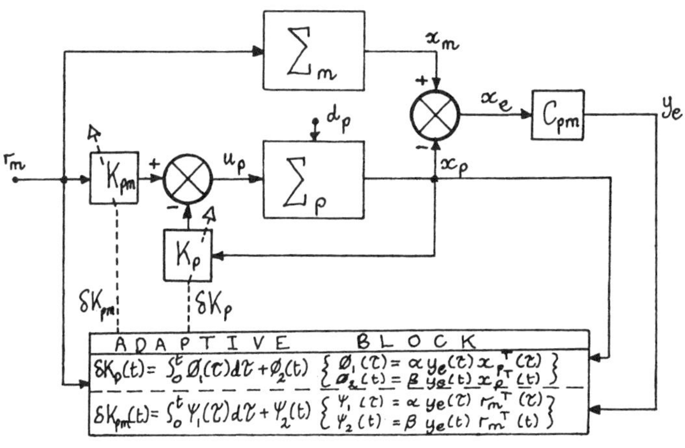

Fig.3. The MRAC Strategy

2.2. *The Discrete-Equivalent MRAC Algorithm*

It should be recognised that the above strategy, although in continuous-time, will most likely be implemented in discrete-time (using hardwired digital electronics or a microcomputer, say). Thus, a set of discrete-time equations equivalent to the above are required. The derivation of such a set, directly in the discrete-time domain, is fraught with controversy at present[6,7], and we take a pragmatic approach which works well in practice. This is based on the *Zero-Order-Hold* filter matching method[14], which yields the required equations in discrete time:

$$u_p(k) = -[K_p - \delta K_p(k)]x_p(k) + [K_{pm} + \delta K_{pm}(k)]\ r_m(k)$$

$$\delta K_p(k) = \delta K_p(k-1) + \beta y_e(k)\ x_p^T(k) - \gamma y_e(k-1)\ x_p^T(k-1)$$

$$\delta K_{pm}(k) = \delta K_{pm}(k-1) + \beta y_e(k) r_m^T(k) - \gamma y_e(k-1) r_m^T(k-1)$$

where $\gamma = \beta - \alpha\Delta$, $\delta K_p(-1) = 0$, $\delta K_{pm}(-1) = 0$, and the sampling interval, Δ, is chosen to avoid aliasing problems.

These equations form the core of discrete-time MRAC implementations, and, as one can see, they require little in the way of computational effort and storage. This is in marked contrast to alternative methods of adaption such as self-tuning control, plant model inversion techniques and look-up tables.

2.3. *The Decoupled MRAC Algorithm*[10]

This simple idea, when applied in practice, generates decoupled closed-loop system responses. The intention is that a highly coupled open-loop multivariable plant (such as a robot, an aircraft, a process plant, *etc.*) is subject to MRAC so that each degree-of-freedom - in the closed-loop configuration - behaves in a well-defined and autonomous way. The strategy, that is only briefly outlined here, generates decoupled control equations which can be implemented by separate processors (or by a concurrent operating system) - with obvious advantages in terms of parallel processing.

The plant consists of m degrees-of-freedom, each described by the state equation

$\dot{x}_{pi}(t) = A_{pi}x_{pi}(t) + B_{pi}\ u_{pi}(t) + d_{pi}(t)$; $i = 1, ---, m$, where d_{pi} contains any coupling components, as well as disturbances of the type previously described. If one chooses the reference model to be decoupled, *viz*:

$\dot{x}_{mi}(t) = A_{mi}\ x_{mi}(t) + B_{mi}\ r_{mi}(t)$; $i = 1, ---, m$, then it turns out that the LMRC and adaptive components of the control equation are also decoupled - both in continuous and discrete-time. Moreover, the *synthesis* of the MRAC strategy is decoupled, too.

3. MRAC OF ROBOTS[8-10]

Fortunately, one can model the vast majority of robots with prismatic and/or revolute joints by the state-space equations given in Section 2.3. Thus, on defining a decoupled reference model for a given robot one can use the resulting decoupled MRAC equations to yield very robust decoupled control for each link of the machine. Details of such strategies are given by Stoten[8-10] and some simulation results of a highly non-linear and coupled class 4 robot are shown in Fig.4.

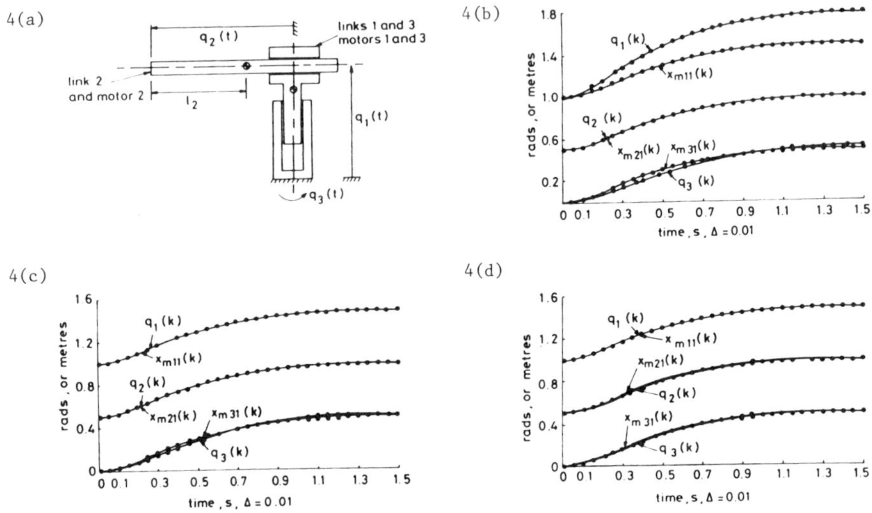

Fig.4. Class 4 Robot: MRAC

4(a) *Schematic* 4(b) *LMRC only* 4(c) *MRAC* 4(d) *MRAC with 50% increase in mass of link 2.*

4. MRAC OF OTHER MACHINES AND PROCESSES

The ability to model machines and processes by the (non-linear) state-space equation shown in Section 2 is well-known. Thus, following the largely qualitative arguments presented in this brief paper, the reader can use his/her own imagination in seeing where the corresponding MRAC strategies can be implemented.

The author wishes to stress the importance of the word *implementation*, since this was the prime motivation behind the work described:- *implementation* of better control systems. Included in the list of the author's implementations of the MRAC strategy are the control of a robot, a tyre-tread extrusion line, a fatigue testing machine, a grinding machine incorporating force feedback and various test/student rigs. In each case, the robustness of the MRAC strategy over a more conventional control policy (*e.g.* PID control) was marked.

We conclude this section with some details of a laboratory-scale implementation of MRAC. The multivariable plant is shown in Fig.5, consisting of two coupled reservoirs, where the problem is to achieve independent control of the liquid levels in each reservoir.

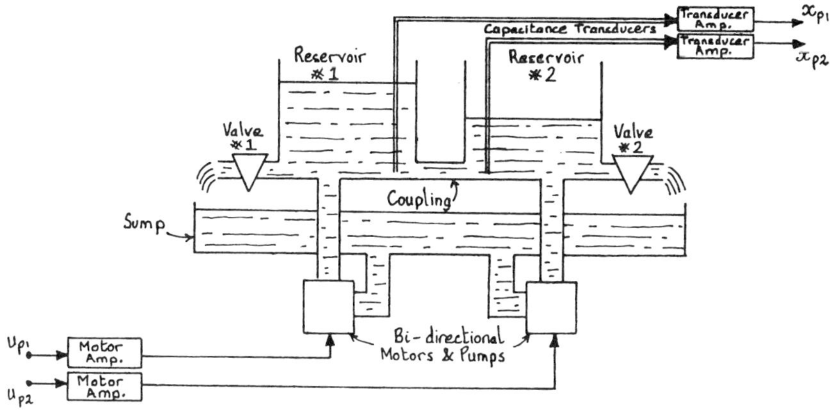

Fig.5. The Multivariable Reservoir

This two degree of freedom system is highly coupled, non-linear and subject to parameter disturbances (mainly the valve settings). A linearised two-state model of the system is given by the standard state-space equation, with

$$x_p = \begin{bmatrix} x_{p_1} \\ \hline x_{p_2} \end{bmatrix} = \begin{bmatrix} \text{measured level in reservoir \# 1} \\ \hline \text{measured level in reservoir \# 2} \end{bmatrix}; \quad A_p = \begin{bmatrix} -0.02265 & 0.01735 \\ 0.01735 & -0.02265 \end{bmatrix};$$

$B_p = 0.0167\ I_2$ and d_p is unknown. The reference model is chosen so that the closed-loop characteristics are critically damped, with a settling time $\simeq$ 100s, so that Erzberger's conditions are satisfied, and so that there is unity steady-state gain between r_m and x_m. Thus, $A_m = -0.04I_2$, $B_m = 0.04I_2$ and hence $K_p = 1.039 \begin{bmatrix} 1 & 1 \\ 1 & 1 \end{bmatrix}$, $K_{pm} = 2.395I_2$. In the synthesis of the adaptive loop, $Q = I_2$, so that $C_{pm} = 0.2088I_2$, and $\alpha = 0.02$, $\beta = 0.2$.

The controller was implemented in discrete time on a BBC-B microcomputer programmed in BASIC, operating with a sampling interval $\Delta = 1.0$s: see Fig.6.

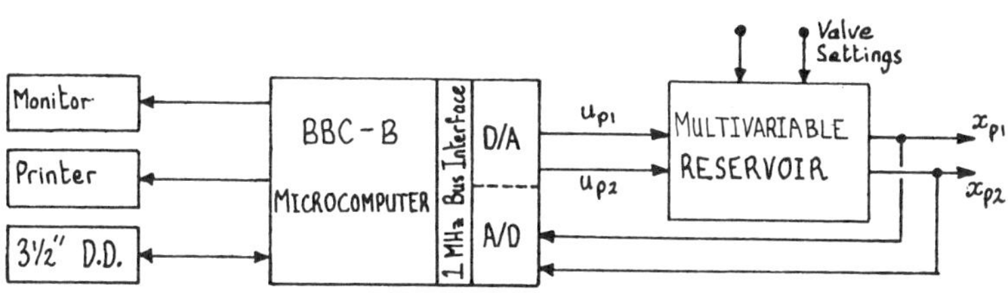

Fig.6. Closed-loop Configuration

Typical recorded results, following full closure of both valves, are shown in Fig.7. These results clearly show the extremely good control that is achieved with MRAC, compared with a conventional linear control system.

7(a)

7(b)

7(c)

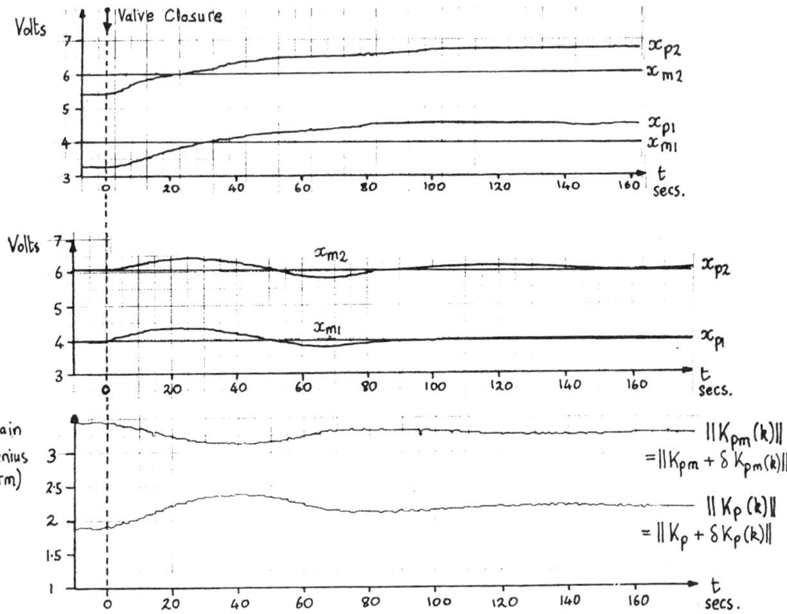

Fig.7. Results

7(a) *Linear State Feedback Tracker*; 7(b) *MRAC*

7(c) *Frobenius Norms of $K_p(k)$, $K_{pm}(k)$ under MRAC*

5. CONCLUSIONS

A Model Reference Adaptive Control (MRAC) algorithm has been described, particularly from the point-of-view of digital implementations. The algorithm is applicable to dynamic systems which can be modelled in state-space form, including systems such as robots, processes and manufacturing machines. The MRAC strategy is particularly useful when unknown, and unmeasurable, dynamical effects would normally cause a degradation in performance and uncoordinated response. In such cases, implementation studies have shown that MRAC is far superior to linear control policies.

In addition, the MRAC strategy is particularly economical in terms of computing power and storage requirements: the implemented equations are simple recursions (similar to digital P+I equations) which only require 'last-step' memory. In terms of the controller *synthesis*, the algorithm is not computationally burdensome, especially in its decoupled guise.

Finally, one recognises that in a manufacturing environment, new technologies must fit into existing or developing frameworks. The MRAC algorithm is of such a compact nature that it readily integrates with supervisory control and PLC (*sic*) systems, programmed in the form of a high-language macro. Indeed, the final form of the equations constitute little more than the equation of a PID controller.

7. REFERENCES

1. Hang, C C and Parks, P C. 'Comparative Studies of Model Reference Adaptive Control Systems'. *IEEE Trans Auto Cont* 18, No. 5, pp. 419-428 (1973).

2. Pierson, B L (ed). 'Special Issue on Self-Tuning Control'. *Opt Cont App and Meth.*, 3, No 4 (1982).

3. Wittenmark, B and Åström, K O. 'Practical Issues in the Implementation of Self-Tuning Control'. *Automatica* 20, No 5 (1984).

4. Landau, I D. 'Adaptive Control - The Model Reference Approach'. Dekker, NY (1979).

5. Donalson, D D and Leondes, C T., 'A Model Referenced Parameter Tracking Technique for Adaptive Control Systems'. *Trans IEEE on App and Industry*, 82, pp 241-262 (1963).

6. Chen, Z J and Cook, P A. 'Robustness of Model Reference Adaptive Control Systems with Unmodelled Dynamics'. *Int Journ Cont*, 39, No 1, pp 201-214 (1984).

7. Bundell, G A. 'Robust Decentralised Adaptive Control in the Manipulator Control Problem'. PhD Thesis, Cambridge University (England) (1985).

8. Stoten, D P. 'The Adaptive Control of Manipulator Arms'. *Mech and Mach Theory*, 18, No 4, pp 283-288 (1983).

9. Stoten, D P. 'Discrete Adaptive Control of a Manipulator Arm'. *Opt Cont App and Meth*, 3, No 4,pp 423-433 (1982).

10. Stoten, D P., 'A Unified Approach to the Decoupled Discrete Adaptive Control of Manipulators'. *IEE Proc.*, 132, Pt D, No 4, pp 151-157 (1985).

11. Winsor, C A and Roy, R J. 'The Application of Specific Optimal Control to the Design of Desensitised Model Following Control Systems'. *IEEE Trans Auto Cont.*, 15, No 3, pp 326-333 (1970).

12. Popov, V M. 'Hyperstability of Automatic Control Systems'. Springer, NY, (1973).

13. Rothschild, D and Jameson, A. 'Comparison of Four Numerical Algorithms for Solving the Lyapunov Matrix Equation'. *Int Journ Cont*, 11, No 2, pp 181-198 (1970).

14. Franklin, G F and Powell, J D. 'Digital Control of Dynamic Systems'. Addison-Wesley, Ma., (1980).

Persistently Exciting Trajectories for Estimation[1]

Pradeep K. Khosla

Department of Electrical and Computer Engineering, The Robotics Institute, Carnegie-Mellon University, Pittsburgh, PA 15213, USA

Abstract

The model-based manipulator control methods achieve complete decoupling and linearization with the assumption that the dynamic model is exactly known. This assumption has been satisfied recently and both symbolic and numerical algorithms to estimate the dynamics parameters have been proposed. In deriving these algorithms the lack of sensing the full torque/force vector at each joint leads to the unidentifiability of some of the dynamics parameters. Moreover, the maximum number of parameters that may be estimated depends on the trajectory used for estimation and the kinematic structure of the manipulator. In this paper, we address the problem of generating persistently exciting trajectories for parameter estimation. To achieve our goal, we propose an algorithm that categorizes the dynamics parameters of a manipulator into three classes: uniquely identifiable, identifiable in linear combinations only and unidentifiable. Since the proposed algorithm categorizes the parameters only on the basis of the kinematic configuration of the manipulator, it allows one to determine if a chosen trajectory (for estimation) is persistently exciting. This knowledge also helps in making the estimation procedure more robust.

---

[1]This research is based upon the work supported by National Science Foundation under Grant ECS-8320364.

1. Introduction

The model-based robot control schemes such as the *computed-torque* [13] and *resolved-acceleration* [11] controllers incorporate the complete dynamic model of the manipulator in the control law computation. The fundamental assumption in these model-based schemes is that the robot dynamics are modeled accurately. To alleviate this strong assumption, we have proposed both symbolic [16] and numerical [6] algorithms to estimate the dynamics parameters of a manipulator. Furthermore, we have also experimentally implemented the identification algorithms and estimated the dynamics parameters of the six degrees-of-freedom CMU Direct-Drive Arm II [7]. Other researchers have also independently proposed algorithms to estimate the dynamics parameters of a manipulator [17, 14, 1]. The estimated parameters from the identification algorithms have been successfully used to evaluate the effect of dynamics compensation in model-based manipulator control schemes [9, 10, 2].

The derivation of the above mentioned identification algorithms is based on the assumption that the input (joint torques) - output (joint position, velocity and acceleration) data are available. Further, in deriving these estimation algorithms it has been noted that the lack of sensing the full force/torque vector at each link leads to the unidentifiability of some of the dynamic link parameters. For example, Khosla and Kanade [8] noted that for the three degrees-of-freedom cylindrical manipulator only three parameters could be identified uniquely, three more parameters could be identified as a linear combination and the remaining twenty four parameters (out of a possible total of thirty) could not be identified at all. Similar observations have also been made for the six degrees-of-freedom CMU DD Arm II [7] and the Stanford manipulator. Typically, only those dynamic parameters that affect the force/torque equation of at least one joint of a manipulator can be identified. Further, all the dynamic parameters of a manipulator can be classified into *three categories*: uniquely identifiable, identifiable in linear combinations, and unidentifiable. The category to which a dynamics parameter belongs is, in general, dictated by the kinematic structure of the manipulator and the input trajectory used for estimation.

A straightforward method to categorize the dynamics parameters is to derive the dynamics equations of a manipulator symbolically and classify the parameters into one of the three categories manually. While it is possible to do this for manipulators having at most two or three degrees-of-freedom, extending it to manipulators with six degrees-of-freedom becomes a formidable task. For example, the symbolically expanded dynamics equations of a six degrees-of-freedom rotary manipulator occupy about 250 printed pages [15]. While it is conceivable that the above method can be used to categorize the parameters, it will not allow us to determine if a chosen trajectory (for identification) is persistently exciting.

In this paper, we propose an algorithmic procedure to categorize the dynamics parameters of an N degree-of-freedom manipulator. We also show that by using our formulation the task of determining whether a given trajectory is reduced to one of finding the rank of a matrix. Thus the proposed algorithm not only categorizes the parameters but it also allows us to determine if the chosen input trajectory (for identification) is persistently exciting.

Specifically, we address the following problem: Given the kinematic description of an N degree-of-freedom manipulator, determine

- the number ν of the dynamic parameters of the manipulator affect the link force/torque of at least one joint.
- which of these ν dynamic parameters can be identified independently.
- which of the ν dynamic parameters occur as linear combinations only and identify these combinations.

The proposed algorithm thus predicts the maximum number of parameters that may be identified and further categorizes them in different classes. This knowledge is extremely important for choosing persistently exciting trajectories for estimating the dynamics parameters.

To accomplish our objective, we reformulate the N-E dynamic robot model from the parameter identification point-of-view. We then apply customization procedure [5] to this reformulation and introduce our algorithm. Recognizing the parameters that can be estimated allows us to reformulate the dynamic parameter estimation algorithm in terms of these identifiable parameters only. Such a reformulation permits us to determine persistently exciting trajectories, increases the robustness of the identification procedure, and also improves the efficiency of the *inverse* dynamics formulation.

This paper is organized as follows: In Section 2, we review the N-E dynamic formulation and reformulate it from the parameter estimation point-of-view. In section 3, we describe the customization procedure and in Section 4, we inroduce our algorithm to find out the identifiable parameters in the dynamic robot model. In Section 5, we describe a numerical technique for categorization and also show it's application determining the persistently exciting characteristics of a ggiven trajectory. We demonstrate the use of our algorithm by presenting an example in Section 6 and finally, in Section 7, we discuss our results. The contributions of our paper are summarized in Section 8.

2. Reformulation of the N-E Dynamic Robot Model

Two formulations have been used to model robot dynamics [18, 3, 12]: closed-form Lagrange formulation and recursive Newton-Euler formulation. While the former leads to physical insight, the latter is computationally more efficient and suited for real-time control applications. We base the development of our algorithm, to categorize the unidentifiable and the identifiable parameters, on the Newton-Euler formulation. We first summarize the O (N) recursive Newton-Euler dynamic equations and then cast it into a form suitable for the development of our algorithm.

2.1. Newton-Euler Formulation

The Newton-Euler formulation shown in equations (1)-(9) computes the inverse dynamics (ie., joint torques/forces from joint positions, velocities, and accelerations) based on two sets of recursions: the *forward* and *backward* recursions. The forward recursions (1)-(3) transform the kinematic variables from the base to the end-effector. Since efficient implementations of the inverse dynamics do not compute the forward recursions for the linear velocities of the links, these have not been outlined. The initial conditions (for i=0) assume that the manipulator is at rest in the gravitational field. The backward recursions (4)-(9) transform the forces and moments from the end-effector to the base, and culminate with the calculation of the joint torques/forces.

$$\omega_{i+1}=\begin{cases} \mathbf{A}_{i+1}^T[\omega_i + \mathbf{z}_o\dot{\theta}_{i+1}] & \text{rotational} \\ \mathbf{A}_{i+1}^T\,\omega_i & \text{translational} \end{cases} \tag{1}$$

$$\omega_{i+1}=\begin{cases} \mathbf{A}_{i+1}^T[\dot{\omega}_i + \mathbf{z}_o\ddot{\theta}_{i+1} + \omega_i\times(\mathbf{z}_o\dot{\theta}_{i+1})] & \text{rotational} \\ \mathbf{A}_{i+1}^T\,\dot{\omega}_i & \text{translational} \end{cases} \tag{2}$$

$$\dot{\mathbf{v}}_{i+1}=\begin{cases} \mathbf{A}_{i+1}^T\dot{\mathbf{v}}_i + \dot{\omega}_{i+1}\times\mathbf{p}_{i+1} + \omega_{i+1}\times(\omega_{i+1}\times\mathbf{p}_{i+1}) & \text{rotational} \\ \mathbf{A}_{i+1}^T[\dot{\mathbf{v}}_i + \mathbf{z}_o\ddot{d}_{i+1} + 2\omega_i\times(\mathbf{z}_o\dot{d}_{i+1})] & \\ \quad + \dot{\omega}_{i+1}\times\mathbf{p}_{i+1} + \omega_{i+1}\times(\omega_{i+1}\times\mathbf{p}_{i+1}) & \text{translational} \end{cases} \tag{3}$$

$$\omega_0 = \dot{\omega}_0 = \mathrm{v}_0 = \mathbf{0}$$

$$\dot{\mathrm{v}}_0 = [g_x \; g_y \; g_z]^T \qquad \text{gravitational acceleration}$$

$$\dot{\mathrm{v}}_i^* = \dot{\omega}_i \times \mathrm{s}_i + \omega_i \times (\omega_i \times \mathrm{s}_i) + \dot{\mathrm{v}}_i \tag{4}$$

$$\mathrm{F}_i = m_i \dot{\mathrm{v}}_i^* \tag{5}$$

$$\mathrm{N}_i = \mathrm{I}_i \dot{\omega}_i + \omega_i \times (\mathrm{I}_i \omega_i) \tag{6}$$

$$\mathrm{f}_i = \mathrm{A}_{i+1} \mathrm{f}_{i+1} + \mathrm{F}_i \tag{7}$$

$$\mathrm{n}_i = \mathrm{A}_{i+1} \mathrm{n}_{i+1} + \mathrm{p}_i \times \mathrm{f}_i + \mathrm{N}_i + \mathrm{s}_i \times \mathrm{F}_i \tag{8}$$

$$\tau_i = \begin{cases} \mathrm{n}_i^T (\mathrm{A}_i^T \mathrm{z}_o) & \text{rotational} \\ \mathrm{f}_i^T (\mathrm{A}_i^T \mathrm{z}_o) & \text{translational} \end{cases} \tag{9}$$

f_{N+1}: external force at the end-effector.
n_{N+1}: external moment at the end-effector.

In the above formulation, the six elements of the symmetric inertia tensor I_i, the three elements of the center-of-mass vector s_i, and the mass m_i constitute the ten dynamics parameters of link i. Thus an N link manipulator has $10N$ dynamics parameters to be estimated. The properties of the Newton-Euler model for parameter identification have been developed and are presented, with proofs, in [8]. Other properties of the dynamic robot model are presented in [19, 20]. For expository convenience, we outline the properties (for identification) without the proofs and these properties are:

1. The Newton-Euler model is linear in the classical link inertia tensors $\mathbf{I}_i$.

2. For rotational joints, the Newton-Euler model is nonlinear in the center-of-mass vectors $\mathbf{s}_i$.

3. The Newton-Euler model is nonlinear in the kinematic parameter vectors $\mathbf{p}_i$.

4. The dynamics equations of links i+1 through N are independent of the mass m_i and the classical inertia tensor $\mathbf{I}_i$ of link i.

It has been noted that if the Newton-Euler model in (1)-(9) is reformulated such that the

Table 1: Kinematic and Dynamic Parameters

m_i	Total mass of link i
τ_i	Joint torque/force at joint i
ω_i and $\dot{\omega}_i$	Angular velocity and acceleration of the i-th coordinate frame
v_i and $\dot{\mathrm{v}}_i$	Linear velocity and acceleration of the i-th coordinate frame
v_i^* and $\dot{\mathrm{v}}_i^*$	Linear velocity and acceleration of the center-of-mass of link i
F_i and N_i	Net force and moment exerted on link i
f_i and n_i	Force and moment exerted on link i by link $i-1$
p_i	Position of the i-th coordinate frame with respect to the $(i-1)$-th coordinate frame: $\mathrm{p}_i=[a_i \;\; d_i\,sin\alpha_i \;\; d_i\,cos\alpha_i]^T$
s_i	Position of the center-of-mass of link i: $\mathrm{s}_i=[s_{ix} \;\; s_{iy} \;\; s_{iz}]^T$
z_o	$=[0 \;\; 0 \;\; 1]^T$
A_i	Orthogonal rotation matrix which transforms a vector in the i-th coordinate frame to a coordinate frame which is parallel to the $(i-1)$-th coordinate frame: $\mathrm{A}_i=\begin{bmatrix} cos\theta_i & -cos\alpha_i\,sin\theta_i & sin\alpha_i\,sin\theta_i \\ sin\theta_i & cos\alpha_i\,cos\theta_i & -sin\alpha_i\,cos\theta_i \\ 0 & sin\alpha_i & cos\alpha_i \end{bmatrix}$ for $i=1,2,\cdots,N$, where $\mathrm{A}_{N+1} \triangleq \mathrm{E}$.
I_i	Classical inertia tensor of link i about the center-of-mass of link i (and parallel to the i-th coordinate frame); with principal inertias I_{ixx}, I_{iyy} and I_{izz}; and cross-inertias I_{ixy}, I_{ixz} and I_{iyz}.

link inertia tensors are expressed about the link coordinate frames instead of the center-of-mass frames then the resulting model is linear in all of the dynamic parameters. If $\mathbf{I}_i$ is the classical link inertia tensor about the center-of-mass of link i, the corresponding inertia tensor $\mathbf{I}'_i$ about the link i coordinate frame C_i is computed according to the parallel-axis theorem or Steiner's law:

$$\mathbf{I}'_i=\mathbf{I}_i+m_i(\mathbf{s}_i^T\mathbf{s}_i\mathbf{E}-\mathbf{s}_i\mathbf{s}_i^T) \tag{10}$$

where $\mathbf{E}$ is the 3×3 identity matrix.

Upon substituing (4)-(6) in (8), we obtain:

$$\mathbf{n}_i=\mathbf{A}_{i+1}\mathbf{n}_{i+1}+\mathbf{p}_i\times\mathbf{f}_i+\mathbf{I}_i\dot{\omega}_i+\omega_i\times(\mathbf{I}_i\omega_i)+m_i\mathbf{s}_i\times[\dot{\omega}_i\times\mathbf{s}_i+\omega_i\times(\omega_i\times\mathbf{s}_i)+\dot{\mathbf{v}}_i] \tag{11}$$

In order to facilitate our reformulation, we introduce the following vector identities:

$$\mathbf{a}\times(\mathbf{b}\times(\mathbf{b}\times\mathbf{a}))=\mathbf{b}\times[\mathbf{a}^T\mathbf{a}\mathbf{E}-\mathbf{a}\mathbf{a}^T]\mathbf{b} \tag{12}$$

$$\mathbf{a}\times(\mathbf{b}\times\mathbf{a})=[\mathbf{a}^T\mathbf{a}\mathbf{E}-\mathbf{a}\mathbf{a}^T]\mathbf{b} \tag{13}$$

where $\mathbf{a}$ and $\mathbf{b}$ are 3×1 vectors and $\mathbf{E}$ is the 3×3 identity matrix.

Substituting the above identities (12), (13) and (10) in the right hand side of (11), we obtain:

$$\mathbf{n}_i=\mathbf{A}_{i+1}\mathbf{n}_{i+1}+\mathbf{p}_i\times\mathbf{f}_i+\mathbf{I}'_i\dot{\omega}_i+\omega_i\times(\mathbf{I}'_i\omega_i)+m_i\mathbf{s}_i\times\dot{\mathbf{v}}_i \tag{14}$$

where $\mathbf{I}'_i$ is the inertia of link i expressed about the link coordinate frame. Henceforth, we will use (14) instead of (8) in our development.

As the next step in our reformulation, we define a six by one vector $\mathbf{g}_i$ as

$$\mathbf{g}_i=[\,\mathbf{f}_i \quad \mathbf{n}_i\,]^T \tag{15}$$

and combine (7) and (14) in a single matrix-vector equation to obtain:

$$\mathbf{g}_i=\mathbf{R}_{i+1}\mathbf{g}_{i+1}+\mathbf{G}_i \tag{16}$$

where $\mathbf{R}_{i+1}$ is the six by six *pseudo-rotation* matrix

$$\mathbf{R}_{i+1}=\begin{bmatrix}\mathbf{A}_{i+1} & 0\\ [\mathbf{p}_i\times]\mathbf{A}_{i+1} & \mathbf{A}_{i+1}\end{bmatrix}$$

and $\mathbf{G}_i$ is defined as:

$$G_i = \begin{bmatrix} F_i \\ N_i' \end{bmatrix}$$

with N_i' being:

$$N_i' = I'_i\dot{\omega}_i + \omega_i \times (I'_i\omega_i) + m_i s_i \times \dot{v}_i + m_i p_i \times [\dot{\omega}_i \times s_i + \omega_i \times (\omega_i \times s_i) + \dot{v}_i]$$

The actuating torques are chosen from the vector g_i according to the following equation:

$$\tau_i = g_i^T \begin{bmatrix} A_i^T & 0 \\ 0 & A_i^T \end{bmatrix} \begin{bmatrix} \sigma_i z_0 \\ (1-\sigma_i) z_0 \end{bmatrix} \qquad (17)$$

where $\sigma_i=1$ for a *i-th* rotary joint and $\sigma_i=0$ for a *i-th* translational joint.

The complete dynamic robot model is given by the forward recursions in (1)-(3) and the backward recursion in (16) and the selection of torques according to Equation (9). In the next section, we describe our customization procedure which is used to derive the algorithm to determine the identifiable parameters of the dynamic manipulator model.

3. Customization Procedure

The concept of *customizing* the N-E dynamic equations of a manipulator to reduce the computational requirements was proposed in [5]. The customization procedure guarantees that no two mathematically equivalent expressions are denoted by the same variable name. We use this property of the procedure to derive our algorithm to categorize the dynamic parameters of a manipulator.

In order to develop a systematic procedure for *customization* of the Newton-Euler algorithm, we adopt the convention proposed in [5]:

The non-zero elements of a vector or a matrix are denoted by subscripted variables, and the zero and unity elements by a 0 and 1, respectively. We propagate the non-zero elements as variables and the zero elements as zeros.

We illustrate our approach by developing the angular velocities in (1) and linear accelerations in (4). The initial conditions for the forward recursions are listed in Table 2-3 and the gravitational acceleration of the manipulator base is:

$v_0=[0 \;\; 0 \;\; -g]^T$.

By using (1 and the initial conditions, we formulate the angular velocities of the links. The angular velocity ω_1 of link 1, expressed in terms of the angular velocity of link 0 is:

$$\omega_1 = \begin{bmatrix} 0 \\ 0 \\ \dot{\theta}_1 \end{bmatrix} = \begin{bmatrix} 0 \\ 0 \\ \omega_{1z} \end{bmatrix}$$

Since the N-E algorithm is recursive and the angular velocity ω_2 is expressed in terms of the angular velocity ω_1, we pass the fact that two of the elements of ω_1 are zero (according to our convention) to the computation of ω_2. The angular velocity ω_2 is:

$$\omega_2 = \begin{bmatrix} 0 \\ \omega_{1z} + \dot{\theta}_2 \\ 0 \end{bmatrix} = \begin{bmatrix} 0 \\ \omega_{2y} \\ 0 \end{bmatrix}$$

Upon continuing the recursion, we compute the angular velocity vector ω_3 according to:

$$\omega_3 = \begin{bmatrix} S_3\omega_{2y} \\ -\dot{\theta}_3 \\ C_3\omega_{2y} \end{bmatrix} = \begin{bmatrix} \omega_{3x} \\ \omega_{3y} \\ \omega_{3z} \end{bmatrix}$$

where $C_3=\cos(\theta_3)$ and $S_3=\sin(\theta_3)$.

Since ω_3 is a full vector, ω_4,ω_5 and ω_6 will also be full vectors. If we had not passed zero elements as zero, we would have required nine variable names to specify all the components of the three angular velocity vectors. Our approach thus creates the minimum number of variables required to specify the vectors uniquely. Using this approach we can customize the *kinematic matrix* of each link of the manipulator to determine which of the inertial parameters of link i will affect its torque/force equation. This idea is further explained in the next section.

4. Categorization of the Dynamic Parameters

To develop our algorithm, we first expand the recursions in Equation (16) to obtain explicit expressions for the vector g_i. And as the next step we decompose the vector G_i as the product of matrix K_i, which is a function of the kinematic parameters and the desired trajectory, and a vector ϕ_i whose elements are the dynamics parameters of the manipulator.

Assuming that the vector of externally applied forces and moments is zero, and expanding the recursions in (16), we obtain:

$$\begin{bmatrix} g_1 \\ g_2 \\ g_3 \\ g_4 \\ g_5 \\ g_6 \end{bmatrix} = \begin{bmatrix} K_1 & R^1{}_2K_2 & R^1{}_3K_3 & R^1{}_4K_4 & R^1{}_5K_5 & R^1{}_6K_6 \\ 0 & K_2 & R^2{}_3K_3 & R^2{}_4K_4 & R^2{}_5K_5 & R^2{}_6K_6 \\ 0 & 0 & K_3 & R^3{}_4K_4 & R^3{}_5K_5 & R^3{}_6K_6 \\ 0 & 0 & 0 & K_4 & R^4{}_5K_5 & R^4{}_6K_6 \\ 0 & 0 & 0 & 0 & K_5 & R^5{}_6K_6 \\ 0 & 0 & 0 & 0 & 0 & K_6 \end{bmatrix} \begin{bmatrix} \varphi_1 \\ \varphi_2 \\ \varphi_3 \\ \varphi_4 \\ \varphi_5 \\ \varphi_6 \end{bmatrix}$$

where

$R^i{}_j = R_{i+1}R_{i+2}......R_j$

and the parameters of link i that affect the torque of link j (for $i > j$) depend on the structure of the matrix $R^i{}_jA_j$.

Further, the vector G_i may be written as:

$$G_i = \begin{bmatrix} \dot{v}_i & [\dot{\omega}_i x] + [\omega_i x][\omega_i x] & 0 \\ [p_i x]\dot{v}_i & [-\dot{v}_i x] + [p_i x][\dot{\omega}_i x] + [p_i x][\omega_i x][\omega_i x] & [.\dot{\omega}_i] + [\omega_i x][\omega_i x] \end{bmatrix} \begin{bmatrix} m_i \\ m_i s_i \\ I_i \end{bmatrix}$$

and the identifiability of the dynamics parameters of link i depends solely on the properties of the rows (or columns) of the kinematic matrix K_i.

As the next step in our development, we divide the matrix K_i into submatrices as follows:

$$K_i = \begin{bmatrix} K_{i11} & K_{i12} & K_{i13} \\ K_{i21} & K_{i22} & K_{i23} \end{bmatrix}$$

where the submatrices K_{i11} and K_{i21} are of dimension 3×1, K_{i12} and K_{i22} are of dimension 3×3, and K_{i13} and K_{i23} are of dimension 3×6. In general, all the submatrices are full except K_{i13} which is always the null matrix. We note that the mass m_i of link i is multiplied only the submatrices K_{i11} and K_{i21}, the parameter vector $m_i s_i$ is multiplied only by the submatrices K_{i12} and K_{i22}, and the vector of inertial parameters is multiplied by the null matrix K_{i13} and K_{i13}. Thus to determine which of the dynamics parameters of link i affect the torque/force of link i, we analytically analyze the submatrices of the matrix K_i. For a given kinematic configuration of a manipulator, the submatrices may or may not be full and this may be determined by using the customization procedure outlined in Section 3. However, in our subsequent analysis we will assume that the submatrices as full and examine the general case.

4.1. Analysis of the Submatrices of K_i

In analyzing the submatrices of the matrix K_i our aim is to search for all the elements in one row which are mathematically equivalent. This in turn implies determining which of the dynamics parameters of link i occur as linear combination in the torque equation. Also the dynamics parameters associated with zero elements of the K_i matrix will not affect the torque equation of link i and thus constitute the unidentifiable parameters. Since this procedure is symbolic it is independent of the chosen trajectory and identifies the maximum number of dynamics parameters that can affect the torque equation of link i.

Further, we assume that our manipulator may have both translational and rotaional joints and that the link angle α_i may take values of 0, 90 and −90 degrees. Then according to the algorithm in Section 2, we need to restrict our search only to the third, fifth and sixth rows of the matrix K_i. It must be noted that our assumptions are satisfied by most of the available manipulators and our analysis thus has wide applicability.

Since the submatrices K_{i11} and K_{i21} have only one column the problem of searching for linearly dependent rows is trivial. To continue with the development of our procedure we symbolically expand the submatrix K_{i21} as:

$$K_{i12} = \begin{bmatrix} -\omega_y^2 - \omega_z^2 & -\dot{\omega}_z + \omega_x\omega_y & \dot{\omega}_y + \omega_x\omega_z \\ \dot{\omega}_z + \omega_x\omega_y & -\omega_z^2 - \omega_x^2 & -\dot{\omega}_x + \omega_y\omega_z \\ -\dot{\omega}_y + \omega_x\omega_z & \dot{\omega}_x + \omega_y\omega_z & -\omega_z^2 - \omega_x^2 \end{bmatrix}$$

where the subcript i has been omitted for the sake of clarity. We note that all of the nine terms in the submatrix above are, in general, unique and hence none of the associated dynamics parameters from the vector $m_i s_i$ can occur in linear combinations for a prismatic joint. Similarly, by expanding the submatrix G_{i22} it can be shown that none of the dynamics parameters included in the vector $m_i s_i$ can occur in linear combinations in the torque equation of the $i-th$ rotary joint.

The submatrix K_{i13} multiplies the inertial elements in the vector I_i' for translational joints. And, since K_{i13} is a null matrix it may be concluded that if the $i-th$ joint is translational, the inertial elements of link i do not affect the torque/force equation of link i. This conclusion could have also been drawn from intuitive arguments. If the $i-th$ joint is rotational then the submatrix K_{i23} multiplies the vector I_i' in the torque/force equation of link i. Upon expanding K_{i23}, we obtain:

$$K_{i23} = \begin{bmatrix} \dot{\omega}_x & \dot{\omega}_y-\omega_x\omega_z & \dot{\omega}_z-\omega_x\omega_y & -\omega_y\omega_z & -\omega_z^2+\omega_y^2 & \omega_y\omega_z \\ \omega_x\omega_z & \dot{\omega}_x+\omega_y\omega_z & \omega_z^2-\omega_x^2 & \dot{\omega}_y & \dot{\omega}_z-\omega_x\omega_y & -\omega_x\omega_z \\ -\omega_x\omega_y & \omega_x^2-\omega_y^2 & \dot{\omega}_x-\omega_y\omega_z & \omega_x\omega_y & \dot{\omega}_y+\omega_x\omega_z & \dot{\omega}_z \end{bmatrix}$$

If the twist angle α_i of the revolute joint i is 90 degrees, then the second row of K_{i23} multiplies the vector I_i'. Searching for symbolically equivalent expressions in the second row, we find that only the linear combination $I_{ixx}'-I_{izz}'$ affects the torque/force equation of joint i and can be estimated. If the twist $\alpha_i=-90$ degrees, then the linear combination $I_{izz}'-I_{ixx}'$ can be estimated. However, if the twist angle is 0 degrees, then the third row of K_{i23} multiplies the vector I' and it is observed that the linear combination $-I_{ixx}'+I_{iyy}'$ can be identified.

The above discussion has served to provide us with a technique of determining which dynamics parameter of link i can be identified from the torque/force equation of link i. One of the properties of the dynamics formulation states that the parameters of link j do not affect the torque/force equation of link i where $i>j$. This suggests that if a dynamics parameter of link i is either unidentifiable or occurs in a linear combination in the torque/force equation of link i, then it may be possible to identify such parameters from the torque/force equation of link j where $j<i$. The complete set of identifiable dynamics parameters is formed by evaluating all combinations and is increasingly complex to be presented here. However, we do present a computer implementable algorithm that allows us to determine the complete set of dynamics parameters for a given manipulator configuration.

Let S_{ij} denote a set of dynamics parameters of link i that can be uniquely estimated from the torque/force equation of link j. Similarly, let L_{ij} and U_{ij} denote the sets of parameters that occur in linear combinations and those that cannot be identified, respectively. Also, let $elem(\cdot)$ denote a function that chooses any single element from the set "$\cdot$". For example $elem(U_{jj})$ selects an element of the set U_{jj}. Then, the following algorithm allows us to classify the dynamics parameters of a manipulator given the kinematic description:

```
for (i=N; i<=1; i--)
 {
  Compute S_ii
  Compute L_ii
  Compute U_ii
  for(j=N; j<i; j--)
   {
    Compute S_ji
    Compute L_ji
    Compute U_ji
    if elem(U_jj) in S_ij
     then delete elem(U_jj) from U_jj
    if elem(L_jj) in S_ij
     then delete elem(L_jj) from U_jj
```

Let S be the set consisting of all the parameters of the manipulator that can be identified uniquely, L be the set of the parameters that can be identified only in linear combinations and U be the set of parameters that cannot be identified at all. Then,

$S_i = \cup_j S_{ij}$
$L_i = \cup_j L_{ij}$
$U_i = \cup_j U_{ij}$

where S_i is the set of parameters of link i that can be identified uniquely, L_i is the set of parameters of link i that can be identified in linear combinations, U_i is the set of parameters of link i that cannot be identified and the subscript j takes the values from 0 to N. Further, the sets S, L and U are obtained from the following relationships

$S = \cup_i S_i$
$L = \cup_i L_i$
$U = \cup_i U_i$

where the subscript i takes the values from 0 to N and N is the number of degrees-of-freedom of the manipulator.

5. Numerical Procedure for Categorization

Earlier in this paper we had alluded to the fact that the category to which a parameter belongs is a function of both the trajectory and the kinematic configuration of the manipulator. Based on the assumption that the trajectory is persistently exciting, we proceeded to outline a symbolic procedure to categorize the parameters. However, we did not address the problem of choosing persistently exciting trajectories nor did we present a method of categorizing the parameters if the trajectory is not persistently exciting.

The symbolic procedure of categorization presented in the previous sections assumes that the trajectory is persistently exciting. However, in practice a trajectory is predetermined and usually nothing is known about its persistently exciting characteristics. In such a circumstance, we can categorize the parameters based on a numerical analysis of the matrix-vector formulation in (???). This means that instead of searching symbolically for mathematically equivalent expressions in the rows of the K_i matrices, we determine the linear dependencies of the columns numerically. This can be achieved through the use of standard matrix analysis [4].

6. Categorized Parameters for the CMU DD Arm II

Based on the symbolic and the numerical procedures outlined in the previous sections we have categorized the parameters of the CMU DD Arm II. We first used the symbolic procedure to determine the complete set of parameters of the CMU DD Arm II. We then selected a trajectory for the numerical estimation of the parameters and also proceeded to classify the parameters based on this trajectory. The categories determined were the same as the ones which for a complete set thus suggesting that the chosen trajectory was indeed persistently exciting.

The complete set of dynamics parameters as determined from the kinematics and confirmed by the numerical procedure is:

$$S_{66} = \{m_6 s_{6x}, m_6 s_{6y}, I'_{6xy}, I'_{6xz}, I'_{6yz}, I'_{6zz}\}$$
$$L_{66} = \{I'_{6xx}, -I'_{6yy}\}$$
$$U_{66} = \{m_6, m_6 s_{6z}\}$$

$$S_{65} = \{m_6, m_6 s_{6z}, I'_{6xx} I'_{6yy}\}$$

$$S_{55} = \{m_5 s_{5x}, m_5 s_{5z}, I'_{5xy}, I'_{5xz}, I'_{5yy}, I'_{5yz}\}$$
$$L_{55} = \{I'_{5xx}, -I'_{5zz}\}$$
$$U_{55} = \{m_5, m_5 s_{5y}\}$$

$$S_{54} = \{m_5 s_{5y}, I'_{5xx}, I'_{5zz}\}$$

$$S_{44} = \{m_4 s_{4x}, m_4 s_{4z}, I'_{4xy}, I'_{4xz}, I'_{4yy}, I'_{4yz}\}$$
$$L_{44} = \{I'_{4xx}, -I'_{4zz}\}$$
$$U_{44} = \{m_4, m_4 s_{4y}\}$$

$$S_{43} = \{m_4 s_{4y}, I'_{4xx}, I'_{4zz}\}$$
$$L_{43} = \{m_5, m_4\}$$

$$S_{33} = \{m_3 s_{3x}, m_3 s_{3z}, I'_{3xy}, I'_{3xz}, I'_{3yy}, I'_{3yz}\}$$
$$L_{33} = \{I'_{3xx}, -I'_{3zz}\}$$
$$U_{33} = \{m_3, m_3 s_{3y}\}$$

$$S_{32} = \{m_3, m_3 s_{3y}, I'_{3xx}, I'_{3zz}\}$$

$$S_{22} = \{m_2, m_2 s_{2x}, m_2 s_{2z}, I'_{2yy}\}$$
$$L_{22} = \phi$$
$$U_{22} = \{m_2 s_{2y}, I'_{2xx}, I'_{2xy}, I'_{2xz}, I'_{2yz}, I'_{2zz}\}$$

$$S_{11} = \{m_1, m_1 s_{1x}, I'_{1zz}\}$$
$$L_{11} = \phi$$
$$U_{11} = \{m_1 s_{1y}, m_1 s_{1z}, I'_{1xx}, I'_{1xy}, I'_{1xz}, I'_{1yy}, I'_{1yz}\}$$

The categorization sets above have been formed only on the basis of the kinematic structure and are hence applicable to any manipulator having the same kinematic structure of the CMU DD Arm II. For a manipulator with a different kinematic structure, we can categorize the parameters by using the algorithm presented in this paper.

7. Discussion

Since the categorization of parameters is a function of the trajectory it is possible to change the classification of the parameter by selecting a trajectory appropriately. For example, consider the trivial trajectory in which no joints are moving in which case it is not possible to identify any parameters. Such a choice of a trajectory results in all parameters being classified as unidentifiable. This suggests that if the dynamics parameters are categorized based only on the kinematic configuration then the choice of a trajectory may re-classify the parameters such that a parameter belonging to a particular category (for persistently exciting trajectories) may be classified as belonging to a *lower* category for the given trajectory. That is, if a dynamics parameter is uniquely identifiable for a persistently exciting trajectory then for a given trajectory this parameter may be re-classified as uniquely identifiable, identifiable in linear combinations or unidentifiable. Similarly a parameter belonging to the class of identifiable in linear combinations may either be retained in the same category or degraded to an unidentifiable parameter.

The unidentifiable parameters are solely determined the kinematic configuration. This implies that for a given kinematic configuration, there exists no trajectory (under the assumption that only the applied link torques/forces are measurable) which can make an unidentifiable parameter identifiable either uniquely or in linear combinations. Similarly, the parameters that are identifiable in linear combinations are also dictated by the kinematic configuration implying that it is not possible to choose a trajectory that will change the category of the parameters to uniquely identifiable. In short, if the complete set of dynamics parameters has been established then there does not exist any choice of trajectory which will serve to upgrade the category to which a parameter belongs. However, as we observed before, the reverse is not true. This suggests that a uniquely identifiable parameter can be made made to belong to the category of identifiable in linear combinations or unidentifiable by a suitable choice of the trajectory.

8. Summary

In this paper, we have presented both the symbolic and the numerical algorithms to categorize the dynamics parameters of a manipulator. Our reformulation of the Newton-Euler algorithm also resulted in a numerical formulation of the identification procedure presented in Section 2. The proposed algorithm allows us to determine persistently exciting trajectories for parameter estimation, makes the identification procedure more robust, and also reduces the computational requirements of the inverse dynamics formulation.

References

[1] An, C. H., Atkeson, C. G. and Hollerbach, J. M.
Estimation of Inertial Parameters of Rigid Body Links of Manipulators.
In *Proceedings of the 24-th CDC.* Florida, December, 1985.

[2] An, C. H., Atkeson, C. G. and Hollerbach, J. M.
Experimental Determination of the Effect of Feedforward Control on Trajectory Tracking Errors.
In Bejczy, A. K. (editor), *Proceedings of 1986 IEEE Conference on Robotics and Automation*, pages 55-60. IEEE, San Francisco, CA, April 7-10, 1986.

[3] Brady, M., et al. (editors).
Robot Motion: Planning and Control.
MIT Press, Cambridge, MA, 1982.

[4] Anon.
IMSL Library of Subroutines.
IMSL, Houston, Texas, 1984.

[5] Kanade, T., Khosla, P. K. and Tanaka, N.
Real-Time Control of the CMU Direct Drive Arm II Using Customized Inverse Dynamics.
In Polis, M. P. (editor), *Proceedings of the 23rd IEEE Conference on Decision and Control*, pages 1345-1352. Las Vegas, NV, December 12-14,, 1984.

[6] Khosla, P. K.
Estimation of Robot Dynamics Parameters: Theory and Application.
In *Proceedings of the Second International IASTED Conference on Applied Control and Identification.* ACTA Press, Los Angeles, CA, December 10-12, 1986.

[7] Khosla, P. K.
Real-Time Control and Identification of Direct-Drive Manipulators.
PhD thesis, Department of Electrical and Computer Engineering, Carnegie-Mellon University, August , 1986.

[8] Khosla, P. K. and Kanade, T.
Parameter Identification of Robot Dynamics.
In Franklin, G. F. (editor), *Proceedings of the 24-th CDC*, pages 1754-1760. Florida, December 11-13, 1985.

[9] Khosla, P. K. and Kanade, T.
Real-Time Implementation and Evaluation of Model-Based Controls on CMU DD ARM II.
In Bejczy, A. K. (editor), *1986 IEEE International Conference on Robotics and Automation.* IEEE, April 7-10, 1986.

[10] Khosla, P. K. and Kanade, T.
Experimental Evaluation of the Feedforward Compensation and Computed-Torque Control Schemes.
In Stear, E. B. (editor), *Proceedings of the 1986 ACC.* AAAC, Seattle, WA, June 18-20, 1986.

[11] Luh, J. Y. S., Walker, M. W. and Paul, R. P.
Resolved-Acceleration Control of Mechanical Manipulators.
IEEE Transactions on Automatic Control 25(3):468-474, June, 1980.

[12] Luh, J. Y. S., Walker, M. W. and Paul, R. P.
On-Line Computational Scheme for Mechanical Manipulators.
Journal of Dynamic Systems, Measurement, and Control 102(2):69-76, June, 1980.

[13] Markiewicz, B. R.
Analysis of the Computed-Torque Drive Method and Comparision with the Conventional Position Servo for a Computer-Controlled Manipulator.
Technical Memorandum 33-601, Jet Propulsion Laboratory, Pasadena, CA, March, 1973.

[14] Mukerjee, A. and Ballard, D. H.
Self-Calibration in Robot Manipulators.
In Fu, K. S. (editor), *Proceedings of the IEEE International Conference on Robotics and Automation,* pages 1050-1057. St. Louis, Missouri, March 25-28, 1985.

[15] Murray, J. J. and Neuman C. P.
ARM: An Algebraic Robot Dynamic Modeling Program.
In Paul R. P. (editor), *Proceedings of the First International IEEE Conference on Robotics,* pages 103-114. Atlanta, GA, March 13-15,, 1984.

[16] Neuman, C. P. and Khosla, P. K.
Identification of Robot Dynamics: An Application of Recursive Estimation.
In Narendra, K. S. (editor), *Advances in Adaptive Systems Theory.* Plenum Publishing Corporation, New York, 1985.

[17] Olsen, H. B. and Bekey, G. A.
Identification of Robot Dynamics.
In *Proceedings of the IEEE Conference on Robotics and Automation,* pages 1004-1011. IEEE, San Francisco, CA, April, 1986.

[18] Paul, R. P.
Robot Manipulators : Mathematics, Programming and Control.
MIT Press, Cambridge, MA, 1981.

[19] Tourassis, V. D.
Robot Control Based on the Properties of the Dynamics Model.
Technical Memorandum TMR-11, University of Rochester, Production Automation Project, Department of Electrical Engineering, Rochester, NY 14627, February, 1987.

[20] Tourassis, V. D. and Neuman, C. P.
Properties and Structure of Dynamic Robot Models for Control Engineering Applications.
Mechanism and Machine Theory 20(1):27-40, 1985.

The Rational Fractional Approach: A Tool for Computer Aided Feedback System Design

Ashok Iyer

Department of Computer Science and Electrical Engineering, University of Nevada, Las Vegas, Las Vegas, NV 89154, USA

1 Introduction

In 1976, Youla, Bongiorno and Jabr published two, now classical, papers in which a complete parameterization of a set of stabilizing compensators for a multivariate feedback system was obtained [11],[12]. In the ensuing years, this work, which is often termed the YBJ theory, has led to the development of an entirely new approach to the feedback system design problem. Indeed, their stabilization theory has been extended to include:

1. The tracking and disturbance rejection problem. [6],[7],[8]
2. Robust design algorithms. [7],[8]
3. Design with a proper or stable compensator. [7],[8]
4. Transfer function design. [7],[8]
5. Pole placement. [4],[7],[8]
6. Simultaneous stabilization. [7],[8],[9]

The key to our theory is a three step design philosophy.

1. Stabilization.
2. Achievement of Design Constraints.
3. Optimization of system performance.

First and foremost, a feedback system must be stable and, as such, the first step in the design process is the parameterization of all stabilizing compensators for the given plant. Once the stabilizing compensators have been characterized, the second step is to choose a subset of compensators which also achieve the design constraints, such as tracking, pole placement etc. Finally, if any design latitude exists, it may be used to optimize some measure of system performance; sensitivity, energy, consumption, etc.

The purpose of the present paper is to lay the foundations for the development of an "interactive" package for the computer aided design of linear control systems on the above described techniques. Section 2 gives a brief overview of the Rational fraction approach.The third section gives an example of how this design technique enhances the design procedure and the last section discusses some conclusions.

2 Rational Fraction Approach

This section deals with a brief overview of the rational fraction approach to feedback system design. Readers interested in the details should see references [2],[7],[8].

2.1 Mathematical Preliminaries

The system will be described by a rational function

$$r(s) = \frac{p(s)}{q(s)} \tag{1}$$

Such a system is said to be *stable* if its poles lie in the (strict) left half plane.

A *fractional representation* for $r(s)$ is a factorization of $r(s)$ in the form

$$r(s) = \frac{n_r(s)}{d_r(s)} \tag{2}$$

where both $n_r(s)$ and $d_r(s)$ are stable and $d_r(s) \neq 0$. If one is given a *polynomial fractional representation* of $r(s)$ one can take

$$n_r(s) = \frac{p(s)}{m(s)} \tag{3}$$

and

$$d_r(s) = \frac{q(s)}{m(s)} \tag{4}$$

where $m(s)$ is any Hurwitz polynomial such that the order of $m(s)$ equals the order of $r(s)$.

The fractional representation $r(s) = n_r(s)/d_r(s)$ is *coprime* if there exist stable rational functions, $u_r(s)$ and $v_r(s)$, such that

$$u_r(s)n_r(s) + v_r(s)d_r(s) = 1 \tag{5}$$

2.2 Stabilization

The basic feedback system we deal with is shown in Figure 1. For the stabilization we assume a coprime fractional representation for the plant is given in the form

$$p(s) = \frac{n_p(s)}{d_p(s)} \tag{6}$$

where $n_p(s)$ and $d_p(s)$ are stable and $d_p(s)$ is not identically zero and there exists stable $u_p(s)$ and $v_p(s)$ such that

$$u_p(s)n_p(s) + v_p(s)d_p(s) = 1 \tag{7}$$

In this single variable setting every plant admits such a representation and hence (6) and (7) are valid without loss of generality. The compensators, represented in the form

$$c(s) = \frac{n_c(s)}{d_c(s)} \tag{8}$$

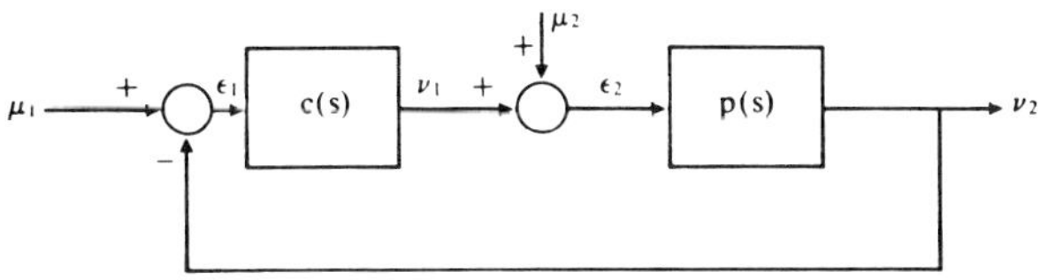

Figure 1: Basic feedback system

where $n_c(s)$ and $d_c(s)$ are stable and $d_c(s)$ is not identically zero, that stabilize the system is given by

$$c(s) = \frac{[-w(s)d_p(s) + u_p(s)]}{[w(s)n_p(s) + v_p(s)]} = \frac{n_c(s)}{d_c(s)} \tag{9}$$

where $w(s)$ is any arbitrary stable rational function and $[w(s)n_p(s) + v_p(s)]$ is not identically equal to zero.

The entire design variable is a parameter of an arbitrary stable rational function $w(s)$. The input-output transfer function with the above compensator is given by

$$h_{\nu_2\mu_1}(s) = -w(s)n_p(s)d_p(s) + u_p(s)n_p(s) \tag{10}$$

This stabilization theory has been extended to solve various classical control problems, such as tracking, disturbance rejection etc. [2],[3],[5],[6],[7],[8]. In the following section a simple example is used to illustrate this powerful design technique.

3 An Example

In this section a very simple second order single variate servomechanism problem is used to illustrate the powerful rational fraction design technique. The polynomial fraction representation of the plant is given by

$$p(s) = \frac{(s+1)}{(s^2-4)} \tag{11}$$

The rational fraction representation of the plant is obtained very easily as shown below

$$p(s) = n_r(s)d_r^{-1}(s) = \left[\frac{(s+1)}{(s+2)^2}\right]\left[\frac{(s-2)}{(s+2)}\right]^{-1} \tag{12}$$

while the coprime equation is obtained via

$$\left[\frac{16}{3}\right]\left[\frac{(s+1)}{(s+2)^2}\right] + \left[\frac{(s+.66)}{(s+2)}\right]\left[\frac{(s-2)}{(s+2)}\right] = u_p(s)n_p(s) + v_p(s)d_p(s) = 1 \tag{13}$$

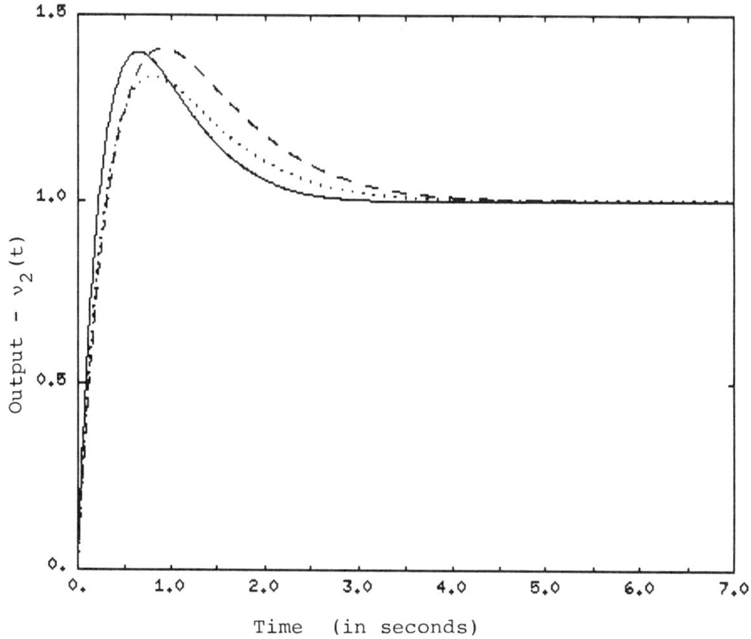

Figure 2: Time Responses

The input-output transfer function as a function of our parameter $w(s)$ is

$$h_{\nu_2\mu_1}(s) = -\left[\frac{(s+1)(s-2)}{(s+3)^2}\right] w(s) + \left[\frac{16(s+1)}{3(s+2)^2}\right] = -n_p(s)d_p(s)w(s) + n_p(s)u_p(s) \quad (14)$$

By the final value theorem the feedback system will asymptotically track a step input if $h_{\nu_2\mu_1}(0) = 1$. Evaluating (14) at $s = 0$ and setting it equal to one yields

$$h_{\nu_2\mu_1}(0) = \frac{1}{4}w(0) + \frac{4}{3} = 1 \quad (15)$$

implying that $w(0) = -4/3$.

Thus any arbitrary $w(s)$ such that $w(0) = -4/3$ will provide a compensator that stabilizes the system and track a unit step function. Let us choose three arbitrary $w(s)$, namely

$$w(s) = -\frac{4}{3} \quad (16)$$

$$w(s) = -\frac{4}{3(s+1)} \quad (17)$$

$$w(s) = -\frac{2(s+2)}{3(s+1)} \quad (18)$$

$w(s)$	$-\frac{4}{3}$	$-\frac{4}{3(s+1)}$	$-\frac{2(s+2)}{3(s+1)}$
Rise Time (sec)	0.1765	0.1831	0.1943
Peak Time (sec)	0.7000	0.7831	0.8000
Overshoot (%)	38.7800	41.2100	44.4200
Settling Time (sec)	2.1000	2.8000	2.7000

Table 1: System Performance

Figure 2 displays the time response with each of the three compensators (16 - solid, 17 - dashed, 18 - dotted) and Table 1 gives the system performances. Both from Figure 2 and Table 1 it is obvious that the compensator given by (16) has the best performance. Of course, if we need to add additional constraints, such as pole locations for the compensators, or other design criteria we could add appropriate constraints to $w(s)$.

4 Conclusions

In this paper, a new approach to computer aided control system design was proposed. Although, the example presented here is a single variate example, the rational fraction theory extends to multivariate systems also [2], [10]. The single variate case is a stepping stone for the ultimate development of a interactive multivariate control system design package.

Although one would ideally like to develop a fully automated computer aided control system design package, experience in others areas, especially in computer-aided circuit design and computer aided machine design, suggests an interactive approach. Indeed, in a practical design environment is is impossible to fully specify all the constraints which one would like to satisfy and/or the trade off between constraints when they cannot be simultaneously satisfied. As such, in the real world of CAD, a minimal set of "hard constraints" are specified and then interactively "tune" the design within the constrained set to achieve the desired trade off among the various "soft constraints" and/or various "intangibles." With this in mind, research work is directed at the development of efficient techniques for interactively tuning a system within the "hard constraints" set defined, by using our analytic theory. To this endeavor, we would like to specify a set of "knobs", i.e., design parameters, which the system designer can "tune" with the CAD package interactively, displaying the effects on the time and frequency responses of interest. Of course, these "knobs" must be matched to the designer's intuition if he is to use them effectively.

5 References

1. Chua, O., and Lin, P. M., "Computer Aided Analysis of Electronic Circuits: Algorithms and Computational Techniques", Prentice Hall, 1975.

2. Desoer, C.A., Liu, R.W., Murray, J. and Saeks, R., "Feedback System Design: The

Fractional Representation Approach to Analysis and Design", IEEE Transactions on Automatic Control, Vol. AC-25, No.3, pp399-412, 1980.

3. Iyer, A., Ph.D. Dissertation, Department of Electrical Engineering, Texas Tech University, 1982.

4. Iyer, A. and Saeks, R., "Feedback System Design: The Pole Placement Problem", International Journal of Control, Vol. 39, No. 3, pp 455 - 472, 1984.

5. Melsa, L. J., and Jones, K. S., "Computer Programs for Computational Assistance in the study of Linear Control Theory", McGraw Hill, 1973.

6. Saeks, R., and J.J. Murray, "Feedback System Design: The Tracking and Disturbance Rejection Problems," IEEE Transaction on Automatic Control, Vol. A-26, pp. 203-217, 1981.

7. Saeks, R., Murray, J.J., Chua, O., Karmokolias, C., and A. Iyer, "Feedback System Design: The Single-Variate Case, Part 1," Circuits, Systems and Signal Processing, Vol. 1, No. 2, pp 137-169, 1982.

8. Saeks, R., Murray, J.J., Chua, O., Karmokolias, C., and A. Iyer, "Feedback System Design: The Single-Variate Case, Part 2," Circuits, Systems and Signal Processing, Vol. 2, No.1, pp 3-33, 1983.

9. Saeks, R., and J. Murray, "Fractional Representation, Algebraic Geometry and Simultaneous Stabilization Problem, " IEEE Transactions on Automatic Control, Vol.AC-27, pp 895 - 903, 1982.

10. Vidyasagar, M., "Control System Synthesis A Factorization Approach," The MIT Press, Cambridge, Massachusetts, 1985.

11. Youla, D.C., Bongiorno, J.J., and H.A. Jabr, "Modern Wiener-Hopf Design of Optimal Controllers, Part I," IEEE Transaction on Automatic Control, Vol. AC-21, pp. 3-15, 1976.

12. Youla, D.C., Bongiorno, J.J., and H.A. Jabr, "Modern Wiener-Hopf Design of Optimal Controllers, Part II," IEEE Transactions on Automatic Control, Vol. AC-21, pp. 319-338, 1976.

Direct and Inverse Dynamic Modelling for Complex Kinematic Structure Robots

M. Becquet

Institute of Applied Mechanics (CP 165), University of Brussels, 1050 Brussels, Belgium

Abstract : The subject of this paper is to establish a general approach for the direct and inverse dynamic modelling of robot with complex kinematic structure. Methods for dynamic modelling of simple kinematic structures are used in combination with holonomic constraints computed automatically in an analytical way. Two examples are considered. One is compared with experimental results.

1. Introduction

In a general manner the dynamic model of an articulated structure made of rigid links connected together can be written :

$$[M(q)]\{\ddot{q}\} + \{V(q,\dot{q})\} + \{G(q)\} + \{F(\dot{q})\} - \{\tau\} = 0 \qquad (1)$$

with $[M(q)]$: the mass matrix

$\{q\}$, $\{\dot{q}\}$ and $\{\ddot{q}\}$: the joint coordinates, speeds and accelerations

$\{V(q,\dot{q})\}$: velocity terms vector

$\{G(q)\}$: gravity terms vector

$\{F(\dot{q})\}$: friction terms vector

$\{\tau\}$: actuators generalized forces.

The system of equations (1) is true for any articulated structure. The main problem is how to establish such a system of equations.

If the articulated structure does not present closed loops in its kinematic structure, what we will call a simple kinematic structure, we may use several methods to build up the system of equations (1) (See [1], [3] and [9]).They are based on recursive computation from bottom to top of each branch of the structure tree. By using recursivity, computation effort is greatly reduced. If the structure presents closed loops, what we will call a complex kinematic structure, we cannot use directly those methods because there is no tree like kinematic structure. The problem is now reduced to transform a complex kinematic structure in a simple one. To solve this probleme we may modify the structure by virtually cut open all the closed loops of the complex structure. We have then to add holonomic constraints to this modified structure so that its dynamic behaviour is the same as the original complex structure (see [3],[5] and [6]).

2. Inverse Dynamic Modelling.

Because mass, velocity, and gravity properties are the same for the original and modified structures and by assuming that the joints that were cut are frictionless the dynamic model of the modified structure is given by :

$$[M(q)]\{\ddot{q}\} + \{V(q,\dot{q})\} + \{G(q)\} + \{F(\dot{q})\} + [\partial\phi/\partial q]^T\{\lambda\} - \{\tau\} = 0 \qquad (2)$$

with : $[\partial\phi/\partial q]$: the Jacobian matrix of the holonomic constraints vector $\{\phi\}$

$\{\lambda\}$: Lagrange multipliers vector.

This leads for the complex structure actuators generalized forces to :

$$\{\tau^c\} = \{\tau\} - [\partial\phi/\partial q]^T \{\lambda\} \qquad (3)$$

In a simple kinematic structure each component of $\{\tau\}$ is related to an active joint, so we don't have any component of $\{\tau\}$ which is identically equal to zero. In the original complex structure some components of $\{\tau^c\}$ are related to passive joints, in other words joints without actuators : those components are identically equal to zero. By selecting in (3) the lines corresponding to those components we have :

$$\{0\} = \{\tau'\} - [\partial\phi/\partial q']^T \{\lambda\} \qquad (4)$$

Because we have as many remaining passive joints (remember that for each loop one passive joint was cut) as Lagrange multipliers $\{\lambda\}$ and as holonomic constraints components, we may compute from (4) the values of the Lagrange Multipliers $\{\lambda\}$. It could be shown that $[\partial\phi/\partial q']$ is a regular matrix (see [3]) if the robot configuration is regular : in other words if the robot is not on one of his workspace limits. By feeding $\{\lambda\}$ in equation (3) we may compute $\{\tau^c\}$, the original structure actuators generalized forces.

To achieve this goal we have to compute :

1- $\{\tau\}$: the generalized forces vector of the simple kinematic chain structure obtained by cutting all the mechanical loops of the original complex kinematic structure.

2- $[\partial\phi/\partial q]^T$: the Jacobian matrix of the holonomic constraints vector $\{\phi\}$ - see section 4.

3- $\{\lambda\}$: the Lagrange multipliers vector.

To perform the first step of this computation scheme we have made the choice of the NEWTON-EULER method which is optimal in terms of number of operations for the inverse dynamic problem (see [4]).

3. Direct Dynamic Modelling

If we take back equation (1) applied to a simple kinematic structure we will have :

$$[M(q)]\{\ddot{q}\} + \{V(q,\dot{q})\} + \{G(q)\} + \{F(\dot{q})\} - \{\tau\} = 0 \qquad (5)$$

If the structure that we have to examine is a complex kinematic structure we may apply equation (5) by modifying the complex structure to obtain a simple kinematic structure. By using equation (3) we may substitute $\{\tau\}$ by $\{\tau^c\}$

$$\{\tau^c\} = \{\tau\} - [\partial\phi/\partial q]^T \{\lambda\} \tag{3}$$

which is the input data for the direct dynamic problem. Equations (3) and (5) give :

$$[M(q)]\{\ddot{q}\} + \{V(q,\dot{q})\} + \{G(q)\} + \{F(\dot{q})\} - \{\tau^c\} + [\partial\phi/\partial q\ ]^T \{\lambda\}) = 0 \tag{6}$$

on which we have to add the holonomic constraints :

$$\{\phi\} = 0 \tag{7}$$

Differentiating (7) twice, gives :

$$\{\dot{\phi}\} = [\partial\phi/\partial q]\{\dot{q}\} = 0 \quad (8) \quad , \quad \{\ddot{\phi}\} = [\partial\phi/\partial q]\{\ddot{q}\} + \{\Sigma\Sigma(\partial^2\phi/\partial q\partial q)\dot{q}\dot{q}\} = 0 \tag{9}$$

Finally equations (6) and (9) may be written in the form :

$$\begin{bmatrix} [M] & , -[\partial\phi/\partial q]^T \\ [\partial\phi/\partial q], & 0 \end{bmatrix} \begin{Bmatrix} \ddot{q} \\ \lambda \end{Bmatrix} = \begin{Bmatrix} \{\tau^c\} - (\{V\} + \{F\} \quad \{G\}) \\ - \{\Sigma\Sigma(\partial^2\phi/\partial q\partial q)\dot{q}\dot{q}\} \end{Bmatrix} \tag{10}$$

The system of equations presented by (10) is the direct dynamic model of a complex kinematic structure considered as a simple kinematic structure added to the holonomic constraints twicely differentiated. The left hand side matrix of (10) is regular if the robot is in a regular configuration (see [7]).We may solve equation (10) to express $\{\ddot{q}\}$ and then proceed to time integration. If we use those equations for time integration we may see that the results are unaccurate if time integration is performed on a large time period. The reason is that we are matching the secondary differentiated holonomic constraints and not the constraints themselves (see [2], [3] and [7]).

At one time step we may have : $\{\ddot{\phi}\} = 0$, $\{\dot{\phi}\} = \epsilon$ and $\{\phi\} = \delta$ (11)

and at the next one we will have $\{\ddot{\phi}\} = 0$, $\{\dot{\phi}\} = \epsilon$ and $\{\phi\} = \delta + \epsilon\Delta t$ (12)

It is well known that equations like equation (9) are unstable. To avoïd this behaviour we have to add stabilization terms. Instead of using equation (9) in equation (11) it is more suitable to use

$$\{\ddot{\phi}\} + 2\alpha\ \{\dot{\phi}\} + \beta^2\ \{\phi\} = 0 \tag{13}$$

which present a damped oscillatory set of equations for the holomonic constraints. If we choose $\alpha = \beta$, critical damping is achieved. Equation (11) became then :

$$\begin{bmatrix} [M] & , -[\partial\phi/\partial q]^T \\ [\partial\phi/\partial q], & 0 \end{bmatrix} \begin{Bmatrix} \ddot{q} \\ \lambda \end{Bmatrix} = \begin{Bmatrix} \{\tau^c\} - (\{V\} + \{F\} + \{G\}) \\ -\{\Sigma\Sigma(\partial^2\phi/\partial q\partial q)\dot{q}\dot{q}\} - 2\alpha\ [\partial\phi/\partial q]\{\dot{q}\} - \beta^2\{\phi\} \end{Bmatrix} \tag{14}$$

By using equation (14) and by making an appropriate choice for $\alpha = \beta$, we can solve accuratelly the direct dynamic problem.

To prepare time integration we have to compute :

1- $[M]$, $\{V\}$, $\{F\}$ and $\{G\}$:	by using a dynamic modelling method for a simple kinematic structure
2- $\{\phi\}$, $[\partial\phi/\partial q]$ and $\{\Sigma\Sigma(\partial^2\phi/\partial q\partial q)\dot{q}\dot{q}\}$:	the holonomic constraints and their first and second differentiated form.
3- $\{\ddot{q}\}$:	the joint acceleration vector.

To perform the first step of this computation scheme we have used the method of the General Theorems which is optimal in terms of number of operations for the direct dynamic problem (see [1] and [3]).

4. Holonomic Constraints Computation

For each closed loop we have to virtually cut a passive joint to create two branches. On each branch end we may locate a frame. By using those two frames we may express holonomic constraints. Three types of constraints must be considered. If the joint virtually cut is a rotational joint the two frames origines are located at a constant distance. If the joint virtually cut is a prismatic joint the two frames have a relative constant orientation. The last constraint is obtained by using the DENAVIT and HARTENBERG frame location rules which lead to a same direction for the {|z} axis of each branch ends frame.

Each of those constraints has the following typical form :

$$\{\phi\} = \prod_{i=1}^{r} [{}^{i}_{i+1}T]\{{}^{r+i}P\} \qquad (15)$$

with : $[{}^{i}_{i+1}T]$: the homogenous transformation matrix between the (i) and (i+1) frames

$\{{}^{r+1}P\}$: a constant vector in the (r+1) frame.

Each matrix $[{}^{i}_{i+1}T]$ only depends on one joint coordinates q_{i+1} and has the following expression.

$$[{}^{i}_{i+1}T] = \begin{bmatrix} \cos\theta_{i+1}, & -\sin\theta_{i+1}, & 0\ , & ai \\ \cos\alpha_i.\sin\theta_{i+1}, & \cos\alpha_i\cos\theta_{i+1}, & -\sin\alpha_i, & -\sin\alpha i\ d_{i+1} \\ \sin\alpha_i\ \sin\theta_{i+1}, & \sin\alpha_i\cos\theta_{i+1}, & \cos\alpha_i, & \cos\alpha_i\ d_{i+1} \\ 0 & 0 & 0 & 1 \end{bmatrix} \qquad (16)$$

with: a_i : distance between $\{|Z_i\}$ and $\{|Z_{i+1}\}$

α_i : angle between $\{|Z_i\}$ and $\{|Z_{i+1}\}$

θ_{i+1} : angle between $\{|X_i\}$ and $\{|X_{i+1}\}$

α_{i+1} : distance between $\{|X_i\}$ and $\{|X_{i+1}\}$

If the joint is a rotational one, the joint coordinate is θ_{i+1}.

If the joint is a prismatic one, the joint coordinate is d_{i+1}.

By using a joint type flag S_{i+1} such that $S_{i+1} = 1$ if the joint is rotational one and $S_{i+1} = 0$ if it is a prismatic one we have :

$$\partial[{}^{i}_{i+1}T]/\partial q_{i+1} = [{}^{i}_{i+1}T]\ [Q_{i+1}] \text{ with } [Q_{i+1}] = \begin{bmatrix} 0 & -S_i & 0 & 0 \\ S_i & 0 & 0 & 0 \\ 0 & 0 & 0 & (1-S_i) \\ 0 & 0 & 0 & 0 \end{bmatrix} \qquad (17)$$

so:

$$\partial\{\phi\}/\partial q_{l+1} = \left(\prod_{i=1}^{l} [{}^{i}_{i+1}T]\right) . [Q_{l+1}] . \left(\prod_{j=l+1}^{r} [{}^{j}_{j+1}T]\right) \{{}^{r+1}P\} \tag{18}$$

and:

$$\partial^2\{\phi\}/\partial q_{l+1}\partial q_{n+1} = \left(\prod_{i=1}^{l} [{}^{i}_{i+1}T]\right).[Q_{l+1}] . \left(\prod_{j=l+1}^{n} [{}^{j}_{j+1}T]\right) . [Q_{n+1}] \left(\prod_{p=n+1}^{r} [{}^{p}_{p+1}T]\right) . \{{}^{r+1}P\} \tag{19}$$

By using equations (15),(18) and (19) we may compute on line $\{\phi\}$, $[\partial\phi/\partial q]$ and $\partial^2\{\phi\}/\partial q_{n+1}\partial q_{l+1}$

5. Examples

We will shortly illustrate two cases, one for the inverse dynamic problem and one for the direct dynamic problem. For the inverse one we have computed the driving forces in the two ball screw units of a complex kinematic structure robot : the ASEA IRB6/2. For a straight line motion at 300 mm/sec with a starting and ending point shown on figure 1, we have the upper arm driving unit force shown on figure 2. Solid line presents experimental measure and doted line computation result for the going motion. A little phase drift appears due to joint trajectory measurements but the average level error is less than 3 N.

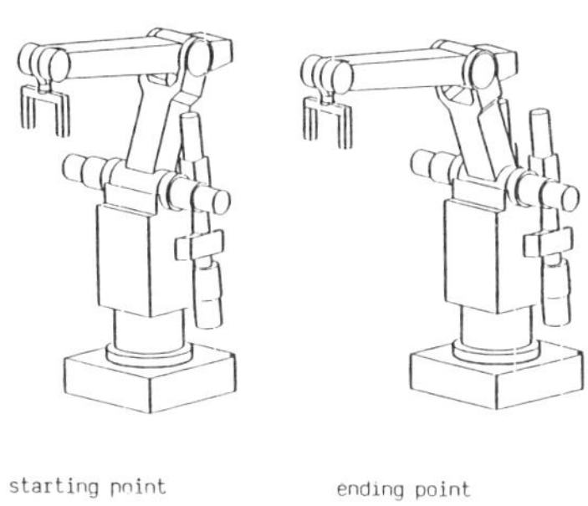

fig.1.

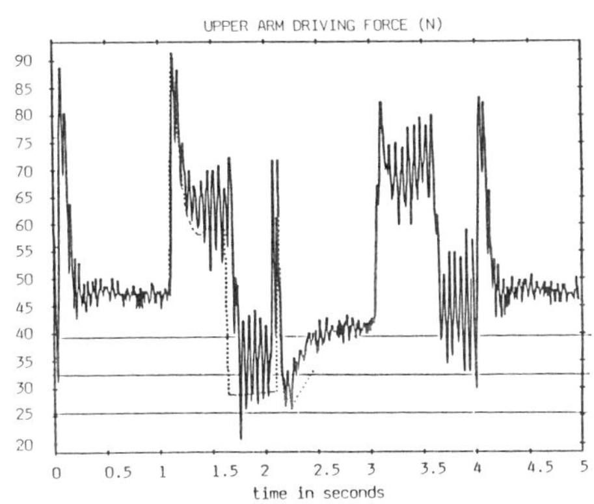

fig 2.

For the direct dynamic problem we have a four-bar mechanism taken from [8] (see fig 3) considered as made of rigid links.
The input torque on the first joint is 1.5 10^{-2} NM. On figure 4 we show the joints position, speed and acceleration computed with a step lenght of 0.04 sec with stabilization term's. By taking a step lenght of 10^{-3} sec as a reference, the maximum error appears on $\ddot{q}_2$ for t = 1.3 sec and its value is of 3.9 %. Without stabilization the maximum error appears on $\ddot{q}_1$ for t = 1.3 sec and its value is of 44.3 %. For a step lenght of 0.025 sec with stabilization the error value is only of 0.7 %. Time integration process of the dynamic model with stabilization terms appears as very accurate.

L_1 = 0.12 m h = 0.024 m

L_2 = 0.264 m b = 0.006 m

L_3 = 0.252 m ρ = 2732 Kg/m^3

O_1O_4 = 0.24 m

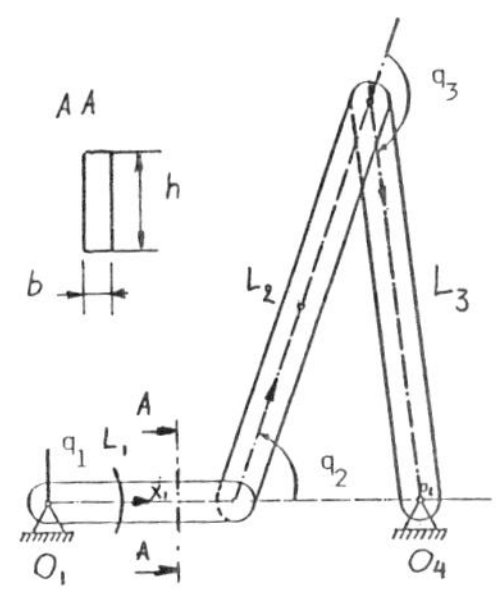

fig 3.

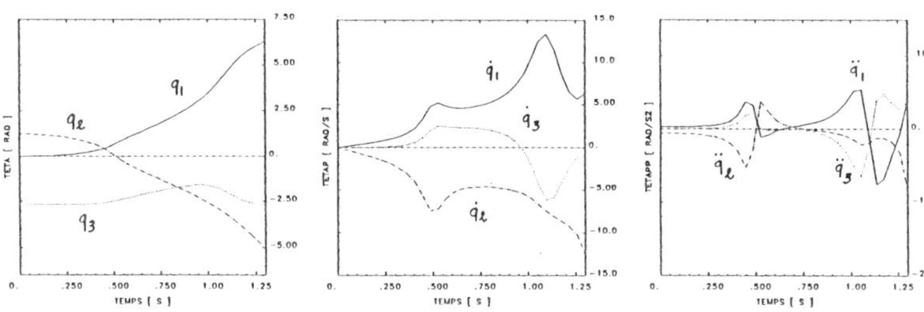

fig 4.

6. Conclusions

We have shown how to establish both direct and inverse dynamic model of complex kinematic structure robots. Each model is based on the transformation of a complex kinematic structure into a simple kinematic structure for which we can use classical dynamic modelling methods. The necessary holonomic constraints and their first and second differentiated form that we have to add at the modified structure can be automatically computed. We have shown for each dynamic problem, an example and brief comments of the results. By using those dynamic modelling methods in a parallel processing scheme, real time dynamic control can be achieved.

References

[1] - BAUDOUIN, Dynamique directe et inverse des manipulateurs en boucle cinématique ouverte : application au robot IRI MSO, travail de spécialisation en automatique robotique, Université Libre de Bruxelles 1986.

[2] - BAUMGARTE, Stabilization of constraints and integrals of motion in dynamical systems, Computer methods in applied mechanics and engineering, North-Holland ed., 1972, pp. 1-16.

[3] - BECQUET M, Validation théorique et expérimentale de modèles dynamiques direct et inverse de mécanismes articulés à structure cinématique complexe, application à la robotique, Thèse de Doctorat en Sciences Appliquées, Université Libre de Bruxelles, 1987.

[4] - HOLLERBACH, Robot motion, planning and control, MIT Press, 1984, pp 51-87.

[5] - LUH et YUAN, Computation of Input generalized forces for robots with closed kinematic chain mechanisms, IEEE Journal of robotics and automation, Vol RA-1, n° 2, June 1985.

[6] - MEGAHED et RENAUD, Dynamic modelling of robot manipulators containing closed kinematic chains, Proc. Advanced software in robotics, Liège, May 1983.

[7] - NIKRAVESH, some methods for dynamic analysis of constrained mechanical systems : a survey, NATO ASI series, vol F9, computer aided analysis and optimization of mechanical system dynamics, Springer Verlag ed., 1984.

[8] - SONG et HAUG, Dynamic Analysis of planar flexible mechanisms, Computer methods in applied mechanics and engineering, North-Holland ed., 1980, pp 359-381.

[9] - VUKOBRATOVIC, Dynamics of manipulation Robots, Scientific fundamentals of Robotics, Vol 1, Springer Verlag, ed., 1982.

Computer Simulation of Manipulator Dynamics Using Different Control Laws

A. Barraco*, B. Cuny*, A. Hoffman†, P. Jamet† M. Lepareux†, and H. Bung†

*ENSAM—Laboratoire de Robotique—151 Bd de l'Hôpital 75013 Paris, France
†CEA—CEN de SACLAY—DEMT/SMTS/LAMS—9191 Gif–Sur–Yvette Cédex, France

0 - INTRODUCTION

PLEXUS is a finite element computer code developed at the CEA — Commissariat à l'Energie Atomique, i.e. French Nuclear Energy Agency — since 1977. Its use was originally centered on mechanical and thermal analyses of metal or concrete bodies, especially when large displacements or fast dynamics phenomena such as impacts were involved. Appendix A.1 recalls the explicit time integration scheme used which is based on an updated Lagrangian formulation, [1], [2], [3].

Recently, there appeared a new and increasing request for design, analysis and simulation tools of spatial mechanisms used in robotics. The first generation robots were indeed built of overly massive and stiff members, empirically designed to ensure a good positioning accuracy for their end effectors. This induced the use of very large motors while the payload versus dead weight ratio stood dramatically low. In order to correct these deficiencies, it has been considered lately to use lighter and therefore more flexible members, while requiring at the same time that a high precision be kept. Since such an optimization process cannot be based on trial and error experiments, it will succeed only if efficient computer codes are used to test changes in structural, motoring or controller components, and quantify their influence.

PLEXUS and its intrinsic potentialities being a good basis to start from, [4], we achieved three developments of this code towards robotics applications. In section I, a rigid body model is presented. We first justify its introduction into PLEXUS and show how a rigid body can be created from and replace the initial realistic node mesh of a mechanical part. In section II, kinematical connexions between adjacent members moving in space are treated using the PLEXUS capability to impose linear relations between nodal point degrees of freedom, named d.o.f. in the following. Specific connexions at robot joints are identified and their treatment as new two-node finite elements is detailed. In section III, the problem of robot control is investigated, first through P.D. control laws directly applied as actuating efforts, and then by correctly modeling the electro-mechanical behavior of direct current electrical motors connected to gear trains, for which experimental data are available. Numerical results are then compared to in-situ measurements. In section IV, the presently achieved developments are discussed and the future evolutions are presented.

I - MODELING RIGID BODIES IN PLEXUS

I.1 - Necessity of a rigid body model

Most robots in activity today are built of over-dimensioned members that can be assumed rigid. As for future robots, though the change of their member constituting material and cross-section dimensions will make them more flexible, some links will still remain sturdy enough to be considered rigid, at least for a first stage of analysis. Therefore the computing tool any robot conceiver will want to have at his disposal must be able to mix rigid and deformable solids in a common analysis, a constraint to which much research work has already been devoted, [5].

Furthermore, PLEXUS uses a conditionally stable explicit numerical integration algorithm, which implies that a minimum time increment value Δt_{crit} be satisfied if calculation results are to be reliable, [6]. Since this lower limit value inversely varies with finite element stiffness, the presence of rigid parts in an otherwise flexible structure will greatly increase computer time and cost, without any benefit in the analysis of the deformable parts. The rigid body model will thus solve this time step problem and in parallel reduce the too large number of d.o.f. issued from a realistic discretization of any rigid structural part.

I.2 - Rigid body parameters and treatment

Let R_o be an arbitrary reference frame in space . From a mechanical standpoint, a rigid solid S is defined by :

- the position in R_o of its center of gravity G_s, and by the value M_s of its mass,
- the orientation with respect to R_o of a frame R_s attached to S in G_s, and by the six components I_{jk} of its inertia tensor I in this frame (reduced to the three diagonal terms I_p when R_s is the principal frame).

From a kinematical point of view, the six d.o.f. of the unconstrained solid S are the three coordinates of G_s in R_o plus three rotation angles specifying the orientation of R_s in R_o.

In order to generate these rigid body quantities for a solid primarily subdivided into 3-D finite elements, the "SOLID" command was developed through which the PLEXUS user specifies first the elements E_k, and hence their node points N_i and related lumped masses m_i, which are to be aggregated into a rigid body S, and secondly the border nodes P_j described here below. The $N_i(m_i)$ nodes are then automatically distributed around G_s at distances computed so that the principal axes and moments of inertia are conserved, figure 1. The twelve d.o.f. of this set of four points are then reduced to the required six by imposing that distances A_iA_j be constant, i.e. :

$$\frac{d}{dt}(\overrightarrow{A_iA_j}^2) = 0 \qquad \Leftrightarrow \qquad \overrightarrow{A_iA_j} \cdot (\vec{v}_j - \vec{v}_i) = 0 \tag{1-1}$$

Since six such scalar equations are written for each rigid part, equidistance conditions when all rigid solids have been handled are collected in matrix form as :

$$[C] \cdot \{v\} = \{0\} \tag{1-2}$$

where C_{ij} coefficients are functions of the coordinates of points A_i and A_j. This type of constraint already exists in PLEXUS which is designed to manage prescribed linear relations between nodal d.o.f., i.e. equations of type (1-2) where $\{v\}$ is replaced by $\{u\}$. In appendix A.2, connexion forces are shown to be induced by such geometrical and kinematical relations : these complementary forces $\{F_c\}$ are introduced into the system's dynamics equations to modify nodal point accelerations, and hence movements, in such a way that all constraints between d.o.f. or velocities are actually satisfied.

As regards frontier nodes P_j, they simply are points arbitrarily chosen from the original finite element discretization and through which the created rigid body S is connected with zero, one or more possibilities of relative motion, to the adjacent structural parts, be they rigid or flexible. The program automatically completes the matrix system (1-2) with scalar equations expressing that distances between any border point P_j and three out of the four solid points A_i also remain constant.

I.3 - Validating examples of the rigid body model

3-D meshed mechanisms with planar motions supplied numerical results that were used to check how correctly the new model behaved with regard to the following treatments : substitution of the four point masses to the initial discretization node points, handling of the kinematics and dynamics of these points held fixed distances apart, return of rigid body results to the original mesh points for numerical or graphical interpretation. Figure 2 shows a three link manipulator with its initial 3-D finite element subdivision : this perspective plot comes from a rigid body analysis of the movement for controlled rotations at the joints. The dynamics analysis of a triple pendulum subjected to gravity was also achieved . Results of both PLEXUS and PAMTIN, a rigid body analysis code developed in our laboratory, [7], and based on Wittenburg's formalism, [8], agreed very well and were compared to experimentally measured data. A time step reduction ratio of about 100 is reached when the rigid instead of deformable member hypothesis is used in PLEXUS.

II - MODELING MECHANICAL MEMBER CONNEXIONS IN PLEXUS

In order for PLEXUS to become a real computer code for mechanisms, the different types of substructure linkages used robotics had to be implemented and made easily available to users, in the form of a library of kinematical models of robot joints. Three most commonly encountered robot connexions are presently implemented : the prismatic, revolute and spherical joints. Their specific treatment is now detailed.

II.1 - Mechanical robot connexion model

A joint between two structural components A and B is treated as a new type of two-node finite element : each of these nodes L_A and L_B has six d.o.f., can possess mass and inertia properties, and is rigidly attached to one of the two components incident on that joint. This last condition is achieved by imposing again that distances between point L_A (resp. L_B) and a limited number of "vicinity" nodes, arbitrarily taken from the original node mesh of component A (resp. B), stay constant, figure 3, thus adding extra equations to the matrix system (1-2). The primary use of these vicinity nodes is to provide point L_A with six d.o.f. when they themselves only have three d.o.f.. However, since a local rigidification is thus enforced on component A (resp. B), appropriate care is required in the choice of vicinity points.

II.2 - Revolute joint, prismatic joint and spherical joint treatment

Depending on the connexion type, some of the six possible relative movements between nodes L_A and L_B must be prevented. For spherical joints, their relative rotation is free, whereas the three translational d.o.f. of the two nodes are identical, which merely yields three equations of type (1-2). For both revolute and prismatic joints, the single relative d.o.f. takes place about or along a $\vec{V}$ axis, the initial direction of which is specified in R_o during user's input. The two distinct problems of connexion relations and axis updating are then handled as follows.

Let $d\vec{u}_A$ and $d\vec{\theta}_A$ (resp. $d\vec{u}_B$, $d\vec{\theta}_B$) denote position and orientation change vectors at node L_A (resp. L_B) during a given time step. For a revolute (resp. prismatic) joint, displacement increments $d\vec{u}_A$ and $d\vec{u}_B$ (resp. rotations $d\vec{\theta}_A$, $d\vec{\theta}_B$) are identical, and so are the projections on a plane normal to $\vec{V}$ of rotation changes $d\vec{\theta}_A$ and $d\vec{\theta}_B$

(resp. translations $\vec{du}_A$, $\vec{du}_B$). Hence five scalar relationships of type (1-2) are written, with C_{ij} coefficents updated after each time step since they involve vicinity node coordinates and $\vec{V}$ axis components. The updating of the $\vec{V}$ axis direction from $\vec{V}_n$ at time t_n to $\vec{V}_{n+1}$ at t_{n+1} is then obtained by :

$$\vec{V}_{n+1} = \vec{V}_n + \left(\frac{1}{2} \cdot (\vec{d\theta}_A + \vec{d\theta}_B) \right) \wedge \vec{V}_n \qquad (2\text{-}1)$$

where the averaged rotation increment of the L_AL_B connexion element is assumed small enough over the Δt time step to be handled as an instantaneous rotation vector.

II.3 - Validating examples of robot joint models

Two spatial mechanisms with 3-D motions are presented, chosen among different test cases that yielded confidence in the treatment of robot joints. On figure 4, a double pendulum made up of two co-planar triangles is shown in its static equilibrium position : the large triangle is hinged about a fixed vertical axis on its left side, while the small triangle is free to rotate about a horizontal axis normal to the first triangle at its right end. An angular velocity step is applied about the vertical hinge so that the small triangle is centrifuged towards a radial equilibrium position. PLEXUS results are smartly compared to the analytical rigid body solution as regards the vertical displacement of the small triangle's lower end.

On figure 5.a is sketched a whole family of anthropomorphic manipulators with a closed parallelogram loop. The dimensions and inertial data for all members were taken from a real structure. The dynamical behavior of this robot was analyzed using several computer codes involving models of rigid and/or deformable solids, [10], [11]. Three linearly varying driving torques were input at the three motorized joints during two seconds. The initial subdivision of the five members into beam elements and added joint masses, was reduced using the rigid body model. On figure 5.b, the $\vec{Y}$ and $\vec{Z}$ displacements of the robot end point E are drawn from PLEXUS and MESAVERDE which agree very well. Other codes also supplied data not reported here.

III - SIMULATION OF ROBOT CONTROL IN PLEXUS

PLEXUS being at this point capable of carrying out dynamic analyses of robots, the last point to be developed was to integrate control strategies as they actually occur in monitoring robot movements. This phase is presently being implemented and numerical results will be discussed after briefly recalling some theoretical ideas.

III.1 - Robot control principles and implementation in the code

Although advanced control concepts are progressively introduced into robot controllers, such as feedforward or computed torque controls which somehow account for robot links dynamics, [12], most industrial systems used today simply consist of independent position, and rarely velocity, servo loops which control each joint angular or linear d.o.f. separately. In these proportional-derivative controllers, constant control gains K_p and K_v are used while position and velocity feedback data are delivered by encoders and tachometers.

Though more complicated control systems can be easily simulated, involving integral or double-derivative models with variable gains, developments in PLEXUS have so far been limited to P.D. control simulation. The user must input for each independently controlled robot joint the two constant gains K_p and K_v, and the either analytical or tabulated prescribed variation $q_r(t)$ from which $\dot{q}_r(t)$ is automatically calculated. The control force/torque generated at the joint from time t_n to $t_{n+1} = t_n + \Delta t$ is then :

$$F_n / C_n = -K_p \cdot (q_n - q_r(t_n)) - K_v \cdot (\dot{q}_n - \dot{q}_r(t_n)) \qquad (3\text{-}1)$$

in which equation position and velocity control terms act as springs with stiffness values K_p and K_v and with varying equilibrium positions q_r and $\dot{q}_r$. These force/torque values are then supplied to the system's dynamics equations which give rise to nodal position and orientation increment vectors $d\vec{u}$ and $d\vec{\Theta}$ satisfying all d.o.f. constraints. Depending on the prismatic or revolute type of the joint, its generalized coordinate q and velocity $\dot{q}$ are updated at time t_{n+1} using the following equations :

$$\left.\begin{array}{l} dq = (d\vec{u}_B - d\vec{u}_A) \cdot \vec{V} \\ dq = (d\vec{\Theta}_B - d\vec{\Theta}_A) \cdot \vec{V} \end{array}\right\} \qquad q_{n+1} = q_n + dq \quad \& \quad \dot{q}_{n+1} = \dot{q}_n + dq/\Delta t \qquad (3\text{-}2)$$

III.2 - Validating examples of P.D. control model

A single beam shown on figure 6, is hinged at its origin and carries a lumped mass M at its end. Reference rotations of amplitude $\pi/2$ are successively prescribed in the shapes of a step, a linear ramp and a parabolic variation. The beam is first assumed to be rigid and have a negligible mass compared to M : its dynamic behavior, expressed as : $C(t) = I \cdot \ddot{\Theta}(t)$, can be analytically integrated twice after replacing the driving torque C(t) by expression (3-1). PLEXUS results for different couples of gain values K_p and K_v are in complete agreement with the corresponding analytical solutions. Both show that P.D. control substantially differs from the pure mechanical solution when velocity, acceleration and torques are considered. The same beam is then treated as a real flexible space frame member : figure 6 superimposes the curves of the reference parabolic angular variation at the beam's origin, of the rigid body response, and of the flexible body responses at its origin and end nodes. The chosen K_p and K_v values yield a pseudo-periodic response in all cases, but most important is the influence of the beam's flexibility which induces vibrations and hence a variable angular response along its length.

The spatial anthropomorphic manipulator previously investigated in paragraph II.3 is presently studied : it has been subjected to independent P.D. joint control at its three link connexions, with reference signals $\Theta_{ir}(t)$ prescribed as ramps. Numerical and graphical results for the rigid body hypothesis are plotted on figure 7 and presently checked against the coupled nonlinear dynamics equations of this robot.

III.3 - Realistic modeling of a D.C. electrical motor control

The preceding robot controller model is not correct since it bypasses the electromechanical behavior of the D.C. motor and directly relates the actuating efforts to position and velocity tracking errors through equation (3-1). Instead, a D.C. motor shows mechanical properties, mass, inertia, friction, that must be accounted for, while its electrical equations relate the generated torque to the input current tension V_a which is really the quantity monitored by P.D. control, see Appendix A.3. Measurements are thus presently under way on one of the motorized revolute joints of the experimental MAC robot developed at the ENSAM Robotics Laboratory, [16]. Figure 8 sketches the experimental set up from which real data are to be obtained. Besides checking values already available in technical notices, this work is aimed at characterizing the actual controller parameters and at measuring friction dissipated torques for different angular velocities and different loads mounted on the output axis of the gear train. Once available, these data will be supplied to PLEXUS so that comparisons can be carried out for different reference signals $\Theta_r(t)$ between experimental and simulation results, namely : torques generated on the rotor and available on the output axis, variations of angular position, velocity and acceleration, and finally member strains and stresses at different points. The full three rotation manipulator will later be simulated, since experimental data for the other two motorized revolute joints, of a simular constitution, can readily be obtained.

IV - CONCLUSIONS

Applying to robots or other spatial mechanisms the developments described in this paper opens new analysis prospects for the better knowledge of their mechanical behavior. Other models of robot controllers and of mechanical actuators can easily be implemented. The calculation of element stresses for various member shapes and different constituting materials will first allow comparisons with experimental data, and then authorize to estimate how influent parameter changes are, when experimental values are no longer available. In particular, the consequences of the rigid / deformable body hypothesis for robot links can now be rapidly quantified, especially in terms of positioning precision and stabilization for flexible robots. Furthermore, the possibility of changing reference and control variables, e.g. replacing joints generalized coordinates $\{q\}$ by end effector cartesian coordinates $\{X\}$ and thus bypassing the $\{q\} = F^{-1}(\{X\})$ transformation step, will permit to simulate and measure the effects of such new control models on manipulator performances, especially when highly flexible links are involved . An experimental platform of a single flexible aluminium beam is now being instrumented to provide experimental data for comparison.

ACKNOWLEDGEMENTS

This work has been supported by DGRST grants N° 83-S-1066 and 85-P-1454 from the French Ministry for Research and Technology. This support is gratefully acknowledged. The competence and friendship of our colleagues at the CEA must also be recognized since they permitted a fruitful collaboration.

APPENDICES

A.1- Explicit solution algorithm of the updated Lagrangian type

Let $\{x\}$, $\{u\}$, $\{v\}$ and $\{\gamma\}$ denote nodal coordinate, displacement, velocity and acceleration vectors ; $\{\sigma\}$ and $\{\varepsilon\}$ internal element stress and strains vectors ; $\{F_e\}$ and $\{F_i\}$ external and internal nodal force vectors ; and [M] the diagonal lumped mass matrix of the system. All quantities at time t_n, with subscript n, are assumed to be known. The successive steps to calculate quantities at time : $t_p = t_n + \Delta t$, with p = n+1, are :

α - new coordinate vector : $\{x_p\} = \{x_n\} + \Delta t \cdot \{v_n\} + \Delta t^2/2 \cdot \{\gamma_n\}$

β - strain vector variation : $\{\Delta\varepsilon\} = [B^k] \cdot \{\Delta u^k\}$ for any element k.

γ - new stress vector : $\{\sigma_p\} = \{\sigma_n\} + \{\Delta\sigma(\{\sigma_n\}, \{\Delta\varepsilon\}, \dots)\}$

δ - internal force vector : $\{F_{i\,p}\} = \int_k [B^k]^T \cdot \{\sigma_p\} \cdot d\Omega_k$

ε - new acceleration and velocity vectors, after matrix assembly :

$$\{\gamma_p\} = [M]^{-1} \cdot (\{F_{e\,p}\} - \{F_{i\,p}\})$$

$$\{v_p\} = \{v_n\} + \Delta t/2 \cdot (\{\gamma_n\} + \{\gamma_p\})$$

then go back to α for the next time step.

A.2- Implicit treatment of relations between d.o.f.

Any linear coupling relation : $[C] . \{u\} - \{D\} \geqslant 0$ can be handled by PLEXUS, the inequality conditions standing for one-sided contacts. To compute the required complementary forces $\{F_{c\,p}\}$ to be added to the dynamics equations at time t_p, these relations are derivated twice into : $[C_p] . \{\gamma_p\} = \{S_p\}$ (1)
the R.H.S. vanishing when [C] and [D] are time independent.

Introducing Lagrange's multipliers $\{\lambda\}$ into $\{F_c\}$: $\{F_c\} = [C]^T . \{\lambda\}$ (2)
the system's dynamics governing matrix equation becomes :

$$[M] . \{\gamma_p\} = \{F_{e\,p}\} - \{F_{i\,p}\} + [C_p]^T . \{\lambda_p\} \quad (3)$$

Accelerations $\{\gamma_p\}$ are solved for and put back into (1) that is rewritten as :

$$[C_p] . [M]^{-1} . [C_p]^T . \{\lambda_p\} = \{S_p\} - [C_p] . [M]^{-1} . (\{F_{e\,p}\} - \{F_{i\,p}\}) \quad (4)$$

The $[H] = [C].[M]^{-1}.[C]$ matrix is symmetrical and can be inverted if the coupling relations are compatible : the $\{\lambda_p\}$ multipliers are then found, leading to connexion forces $\{F_{c\,p}\}$ and finally accelerations $\{\gamma_p\}$ by means of (2) and (3).

Concerning any rigid body S, the constant distance conditions (1-2) to hold between points belonging to S read at time t_p : $[C_p] . \{v_p\} = \{0\}$ (5)
where C_{ij} coefficients, functions of nodal point coordinates $\{x_p\}$, are known at t_p. When $\{v_p\}$ is replaced by the expression given in A.1.ε, this becomes :

$$[C_p] . \{\gamma_p\} = - 2/\Delta t . [C_p] . (\{v_n\} + \Delta t/2 . \{\gamma_n\}) \quad (6)$$

leading back to equation (1) above with the appropriate expression for $\{S_p\}$.

A.3 - Electrical and mechanical equations for a D.C. electrical motor

Let R and L denote the resistance and self inductance of the coils ; I and V_a the input current intensity and tension ; C_m the torque generated by the motor ; N the speed reduction ratio ; Ω_m the rotor's angular velocity ; E the counterelectromotive force. The governing electrical equations are :

$$V_a = R . I + L . dI/dt + E \quad ; \quad E = K_E . \Omega_m \quad ; \quad C_m = K_m . I$$

where K_E and K_m are constant coefficients usually taken equal. The mechanical equation is :

$$N . C_m = J_{eq} . \ddot{\theta}_m + f . \dot{\theta}_m + C_f + C_a$$

in which J_{eq} is the total moment of inertia of rotating parts reflected to the output axis, $\dot{\theta} = N.\Omega_m$ the output angular velocity, f the viscous friction coefficient, C_f the dry friction torque and C_a the remaining available torque used to move members. The analogical angular velocity control loop is considered to be linear with a saturation level, i.e. :

$$V_a = - K_a . (\Omega_m - \Omega_{mr}) \quad \text{with} \quad V_a \leqslant V_{a\,max}$$

Ω_{mr} being the reference angular velocity of the motor.

REFERENCES

[1] - Système CASTEM - Programme PLEXUS - Manuel d'utilisation - CEA - Update 01/87

[2] - M. LEPAREUX, H. BUNG - Système CASTEM - Programme PLEXUS - Présentation résumée des possibilités - DEMT/SMTS/LAMS/83-84 Report - Spring 83

[3] - Calcul et Analyse des Structures en Mécanique et Thermique - Fiches de Validation - DEMT/SMTS/LAMS/82-75 Report - Autumn 82

[4] - A. HOFFMANN, M. LEPAREUX - Le calcul des robots : une approche à l'aide de CASTEM - DEMT/SMTS/82-83 Report - Spring 83

[5] - G. BIANCHI, W. SCHIEHLEN - Dynamics of Multibody Systems - IUTAM/IFToMM Symposium - Udine - SPRINGER-VERLAG - 85

[6] - K.J. BATHE, E.L. WILSON - Numerical Methods in Finite Element Analysis - PRENTICE HALL - 76

[7] - F. SAKHITAB - Contribution à la simulation du comportement dynamique des mécanismes tridimensionnels. Application à la conception de robots manipulateurs - Thèse de Docteur Ingénieur - 6/84

[8] - J. WITTENBURG - Dynamics of systems of rigid bodies - B.G. TEUBNER - Stuttgart - 77

[9] - H. BUNG, M. LEPAREUX - Dynamique des robots - Mécanismes spatiaux - Articulations - DEMT/86-201 Report - 6/86

[10] - J. WITTENBURG, U. WOLZ - The program MESA VERDE for robot dynamics simulations - 3rd Int. Symposium of Robotics Research - Gouvieux - 10/85

[11] - A. BARRACO - Dynamique des Systèmes Mécaniques Déformables - Thèse de Doctorat d'Etat - Paris - 6/83

[12] - C.H. AN, C.G. ATKESON, J.M. HOLLERBACH - Experimental determination of the effect of feedforward control on trajectory tracking errors - IEEE Int. Conf. on Robotics and Automation - San Francisco - 4/86

[13] - W. KHALIL - Contribution à la commande automatique des manipulateurs avec l'aide d'un modèle mathématique des mécanismes - Thèse de Doctorat d'Etat - Montpellier - 10/78

[14] - P. COIFFET - Les robots : modélisation et commande - Tome 1 - Hermès 81

[15] - P. ANDRE, J.M. KAUFMANN, F. LHOTE, J.P. TAILLARD - Les robots : constituants technologiques - Tome 4 - Hermès 83

[16] - T. KIRATI - Etude et réalisation d'un robot modulaire allégé utilisant les matériaux composites - Thèse de Docteur Ingénieur - Paris - 4/86

FIGURES

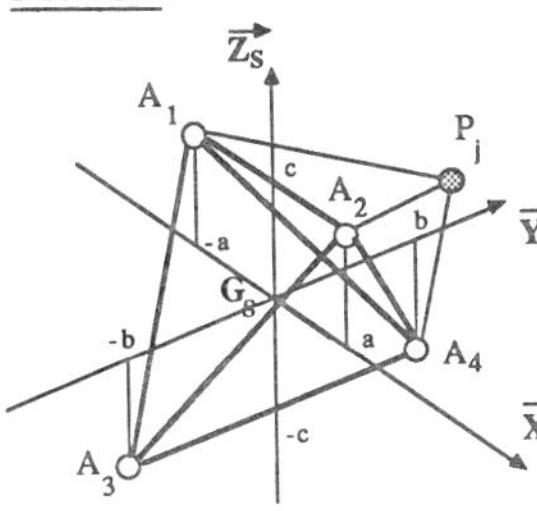

$$a = \sqrt{\frac{Jy + Jz - Jx}{M_s}} \qquad b = \sqrt{\frac{Jz + Jx - Jy}{M_s}} \qquad c = \sqrt{\frac{Jx + Jy - Jz}{M_s}}$$

$$m_i = \frac{M_s}{4} \qquad i = 1, 4$$

Figure 1: Rigid body model with four point masses $A_i(m_i)$ and border nodes P_j held fixed distances apart

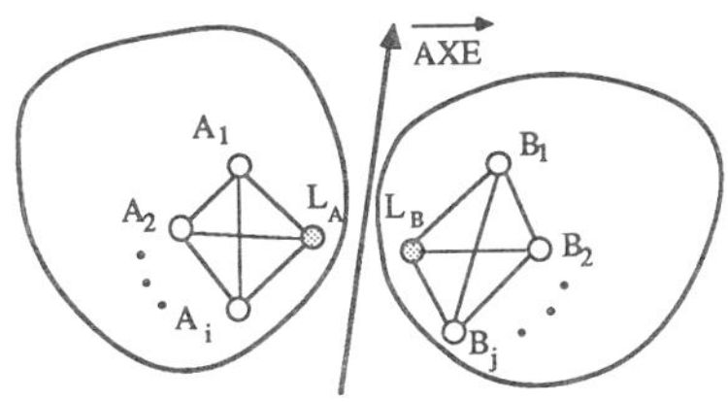

Figure 3: Definition of kinematical connexion element L_AL_B between two bodies

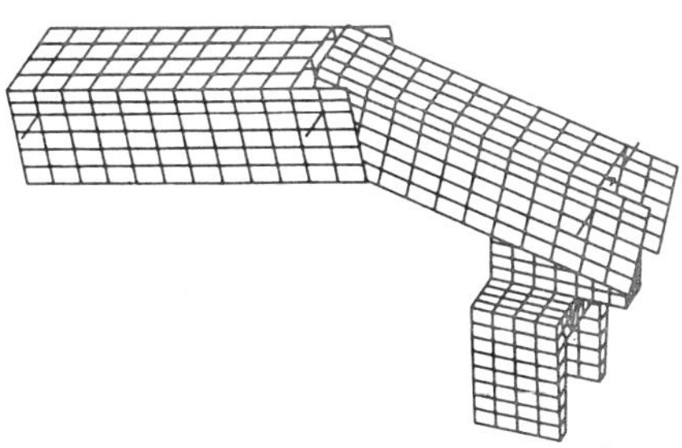

Figure 2: Rigid body model application to manipulator with 3 revolute joints : displaced 3-D mesh

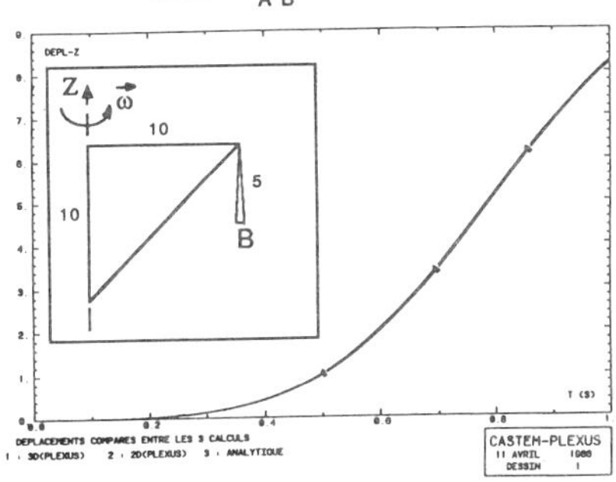

Figure 4: Double triangle pendulum definition and vertical displacement at point B

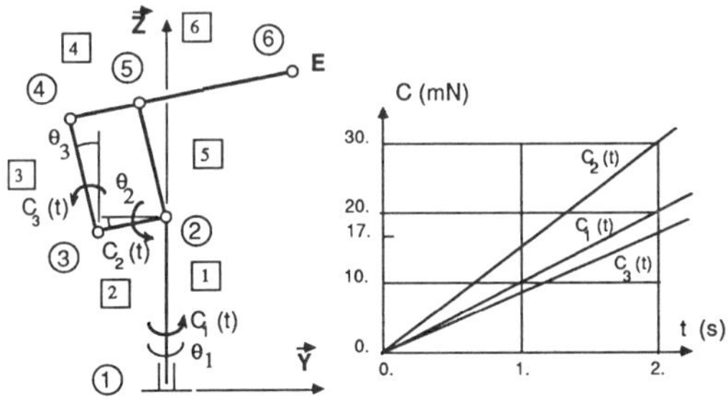

Figure 5-a: Definition of balanced anthropomorphic manipulator and of input torques at the three revolute joints

Members □ masses : 4.468 - 0.705 - 1.269 - 1.069 -1.560 - 2.266 (kg)
lengths : 0.7 - 0.25 - 0.45 - 0.25 - 0.45 - 0.67 (m)

Nodes ○ masses : 0.0 - 10.80 - 2.84 - 4.86 - 5.40 - 7.40 (kg)

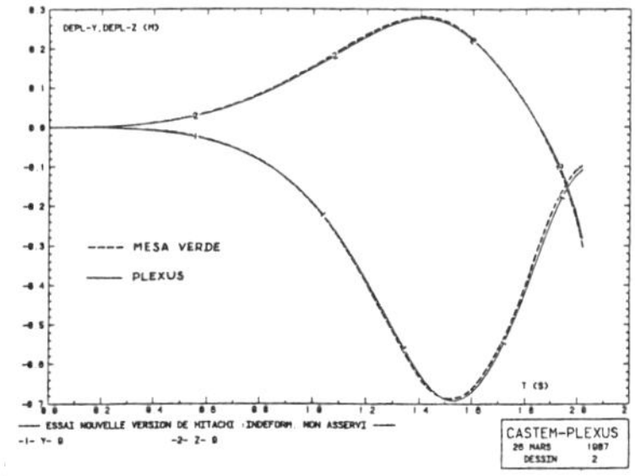

Figure 5-b: Output Y and Z displacements at end point E of manipulator

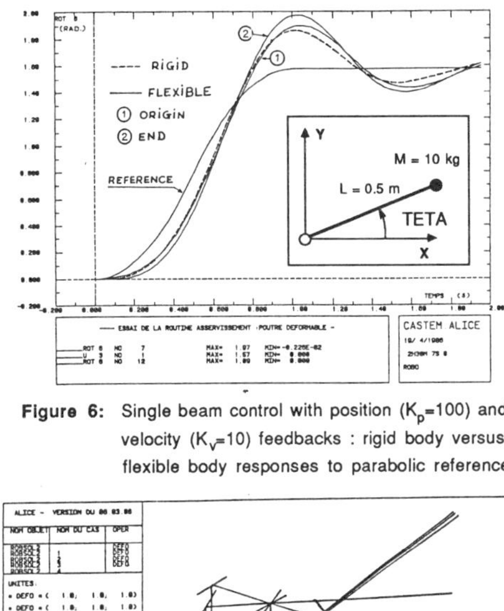

Figure 6: Single beam control with position (K_p=100) and velocity (K_v=10) feedbacks : rigid body versus flexible body responses to parabolic reference

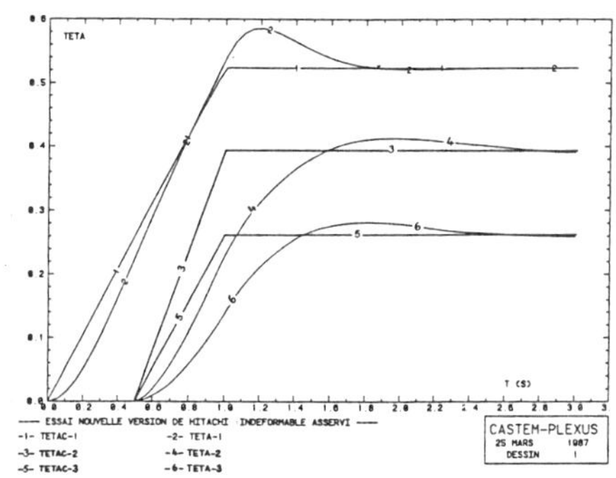

Figure 7-a: Independent joint control (position K_p=100 and velocity K_v=10 feedback) for the anthropomorphic robot : responses to ramp inputs

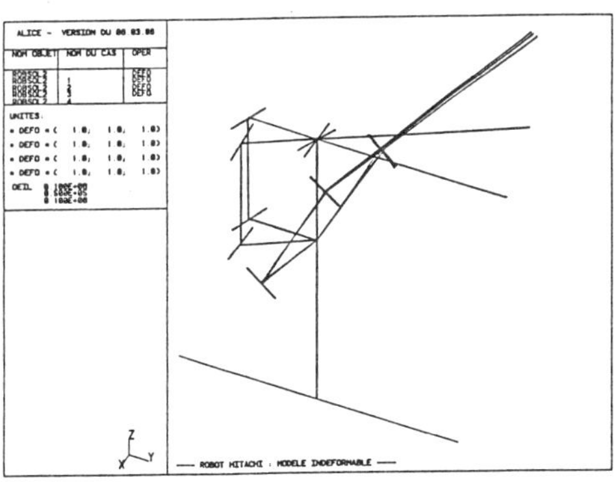

Figure 7-b: Configurations in space of anthropomorphic manipulator at times : 0. , 0.5 , 1. ,1.5 and 2. sec.

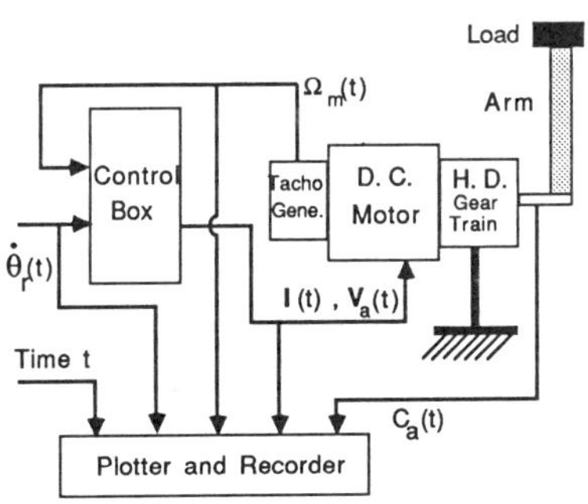

Figure 8: Experimental setup for controlled motor parameters measurement

Computer-Assisted Robot Modeling for Real-Time Dynamic Control

Sandeep Gupta

Vigyan Research Asociates, Inc., Hampton, VA 23666-1325, USA

INTRODUCTION

A major difficulty in the real-time implementation of dynamic controllers for robot arms has been the large computational time required for the solution of robot dynamics. Recursive techniques have been applied successfully for efficient computation of the actuator torques [1,2]. However, the matrix structure of closed-form Lagrange-Euler formulation is very attractive from the control viewpoint. This paper presents an approach for efficient real-time evaluation of the dynamic matrices for robot arms based on the Lagrangian formulation.

General formulations for robot dynamics like Lagrange-Euler formulation [3] contain many parameters which are constant for a specific robot arm, e.g. the masses and moments of inertia of the links, the link lengths, etc. Frequent solution of these equations required for real-time control of robot arms causes repetition of many mathematical operations involving the constant parameters. The computational content of control algorithms can be reduced by performing these computationally intensive operations during preliminary processing by using numerical and symbolic manipulations [4,5]. The inverse dynamic solution then requires only those computations that are necessitated by changes in the system variables, eliminating the duplication of computational effort in repetitive solutions. The minimal computation needed for the solution of this simplified model is feasible for real-time control. Further, the independent calculations required for model evaluation make this algorithm amenable to parallel computation.

MODEL DEVELOPMENT

An efficient robot model is developed here by applying the symbolic polynomial approach [4,5] to the Lagrangian formulation of robot dynamics [3]. This discussion is limited to robot arms with revolute joints, and can easily be extended to include prismatic joints. The model is developed with respect to some reference kinematic configuration chosen as the home position. The dynamic elements are obtained as functions of joint coordinates relative to the home position.

Kinematics: The kinematics of a robot arm are described using the Denavit-Hartenburg convention. The homogeneous transformation of link i coordinates to link i-1 coordinates is given by

$$A_i = \begin{bmatrix} c\theta_i & -c\alpha_i \; s\theta_i & s\alpha_i \; s\theta_i & a_i \; c\theta_i \\ s\theta_i & c\alpha_i \; c\theta_i & -s\alpha_i \; c\theta_i & a_i \; s\theta_i \\ 0 & s\alpha_i & c\alpha_i & d_i \\ 0 & 0 & 0 & 1 \end{bmatrix} \quad (1)$$

where θ_i, d_i, a_i and α_i are the joint and link parameters defined by the Denavit-Hartenburg representation to describe the kinematic relationship between link i-1 and link i.

The joint parameter, θ_i, includes the effect of rotation of the ith joint. Rewriting, θ_i as

$$\theta_i = \theta_{hi} + q_i \quad (2)$$

where θ_{hi} is the rotation about z_{i-1} axis required to define the home position, and

q_i is the joint coordinate or joint angle relative to the home position.

This is done to separate the constant part of the transformation, θ_{hi}, from the variable joint coordinate, q_i, which is expected to change in each solution iteration. The homogeneous transformation matrix, A_i, then becomes

$$A_i = \begin{bmatrix} c\theta_{hi}cq_i - s\theta_{hi}sq_i & -c\alpha_i s\theta_{hi}cq_i - c\alpha_i c\theta_{hi}sq_i & s\alpha_i s\theta_{hi}cq_i + s\alpha_i c\theta_{hi}sq_i & a_i c\theta_{hi}cq_i - a_i s\theta_{hi}sq_i \\ s\theta_{hi}cq_i + c\theta_{hi}\ sq_i & c\alpha_i c\theta_{hi}cq_i - c\alpha_i s\theta_{hi}sq_i & -s\alpha_i c\theta_{hi}cq_i + s\alpha_i s\theta_{hi}sq_i & a_i s\theta_{hi}cq_i + a_i c\theta_{hi}sq_i \\ 0 & s\alpha_i & c\alpha_i & d_i \\ 0 & 0 & 0 & 1 \end{bmatrix} . \tag{3}$$

Elements of the matrix A_i above are polynomials of the sines and cosines of the joint coordinate, q_i. The coefficients of these polynomials involve constant trigonometric functions of angles α_i, θ_{hi} and distance parameters a_i, d_i, which can be precomputed.

For example the element, $A_i(2,2)$, is written as

$$A_i(2,2) = a_1\ cq_i + a_2\ sq_i \tag{4}$$

where a_1, a_2 are constant coefficients given by

$$\begin{aligned} a_1 &= c\alpha_i\ c\theta_{hi} \\ a_2 &= -c\alpha_i\ s\theta_{hi}. \end{aligned} \tag{5}$$

The transformation from the coordinates of link i to the base coordinates, is obtained as

$$T_i = A_1\ A_2,\ldots,A_i. \tag{6}$$

Note that the mulitplication of matrices A_i in eq. (6) above require polynomial additions and mulitplications rather than numerical operations. Consequently, the elements of the resulting transformation matrices, T_i, are polynomials of cq_j and sq_j $(j=1,\ldots,i)$.

For notational convenience, the derivatives of T_i with respect to the joint coordinates, are written as

$$U_{ij} = \frac{\partial T_i}{\partial q_j} \tag{7}$$

$$U_{ijk} = \frac{\partial^2 T_i}{\partial q_j \, \partial q_k} . \tag{8}$$

Dynamics: The Lagrangian equations of motion for an n degrees of freedom robot arm are written in matrix form as

$$\tau = H(q)\ddot{q} + C(q,\dot{q}) + G(q) \tag{9}$$

where τ is n×1 actuator torque vector,
q, $\dot{q}$, $\ddot{q}$ are n×1 column vectors for joint coordinates, joint velocities, and joint accelerations, respectively.

$H(q)$ is an n×n inertial matrix, whose elements are given by

$$H(i,j) = \sum_{k=\max(i,j)}^{n} \mathrm{Tr}(U_{kj} J_k U_{ki}^T), \qquad \text{for} \quad i,j = 1,\ldots,n. \tag{10}$$

$C(q,\dot{q})$ is an n×1 column vector representing the centripetal and Coriolis component of the dynamic torques acting on the robot arm.

$$C(q,\dot{q}) = \begin{bmatrix} \dot{q}^T C_1(q)\dot{q} \\ \dot{q}^T C_2(q)\dot{q} \\ \vdots \\ \dot{q}^T C_n(q)\dot{q} \end{bmatrix} \tag{11}$$

where $C_i(q)$ are n×n matrices such that $\dot{q}^T C_i(q)\dot{q}$ gives the centripetal and Coriolis forces acting at the ith joint.

Elements of the matrix, C_i, are

$$C_i(j,\ell) = \sum_{k=\max(i,j,\ell)}^{n} \mathrm{Tr}(U_{kj\ell} J_k U_{ki}^T), \quad \text{for } i,j,\ell = 1,\ldots,n. \tag{12}$$

Finally, $G(q)$ is a n×1 gravity column vector given by

$$G_i(q) = -\sum_{k=i}^{n} m_k \, g^T \, U_{ki} \, \bar{r}_k, \qquad \text{for} \quad i = 1,\ldots,n. \tag{13}$$

All variables in eqs. (10), (12), and (13) have constant known numerical values, except the derivative matrices U_{ij} and U_{ijk}, which are written in terms of the joint variables. Substituting the symbolic polynomial form of the derivative matrices and the explicit numerical values of other parameters, the expressions for the elements of the dynamic matrices are simplified to polynomials of the sines and cosines of the joint coordinates.

Example: Consider the development of a symbolic polynomial model for the PUMA arm shown in Fig. 1. The home position at which all joint coordinates are defined to be zero, is shown in this figure. A 3 degrees of freedom model is considered here for simplicity. A 6 dof model will follow exactly the same procedure - the only difference being that more symbolic variables will be involved with the larger number of joint coordinates; and in general, the polynomials of the dynamic elements will be larger. The kinematic parameters used in this example are shown in Table 1, and estimated values of the dynamic parameters are given in Table 2.

Consider the derivation of element H(1,1) of the inertial matrix,

$$H(1,1) = Tr(U_{11}J_1U_{11}^T) + Tr(U_{21}J_2U_{21}^T) + Tr(U_{31}J_3U_{31}^T). \tag{14}$$

The product matrix $U_{11}J_1$ is

$$U_{11}\ J_1 = \begin{bmatrix} 0.28\ cq_1 & 0 & 4.22\ sq_1 & 7.5\ sq_1 \\ 0.28\ sq_1 & 0 & -4.22\ cq_1 & -7.5\ cq_1 \\ 0 & 0 & 0 & 0 \\ 0 & 0 & 0 & 0 \end{bmatrix} \tag{15}$$

and

$$U_{11}^T = \begin{bmatrix} cq_1 & sq_1 & 0 & 0 \\ 0 & 0 & 0 & 0 \\ sq_1 & -cq_1 & 0 & 0 \\ 0 & 0 & 0 & 0 \end{bmatrix} \tag{16}$$

This gives

$$Tr(U_{11}\ J_1\ U_{11}^T) = 4.5. \tag{17}$$

Similarly, the second trace term is

$$\mathrm{Tr}(U_{21}\ J_2\ U_{21}^T) = -54.277\ s^2q_2 + 9.5\ c^2q_2 + 1.363 \tag{18}$$

and the third trace term is

$$\begin{aligned}\mathrm{Tr}(U_{31}\ J_3\ U_{31}^T) = {} & 0.37\ s^2q_2s^2q_3 + 26.72\ cq_2sq_2cq_3sq_3 + 0.37\ c^2q_2c^2q_3 \\ & + 0.432 + 13.73\ s^2q_2c^2q_3 + 13.73\ c^2q_2s^2q_3 \\ & + 42.157\ s^2q_2cq_3 + 42.157\ cq_2sq_2sq_3 + 0.74\ s^2q_2.\end{aligned} \tag{19}$$

Adding the three terms as in eq. (14) gives

$$\begin{aligned}H(1,1) = {} & 6.295 - 53.537\ s^2q_2 + 9.5\ c^2q_2 + 0.37\ s^2q_2s^2q_3 \\ & + 26.72\ cq_2sq_2cq_3sq_3 + 0.37\ c^2q_2c^2q_3 + 13.73\ s^2q_2c^2q_3 \\ & + 13.73\ c^2q_2s^2q_3 + 42.157\ s^2q_2cq_3 + 42.157\ cq_2sq_2sq_3.\end{aligned} \tag{20}$$

Polynomials for other elements of the dynamic matrices, H, C_1, $C_2,\ldots,C_n$ and G, are derived in a similar manner.

As can be seen from the example, deriving the symbolic polynomials by hand is tedious and error-prone. Also, modification of the model would require repetition of the whole effort. Computer generation of these polynomials is highly desirable. Software for the numerical and symbolic manipulations required for automatic generation of these polynomials has been developed.

CONCLUSION

Dynamic matrices for the robot arm at any point along its trajectory are obtained by the evaluation of its symbolic polynomials at the joint coordinates of that position. The polynomials are identical for many elements, and frequently, larger polynomials of elements associated with proximal links include the smaller polynomials of the remote links; therefore, computing the same term twice should be avoided in the model implementation. The exact number of operations required for the solution would depend on the robot structure, choice of the home position and the specific implementation of the model. However, in general, the computational load of this algorithm is far lesser than that for recursive techniques. (Estimated computational content of the 3 dof PUMA arm model : 70 multiplications and 55 additions).

Real-time computation of the dynamic matrices is very useful in the implementation of sophisticated control algorithms developed using modern state-space control methods.

Finally, the polynomials for different dynamic elements can be evaluated independent of each other. Thus, a suitable number of co-processors can be used with good load balancing and low idle times in a parallel computing configuration for fast computation of the inverse dynamic solution.

REFERENCES

[1] Luh, J. Y. S., Walker, M. W. and Paul, R. P. C.: "On-Line Computational Scheme for Mechanical Manipulators," ASME J. Dyn. Sys. Meas. and Control, Vol. 102, No. 2, 1980, pp. 69-70.

[2] Hollerbach, J. M.: "A Recursive Formulation of Lagrangian Manipulator Dynamics," IEEE Trans. Systems, Man., Cybernet., SMC-10, 1980, pp. 730-36.

[3] Lee, C. S. G.: "Robot Arm Dynamics," Tutorial on Robotics, IEEE Computer Society, 1983, pp. 93-102.

[4] Gupta, S.: "The Symbolic Polynomial Technique for Modeling and Real Time Control of Robot Arms," Master's Thesis, University of Virginia, January 1987.

[5] Townsend, M. A. and Gupta, S.: "Automated Modeling and Rapid Solution of Robot Dynamics Using the Symbolic Polynomial Technique," ASME J. Mechanism, Transmission and Automation in Design (submitted for review).

Table 1. Kinematic Parameter of PUMA Arm

Link i	θ_i	d_i	a_i	α_i
1	-90.0	0.0	0.0	90.0
2	90.0	0.25	0.86	0.0
3	90.0	0.0	0.0	90.0

Table 2. Estimated Dynamic Parameters of PUMA Arm

Link i	M_i	I_{ixx}	I_{iyy}	I_{izz}	$\bar{x}_i$	$\bar{y}_i$	$\bar{z}_i$
1	50.0	4.5	4.5	0.56	0.0	0.0	0.15
2	109.0	10.8	26.9	35.1	-0.43	0.0	0.0
3	57.0	14.1	14.1	0.74	0.0	0.0	0.43

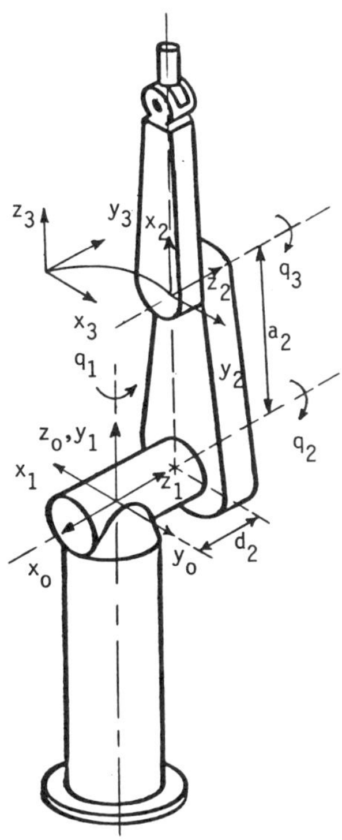

Fig. 1. PUMA Robot Arm

Optimal Distributive Relation Between Fingertip Forces for the Active Robot Hand to Manipulate an Object

Guo Gongliang and Zhang Qi-xian

Research Division for Mechanisms and Robotics, Beijing Institute of Aeronautics and Astronautics, Beijing, The People's Republic of China

In this paper, the dynamic equations are established for the active robot hand to manipulate an object. According to the hand design features and certain assumptions, we outline the geometrical and frictional conditions that must be satisfied for the hand to manipulate an object, and propose the objective performance functional of the optimal distributions of its fingertip forces. Further, we adopt the integral SUMT of the closed-region-functional extreme value to determine the force functional extreme value to determine the force function for each fingertip of the hand. Finally, an example is given for the A3-A3-A3 kind of the active hand(The hand is composed of three fingers, each with 3 degrees of freedom, while A3 signifies that the fingertip contact with the object is a point contact with friction) to make the pen draw a planar curve.

Introduction

Recently, the chief study of the robot hand concentrates on the following aspects:(1) originated from the general functions of human hand being studied and simulated, the pricipal function requirements of the robot hand are investigated in order to select and design the proper hand mechanisms(9-14); (2) for realizing motion with specified loci and velocities, the kinematic equations and the dextrous working space of the finger mechanisms are analyzed, and the suitable force/position control technique is studied(9,10); (3) the Coulomb frictional condition between the fingertip and the object manipulated and the stiffness, strength, and stability for the hand to grasp an object are studied with the view of controlling the internal gripping forces and the small displacement of the hand/object(9,10).

One of the important questions for study is how to make the efficiency of the actuators be optimal, as they are set up within the small space of a robot hand. On account of this, the present paper only studies how to determine the optimal relations between the fingertip forces for making the hand energy consumed be minimal, in the case of its dynamic equations being taken into consideration. Based on the hand design features, geometrical and frictional conditions being considered, the paper proposes an objective performance functional of the optimal distributions of the fingertip forces for manipulating objects, and then adopts the optimization method of the integral SUMT of the closed-region-functional extreme value to determine the force function for each fingertip of the hand(1-6). As an example, the A3-A3-A3 kind of the hand is taken to manipulate a pen to draw a planar curve and the optimal solutions of the force function for each fingertip are obtained(7).

Objective Performance Functional of the Optimal Distributions of the Fingertip Forces

If the robot hand manipulates an object moving from A to B, its energy consumed in the course of the whole motion is needed to be minimized. Thus it is necessary to establish an objective performance functional with satisfied dynamic equations, geometrical and frictional conditions.

Notations and Assumptions:

(1) As shown in Fig.1, OXYZ is the base coordinate system; $C\xi\eta\zeta$ is the body coordinate system fixed to the moving object, whose mass center is taken as the origin, while the ξ, η, ζ axes are the major, middle and minor axes of the object center inertia symmetry, respectively.

(2) The object is assumed to be rigid with the surface equation $H(\xi,\eta,\zeta)=0$, and the following moments of inertia:

$$J_{\xi\xi} = \int_m (\eta^2+\zeta^2)\,dm \quad , \qquad J_{\xi\eta} = 0$$

$$J_{\eta\eta} = \int_m (\zeta^2+\xi^2)\,dm \quad , \qquad J_{\eta\zeta} = 0$$

$$J_{\zeta\zeta} = \int_m (\eta^2+\xi^2)\,dm \quad , \qquad J_{\zeta\xi} = 0$$

where m is the mass of the object.

(3) The hand is composed of several fingers, and the position vector of each finger contact with the object is described as ${}^{B}\underline{P}_i\,(i=1,2,\cdots,m)$ in the body system.

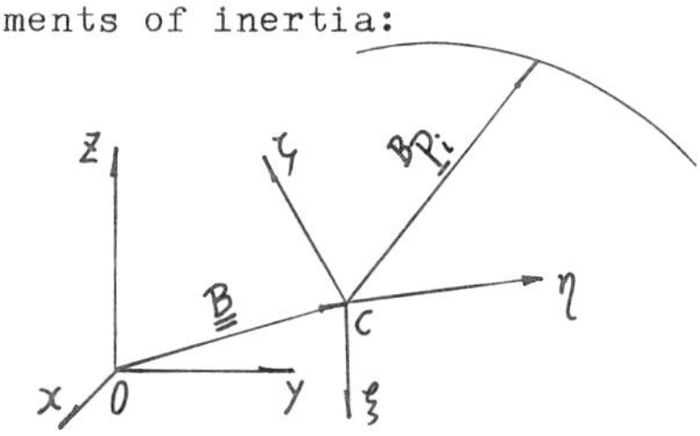

Fig.1

(4) The fingertip exerts a force $\underline{F}_i$ on the object along its internal normal direction and another force $\underline{f}_i$ tagent to the object surface.

(5) For the specified initial state $\underline{X}(0)=\underline{X}_o$, final state $\underline{X}(T_f)=\underline{X}_f$, and other conditions, the kinematic equations of the object can be represented as

$$\begin{array}{ll} X_c=h_1(t) & \psi=h_4(t) \\ Y_c=h_2(t) & \theta=h_5(t) \\ Z_c=h_3(t) & \varphi=h_6(t) \end{array} \qquad (1)$$

where (X_c,Y_c,Z_c) are the base coordinates of the mass center, and (ψ,θ,φ) are the Eular angles about the (X,Y,Z) axes of the base system.

(6) A vector is represented as either an alphabet with an underline

or as a column of homogeneous coordinates; Matrix is represented as an alphabet with two underlines or as a 4x4 matrix. Their reference systems can be indicated by using the left superscripts; moving body system as "B", and base system as "O" but usually omitted.

Dynamic Equations:

According to the relation of the homogeneous coordinate transformation, the transformation matrix from base (X,Y,Z) to moving body (ξ,η,ζ) can be represented as

$$\underline{\underline{B}}=\begin{bmatrix} \cos\psi\cos\theta\cos\varphi-\sin\psi\sin\varphi, & -\cos\psi\cos\theta\sin\varphi-\sin\psi\cos\varphi, & \cos\psi\sin\theta, & X_c \\ \sin\psi\cos\theta\cos\varphi+\cos\psi\sin\varphi, & -\sin\psi\cos\theta\sin\varphi+\cos\psi\cos\varphi, & \sin\psi\sin\theta, & Y_c \\ -\sin\theta\cos\varphi & , \sin\theta\sin\varphi & , \cos\theta & , Z_c \\ 0 & , 0 & , 0 & , 1 \end{bmatrix} \quad (2)$$

From the kinetic theorem of the mass center, we have

$$m\underline{a}=\sum_{i=1}^{m}(\underline{F}_i+\underline{f}_i)+m\underline{G} \quad (3)$$

or

$$m\underline{a}=\underline{\underline{B}}\sum_{i=1}^{m}({}^{B}\underline{F}_i+{}^{B}\underline{f}_i)+m\underline{G} \quad (4)$$

where $\underline{F}_i$ and $\underline{f}_i$ are the normal and tangential forces exerted from the fingertip on the object, $\underline{a}$ is its mass center acceleration and $\underline{G}$ is the gravitational acceleration.

Similarly, from the theorem of the moment of momentum of the object with respect to its mass center, we obtain

$$\underline{\underline{J}}\cdot\underline{\varepsilon}+\underline{\underline{J}}_{ij}\cdot\underline{w}_{ij}=-\sum_{i=1}^{m}(({}^{B}\underline{F}_i x)+({}^{B}\underline{f}_i x))\cdot{}^{B}\underline{P}_i \quad (5)$$

where

$$\underline{\underline{J}}=\begin{bmatrix} J_{\xi\xi} & & & \\ & J_{\eta\eta} & & \\ & & J_{\zeta\zeta} & \\ & & & 1 \end{bmatrix} \qquad \underline{\varepsilon}=\begin{bmatrix} \dot{w}_1 \\ \dot{w}_2 \\ \dot{w}_3 \\ 0 \end{bmatrix}$$

$$\underline{\underline{J}}_{ij}=\begin{bmatrix} J_{\zeta\zeta}-J_{\eta\eta} & & & \\ & J_{\xi\xi}-J_{\zeta\zeta} & & \\ & & J_{\eta\eta}-J_{\xi\xi} & \\ & & & 1 \end{bmatrix} \qquad \underline{w}_{ij}=\begin{bmatrix} w_2w_3 \\ w_3w_1 \\ w_1w_2 \\ 0 \end{bmatrix}$$

$$({}^{B}\underline{F}_i)+({}^{B}\underline{f}_i)=\begin{bmatrix} 0 & ,-F_{i\zeta}-f_{i\zeta} & ,F_{i\eta}+f_{i\eta}, & 0 \\ F_{i\zeta}+f_{i\zeta}, & 0 & ,-F_{i\xi}-f_{i\xi}, & 0 \\ -F_{i\eta}-f_{i\eta}, & F_{i\xi}+f_{i\xi}, & 0 & , 0 \\ 0 & , 0 & , 0 & , 0 \end{bmatrix}$$

$w_1=\dot{\psi}\sin\theta\sin\varphi+\dot{\theta}\cos\varphi$, $w_2=\dot{\psi}\sin\theta\cos\varphi-\dot{\theta}\sin\varphi$, $w_3=\dot{\psi}\cos\theta+\dot{\varphi}$

Geometrical Conditions:

In order to determine the directions for the fingertip to exert forces on the object, the directions of the internal normal line and the tangential surface at the point of contact must be known.

The direction number of the normal line at the point P_i of the object can be expressed as

$$^B m_i = \frac{\partial H(\xi,\eta,\zeta)}{\partial \xi}\bigg|_i , \quad ^B n_i = \frac{\partial H}{\partial \eta}\bigg|_i , \quad ^B l_i = \frac{\partial H}{\partial \zeta}\bigg|_i$$

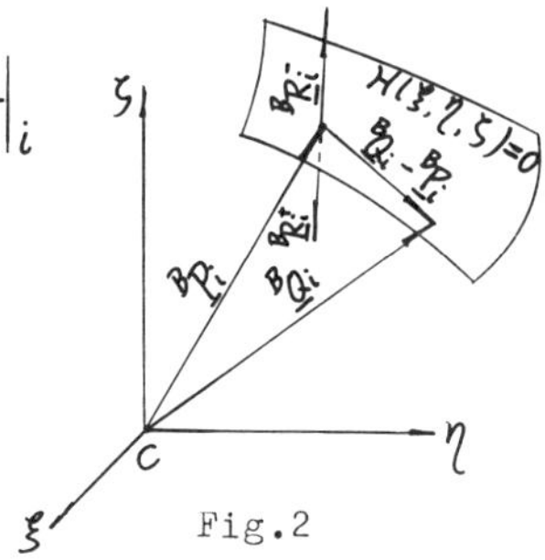

Fig.2

If another point Q_i is taken very close to the point P_i, we have the vector

$$^B\underline{Q}_i - {}^B\underline{P}_i = \{\Delta\xi_i, \Delta\eta_i, \Delta\zeta_i\}.$$

Thus, the vector of the internal normal line at P_i is represented as

$$^B\underline{R}_i^+ = \{s_i \cdot {}^B m_i \ , \ s_i \cdot {}^B n_i \ , \ s_i \cdot {}^B l_i\}$$

where the parameter s_i is taken as +1 or -1.

Due to the angle between vectors $^B R_i^+$ and $^B\underline{Q}_i - {}^B\underline{P}_i$ being always an acute angle, we have

$$s_i(\Delta\xi_i \cdot {}^B m_i + \Delta\eta_i \cdot {}^B n_i + \Delta\zeta_i \cdot {}^B l_i) \geqslant 0 \tag{6}$$

From the inequality (6), the s_i can be given +1 or -1, therefore the vector $^B\underline{R}_i^+$ can be easily determined.

Because the force vector $^B\underline{F}_i$ is parallel to $^B\underline{R}_i^+$, and orthogonal to $^B\underline{f}_i$, we have

$$^B\underline{F}_i = K_i(t) \cdot {}^B\underline{R}_i^+ \quad , \quad K_i(t) \geqslant 0 \ , i=1,2,\cdots,m. \tag{7}$$

$$({}^B\underline{F}_i)^T \cdot ({}^B\underline{f}_i) = 0 \ , \ i=1,2,\cdots,m \tag{8}$$

Frictional Conditions:

The fingertip can exert a force on the object along the tangential surface direction, because their contact is a point contact with friction. If the coefficient of frictin of their contact is assumed as μ , and their contact conditions are ensured against change during the object moving, the relations between the normal force $^B\underline{F}_i$ and the tangential force $^B\underline{f}_i$ must satify the following condition

$$\left\| {}^B\underline{f}_i \right\|_2 \leqslant \mu \cdot \left\| {}^B\underline{F}_i \right\|_2 \quad , \quad (i=1,2,\cdots,m) \tag{9}$$

The Objective Performance Functional:

In order to make the hand energy consumed be as small as possible during manipulating the object from the initial state $\underline{X}_0$ to the final $\underline{X}_f$, we have to use an energy criterion which is to minimize the objective performance functional

$$V(\underline{F}(t)) = \int_0^{T_f} \sum_{i=1}^{m} (({}^B\underline{F}_i)^T \cdot ({}^B\underline{F}_i) + ({}^B\underline{f}_i)^T \cdot ({}^B\underline{f}_i))dt \tag{10}$$

Substituting equation (7) into equation (10), we have

$$V = \int_0^{T_f} \sum_{i=1}^{m} (k_i^2 \cdot K_i^2(t) + ({}^B\underline{f}_i(t))^T \cdot ({}^B\underline{f}_i(t)))dt \tag{11}$$

where $k_i^2 = ({}^B\underline{R}_i^+)^T \cdot ({}^B\underline{R}_i^+)$, $i=1,2,\cdots,m$.

The Integral SUMT of the Closed-region-functional Extreme Value

Now, the problem of optimal distributions of the fingertip forces for the hand to manipulate an object may be described as follows: to find such functions $K_i(t)$ (that is $\underline{F}_i$ and $\underline{f}_i$ of the fingertip forces as to make the objective performance functional equation (11) be of minimal value V_{min}. Clearly, in finding K_i and $\underline{f}_i$, they have to be satisfied with the equality constraint conditions

$$\begin{aligned} &\underline{\underline{B}}\cdot\sum_{i=1}^{m}({}^{B}\underline{F}_i+{}^{B}\underline{f}_i)+m\underline{G}-m\underline{a}=0 \\ &\sum_{i=1}^{m}(({}^{B}\underline{F}_i x)+({}^{B}\underline{f}_i x))\cdot{}^{B}\underline{P}_i+\underline{\underline{J}}\cdot\underline{\varepsilon}+\underline{\underline{J}}_{ij}\cdot\underline{w}_{ij}=0 \\ &K_i(t)\cdot({}^{B}\underline{R}_i^{+})^{T}\cdot({}^{B}\underline{f}_i)=0\ ,\quad (i=1,2,\cdots,m) \end{aligned} \tag{12}$$

and the inequality constraint conditions

$$K_i(t)\geqslant 0\ ,\quad \mu\|{}^{B}\underline{F}_i\|_2-\|{}^{B}\underline{f}_i\|_2\geqslant 0\ ,\ (i=1,2,\cdots,m) \tag{13}$$

the above problem can be solved by using the variational method of the functional extreme value and the integral SUMT of the optimal control.

Conditional Extreme Value and the Initial State for Iteration:

If the inequality constraint conditions (13) are not included for the moment, the functional extreme value can be obtained easily by the method of lagrangian multipliers. Here, the new objective performance functional is formed as

$$\begin{aligned} V_1=\int_0^{T_f}\Big(&\sum_{i=1}^{m}(k_i^2\cdot K_i^2(t)+({}^{B}\underline{f}_i(t))^{T}\cdot({}^{B}\underline{f}_i(t)))+\underline{\lambda}_1(t)\cdot(\underline{\underline{B}}\sum_{i=1}^{m}({}^{B}\underline{R}_i^{+}\cdot K_i(t)+ \\ &{}^{B}\underline{f}_i(t))+m\underline{G}-m\underline{a}(t))+\underline{\lambda}_2(t)(\sum_{i=1}^{m}(K_i(t)\cdot({}^{B}\underline{R}_i^{+}x)+({}^{B}\underline{f}_i x))\cdot{}^{B}\underline{P}_i+ \\ &\underline{\underline{J}}\cdot\underline{\varepsilon}+\underline{\underline{J}}_{ij}\cdot\underline{w}_{ij})+\sum_{i=1}^{m}(\lambda_{3i}(t)\cdot K_i(t)\cdot({}^{B}\underline{R}_i^{+})^{T}\cdot({}^{B}\underline{f}_i(t)))\Big)dt \end{aligned} \tag{14}$$

where the lagrangian undermultipliers are represented as

$$\lambda_1(t)=(\lambda_{11}(t),\lambda_{12}(t),\lambda_{13}(t),\ 0\ )^{T}$$

$$\lambda_2(t)=(\lambda_{21}(t),\lambda_{22}(t),\lambda_{23}(t),\ 0\ )^{T}$$

$$\lambda_{3i}(t)\ ,\ i=1,2,\cdots,m$$

If the equation (12) and (13) are simultaneously taken into consideration, the functions $K_i^{(0)}(t)$ and ${}^{B}\underline{f}_i^{(0)}$ can be obtained analytically or numerically by the method of variation. There are two possible results. If the results are satisfied with the inequality (13), they can serve as the solution of the original problem. On the contrary, the solution will have to be developed by using Taylor's series(In engineering, they can usually converge to the original function.). Thus, we have

$$K_i^{(0)}=\sum_{j=0} a_{ij}^{(0)}\cdot t^j \quad , \quad {}^B\underline{f}_i^{(0)}(t)=\sum_{j=0} \underline{b}_j^{(0)}\cdot t^j \quad (i=1,2,\cdots,m) \tag{15}$$

The coefficients $a_{ij}^{(0)}$ and $\underline{b}_j^{(0)}$ of the first n+1 term of the series (14) may be taken as the initial state for iteration of the integral SUMT.

Integral SUMT Method:

If both the equality and inequality constraint conditions (12) and (13) are taken into consideration, according to the numerical method of the optimal control, the terms of the integral SUMT are composed of

$$\int_0^{T_f} w_1 dt=\int_0^{T_f} r_k(t)\cdot((\underline{B}\cdot\sum_{i=1}^{m}({}^B\underline{R}_i^+\cdot K_i(t)+{}^B\underline{f}_i(t))+m\underline{G}-m\underline{a}(t))^T\cdot(\underline{B}\cdot\sum_{i=1}^{m}({}^B\underline{R}_i^+\cdot K_i+$$

$${}^B\underline{f}_i \quad +m\underline{G}-m\underline{a})+(\sum_{i=1}^{m}(K_i\cdot({}^B\underline{R}_i^+x)+({}^B\underline{f}_i x))\cdot{}^B\underline{P}_i+\underline{\underline{J}}\cdot\boldsymbol{\varepsilon}+\underline{\underline{J}}_{ij}\cdot\underline{\boldsymbol{W}}_{ij})^T$$

$$\cdot(\sum_{i=1}^{m}(K_i\cdot({}^B\underline{R}_i^+x)+({}^B\underline{f}_i x))\cdot{}^B\underline{P}_i+\underline{\underline{J}}\cdot\boldsymbol{\varepsilon}+\underline{\underline{J}}_{ij}\cdot\underline{\boldsymbol{W}}_{ij})+\sum_{i=1}^{m}(K_i\cdot({}^B\underline{R}_i^+)^T\cdot({}^B\underline{f}_i))^2)\cdot dt \tag{16}$$

$$\int_0^{T_f} w_2 dt=\int_0^{T_f} M_k(t)\cdot(\sum_{i=1}^{m}(\min^2(0,K_i(t))+\min^2(0,\mu\|{}^B\underline{F}_i\|_2-\|{}^B\underline{f}_i\|_2\ ))\cdot dt \tag{17}$$

If we assume that

$$W=\sum_{i=1}^{m}(k_i^2\cdot K_i^2(t)+({}^B\underline{f}_i(t))^T\cdot({}^B\underline{f}_i(t)))+w_1+w_2 \tag{18}$$

then, the new objective functional can be adopted as

$$V_2=\int_0^{T_f} W\cdot dt \tag{19}$$

The equation (19) can be solved by the numerical method, provided that its solution can be represented as the series

$$K_i(t)=\sum_{j=1} a_{ij}\cdot t^j \quad , \quad {}^B\underline{f}_i(t)=\sum_{j=1} \underline{b}_j\cdot t^j \quad ,i=1,2,\cdots m \tag{20}$$

If the sum of the first n+1 terms of the series (20) is taken as the approximate value of the functional (19), we get

$$V_2=\int_0^{T_f} W(t,a_{ij},\underline{b}_j,\ i=1,2,\cdots,m;\ j=1,2,\cdots,n+1)\cdot dt \tag{21}$$

If the integral equation (21) is solved by the Newton-Cotes method, we have

$$V_2\approx T_f\cdot\sum_{k=0}^{p} C_k^{(p)}\cdot W(t_k,a_{ij}^{(k)},\underline{b}_j^{(k)}) \tag{22}$$

where

$$C_k^{(p)}=\frac{1}{T_f}\int_0^{T_f}\prod_{\substack{j=0\\ j\neq k}}^{p}\frac{t-t_j}{t_k-t_j}dt \quad , \quad t_k=k\cdot h \quad , \quad h=T_f/p$$

Up to now, V_2 has been described as a general multivariable function and the problem of the functional (19) has been turned into the one of the multivariable function. Its solution can be attained by the method of non-linear programming for the unconstrained multidimensional problem, where the initial state for iteration can be determined by the formula (15) and the functions of penalty constants $r_k(t)$ and $M_k(t)$ have to be satisfied with

$$r_k(t)=\beta_1 \cdot r_{k-1}(t)\ ,\quad \beta_1 \geqslant 1 \quad ; \quad M_k(t)=\beta_2 \cdot M_{k-1}(t)\ ,\quad \beta_2 \geqslant 1$$

Example:

Determine the optimal function $F_i(t)$ of the fingertip force for the A3-A3-A3 kind of the hand, if it is required to manipulate the pen to draw a planar sine curve.

Given $r=1cm$, $m=1kg$, $J_c=0.5kg\cdot cm^2$, $\mu=0.1$ as shown in Fig.3. The planar kinematic equations of the pen are supposed to be

$$X=\sin(10t)\ ,\ Y=0.1t\ ,\ \varphi=0\ ,\ t\in(0,36) \tag{23}$$

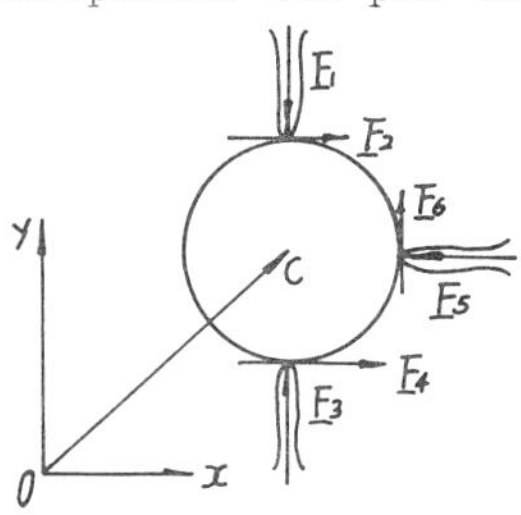

Fig.3

(1) Dynamic Equations

Since

$$\begin{cases} m\ddot{X}_c=F_x=F_2+F_4-F_5 \\ m\ddot{Y}_c=F_y=F_3+F_6-F_1 \\ J_c\ddot{\varphi}=M_c=F_4+F_6-F_2 \end{cases},$$

we have

$$\begin{cases} F_1=F_3+F_6 \\ F_2=F_4+F_6 \\ F_5=2F_4+F_6+100\sin(10t) \end{cases} \tag{24}$$

(2) Geometrical and Frictional Conditions

$$\begin{cases} F_1(t)\geqslant 0\ , \\ F_3(t)\geqslant 0\ , \\ F_5(t)\geqslant 0\ , \end{cases} \quad \begin{cases} |F_2|\leqslant|\mu F_1| \\ |F_4|\leqslant|\mu F_3| \\ |F_6|\leqslant|\mu F_5| \end{cases} \tag{25}$$

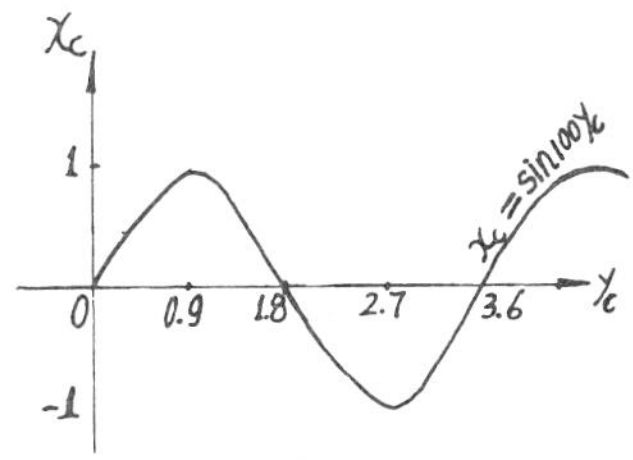

Fig.4

(3) Objective Functional

$$\min_{\forall F_i(t)} V=\min\int_0^{36}\sum_{i=1}^{6}F_i^2(t)\cdot dt \tag{26}$$

Substituting equations (24) into equation (26), we have

$$V=\int_0^{36}((F_3+F_6)^2+(F_4+F_6)^2+F_3^2+F_4^2+(2F_4+F_6+100\sin(10t))^2+F_6^2)\cdot dt \tag{27}$$

(4) Numerical Solution

If we substitute the following equations

$$F_3=a_0+a_1 t\ ,\quad F_4=b_0+b_1 t\ ,\quad F_6=c_0+c_1 t \tag{28}$$

into the equation (27), and take the constrained conditions (25) into consideration, we get the following solutions

$$\begin{cases} F_1=60.63+16.72t \\ F_2=5.49+1.69t \\ F_3=60.40+16.72t \end{cases}, \quad \begin{cases} F_4=5.26+1.69t \\ F_5=10.75+3.38+100\sin(10t) \\ F_6=0.233 \end{cases} \qquad (29)$$

Table 1 gives out the results of this optimal problem, in which $V_{min}=1.19\times10^7$ is much smaller than the others.

Conclusion

1. In the course of moving the hand to manipulate an object, the criterion of the objective performance functional of the fingertip force functions is proposed along with considerations of dynamic equations of the object, and the geometrical and frictional conditions of the hand.
2. In this paper, based on the classical variation, the maximum principle of Pontryagin and the dynamic programming theory of Bellman, the penalty functional terms and the numerical integration are introduced to develop the integral SUMT of the closed-region-functional extreme value.

Table 1 Data of Optimization Process

Variables	a_0	a_1	b_0	b_1	c_0	c_1	V
Initial	201.00	201.00	10.00	10.00	1.00	1.00	1.38×10^9
Feasible	63.654	21.026	5.332	2.069	0.02	0.0	7.6×10^7
Optimal	60.496	16.721	5.257	1.69	0.23	0.0	1.2×10^7
Final	60.40	16.72	5.26	1.69	0.23	0.00	1.2×10^7

References

1. Cicala, P., An engineering approach to the calculus of variations. Turin, Italy, Levrotto and Bella, 1957.
2. Bolza, O., Lectures on the calculus of variations, Chicago, The University of Chicago Press, 1904.
3. Sage, A .P., White, C.C., Optimum systems control, Englewood clif, New Jersey, Prentice-hall, Inc, 1977.
4. Athans,M.,and Falb,P.L., Optimal control, New York, McGraw-hill,1966.
5. Bellman, R., Dynamic programming, Princeton, N.J., Princeton University Press, 1957.
6. Gong Xi-Fang, The calculus of the optimal control problems, Beijing,China, Science Press(in Chinese), 1979.
7. Guo,G.L., The research report of BIAA, BH-2275(in Chinese),1986.
8. Fox, R.L., Optimization method for engineering design, Mass., A ddison-Wesley, Reading, 1972.
9. Matthew T. Mason, Kenneth Salisbury, J., Robot hands and the mechanics of manipulation, Mass., The M.I.T. Press, 1985.
11. Okada, T., IEEE Trans. on System, man, and Cybernetics, Vol.SMC-9 No.2, 1979.
12. Hanafusa, H. and A sada, H., Proc. 7th Int. Symp. on Industrial Robots, Tokyo, 1977.

A PID Sliding Mode Controller for a Robotic Link

Romano M. DeSantis

Electrical Engineering Department, Ecole Polytechnique de Montreal, Montreal, Canada

Abstract

A classical PID controller modified via the parallel addition of an on-off switching controller is presented. The ensuing controller is conceived so as to behave as an element of the recently proposed sliding mode class; its design is developed by combining toghether the classical transfer function techniques with more recent variable structure systems ideas. Theoretical considerations reinforced by simulations indicate a potential for a significant improvement over the dynamic performance of currently available PID robotic link controllers.

Introduction

The large and rapid variations of the manipulator parameters values, the highly complex nature of the intercouplings, the elasticity of the joints and the flexible structure of the manipulator, the effect of the nonlinearities, the uncertainty of the model etc.. are all factors which make it difficult to obtain a satisfactory control performance when in the presence of the usually employed classical PID controllers. As a consequence, these controllers, though adequate in many current industrial applications are often inadequate in a number of more advanced applications where a much higher degree of both transient and steady state performance may be required and where, at the same time, greater and faster plant parameters variations toghether with stronger intercoupling forces are expected. It becomes very hard in these cases to attain a satisfactory balance between a good transient and steady state performance toghether with an acceptable level of performance robustness.

Recently, this situation has been partially alleviated by inserting the classical PID servo-controller in an adaptive control loop [As.1, Du1, Ko.1]. While effective to reduce the negative effect of slow variations in the manipulator parameters, this approach is, however, less effective when directed at compensating for rapidly changing parameters or high frequency external perturbations. To overcome this difficulty, a more recent line of research has focused on the replacement of the classical PID with sliding mode controllers [Ut.1, As.1]. This new type of controllers, though characterized by a much higher robustness to parameters variations and external perturbations, are afflicted, however, by a tendency to high control levels, chattering generation, and oscillations. Recent improvement via the application of an appropriate "boundary layer" technique coupled with the subsequent adoption of an adaptive strategy [As.1] offers only a partial solution to this problem in that it imposes an often unacceptable trade-off between quality of transient performance on the one hand, and steady state disturbance rejection capabilities on the other.

The present paper discusses a modified PID sliding mode controller with the objective of removing this latter difficulty. The novelty with respect to the available literature is in that our sliding mode controller is developed with a view at complementing rather than at replacing the action of the classical counterpart. Accordingly, we construct it by inserting in parallel with the usual PID controller a nonlinear on-off switching compensator (figure 1). The original PID component of the issuing controller is designed using standard transfer function techniques; the switching component is designed using sliding mode systems techniques [Ut.1, As.1].

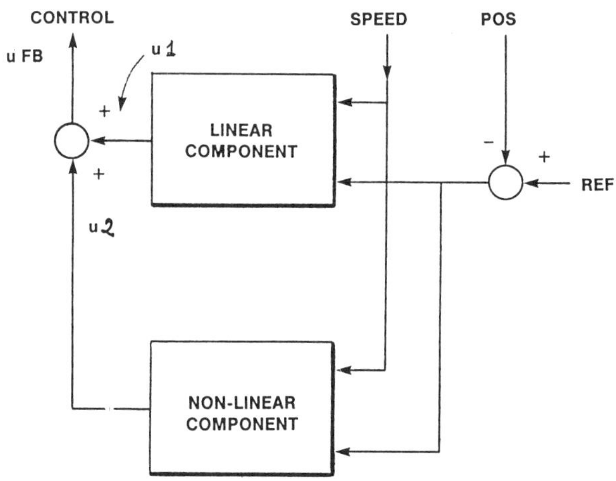

Figure 1 : i) Typical configuration of the classical PID controller; ii) proposed PID-SLM controller configuration.

The Robotic Link Model

For the purposes of the present development it will be sufficient to model the dynamics of a robotic link in terms of the following simple equations

$$\dot{x}1 = x2$$
$$\dot{x}2 = -(\alpha/J)*x2 + (1/J)[uTOT + W]$$

where: x1 and x2 represent respectively the link position and speed ; α and J represent the link viscosity coefficient and moment of inertia; W represent the compounded effect of gravity, nonlinear and intercoupling forces, plus the influence of friction forces and of an only partially predictable load ; uTOT is the control action. Introducing the notation

$$KM = 1/J \text{ and } \tau = \alpha/J,$$

with KM and τ representing the estimated values of the link static gain and time constant, one has

$$\dot{x}1 = x2$$
$$\dot{x}2 = -x2/\tau + (KM/\tau)*(uTOT + W)$$

The Design of the PID Sliding Mode Controller

The control uTOT,

$$uTOT = uFF + uFB,$$

can be usually seen as given by the sum of a feedforward action, uFF, directed at neutralizing as much as possible the predictable part of W, plus a feedback control action, uFB, directed at having the actuator follow as closely and as rapidly as possible the set-point, REF, in spite of the presence of the residual perturbation ΔW = uFF+W. The servo actuator dynamics can then be modeled in terms of

$$\dot{x}1 = x2$$
$$\dot{x}2 = -x2/\tau + (KM/\tau)*(uFB + \Delta W).$$

Usually, the uFB control is obtained by employing a classical PID controller. A typical configuration of such a controller is represented by part i) of figure 1. In what follows we consider a modification of this configuration by inserting in parallel to the classical controller the nonlinear on-off controller indicated by part ii) in figure 1. The feedback action of the overall controller , uFB, can then be seen as given by

$$uFB = u1 + u2,$$

where u1 and u2 denote the respective contributions of the PID component and of the nonlinear component.

The new controller will be designed by viewing the PID linear component as having the function to provide the control needed to assure a satisfactory performance under the hypotheses that the values of KM and τ are perfectly known and that the disturbance is stationary. Such a control corresponds to that which would be generated by the classical PID controller if acting alone. The switching component will be viewed as having the function to generate a control action primarily directed at reducing the influence of transient perturbations, fast parameters variations, and PID controller gain tuning offsets caused by an inaccurate estimation of the KM and τ values.

The design of the PID component: The action of the controller PID component is described by

$$u1= -K1*(x1-REF)-K2*x2 -K3*(x3-K4*REF)$$

where x3 is the integral of the position error

$$x3 = \int_0^t (x1 - REF)dt$$

and K1, K2, K3 and K4 denote, respectively, the proportional, derivative, integral and feedforward gain. The values of these gains are determined by simply ignoring the presence of the switching component. Under this assumption, using Laplace transform notations, one has

$$x1(s) = F1(s)*REF(s) + F2(s)*\Delta W(s)$$

with

$$F1(s)=[s(KM*K1+KM*K3*K4)+K3*KM]/[\tau*s^3+(1+KM*K2)*s^2+KM*K1*s+KM*K3]$$
$$F2(s)=s*KM/ [\tau*s^3+(1+KM*K2)*s^2+KM*K1*s+KM*K3]$$

Following standard practice, we determine the PID gains by imposing that the poles and the zero of the F1(s) transfer function have assigned values p1, p2 and p3, and z = p3. This leads to:

$$K1=(p1*p2+p3*p1+p2*p3))*\tau_{nom}/KM_{nom};$$
$$K2=-((p1+p2+p3)*\tau_{nom}-1)/KM_{nom}; \qquad (1)$$
$$K3=-p1*p2*p3*\tau_{nom}/KM_{nom};$$
$$K4=(p1+p2)/p1*p2$$

where τ_{nom} and KM_{nom} represent the nominal values of τ and KM. It follows that under nominal plant parameters values one has

$$F1(s) = 1/(1+a1*s+a2*s^2)$$
$$F2(s) = s/[K3*(1+a1*s+a2*s^2)*(1-s/p3)]$$

with

$$a1 = -1/p1 -1/p2, \text{ and } a2 = 1/p1*p2.$$

The Design of the Switching component: The controller switching component is designed by following the steps which are typical of the sliding mode systems approach [Ut.1], namely: - the characterization of the desired system dynamic behavior in terms of an assigned sliding mode; - the choice of an appropriate controller structure so as to guarantee the existence and the stability of the sliding mode; - the determination of the controller parameters which will steer the state of the plant to the sliding mode. Coherence with the linear component design suggests that the desired system dynamics be characterized in terms of the "sliding mode" function:

$$slm = a1*x1 + a2*x2 + x3$$

This function has the property that, if one builds the switching component so as to attain the "sliding mode condition"

$$slm=0,$$

then, working once again with Laplace notations, it would follow

$$F1(s) = 1/(1+a1*s+a2*s^2)$$

$$F2(s) = 0.$$

Note that, with x1, x2 and x3 satisfying the sliding mode condition, the servo controller performance is no longer affected by the presence of ΔW; the value of F1(s) is no longer influenced by potential discrepancies between actual and nominal KM and τ values.

In line with the variable structure systems approach [Ut.1], we pursue this result by considering the sliding mode stability condition

$$\dot{slm}*slm < 0$$

This condition guarantees that $\lim_{t\to 0} slm(t)^2 = 0$, hence the convergence to zero of the sliding mode function, and therefore the attainement of the desired dynamic behavior.

We first rewrite the plant equations in terms of τ_{nom} and KM_{nom},

$$\dot{x}1 = x2$$

$$\dot{x}2 = -x2/\tau_{nom} + (KM_{nom}/\tau_{nom})*u1 + KM/\tau*u2 + PERT$$

where: PERT represents the input equivalent overall perturbation,

$$PERT = (KM/\tau)*\Delta W + (1/\tau_{nom}-1/\tau)*x2 + (KM/\tau-KM_{nom}/\tau_{nom})*u1,$$

and

$$u1= -K1*(x1-REF)-K2*x2 -K3*(x3-K4*REF)$$

Using the gain values (1), this gives

$$\dot{x}2 = -(p1*p2+p1*p3)+p2*p3)*x1 +(p1+p2+p3)*x2 +p1*p2*p3*x3+p1*p2*REF + KM/\tau*u2 + PERT$$

We consider now

$$\dot{slm} = a1*x1 + +a2*x2 + x1 - REF,$$

and, implementing the appropriate substitution,

$$\dot{slm} = p3*slm + a2*[(KM/\tau)*u2+PERT],$$

hence

$$\dot{slm}*slm=p3*slm^2+ a2*[(KM/\tau)*u2+PERT]*slm$$

It follows that to satisfy the sliding mode stability condition

$$\dot{slm}*slm < 0$$

it is sufficient to impose,

$$SIGN[(KM/\tau)*u2+PERT] = -SIGN[slm]$$

This can be done by simply considering a on-off switching control of the type

$$u2 = -M*SGN[slm]$$

with

$$M > |PERT|*\tau/KM.$$

A reasonable choice of M can be easily made on the basis of an upper bound estimate of $PERT*\tau/KM$.

An Application Example.

To illustrate the dynamical behavior of the proposed controller, we consider the test bench described in [De.1]. The actuator is an Electro-Craft Model M-100 servo-motor equipped with a separate moving coil tachometer. The other system components are as indicated in table 1; the measured speed drive open loop parameters as listed in table 2.

Table 1: TEST BENCH SYSTEM COMPONENTS

-Micro-computer: IBM PC, 512 Kbyte, co-processor 8087;
-D/A A/D Converters: Lab Master card, Techmar Inc.;
-DC Motor with built in tachometer: Electrocraft model M-100;
-Flywheel: home cylinder, .01 m dia, 1.18 kg;
-Reduction gear: 80:1;
-Amplifier: Kepco, Bipolar Power Amplifier.

Table 2: OPEN LOOP PLANT PARAMETERS VALUES

-Amplifier gain (μ): 4;
-DC Motor Static Gain (KM): 3 rad/volt*sec;
-DC Motor + Flywheel time constant (τ): .16 sec;
-Tachometer gain (KS): .02 volt*sec/rad;
-Potentiometer gain (KP) = .06 volt/rad;
-Sampling period (T): .005 sec.

Table 3: PID-SLM CONTROLLER GAINS

K1 = 6.25;	K2 = 0.4;	K3 = 26;	K4 = -.16;
a1 = 1/6.5;	a2 = 1/13^2	M = .6;	ϵ = . 1.

Taking into account that KM*μ = 12 rad/volt*sec and τ=.160 sec, we design a PID-SLM controller by choosing p1 = p2 = p3 = -13 , and z = p3. From equations (1) we obtain : K1 = 6.25, K2= 0.4, K3 = 26, and K4 = -.16. For the sliding mode component we obtain a1 = 1/6.5, and a2 = 1/13^2; the value of M is selected as equal to 5. To avoid chattering, we introduce the threshold u2=0 if $|slm| <$ .1.

The following simulation experiments illustrate the response of this PID-SLM controller to a step input under nominal and perturbed operating conditions; this response is compared with that obtained when applying the classical PID (M=0).

Experiment N.1. Under nominal operating conditions, (KM =12 rad/volt*sec; τ= .16 sec; sampling time T=.001 sec), we apply a step input with an amplitude REF = 5 rad. The simulated response of the system in correspondence to the classical PID controller (M=0), is in this case identical to that of the PID-SLM controller (figure 2).

Experiment N.2. We repeat experiment N.1 by now posing REF=0 and by applying a step disturbance with an amplitude PERT=750 rad/sec^2. As illustrated in figure 3, the influence of this disturbance in the classical PID controller is considerably larger than in the PID-SLM controller.

Experiment N.3. We repeat experiment N.1 by now varying the servo-motor time constant and static gain actual values. Figure 4 illustrates the effect over the system output in the case of a + 300% variation in the time constant and -50% variation in the static gain (τ=.48 sec; KM = 6 rad/sec/volt). Figure 5 concerns the case of a -65% variation in the time constant and a + 100% variation in the gain. One can clearly see that the PID-SLM controller has a considerably smaller sensitivity to these parameters variations than the classical PID.

Experiment N.4. Figure 6 shows the behavior of the system under a sampling time variation in the experiment N.1 operating conditions (T=.02 sec). It clearly indicates that the sampling period does not play any significant role in the performance of the controllers under investigation.

Experiment N.5. We change experiment N.1 operating conditions by introducing a large delay in the control (delay=.030 sec); the issuing results are reported in figure 7. One can observe that while this large delay has a deteriorating effect in the case of the classical PID controller it does have a catastrophic effect on the PID-SLM controller. Clearly the new controller is somewhat weaker under this performance aspect. One notices on the other hand that this inconvenience is quite tolerable in that the PID-SLM controller is capable of withstanding reasonable delay time values.

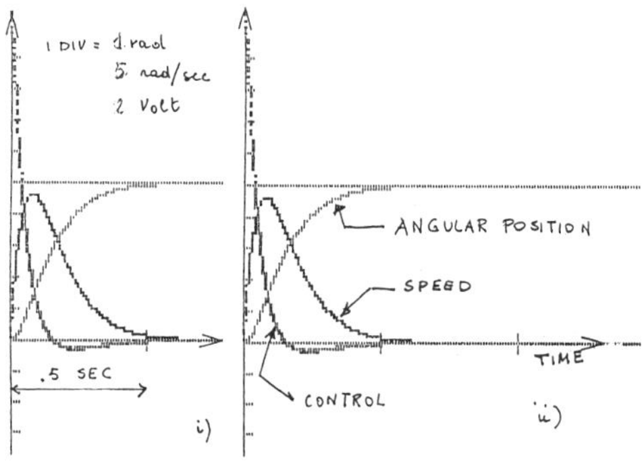

Figure 2: Step input response under nominal parameters values: τ = .16 sec; KM = 12 rad/volt*sec: i) classical PID; ii)PID-SLM controller.

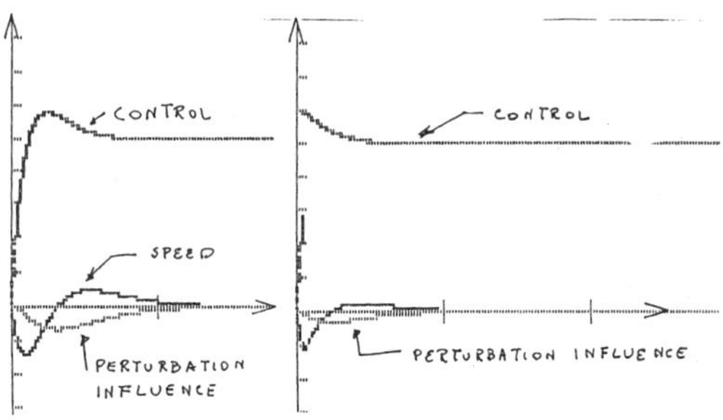

Figure 3: Step disturbance response under nominal parameters values: i)classical PID; ii)PID-SLM compensator.

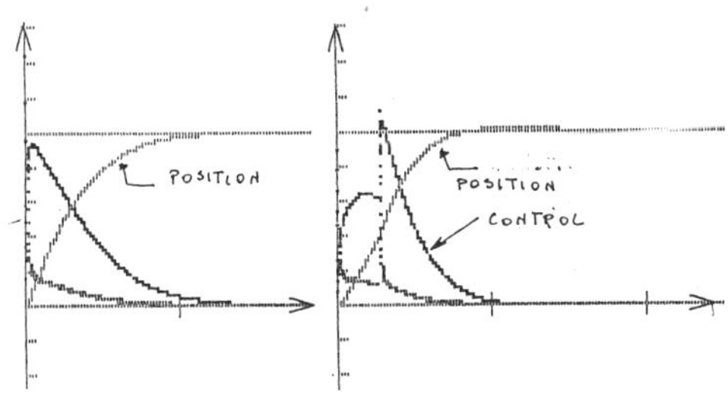

Figure 4: Step input response under perturbed parameters values: τ = .48 sec; KM = 6 rad/volt*sec: i)classical PID; ii)PID-SLM compensator.

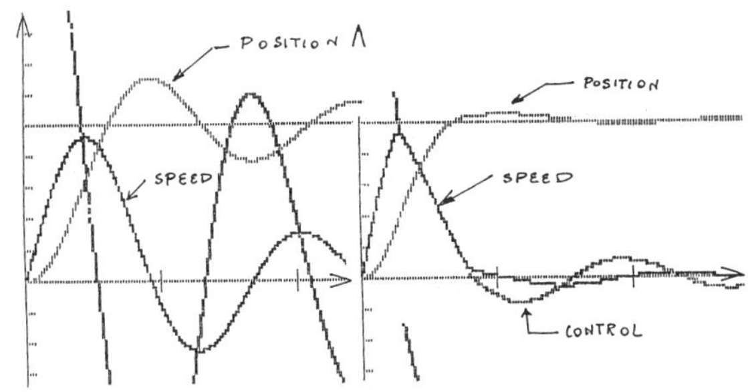

Figure 5: Step input response under perturbed parameters values: τ = .05 sec; KM = 24 rad/volt*sec: i)classical PID; ii)PID-SLM compensator.

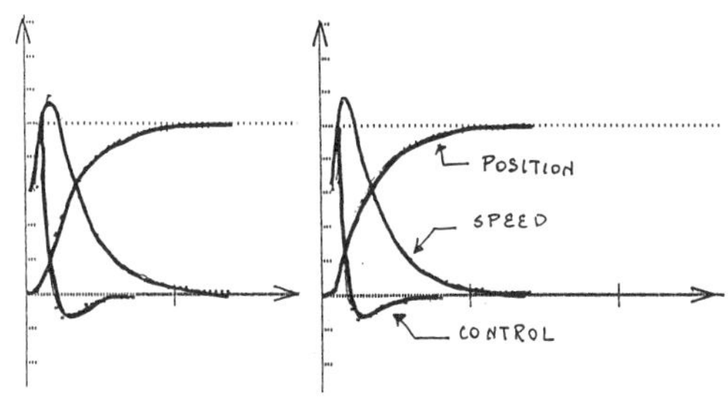

Figure 6: Step input response under a large sampling period : i)classical PID; ii)PID-SLM compensator.

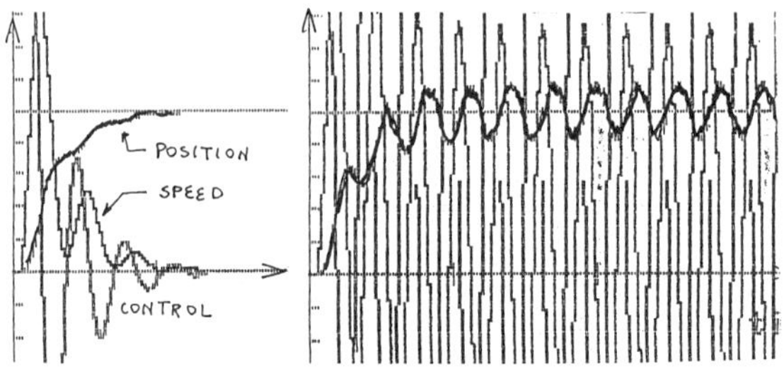

Figure 7: Step input response under a large dead time (delay =.03 sec): i)classical PID; ii)PID-SLM compensator.

Closure

The PID sliding mode (PID-SLM) servo-controller discussed in this paper, a natural modification of the classical PID controller, allows one to complement the wellknown transient and steady state properties of the calssical PID with the superior performance robustness to parameters variations and external perturbations of the sliding mode controller. In analogy with the sliding mode controllers already vailable in the technical literature, care must be taken to avoid the potential presence of persistent chattering and transient high frequency oscillations. Given the special structure of our PID-SLM controller, this is, however, a particularly easy task: the chattering can be eliminated by introducing a small threshold on the output of the switching component without altering the basic properties of the controller; transient oscillations can be reduced to a minimum by the eventual application of self-tuning of the PID component parameters. Contrary to what happens in sliding mode controllers with integral action already reported in the technical literature (for example [Sl.1]), these corrective actions no longer imply a trade off between transient behavior performance and steady state disturbance rejection capabilities.

References

[As.1] Asada,H., Slotine,J.J., Robot Analysis and Control, Addison Wesley, 1986.

[DC.1] DC MOTORS SPEED CONTROLS SERVO SYSTEMS, An engineering handbook by ELECTRO-CRAFT CORPORATION, Pergamon Press 1977.

[De.1] DeSantis, R.M., Hurteau, R., Authie, G., A Real-Time Computer Control Demonstrator for Use in a University Control Laboratory, IEEE Transactions on Education, Vol. E-25, NO.1, Feb 1982.

[Du.1] Dubowsky, S., Kornbluh, R., On the Development of High Performance Adaptive Control Algorithms for Robotic Manipulators, The Robotic Research: The 2nd International Symposium, MIT Press 1985.

[Ko.1] Koivo, A.J., Guo,T.H, Adaptive Linear Controller for Robotic Manipulators, IEEE Transactions on Automatic Control, AC-28, pp 162-171, 1983.

[Sl.1] Slotine, J.J.E., Coetsee, J.A., Adaptive Sliding Controller Synthesis for Nonlinear Systems, Int. J. Control, 1986.

[Ut.1] Utkin, V.I., Sliding Modes and their Application to Variable Structure Systems, MIR Publishers, Moscow, 1978.

TRAJECTORY PLANNING

An Intelligent Path Planner in 2D and 3D with Rotation

K.L. Muck and C.B. Besant

Imperial College, Dept. of Mech. Eng., Exhibition Road, London SW7 2AZ, UK

Summary

A fast and efficient path planner that produces paths that are deterministic in path properties have been proposed. Search space can be updated constantly. The workspace is hierarchically decomposed into convex free space polygons. Two levels of path construction is employed. At the lower level, local path planner guided by rules of motion based on theories of convex bodies and constrains of motion to construct safe direct local paths while at higher level global path planner make use of rich accurate information to construct a short global path.

1 Introduction

In most cases path planning has been formulated as a purely heuristic search problem in some abstract space, eg. configuration space approach [4 , 5 , 7 , 8]. There are many short falls associated with this class of approach. The abstract search space are computationally expensive to generate. The search space are usually too large to be handled efficiently by any search technique. Various techniques have been attempted to cut down the search space (slicing of configuration space), leaping forward in search space (dividing C space into connected convex region) and hypothesizing plausible channels, in order to improve the efficiency of the method. However they are proved to be either too clumsy or oversimplifying the search space and tend not to be reliable in 3D path planning. Uncertainties associated, world model, robot motions and movement of people and other robots in and out of the work space will make any preplanned path unworkable unless the search space can be updated constantly. Updating the abstract space will be too difficult. Representing free space of workspace as connected cones or convex polygons belongs to another class of approach [1 , 2 , 3 , 6 , 9]. Collision free motion consists of collision free motion within and between cones or polygons. It is potentially fast and pruduces safer paths. Updating the search space is easy and efficient. Some methods tends to over- simplify the workspace and overconstrain the motion of a robot. As a result, legitimate paths may be missed in a crowded enviroment. Although the author's method belongs to the second class of approach, it shows that a near complete solution is possible and yet still retains the desirable properties of this class of approach.

2 Work Space and Robot Modelling

Complex enviroment like a factory floor or an office block is built up of small independent workspaces called unit workspace. They are linked together by corridors, stairs and lifts etc. A room is a typical unit workspace. Obstacles and robots are represented as convex polygons and boundaries of a workspace are represented as lines. A robot has a local frame of reference at the centre of rotation where the Y axis passes through the mid point of the widest width while the X axis is orthogonal to the

·minimum width of a robot . Motion of robot with + ve X axis pointing roughly in the direction of motion is + ve **mode of motion** and the opposite is - ve **mode of motion** as shown in fig. 1b. There is always uncertainty associated with robot heading called **rotational uncertainty** . A robot is assumed to have three degrees of freedom . Motion of a robot can be described as a combination of pure rotational motion about its centre of rotation and a pure translation motion through the centre of rotation .

3 Generation of Free Space Polygons

Free space of a workspace is represented explicitly as **free space polygons (fsp)** . They are defined as convex polygons of triangular or quadrilateral shapes and boundaries of a fsp are either common boundaries with obstacles or common boundaries with neighbouring fsps (**pseudo boundary**) shown as a dotted line in fig . 2a . A cluster is defined as a group of obstacles (no less than one) and its minimum convex hull include no other obstacle other than obstacles of its own group . A cluster can be treated as a subunit workspace . If there are N clusters with more than one element in each cluster then there will be N+1 distinct groups of fsps . The (N + 1) th group of fsps represent free space among clusters and between clusters and boundaries of a workspace . **Obstacles connected graph** consists of a list of centre obstacles and its three closest obstacles . Segmentation of free psace of a workspace is as follows : All potential clusters are formed . Free space of each cluster is segmented and its minimum convex hull obtained . If there is interference between convex hulls of clusters the affected clusters are combined and resegmented until no interference occurs. Treat each cluster as a large obstacle and segment the free space between clusters followed by segmentation of free space between the convex hull of all clusters and boundaries of a workspace as shown in fig . 2 . An efficient cluster formation algorithm operating on obstacles connected graph can be implemented based on the facts that : an obstacle can belong only to one cluster and the obstacle is a neighbour of another obstacle which in turn must be the neighbour of the obstacle itself . Algorithm for triangulating free space of a cluster :

1) pick a boundary of an obstacle and call it an open boundary .
2) put an open boundary on current state .
3) pick the nearest available vertex facing the current boundary from three nearest facing obstacles .
4) form a free space triangle , put the pseudo boundaries of the newly formed triangle on the list OPEN if the boundaries are not already on the list OPEN . If a newly formed boundary is already on the list OPEN then delete the new and old one from the list.
5) repeat 2 -> 5 .

The process stops when no obstacle is facing any open boundary . The open boundaries and the boundaries of the obstacles form a minimum convex hull of the cluster as shown in fig 2 . Channel transformation combines connecting triangles with solid boundaries to form convex quadrilateral polygons called channels . The remaining groups of connected triangles is decomposed into small number of convex quadrilateral and triangular polygons . Small triangles is deliberately formed during triangulation by choosing the nearest vertex and during decomposition of connecting triangles . It is believed that polarisation of the sizes of polygons will make fsps that do not allow a robot to pass more outstanding. It also represents the local geometry more closely . A pointer is assigned to each fsp indicating the group it belongs to . This is useful in helping to reduce the list of objects to be searched during generation of connected obstacles graph , connected free space graph and during global path searching .

4 Free Space Graph and Local Path

Connected free space graph consists of a list of connected free space nodes . Each node has a **centre** fsp (cfsp). The immediate and second immediate neighbours of a centre fsp are called **primary** fsps (pfsp) and **secondary** fsps (sfsp) respectively . A list of **non sucessors** is also associated with the cfps which are primary fsps having maximum width that are less than the minimum width of the robot , or having only one secondary fsp and the maximum width of the secondary fsp is less than the minimum width of the robot . Each node represents a patch of area in the enviroment . The boundaries between cfsp and pfsp are **starting boundaries** and boundaries between pfsp and sfsp are **ending boundaries** . Collision free paths (+ve and -ve paths) from starting to ending boundaries are called **local paths** , see fig. 3 . The following concepts are important in local path construction . A **point** is a position vector with respect to the global coordinate . A **line state** is a line segment on the boundary or within a fsp . The end points of a line state are the nearest collision free point within a fsp . Line states that are on the boundaries of a fsp are called **boundary line states** and **interia line states** are line states that are within a fsp . There are usually multiple ranges of collision free rotational motion associated with a point about the centre of rotation . These are simply called **Ranges of motion** . A set of **constrains of motion** is associated with each range of motion . It is a subset of the vertices and boundaries of a node . A **configuration vector** (**CV**) is defined as a point with all its ranges and associated constraint sets . For example ,

$CV = [p , \{ (R1 , C1) , (R2 , C2) , ... (R_n , C_n) \}]$, where R and C are ranges and associated constrain sets respectively see fig . 4 . A **local path configuration point** (**LPCP**) is a subset of CV consisting of a point, a range and its associated constrain set [p , (R1 , C1)] . An **approach vector** (**AV**) AB is vector from point A to point B . A LPCP B is said to be **visible** from A if the approach vector AB is within the range of LPCP A . A **motion vector** (**MV**) of a robot AB is a collision free translation from A to B . Angle between two sucessive MVs is called **angular displacement** (**AMV**) . A motion vector is considered safe if the angle between the motion vector and the nearest range limit (**min . difference range**) of the two LPCPs are larger than the **rotational uncertainty** of a robot . A **local path** is a list of LPCPs from the starting line state to the ending line state where neighbouring LPCPs are linked by safe motion vectors . See Fig . 5 .

4.1 Rules of motion between LPCPs :

The following theorems and rules provides the theoretical basis of an efficient implementation of a local path constructor .

Theorem of convex bodies : *A simple closed curve (surface) T bounds a convex body K iff every line through a point interior to T intersects T in exactly two points .*

Theorem of Direct Passage : *There exists a collision free path between LPCPs on boundary line states if a) there is an overlapping range r' between the two LPCPs and b) the approach vector is within r' or (180° + r')* . Also applicable for any two interior line states .

Theorem of Identical Constrain Sets : *If two LPCPs within a convex fsp are having identical sets of constrains , then there exist a collision free path between the LPCP if there is an overlapping range between the two LPCPs* . Case 1 : The identical constrain set consists of two line segments , eg. two LPCPs within a long thin channel . Case 2 : While two LPCPs are constrained by a single line boundary . There is always an overlapping range between the two LPCP and collision free translation is possible in the overlapped range. . In general this theorem is applied to any two LPCPs with identical constrain sets within a fsp .

Theorem of Compatible Constrain sets : *A collision free path between two LPCPs is possible If the intersection of the two constrain sets is non empty , and overlapping range exists between the two LPCPs .* Implication : LPCPs (a , b , c) , if (a intersect b) and (b intersect c) are non empty then a -> c via b is possible .

Theorem of Impossible Path : *If a CV on line state A has say LPCPs p1 and p2 where p1 and p2 are mutually exclusive , it is impossible for LPCP q on line state B to proceed to p1 if the constrain set of q is compatible or identical to p2 .*

Note : All theorems breaks down if the two LPCP is not in the same fsp and robot model is assumed to be regular . **Safe Path :** A path is considered as safe if the minimum distance between a robot and an obstacle and the min. difference range is larger than the positional and the rotational uncertainty of a robot respectively at all LPCP on a path .

4.2 Construction of Local Paths :

There are infinite number of local paths possible between two line states . A pair of direct paths (+ ve and -ve) are chosen . This is because direct path is among the shortest path available and the algorithm for it is relatively easy . Complex path can be efficiently constructed during global path searching when necessary . An **independent line state** is a line state where all the CVs on the line state has compatible constrain sets , ie no more than two distinct elements that make up the constrain sets of all CVs on the line state . **Multiple ranges line state** there are more than two distinct elements that make up the constrain sets of all CVs . LPCPs of a CV may be mutually exclusive . Theorem of compatible constrain set shows that collision free path exists between LPCPs of an independent line state . Theorem of impossible path also shows that collision free path is not possible between two line states if either one is a multiple range line state and the two LPCPs are mutually exclusive . It is therefore important to identify the ending line state and decide a suitable constrain sets for the potential ending LPCP , see fig . 5 . Once a constrain set is decided , the line state is stepwise refined until a LPCP compatible with the chosen constrain set is found . The range of that LPCP may be enlarged by rotating the robot to the range limit that has to be enlarged and move the robot in the allowed direction along the line state until a suitable range is found . A **suitable** LPCP on a line state is a LPCP where the ending LPCP is visible or as closed to visible as possible from it .

Algorithm for Local Path Construction

1) if safe path between the current starting and ending LPCP is possible then return the two LPCP as a safe path , else ,

2) generate a CV and hence its LPCPs which is visible from the current starting LPCP ,

3) if no safe path between the current CV and the starting LPCP then step 5 else step 4 ,

4) locate a suitable LPCP on the current line state . If a safe path is possible between the starting LPCP and the current LPCP then step 6 else treat the chosen LPCP as ending LPCP and start step 5.

5) generate another mid . line state between the starting line state and the current line state and repeat step 2 until a safe path between the starting and mid . line state is found then step 3 or declare the path not passable if the distance betwen line state is less than a critical value .

6) put the current line state and LPCP as the starting line state and LPCP and start the next phase of

local path construction towards the ending line state repeat step 1.

A complete local path between the starting and target LPCPs is constructed by connecting all the sub paths together. Note that intermediate line states and LPCPs are only generated when required. Hence in a space enviroment, local path searching which is by far the most computationally expensive algorithm of the path planner can be fast while in a crowded environment, more **detail** paths are generated when required. The ending LPCP of a node is then used as a starting LPCP for its sucessors. Each boundary line state has no more than two starting LPCPs or ending LPCPs see fig. 6 for local path generation.

5 Global Path Searching

The geometry of a connected free space graph node has been discribed in section 3. **Path length** and **normalised path safety** of each node is obtained from the corresponding local paths. Path length (**PL**) is the sum of L and R. Where L is the sum of all | MV | and $R = (r / 180^{o}) \cdot (L)$, r is the sum of all absolute value of angular displacement of a path. **Normalised Path safety** is the average difference between the min. difference range and the rotational uncertainty of a robot. An availability status (binary or function of time) and uncertainty function represent the availability and physical conditions of an area of the floor are added to each node. Hence a node A is := [status, uncertain function, geometry of the node, { geometry of local paths and its properties }]. A* search technique is used in global path planning. The heuristic function H is defined as

$H = s + e + h1 + h2 + h2'$. Where s and e are the local path from given configurations to the starting or ending nodes respectively, h1 is the total path length of a current global path between the starting and the current node, h2 is the estimated euclidean distance between the current and ending node and h2' is only added to H when the mode of the current node is not compatible with the ending node chosen. It is simply equal to the path length of the current node and deleted when the current path becomes compatible. To start global path searching, local paths leading to the starting boundaries of the starting and ending nodes are constructed. There is only one possible starting node but usually more than one ending node (convension of a node used). Hence h2 is the minimum estimated value from the current node to an ending node.

A* Search Algorithm For Global Path Searching :

1) if the starting node is identical to one of the ending nodes then step 1*, else put the starting node on current state,

2) if the current node is an ending node then proceed to 2*,

3) open all the relevant successors of the current node, if the following conditions of the current node is true, then put the current node on a list CLOSED.

a : successor is available, indirect successor is generated only when no direct successor is possible,

b : status of the current node is favorable for generating successor, status may be time dependent,

c : the local path from the current node to the successor is safe after uncertainty factor of the node is added to the path,

d : not a parent node of the current node, prevent self repeating,

e : not fully generated,

f : not in the list CLOSED, if in CLOSED, all its successors must not be in CLOSED or OPEN, g : if the successor is in the correct mode of motion , this condition is optional , applied only when robot path has to be in a particular mode of motion .

4) update the heuristic function H of each newly generated node .

5) assign a pointer to each successor indicating its parent node .

6) place each successor node on the list OPEN in an accending order of H .

7) put the upper most node on OPEN on current status and repeat step 2 -> 7.

Step 1* : Construct a local path between the two configuration points and declare a path is found . If no path is possible between the two configuration points , then step 2 .

Step 2* : trace back the history of the current path from the current node until the starting node . If the mode of motion is compatible , then a global path is found , else intitiate an indirect motion at a suitable place along the path to correct the mode . If no suitable place is found then declare the path invalid . Generate the remaining successor apart from the successor that contains the ending configuration point and proceed to step 4. Declare no solution is possible if all opened nodes are exhausted . Note : An indirect successor is formed when no direct path from the starting line state to the ending line state of a successor is possible but a path via the line state of another successor is possible . Mode changes is always associated with an indirect path . Indirect successor is redundent if direct successor is available . See fig . 7 for global path searching .

6 Conclusions

Three major issues of path planning has been discussed in this paper . Firstly , large amount of information of a work space and efficiency can be achieved by explicitly representating and hierarchically decomposing complex work space into unit work space , clusters and finally connected free space polygons . Secondly theories of motion within a fsp based on properties of convex bodies and constrains of motion provides the theoretical basis for an efficient local path planner to be implemented . Safe and direct local paths can be obtained . Accurate and useful information is also derived from the structure of a connected free space graph as well as from local paths of each node of the graph . Avaliability status and uncertainties function of each node help the planner to keep up with the changing external world . Global paths obtained are deterministic . By changing the weight of individual functions of the heuristic function H or generating only the safest available successor or choosing successor that are in a chosen mode of motion will result in global paths with very different characteristics . All the concepts discussed in this paper can be extended to 3D path planning . However segmentation of free space into convex polyhedra and the constrain sets associated with a CV in 3D will be much more complex .

References

[1] Brooks , R. A . " Solving the Find - Path Problem by Good Representation of Free Space , " * SMC-13 83

[2] Brooks , Rodney A . , " Planning Collision Free Motions for Pick And Place Operations ", Proceeding of First International Symposium of Robotics Research MIT Press. 1983

[3] Brooks , R. A . ," Aspects of Mobile Robot Visual Map Making ", Proc. 2nd Int. Sym. of Rob Re. 85 MIT

[4] Brooks , Rodney A . and Lozano - Perez , Tomas " A sub- division algorithm in Configuration Space for Find - Path with Rotation , " IEEE Trans. Sys . , Man and Cybernetics SMC-13 Mar. / April 1983

[5] Donald Bruce R. " Hypothesizing Channels Through Free Space in solving the Fing Path Problem " MIT A . I . Memo No . 736 . June 1983 .

[6] Giralt et al , " An intergrated Navigation and Motion Control System for Autonomous Multisensory Mobile Robot " Proc . of 1st . Int . Sym. of Robotics Research MIT Press. 1983 .

[7] Lozano-Perez , T " Spatial Planning : A configuration Space Approach , " IEEE Trans. on Comp Feb 83

[8] Lozano - Perez , T. " Automatic Planning of Manipulator Transfer Movements , " * SMC - 11 81

[9] Van Duc Nguyen , " The Find Path Problem in Plane , " MIT A. I. Memo 760 1984

* IEEE Trans. on Sys . , Man and Cyb.

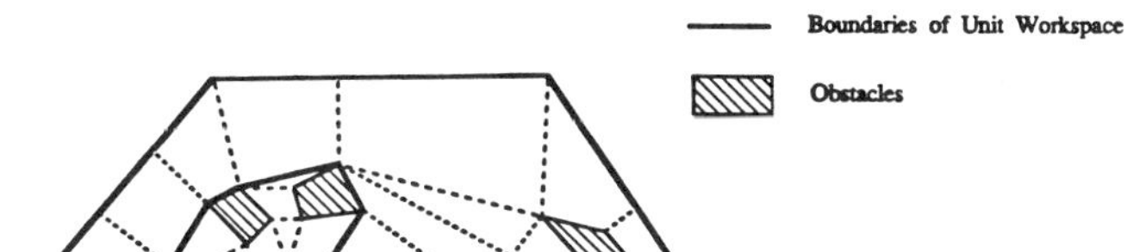

Fig. 1a : A typical Unit Work Space.
The diagram also shows that the free space of the enviroment is represented by free space polygons.

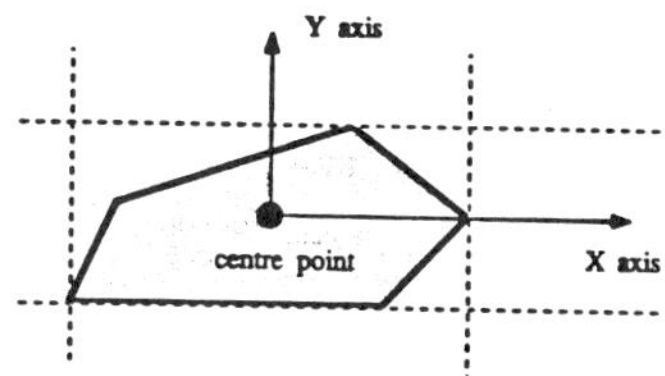

Fig. 1b : A Typical Robot.

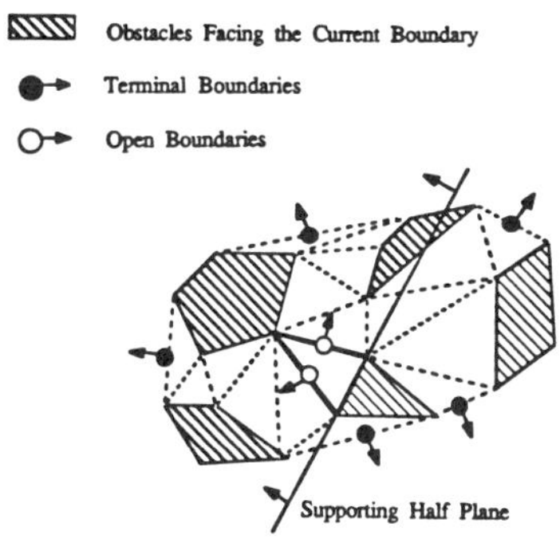

Fig. 2a : Free space of a cluster being segmented into trianglular free space polygons.

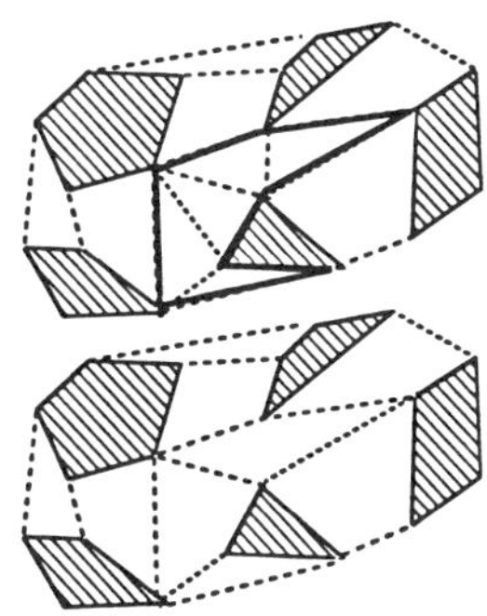

Fig. 2b : Channel Formation and Transformation of the Remaining Free Space Polygons.

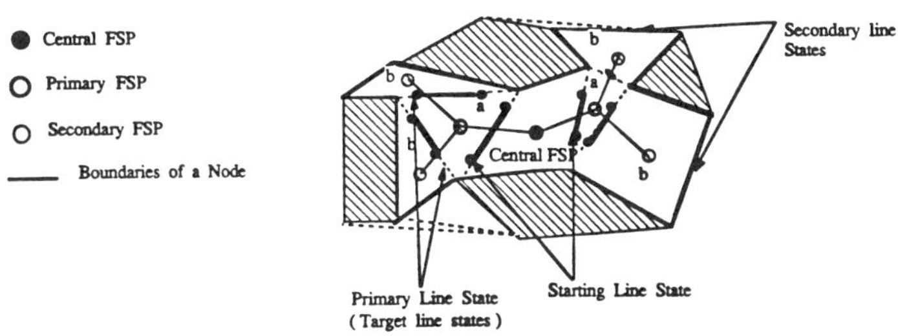

Fig. 3 : A Typical Connected Free Node.

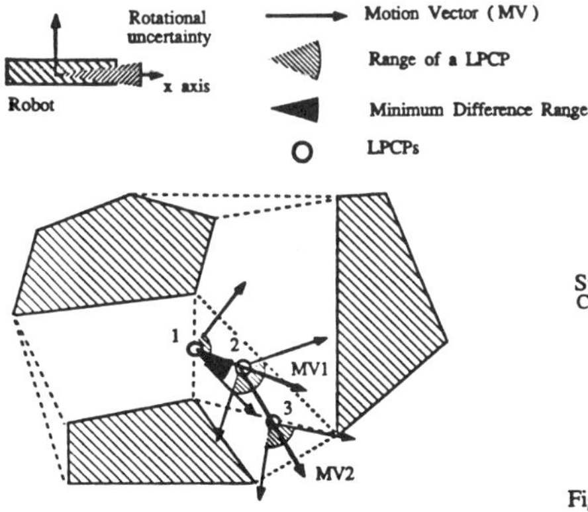

Fig. 5 LPCPs and Motion between LPCPs

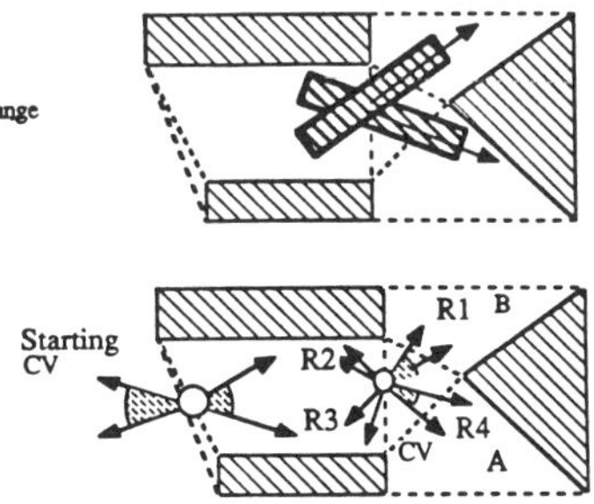

Fig. 4 : A Typical CV on Multiple ranges Line States.
R1 and R2 are mutually exclusive and R1 -> B and R4 -> A only

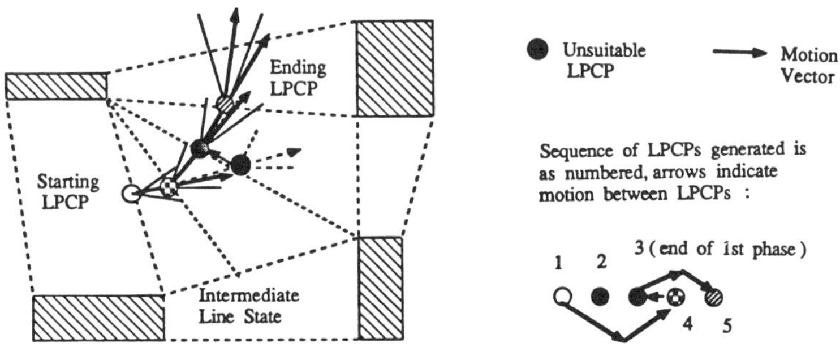

Fig. 6 : Local Path Construction.

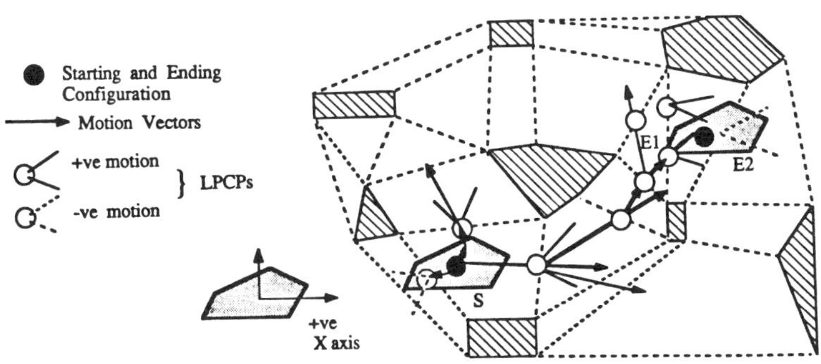

Fig. 7 Global Path Searching, E1 and E2 are the two Ending Nodes

Robot Path Adaptation Based on Axis Drive Current Monitoring Including Discrimination Between Collision and Acceleration Transients

D.A. Sanders, F. Naghdy, J. Billingsley, D. Robinson, and A. Collie

Robotics Laboratory, Portsmouth Polytechnic, Portsmouth, Hampshire, England, PO1 3DJ

A. SYNOPSIS

This paper reports on a technique expected to increase the speed of operation and working life of robot servo motors and machinery. The system was first proposed at the CADCAM 87 Conference in Birmingham England. Sanders 1987. [Ref.1]. Since that paper was prepared changes in the system and techniques have been made and are included.

Unnecessary changes in velocity of robot motors occur as a robot attempts to closely follow the programmed paths produced by CADCAM or teaching packages. These changes in direction reduce the operating speed and efficiency of the robot and may excite resonances in the manipulator.

The described technique uses information from the servo motor currents to adapt the robot paths and minimise current and torque transients. The current transients produced by changes in direction or velocity are similar to those produced by collisions and methods of discrimination are considered and described.

Transients due to contact with hard obstructions have a larger amplitude and slope characteristic than transients associated with changes in direction. These collisions are detectable. Contact with softer objects is more difficult to discriminate. Velocities are monitored and a large error with a low velocity suggests an overload or collision. Any transients not associated with the generation of new demand signals are regarded as collisions. Some definitions are necessary and are included in Appendix ONE.

B. INTRODUCTION (A REVIEW)

B.i Proliferation and Development of Modern Robot Programming.

Low cost computing is allowing manufacturers to run Computer Aided Design packages for the first time. This new computing power, with the introduction of robot systems, increases productivity and flexibility. CAD systems are communicating directly with Computer Aided Manufacture systems to form CADCAM systems. Research into automatic inspection and testing, Billingsley 1983. [Ref.8] combined with CADCAM systems will form CADMAT(Computer Aided Design, Manufacture and Testing) systems Scott 1984, [Ref.3] and Craftsman robot systems [Ref.7 and 8]. Robot programs are created off-line by these systems for later use on the shop floor.

B.ii Are off line programs becoming too complex?

The work-piece handling tasks successfully robotised to date are generally suited to first generation robots working in isolation from the rest of the production line. Many first generation operations can be completed by simple point to point movement with relatively uncoordinated joint control between the points. These robot tasks will remain economic even as future generations and further developments become commercially available and CADCAM systems are installed. The end effector paths produced by off-line programming or CADCAM may be too specific and therefore the joint trajectories more complex than is required for these simple tasks.

B.iii Computer aided systems and off-line teaching

Recent commercial systems such as GRASP, developed at Nottingham University, England, allow dynamic modelling of robots and machine tools within flexible assembly systems. Cell layout can be improved by testing various configurations and running different robot programs to optimise the cell design and product construction sequence. The programs produced can then be used through post-processors to directly program the robots on the factory floor.

Within the computer, complicated functions of space and time are decoupled from the operator and only simple descriptions of desired motion are considered. Positions calculated by off-line computer or CAD systems are in a coordinate frame related to the joint variables by some homogeneous transform. CAD systems tend to use objective level programming relating the end effector to an object in the cell. This type of programming is easier for operators to visualise and allows the same path program to be used with different robots.

The software decides the path to arrive at a destination (including duration and velocity profile). The paths produced pass through via points where joint velocities may change unnecessarily in trying to follow the preprogrammed path.

Teaching by following or showing produces a continuous path control that appears simple to the human operator but which may include unnecessary corners and via points. Trajectories and paths may be further complicated in this case by physical or safety restrictions for human teachers within the robots working volume.

B.iv Adapting the preprogrammed path

The sequence of moves produced by computer or with teaching pendants usually consists of straight line and simple curved trajectories in some coordinate frame. The paths may include sharp corners and the trajectories unnecessary or irrelevant changes in velocity. During a sequence of moves, the servo motors experience large current transients and torques as the motors start, stop and change direction.

The trajectories produced by integrated systems may involve excessive complex computation and typical calculation theory has been discussed by the authors. Sanders, 1987. [Ref.**1**]. Angular joint velocities are calculated from required cartesian velocities. Strategies for selecting the joint velocities at via points can be complicated. A simple strategy considers zero velocities at via points and the proposed system will use this case.

For the first generation robot tasks discussed in B.ii, precision through via points is not always necessary and may make the system less efficient. The adaptive path control proposed can be used to minimise current and torque transients in the robot motors, providing vital trajectories and destinations are defined (see Appendix One.)

The proposed software controller switches from the optimum look-up table to other selected look-up tables, adapting the state space description of the controller. The path is adapted accordingly:-

1. A joint stops and restarts in the same direction without a vital destination or trajectory between the via points. Control for this changes to a look-up table which slows but never reaches a zero velocity. A small output is preset for zero error. Arrival is signalled just before the destination is passed and a new joint target and look-up table is selected.

2. A joint stops and restarts as above but in the opposite direction. Control for these switches to a look-up table with a low gain characteristic. Arrival at the joint destination is signalled immediately and the joint moves at a lower velocity, never reaching the destination.

C. FORCE SENSING AND COLLISION DETECTION

It is possible to use software calculation with information from joint motor currents for force sensing in addition to path modification, Naghdy 1987. [Ref.**10**]. The motor current varies as the square of motor torque. Manipulator forces are transmitted to the joints and the motors try to overcome these forces. Signals for collision and change of direction are similar and typical results from a push, (figure one a), and a move (figure one b) are shown below. The first difference is taken to highlight the transients.

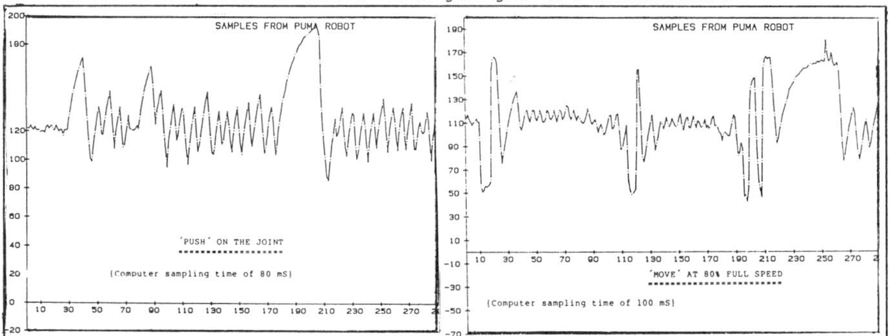

(a) (b)

Results from a push and a move, taking first difference.

FIGURE ONE

Forces exerted in cartesian space can be related to forces in the joint variables by the Jacobian matrix, Coiffet 1983.[Ref.**9**].

$$|\mathbf{F}_t| = [\mathbf{T}+\mathbf{Q}] \;.\; [J]^{-1},$$

where:- $\mathbf{F}_t$ = vector of cartesian forces
$\mathbf{T}$ = vector of joint variables
$\mathbf{Q}$ = vector of external forces.

The current in the motors may be used to represent the joint forces and the end effector force calculated using algorithms derived from the above. Work is still required to determine the optimum sampling time for the motor currents for the system. The amplitude of the transients is monitored and compared to a preset limit. Transients exceeding this limit are regarded as collisions. The error demand value is compared with changes in absolute joint position. A small change in position associated with a large error demand is suggested as a collision. Abnormal and unexpected forces not associated with a marker from the master computer are regarded as collisions.

D. METHOD

Multi-axis robot movements are modelled, concentrating on and controlling the movement of a robot base. Current to the base motor is sampled and stored. Detailed analysis of actuator current is difficult and high levels of noise may be present. Current transients are detected by considering the level and gradient of consecutive samples following a new destination marker signal. Actuator moves are signalled as new manipulator destinations are simulated and a new actuator destination generated.

The velocity of the joint is calculated from the monitored absolute positions. Any velocities approaching zero are then compared to the demand error signal. The sampled data is analyzed and information on adaptable trajectories is considered. Transients not associated with a new destination signal are assumed as collisions. The joint trajectories are adapted by changing the controller look-up table. The simulated robot paths are then adapted to minimise some current torque transients in the motor circuits. This is achieved by :-

1. Running the actuator motor at low speed instead of stopping at non vital points in trajectories which would normally mean the motor stopped and restarted in the same direction (Irrelevant stops). This is detected by two consecutive current transients in opposite directions.

2. Replacing the look-up table with a table of low gain. This slows the actuator so that non vital destinations are never reached. Velocity is low so that current transients are reduced when the joint stops and restarts in the opposite direction (irregular stops). This is detected by two consecutive current transients in the same direction.

E. SYSTEM

E.i General

The proposed system uses two Apricot microcomputers to control a prototype robot base joint and sample actuator current from the Billingsley dc Servo Amplifier. Algorithms are applied to the data and as the robot simulation repeats a sequence of moves the activator motion is adapted. A high quality brushed d.c. servo motor is used for reasons described by Rogers 1986. [Ref.**4**].

E.ii The system hardware

As shown in figure two, the basic single axis controller consists of a 16 bit 80286 Apricot Xen microcomputer and interfacing circuitry to a prototype robot base joint. The main computer controls the base via D/A circuits based on a non linear amplifier/mixer stage with multiple tacho feedback paths feeding a Billingsley dc Servo Amplifier, Sanders 1986.[Ref.**6**].

An Electro-Craft Corporation motor and gearbox are used to drive the actuator via a large helical spur gear. The motor has an integral tacho-generator and optical encoder attached to the motor shaft. The pulse-train from the incremental encoder is converted to absolute position using an LS 2000-based decoder circuit. Velocity is monitored for collision detection by comparing absolute positions from the decoder. This data is compared with the demand error signal.

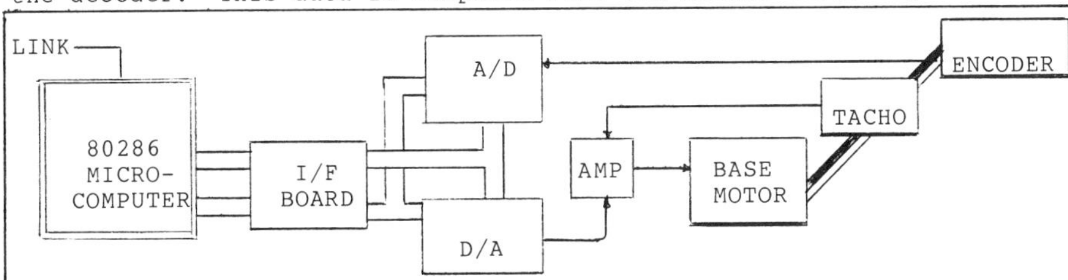

Block diagram of the single axis controller.
FIGURE TWO

The tacho-generator supplies an analogue voltage to two feedback loops, one low gain linear loop for high velocities and one high gain non-linear loop for low velocities and small demand signals. This strategy allows loose control during fast movements and tight control close to the destination.

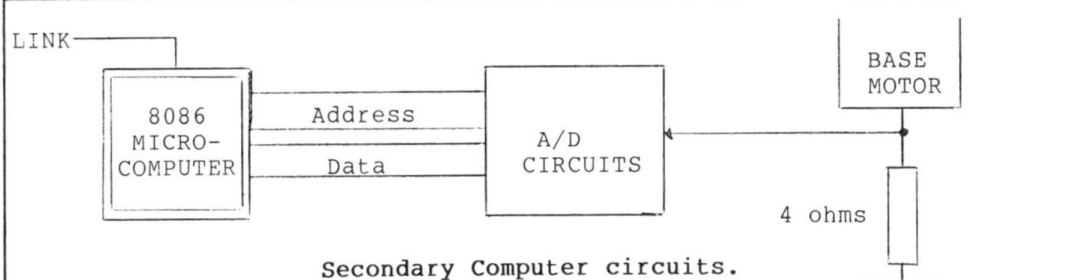

Secondary Computer circuits.
FIGURE THREE

The second computer is attached to a multiplexed six channel A/D circuit based on an R8574. This computer samples the motor currents using the voltage across a small resistance relative to the motor (4 ohms), as shown above in figure three. The analogue signal from the resistor is passed through a simple filter and is digitally filtered by the sampling system, Naghdy 1986. [Ref.**5**].

The two microcomputers communicate via a parallel interfaces based on Zilog 8536 C.I.Os.

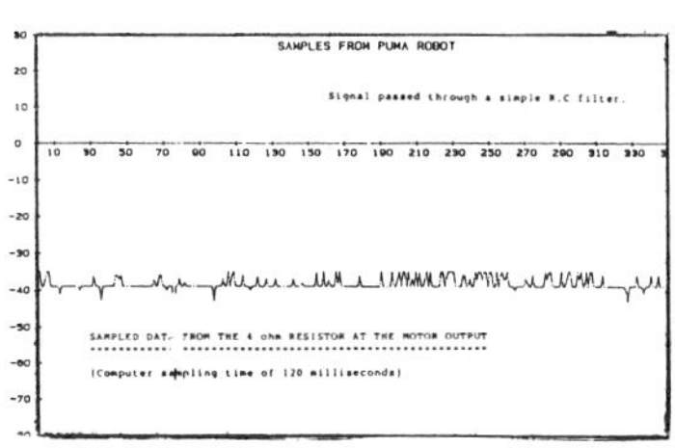

Typical motor current samples.
FIGURE FOUR

E.iii System Software

1 GENERAL

The main program, communication routines and man/machine interfacing is written in ´C´. Time critical routines are in assembly language. Assembly routines are linked by a common variable syntax. ´C´ is used for its portability, allowing development on off-line IBM and VAX computers.

2 MAIN COMPUTER (see figure five)

The master program and various non-linear control algorithms in sub-processes are loaded into the main computer. Included is the optimum solution developed by non linearising the experimentally achieved, critically damped control algorithm for the closed loop system. These are used to produce look-up tables in memory.

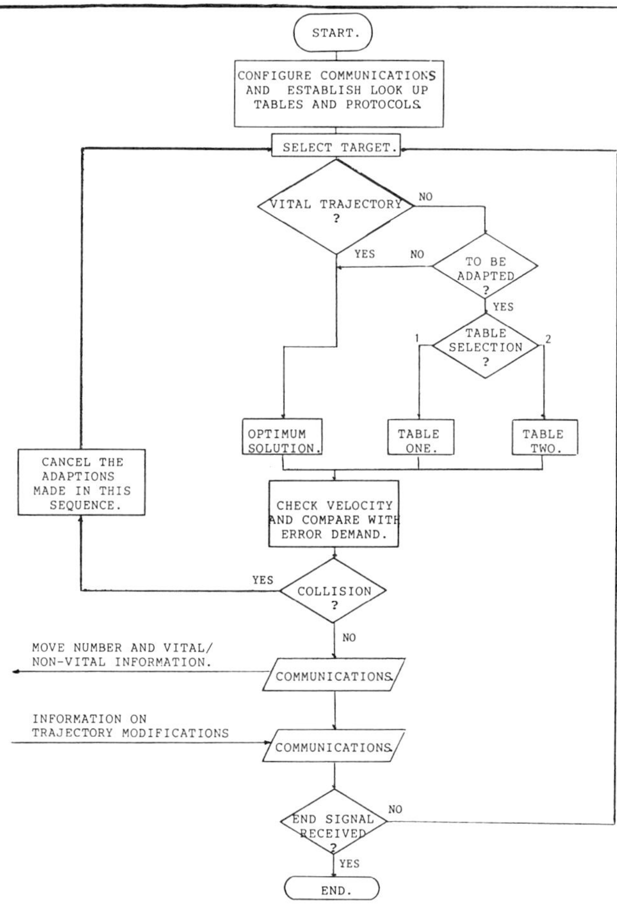

Flow chart of the main computer program

FIGURE FIVE

A repeated sequence of robot moves is simulated. During the first sequence the optimum solution is used for actuator control. As each joint angle target is passed, the master computer signals the second computer. The information transferred includes detail of vital trajectories and destinations in the form of an Ascii character and the move number as a hex integer, starting with move 0 at time, t=0.

V = vital **N** = non-vital.

A typical two bytes of data would be **"08 V"** if the eighth move was a vital trajectory for the joint.

If consecutive actuator trajectories are "non vital", information on the type of non vital change is passed back from the secondary computer and the program is modified in each sequence by passing control to the relevant look-up table for that type modification. The new look up table is then used between the via points.

Velocity is calculated in the software by sampling changes in position and is used to consider suspected collisions by comparison with the error demand value.

3 SECONDARY COMPUTER (see figure six)

The second computer samples the D.C current from the Servo Amplifier driving the actuator (Up to six joints can be monitored). The sampled data is stored for calculation and analysis, and later printing or display. The transients associated with non-vital trajectories are considered. If two consecutive transients are non-vital, the relevant change of look-up table for the type is signalled to the main computer. This data is transmitted via the parallel link with the information as two bytes. The move number and type of non vital transients. Once a move is signalled by the main computer, transients are identified by considering the relative level and gradient of four consecutive samples.

If

$\mathbf{i}_{(n+2)} - \mathbf{i}_{(n-1)}$ $\mathbf{i}_{(n+1)} - \mathbf{i}_{(n-1)}$

and $\mathbf{i}_n - \mathbf{i}_{(n-1)}$ are the same sign,

then:-

$$\mathbf{i}_{(n+2)} - \mathbf{i}_n = +/- \mathbf{t.g}$$

where:-

$\mathbf{i}$ = instantaneous current.
$\mathbf{g}$ = gradient over 4 samples.
k = gradient constant.
$\mathbf{t}$ = sampling time.
n = sample number.

If $\mathbf{g}$ is greater than kt, a transient is detected. The sign of $\mathbf{g}$ is then considered to give the direction of the actuator drive. Considering two consecutive transients, the relevant change is signalled. If noise is introduced into the system and a reading cannot be taken because the signs of the gradient change during the sampling periods, a recalculation would take place in the next pass.

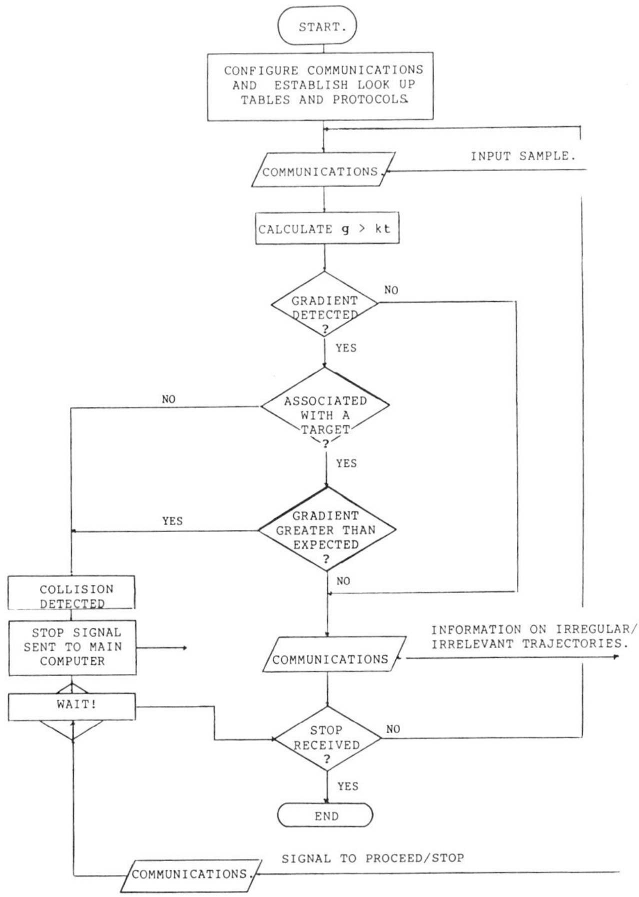

Flow chart of the Secondary computer program.
FIGURE SIX

Software calculation using statistical information from motor currents being recorded, Naghdy 1985. [Ref.5] may be used to detect a collision. The amplitude of the transient is monitored. If a transient gradient is detected which is not associated with a new destination signal from the master computer a collision may be assumed. In a working system, specific external state sensors (vision, proximity, taction, force) or software modelling of the environment may be used to complement the transient calculations in order to detect collision or danger.

F. CONCLUSIONS

The paths produced from modern auto-programming methods can be adapted by considering individual joint motions, but high accelerations may be produced at the end effector. If this is not acceptable, accelerations must be monitored and the software adapted accordingly.

Some active compliance may be achieved by considering the joint position errors and the joint forces discussed earlier. This active compliance may not be sufficient for difficult assembly tasks, but could aid specialised remote centre compliance devices or robots such as the I.B.M Selective Compliance Assembly Robot Arm.

The robot is provided with a degree of autonomy and the result is similar to human workers adapting a repetitive task to make movements easier and less tiring. The effect is similar to the addition of parabolic blends to multi-segment linear paths, Snyder 1985. [Ref.**11**] when via points may not actually be reached.

APPENDIX ONE (Definitions)

End effector destinations where orientation or configuration are important are defined as ´**vital destinations**´. Manipulator and end effector movements are defined as ´**vital trajectories**´ and ´**vital paths**´, when the approach vector needs to be defined between via points or the move is to a vital destination.

Manipulator movements to non-vital destinations or via points may have their trajectories modified by adapting individual actuator control algorithms within software to remove motor current transients. The distinction between vital and non-vital paths and trajectories is necessary within the software as only non-vital paths and trajectories can be modified while preserving the integrity of the robot task.

The terms trajectory and path are as defined by Brady 1983 [Ref.**12**]. **Trajectory** is the time sequence of intermediate configurations of a robot arm between source and destination. **Path** is the space curve traced by the end effector. Trajectories include via points as defined by Craig 1986. [Ref.**2**] and are the programmed intermediate points between the source and destination.

REFERENCES

1. **SANDERS D,** NAGHDY F and BILLINGSLEY J. **1987**. "Adaptive path control to minimise current and torque transients in motors using information from the servo amplifiers connected to robot joints". Proceedings of CADCAM 87 Conference, Birmingham, UK.

2. **CRAIG 1986**. "Introduction to robotics, mechanics and control" A/W.

3. **SCOTT P.B. 1984**. "The Robotics Revolution". Bell & Bain. UK.

4. **ROGERS C.G. 1986**. "Progress in the design of electrical drives for modular robotic systems". Proceedings of IMechE C363/86. UK.

5. **NAGHDY F** et al´.1986. "Robot force sensing using stochastic monitoring of the actuator torque". Portsmouth Polytechnic. UK.

6. **SANDERS D** and BILLINGSLEY J. **1986** "Servo Motor Power Amplifier". Electronic Engineering. UK, Aug 1986. UK.

7. **NAGHDY F** et al´. **1985**. "Craftsman Computer Integrated Manufacture". Quality Assurance UK, vol 11. June 1985.

8. **BILLINGSLEY J** et al´. **1983**. "The Craftsman Robot" Electronics and Power, UK, December 1983.

9. **COIFFET P. 1983**. "Interaction with the environment" Prentice/Hall.

10. **NAGHDY F** and WU. **1987**. "Interpretation of wave-forms generated by robot actuator torques." 1987 International symposium on cybernetics.

11. **SNYDER W.E. 1985**. "Computer interfacing and control of industrial robots". Prentice Hall International. USA.

12. **BRADY. 1983**. "Basics of Robot Motion, Planning and Control". Robot Motion. pp 1-50. MIT Press. UK.

Off-Line Integration Techniques for Robot Path Planning

Xuguang Wang and W. Edward Red

System Automation Laboratories, Brigham Young University, Provo, UT 84602, USA

ABSTRACT

Off-line programming techniques are not developed to the stage of practical application because of robot inaccuracy, difficulty in precise robot calibration,and difficulty in interfacing to CAD data bases. In this paper we present a practical approach to off-line programming of robots used in microelectronics assembly by integrating an off-line simulation program with data base preprocessing, robot calibration and robot inaccuracy correction. The calibration algorithms and the integration techniques described in this paper were verified in an environment of microelectronics assembly using a cylindrical robot.

INTRODUCTION

The off-line programming techniques developed apply data driven, relative methods instead of conventional absolute methods which assume that the workcell and robot tool geometry are known relative to the robot base frame for any robot configuration. Global off-line techniques fail because robots cannot be commanded to move to global positions in space accurately due to joint misalignment, joint compliance, gear train errors, etc. The relative off-line techniques described in this paper require a vision system to locate parts on which the robot operates relative to the vision camera frame using relative joint transformations. Once these parts are located in the robot base frame, inverse kinematics solutions determine robot joint values that move the robot tools (camera, gripper, etc.) to the desired operation points. To successfully implement these relative techniques, robot calibration, off-line simulation and integration, conformed path deviation, and systems communication methods have been developed to integrate the activities of the workcell shown in Figure 1.

Calibration techniques can be used to improve robot accuracy through software correction rather than by methods which change the mechanical structure or design of a robot. Two levels of robot calibration, namely off-line calibration and on-line calibration, are developed. Off-line calibration techniques are used to precisely locate the terminal control frames (TCF) for end-effectors (gripper, camera, etc.) attached to the robot distal (last) link. Robot teach terminal readout is used to determine the end-effector TCF location. On-line calibration is used to position the robot TCF relative to the parts in the workspace and thereby compensate for robot inaccuracy in localized regions of the robot's workspace. Since the calibration methods use localized reference features, the robot inaccuracy is less than the desired tolerances required for assembly. In addition statistical averaging of the calibration data compensates for some of the robot position errors.

SMART (Simulation of MAnipulator and Robot Tasks), a robot workcell simulation software package, has been modified for off-line planning of both calibration tasks and working tasks. The calibration path is a set of robot configurations which, when integrated, downloaded to the robot controller and executed, will cause the robot to move the vision system camera to near the desired calibration feature for part location. The working path is a set of robot configurations, speed settings, pick/place commands, etc. which, when integrated, downloaded to the robot controller, and executed after proper path deviation to account for calibration error as measured by the vision system, will cause the robot to move the robot tool to the desired operation points.

Due to robot inaccuracy and workcell geometry uncertainties, the off-line path planned by computer simulation must be modified to account for errors measured by vision system or other sensors. Conformed path deviation algorithms are developed which use appropriate control points in the pre-planned paths to perturb the pre-processed working paths, and yet maintain similar Cartesian path shape.

CALIBRATION METHODS

The robot calibration methods discussed in this paper use two levels of robot calibration, namely, 1. off-line calibration for the TCF locations of end-effectors and sensors attached to the distal link of the robot; 2.on-line calibration for properly positioning the robot TCF relative to parts.

Calibration for TCF Position

Figure 2 graphically defines the relative frames used extensively in this TCF calibration method, including the manipulator global frame $(XYZ)_m$, the end-effector or vision TCF frame $(XYZ)_T$, and the robot terminal frame $(XYZ)_e$. The calibration operation proceeds as follows, Figures 2 and 3:

STEP 1 - Attach a pointer to the robot distal link such that the pointer is aligned with the z-axis of the robot terminal frame, and move the robot arm to a position such that the tip of the pointer is at the selected point in the workspace, Figure 3(b). The pose of the robot terminal frame $(XYZ)_e$ at this particular robot arm position can be obtained from the robot teach terminal.

STEP 2 - Take the pointer off and put the gripper and camera on the robot distal link.

STEP 3 - Move the robot arm such that the TCF origin of the end-effector (grip or camera) is at the same selected point as the tip of the pointer used in step 1, Figure 3(c). The pose of the robot terminal frame $(XYZ)_e$ at this particular robot arm position can also be obtained from the robot teach terminal.

STEP 4 - Find the TCF position using the information obtained in steps 1 and 3.

To accomplish step 4 the related equations can be derived by merging the configurations of Figure 3(b) and 3(c). From this sequence of steps it is possible to generate a set of equations which can be solved for X and Y, the TCF origin position coordinates in the robot distal link frame:

$$X = (X_2 - X_1) S_2 - (Y_2 - Y_1) C_2 \qquad (1)$$

$$Y = (X_2 - X_1) C_2 + (Y_2 - Y_1) S_2 \qquad (2)$$

where $S_i = \text{Sin}\,\theta_i$, $C_i = \text{Cos}\,\theta_i$; and $X_1, Y_1, X_2, Y_2, \theta_1, \theta_2$ are the robot teach terminal readout.

Due to the robot inaccuracy from sources such as visual inspection, robot encoder and decoder errors, gear backlash, controller error, etc., the same calibration procedure is conducted for a number of randomly selected workspace locations. The statistical average is then used to determine the best TCF transformation matrix.

Calibration for TCF Orientation

The previous method only finds the origin coordinates of the end-effector TCF in robot distal link coordinates, but not the orientation of the end-effector with respect to the robot distal link. This information is required when assembly requires both position and orientation. It is not difficult to define the TCF orientation once the position of the end-effector is known by applying the following steps:

STEP 1 - Using the TCF position coordinates found by the previous calibration method, find the positions of two arbitrarily selected, but marked spots p and q. This can be done by moving the robot to a position such that the center of the end-effector TCF is at the arbitrarily selected spot p, and then read the current robot position from the robot teach terminal. The position of q in robot global coordinates can be found similarly.

STEP 2 - Find the orientation of vector **pq** = **p** - **q** in the robot global frame, angle α, using the information obtained in step 1, Figure 4.

$$\alpha = \arctan(Y_q - Y_p , X_q - X_p) \qquad (3)$$

where X_p, Y_p, X_q, Y_q are the position coordinates of p and q in the robot global frame.

STEP 3 - Move the robot arm to a position such that the orientation of vector **pq** relative to the end-effector TCF (vision camera), angle γ as shown in Figure 4, can be obtained; then read the current robot position from the robot teach terminal. For the vision camera angle γ can be determined from (4). For other end-effectors that are not axisymmetric align the end-effector's TCF X-axis with vector **pq** such that $\gamma = 0$.

$$\gamma = \arctan(Y_{qv} - Y_{pv} , X_{qv} - X_{pv}) \qquad (4)$$

where:

X_{pv} , Y_{pv} = position of p in vision screen coordinates.

X_{qv} , Y_{qv} = position of q in vision screen coordinates.

From this sequence a set of equations can be generated, and TCF orientation, β , can be determined by:

$$\beta = \arctan(\cos(\alpha - \theta - \gamma), \sin(\alpha - \theta - \gamma)) \qquad (5)$$

The TCF orientation calibration procedures introduced above can be used for any end-effector attached to the robot distal link. Statistically averaging for the best TCF orientation can be conducted using the procedures previously discussed.

On-Line Robot Calibration

On-line robot calibration is conducted at run-time. By using a vision camera, the calibrated TCF geometry infomation, and the workpiece geometry information from a preprocessed database, on-line calibration will determine the relative locations of workpieces in the workspace, allowing us to correct rigid body and robot position errors which are detailed in a companion paper[6].

The relative approaches developed use the calibration features shown in Figure 2 and a vision system to capture the relative geometry information between parts and the robot end-effectors. Using the relative frames shown in Figure 2, the component pose with respect to the robot global frame can be represented by the transform equation (6).

$$\mathbf{T}_{m}^{c} = {}^{c}\mathbf{T}_{m}^{e}\,\mathbf{T}_{e}^{v}\,\mathbf{T}_{v}^{f}\,\mathbf{T}_{f}^{c} \tag{6}$$

where:

$\mathbf{T}_{m}^{c}$ = transformation from the component frame to the robot global frame.

${}^{c}\mathbf{T}_{m}^{e}$ = transformation from the robot terminal frame to the robot global frame when the vision camera acquires the calibration feature.

$\mathbf{T}_{e}^{v}$ = transformation from the vision camera frame to the robot terminal frame, as determined by the off-line robot calibration.

$\mathbf{T}_{v}^{f}$ = transformation from the calibration feature frame to the vision camera frame, as determined by the image analysis [6].

$\mathbf{T}_{f}^{c}$ = transformation from the component frame to the calibration feature frame, and determined from pre-processed workpiece geometry data bases.

DATA BASE INTEGRATION

In a typical manufacturing process parts are made or assembled using relative rather than global methods[3]. Often, robot accuracy is defined as the robot's ability to move to the designated assembly point relative to some abstract global reference frames. The accuracy of most modern robots is usually insufficient for off-line programming without the use of real time sensor systems. Here we present methodologies which will compensate for inaccuracy of robot using local relative information as determined by sensor systems used discretely in time. Referring to Figure 5 which shows the information flow and data base structure, the general procedures follow:

STEP 1 - Move the robot arm to a predetermined position using off-line planned moves such that the camera is approximately over the calibration feature.

STEP 2 - Process the vision system image for the transformation matrix which relates the part calibration feature frame relative to the vision system camera frame.

STEP 3 - Find the TCF pose in the robot global frame for exact pick/place operation using the transformation matrix obtained in step 2, calibrated gripper and vision system camera TCF information, and the workpiece geometry information pre-processed off-line.

STEP 4 - Find the joint coordinates for the robot joints corresponding to the TCF pose found in step 3 using robot inverse kinematics.

STEP 5 - Modify the working path to maintain path conformity using control points.

STEP 6 - Use the modified working path to carry out the pick/place operations.

Off-Line Path Planning

The task off-line procedures require only the approximate positions of the workspace elements. Both calibration and working paths are then planned off-line using the capabilities of SMART, Figure 6. During the calibration off-line path planning, the user enters a set of generalized displacement matrices in global coordinates to describe a set of desired calibration poses for the vision system TCF. Figure 7 illustrates the calibration path segments as they are connected to form a path in Cartesian space.

During the working path simulation planning, the gripper TCF is used as the robot end-effector TCF and the user enters displacement matrices to describe a set of desired poses of the gripper TCF relative to the components on which the gripper acts. Figure 8 illustrates the concatenated working path segments in Cartesian space and the critical path nodes.

After planning the calibration and working paths, the planned path segments are filed for later use in path segment integration and path deviation. The path file contains the number and sequence of path configurations, pick/place activation flags, and other information relating to robot speed settings.

Path Segment Integration

Given an off-line planned program, the pre-planned tasks are integrated and converted to the control languages used in workcell system controllers such as those in the vision system or robot. Integration software has been developed that will allow the user to combine the pre-planned path segment files by adding or deleting path segments from the total task file or by changing the sequence of path segments. Additionally, the user can enter commands in the robot control language (DARL) during the integration process.

Workpiece Geometry Data Base

The off-line procedures presented in this paper require workpiece geometry data bases that contain the geometric relationships between the workpiece components and the workpiece calibration features. Modification of modern CAD system could be used to generate and store this information at the parts design or early manufacturing stage. For example, NC machines require a data file containing part dimensions and other relative information for the part being machined. In this project we emulated this processing by using an NC mill as a measuring device to construct a data base of relative information between the workpiece components and calibration features.

PATH DEVIATION

The off-line path planned by computer simulation must be modified to account for calibration errors measured by the vision system or other sensors and to account for geometric modeling uncertainties. There are various ways that an off-line path can be modified, but we will objectively maintain the Cartesian path shape or path conformity of critical path segments by deviating the path while maintaining consistent relative Cartesian motions of the robot end-effector terminal control frame.

It is impractical to deviate the entire assembly task path to maintain the path conformity. Path segments near critical points such as pick and place locations are candidates for conformed path segment deviation by specifying three control points, 1.*Control Pose Point (CPP)* = critical point (robot configuration) where calibration error affects the effectiveness of the robot end-effector interaction with the part ; 2.*Control Point Next (CPN)* = point after which there will be a critical point (CPP). The Cartesian shape of the deviated path segment between CPP and CPN must conform to the shape of the Cartesian path planned by simulation ; 3.*Control Point Last (CPL)* = point before which there has been a critical point (CPP). Path conformity after path deviation must be maintained between CPP and CPL.

In a Cartesian workspace the task of modifying a path segment to maintain path conformity can be described as deviating each off-line planned pose of an end-effector TCF between the control points exactly the same amount of translational and rotational error detected during calibration. In other words, the displacement matrix of the end-effector TCF for every node on the simulated path segment is post multiplied by the same error correction transform matrix to get a new displacement matrix from which the conformed deviated path segment results, Figure 9.

MICROELECTRONICS ASSEMBLY DEMONSTRATION

The practical off-line programming techniques developed are applied to a microelectronics assembly task, which can be described by following steps, 1. pick microelectronics chips of different sizes from a component feeder; 2. avoid workspace collisions during transitional paths; 3. insert microelectronics chips into designated IC board locations.

The task has been accomplished successfully and the off-line programming techniques developed have been shown to improve the robot position accuracy to ±0.1 mm (±0.004") within small areas of the workspace using rigid body error correction. For areas larger than 25 $(\text{inch})^2$ in the workspace, robot inaccuracy must be corrected in software to meet these position requirements [6].

DISCUSSION AND CONCLUSIONS

To successfully implement the off-line programming techniques, precise TCF calibration, data base integration and robot inaccuracy correction methods are necessary. The TCF calibration methods developed successfully determine the TCF pose of the end-effectors attached to the robot distal link. It is shown that the TCF position accuracy is determined within ±0.05 mm (±0.002") and the TCF orientation accuracy is determined within ±0.15°, near the camera resolution as determined by the vision screen pixel size.

The flexible data base integration techniques developed integrate calibration paths, working paths, and robot and vision system controller commands into a total task file. Control points integrated into the total task file are used at run time to carry out the conformed working path deviation to realize collision free operations.

REFERENCES

[1] Wang, Xuguang , "Off-Line Integration Techniques for Robot Path Planning," Master's Thesis, Brigham Young University, April, 1987.
[2] Roth, Z., B. W. Mooring, and B. Ravani, "Robot Precision and Calibration Issues in Electronic Assembly," Proceedings of Southeon 86, Record 21, March 18-20, 1986.
[3] Red, W. E., "Robotic Component Insertion and SMD Placement in A Microelectronics Low Torlerance Environment," Research and Development Prospectus, 1986.
[4] Smith, Randall C., D. Nitzan, "A Modular Programmable Assembly Station," Proceedings of 13th International Symposium on Industrial Robots and Robots 7, April 17-21, 1983.
[5] Paul, R. P., Robot Manipulators: Mathematics, Programming, and Control, The MIT Press, 1982.
[6] Turner, Ed, "A Vertical Assembly Methodology That Accounts for Robot Inaccuracy," Proceedings of Second International Conference on Robotics and Factories of the Future, July 28-31, 1987.
[7] Whitney, D. E., C. A. Lozinski, and J. M. Rourke, "Industrial Robot Calibration Method and Result," Proceedings of the ASME Computers in Engineering Conference, Las Vegas, August 1984.

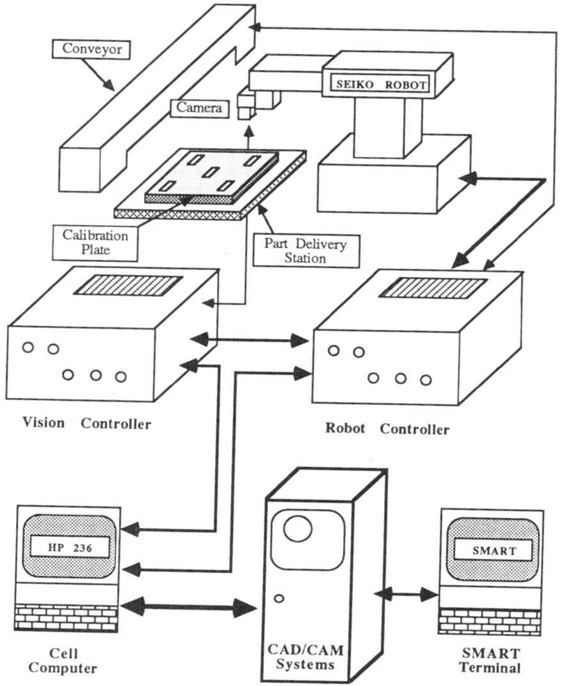

Figure 1 The Integrated Workcell

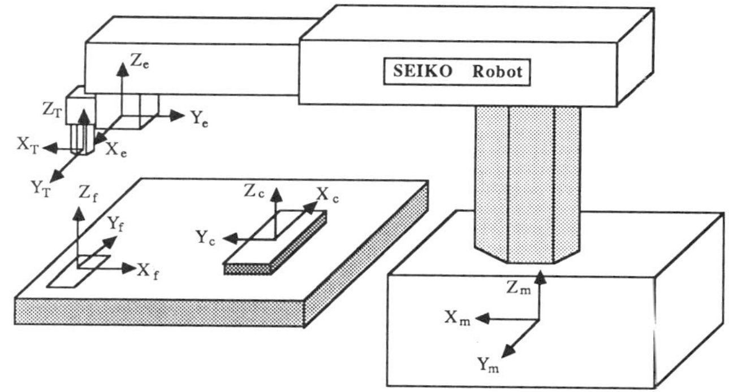

$(XYZ)_f$ = calibration feature frame $(XYZ)_c$ = component frame $(XYZ)_T$ = vision camera frame
$(XYZ)_e$ = robot terminal frame $(XYZ)_m$ = robot global frame

Figure 2 Relative Frames in the Workcell

NOTE: ● stands for an arbitrarily selected location which is a designated and physically marked location in the workspace.

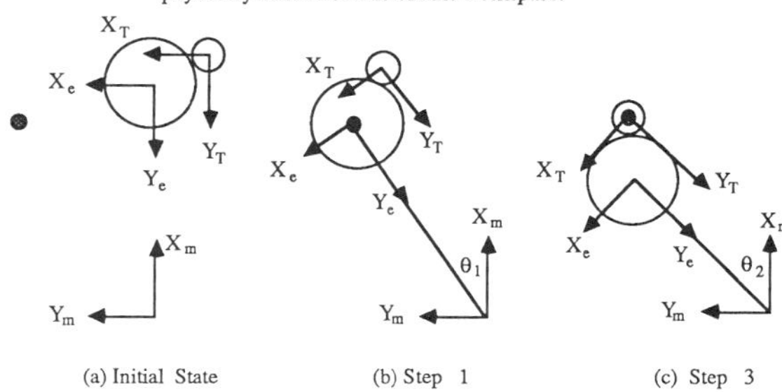

Figure 3 TCF Position Calibration

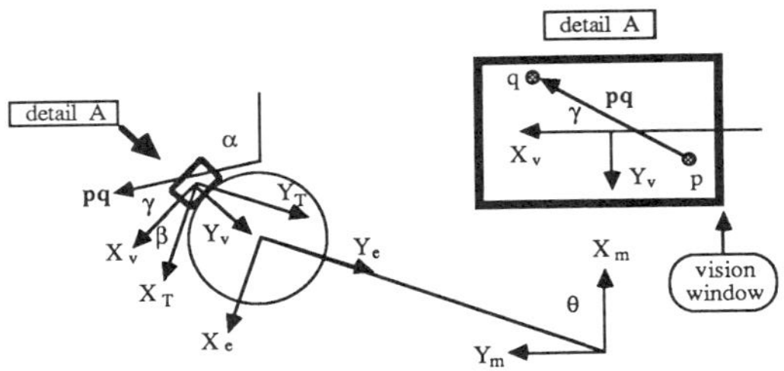

Figure 4 TCF Orientation Calibration

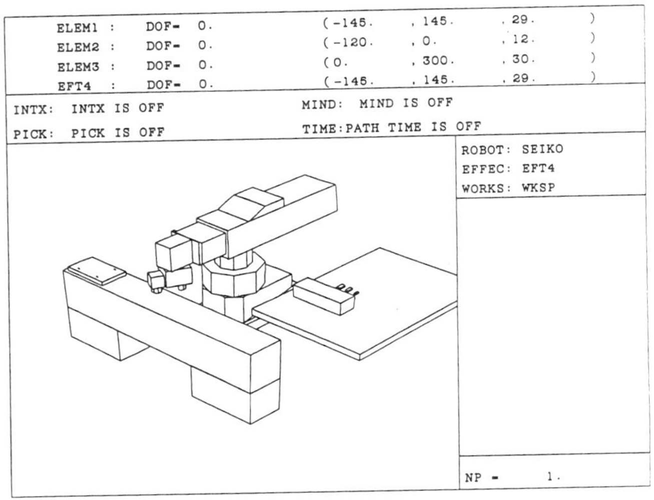

Figure 6 Simulation Model

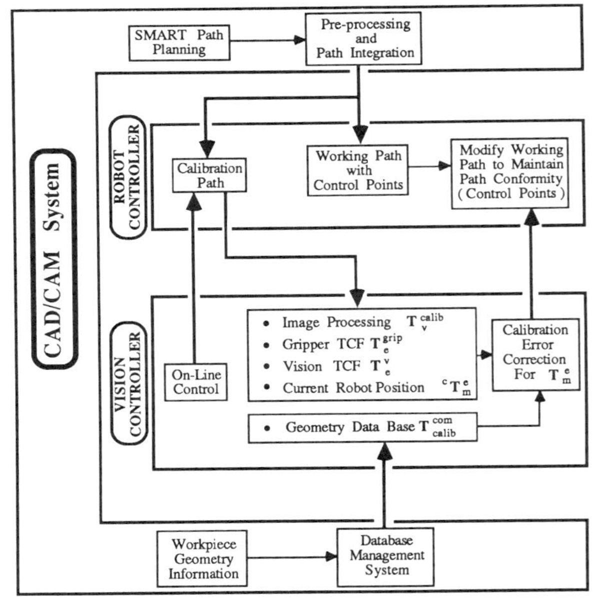

Figure 5 Off-Line Programming Data Base Structure

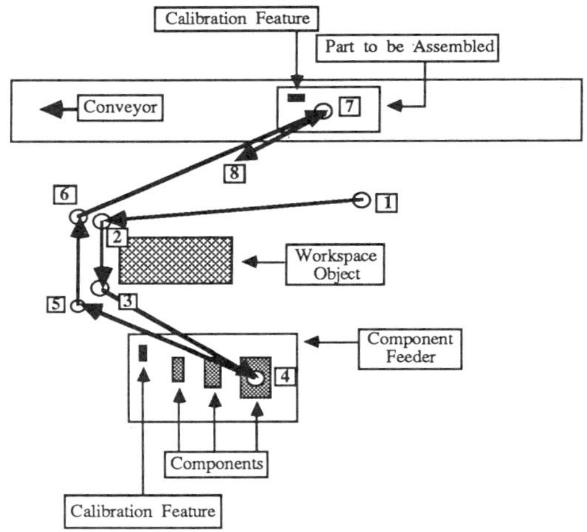

1. Gripper TCF (Home Position)
2. Obstacle Avoidance (Pick Off)
3. Obstacle Avoidance (Pick Off)
4. Pick Component
5. Obstacle Avoidance (Pick On)
6. Obstacle Avoidance (Pick On)
7. Place Component Picked
8. Next Operation

Figure 8 Working Path Segment in Cartesian Space

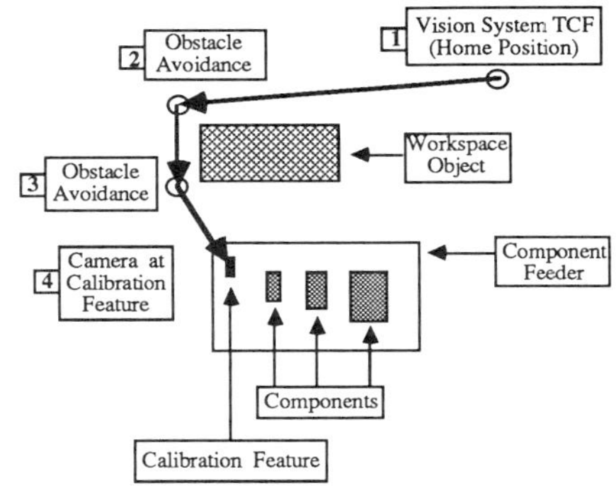

Figure 7 Calibration Path Segment in Cartesian Space

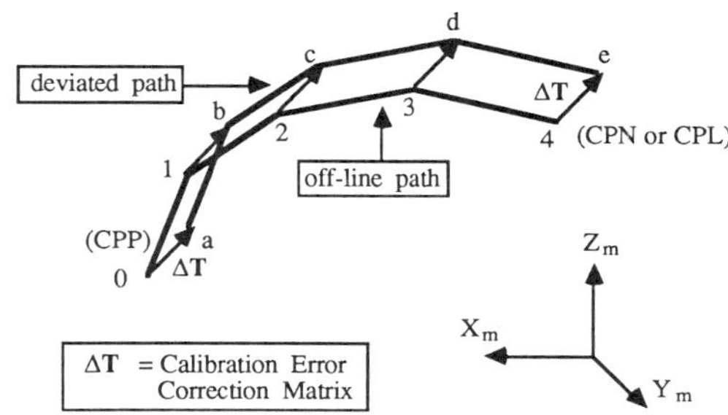

Figure 9 Path Segment Deviation

Collision Detection Using Sphere Approximations

David W. Sandberg and Richard B. Wodtli

Oregon State University, Corvallis, OR, 97331, USA

Abstract

Collision detection is important in robotics and mechanical design. Objects can be represented for efficient collision detection by a hierarchy of approximations formed by the union and intersection of spheres. The approximations are encoded in a binary tree of spheres. This representation is superior to representation as collections of faces (because pairwise testing of components is not required), and to representation as octrees (because spheres are easier to translate and rotate than are cubes). Algorithms are given for building sphere representations and for detecting collisions between objects.

1 Introduction

An important problem both robotics and mechanical design is to determine if a moving object collides with an object in its environment. One method for solving this problem is to model the objects as collections of faces[4,8]. If a face from the moving object intersects with a face from another object, then the objects intersect. This algorithm must test the faces pairwise, which slows the algorithm down as the number of faces increases. If the moving object contains m faces and the environment n faces then nm pairs must be tested. To quickly rule out many possible face intersections, each object is often enclosed in a sphere; if the spheres around the objects do not intersect, then the objects do not intersect.

The sphere enclosing the object can be viewed as an approximation to the volume of the object. A single sphere is a poor approximation to the volume of a long, thin object. Better approximations can be formed by using the union and intersection of more than one sphere. In this paper, we present an algorithm for collision detection that uses a hierarchy of sphere approximations to quickly locate the area of intersection if it exists.

Octrees, a three-dimensional form of quad-trees using cubes, have also been used for collision detection[1,2]. Octrees also contain a series of approximations to the object being represented, so the area of intersection can be detected quickly. Unfortunately, octrees are difficult to rotate and translate. Spheres are easy to rotate and translate.

Other primitive objects besides spheres and cubes could be used to approximate objects[5,6,12,13]. Half-spaces could be used, but spheres are better since they divide the space into a bound piece and an unbound piece, whereas a minimum of four half-spaces are needed to obtain a bounded volume. More complicated objects such as ellipsoids or cylinders could be used, but the test for intersection of such objects is more complicated.

Collision detection can be done as a by-product of modeling solids as intersections, unions, and differences of primitive objects, although this approach does not lead to fast algorithms[12].

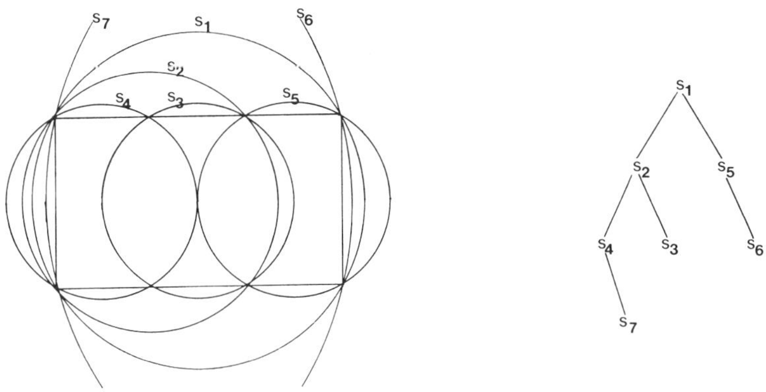

Figure 1: Approximating a cylinder with spheres. The side of the cylinder is represented by the rectangle.

Most of the work in modeling objects has been done with surfaces, not volumes. For collision detection volume seems to be more appropriate. Spheres have been used to model objects for other purposes, such as shape extraction[9], animation[3], and rendering[10]. An algorithm for detecting intersections between individual spheres is given in [7], but it does not deal with objects represented as collections of spheres.

2 Approximating Objects with Spheres

Any object can be approximated by the union and intersection of spheres to within a given error factor. For collision detection, we require that no point further than a distance e from the object is in the approximation and that all points within the object are within the approximation.

Figure 1 shows how a cylinder can be approximated by a collection of spheres. A crude approximation is given by the sphere s_1. A better approximation, which is poor at the ends of the cylinder, is the union of s_3, s_4, and s_5. The best approximation given in Figure 1 is the intersection of s_7 and s_6 and the union of s_3, s_4, and s_5. This one contains all the points within the cylinder and no points outside the cylinder further than distance e from the surface of the cylinder.

This collection of approximations can be represented by a binary tree whose nodes are labeled with spheres. The volume described by a tree T, $\sigma(T)$, is defined as $\sigma(T) = S_T \cap (\sigma(T_l) \cup \sigma(T_r))$ where S_T is the sphere labeling the root, T_l is the left subtree, and T_r the right subtree. If the left subtree, the right, or both are empty, the expression is $S_T \cap \sigma(T_r)$, $S_T \cap \sigma(T_l)$, or S_T, respectively. The tree in Figure 1 represents the volume $s_1 \cap ((s_2 \cap ((s_4 \cap s_7) \cup s_3)) \cup (s_5 \cap s_6))$. S_T can be viewed as an approximation to $\sigma(T)$. The tree in Figure 1 also represents at least three other approximations, s_1, $s_1 \cap (s_2 \cup s_5)$, and $s_1 \cap (s_2 \cap (s_4 \cup s_3) \cup (s_5 \cap s_6))$.

The intersection of the two volumes represented by the trees T and Q is calculated by $\sigma(T) \cap \sigma(Q) = S_T \cap S_Q \cap ((\sigma(T_l) \cap \sigma(Q_l)) \cup (\sigma(T_l) \cap \sigma(Q_r)) \cup (\sigma(T_r) \cap \sigma(Q_l)) \cup (\sigma(T_r) \cap \sigma(Q_r)))$. If $S_T \cap S_Q$ is empty, there is no need to compute the rest of the expression.

3 Approximating with Intersections of Spheres

Bounded convex polyhedra are easy to approximate by intersecting spheres. Each face of the object is associated with a sphere that rises no more than a distance e above that face, and encloses the entire object. The sphere tree representation of this set of intersecting spheres is a degenerate binary tree, that is, each internal node of the tree has one son. The set of approximations represented by this tree can be improved by adding the smallest sphere that encloses the convex object as a new root to the tree. A careful ordering of the other nodes in the tree may also improve the approximations.

The radius of the sphere used to approximate a face is given by $(d^2+e^2)/2e$ where d is one half the maximum distance between any two vertices of the face being approximated. Thus, the center of a sphere may be a considerable distance from the object being approximated, which may cause errors since computers use finite representations of numbers.

Another disadvantage of this technique of approximating a convex object with many faces is that a test against every face must be done before it is known that two objects intersect. Performance can be improved by making the unused son of the root be a crude approximation to the object formed from unions of spheres. Then the only time all the tests against the faces need to be done is when the two objects just miss intersecting or just barely intersect.

4 Approximation Using Unions

Any volume can be approximated by a union of spheres. A simple way to form a series of approximations to an object represented as a union of spheres is by combining pairs of spheres that are neighbors into a single sphere that encloses both spheres.

If the volume has a sharp edge, a proliferation of very small spheres at the edge occurs as the accuracy of representation increases. This causes the parts of the tree representing the edge to become very large, and intersection detection for parts of the object near those areas becomes inefficient. Figure 2 shows how a cylinder can be approximated as a union of spheres. Rings of spheres are used to approximate the edges of the cylinder. The proliferation of small spheres is avoided in Figure 1 by using an intersection with a sphere that approximates the end of the cylinder.

5 Building Sphere Trees

This section presents an algorithm for building a sphere tree from a solid. We assume that it is possible to determine if a sphere is totally contained in the solid, partially contained in the solid, or totally outside the solid. The first step is to find a cube that totally encloses the object. We will then pack this solid with a face-centered packing of spheres of radius 1/2 the length of an edge of the cube. A face-centered packing puts spheres on each corner and in the center of each face of the cube. Unlike packing atoms in a metal, these spheres will

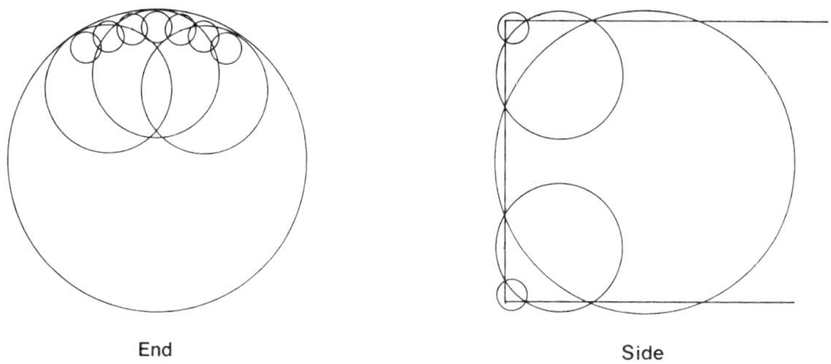

Figure 2: Packing of Spheres near the end of a cylinder.

overlap and fill all space within the cube. To refine this approximation the cube is replaced with a cubical arrangement of 27 nonoverlapping cubes with side one half the length of the original cube. The arrangement has the same center as the original cube. Each of the spheres will contain one of the smaller cubes. If the sphere is totally contained in the solid or totally outside the solid, then the small cube within the sphere will need no further subdivision. Otherwise, the small cube is packed with spheres and then subdivided again. Small cubes not within a sphere are always subdivided. This process continues until the spheres are small enough to achieve the desired degree of precision. All spheres that are totally within the object form a set of spheres whose union approximates the solid. Spheres of the minimum size that are neither inside nor outside the solid are included in this set. Any sphere that is entirely contained in another should be removed from the set. A tree can be formed from this set by pairing the smallest spheres with their smallest, nearest neighbor.

As mentioned above, a pure union of spheres produces many small spheres near an edge. The algorithm above also creates many small spheres near the surface of an object. Many of these spheres can be removed if the algorithm is modified to take advantage of intersections. Whenever a sphere is neither in the object nor outside the object and the part of the object's surface that is within the sphere can be modeled with an intersection of one or two spheres, then instead of subdividing the cube within the sphere, a subtree of spheres is used to model the part of the solid within the cube.

6 References

[1] Ahuja, N., Chien, R. T., Yen, R. and Bridwell, N. Interference detection and collision avoidance among three dimensional objects. *Proceedings of the First National Conference on Artificial Intelligence* Stanford University, August 1980, 44-48.

[2] Ahuja, N. and Nash, C. Octree representations of moving objects. *Computer Vision, Graphics, and Image Processing* 26(2), May 1984, 207-216.

[3] Badler, N. I., O'Rourke, J., and Toltzis, H. A spherical representation of a human body for visualizing movement. *Proc. IEEE* 67(10), October 1979, 1397-1403.

[4] Boyse, J. W. Interference detection among solids and surfaces. *Commun. ACM* 22(1), January 1979, 3-9.

[5] Braid, I. C. The Synthesis of Solids Bounded by Many Faces. *Commun. ACM* 18(4), April 1975, 209-216.

[6] Goldstein, R. A., and Nagel, R. 3-D Visual simulation. *Simulation* 16(1), January 1971, 25-31.

[7] Hopcroft, J. E., Schwartz, J. T., and Sharir, M. Efficient detection of intersections among spheres. *The International Journal of Robotics Research* 2(4), Winter 1983, 77-80.

[8] Maruyama, K. A procedure to determine intersections between polyhedral objects. *International Journal of Computer and Information Sciences* 1(3), 1972, 255-266.

[9] Mohr, R. and Bajcsy, R. Packing volumes by spheres. *IEEE Transactions on Pattern Analysis and Machine Intelligence* 5(1), January 1983, 111-116.

[10] O'Rourke, J. and Badler, N. Decomposition of Three-Dimensional Objects into Spheres. *IEEE Transactions on Pattern Analysis and Machine Intelligence* 1(3), July 1979, 295-305.

[11] Requicha, A. A. G. Representations for Rigid Solids: Theory, Methods, and Systems. *Computing Surveys* 12(4), December 1980, 437-466.

[12] Voelcker, H. B., and Requicha, A. A. G. Geometric modeling of mechanical parts and processes. *Computer* 10(12), December 1977, 48-57.

[13] Woo, T. C. Progress in shape modeling. *Computer* 10(12), December 1977, 40-46.

A Derivation of Robot Control Algorithms From the Lagrange Formalism

M. Cotsaftis* and C. Vibet†

*Professor, UGRA/DPT CEN-FAR BP No 6, 92260 Fontenay aux Roses, France
†Professor, Laboratoire de Robotique Industrielle, 22 Allée J. Rostand 91011 Evry France

Abstract: For automatic derivation of control algorithms, the decoupling method is directly applied to the Lagrange formalism and explicit expressions are given.

Keywords: Decoupling method, Lagrange formalism, control law synthesis

INTRODUCTION Recently a number of papers suggesting new approaches for controlling robots appeared in the litterature. The decoupling method [1] first introduced by Freund in robot control field opened a way to systematically derive the control algorithms needed for an exact linearization with arbitrary choice of dynamics. Related to this problem, the cartesian method [2] takes a more natural approach in using a Cartesian dynamic model of robots. In other methods including the resolved-acceleration control method [3] and inverse kinematic acceleration method [4], specific procedures are proposed to define efficient algorithms to implement in nonredundant robot controllers.

Alternatively, an adaptive control scheme of mechanical manipulators has been studied in ref [5] which estimates parameters on-line. Another method proposes a Cartesain trajectory tracking control system [6] in which the adaptation laws are derived from the model reference adaptive control theory.

In all these methods, the control algorithms are derived from a joint dynamical model of robots. With the aim at developping a direct methodology to find the control laws involved in robots with sensors, this letter introduces a new strategy by applying directly the decoupling method to the Lagrange formalism itself. In the first paragraph the Lagrange formalism is rewritten with a state space representation and a block-diagram is proposed. Next, the decoupling method is directly applied to the new formulation of the Lagrange formalism and explicit control algorithms are given.

I-A NEW LOOK AT LAGRANGE FORMALISM In this paragraph, the Lagrange formalism is viewed as a state space equation and a block-diagram representation is given. This methodology originates in a research to include systematically the initial conditions appearing in nonlinear systems. A new efficient concept based on block-diagram representation and control theory (i.e. the causality principle) was introduced [7,8]. Recently a general method which takes into account initial conditions in Lagrange formalism has been derived on the following way [9]. For a system whose Lagrangian is defined by

$$\mathbf{L} = \mathbf{L}(\mathbf{q},\dot{\mathbf{q}},t), \tag{1}$$

where $\mathbf{q} = col(q_1, q_2, \ldots q_n)$ are the generalized multibody system coordinates, the dynamical equations are deduced from the Lagrange formalism:

$$\frac{d}{dt}\frac{\partial \mathbf{L}}{\partial \dot{\mathbf{q}}} - \frac{\partial \mathbf{L}}{\partial \mathbf{q}} = \mathbf{T}, \tag{2}$$

where $\mathbf{T}$ represents the external generalized forces. A reformulation of the Lagrange formalism given in eqn (2), consists in introducing the state variables $\mathbf{X} = col(\mathbf{q},\mathbf{P})$, instead of the usual $\mathbf{X}_o = col(\mathbf{q},\dot{\mathbf{q}})$, where the generalized momentum $\mathbf{P}$ is defined from [9]

$$\mathbf{P} = \frac{\partial \mathbf{L}}{\partial \dot{\mathbf{q}}}(\mathbf{q},\dot{\mathbf{q}},t). \tag{3}$$

By introducing eqn (3) into eqn (2) and with the help of eqn (1), the Lagrange formalism takes the following classical form

$$\dot{\mathbf{X}} = \mathbf{F}(\mathbf{X}) + \mathbf{B}(\mathbf{X})\,\mathbf{U}(t), \tag{4}$$

with

$$\mathbf{F}(\mathbf{X}) = col(\mathbf{p},\frac{\partial \mathbf{L}}{\partial \mathbf{q}}), \quad \mathbf{B}(\mathbf{X}) = \begin{bmatrix} 0 & 0 \\ 0 & 1 \end{bmatrix}, \quad \mathbf{U}(t) = col(0,\mathbf{T}) \tag{5}$$

and where

$$\mathbf{p} = \dot{\mathbf{q}}(\mathbf{q},\mathbf{P},t), \tag{6}$$

is obtained by inverting eqn (3). In eqn (6), $\mathbf{P}$ represents the initial conditions resulting from shocks in the Lagrangian system [10,11], as it appears from integration of eqn (2) during the short time interval of the shock. Now, it is easy to derive a block-diagram representation of the Lagrange formalism as shown in Fig 1. Many useful features of practical significance are depicted by such a representation. In particular, the incidence of any saturation effect on the components of state configuration $\mathbf{q}$ and $\dot{\mathbf{q}}$ is immediatement deduced from this block-diagram.

II-DECOUPLING METHOD AND LAGRANGE FORMALISM In this paragraph, the nonlinear decoupling method is directly applied to the reformulated Lagrange formalism. The specific case of robot controllers is next investigated. Usually, the decoupling method starts from a state space equation as eqn (4) where the dimension of vector and matrix are deduced from eqn(6). Outputs of the system are written as

$$\mathbf{Y} = \mathbf{C}(\mathbf{X}), \tag{8}$$

where $\mathbf{C}(\mathbf{X})$ is a nonlinear vector depending upon $\mathbf{X}$. In robotics, the usual dimension of $\mathbf{Y}$ is (6x1). To obtain decoupling and linearization, the required feedback law is known to be of the following form:

$$\mathbf{U}(t) = \boldsymbol{\alpha}(\mathbf{X}) + \boldsymbol{\beta}(\mathbf{X})\, \mathbf{W}(t), \tag{9}$$

where

$$\boldsymbol{\alpha}(\mathbf{X}) = \begin{bmatrix} \alpha_1 \\ \alpha_2 \end{bmatrix}, \qquad \boldsymbol{\beta}(\mathbf{X}) = \begin{bmatrix} \beta_1 & 0 \\ \mathbf{0} & \beta_2 \end{bmatrix}, \tag{10}$$

are to be determined. The synthesis of control law is obtained by taking the successive derivatives of the ith output $\mathbf{Y}$ with respect t. Because robots are decribed by differential equations of second order, two derivatives are to be computed. One gets first:

$$\dot{\mathbf{Y}}_i = \frac{\partial \mathbf{C}_i}{\partial \mathbf{X}} \mathbf{F}(\mathbf{X}), \quad \text{because} \quad \frac{\partial \mathbf{C}_i}{\partial \mathbf{X}}(\mathbf{X}).\ \mathbf{B}(\mathbf{X}) = 0. \tag{11}$$

To the next order

$$\ddot{\mathbf{Y}}_i = \frac{\partial}{\partial \mathbf{X}} \left[\frac{\partial \mathbf{C}_i}{\partial \mathbf{X}}(\mathbf{X}).\mathbf{F}(\mathbf{X})\right].[\mathbf{F}(\mathbf{X}) + \mathbf{B}(\mathbf{X})\ \mathbf{U}(t)]. \tag{12}$$

Substituting for $\mathbf{U}(t)$ from eqn (9) into eqn (12), $\boldsymbol{\alpha}(\mathbf{X})$ and $\boldsymbol{\beta}(\mathbf{X})$ are obtained by exactly linearizing eqn (12), i.e.

$$\frac{\partial}{\partial \mathbf{X}} \left[\frac{\partial \mathbf{C}_i}{\partial \mathbf{X}} \mathbf{F}(\mathbf{X})\right].[\mathbf{F}(\mathbf{X}) + \mathbf{B}(\mathbf{X})\ \boldsymbol{\alpha}(\mathbf{X})] = 0 \tag{13}$$

and

$$\frac{\partial}{\partial \mathbf{X}} \left[\frac{\partial \mathbf{C}_i}{\partial \mathbf{X}} \mathbf{F}(\mathbf{X})\right] \mathbf{B}(\mathbf{X})\boldsymbol{\beta}(\mathbf{X}) = \mathbf{1}. \tag{14}$$

This gives $$\ddot{\mathbf{Y}}_i = \mathbf{W}_i. \tag{15}$$

By choosing

$$\alpha_2 = [\frac{\partial C}{\partial q}\frac{\partial p}{\partial P}]^{-1} [-\frac{\partial(}{\partial q}\frac{\partial C}{\partial q} p).p] - \frac{\partial L}{\partial q} \qquad (16)$$

and

$$\beta_2 = [\frac{\partial C}{\partial q} \cdot \frac{\partial p}{\partial P}]^{-1} . \qquad (17)$$

The first components α_1 and β_1 are not determined here, because the feedback law is not acting on the variable **p**, and can be taken as zero. However this freedom can be used in the more general situation of Fig 1 to control the pulse effects resulting in non-zero δX. To obtain decoupling and pole assignment, a more general feedback law of the following form

$$U(t) = \alpha(X) + \beta(X) [KQ(X) + \Gamma z], \qquad (18)$$

is used intead of eqn (9). Taking $K = diag(K_1, K_2, \ldots, K_n)$ with $K_i = col(-k_i^o, -k_i^1)$, and $\Gamma = diag(\gamma_1, \gamma_2)$ with $\gamma_j = col(\gamma_j^1, \gamma_j^2, \ldots \gamma_j^n)$, $j = 1,2$; one now gets

$$\ddot{Y}_i = - k_i^1 \dot{Y}_i - k_i^o Y_i + \Gamma_i z. \qquad (19)$$

provided the control law is defined by

$$K_i Q(X) = - k_i^1 \dot{Y} - k_i^o Y \qquad (20)$$

and $\alpha(X)$ and $\beta(X)$ are defined as in eqns(17,18). With these expressions, it is possible to design pole location by choosing the appropriate parameters **K** and Γz as

$$\Gamma z = -KQ(X_d) + \ddot{Y}_d + A'^{-1}(X) f_d \qquad (21)$$

where X_d corresponds to the desired state space inputs, $\ddot{Y}_d$ is the desired value of the acceleration output, $A'(X)$ is the inertia matrix in representation space and f_d is the exerted force by the end-effector on environment. It should be stressed that control algorithms as explicitely given by eqns(10,16,17,2,21), only require the knowledge of the Lagrangian $L(q,\dot{q},t)$ of the system and the outputs $Y = C(X)$. So, it is now possible to define a specific controller without needing a derivation of the system dynamic equations.

REFERENCES

1-Freund E.; Fast nonlinear control with arbitrary pole-placement for industrial robots and manipulators. Robotic Research 1,1 1982 pp 65-78

2-Khatib O, Burdick J. ; Motion and force control of robot manipulators. 1986 IEEE Conf. on Robotics pp 1381-1386.

3-Luh J., Walker W., Paul R.; Resolved-Acceleration control of mechanical manipulators. IEEE Trans. Vol AC-25 No 3, 198Ø pp 468-474

4-Hollerbach J., Sahar G.; Wrist-partitioned inverse kinematics acceleration and manipulators dynamics. 1984 IEEE Conf. on Robotics pp 152-161

5-Craig J. et al.; Adaptive control of mechanical manipulators. 1986 IEEE Conf. on Robotics pp 19Ø-195

6-Seraji H.; An adaptive Cartesian control scheme for manipulators. IEEE Conf. on Robotics pp 157-162

7-Vibet C.; Initial conditions and differential equations. Simulation, Vol. 48, Nb 5, 1987 pp. 21Ø-212. See also: Robots: Principles and Contrôle, Editions Ellipses/Marketing, Paris 1987 (in French)

1Ø-Vibet C.; General method of finding initial conditions. Revised form in IEEE Trans. on Educ. March 1987. See also: The initial conditions problem. To appear in IJEEE, 1987

11-Cotsaftis M.; Impulse effects in polyarticulated systems dynamics. Submitted to Electr. Lett., June 1987

12-Cotsaftis M., Vibet C.; Consistent impulse conditions in Lagrangian system dynamics. To appear in Robotica, 1987

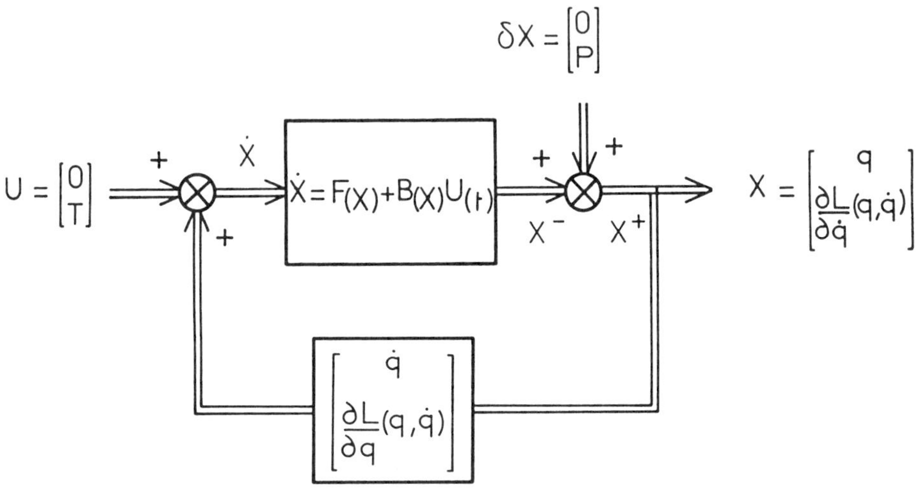

Fig 1: Block-diagram representation of the Lagrange formalism which includes initial conditions

Probabilistic Estimate of Robot End–Effector's Position Error Due to Process Noise

L.C. Chang*, G.R. Schade†, and Y.C. Pao*

*Graduate Student and Professor, Department of Engineering Mechanics, University of Nebraska, Lincoln, NE 68588-0347, USA
†Associate Professor of Mechanical Engineering, University of Nebraska, Lincoln, NE 68588-0347, USA

INTRODUCTION

The expected position of the end effector attached to a robot often needs to be controlled so that it will remain on the planned trajectory [1]. But due to process noise and measurement errors, the end effector will be driven off the expected (nominal) position. Control systems can be designed to guide the robot back to the correct trajectory. However, in order to predict the perturbed state of the end effector, probabilistic characteristics of the disturbing noises must be taken into consideration. Wu [2] has considered the random kinematic errors of the robot joint variables and derived the mathematical expressions for describing the position error of the end effector. He assumed that the errors of the joint variables are independent of each other, and employed a linear model for estimating the perturbed state derived from a Taylor's series expansion.

This paper proposes a probabilistic model for predicting the size and shape of an envelope within which the probability value of the perturbed position of the end-effector is confined can be quantified. The position of the end effector can be perturbed by process and/or measurement noises. Although we have studied both types of noises, in this paper only the process error due to the Gaussian-distributed white noise with zero-mean is considered. Also assumed are that all robot links are rigid, noises affecting the link variables are not correlated, and that perturbations are small so as to allow linear analysis based on a Taylor's series expansion.

[1,3]Graduate Student and Professor, Department of Engineering Mechanics, respectively, and [2]Associate Professor of Mechanical Engineering, University of Nebraska, Lincoln, Nebraska 68588-0347 USA.

DYNAMIC EQUATIONS OF n-LINK ROBOT

For an n-link robot, the dynamic equation can be derived by the Lagrange method using Denavit-Hartenberg notation [1,3]. It can be shown that the generalized driving forces Q_i's are related to the generalized link variables q_i's by the equations

$$Q_i = \sum_{k=1}^{n} D_{ik}\ddot{q}_k + \sum_{k=1}^{n}\sum_{m=1}^{n} H_{ikm}\dot{q}_k\dot{q}_m + G_i, \quad \text{for } i=1,2,\ldots,n \tag{1}$$

where dots denote differentiation with respect to time and the coefficient matrices are

$$D_{ik} = \sum_{j=\max(i,k)}^{n} \mathrm{Tr}([U_{jk}][J_j][U_{ji}]^T), \quad \text{for } i,k=1,2,\ldots,n \tag{2}$$

$$H_{ikm} = \sum_{j=\max(i,k,m)}^{n} \mathrm{Tr}([U_{jkm}][J_j][U_{ji}]^T), \quad \text{for } i,k,m=1,2,\ldots,n \tag{3}$$

and

$$G_i = \sum_{j=i}^{n} -m_j g[U_{ji}]\underline{s}_j, \quad \text{for } i=1,2,\ldots,n \tag{4}$$

In Eqs. (2), (3), and (4), $[U_{ij}]$ is a matrix obtained by partial differentiation with respect to the generalized variable q_j of the matrix $[T_{i\to o}]$ which transforms the coordinate system at the joint i to that at the joint 0, defined by

$$[U_{ij}] = \frac{\partial[T_{i\to o}]}{\partial q_j} = \begin{cases} [T_{j-1\to o}][S_j][T_{i\to j-1}], & \text{for } j\le i \\ 0, & \text{for } j>i \end{cases} \tag{5}$$

$[U_{jkm}]$ is obtained from $[U_{jk}]$ by differentiation with respect to q_m and can be explicitly written as

$$[U_{jkm}] = \frac{\partial[U_{jk}]}{\partial q_m} = \begin{cases} [T_{k-1\to o}][S_k][T_{m-1\to k-1}][S_m][T_{j\to m-1}], & \text{for } j\ge m\ge k \\ [T_{m-1\to o}][S_m][T_{k-1\to m-1}][S_k][T_{j\to k-1}], & \text{for } j\ge k\ge m \\ 0, & \text{for } j<k \text{ or } j<m \end{cases} \tag{6}$$

where $[S_i]$ for prismatic and revolute joint are defined respectively by

$$[S_j] = \begin{bmatrix} 0 & 0 & 0 & 0 \\ 0 & 0 & 0 & 0 \\ 0 & 0 & 0 & 1 \\ 0 & 0 & 0 & 0 \end{bmatrix} \quad \text{and} \quad [S_j] = \begin{bmatrix} 0 & -1 & 0 & 0 \\ 1 & 0 & 0 & 0 \\ 0 & 0 & 0 & 0 \\ 0 & 0 & 0 & 0 \end{bmatrix}. \tag{7,8}$$

In Eqs. (2) and (3), Tr denotes trace of a matrix. In Eq. (4), $\underline{g}$ is a vector equal to $[0\ -g\ 0\ 0]^T$, and $\underline{s}_j=[\bar{x}_i\ \bar{y}_i\ \bar{z}_i\ 1]^T$ is the position vector directed from the origin of the ith coordinate system to the mass center of the link i, with T denoting transpose of a matrix. m_j in Eq. (4) is the mass of link j. $[J_j]$ in Eqs. (2) and (3) is a matrix composed of the second moments of mass of the link j defined by

$$[J_j] = \begin{bmatrix} \int x^2dm & \int xydm & \int xzdm & \int xdm \\ \int xydm & \int y^2dm & \int yzdm & \int ydm \\ \int xzdm & \int yzdm & \int z^2dm & \int zdm \\ \int xdm & \int ydm & \int zdm & \int dm \end{bmatrix} \tag{9}$$

For convenience of further discussion, Eq. (1) is converted into a standard notation of state-variable as a system of first-order equations

$$\dot{\underline{x}} = \underline{F}(\underline{x},t) \tag{10}$$

In order not to overly complicate the discussion about the perturbed position of the robot end-effector's dynamic state, we shall use a 2-link robot but the procedure matter can be extended for a n-link robot.

ANALYSIS OF PERTURBED STATE

For a robot with two revolute links, the state vector $\underline{x}$ consists of four components, the angular displacements θ_1 and θ_2, and angular velocities $\dot{\theta}_1$ and $\dot{\theta}_2$. In state-variable notation, these are denoted as x_i for i=1,2,3,4, respectively. Denote the perturbed state by a vector $\underline{\epsilon}_x$ which can then be expressed in terms of the expected, nominal state $\underline{x}$ by a Taylor's series expansion following the approach of [4]. The dynamic equation for the perturbed state derived in [5] is

$$\dot{\underline{\epsilon}}_x = \begin{bmatrix} \delta\dot{x}_1 \\ \delta\dot{x}_2 \\ \delta\dot{x}_3 \\ \delta\dot{x}_4 \end{bmatrix} = [A] \begin{bmatrix} \delta x_1 \\ \delta x_2 \\ \delta x_3 \\ \delta x_4 \end{bmatrix} + [B] \begin{bmatrix} \delta Q_1 \\ \delta Q_2 \end{bmatrix} = [A]\underline{\epsilon}_x+[B]\delta\underline{Q} \tag{11}$$

where $\delta\underline{Q}$ is the perturbed input-forces vector,

$$[A] = \begin{bmatrix} \partial F_1/\partial x_1 & \partial F_1/\partial x_2 & \partial F_1/\partial x_3 & \partial F_1/\partial x_4 \\ \partial F_2/\partial x_1 & \partial F_2/\partial x_2 & \partial F_2/\partial x_3 & \partial F_2/\partial x_4 \\ \partial F_3/\partial x_1 & \partial F_3/\partial x_2 & \partial F_3/\partial x_3 & \partial F_3/\partial x_4 \\ \partial F_4/\partial x_1 & \partial F_4/\partial x_2 & \partial F_4/\partial x_3 & \partial F_4/\partial x_4 \end{bmatrix} \tag{12}$$

and

$$[B] = \begin{bmatrix} \partial F_1/\partial Q_1 & \partial F_1/\partial Q_2 \\ \partial F_2/\partial Q_1 & \partial F_2/\partial Q_2 \\ \partial F_3/\partial Q_1 & \partial F_3/\partial Q_2 \\ \partial F_4/\partial Q_1 & \partial F_4/\partial Q_2 \end{bmatrix} \tag{13}$$

PROBABILISTIC ANALYSIS OF PROCESS NOISE

The disturbance to the robot dynamic state could be process noises such as vibratory motion at the robot base, wind loading, and so forth, rather than the inaccurate description of the input loads $\underline{Q}$. Let us denote the arbitrary process noise by $\underline{w}$ and rewrite Eq. (11) as

$$\dot{\underline{\epsilon}}_x = [A]\underline{\epsilon}_x + [B]\underline{w} \tag{14}$$

To be more specific, let us assume all components of $\underline{w}$ be stationary Gaussian white noises with means equal to zero. Following the procedure in [6], the solution of Eq. (14) can be obtained as

$$\underline{\epsilon}_x(t) = [\Phi(t,0)]\underline{\epsilon}_x(0) + \int_0^t [\Phi(t,\tau)][B]\underline{w}(\tau)d\tau \tag{15}$$

where $[\Phi]$ is the transition matrix.

The <u>mean</u> and <u>covariance matrix</u> can be, respectively, written as

$$\underline{\mu}_x = E\{\underline{\epsilon}_x\} = [\Phi(t,0)]\underline{\mu}_x(0) \tag{16}$$

and

$$[P] = E\{(\underline{\epsilon}_x - E\{\underline{\epsilon}_x\})(\underline{\epsilon}_x - E\{\underline{\epsilon}_x\})^T\} \tag{17}$$

In [5], it is shown that [P] satisfies the Ricatti equation

$$[\dot{P}] = [A][P] + [P][A]^T + [B][W][B]^T \tag{18}$$

where [W] is the covariance matrix of the process-noise vector $\underline{w}$. For noises that are not correlated, [W] can be assumed to have the form

$$[W] = \begin{bmatrix} W_1 & 0 & \dots & 0 \\ 0 & W_2 & \dots & 0 \\ \cdot & \cdot & & \cdot \\ \cdot & \cdot & & \cdot \\ \cdot & \cdot & & \cdot \\ 0 & 0 & \dots & W_n \end{bmatrix} \quad (19)$$

When it is necessary to find only the steady-state solution for which $[\dot{P}]$ is equal to zero, Eq. (19) is reduced to an <u>algebraic Ricatti equation</u> [7]. The steady-state solution is readily available in [8].

Using the elements in the covariance matrix, it is possible to construct a confidence region and to quantify the asscoiated probability that the perturbed position of the end effector lies within the region. In next section, we will demonstrate the step-by-step procedure of determining such a region for a two-link robot.

PROBABLE ELLIPTICAL ENVELOPE FOR 2-D END-EFFECTOR

Let the coordinates of the perturbed position of the end effector of a two-dimensional robot be denoted as (X,Y), and the means of X and Y be denoted by μ_X and μ_Y, respectively. It should be obvious that the coordinates of the unperturbed, nominal position of the end effector are (μ_X,μ_Y). As a <u>linear</u> model, we can thus apply the Taylor's series expansion to express the coordinates of the perturbed position in terms of those of the nominal position, the lengths and orientations of the two links as

$$X = \mu_X+a_1\delta\theta_1+a_2\delta\theta_2 \quad \text{and} \quad Y = \mu_Y+b_1\delta\theta_1+b_2\delta\theta_2 \quad (20,21)$$

where a_1, a_2, b_1, and b_2 are partial derivatives of X and Y with respect to θ_1 and θ_2 evaluated at the nominal coordinates.

The variances and covariance of X and Y can also be expressed in terms of the coefficients a_i and b_i, and the errors in θ_1 and θ_2, as

$$\mathrm{Var}(X) = a_1^2\mathrm{Var}(\delta\theta_1)+a_2^2\mathrm{Var}(\delta\theta_2)+2a_1a_2\mathrm{Cov}(\delta\theta_1,\delta\theta_2) \quad (22)$$

$$\mathrm{Var}(Y) = b_1^2\mathrm{Var}(\delta\theta_1)+b_2^2\mathrm{Var}(\delta\theta_2)+2b_1b_2\mathrm{Cov}(\delta\theta_1,\delta\theta_2) \quad (23)$$

and

$$\mathrm{Cov}(X,Y) = a_1b_1\mathrm{Var}(\delta\theta_1)+a_2b_2\mathrm{Var}(\delta\theta_2)+(a_1b_2+a_2b_1)\mathrm{Cov}(\delta\theta_1,\delta\theta_2) \quad (24)$$

The variances and covariance of $\delta\theta_1$ and $\delta\theta_2$ are simply the elements of the state variables' error-covariance matrix [P]. Since the process noise $\underline{w}$ is assumed Gaussian-distributed, it can therefore be expected that the perturbed position of the two-link robot end-effector is also

Gaussian-distributed. As both of the coordinates of the end effector's position, X and Y, are involved, the probability density function with bivariate normal distribution [9,10] is hence applicable here. That is

$$f(X,Y) = \frac{1}{2\pi\sigma_X\sigma_Y\sqrt{1-\rho_{XY}^2}}\exp(-C/2) \tag{25}$$

where

$$C = \frac{1}{1-\rho_{XY}^2}\left[\frac{(X-\mu_X)^2}{\sigma_X^2} - 2\rho_{XY}\frac{(X-\mu_X)(Y-\mu_Y)}{\sigma_X\sigma_Y} + \frac{(Y-\mu_Y)^2}{\sigma_Y^2}\right] \tag{26}$$

and σ_X and σ_Y are the standard deviations of X and Y, respectively, and ρ_{XY} is the correlation coefficient between X and Y. Equation (25) describes a bell-shaped surface over the X-Y plane and the value of its right-hand side when C=0 gives the maximum height of the bell. By assigning a positive value for C, Eq. (26) depicts an elliptical contour that is the intersection of the bell-shaped surface and a plane parallel to the X-Y plane.

The probability that the end effector lies within some confidence region R may be computed using the equation

$$\iint_R f(X,Y)\,dX\,dY \tag{27}$$

For the specific elliptical confidence region defined by some positive value of C, Eq. (26), this probability is defined by the χ^2 distribution with two degrees of freedom. For the planar case, the probability is

$$\Pr(\chi^2<C) = 1-\exp(-C/2) \tag{28}$$

Figure 1 shows that the ellipse is to be drawn using the expected, nominal position (μ_X,μ_Y) as center and has values of semi-major and semi-minor axes calculated by the equations

$$r_X = \{C\sigma_X^2\sigma_Y^2(1-\rho_{XY}^2)/(\sigma_Y^2\cos^2\gamma-\rho_{XY}\sigma_X\sigma_Y\sin2\gamma+\sigma_X^2\sin^2\gamma)\}^{1/2} \tag{29}$$

and

$$r_Y = \{C\sigma_X^2\sigma_Y^2(1-\rho_{XY}^2)/(\sigma_Y^2\sin^2\gamma+\rho_{XY}\sigma_X\sigma_Y\sin2\gamma+\sigma_X^2\cos^2\gamma)\}^{1/2} \tag{30}$$

The major axis of the ellipse is at an angle, measured from the X-axis, equal to

$$\gamma = \{\mathrm{Tan}^{-1}(2\rho_{XY}\sigma_X\sigma_Y/(\sigma_X^2-\sigma_Y^2))\}/2 \tag{31}$$

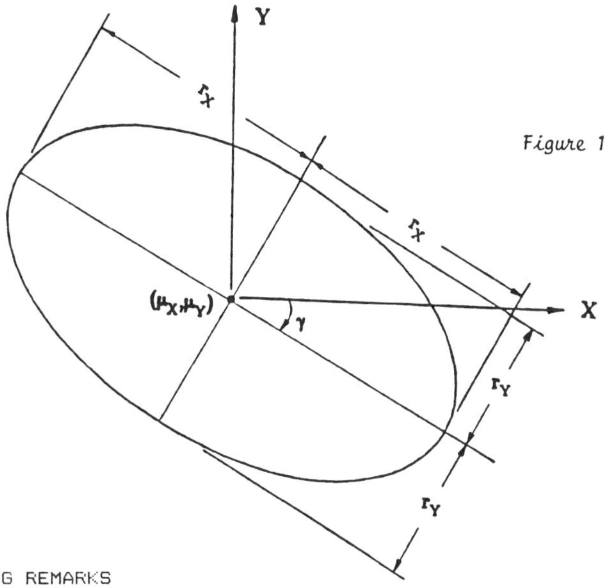

Figure 1

CONCLUDING REMARKS

The proposed probabilistic model enables to predict the envelope within which the perturbed position of the robot end-effector will be confined in response to process noise. The procedure can be summarized as follows:

1. Based on the robot geometric shapes, material properties, and interrelationships among links, calculate matrices [A] and [B] according to Eqs. (12) and (13).
2. Define the covariance matrix, [W] in Eq. (19), of the process noise vector $\underline{w}$.
3. Solve the steady-state Ricatti equation, that is using Eq. (18) but with $[\dot{P}]=0$.
4. Knowing [P], σ_x, σ_y, and ρ_{xy} can be calculated using Eqs. (22)~(24).
5. Choose different values of C to construct the elliptical envelopes using Eq. (26), and calculate the corresponding probability values using Eq. (28).

The envelopes for a two-dimensional robot are ellipses. It should be easily to observe that for a three-dimensional robot, the envelopes for the perturbed position of the end effector would be ellipsoids [5], and the corresponding probability would be distributed χ^2,with three-degrees-of-freedom. In [5], Monte-Carlo simulation [11] has also been applied to verify the proposed model and produces very good agreement in results.

For most robot systems, their dynamic behaviors responding to process noise are unstable. Feedback control designs often are necessary to stabilize the systems, and Kalman filters are effective means for considerations [1,7,12]. For lack of space here, we have to deliberate the consideration of measurement noise and application of Kalman filter for stabilizing robot systems elsewhere [13].

The methodology developed herein can be readily adapted for interactive determination of the envelope by use of microcomputer and peripheral graphics devices [14]. And for further refinement of the probabilistic models, nonrigid links could be considered by application of finite element method [15].

ACKNOWLEDGEMENT

This work was supported, in part, by the General Motors Corporation through a CAD/CAM Fellowship Program and by the Engineering Research Center, University of Nebraska-Lincoln. The authors would like to thank their colleague Professor William L. Brogan for enlightening discussion of the Ricatti equation.

REFERENCES

1. K. S. Fu, R. C. Gonzalez, and C. S. G. Lee, ROBOTICS: Control, Sensing, Vision, and Intelligence, McGraw-Hill, New York, 1987.
2. C. H. Wu, "A Kinematic CAD Tool for the Design and Control of a Robot Manipulator," International Journal of Robotics Research, Vol. 3, 1984, pp. 58-67.
3. J. Denavit and R. S. Hartenberg, "A Kinematic Notation for Lower-Pair Mechanism Based on Matrices," Journal of Applied Mechanics, Vol. 77, 1955, pp. 215-221.
4. M. J. Chung and C. S. G. Lee, "An Adaptive Control Strategy for Computer-Based Manipulators," Technical Report RDS-TR-10-82, Center for Robotics and Integrated Manufacturing, University of Michigan, Ann Arbor, Michigan, 1982.
5. L. C. Chang, "Probabilistic Error Model of Robot End-Effector's Dynamic State," Doctoral Dissertation, Department of Engineering Mechanics, University of Nebraska, Lincoln, Nebraska, 1987.
6. W. L. Brogan, Modern Control Theory, Prentice-Hall, Englewood Cliffs, New Jersey, 1985.
7. F. L. Lewis, Optimal Estimation, John Wiley, New York, 1986.
8. B. D. Anderson and J. B. Moors, Linear Optimal Control, Prentice-Hall, Englewood Cliffs, New Jersey, 1971.
9. G. R. Schade, "A Bivariate Normal Model of Mechanism Coupler - Point Position," Paper No. 82-DET-44, American Society of Mechanical Engineers Publication, 1982.
10. S. L. Meyer, Data Analysis for Scientists and Engineers, John Wiley, New York, 1975.
11. S. R. Ross, Introduction to Probability Models, Academic Press, London, United Kingdom, 1985.
12. R. G. Brown, Introduction to Random Signal Analysis and Kalman Filter, John Wiley, New York, 1983.
13. L. C. Chang, W. L. Brogan, and Y. C. Pao, "Kalman-Filter Feedback Control of Probable End-Effector Position Error Due to Process and Measurement Noises," a manuscript under preparation.
14. Y. C. Pao, Elements of Computer-Aided Design and Manufacturing, CAD/CAM, John Wiley, New York, 1984.
15. Y. C. Pao, A First Course in Finite Element Analysis, Allyn & Bacon, Boston, Massachusetts, 1986.

SENSORS AND VISION SYSTEMS

Methodology and Development of a Software Package for a Low Cost Machine Vision System for Real Time Applications

Helmut T. Zwahlen and Hugo M. Straumann

Industrial and Systems Engineering, Ohio University, Athens, OH 45701-2979, USA

A software package for a low cost machine vision system was developed. The system, which might be used as an educational tool, or for industrial applications, is capable of detecting, locating and identifying simple objects made up by straight lines with no constraints on orientation or location in the viewing area.

INTRODUCTION

According to Groover (1986), machine vision systems need to fulfill two important criteria which currently limit their feasibility in many manufacturing processes. These criteria include the need for being relatively low-cost, typically under $ 30,000, and the need for rapid response time, typically a fraction of a second.

The objective was to develop a software package (BASIC and assembler) for a low cost machine vision system which had to be able to detect, locate and identify objects (made up by straight lines) in any position or orientation within the viewing area in as short a time interval as possible. The primary purpose of this paper is to report on the method which was used to detect an object, find the location, detect the edges, follow the edges and to identify the object.

APPARATUS

The low cost machine vision system (less than $10,000) is comprised of an Apple IIe microcomputer, a General Electric TN2500 digitizing camera, a 4K frame buffer and a monitor. A Teachmover Microbot is used in conjunction with the machine vision system and the Apple IIe microcomputer to pick up, move and place objects with parallel sides in a designated area for demonstration purposes.

The General Electric camera has a pixel array of 244 rows and 256 columns and is capable of detecting and transmitting up to 256 gray scale values (0-255) for each pixel. Since it was not feasible to store and analyze all of the information which the camera is capable of collecting in the Apple IIe computer, a 4K frame buffer was added to the system in order to reduce the amount of pixels and information. This buffer, which was developed by the Welding Engineering Department of the Ohio State University in Columbus, Ohio, is capable of storing the gray scale values for 4000 pixels. One can choose the rows and columns of pixels within the viewing area for which the pixel gray scale values are to be stored. The ability to select certain rows and columns of pixels for storage and analysis is called windowing. The Apple IIe microcomputer, which is equipped with a 6502 microprocessor and 64K Random Access Memory, is then used to process the information which is stored by the frame buffer.

APPROACH

General Software Package

The software package allows the user to execute the machine vision process, store coordinates of new objects into memory, or view and if necessary delete coordinates of objects already stored into memory. The machine vision process encompasses five phases including object detection, object location, edge detection, edge following and object identification. Each of these processes are fairly complex and require a large number of calculations. Since the Apple IIe has a relatively low processing speed (1 MHz) and limited storage capacity (64K RAM) it was necessary to formulate constraints to reduce the processing time. The following constraints were placed upon objects in the viewing area (23 x 18.5 degrees): 1) only one object can be present in the viewing area at any time, 2) the contrast ratio between the object and the background (dark) should be as high as possible, 3) the smallest dimension of the object has to be greater than 1 degree of the visual angle, 4) the largest dimension of the object has to be less than 16 degrees of the visual angle, 5) the ratio between the largest dimension and smallest dimension within an object should not be greater than 7:1, 6) all edges have to be straight lines.

Object Detection

A locating algorithm using the Roberts cross edge gradient is utilized to determine the location of an object's edges. In this edge detection algorithm an element in a matrix is chosen. Let us assume that this matrix element is in row i and column j. In order to calculate the Roberts cross edge gradient R(i,j) for the element (i,j) the following equation is employed:

$$R(i,j) = |G(i,j) - G(i+1,j+1)| + |G(i,j+1) - G(i+1,j)|$$

where G(i,j) represents the gray scale value for element (i,j).

To locate the object the frame buffer accepts first every 4th row and every 4th column of gray scale values such that the entire viewing area can be represented and stored in a low resolution matrix (57 x 57) which requires 3249 bytes of memory. A portion of such a matrix is shown in Figure 1. The Roberts cross edge gradient is then calculated for every element in this matrix yielding a new matrix as shown in Figure 2. Beginning at the upper left hand corner and sweeping downwards the algorithm compares each value against an arbitrary threshold value of 70. If two consecutive values larger than the threshold are not found then it is assumed that there is no object present in the viewing area and a new set of gray scale values of the viewing area is taken. If two or more consecutive values are found which are larger than the threshold value then an object is assumed to be present in the viewing area.

Object Location

Once an object has been detected in the viewing area it is then necessary to determine its approximate location and dimensions as illustrated in Figure 3. When the computer determines that an object is present the row number where the object was first detected is stored. The computer then begins at position 2 in Figure 3 which is the first column of the row following the row in which the object was first detected and scans down the columns as shown in Figure 3 until two additional consecutive values which are larger than the threshold value are found and stores the column number in which these values were found. The computer then proceeds to position 3 in Figure 3 and scans the matrix as shown until two additional consecutive values are found which are larger than the threshold value and stores the row number in which they are found. Finally, the computer begins at position 4 in Figure 3 and scans until it finds a fourth pair of consecutive values which are larger than the threshold value and stores the column number in which they are found.

Edge Detection

Once the object has been located, a new window of gray scale values is recorded which includes a boundary of 8 pixel values around the four points which were found in the last step. The program then calculates the Roberts cross edge gradient for each element in the matrix. This results in another matrix similar to the one shown in Figure 2 with the exception that it only includes values which are inside the window (higher resolution). In order to reduce the amount of data

```
22 22 24 24 24 22 22 23 23 24 22 24 22 22 20 19 20 20 22 22 18 20
24 23 23 24 23 22 22 23 22 24 22 20 24 24 22 20 22 20 18 22 20 18
22 22 21 22 24 23 24 24 25 22 24 22 23 22 23 21 20 20 20 20 20 18
22 24 24 22 23 24 26 22 14490 26 23 23 22 22 20 20 18 19 20 20 20
22 20 23 24 25 23 26 20 14614614712028 22 22 24 20 22 20 19 20 20
20 21 21 22 22 22 24 37 14414214614614413650 22 20 18 20 18 20 22
20 22 22 22 24 22 24 15214814815014814414214213484 24 20 20 19 20
22 24 24 24 24 24 22 15815415315214814414614213813413110826 23 22
22 22 22 25 24 24 28 15816015215214614514213913813813213638 24 20
22 22 20 22 21 23 68 15816015815615014614313913613813440 20 18 20
22 24 20 20 23 24 15215615815615215214614314013613236 20 22 18 20
20 22 22 22 22 24 14815415615415214814814613913837 20 18 24 18 18
18 22 22 22 22 28 14815215415215114814814514030 20 18 18 20 16 17
18 18 22 20 22 93 14414815014814814414414434 20 18 22 21 16 20 16
18 20 20 20 20 14414214614614414414214032 20 20 20 20 18 20 14 18
20 22 24 22 22 13813814214214214213830 22 19 19 20 18 18 18 16 18
18 20 24 20 31 14414214614814714634 20 22 20 20 20 18 18 18 16 18
18 18 20 19 12214514615015215029 20 20 19 20 18 18 18 20 18 16 17
20 18 21 20 14414814815415027 22 22 21 19 18 17 18 20 18 18 16 19
18 16 20 20 14615015015232 20 18 20 18 18 17 18 18 18 18 16 16 16
18 24 16 30 14615215224 20 22 20 18 20 22 18 18 18 18 18 16 16 16
18 18 19 13614415024 22 20 19 18 18 18 18 18 17 15 19 16 16 16 18
16 17 16 14413822 19 20 18 19 20 18 18 17 16 16 14 16 16 18 16 15
18 20 14 13620 19 18 16 18 18 18 18 20 18 18 18 14 16 18 19 18 16
16 18 36 20 21 18 19 18 18 20 17 18 18 18 16 16 16 16 18 16 16 15
18 20 20 18 19 19 19 19 17 20 16 20 17 17 16 17 18 18 16 16 18 16
16 16 16 19 18 20 18 17 16 17 18 18 16 14 16 16 17 14 16 16 16 14
```

Figure 1 Typical Matrix of Pixel Gray Scale Values for a Selected Portion of the Viewing Area Containing a Triangle

```
3 2 1 1 3 0 2 1 3 4 4 2 4 4 3 3 2 4 4 6 0 4
3 3 4 1 2 3 3 4 1 0 4 5 3 1 4 1 2 2 4 2 4 2
2 5 3 3 0 3 4 12318770 5 1 2 1 4 1 2 3 1 0 2 2
4 5 1 4 1 5 10 24858 1772181027 0 4 4 4 5 0 1 0 0
1 3 4 5 4 5 15 2336 5 27 14223014226 6 4 4 3 3 2 2
3 2 1 2 2 4 14111910 10 6 6 10 98 20417670 6 2 1 3 4
4 4 4 2 2 0 26412 11 7 6 8 4 4 12 54 15719594 10 6 4
2 4 3 1 0 6 2666 9 1 10 5 3 7 5 4 9 29 18017 5 4
0 2 5 6 4 49 2204 10 10 12 5 6 7 4 2 10 94 11824 6 2
2 6 2 2 3 17394 4 4 6 6 10 6 7 7 6 10411818 6 4 0
4 4 4 3 3 25210 4 4 6 4 6 5 10 5 10511318 8 10 2 4
6 0 0 0 8 24410 4 4 3 7 0 5 12 11112519 2 8 10 3 3
4 4 2 2 77 1718 8 8 7 7 8 5 11512412 2 7 3 0 3 1
2 4 2 2 195 49 8 6 8 8 6 6 11612614 2 4 5 3 2 4 2
4 6 6 4 240 10 8 8 6 4 6 11412015 2 1 2 2 2 8 6 4
4 6 6 11 229 10 8 10 11 9 11612210 5 2 1 4 0 0 4 4 6
2 6 5 114136 5 8 8 7 12213114 3 3 2 4 2 2 2 4 3 3
2 5 2 22727 5 10 2 1251309 3 3 2 3 1 2 0 2 4 4 4
4 7 1 2508 4 8 12413511 6 5 4 2 0 1 2 2 2 2 3 1
8 4 14 24210 4 12614010 4 0 0 6 5 1 0 0 0 4 0 0 2
6 7 13112412 1301306 3 5 2 2 6 4 1 4 4 3 2 0 2 4
3 4 2452 134 1337 4 2 2 2 0 1 3 3 4 6 3 2 2 3 5
5 7 250126121 4 5 4 1 3 2 2 3 3 4 6 4 2 3 3 3 5
4 20 1061154 0 3 2 2 3 1 2 2 2 4 4 2 4 3 5 3 3
4 18 18 4 3 1 1 2 5 7 5 3 2 3 1 3 4 0 2 2 3 3
6 8 3 0 2 2 3 3 4 3 4 5 4 3 1 2 5 4 0 2 4 6
2 2 1 5 5 6 4 3 3 3 6 6 0 2 0 2 3 2 2 2 1 2
```

Figure 2 Typical Matrix of Roberts Cross Edge Gradient Values Calculated from Pixel Gray Scale Values shown in Figure 1

required to be stored the computer compares each of these values to the threshold value of 70, storing only the positions of the values which are larger than the threshold value as part of the edge list. The program then attempts to link the stored location values together to determine the shape of the object and to filter out local disturbances accepted by Roberts cross edge gradient.

Edge Following

The algorithm developed to follow the object's edges can best be

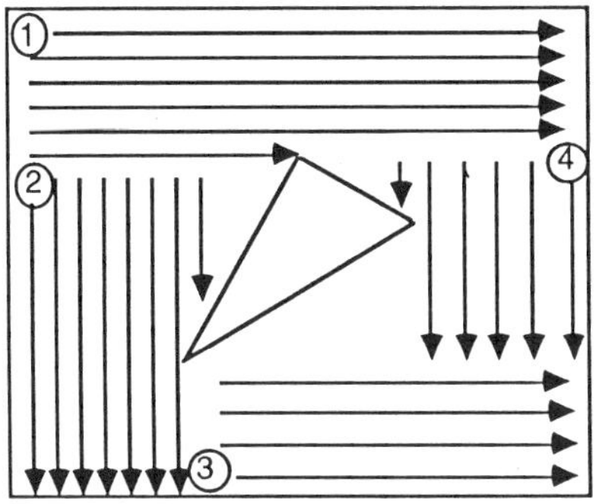

Figure 3 Example of the Algorithm used to Determine the Window in which the Object is Located

described by using an example. The algorithm always begins by storing the recorded edge position nearest the upper left hand corner of the viewing area. Once a starting point has been determined this point may be thought to occupy the center position (position 8) in the 3x3 matrix shown in the top left corner of Figure 4. From this position the algorithm then checks the position diagonally towards the lower right which is depicted as position 4. If this value is not present in the list of edge positions then the algorithm checks position 3, then positions 5, 2, 6, 1, and 7 respectively. As soon as one of these positions are found in the list of edge points this position is stored in the linked edge list and becomes the new center position in a 3x3 matrix while the previous position assumes the position of 0 in this 3x3 matrix. The algorithm then follows the edges of the object rotating the 3x3 matrix such that position 4 is always in a straight line with the two previous positions.

Let us choose the lower left hand corner of the triangle shown in Figure 2 as an example. Assuming that the current position is the position occupied by the Roberts Cross Edge Gradient of 135 shown in the 9x9 matrix in Figure 4, this position would be considered the center position (position 8) in a 3x3 matrix. If we further assume that the position in the 2nd row and 9th column of the 9x9 matrix occupied by 130 has already been established as part of the edge then this position assumes position 0 and would require the 3x3 matrix to be positioned as the 3x3 matrix shown in the middle of the top row of 3x3 matrices is positioned. The algorithm would then look for the position occupied by 140 in the 9x9 matrix (since 140 is greater than 70 this position is present in the edge list), find it is present, store it and advance to this position. This would be repeated storing the positions occupied by 130, 133, 121 and 115 in the linked edge

```
0 1 2    6 7 0    5 6 7
7 8 3    5 8 1    4 8 0
6 5 4    4 3 2    3 2 1

    3 4 5    2 3 4
    2 8 6    1 8 5
    1 0 7    0 7 6
```

6	5	114	136	5	8	8	7	122
5	2	227	27	5	10	2	125	130
7	1	250	8	4	8	124	135	11
4	14	242	10	4	126	140	10	4
7	131	124	12	130	130	6	3	5
4	245	2	134	133	7	4	2	2
7	250	126	121	4	5	4	1	3
20	106	115	4	0	3	2	2	3
18	18	4	3	1	1	2	5	7

Figure 4 3x3 Pixel Matrices Indicating Order of Search Using Positions in the Edge List and 9x9 Pixel Matrix Showing Roberts Cross Edge Gradient Values for Lower Left Hand Corner of Triangle Shown in Figure 2

list, however, once 115 has been stored the algorithm can no longer continue straight in the direction of position 4. Therefore the algorithm checks position 3, can not find this position in the edge list, checks for position 5, finds this position in the edge list, stores the position occupied by 106, advances the center of the 3x3 matrix to the position occupied by 106 and position 0 of the 3x3 matrix to the position occupied by 115 rotating the 3x3 matrix to the direction shown in the 3x3 matrix in the top right of Figure 4. Once again the algorithm checks position 4, finds it not in the list of edge positions, checks positions 3, 5 and 2, finds they are not available, checks position 6, finds it is available and stores the position occupied by 250 in the 9x9 matrix as part of the linked edge list and advances the center of the 3x3 matrix to the position occupied by 250 and position 0 of the 3x3 matrix to the position occupied by 106 rotating the 3x3 matrix to the direction shown in the 3x3 matrix on the bottom left of Figure 4. The algorithm then looks for the position now occupied by position 4 (245) of the 3x3 matrix and finds it is available in the list of edge positions. Again this value is stored, but note that the direction of the 3x3 matrix has changed and maintained its present direction for two consecutive pixel values, therefore, the algorithm returns to the last stored pixel position before the last change in direction occurred (106) and stores it as a corner point and stores the two positions found after the corner point as positions along a new edge. The algorithm then returns to following the edge and stores the positions occupied by 245 and 131 as points along this new edge. However, once the position occupied by 131 is established as the center of the 3x3 matrix and the position occupied by 245 is established as position 0, it is unable to advance to position 4 of the 3x3 matrix since this position is not in the edge list. The algorithm looks for position 3, is unable to find it, looks for position 5, finds it, stores it and advances the 3x3 matrix so that the center position is the position occupied by 242 and position 0 is occupied by 131. Note that the 3x3 matrix has changed its direction as shown in the bottom right 3x3 matrix of Figure 4. Again position 4 is checked but can not be found, position 3 is checked, found, and stored, therefore the 3x3 matrix rotates back to the position shown by the bottom left 3x3 matrix in Figure 4. It should be noted that although the direction of the 3x3 matrix changes in both of these last steps a new direction is not maintained for at least two consecutive positions and therefore the algorithm does not consider these changes in direction as separate edges. Once the 3x3 matrix has advanced to the position occupied by 250 it maintains this direction throughout the rest of the example storing 227 and 114 as positions along the same edge and follows the edge of the object back to the starting point. A linear least squares regression is performed for each edge of the object and new corner points of the object are calculated. However, if the algorithm is unable to follow the edges back to the starting point due to missing adjacent pixel edge positions then a new matrix of gray values is stored and the object detection, location, edge detection and edge following are repeated.

Object Identification

Once the corner points of the object have been calculated the object can be identified. This is done by comparing the coordinates of the object with the coordinates of objects which have been stored in memory. However, one must consider that the object may have an infinite number of possible orientations, therefore, the orientation and location of the object must be the same as was used to store the objects in memory. To accomplish this each side's length is computed using the object's corner coordinates. These values (positive or negative depending upon which way the coordinates are subtracted from each other) are then squared. These values are then ranked and the point of origin is found as determined by the following four rules: 1) If all sides have the same length, and the object has no subshapes, then the origin can be chosen at any corner; 2) If all sides have the same length and the object has a subshape (for example a square hole), then the origin is chosen at the corner with the shortest distance to the subshape; 3) Select the longest side as the base; 4) Select the origin such that the shortest side is closest to the y-axis subject to rule 3. Using the triangle shown in Figures 1 and 2 as an example, one can see the sides have different lengths, thus rules 1 and 2 do not apply. According to rule 3 the bottom side (longest side) will be the base along the x-axis and according to rule 4 the top side (shortest side) will be closest to the y-axis and intersect with the bottom side at the origin (0,0 coordinates).

When the origin has been found the corner points are calculated with respect to this origin. These values are then compared to the coordinates for each object in memory. If the coordinates for the object in the viewing area do not vary by more than a certain tolerance from the coordinates for an object in memory then the object in the viewing area is classified as this object. One of three situations can occur at this stage: 1) the object is identified as an object in memory with parallel sides in which case the steps for each of the 6 motors in the robot are calculated and the robot is instructed to retrieve the object and place it in a designated location with a designated orientation, 2) the object is identified and does not have parallel sides and therefore can not be retrieved by the robots parallel grippers, or 3) the object can not be identified. Regardless of which situation occurs an appropriate message will appear on the screen (triangle example: Robot Retrieval Not Possible!) and the user is asked if he or she would like to repeat the vision process.

CONCLUSIONS

A software package for a low cost machine vision system has been developed to successfully locate and identify simple objects. This package might be used as an educational tool, or might have industrial applications. Reliability tests, which have been conducted using a number of simple objects with straight edges, indicate that the system has a rather low reliability which is dependent on the orientation and the shape of the objects. It appears that some improvements could be made to increase the reliability of the system by expanding the algorithms used. The tests conducted indicate further, that the processing time from detection to identification for a simple object is about 15 seconds. Based upon an analysis of the algorithms used, it appears that this time can not be significantly reduced through simplifications and modifications. Therefore, it appears that microprocessors with a much faster clock speed will have to be used if processing times of less than a second need to be achieved.

REFERENCES

Groover, M.T., Weiss, M., Nagel, R.N., and Odrey, N.G., Industrial Robotics, McGraw-Hill Book Company, 1986, p. 160.

A Sensorized System for Delicate Piece Selection, Handling and Packaging

G. Belforte, F. Bellosta, N. D'Alfio, and T. Raparelli

Department of Mechanics, Polytechnic of Turin, Turin, Italy

Identification and orientation

Automatic handling of fragile objects of different shapes and dimensions requires orientation devices and an adaptive gripper with touch sensitivity.

Packaging of assorted biscuits is here considered. They are presented in a number of heaps, one for any different sort. They have to be grasped one by one and accomodated in the proper plastic case inside a cardboard box. Each case holds several pieces, one upon the other.

A typical package holds many different biscuit shapes.

Biscuits are extracted from heaps by moving belts and gravity channeling and are forwarded randomly to the orientation device.

The assortment is presented in Fig.1. The biscuits are partly axialsymmetric with one leaning surface (a, e, f, h, m), or two equal surfaces (o); in part they have two symmetry planes and one leaning surface (c, g, i, l, n), or two equal surfaces (g); type (d) has one symmetry plane and one leaning surface; type (b) has a line contact with the supporting plane.

A separate orientation device is provided for the different types; all devices are nevertheless conceived in the same way.

The biscuits arrive on a perforated plate through which flows compressed air, providing pneumostatic support (Fig. 2).

Lateral nozzles and mechanical deviators orientate and channel the biscuits.

Discrimination between correct and upside down posture is made by arranging hole density and flow throughput in order to let floating only the correctly positioned biscuits.

If the biscuit does not move, it is reversed after sensing its stay in area G(Fig. 2).

The wide field air jet sensor is described in Fig.3; it has two emitters E_1, E_2, and a receiver R, made by a nozzle plus a tube where the recovered pressure is measured.

The jets from E_1, E_2 interfere and the recovery pressure is slightly negative, but for the case when an object obstructs the E_2 jet; then the E_1 jet impinges on the R nozzle, and the recovery pressure rises.
The sensor signal is forwarded to a pneumatic actuator that pushes the biscuit against a curved wall where it is reversed, aided by three air jets (Fig.2).
The wafer biscuit "g" has two equally suitable contact areas, but they have different texture. Identification is performed by an induction fluidic sensor, that senses air leaks between its flush mounted tip and the biscuit. Leaks depend on surface coarseness.

Grasping and packaging

The same gripper has been used for all the shapes. It must insert biscuits of different sizes into cases that are just a little larger, then it opens very slightly the fingers for avoiding damages to the plastic case walls. The fingers are very thin and their motion is parallel.
The gripper is represented in Fig.4. The fingers are driven by four bar linkages, with rods laying in parallel planes. The finger metallic surface is covered by a membrane forming a poach filled with distilled water; it is the sensitive fingertip, connected with a pressure transducer. Grasping is stopped at a threshold pressure.
Actuation is performed pneumatically (Fig.5) by flat, parallel micro-actuators CA and CC, and by the braking cylinder CF. The CA body is joint to the rod of CC and controls the gripper opening.
The rod of CA actuates both fingers through a slotted link. The actuator CF acts on a wedge that brings to rest the rod of the closure actuator CC.
In this way, gripper closure and opening are independent; the opening width is fixed with respect to the closure position; such position is controlled by the brake, according to the sensor signal. During closure, the actuator CA is powered and its rod is protruding; opening occurs by de-powering such actuator and letting its rod be reset by a spring.
The gripper has been mounted on a five degrees of freedom robot (Fig.6).
The authors thank FESTO s.p.a for granting a fellowship.
This work has been sponsored by the Italian Ministry of Education.

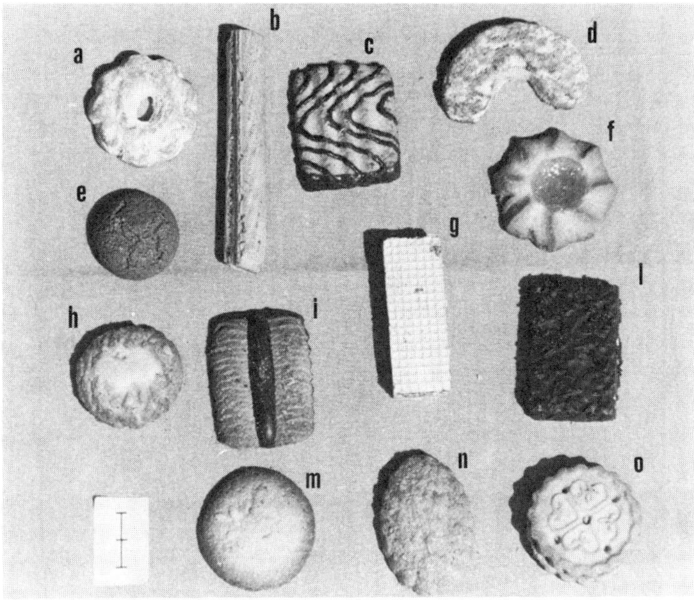

Fig. 1 - Assorted biscuits.

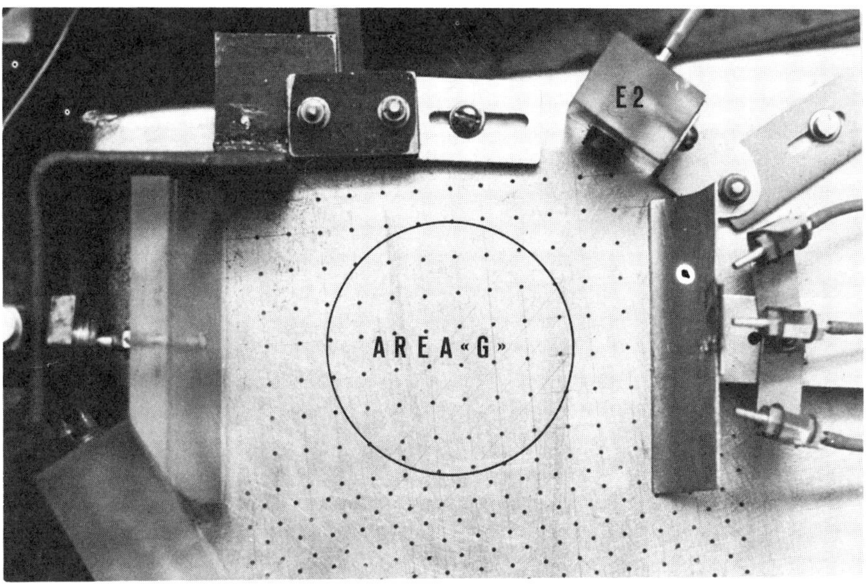

Fig. 2 - Orientation device

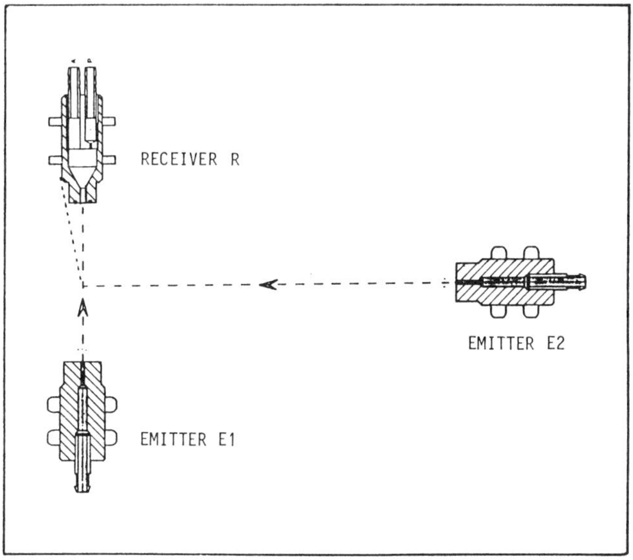

Fig. 3 - Sensor system

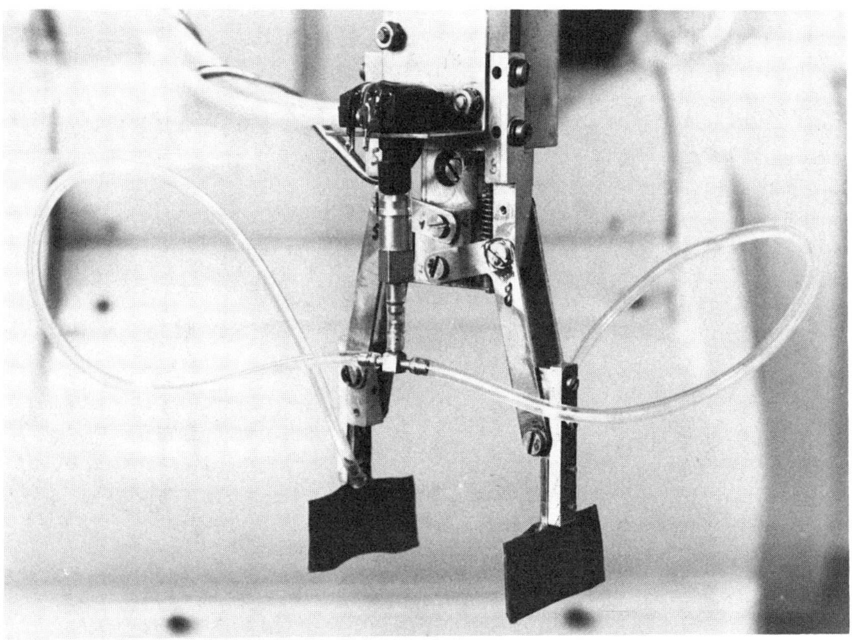

Fig. 4 - Gripper picture

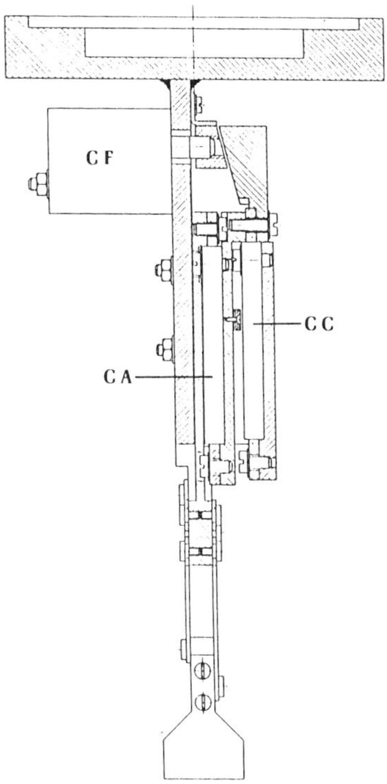

Fig. 5 - Gripper schematics

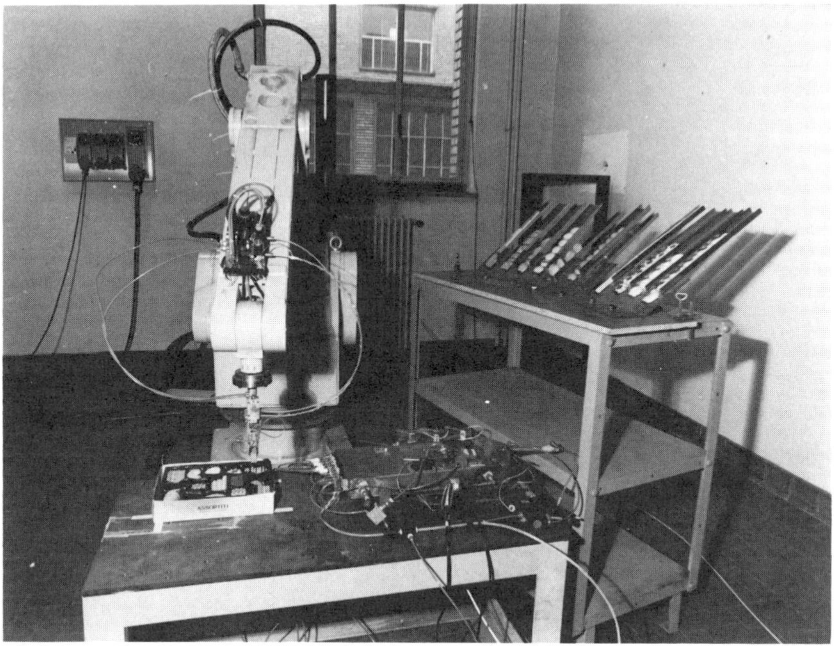

Fig. 6 - Robotic working cell test rig.

Adaptive Visual Control of a Robot Arm Using Tiled Space

T. Fancott, P. Grogono, and H. Polley

Department of Computer Science, Concordia University, Montreal, PQ, Canada H3G 1M8

1. Introduction

Suppose that we have a *robot* consisting of an *arm* with several joints and a *hand*. There is also an object, or *target*, which we want the hand to locate. We have a *visual subsystem* consisting of one or more cameras whose field of view contains both the hand and the target. The problem that we discuss in this paper is how to use the visual information obtained from the cameras to move the hand to the target.

Conventional methods of control of a robot arm require considerable computation for each movement. An arm trajectory is usually specified by a sequence of Cartesian corner points, each of which must be transformed into the corresponding joint coordinates [5]. Controlling the trajectory of the arm thus requires either inverse homogeneous transforms [9] or inverse kinematic computations which are specific to the design of the arm [4]. The latter method has been investigated frequently in recent years due to its promise of greater computational efficiency [2, 3]. Manipulator control along a trajectory has been studied using both pre-computed joint coordinate values [6, 11] and values computed while the arm is in motion [8, 12]. Clearly, the latter is necessary for adaptive control, as described by Sweet in his discussion of a vision-controlled robot welder [10]. Nitzan summarizes the issues of adaptive control as: optimal flexibility, sensor integration, planning of conditional branching, and execution speed [7]. All of these criteria require an approach which minimizes the computational load for each point in the trajectory to allow frequent, in motion, control of end effector trajectory.

A more attractive alternative is to measure the actual position of the hand and to use this information to move the hand towards the desired position. In a system that we built previously [1], a television camera was used to control an arm with three degrees of freedom, allowing the hand to grasp an object lying on a plane surface. We are currently building a system with two cameras that will enable the hand of an arm with five degrees of freedom to grasp an object in three-dimensional space. We anticipate that the system will also be able to pursue a moving object.

Adaptive feedback using visual control has several advantages. The robot itself can be cheaper and simpler because it does not have to be equipped with sensors. The independence of the mechanical and visual systems simplifies the design of a failsafe system.

An important component of our method, especially when working with moving targets, is that we use neither the kinematic equations nor their inverses. We can compute a suitable correction to the arm position from the error in the spatial position. Corrections involve only a small amount of computation and can be made rapidly. Information about the effect of the correction is available from the camera shortly after the correction has been made. Consequently, a first-order approximation of the required movement provides sufficient accuracy.

2. Geometry

The system operates in three-dimensional Euclidean space. Any small object can be represented by its coordinate vector in this space. If the object can be seen by the cameras, it will have a position $\bar{p} = (p_1, p_2, \ldots, p_m)$ in *pixel space*. Since each camera provides the coordinates of a point in its image plane, *m* will usually be equal to twice the number of cameras. The robot hand is driven by *n* motors and its position in *arm space* can be expressed as $\bar{r} = (r_1, r_2, \ldots, r_n)$, in which r_i is the position of the i'th motor. If $m > n$, the cameras are providing redundant information which can be used for error estimation. If $m < n$, there are more motors than dimensions in pixel space and we do not have sufficient information. In practice, the latter situation may arise because the arm has motors for orienting the hand, a problem that we do not address in this paper.

In principle, we could derive a relation between the vectors $\bar{p}$ and $\bar{r}$ describing a point belonging to both pixel and arm space. For our purposes, the existence of this relation is important, but its analytical expression is not. We do, however, assume that, if the gripper is at position $\bar{p}$ in pixel space, the corresponding arm configuration $\bar{r}$ is uniquely determined.

Two observations about the geometry of the system are of interest. The first is that we never actually use Euclidean space. The second is that both pixel space and arm space are discrete. This means that both can be represented as integers, resulting in potential saving of both computation and storage.

3. Moving the Hand

If we assume that the visual analysis can identify the hand and the target, and can therefore provide vectors $\bar{p}_H$ and $\bar{p}_T$ in pixel space, our problem is to move the hand from $\bar{p}_H$ to $\bar{p}_T$. One way of accomplishing this movement is to transform to arm space, obtaining vectors $\bar{r}_H$ and $\bar{r}_T$. The vector $\bar{r}_T - \bar{r}_H$ then defines the movement required from the actuators. This, however, involves the computation of the coordinate transformation that we wish to avoid.

Suppose that $\bar{r}_T - \bar{r}_H$ is small and let $\overline{\Delta r} = \bar{r}_T - \bar{r}_H$. The corresponding vector in pixel space is $\overline{\Delta p}$. We have, as a first-order approximation, $\overline{\Delta r} = J \cdot \overline{\Delta p}$ where J is the Jacobian matrix of arm space with respect to pixel space:

$$J_{ij} = \frac{\partial r_i}{\partial p_j}.$$

The value of J is a function of position. We will write it as J(p), to indicate this dependence. Thus $\overline{\Delta r}$ is a first order approximation to the movement required in arm space. Consequently, we can interpret $\overline{\Delta r}$ as an instruction to move the actuators of the arm. The instructions will be accurate only if $|\overline{\Delta p}|$ is small enough to ensure that the non-linearity of the coordinate system is unimportant. If $|\overline{\Delta p}|$ is large, this assumption does not hold. There is no reason to suppose that the arm movement $\overline{\Delta r}$ would move the hand to the target because the transformation is non-linear. We hypothesized, however, that the *direction* indicated by $\overline{\Delta r}$ would be approximately correct. Thus $\overline{\Delta r}$ is used to move the arm through a small distance, after which $\bar{p}_H$ is measured again and the process repeated.

4. The Algorithm

In this section, we describe the guiding algorithm itself. The only point of interest not discussed above is the selection of the stepper motor commands. We define a constant, M, the *frame size*, that determines the largest number of steps that can be issued to a motor during a single iteration. Criteria for choosing the frame size are given in the discussion that follows. The other parameter used in the algorithm is the tolerance, ε, within which the hand is deemed to have reached the target.

1. Use camera data to determine the positions, $\bar{p}_H$ and $\bar{p}_T$, of the hand and the target in pixel space.
2. If $|\bar{p}_T - \bar{p}_H| < \varepsilon$, the hand has reached the target and the algorithm terminates.
3. Find the tile that contains $\bar{p}_H$ and look up the value of $J(\bar{p}_H)$.

4. Compute $\Delta r = J(\bar{p}_H)(\bar{p}_H - \bar{p}_T)$.
5. Let $\Delta r = (\Delta r_1, \Delta r_2, \ldots, \Delta r_n)$ and $\lambda = \frac{M}{\max(|\Delta r_i|)}$.

   For i := 1, 2, ..., n, send $round(\lambda \Delta r_i)$ to motor i.
6. Allow time for the arm to move and go back to step 1.

5. Computing the Jacobian

The computations that we have described are straightforward provided that we can rapidly obtain values for $J(\bar{p})$. The Jacobian is, of course, a function of position, and we require its value at the position of the hand. Approximate values of the Jacobian are acceptable because visual feedback prevents errors accumulating.

Our method is to divide pixel space into small, disjoint, contiguous regions and to store the Jacobian at the center of each region in an array. We refer to the regions as *tiles* and to the process of selecting them as *tiling pixel space*. The system determines the tile that currently contains the hand position and uses the Jacobian that it finds there in the calculation described above. The Jacobian for a particular tile can be computed analytically from the geometry of the visual subsystem and the robot arm, or it can be estimated by the system. For either of these methods, there are several ways of organizing the computation.

First, we could compute the Jacobian for every tile in pixel space before using the robot. This requires a large amount of computation, or a lengthy training period, and a large amount of storage. Alternatively, we could evaluate the Jacobian for the tiles in pixel space which we expect the hand actually to visit. For some applications, this subspace might be considerably smaller than the space of all accessible tiles. Finally, we could compute the Jacobian in real time; each time the arm is moved, the system would obtain an approximation for the Jacobian at the new position. If we compute the Jacobian analytically, this is the conventional *differential method* of robot control. There is not enough time to estimate the Jacobian by making small arm movements during the motion, although this method might be feasible for a training session.

In self-calibration mode, the system moves the arm to a known position and locates the hand using the cameras. Thus $\bar{r}$ and $\bar{p}$ are known. Then one pulse is sent to stepper motor 1. Since $\overline{\Delta r_j}$ is zero for $j \neq 1$, the system can obtain an approximate value for $\frac{\partial p_i}{\partial r_1}$ by measuring Δp_i and using the fact that

$\frac{\partial p_i}{\partial r_1} \cong \frac{\Delta p_i}{\Delta r_1}$ for i = 1, 2, ..., *n*. The arm is moved back to its original position, and the measurement is repeated for the other motors. The matrix obtained is then inverted to obtain the Jacobian matrix at the position $\overline{p}$.

The use of a table of computed values is a standard method for decreasing the time requirement of a computation at the expense of storage. A three-dimensional array of values of the Jacobian would require excessive space in some applications. It is feasible for our system because we can tolerate small errors and consequently we can use a relatively coarse tiling. Nevertheless, it may be wasteful and unnecessary to compute the Jacobian for every tile that the arm can physically reach. If it is known that the arm only makes certain movements during a particular process, we can evaluate the Jacobian for points near the expected motion. Sparse array techniques can be used to store the required values.

6. Discussion

Since the algorithm employs relatively crude approximations, it is important to analyze its effectiveness in controlling the arm. There are three issues to address. The first is convergence: does the hand always approach, and eventually reach, the target? The second is accuracy: is the final position of the arm sufficiently close to the target? Finally, there is the question of stability: are there situations in which the arm oscillates instead of approaching the target?

We have not undertaken a complete mathematical analysis of the method. This would involve the geometry of the visual system, the geometry of the arm, an error analysis within each geometry, and an analysis of the approximations that we use. We have chosen instead to simulate the movement of the arm under a variety of conditions. The simulation is described in section 7. We now discuss the reasons that we had for believing that the method would work before we simulated it.

The product $J(\overline{p})\overline{\Delta p}$ is a first-order approximation to the required motor movement $\overline{\Delta r}$. This approximation is useful if $|\overline{\Delta p}|$ is small enough to ensure that the coordinate system is approximately linear in the region of interest. As explained above, however, we actually use this approximation with large values of $|\overline{\Delta p}|$ to obtain an estimate of the direction of $\overline{\Delta r}$. This direction will be approximately correct in Euclidean space if the transformation from pixel space to Euclidean space is almost linear. Provided that this condition is satisfied, we can guarantee that the hand will at least move towards the target even if the transformation to arm space is highly non-linear.

The effectiveness of the method depends on two parameters. The first is the size of the tiles. If the tiles are large, the values obtained for $J(\bar{p})$ may be so inaccurate that guidance fails. If the tiles are small, a large amount of storage will be required for values of $J(\bar{p})$. The other parameter is M, the frame size or, in other words, the maximum number of pulses that may be sent to a stepper motor during each iteration. If this number is too small, the arm will move slowly and jerkily, with only one motor being stepped at a time. A large value of the frame size will produce smoother motion, with several motors runing simultaneously, but guidance may be inaccurate and the hand may overshoot the target.

In practice, these parameters are related, as can be seen by considering extreme examples. If the arm typically moves over many tiles during a single iteration, the tiles are unnecessarily small. On the other hand, if the arm usually makes many moves within a single tile, most of the computation is redundant. These considerations suggest that the maximum motion for an iteration should be of the same order of magnitude as the tile size. (It is impossible to specify a precise relationship because of non-linearity.) We have found that this rule of thumb works well in practice. In our simulations, the hand moves smoothly towards the target and does not overshoot. We have also found that by setting the parameters inappropriately, we can obtain the varieties of poor performance that we predicted.

7. The Simulation

We decided to simulate the method before testing it with expensive hardware. A simulation is quicker to set up, easier to change, and more convenient for experiments. It is often simpler to obtain a detailed performance analysis from a simulation than from physical machinery. The simulation has been sufficiently succesful for us to plan the construction of an experimental system.

The simulation consists of a program written in C and executed on a Sun workstation. The program has two components: one component executes the algorithm described above and the other simulates the system controlled by the algorithm and displays the results graphically. The algorithm is provided with simulated visual information giving the observed positions of the hand and the target; its output consists of commands to the motors. The simulator interprets the motor commands and provides appropriately updated visual information. In the current version of the program, the Jacobian is computed for each tile in pixel space before the simulation is started. The parameters mentioned above, namely the tile size and the maximum motor movement, are set by the user before the simulation begins.

We have found that the simulated arm succesfully locates the target in a wide variety of situations. As we expected, if the tiles are too coarse, or the value of the frame size, M, is set too high (so that a motor may execute many steps without receiving feedback), the hand fails to reach the target, usually going into some kind of oscillation. When these parameters have appropriate settings, the hand reaches the target unless the initial conditions are pathological.

8. Further Research

We have demonstrated that our algorithm is both efficient and effective in a variety of conditions. Nevertheless, there are a number of directions in which we perceive the need for future research.

Although we have claimed that our system is self-calibrating, we have not actually constructed a robot that calibrates itself. Both theoretical and practical work is required here. In particular, it would be interesting to investigate "on-line" self-calibration, in which the system continuously calibrates itself as it operates.

We have claimed that the guidance system should work for moving targets. Although we see no reason why this should not be true, provided that the targets do not move too fast, we have not yet experimented with moving targets.

We are using a classical positional control system. For robots that are designed to move heavy objects rapidly, velocity feedback and velocity control must be considered.

It is easy to say that a robot using our algorithm is behaving in a human-like way, using its cameras and arm as we use our eyes and limbs. When we perform a habitual mechanical task, however, we use visual feedback for confirmation only. Perhaps a robot could do the same. Very little work has been done with hybrid, open/closed-loop systems of this kind.

Acknowledgements

The research described in this paper was supported in part by the Natural Sciences and Engineering Research Council of Canada.

References

1. Brochu, B. *Implementation of a Pick-and-Place Robot with Manipulator and Camera.* M. Comp. Sc. Report, Department of Computer Science, Concordia University, August 1985.
2. Crochetiere, W.J. Locating the wrist of an elbow-type manipulator. *IEEE Trans. Systems, Man and Cybernetics, SMC-14,* 3, 497-499 (1984).
3. Elgazzar, S. Efficient kinematic transformation for the PUMA 560 robot. *IEEE Journal of Robotics and Automation, RA-1,* 3 (1985).
4. Featherstone, R. Position and velocity transformations between robot end effector coordinates and joint angles. *International Journal of Robotics Research, 2,* 2 (1983).
5. Luh, J.Y.S. Approximate joint trajectories for control of industrial robots along Cartesian path. *IEEE Trans. Systems, Man, and Cybernetics, SMC-14,* 3, 444-450 (1984).
6. Luh, J.Y.S. and C.S. Lin. Optimum path planning for mechanical manipulators. *ASME Trans. J. of Dynamic Systems, Measurement and Control, 103,* 2, 142-151 (1951).
7. Nitzan, D. Development of intelligent robots: achievements and issues. *IEEE J. Robotics and Automation, RA-1,* 1, 3-13 (1985).
8. Paul, R.P. Manipulator Cartesian path control. *IEEE Trans. Systems, Man and Cybernetics, 9,* 11, 702-711 (1979).
9. Paul, R.P. *Robot Manipulators: Mathematics, Programming and Control.* MIT Press, 1981.
10. Sweet, L.M. Sensor-based control systems for arc-welding robots. *Robotics and Computer Integrated Manufactoring, 2,* 2, 125-133 (1985).
11. Taylor, R.H. Planning and execution of straight line manipulator trajectories. *IBM J. Research and Development, 23,* 4, 424-436 (1979).
12. Whitney, D.E. Resolved motion rate control of manipulators and human prostheses. *IEEE Trans. Man-Machine Systems, 10,* 2, 47-53 (1969).

Dynamic Six Components Measurement of Robot Precision

Ario Romiti and Terenziano Raparelli

Department of Mechanics, Polytechnic of Torino, Turin, Italy

An analysis of the robot dynamic precision by measurement of the six spatial coordinates of the geometrical error is undertaken.
The sensor equipment on a robot hand has been designed, built and tested. The measurement sensitivity is in the 0.01 mm range. The measuring occurs with respect to a fixed straight trajectory, that is materialized by a bar. The robot hand brings a set of five proximity sensors, that give continuous informations about position and orientation of the hand during the motion.
The hand centre position along the bar and its speed are measured by linear sensors.
After comparison of many different proximity sensors, inductive sensors showing very good performances were choosen and they were individually calibrated.
A test of the method has been performed on an anthropomorphic robot having six degrees of freedom. A straight, ferromagnetic, one meter long bar is provided, that has two perpendicular plane faces.
A frame supported by the hand bears the five proximity sensors. The tip ends of three sensors are on a plane that is perpendicular to the plane of the two other sensor tips.
The linear sensor is a wire actuated potentiometer. The parallelism error is negligible. The same wire actuates a tachometric dynamo for speed measurement.
The proximity sensor location is characterized by the tip centre point. The tension - distance diagrams for each sensor have been approximated by algebraic polinomials of ninth degree. The maximum discrepancies between experimental and calculated values are less than about 2,5 μm. Fig.1 shows the sensitivity curve; the aspect ratio sensitivity is shown in Fig.2.
The measurement requirements of the longitudinal displacement are less strict, so that a one per thousand end of scale repeatability is satisfactory.
The faces of the bar are honed plane with a 0.01 mm precision, and they are perpendicular with a tolerance of 0.006 degrees.

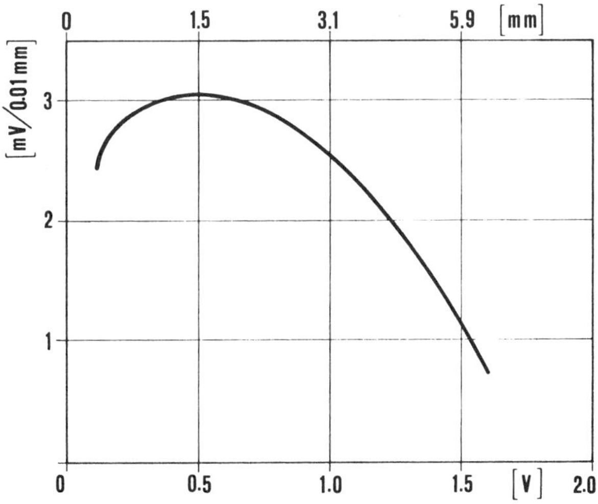

Fig. 1 - Proximity sensor sensitivity to distance.

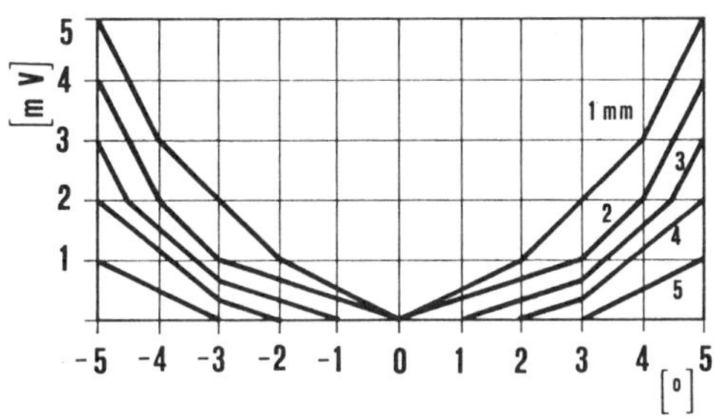

Fig. 2 - Proximity sensor angular sensitivity.

Hand coordinate evaluation

Fig.3 represents the coordinate frame x, y, z fixed to the bar; the frame u, v, w is fixed to the robot hand. The unit vectors of the axes of the first frame are $\underline{i}$, $\underline{j}$, $\underline{k}$, and those related to the second frame are $\underline{r}$, $\underline{s}$, $\underline{t}$. The origins of the fixed and of the moving frames are respectively O and C. The proximity sensor tip centre points are called $S_i (i=1,....5)$.

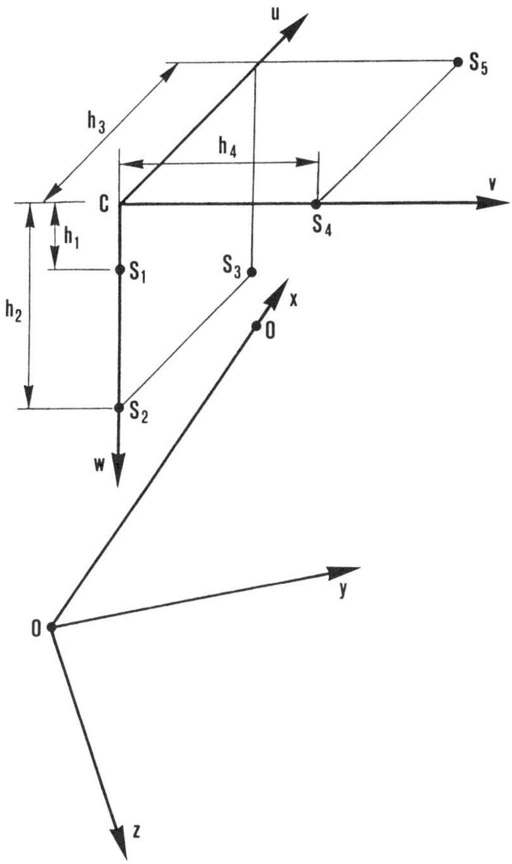

Fig. 3 - Coordinate frames.

The points S_1, S_2, S_3 are on the plane v=0; the points S_4, S_5 are on the plane w=0.

The moving coordinates of the points S_i are:
$S_1\ (0,0,h_1)$; $S_2\ (0,0,h_2)$; $S_3\ (h_3,0,h_2)$ $S_4\ (0,h_4,0)$; $S_5\ (h_3,h_4,0)$.
The S_1 , S_2 , S_3 sensor axes are parallel to v; the S_4 , S_5 sensor axes are parallel to w.
Corresponding to each point S_i there is a point P_i that lays on the opposit bar face on the prolongation of the S_i sensor axis. The distance between S_i and P_i is called d_i.
The points P_1 , P_2 , P_3 are on the plane y=0; the points P_4 ,P_5 are on the plane z=0.
The coordinates u_i , v_i , w_i of points P_i are:
$P_1(0,d_1,h_1)$; $P_2(0,d_2,h_2)$; $P_3(h_3,d_3,h_2)$; $P_4(0,h_4,d_4)$; $P_5(h_3,h_4,d_5)$.
The components j_u, j_v, j_u with respect to the moving frame of the unit vector of axis y, normal to the plane P_1 , P_2 , P_3, may be easily computed.
The components of unit vector $\underline{k}$ of axis z, normal to y and to the straightline P_4 P_5, are computed in their turn.
The unit vector $\underline{i}$ of axis x is then calculated by cross product of $\underline{j}$ and $\underline{k}$.
The point where the x axis intersects the plane u=0 is named 0_o. The points P_1 , P_2 , P_4 are on the plane x=0. The straightline connecting P_1 and P_2 corresponds to the intersection of the planes y=0 and u=0. The intersection of the plane z=0 , u=0 is a straightline through the point P_4 perpendicular to the straightline connecting P_1 and P_2. The intersection of these two straightlines is the point 0_o; its coordinates v_o, w_o may be easily computed.
The coordinates x_c, y_c, z_c of point C in the fixed frame x, y, z with origin in 0 are $y_c = -(v_o j_v + w_o j_w)$; $z_c = -(v_o k_v + w_o k_w)$.
The wire of the potentiometric sensor that measures the run length is hooked at the robot hand in the point C, so that one can put $x_c=\sigma$, that is the measured lenght.
The moving frame unit vectors are obtained by transposing the orientation matrix of the fixed frame, evaluated with respect to the moving axes.

The robot hand trajectory

The values y_c , z_c , $\underline{r}$, $\underline{s}$, $\underline{t}$ may be evaluated continuously by the proximity sensor signals as functions of the run lenght σ. The robot dynamic errors are given by the deviations of such values with respect to the corresponding programmed values.
When programming a robot by manual teaching, the hand is brought successively in two positions at the two ends of the reference bar.
The hand is there oriented to let the planes v=0 and w=0 be approximately parallel, respectively, to the planes y=0 and z=0 of the bar faces.

A linear interpolation between the two end points is set; the hand centre is then programmed to run along a straightline, and the moving frame to carry out a constant speed rotation.

The angle traveled during the rotation is of the same order of magnitude of the angular errors, so that it cannot be neglected for measuring purpose.

Call $\underline{r}'$, $\underline{s}'$, $\underline{t}'$ the mobile frame unit vectors in the initial position, and $\underline{r}''$, $\underline{s}''$, $\underline{t}''$ the same vectors in the final position.

A 3x3 matrix $H = \underline{r}\ \underline{s}\ \underline{t}$ may be defined.

The orientation change between the initial condition represented by a matrix H' and the final condition represented by a matrix H'' is given by the matrix $(H')^{-1} H'' = F(L)$, that is a function of the total lenght of stroke L.

The total angle of rotation, φ, and the unit vector $\underline{n}$ of the rotation axis are given by the expressions:

$$\cos \varphi = \frac{1}{2} (\underline{r}'.\underline{r}''+\underline{s}'.\underline{s}''+\underline{t}'.\underline{t}'' - 1) \tag{1}$$

$$\underline{n} = \frac{1}{2\cdot\sin\varphi} |(\underline{t}'.\underline{s}''-\underline{s}'.\underline{t}''),(\underline{r}'.\underline{t}''-\underline{t}'.\underline{r}''),(\underline{s}'.\underline{t}''-\underline{t}'.\underline{s}'')|^T \tag{2}$$

Consider now an intermediate run lenght σ. The interpolation between the end positions occurs uniformly; the matrix H^* representing the hand frame orientation for $x = \sigma$ may be written as $H^* = H' F(\sigma)$.

In order to find $F(\sigma)$, one has first to consider the column vectors of the matrix $F(L)$, that may be written in the form $F(L) = \underline{p}_1''\ \underline{p}_2''\ \underline{p}_3''$

The hand orientation defined by $F(\sigma)$ corresponds to a rotation from the final position of an angle $-(1 - \sigma/L)\varphi$ around an axis having $\underline{n}$ as unit vector. One can put $F(\sigma) = \underline{p}_1\ \underline{p}_2\ \underline{p}_3$. All the vectors $\underline{p}_i$ are rotated around $\underline{n}$ of the same angle with respect to the corresponding vectors $\underline{p}_i''$. One obtains:

$$\underline{p}_i = \{1 - \cos(1-\sigma/L)\varphi\}(\underline{n}.\underline{p}_i'')\underline{n} + \underline{p}_i''\cos(1-\sigma/L)\varphi - (\underline{n} \times \underline{p}_i'')\sin(1-\sigma/L)\varphi \tag{3}$$

One can then find $F(\sigma)$ and the matrix H^* that defines the moving frame corresponding to the run lenght σ.

Evaluation of the dynamic errors

Call $\underline{r}^*$, $\underline{s}^*$, $\underline{t}^*$ the moving frame unit vectors, given by the matrix H^*, according to the programmed motion, and $\underline{r}$, $\underline{s}$, $\underline{t}$ the moving frame unit vectors

that are evaluated through the measurements at the run lenght σ.
The total angular error δ when $x = \sigma$ is given by the relation:

$$\cos \delta = \frac{1}{2} (\underline{r}.\underline{r}^* + \underline{s}.\underline{s}^* + \underline{t}.\underline{t}^* - 1) \quad (4)$$

The linear error may be related to the deviations from the trajectory of the wrist centre Q, that has known coordinates in the moving frame.
If $\underline{Q}$, $\underline{C}$ are the point vectors of Q and C in the fixed frame, and ${}^m\underline{Q}$ is the point vector of Q in the moving frame, one has $\underline{Q} = \underline{C} + H\,{}^m\underline{Q}$.
The components of $\underline{C}$ are obtained by elaboration of the sensor measurement at each run lenght $x_c = \sigma$. The values $\underline{Q}^*$ of the programmed trajectory are given by: $\underline{Q}^* = \underline{C}^* + H^*\,{}^m\underline{Q}$.
The components $y_c{}^*$ and $z_c{}^*$ of $\underline{C}^*$ are obtained by linear interpolation along the straightline for $x_c{}^* = \sigma$.
The total linear error ε is:

$$\varepsilon = \{(y_q - y_q{}^*)^2 - (z_q - z_q{}^*)^2\}^{\frac{1}{2}} \quad (5)$$

Fig.4 shows the diagrams of the errors δ and ε as functions of the run lenght (L = 1m) at full speed.

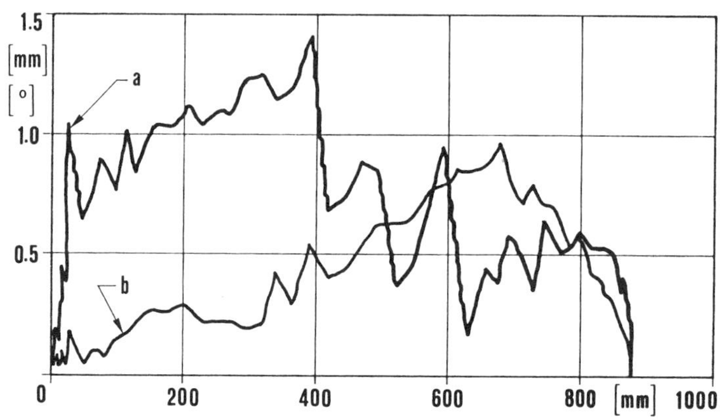

Fig. 4 - a) Total displacement error; b) Total angular error.

The authors thank Mr. Mario Pagani for his help in experiments and data elaboration. This work was sponsored by the Italian Ministry of Education

Tactile Pattern Recognition of Polyhedra on a Plane

Swaminathan Mahalingham*, Ramamoorthy Ganesh†, and
Dr. Suren N. Dwivedi‡

*Graduate Research Assistant, Department of Mechanical Engineering, and Engineering Science, University of North Carolina at Charlotte, Charlotte, NC 28223, USA
†Graduate Fellow, Department of Electrical and Computer Engineering, Syracuse University, Syracuse, NY, USA
‡Associate Professor, Department of Mechanical Engineering and Engineering Science, University of North Carolina at Charlotte, Charlotte, NC 28223, USA

ABSTRACT:

An approach to identify a polyhedral object as an element of a set of predefined objects and to determine its orientation with reference to a fixed frame utilizing tactile sensors is presented. The object is defined in one frame, the Object Frame (OF) as a polygon and local data from the sensors is obtained in another frame, the Actual Hand Frame (AHF). Correlation of the data obtained from the two frames would constitute the main problem dealt with in this paper. This approach successfully determines the position of the sensors even when more than one, but not all, contact points are either on the same or parallel edges. This approach uses three sensors for recognition and localization.

INDEX TERMS: Tactile Sensors, Tactile Pattern Recognition, Pruning.

I. OBJECT DEFINITION

For a given stable state, an object is defined by the polygon obtained by the intersection of the sensor plane with the polyhedral object at a given height from the necessarily parallel support plane, on which the object rests. This height should be the same for all the objects. Assuming a convenient frame for object definition, either the edges and vertices are sequentially defined by their equations and co-ordinates respectively, or the vertices are defined and the formula for obtaining the equation of the line incorporated in the program. Additional data to be provided include the angle (α_{ij}) and minimum and maximum distances ($dmin_{ij}$ and $dmax_{ij}$) between two sides, for all combinations of two sides, for every object, where $i = 1, N$; $j = 1, N$; $i \neq j$, N = number of sides of the OF polygon. These values are stored in arrays. For convenience these will be referred to as the α, dmin and dmax arrays.

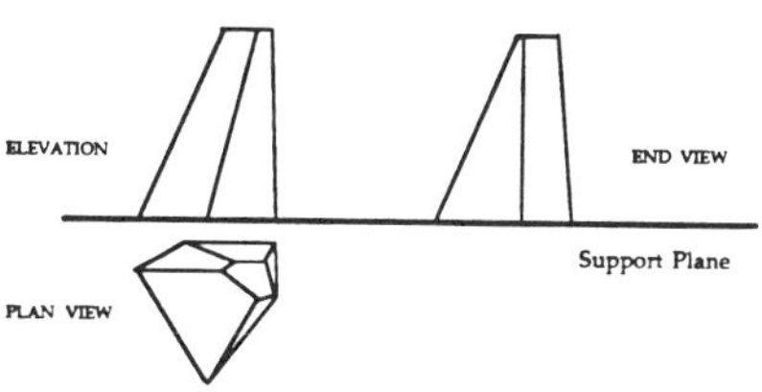

Fig.1: OBJECT TO BE DEFINED

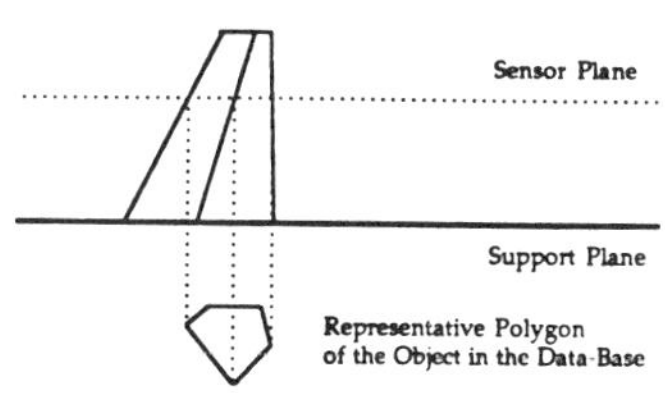

Fig.2: OBJECT REPRESENTATION

II. OBJECT RECOGNITION

1. PRUNING: As a solution, it is to be determined that in the sample space Ω at least one $O \in \Omega$ occurs, i.e., $\cup O$ where $\Omega = \{O(1), O(2), \ldots O(N)\}$, N being the number of recognizable objects. The elements of Ω are stored internally as a vector in the suggested form in the object definition.

The data from the three sensors may be represented as

$$S(i) = x(i),\ y(i),\ \psi(i),\ i = 1,3$$

where x, y are the coordinates of the sensor position and the direction of the edge normal, all with respect to the AHF. These data define the equations of three straight lines. Each of the internal angles formed by the intersection of these lines $\beta(i)$, i = 1,3 (or non intersection in the case of parallel edges where β is considered to be zero), are compared with the elements of the α-array of each object and three sets A_1, A_2 and A_3 are formed, for one object at a time, where A_i is a set comprising of ordered pairs of sides, the α value of which is equal to the β_i value.

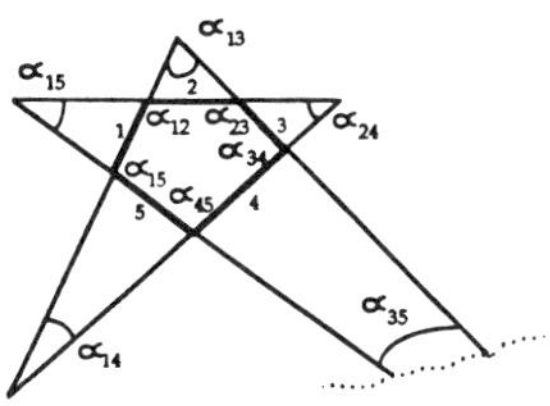

Fig.3: DEFINITION OF α_{ij}

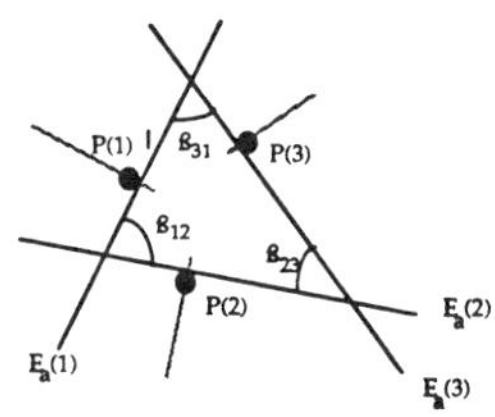

Fig. 4: DEFINITION OF β_{ij}

As an illustration

$$A_1 = \{(1,5),\ (2,4)\ldots\}$$
$$A_2 = \{(4,3),\ (6,2)\ldots\}$$
$$A_3 = \{(2,3),\ (5,6)\ldots\}$$

might represent one such possibility. For a locally legal solution, for any α_{ij} in A_1, there must occur one α_{jk} in A_2/A_3 and an α_{ki} in A_3/A_2. If this condition is not satisfied for any α_{ij} in A_1, the possibility of the object under consideration as a solution is pruned. The procedure is repeated for all objects and a set of all locally legal objects is formed. In the above illustration, this condition is satisfied by (2,4); (4,3) and (2,3) hence rending this object a locally legal solution.

This set is further pruned using the dmin and dmax criteria, wherein, the Euclidean distance between any two of the contact points should lie within the permissible range as delimited by dmin and dmax for the corresponding sides, and should do so for all the three pairs of contact points, i.e., for S(1)-S(2); S(2)-S(3) and S(3)-S(1) combinations. This further reduces the number of objects to be tested.

2. RECOGNITION

In a general case, the AHF and OF are rotated and linearly translated with respect to each other.

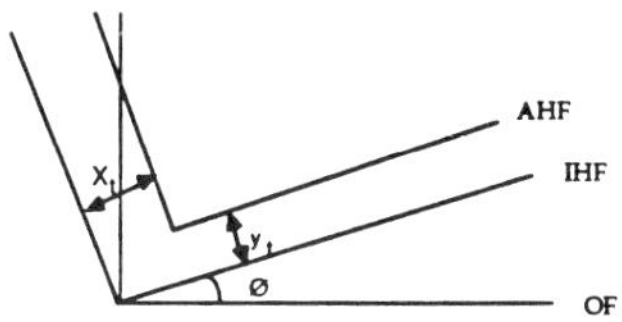

Fig.5: RELATIVE FRAME ORIENTATIONS

Hence, the equations in the AHF after a translation and rotation, the values of which are to be determined, should correspond to the equations in the OF. For a correct assumption, the three AHF equations transformed should be found in the OF, and should satisfy necessary criteria.

Proceeding in a reverse direction, for a correct assumption, the OF edge equations should transform to the AHF equations. This approach is followed here.

The assumptions are made sequentially. The first object in the existing vector is assumed to be the object sensed by the tactile sensors. Hence the geometric model of the object is retrieved from the data bank to be manipulated. The next assumption made is that the first sensor S(1) is on the first legal edge. Hence the assumption is that $E_a(1) \leftrightarrow E_o(n)$; where $E_a(1)$ denotes the edge on which S(1) is present, which is given by the equation in the AHF constructed by the data given by S(1), i.e., coordinates of P(1), and $\psi(1)$; and $E_o(n)$ denotes the first legal edge equation.

Here the concept of the Ideal Hand Frame (IHF) is introduced. As mentioned previously, the skewing of the AHF and OF may be considered as a combination of a rotation and translation between the two. An intermediate frame is introduced which can be considered to have undergone a rotation only with respect to the OF, and only a linear translation of which would give the AHF. This intermediate frame is termed as the IHF. A point worth noting is that the slopes read off in the AHF would be undistorted in the IHF.

3. ROTATION

The next step is to define the object model in the IHF. This can be done by following either of the two methods described below; (i) The Conventional approach and (ii) The Complex Plane (Conformal Mapping) approach.

For the assumed $E_a(i) \leftrightarrow E_o(n)$ correspondence, the rotation of the IHF with respect to the OF, Ø, is given by the difference of the slopes of these two edges in their respective frames. If Ø(n) be the slope of $E_a(1)$ in the OF, and $\psi(1)+\pi/2$ being the slope of $E_a(1)$ in the IHF, then $\emptyset = [\ \psi(1)+\pi/2-\emptyset(n)$. Proper signs would be considered in the program and poses no problem, throughout the following derivations.

With Ø known, transformation of all the equations in the OF to the IHF can be realized.

Conventional Approach

The edge $E_o(n)$ may be

(i) Parallel to the X axis; (ii) Inclined to both the axes; (iii) Parallel to the Y axis.

(i) Parallel to the X axis

Eqn. in the OF y = k

$m = \tan[\psi(N) + \pi/2]$

$= \tan[\psi(n) + Ø] = \tan Ø$ as $\psi(N)=0$

when parallel to the X axis

$C_i = k\sec Ø$

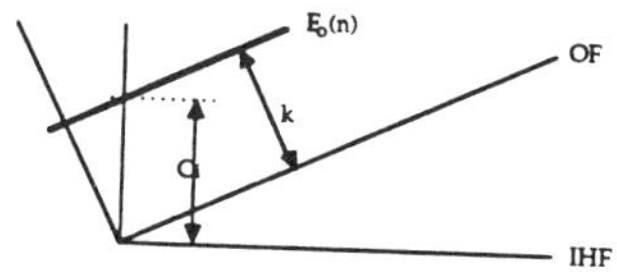

Fig. 6: PARALLEL TO OF X-AXIS

(Convention followed is anti-clockwise rotation of the IHF with respect to OF is positive).

∴ Eqn in the IHF is

$y = \tan[\psi(N)+Ø]x + k\sec Ø$

$y = (\tan Ø)x + k \sec Ø$

(ii) Inclined to both the axes: Eqn in the OF: $y = mx + C_o$

$C_t \cos[\psi(N)+Ø] = C_o \sin Ø$ or $C_t = \frac{C_o \sin Ø}{\cos[\psi(N)+Ø]}$

$C_i = C_t \sin[\psi(N)+Ø] + C_o \cos Ø$

$= \frac{C_o \sin Ø}{\cos[\psi(N)+Ø]} \sin[\psi(N)+Ø] + C_o \cos Ø$

$= C_o[\tan(\psi(N)+Ø) \sin Ø + \cos Ø]$

∴ Eqn. in the IHF would be

$y = \tan[\psi(N)+Ø]x + C_o[\tan[\psi(N)+Ø]\sin Ø + \cos Ø]$

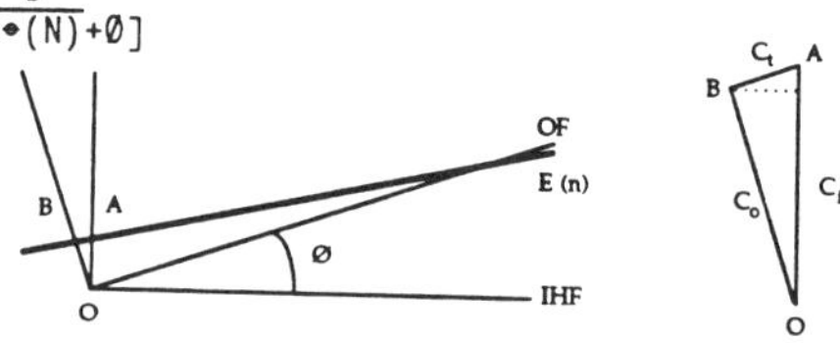

Fig.7: $E_o(n)$ NON-PARALLEL TO BOTH OF AXES

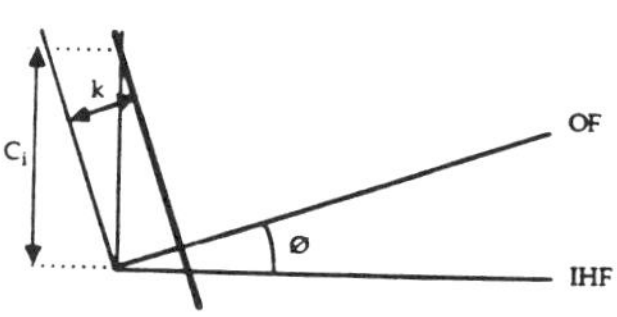

Fig.8: $E_o(n)$ PARALLEL TO OF Y-AXIS

(iii) Parallel to Y axis Eqn in OF: $x = k$

$$C_i = k \sec(\pi/2 - \emptyset)$$

$$= k \operatorname{cosec}(\emptyset)$$

$\therefore$ Eqn. in the IHF

$$y = \tan[\bullet(N)+\emptyset]x + k\operatorname{cosec}\emptyset$$

Further, two subcases may arise for case (ii) previously dealt with; (i) The transformed edge E(N) may be parallel to the X axis in the IHF (ii) may be parallel to the Y axis in the IHF.

(i) When parallel to the X axis of the IHF.

Eqn. in OF: $y = mx + C_0$

From fig. $k = C_0 \cos \emptyset$

Eqn. of this line would be

$y = k = C_0 \cos \emptyset$ in the IHF.

(ii) When parallel to the Y axis of the IHF:

$$k = C_0 \cos(\pi/2 - \emptyset)$$

$$= C_0 \sin \emptyset$$

Eqn. of this line in the IHF

would be $x = k = C_0 \sin \emptyset$

Using these transformation formulae, all the equations in the OF can be transformed into the IHF equations.

Complex Plane Method:

Here, all the co-ordinates in the OF of the vertices are first transformed into complex numbers by taking x coordinate as the real part and y the imaginary part of the complex number. Hence, if x(N), y(N) are the coordinates of a vertex, we transform it into Z(N) = x(N) + iy(N). Once rotation of the frames has been determined for each assumption, conformal mapping of the complex variables is performed.

The mapping of the z plane onto the w plane by means of the equation

$$w = Bz + C$$

is a translation accompanied by rotation.

First considering only rotation, $w = Bz$

Where B is a complex, in the polar form, $B = be^{i\emptyset}$

b = magnification factor, $\emptyset$ = rotation.

We assign a value of 1 to b, hence $B = e^{i\emptyset}$

Once rotation $\emptyset$ is determined, B can be determined. The rest of the vertices can then be obtained by the transformation.

$$W(N) = Bz(N)$$

The equation of the edges in the IHF are then obtained by finding the equations of the lines joining the vertices.

After rotation another criterion for pruning arises, as, upon rotation, there should be at least one edge each having an inclination of $\psi(2)+\pi/2$ and $\psi(3)+\pi/2$ respectively in the IHF. Invalid object models are rejected.

4. TRANSLATION:

To determine the translation between the IHF and AHF, the coordinates of one and the same point in both the frames are to be known. This approach necessitates the subdivision of this problem into two parts: (i) Only one line with slope corresponding to either $\psi(2) + \pi/2$ or $\psi(3) + \pi/2$ found: Translation in this case is very easily determined.

If the coordinates of the point 'p'be x_a, y_a and x_i, y_i in the AHF and IHF respectively then the translations x_t and y_t are given by $x_t = x_i - x_a$ and $y_t = y_i - y_a$.

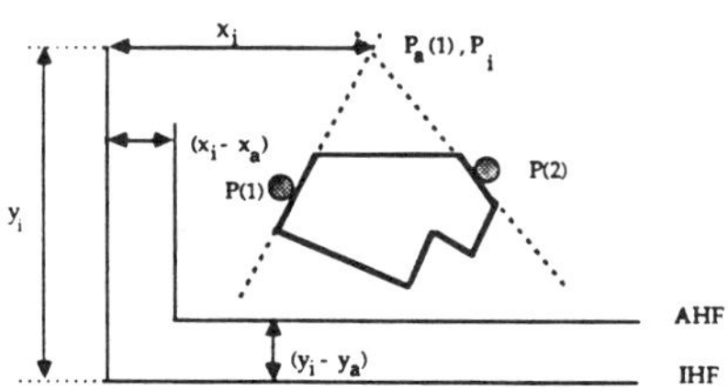

Fig.9: SINGLE SLOPE-TO-SLOPE CORRESPONDENCE

(ii) When more than one edge is found in the IHF with slope equal to $\psi(2) + \pi/2$ and $\psi(3) + \pi/2$ both.

In this case, correspondence of points in the two frames becomes ambiguous. The point of intersection of $E_a(1)$ and $E_a(2)$ as an example, might correspond to any one of the points of intersection of $E_i(n)$, the legal edge being considered as an assumption, with the parallel edges having a slope equal to $\psi(2) + \pi/2$.

Each assumed correspondence gives a different transformation. An example is shown as as illustration, for which two possible solutions exist.

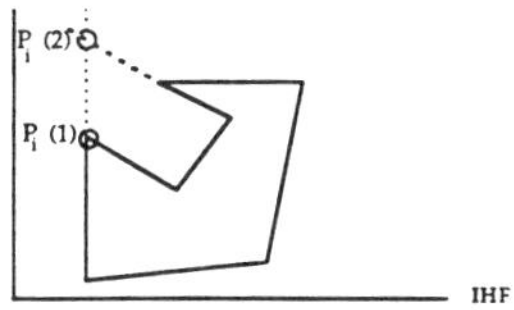

Fig. 10: MULTIPLE SLOPE-TO-SLOPE CORRESPONDENCE

Hence two sets of values of x_t, y_t are obtained for the assumption $p_i(1) \leftrightarrow p_a$ and $p_i(2) \leftrightarrow p_a$.

Two possible solutions to the problem of ambiguity exist, which seem to be highly efficient.

(i) For an assumption that P(1) is on one of the legal edges, the end points of that edge are known in the IHF. The calculated distance between the sensors, and the slope of the line joining the two, P(1) and P(2) are invariant with the frame translation.

A parallelogram with one side equal to the length of the assumed legal edge and the other equal to the distance between P(1) and P(2), is constructed with one side coinciding with $E_i(n)$, the assumed legal edge.

The edge of the polygon, which is cut by that side of the parallelogram which is parallel to the edge $E_i(n)$, is the edge on which the second sensor lies on. Once the edge is identified, the determination of translation will proceed along the same lines as described before. If ambiguities further persist, both for P(1)-P(2)and P(1)-P(3) combinations, the same procedure applied to P(2)-P(3) combination solves the problem.

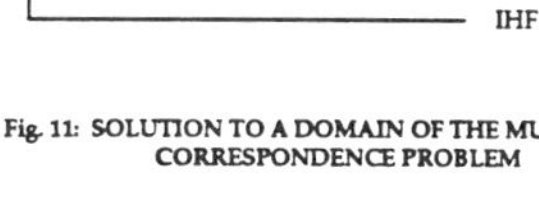

Fig. 11: SOLUTION TO A DOMAIN OF THE MULTIPLE CORRESPONDENCE PROBLEM

(ii) Another solution would be to make sure that P(2) does not lie within or outside the region enclosed by the polygon, for every assumption. This is done by transforming the polygon defining equations from the IHF to the AHF, thus defining the region enclosed by the polygon in the AHF. To determine whether P(2) is within or without, any straight line drawn from the point P(2) and proceeding in a particular direction with a constant sense, is drawn. An odd no. of intersections ensures that the point is within the polygon. For the correct assumption, P(2) should lie exactly on the assumed edge, E(n).

5. CONCLUSION:

A simplistic approach to recognize and locate a polyhedral object on a plane has been presented, the salient features being the ability to localize the object when more than one, but not all, sensors are either on the same side or parallel sides, and simplifying the heuristics by using effective pruning techniques for discarding incorrect object model solutions in the initial stages. This approach however fails to arrive at a unique solution when more than one object corresponds exactly to the AHF data for a given configuration.

Extension of the algorithm to the representation and tactile recognition of 3-D objects is fairly obvious. We wish to point out that for 3-D object recognition, numerical analysis techniques such as polynomial fitting would be required to generate the equations of the surface and this would necessitate two important factors:

(1) Increase in number of sensor points

(2) Ability to displace individual sensor points to take the shape of the object.

This would allow the construction of a partial surface of the object which will lead to recognition and localization.

Sensor Fusion: The Quest for Autonomous Robots

Bruce Moxon

BBN Advanced Computers, Inc.

Abstract

"Sensor fusion" is the process that merges environmental information derived from multiple sensor sources to eliminate ambiguity inherent in the transducer measurements, thus providing a more accurate perceived world model. This perceptual process is critical to the development of intelligent, autonomous systems capable of reacting to and interacting with their environments. The sensor fusion process comprises data acquisition and analysis tasks, symbolic abstraction of the raw data, and subsequent knowledge based interpretation and correlation. These complex processes are employed to build consistent world models that are used in the planning and control of the robot system. Powerful parallel processing computer architectures are required to support the immense computing requirements called for by this enabling technology. The Butterfly™ Parallel Processor is presented as an experimental environment in which to carry out research and development of sensor fusion software architectures.

Introduction

Fully autonomous robots, capable of functioning in unconstrained, even hostile environments, have been proposed as solutions to many sociological and economic problems facing our modern society. Intelligent robots play a key role in many of the systems evolving for manufacturing, the handling of toxic materials, industrial security, defense, and space exploration. These ideas are as old as the genre of science fiction. Yet robots currently employed in the manufacturing automation tasks of even our most advanced factories have nowhere near the capabilities exhibited in recent science fiction cinema.

In fact, until very recently, robot technology had been applied to only the simplest human tasks: those satisfied by a sequence of preprogrammed, repetitive actions in a tightly constrained and unchanging environment. Many of today's industrial robots are "taught" what to do by a human operator, who guides a robot arm through a sequence of motions that are mimicked time and time again in a mindlessly dutiful performance.

This discrepancy between the real and imagined capabilities of robots is due almost entirely to the lack of sophistication in the sensory processing systems of these machines. Over the years, tremendous advances have been made in some aspects of robot technology, including the evolution of control theory for dealing with complex kinematics and dynamics, the development of smaller, more powerful servo mechanisms, the development of sophisticated sensing devices, and the origination of capable AI-based planners. But the ability to quickly and intelligently use the information generated by sophisticated sensors has not kept pace -- for both technological and economic reasons.

Simple sensing devices, such as position encoders and proximity sensors, have been well integrated into existing robot systems. But these sensing devices provide minimal control capabilities, making the systems applicable to only a small range of problems. More aggressive systems have attempted to use vision, ranging, tactile, or force sensing. However, most of these systems have been designed in rather ad hoc manners in order to economically perform a set of very specific tasks (e.g. painting, welding, or windshield placement). The integration of advanced multi-modal sensor systems to support more fully autonomous operation places extreme demands on the associated computing environment. The data from multiple sensors must be integrated or "fused" to provide accurate information about the state of the environment. Based on this world model, an intelligent planning module must then establish a course of action to be carried out by the effectors. Sensor based monitoring is required to verify proper execution of the plan and to react to any unexpected situations. This scenario is repeated perhaps tens or hundreds of times a second in order to ensure smooth, continuous operation.

Tremendous amounts of computational power are required to accommodate the real-time requirements in both the signal and symbol processing domains. Until recently, such processing was available only in the most expensive supercomputers. This high cost made it impractical to develop sophisticated sensory systems for autonomous robot projects.

But the development of high performance, massively parallel computers based on inexpensive microprocessor technology is beginning to change this situation. With the availability of these systems, robotics applications are taking a more aggressive posture, incorporating multiple sensors into fully integrated robot systems. In fact, work is already underway in the development of multiprocessor based sensing and control systems for autonomous operation.

DARPA's Autonomous Land Vehicle (ALV) project, for example, is using laser ranging, video, and ultrasonic sensing for navigation and obstacle avoidance. Researchers in the Stanford Artificial Intelligence Laboratory robotics group are developing a multiprocessor based system called NYMPH to implement force and tactile based control of the Stanford/JPL hand [Chen86]. Researchers at MIT are using a multiprocessor architecture to control the Utah/MIT hand [Siegel85]. The Robotics Systems Research and System Architectures Research groups at AT&T Bell Laboratories have collaborated on a number of projects employing a distributed multiprocesser environment for robot control [Gaglianello86] [Andersson86].

To see why parallel processing has sparked new interest in robot sensor fusion, we first present a discussion of a structured, autonomous system software architecture. The fundamental characteristics of this software architecture are then used to formulate "computational requirements" which must be met by a supportive software/hardware architecture. We then present the Butterfly Parallel Processor as an environment that supports these requirements.

Autonomous system architecture

An autonomous robotic system is composed of four major components: the sensor unit, the effector unit, the planning and control unit, and the world model, or world database. Figure 1 illustrates the four components and their interactions. This architecture is consistent with other robot architectures presented in the literature, e.g. [Albus84].

The sensor unit transforms the sensor signals into high level world models which symbolically describe the current state of the robot's environment. In this process, information from multiple sensor sources must be fused into a consistent perceptual "view". The sensor unit provides a data driven flow of information which allows the system to recognize and react to unforseen changes in the environment.

The planning and control unit accepts high level directives, i.e. tasks, from an intelligent "operator" (human or machine), analyzes the requests in conjunction with the current state of the world, and generates symbolic plans of action to be carried out by the effector unit. Goal-directed reasoning is applied by the planner to support efficient and effective reactions to changes in the environment in accordance with the prescribed objectives.

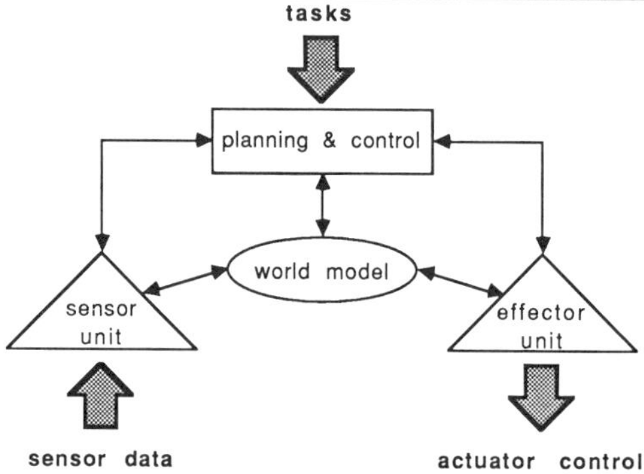

Figure 1. Autonomous system architecture

The effector unit recursively decomposes the symbolic plans into simpler and simpler tasks until such tasks can be satisfied by signal perturbations that drive actuators. The effector unit can be used to modify the environment, either in direct response to a task request (e.g. moving a part into an assembly position), or to obtain more information about its environment (e.g. rotate a part so the vision system can have a different view). Additionally, the plans may call for modification of sensor parameters, such as device sensitivity or sensor data processing algorithm selection. For example, stiffness matrix coefficients might be modified to change the characteristics of compliant motion.

The effector unit might also be capable of dynamically changing the allocation of sensor processing resources in response to a directive to focus attention on one or more sensing modalities. The ability to tune in or out various sensory devices, and to redistribute sensor processing power where required will provide autonomous systems with the adaptability necessary for autonomous operation. This adaptability also conserves computing resources, as sensors can be directed to simply verify predictions in portions of scenes, rather than requiring the system to rederive the entire perceived world model every "frame". To distinguish between actuators which directly modify their environment (e.g. servos), and those which modify only the perception of the environment, we define the concept of **virtual actuators:** system components which effect changes on sensor or processing parameters or algorithms.

The final component in an autonomous robotic system, the world model, actually consists of two parts. The **long term world model** maintains constant or slowly changing information, such as 3-d models of known objects, models of sensor behavior in response to environmental stimuli, and kinematics and dynamics models. The **short term world model** maintains the currently "believed" state of the environment (rather than just the perceived state). Hence, it maintains information about the states and properties of all of the objects that it has encountered. As new objects are sensed, they are placed into the world model. These objects remain in the world model, regardless of whether or not they are "visible", until their absence can be verified, or until they are no longer of consequence to the task at hand.

Much attention has recently been given to the planning and control tasks, and to the plan decomposition tasks of the effector unit, including work in task level planning [Lozano-Perez83]. Less frequently discussed are the requirements of the sensor unit. It is within this "sensor fusion pyramid" and in the interactions between the sensor unit, the planning and control unit, and the world model, that most of the intense computation takes place. In the following section, we explore the structure and function of the sensor unit, and thereafter present the computational characteristics of an autonomous system architecture employing sensor fusion techniques.

Sensor fusion software architecture

The sensor processing unit depicted in figure 1 has the job of acquiring and interpreting data from various sensor sources to provide a perceived world model. The sensor fusion process can be subdivided into 4 major levels: data acquisition, data analysis, data abstraction, and data correlation and interpretation. The sensor fusion processing hierarchy is illustrated in figure 2. Data acquisition and data analysis are functions often referred to as signal or image processing, while data abstraction, correlation, and interpretation represent capabilities supported by pattern recognition and artificial intelligence techniques. Each of the four levels of processing obtains data from "lower" level subprocesses, and generates information for "higher" level processes, perhaps under some external control. These levels are present across all sensing devices, though the complexity of the respective processes may differ significantly. The complexity of the data analysis level for visual sensing, for example, is much greater than that for position, velocity, or acceleration determination from the servo shaft sensors.

Data acquisition is performed by the sensing devices themselves. These devices are transducers, generating electronic signals from the physical characteristics of the environment. Control at the data acquisition level is typically exercised by adjusting the sensitivity of the sensor using, for example, gain control parameters or signals. Data may be acquired from any of a number of sensing devices. Sensing modalities that have been proposed for the development of various autonomous robotics projects include: **joint position, velocity, acceleration, and torque**; **vision** (monochrome, color, infrared, radar); **range** (laser, ultrasonic); **force and moment**; and **tactile** sensing.

Data analysis is performed by a set of prescribed numerical algorithms, and is typically carried out in a digital computer. Algorithms at this level tend to be strictly numeric in nature, that is they massage the data according to some mathematically prescribed formula. Control at this level is exercised by appropriate selection of algorithms and their parameters. Examples of data analysis

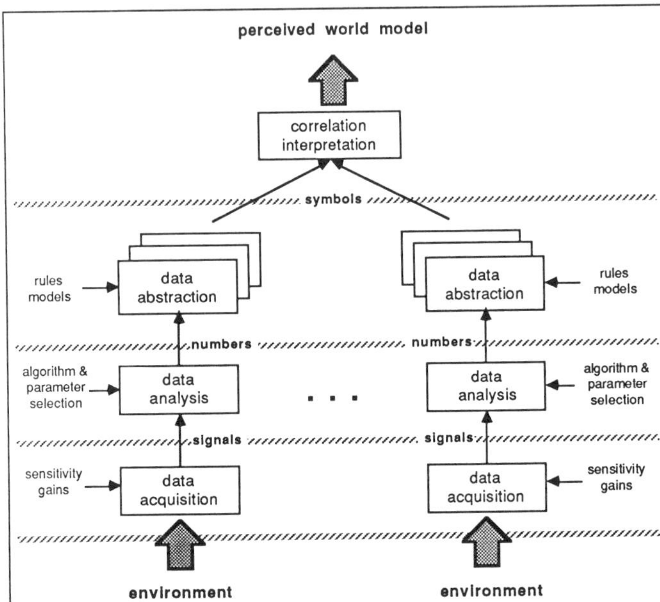

Figure 2. The sensor fusion pyramid

functions include low level image processing techniques, such as image enhancement, generation of multiple resolution images, edge detection, and simple region formation.

The **data abstraction** process performs the transformation from the numerical data generated by the sensors to symbolic data which will be used by the "higher" level processes. The abstraction process transforms information about the environment into more efficient representations for higher level processing. These representations more closely resemble those of the world model, and hence facilitate comparison (matching) with various aspects of the model. Additionally, the abstraction process reduces the amount of data required to describe the physical phenomena in the environment, thus facilitating more efficient communication between system subprocesses.

The abstraction process itself might be decomposed into a network of interconnected processes, each of which operates at a different resolution of the data. In the processing of visual information, for example, it might be useful to obtain both very coarse location information for collision avoidance as well as more detailed information for determining an appropriate grasp position on a part.

The **interpretation and correlation** process takes the symbolic descriptions of the environment produced by the various abstraction modules and attempts to generate a plausible world model which could account for all of the sensed data across all of the sensors. As sensor generated information is contaminated by noise and is inherently ambiguous, this process must not only interpret the information from each source, but must correlate information from the various sensing modalities in order to disambiguate information which might be present in only one "view". This requires multiple hypothesis generation, and reasoning with uncertainty about the quality or credence of the information from the various sources.

A model-based matching scheme provides the interface between the sensor unit and the world model. This matching process includes a discriminating controller to interpret the sensed information and update the world model. Figure 3 illustrates a representative model-based matching scheme. In the matching process, the sensed information is compared with model-based "predictions" of what the system expects to encounter. These predictions are, in essence, forward simulations that are carried out within the world model and used to generate environment expectations. The simulations are computed in parallel with the driving of the physical actuators, and in fact can be triggered by the

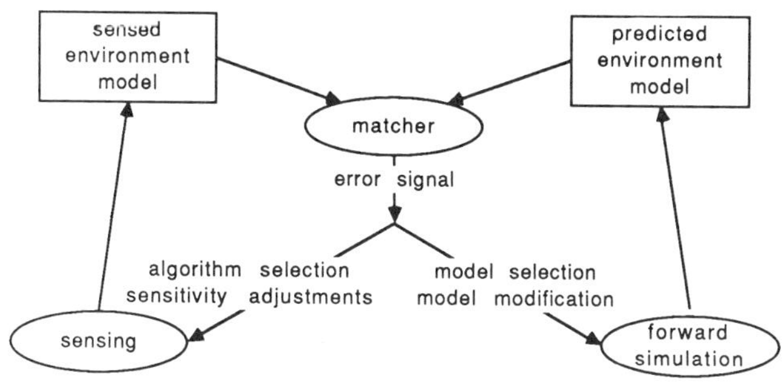

Figure 3. Model-based matching scheme

same internal signals. This real-time simulation enables the prediction to be compared with the sensed environment in the next sensing "frame". The results of the comparison can be thought of as "error signals". These error signals can be used to perform model selection or modification, and to direct further processing of sensed data via sensitivity controls and algorithm selection. This large feedback loop between the sensor and effector system provides the robot sensory system with an adaptive capability.

Software Architecture Characteristics

The autonomous system software architecture described above has a number of important characteristics that affect its implementation on a computer system. Some of these characteristics are outlined below:

High degree of parallelism

The hierarchical sensor fusion architecture exhibits a high degree of parallelism, both across sensor modalities, and within the processing of a single sensor "spectrum". Multiple resolution "views" of the same sensor data can be thought of as providing information from a number of **virtual sensors**. The data from each of these virtual sensors is processed in parallel to generate a single world model at various levels of abstraction. Additionally, multiple hypothesis generation must be supported at the higher levels of the sensor fusion pyramid. These concepts fall into the category of **macro-level parallelism**, i.e. explicit parallelism which reflects the structure of the problem. In addition, there is **micro-level parallelism**, i.e. parallelism which can be derived at the algorithm level, and which can be exploited to reduce the processing time of the data analysis phase.

Large, central model database

The architecture relies on a large, globally accessible model database to support all aspects of processing. This includes 3-D models for recognition tasks, sensor models to support forward simulation, and complex kinematics and dynamics models to facilitate both simulation and task planning. Additionally, the model database needs to support multiple hypotheses that can account for the perceived world model.

Low latency , high bandwidth interprocess communication

The perceptual process requires much interaction between the various sensor processing components and the world model. The tasks of recognition and correlation require a substantial amount of communication bandwidth during the hypothesis matching procedure. Additionally, interprocess communication may be required to insure consistency between different levels of abstraction among a family of virtual sensors.

Mixture of data driven and goal directed processing

Data driven processing is predominant in the sensor fusion pyramid, as information from the sensors is processed to generate a perceived world model. Goal directed processing is used in plan generation, and in focusing attention of the sensing unit to specific areas of the scene.

Flexible application of processing resources

Concepts such as focus of attention, and prediction based verification require that the software support flexible application of processing resources under adaptive control. When an autonomous system is first powered up, it must spend a great deal of effort attempting to understand its environment. Thereafter, it can proceed by prediction based verification of the environment. Significant deviations from these predictions cause the system to apply appropriate resources to the perceived changes in an attempt to understand the deviation. These processing resources must be transferrable from one sensor modality to another and from one level of abstraction to another in order to achieve the required flexibility.

Range of numeric and symbolic processing

Additionally, such an architecture must support a wide range of numeric and symbolic processing tasks. Sensor data analysis processes, forward simulation and geometric transformation processes all require significant amounts of numeric processing power. Symbolic processing is required by the planning and control unit to implement goal-directed skeletal planning, by the sensor fusion pyramid in combining uncertain information from multiple sensor sources, and by the world model maintenance process in performing match and update tasks.

Computing Environment Requirements

These characteristics imply that a computing environment capable of supporting research and development activities in sensor fusion and autonomous robot control must meet the following related requirements:

[1] Efficient numeric and symbolic processing
[2] Parallel execution of diverse programs
[3] Dynamic load balancing
[4] High interprocess communications bandwidth
[5] Facilities to support rapid access to a large, global database

In the next section, we present the Butterfly Parallel Processor as a computing environment which meets these requirements.

The BBN Butterfly™ Parallel Processing Computer

Hardware

The Butterfly Parallel Processor is a tightly coupled, homogeneous, multiple instruction multiple data (MIMD) multiprocessor. The system supports up to 256 "processor nodes" interconnected through BBN's Butterfly switch. The switch is a multi-stage, synchronous interconnect developed from BBN's packet switching technology. It is constructed from sets of 4x4 switching elements arranged in a topology similar to that of the FFT Butterfly. The Butterfly switch supports interprocessor communication rates of 32 Mbits per second. This switch bandwidth scales linearly with the number of processor nodes, so a 16-processor system supports 512 Mbits/sec of communication bandwidth and a 256-processor system supports 8 Gbits/sec.

The switch is used to interconnect the homogeneous processing nodes. Butterfly processing nodes are equipped with a Motorola MC68020 microprocessor, an MC68881 floating point coprocessor, and 1 or 4 MBytes of main memory. Additionally, each processor node contains a microcoded Processor Node Controller (PNC), an auxilliary processor which provides the interface to the Butterfly switch. The PNC performs the virtual to physical address translation, packaging up appropriate switch messages for remote memory references. Because the PNC is involved in every memory reference, whether local or remote, Butterfly memory appears as a single large global memory to application programs. Hence the Butterfly presents a shared memory model through global access to the distributed memories. The PNC also provides for an extended instruction set, in order to support various parallel processing operations. These operations include atomic test and set operations, queueing operations, and event mechanisms, all used to provide efficient communication and syncronization between processors. Additionally, the Butterfly supports communication with both Multibus and VMEbus based devices. This opens up the Butterfly architecture to a wide variety of devices for robot sensing and control. The Butterfly can be configured in single processor increments, from 3 to 256, providing tremendous flexibility and scalability. A fully configured machine supports 256 processors with 1 Gigabyte of memory, 8 Gigabit/sec of communication bandwidth, and 256 Mips / 64 MFlops of processing power.

Software

The Butterfly uses an object-based operating system called Chrysalis to perform all process and memory management operations and to provide interprocess communication and synchronization. Chrysalis is a layered operating system, containing an operating system kernel and a number of operating system libraries. Process management is performed by a real-time multiprocess scheduler which runs on each processor node. Chrysalis implements a segmented virtual memory management system, with the address translation mechanism implemented in PNC microcode. Interprocess communication is supported by a number of operating system primitives, including shared memory atomic operations, dual queues (data queues which are interlocked at the PNC level), and events. These primitives are used to develop more abstract interprocess communication and synchronization mechanisms, including locks, semaphores, monitors, and servers.

The Butterfly is programmed in either C, FORTRAN, or LISP. Interprocess communication is performed via calls to Chrysalis routines and to parallel programming methodology libraries. Three basic parallel processing models have been developed thus far for applications written in C or FORTRAN. The **Communicating Sequential Processes (CSP)** approach is supported by the Chrysalis process management and interprocess communication primitives. A library called the **Uniform System** has been developed to implement a single program multiple data (SPMD) approach to parallel computation. The Uniform system supports the notion of a set of homogeneous workers that cooperate in the execution of a number of tasks. A task queue is established that processors use to acquire tasks as they become free and need work. This model insures dynamic load balancing across the resources of the machine. The University of Rochester has developed a **structured message passing library, SMP**, for the Butterfly [LEBLANC86]. SMP provides a flexible message passing approach, including message buffering, blocking of processes, and dynamically created process families which communicate according to predefined interconnection topologies.

Butterfly Lisp is implemented as a Common Lisp compatibility package on top of the language MultiLisp, a Scheme variant devloped at MIT. Butterfly Lisp uses the **future** construct to support an implicit synchronization and communication model. This presents yet a fourth parallel processing model available to applications programmers. An object-oriented programming methodology based on an extended version of CommonLoops has also been developed. This methodology uses the parallelism available in Butterfly Lisp to support a **Parallel Expert Systems Execution Environment** [QUAYLE86]. Additionally, researchers at the Rockwell Science Center [CHIU86] have proposed an object-oriented, icon based **Function Network Programming Environment** for developing and organizing complex, asynchronous systems using a "data-flow" model of computation.

The Butterfly Parallel Processing Computer is ideal for experimenting with complex sensor fusion architectures such as the one proposed earlier. It combines powerful numeric processing capabilities with a flexible, software partitionable architecture. This enables support of both numeric and symbolic processing in a single, multitasking environment. The homogeneous processor architecture supports dynamic load balancing at both the operating system and application levels. The high communications bandwidth of the Butterfly Switch, coupled with a large shared memory architecture, supports rapid interprocess communication and global access to shared resources. The Butterfly also provides Multibus and VMEbus support of external devices, including frame grabbers, low-level image processing subsystems, robot controllers, and other sensors. All of these capabilities exist in a single, unified development environment that supports C, FORTRAN, LISP, and higher level artificial intelligence development tools.

Conclusions

The ability to fuse information from diverse sensor sources is critical to the development of fully autonomous robotic systems. This process is data and compute intensive, and until now, research in this area has been somewhat limited by the unavailability of appropriate computing power. The recent development of highly parallel multiprocessing architectures now provides the computing capabilities necessary to advance autonomous system technology.

In this paper, we presented a high level software architecture for the development of intelligent, autonomous robots. We investigated the structure of the sensor fusion pyramid and presented some computing environment system requirements to support such a software architecture. The BBN Butterfly, because of its ability to meet these requirements, was proposed as a multiprocessor environment in which to carry out further autonomous system and sensor fusion research.

In particular, we would like to see the development of a blackboard based control system to integrate the sensor unit, the effector unit, the planning and control unit, and the world model database. Blackboard based systems have been used in a number of complex real-time signal processing tasks, including CMU's Sidewalk Navigation System [Goto86]. These systems support a parallel, asynchronous computation model in conjunction with a globally shared database, and appear to match the requirements of an autonomous system software architecture quite well. We are open to, and encourage, joint efforts in these and other related research areas.

References

Andersson, R. L., "Living in a Dynamic World", *Proc. 1986 Fall Joint Computer Conference*, November 1986, pp. 97-104.

BBN Laboratories, Inc., "Butterfly™ Parallel Processor Overview", Report No. 6148, BBN Laboratories, Inc. , March 1986.

BBN Laboratories, Inc., "The Uniform System Approach to Programming the Butterfly™ Parallel Processor", Report No. 6149, June, 1986.

BBN Laboraties, Inc., "Parallel Expert Systems Execution Environment: A Functional Specification (Version 1.0)", Report No. 6225, BBN Laboratories Inc., August 1986.

Chen, J. Bradley, Ronald Fearing, Brian Armstrong, and Joel Burdick, "NYMPH: A Multiprocessor for Manipulation Applications," *Proc. 1986 IEEE Conference on Robotics and Automation*, April 1986, pp. 1731-1736.

Chiu, Stephen L., David J. Morley, and Jim F. Martin, "Sensor Data Fusion on a Parallel Processor", *Proc. 1986 IEEE International Conference on Robotics and Automation*, April 1986.

Gaglianello, Robert D., and Howard P. Katseff, "A Distributed Computing Environment for Robots," *Proc. 1986 IEEE Conference on Robotics and Automation*, April 1986, pp. 1890-1896.

Goto, Y., K. Matsuzaki, I. Kweon, T. Obatake, "CMU Sidewalk Navigation System: A Blackboard-Based Outdoor Navigation System Using Sensor Fusion with Colored-Range Images," *Proc. 1986 Fall Joint Computer Conference*, November 1986, pp. 105-113.

Kent, Ernest W., and James S. Albus, "Servoed world models as interfaces between robot control systems and sensory data", *Robotica*, Vol 2, 1984, pp. 17-25.

Leblanc, T. J., Gafter, N. M., and Ohkami, T., "SMP: A Message-Based Programming Environment for the BBN Butterfly", Butterfly Project Report No. 8, University of Rochester Computer Science Department, Rochester. NY, July 1986.

Lozano-Perez, Tomas, "Robot Programming", *Proceedings of the IEEE*, Vol. 71, No. 7, July 1983.

Ozunger, F. and M. L. Kao, "A Reconfigurable Multiprocessor Architecture for Reliable Control of Robotic Systems," *1985 IEEE International Conference on Robotics and Automation*, March 1985, pp. 802-806.

Siegel, David M., Sundar Narasimhan, John Hollerbach, David Kriegman, and George Gerpheide, "Computational Architecture for the Utah/MIT Hand," *Proc.1985 IEEE International Conference on Robotics and Automation*, March 1985, pp. 918-924.

A Simple Vision System for an Educational Robot

J.B. Cheatham and Chris Wu

Mechanical Engineering and Materials Science Department, Rice University, P.O. Box 1892, Houston, TX 77251-1892, USA

Abstract

An inexpensive computer vision system has been interfaced with a small educational robot to provide vision data for control of a robot in simple pick-and-place operations. The equipment, procedures, and limitations are described in this paper. For miniminal cost, educators and engineers can have a system for demonstrating how machine vision and a robot can be made to work together in an elementary fashion.

Introduction

Robot vision systems can be expensive, complex, and difficult to understand and work with. In order to give undergraduate engineering and computer science students an introduction to the use of vision with robotics applications, the inexpensive system described in this paper has been developed at Rice University. One important objective of this effort was to keep the procedures as simple as possible in order to emphasize principles and to permit the students to apply the system to a laboratory experiment in a minimum amount of time. After achieving success in the limited experimental environment the students may then be interested in studying more complex robotic vision systems and procedures. The purpose of this paper is to describe what has been done with the system, and thereby to help others in developing similar systems for introducing machine vision to robotics using a personal computer such as an Apple II or an IBM PC.

Equipment used to perform the task of providing image data to the robot is described. The formation of binary or black and white images on the computer monitor is described and the computer programs used to implement the interaction between the camera system and the robot are summarized.

Description of Equipment

The robot used in these experiments is a Microbot Minimover-5 manipulator with plug-in card for interfacing the arm to a microcomputer. This small tabletop robotic manipulator has five revolute joints and a parallel jaw gripper controlled by stepper motors using a 64K Apple II computer. The interface card allows the user to control the stepper motors manually from the computer keyboard. It also supports an interactive command interpreter called ARMBASIC that enables the user to program the robot in BASIC. A BASIC program is written to move 3 blocks from known or taught locations to prescribed positions. A built-in sensor using a simple microswitch activated by tension in the end effector drive cable can be used to determine the presence of an object as well as the size of an object upon which the hand closes. Thus, when an object is not present in the proper location, a signal is sent to detect the absence of the object and the arm moves back to the home position. A program is then called to activate the camera system.

The MicroDcam is a small CCD camera with a binary digitizer for the Apple II computer (Figure 1). It is a low-cost all digital imaging system. All necessary hardware, software and optics were furnished in one package with the camera. This system provides a resolution of 128 x 256 x 1. The digital picture output is stored in computer memory for further processing. The system allows the user to digitize images of various sizes with a choice of either fixed exposure time or automatic exposure time adjustment. The system is used to produce a binary image of a circular disk used in the robot pick-and-place program.

Figure 1. Equipment Set Up of the Vision Guided Robotics System

Problem Statement

The problem addressed is one of having a camera system provide the data for locating a circular black disk on a white background relative to the robot. Once the disk location has been stored in memory, it is then desired to have the robot move its end effector to the object, pick it up, and transport it to a specified location. In order to achieve this, several subtasks are required.

The camera system provides a digitized image that can be displayed on the computer monitor and from which data are stored in the computer memory. In dealing with binary images, one need only work with ones and zeros representing either white or black pixels (picture elements). In order to be able to compute the center or centroid of the black image on a white background, it is necessary that one know the locations in computer memory corresponding to the display on the monitor or CRT. Once the black and white pixels have been identified, a simple algorithm can be written to compute the centroid and other geometric properties. Details of each of these operations are described below.

Binary Vision System

Industrial vision systems and more complex vision systems for personal computers work with many graylevels. However, there is a distinct advantage in working with only two levels of light intensity, black and white, or a so called binary vision system. The output picture contains the silhouette of the viewed object. Two-dimensional geometric properties such as the first and second order moments, area, perimeter, and maximum radius can be easily measured from the binary image. Furthermore, it can be shown [1], [2], [3] that transformations of these geometric properties provide invariant measurements that can be used as discriminators to identify the viewed objects from a set of library models. This paper presents discussions of how these measurements can be obtained from a simple binary vision system and fed to the robot arm control program.

Binary Image Display

The BASIC program provided by Micromint, Inc., to drive the camera system was modified to obtain a subroutine for acquiring a binary image of size 64 x 128. This output image is displayed on the upper left corner of the screen of the Apple II computer. The video display of the Apple computer uses information in the system RAM to generate its display. In High Resolution Graphics mode, the value of a single memory location (a byte or 8 bits) controls a line of seven pixels. The most significant pixel of a byte is reserved for color information. On the screen a pixel is turned on, or white, when the corresponding bit in display memory has a value of 1. A pixel is turned off, or black, when the corresponding bit in display memory has a value of 0.

In the experiments, the disk used in the pick-and-place task is painted black and placed on a white background. This arrangement results in a clear binary image in which the disk is represented by the black pixels. With the digitized data stored in the RAM, a screen searching subroutine that will scan through the screen row by row searching for the black pixel by peeking the value from the corresponding bit in RAM is implemented. For instance, the status of the upper left pixel can be found by examining the value of the least significant bit of memory address 16384 using a peek command. Details of the screen searching subroutines for the Apple II and IBM-PC are given in the Appendices.

Calculation of Geometric Properties

The equations discussed in this section are valid only if the binary image contains a single black object on a white background. The set up of the screen coordinates, or image space, is with the origin at the upper left corner, the i axis pointing to the right and the j axis pointing downward. Geometric properties are calculated in the image space.

The screen searching algorithm is a one pass, raster by raster process. After each pixel is examined, the value of the pixel, b_{ij}, in the i-th column and j-th row is returned. Where

$$b_{ij} = 1 \text{ for black dot}$$
$$= 0 \text{ for white dot}$$

Thus the area is given by

$$A = \sum_i \sum_j b_{ij}$$

and the geometric center by

$$\bar{i} = \sum_i \sum_j i \times b_{ij}/A \quad \text{and} \quad \bar{j} = \sum_i \sum_j j \times b_{ij}/A$$

If the viewed object is assumed to be elongated, the orientation of the maximum principal axis of area measured counterclockwise from the ordinate is given by

$$\theta = \frac{1}{2} \tan^{-1} \left(\frac{2I_{ij}}{I_j - I_i}\right)$$

where

$$I_i = \sum_i \sum_j j^2 \times b_{ij} \quad \text{and} \quad I_j = \sum_i \sum_j i^2 \times b_{ij}$$

$$I_{ij} = \sum_i \sum_j i \times j \times b_{ij}$$

The perimeter P, topmost pixel TM, bottommost pixel BM, rightmost pixel RM, and the leftmost pixel LP are given by

$$P = \sum_i \sum_j p_{ij}$$

If $b_{ij} = 1$ and $b_{(i-1)j} = 0$ or $b_{(i+1)j} = 0$ or
$b_{i(j-1)} = 0$ or $b_{i(j+1)} = 0$
then $p_{ij} = 1$ else $p_{ij} = 0$

and

$$TM = \min_j (p_{ij} | p_{ij} = 1) \qquad BM = \max_j (p_{ij} | p_{ij} = 1)$$

$$RM = \min_i (p_{ij} | p_{ij} = 1) \qquad LM = \max_i (p_{ij} | p_{ij} = 1)$$

The maximum radius of an object is defined as the radius of the smallest circumscribing circle about the geometric center to the boundary of the object. This quantity is equal to the maximum distance between the geometric center and any of the four outmost pixels. Computer implementation of these equations is given in the Appendices.

Image with Multiple Objects

In the case where multiple objects are present in the image, a much more sophisticated algorithm is needed for extracting the geometric properties of each object in the image. In the pick-and-place task, the problem is simplified by assuming that only a disk is within the field of view to ensure an unobstructed image. However, a modified version of the screen searching routine based on the Label-Merge algorithm can be implemented for extracting the geometric properties from an image containing multiple objects. Details of this algorithm are beyond the scope of this paper. However, helpful information can be found in references [4] and [5].

Robot Movement Routine

Movements of the Minimover-5 robot manipulator are commanded in BASIC using the @STEP statement followed by seven arguments. The first argument defines the speed of motion, and the last six give the magnitudes of rotations of the stepper motors. This command allows the robot to be programmed in the joint coordinate system. To perform a robot movement in which the position and orientation of the robot hand are defined in the world coordinate system, a fixed Cartesian frame, an inverse kinematic transformation from joint to world coordinates is performed. Equations of inverse kinematics for the Minimover-5 and an example BASIC program are given in the Minimover-5 user's manual. A modified version of that BASIC program is implemented to perform the following tasks:

1) Six numbers (x,y,z, pitch, roll, jaw) that define the position, orientation, and jaw opening of the hand at the target location in the world coordinate are input to the program.
2) Perform inverse kinematics to obtain the corresponding joint angles of the robot at the target location.
3) Determine the differences between the joint angles corresponding to the target location and that of the current position.
4) Issue movement commands to move the arm to the target location. A vertical departure and approach path is designed for surface clearance.

Description of Computer Program

Software discussed in this paper is implemented in BASIC, which is the language used in existing programs provided by the manufacturers of the vision and robot systems. Four subroutines are called by the main program to accomplish the vision guided pick-and-place task. The first subroutine acquires the image data and stores the binary numbers in computer memory, and the second subroutine interprets these numbers in memory in terms of black and white pixels and computes the geometric center in image space. Then the third subroutine moves the arm to a prescribed target position, and the final subroutine generates a graphics display of the disks and messages on the screen for monitoring the system.

The execution cycle of the main program consists of three routines: initialization routine, pick-and-place routine, and position recovery routine. Figure 2 shows the flow chart of the execution cycle. In the initialization routine, six parameters are input to the program: the window size in which the black pixels are searched, W_x and W_y, the world coordinates of the image plane origin, O_x and O_y, and the world coordinates of the calibration disk's geometric center, D_x and D_y. After the calibration disk is placed at the specified location, the image acquisition subroutine is called and then the search subroutine. The camera system is set up such that the image plane, i-j plane, is aligned and parallel to the x-y plane in the world coordinates. It is assumed that the missing degree of freedom, the depth from the image plane to the viewed object, is a known quantity. To convert the numbers D_i and D_j, returned from the search subroutine into world coordinates, conversion factors along the x and y directions are given by C_x and C_y, where

$$C_x = (D_x - O_x)/D_i$$

$$C_y = (D_y - O_y)/D_j.$$

After the conversion factors are found, the arm is put into manual mode to teach the home position.

As an example of the use of the pick-and-place routine, three disks in a stack are moved from one location to another based on the rules of the famous Artificial Intelligence game "Towers of Hanoi". If no accident occurs, the disks will be moved one after the other consecutively. The graphics display will monitor the current state of the system and indicate the next disk to be moved, and its destination. If absence of a disk is detected, the recovery routine is called.

In the position recovery routine, the image acquisition and screen searching subroutines are called. If a disk is detected, the numbers generated for its geometric center are converted into the world coordinates. The robot is instructed to move to the disk, pick up the disk and move it to the specified destination. Due to the low resolution of the camera system and the open loop control scheme of the Minimover-5, accumulated position error is compensated for by opening the jaw amply in executing the pick task.

Summary and Conclusions

An inexpensive vision system for an educational robot can be used to provide an introduction to robotic vision. This paper has described how the MicroDcam can be interfaced with a Microbot Minimover using an Apple II computer to permit location of a black disk on a white background, and how the robot can be instructed to retrieve this object and move it to a specified location. This experiment requires that one be able to locate the binary numbers in the computer memory that are associated with black and white picture elements and use these in a simple algorithm to compute the centroid of the area. The simple vision system can also be used to obtain other geometric properties such as area, orientation, perimeter, and maximum radius for advanced vision applications. This exercise points the way for more advanced vision work beyond the scope of this paper. It is believed that by introducing robotic vision in such a simple manner, some of the mystery and difficulty associated with this topic can be eliminated, and that students will be encouraged to proceed to more advanced topics related to robotic vision.

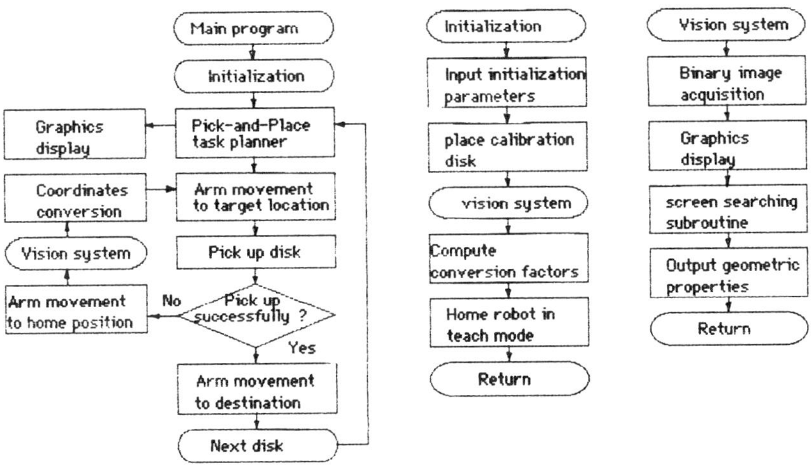

Figure 2. Flow Chart of the Main Program and Subroutines

References

1. S. A. Dudani, K. J. Breeding, and R. B. McGhee, "Aircraft Identification by Moment Invariants", IEEE Trans. on Comp., Vol. C-26, No. 1, Jan. 1977.
2. C. A. Rosen and G. J. Gleason, "Evaluating Vision System Performance", in Robot Vision, edited by Alan Pugh, Springer-Verlag, New York, 1983.
3. Y. F. Wang, M. J. MaGee, and J. K. Aggarwal, "Matching Three-Dimensional Objects Using Silhouettes", IEEE Trans. on PAMI, Vol. PAMI-6, No. 4, July 1984.
4. Berthold K. P. Horn, Robot Vision, McGraw-Hill, New York, 1986.
5. C. K. Wu, "Computer Aided Grain Size Distribution Analysis of Polycrystalline Materials", Master of Science Thesis in Mechanical Engineering, Rice University, May 1986.
6. William F. Luebbert, What's Where in the Apple, Micro Ink, Chelmsford, Mass., 1982.

Appendix A

The subroutine listed here scans the pixels within the window of size nrow by ncol and returns the area and the geometric center of the black pixels. The search pattern is a row by row, top to bottom, left to right process. The binary image is stored in the page 2 in the high resolution graphics.

```
100  REM  SCREEN SEARCHING SUBROUTINE FOR APPLE II COMPUTER
110  REM
120  REM  INPUT ARGUMENTS: NROW,NCOL , THE WINDOW SIZE
130  REM  OUTPUT ARGUMENTS: KOUNT , TOTAL NUMBER OF THE BLACK PIXELS
140  REM   TI,TJ,LOCATION OF THE GEOMETRIC CENTER ON THE IMAGE PLANE
150  REM
160 JJ = 0:TI = 0:TJ = 0:KOUNT = 0
170  FOR L0 = 16384 TO 16464 STEP 40
180  FOR L1 = L0 TO L0 + 896 STEP 128
190  FOR L2 = L1 TO L1 + 7168 STEP 1024
200  IF JJ > NROW GOTO 380
210 II = 0
220  FOR L3 = L2 TO L2 +  INT (NCOL / 7) + 1
230 TP =  PEEK (L3)
240  IF TP > 127 THEN TP = TP - 128
250  IF TP = 127 THEN II = II + 7: GOTO 320
260  FOR L4 = 1 TO 7
270  IF II > NCOL - 1 GOTO 340
280 PV = 1 -  FN MOD(TP):TP =  INT (TP / 2)
290 KOUNT = KOUNT + PV
300 TI = TI + II * PV:TJ = TJ + JJ * PV:II = II + 1
310  NEXT L4
320  POKE L3,0
330  NEXT L3
340 JJ = JJ + 1
350  NEXT L2
360  NEXT L1
370  NEXT L0
380 TI = TI / KOUNT:TJ = TJ / KOUNT
390  TEXT : HOME
400  RETURN
```

Appendix B

The subroutine listed here scans the pixels within the window of size nrow x ncol and returns the following quantities: area, geometric center, orientation, perimeter, four outmost pixels, and the maximum radius. The search pattern is a row by row, top to bottom, left to right process. Binary image is stored in the screen 2 memory.

```
100 REM SCREEN SEARCHING SUBROUTINE FOR IBM-PC
110 REM
120 REM INPUT ARGUMENTS: NROW,NCOL , THE WINDOW SIZE
130 REM OUTPUT ARGUMENTS: KOUNT , TOTAL NUMBER OF THE BLACK PIXELS
140 REM  TI,TJ , LOCATION OF THE GEOMETRIC CENTER ON THE IMAGE PLANE
150 REM  THETA , CCW ANGLE FROM ORDINATE TO THE MAXIMUM PRINCIPAL
160 REM          AXIS
170 REM  P , PERIMETER. MR , MAXIMUM RADIUS
180 REM  TM(0),TM(1) , I,J LOCATION OF THE TOPMOST PIXEL
190 REM  BM(0),BM(1) , I,J LOCATION OF THE BOTTOMMOST PIXEL
200 REM  LM(0),LM(1) , I,J LOCATION OF THE LEFTMOST PIXEL
210 REM  RM(0),RM(1) , I,J LOCATION OF THE RIGHTMOST PIXEL
220 KOUNT =0:TI=0:TJ=0:AI=0:AJ=0:AIJ=0
230 TM(1)=NROW:BM(1)=0:LM(0)=NCOL:RM=0:MR=0
240 FOR J = 0 TO NROW-1
250 FOR I = 0 TO NCOL-1
260 IF (POINT(I,J)) GOTO 380
270 KOUNT = KOUNT + 1
280 TI = TI + I:TJ = TJ + J
290 AI = AI + J^2:AJ = AJ + I^2:AIJ = AIJ + I*J
300 IF (I=0  OR I=NCOL-1 OR J=0 OR J=NROW-1) GOTO 380
310 IF (POINT(I-1,J) OR POINT(I+1,J)) GOTO 330
320 IF (POINT(I,J-1)=0 AND POINT(I,J+1)=0) GOTO 380
330 P = P+1
340 IF (TM(1) > J) THEN TM(0)=I:TM(1)=J
350 IF (BM(1) < J) THEN BM(0)=I:BM(1)=J
360 IF (LM(0) > I) THEN LM(0)=I:LM(1)=J
370 IF (RM(0) < I) THEN RM(0)=I:RM(1)=J
380 NEXT I: NEXT J
390 SCREEN 0
400 TI = TI / KOUNT
410 TJ = TJ /KOUNT
420 THETA = ATN(2*AIJ/(AJ-AI))/2
430 TMR = SQR((TI-TM(0))^2+(TJ-TM(1))^2)
440 IF(TMR>MR) THEN MR=TMR
450 TMR = SQR((TI-BM(0))^2+(TJ-BM(1))^2)
460 IF(TMR>MR) THEN MR=TMR
470 TMR = SQR((TI-LM(0))^2+(TJ-LM(1))^2)
480 IF(TMR>MR) THEN MR=TMR
490 TMR = SQR((TI-RM(0))^2+(TJ-RM(1))^2)
500 IF(TMR>MR) THEN MR=TMR
510 RETURN
```

Visual Robot Guidance for Materials Handling Operations

M.B. Zaremba*, M.A. Poissant†, O.L. Hageniers†

*University of Quebec in Hull, Dept. of Computer Science C.P. 1250, succ. "B", Hull, Quebec, Canada J8X 3X7
†Diffracto Ltd., Windsor, Ontario, Canada N8T 3B7

1. INTRODUCTION.

Machine vision is rapidly becoming an essential and indispensable element of modern manufacturing systems, particularly those incorporating robotics technology. It allows to achieve system flexibility and to greatly enhance robot interaction with its work environment.

One of the most difficult and frequently encountered problems in robotised manufacturing is visual tracking of a moving object. A variety of materials handling operations can be imagined in which the object has to be tracked by the robot before the conditions (such as a sufficiently small and stable value of the tracking error) are established for grasping the object. Another example of a field in which tracking is an essential element of the manufacturing process is visual guidance of arc-welding robots.

The vast majority of machine vision systems used in such applications have been based on image processing [3.], [5.]. From the point of view of real-time control of robots moving at high speed, image interpretation is in many cases computationally expensive and may introduce considerable noise into the control system. The paper presents a robot guidance system that applies a 3-D vision sensor based on photogrammetry technology developed at the National Research Council of Canada [4.], [7.]. A specific feature of the sensor is that it provides six Cartesian coordinates (x, y, z, pitch, yaw, roll) of the targeted object with reference to the camera. The solution time is short enough to prevent significant time constraints in real-time control applications.

Following a description of the photogrammetry vision sensor, the paper discusses the control part of the system. Using specific features of the sensor, a higher level of control can take decisions on selection of the appropriate phase of the robot operation, or supervise target reacquisition after a more significant disturbance. Thus, an extension toward AI-based supervisory control of the materials handling process is provided.

2. PHOTOGRAMMETRY VISION SYSTEM.

Photogrammetry, used originally for cartographic purposes, also found application in other engineering fields. In manufacturing, photogrammetry methods may be used for positioning, calibration, and as a part of control systems. Their big advantage is simplicity (images from a single camera are processed) and short solution time. The operation of the sensor is fast enough for real-time robot control (30-60 Hz).

The 3-D vision sensor, based on photogrammetry technology and developed at the NRCC, provides precise information on both position and orientation of a special target with reference to the optical centre of the camera. The sensor is therefore particularly well suited to situations where the information required relates to the location rather than to the shape of an object.

The sensor concept is based on the projection of a four target array on the image plane of the camera (see Figure 1).

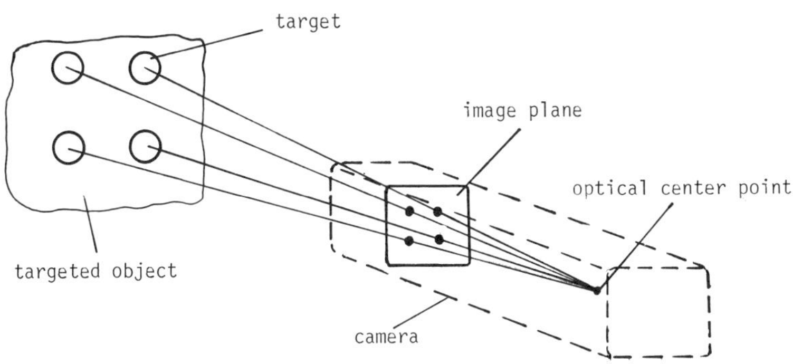

Figure 1. Projection of target on image plane of the camera.

The position and orientation of the target with respect to the camera is obtained directly from the vision system. The solution is in the form of 6 coordinates: 3 linear - x, y, z, and 3 angular - pitch (θ), yaw (ψ), roll (ϕ). The vision system is therefore able to immediately define the transformation matrix S between the target and the camera coordinate systems:

$$S = \begin{bmatrix} \cos\psi \cdot \cos\phi + \sin\psi \cdot \sin\theta \cdot \sin\phi & \sin\psi \cdot \sin\theta \cdot \cos\phi - \cos\psi \cdot \sin\phi & \cos\theta \cdot \sin\psi & x \\ \cos\theta \cdot \sin\phi & \cos\theta \cdot \cos\phi & -\sin\theta & y \\ \sin\theta \cdot \cos\psi \cdot \sin\phi - \sin\psi \cdot \cos\phi & \sin\psi \cdot \sin\phi + \cos\psi \cdot \sin\theta \cdot \cos\phi & \cos\theta \cdot \cos\psi & z \\ 0 & 0 & 0 & 1 \end{bmatrix} \quad (1.)$$

The accuracy of the solution obtained is 0.001" for x, y, z, and 0.1 deg for yaw, pitch and roll [2.].

3. ROBOT GUIDANCE.

3.1. DESCRIPTION OF THE MANUFACTURING PROCESS.

In order to design the robot guidance system and to evaluate its performance, an experimental workcell was set up, which included a Cincinnati Milacron T3 robot equipped with a photogrammetry vision system. The task of the robot was to unload a part from a swinging carrier and deposit it on a belt conveyor. Figure 2 shows the robot arm with a camera mounted on its wrist, pointing at an automotive part.

Figure 2.

The major phases of the robot operation included:

- Part selection,
- Target acquisition,
- Part tracking with the target within the camera field of view (FOV),
- Target reacquisition when the target is partially outside the FOV,
- Approaching and grasping the mechanical part,
- Materials handling sequence,
- Termination of the cycle.

3.2. ROBOT STATIC CONTROL.

As explained in section 2, output signals from the photogrammetry vision system can be used directly to determine desired position and orientation of the end effector with reference to the object to be handled. The information can be used for static control, to perform operations such as positioning and calibration, or for dynamic, robot path control.

A general equation serving as the basis for the calculations is obtained from the kinematic chain:

$$R \cdot E \cdot S \cdot P = R_D \cdot F \tag{2.}$$

with the elements of the equation being 4 x 4 homogeneous transformation matrices [6.] between the following coordinate systems:

R - end effector with respect to the robot world system,
E - vision sensor (camera) with respect to the end effector,
S - target with respect to the vision sensor,
P - handled part with respect to the target,
R_D - desired location of the end effector with respect to the robot world system,
F - handled part with respect to the end effector.

From equation (2):

$$R_D = R \cdot E \cdot S \cdot P \cdot F^{-1} \tag{3.}$$

The desired position of the end effector corresponds to the fourth column of the right-hand matrix, while the desired orientation is obtained from trigonometrical relationships between the elements of the first three columns of the matrix.

3.3. DYNAMIC PATH CONTROL.

During the whole target acquisition and object tracking cycle, the control of the robot must take into account robot arm inertia and time delays introduced by different elements of the system. The design of the tracking system is, thus, a classical control problem.

Because robot characteristics vary within the workspace, resulting in stability problems (particularly severe at higher velocities), there are major advantages to an open-loop scheme with prediction [3.], [8.].

The basis for our consideration of the control algorithm will be equation (2) expressed as a function of time

$$R_D(t) \cdot F = R(t) \cdot E \cdot S(t) \cdot P \tag{4.}$$

The control signal can be calculated on the basis of the predicted location of the handled part. Denoting the right hand side of equation (4) by W(t), which is the matrix describing current location of the tracked part with respect to the world coordinate system, we obtain:

$$R_D(t+t_{pred}) = (W(t) + dW(t_{pred})) \cdot F^{-1} \tag{5.}$$

Now, the problem resolves into correctly calculating W(t) and properly evaluating $dW(t_{pred})$. The calculation of W(t) should fulfill the condition of consistency, which means that the matrices R(t) and S(t) have to be determined for the same time instant. In the system implemented by the authors, the vision sensor sampling instants (t_v), having been accurately determined, were taken as reference. The location of the end effector at that time, $R(t_v)$, can be calculated from the kinematic model of the manipulator, if robot joint angles can be measured. However, this is not always possible. A different approach was adopted by the authors. Current end effector position and orientation were determined from previous control signals making use of experimentally found robot arm dynamics models of the following form:

$$K_i(z) = \frac{Y_i(z)}{X_i(z)} = \frac{b_{0i} + b_{1i} \cdot z^{-1} + b_{2i} \cdot z^{-2}}{1 + a_{1i} z^{-1} + a_{2i} \cdot z^{-2}} \cdot z^{-d_i} \tag{6.}$$

The capabilities of external control of the robot, such as DEPC (Direct External Path Control) communication link for the Cincinnati Milacron, allow direct application of control signals defined in world coordinate system.

When evaluating $dW(t_{pred})$, the choice of the prediction algorithm (ex. polynomial extrapolation, Kalman filters, etc.) should consider the computation time and the dynamics of the motion of the part. A simple polynomial extrapolator was sufficient in the implemented system.

4. TARGET ACQUISITION.

Apart from the main solution, i.e. the position and orientation of the target, the vision system provides additional signals that may be used primarily for target acquisition. The signals fall into three categories:

1.) window errors,

2.) target dot detection,

3.) lighting conditions.

The first two sets of signals are related to the windows generated - at the vision sensor control level - around the four dots of the target. Where there are strong disturbances, the signals - e.g. "Target overlaps bottom edge of window", "Window is outside of X F.O.V.", etc. - can be used by the supervisory robot control level to undertake actions aiming at target reacquisition. Those actions, in the form of elementary robot control signal increments are determined from a decision table. The decision table consists of n conditions B_i (i = 1, 2, ..., n), which are the additional signals provided by the vision system, N rules R_i (i = 1, 2, ..., N) and actions A_k (k = 1, 2, ..., N). The actions are binary sequences of length 6 (six control signals for a robot of six degrees of freedom).

In order to avoid excessive overshoots, the amplitude of the elementary increments have to be calculated taking into account the dynamics of the robot arm and the relations between the dimensions of the target and the camera F.O.V.

The decision table representation provides a fast, efficient solution. However, it lacks flexibility. Currently, at the University of Quebec in Hull work is under way to integrate target acquisition with collision avoidance applying AI techniques for the determination of the trajectory of the end effector.

4. CONCLUSIONS.

The analysis and implementation of the sensory-based robot guidance development presented in the paper demonstrated that the NRCC photogrammetric vision system combined with the hierarchical structure of the system controller and prediction-based tracking provide a real-time, industrially applicable solution for a flexible system with a certain degree of intelligence. The total computation time of the six axes control system was short enough to provide a control signal cycle time of less than about 34 Hz.

The type of system discussed in the paper has vast application potential in manufacturing processes. The photogrammetry vision system may be used for positioning and calibration as well as for dynamic guidance of robots. Other application areas that can be envisaged include vehicle guidance, automatic artillery emplacement, automatic or semi-automatic docking of helicopters, Harriers and light planes, and others.

5. ACKNOWLEDGEMENTS.

The work reported in this paper was supported by the National Research Council of Canada, PILP program office.

6. REFERENCES.

[1.] Baudot, W. and Clermont, G. and Gaspart, P. and Lecoq, P. "Visually Guided Arc-Welding Robot With Self-Training Features", Proceedings of the Robots 7 Conference, Chicago (1983).

[2.] Diffracto. "RPS: Realtime Photogrammetry System". Application note.

[3.] Hunt, A.E. and Sanderson, A.C. "Vision-Based Predictive Robotic Tracking of a Moving Target", Carnegie-Mellon University, The Robotics Institute, Report No. CMU-RI-TR-82-15, (1982).

[4.] Kratky, V. "Real-time Photogrammetric Support of Dynamic Three-Dimensional Control". Photogrammetric Engineering and Remote Sensing, Vol. 45, No. AP9, (1979).

[5.] Nitzan, D. "Development of Intelligent Robots", IEEE Journal of Robotics and Automation, vol. RA-1, No. 1, March (1985).

[6.] Paul, R. "Robot Manipulators: Mathematics, Programming, and Control." MIT Press. Cambridge, Mass., (1981).

[7.] Pinkney, H.F.L. "Theory and Development of an On-Line 30 HZ Video Photogrammetry System for Real-Time Three-Dimensional Control". Proceedings of the ISP Symposium on Photogrammetry for Industry, Stockholm, August (1978).

[8.] Zaremba, M.B. and Poissant, M.A. and Hageniers, O.L. "Robot Target Tracking System". Proc. 9th Conference of the British Robot Association, Stratford-Upon-Avon, May (1986).

A Geometric State Sensor for Robot Workcell Monitoring

Christopher H. Orgill

AI & Robotics Research Group, Dept. of Computer Science UCW Aberystwyth, Dyfed, Wales SY23 3BZ

1. Abstract

A processing scheme is given that provides data at a point along a sensory continuum which is distinct from the region normally occupied by those data provided by the sensing devices more conventionally used in robotic assembly. In contrast to the operation of typical binary sensors such as contact probes and beam-interruption devices, which give an unambiguous datum about a local event in the cell, the sensor described is sensitive to deviations of any component in the workcell volume with an implementation dependent resolution, presenting no other information except that a deviation has occurred.

Where required in the assembly sequence, a signature is derived by a simple off-line teaching procedure. Although the use of the sensor in an analytic role is not expected to be tenable, it provides a cue for the onset of error diagnosis in a cell supervisory computer, operating either in stand-alone mode or integrated with other sensory input.

An argument from the field of statistical communication theory is presented, showing the elicitation of system behaviour from a communication theoretic formalism to be inappropriate for robot domains due to the nature of the knowledge representation. The use of domain specific knowledge is discussed with a view to establishing likely bounds for probabilistic descriptors of the scene.

The A.I technique of 'evidential propositional calculus' is applied for the combination of multiple independent sensor readings.

2. Introduction

Many implementations of industrial robot assembly workcells have a sensory capacity that is composed of a set of measurands which provide highly localised data about the assembly sequence as it progresses. If this capacity is seen as the ability to illuminate a part of a space, such as the hyper-volume proposed by ALBUS 1981 then such measurands can be viewed as 'high-intensity point illumination', where by intensity we give a qualitative expression of the probability that the sensor has correctly performed it's function, and by 'point' we mean the order of magnitude of the sub-volume within the space that is being measured, or 'illuminated'. This is illustrated in figure 1, for some conventional sensors.

We therefore conceived of a need for a simple sensor whose measurand occupies the region of the space of figure 1, corresponding to 'low-intensity diffuse illumination'. Such a sensor presents a measurand which represents a change in workcell state -here a geometric perturbation- which is not constrained to occur in a small sub-volume of the envelope . A specification and preliminary analysis for such a type of device is presented in this paper.

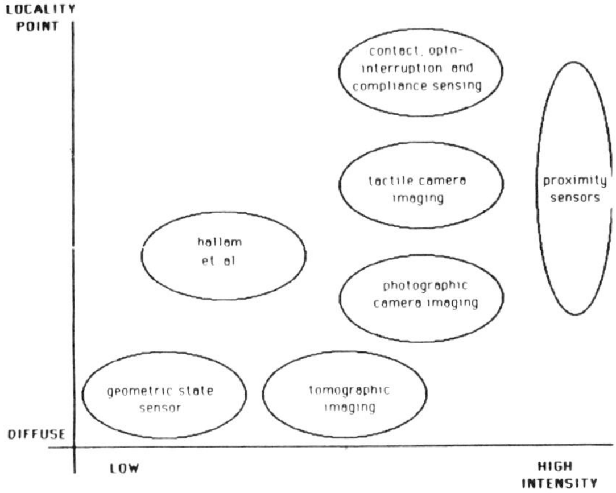

figure 1- A qualitative two dimensional sensor space

All robotic workcells suffer from sensory undercapacity. Indeed, from a mechanistic standpoint, it is impossible to 'know' every parameter of a system. Purely pragmatic considerations force the equiping of workcells with senses to be done on a 'just-suffices' basis for the proposed assembly task. Brooks has addressed the problem of automating the process of optimising where in an assembly task sensing should be performed (BROOKS 1982).

However, with the advent of supervisory systems which are precursors to 'intelligent' supervisors- such as the frame-based AFFIRM (BARNES et. al. 1983) the issue of redefining the criteria for 'adequacy' is raised, with a view to providing additional sensory capacity beyond that which is required solely to provide local monitoring of those parts of the cell that are being acted upon by the known agents. In addition, preliminary work on mechanism identification for the origin of tolerance errors[1], e.g see LEE 1986, has shown that certain types of free-transfer assemblies present considerable problems for conventional point sensing, where it is not possible to use a fixed strategy for the monitoring of geometric deviations . As to the use of highly flexible remote sensors, it is often neither cost-effective, nor practical[2] to consider the continuous use of imaging systems, whether they be camera or tomographic in their mode of operation. Hallam (HALLAM, et. al. 1985) has acheived a system whose measurand can return a 'medium-intensity semi-diffuse' datum about a pre-processed component of a scene, by our prior classification. However, our required sensor function is not the elicitation of structural properties from the scene, it is merely to provide a fast cue to a supervisory system that a task error has occurred. This cue should provide it's knowledge in a formalism that allows simple integration with data provided from task and assembly models, together with the data from the local senses. The knowledge source(KS) thus constructed can be used either in an 'expectation-free' manner, placing heavy reliance on the authenticity of the KS; a technique more easily integrable with earlier generation supervisors; or it can be used in a 'hypothesis-confimation' manner, where it is required to augment or verify assertions posed by the interpretation of other (Boolean) senses.

3. Monitoring the workcell scene

3.1. Introduction

We consider the measurement problem in two stages. We firstly cast the problem in the framework of statistical communications theory. It will be shown that the construction of a classical optimum signal detection computer, although completely general and not requiring off-line teaching, in practice turns out to be ill-suited to this problem domain. The analysis highlights the inherent experimental difficulties in attempting the embodiment of system behaviour as a multivariate vector of conditional probabilities. If we "pre-learn the noise" (c.f following section) ,the problem complexity is drastically reduced. We then consider the use of an artificial intelligence technique for interpretation of the data detected, namely 'evidential propositional calculus'.

Before we consider the first treatment it is necessary to state what we consider to be the principal elements of the measurement system. In the parlance of communication theorists we would speak of the symbol source ,the information channel and the receiver. We identify, respectively; the quantity being measured ;the carrier of the measurand and the measurand itself. With our assumptions about the nature of these elements made explicit we can then proceed to build a methodolgy for analysis.

By carrier we mean the physical property whose vector field creates a complex set of 'object waves' by interaction with the scene objects.

The quantity being measured is defined as follows. We state here that there exists an infinite set of perturbations, any member being interpreted as the difference between the desired workcell state[3] and the (unknown) perturbed workcell state. We shall call the measurement of the desired state noise, as it is the background against which we wish to detect any member of the set of perturbations.

The measurand is that property which, after an arbitrary number of invariant transformations and transductions performed on the quantity being measured , is presented to the input of the detection unit. It is ultimately a boolean sequence, but may represent a sample of a pressure field, or an electromagnetic intensity[4], depending on the implementation.

3.2. The Ideal Observer[5] model: knowledge requirements

3.2.1. Statistical unpleasantness

We complete the formulation of the problem as one of binary detection by stating the probabilities we still require from the system in order to perform an optimum thresholding. We seek to optimise two quantities: the false positive probability[6] and the missed event probability. We compute a threshold which minimises both of these; the maximum likelihood ratio, or MLR, as is usual for problems of this type. For values of this parameter ≥ 1 we decide that a signal is present against the noise, for values of MLR <1 we decide noise alone. As the detector, for practical reasons, is a discrete sampling device, we

do not have an infinite set of measurand values. However, we must still compute a value of the MLR for each member of this set, depending on the chosen resolution of the digitisation. We have[7] the vector quantities: $S(\Theta)=S_j(\Theta_i);j=1\ldots\infty;i:=1\ldots m$, giving our signal vector- we are denoting that the signal is a function of m variables- our measurand is: $V=V_k;k=1\ldots N;N=2^1$; for 1-bit digitisation, and we have the following probabilities:

p,q: are the probabilities that S+N and N will be observed, respectively. In the workcell context this is simply the measured error probability for the event $S_i,N_i;i=1\ldots T$ for T states to be monitored.

$F_1(V|S)$: is the N-dimensional conditional probability distribution describing the chance of measuring V_k given that signal S_j was present.

$w_H(S);H=0,1$: are two probability densities. The sub-script represents the two contesting hypotheses in the detection. H_0 is the hypothesis of no signal and hence w_0 is trivial, reducing to a single value, q. w_1 is the distribution of the rest of the S.

Given that this knowledge of the assembly task is available, we can then compute the MLR, which is given by :

$$MLR_1(V)=\frac{p\langle F_1(V|S)\rangle_S}{qF_1(V|0)} \qquad 1$$

where ;

$$p\langle F_1(V|S)\rangle_S=p\int_\Theta w_1(\Theta)F_1(V|S(\Theta))d\Theta \qquad 2$$

and ;

$$qF_1(V|0)=q\int_\Theta w_0(\Theta)F_1(V|S(\Theta))d\Theta \qquad 3$$

3.2.2. Some pragmatism

We now address the problem of evaluating the necessary probabilities, easiest first !. p and q are straightforward. It may be possible accurately to estimate these, a priori, using the proposed extensions to current solid modelling systems (see discussion in REQUICHA 1984) which can predict, by Monte Carlo simulation, the expected tolerance locus for a particular i in T, and hence the failure probability. If the sensor is retrofitted, then the statistic can be extracted from the production log. The latter approach is more desireable, as it will include the contributions from the tails of the p.d.f $w_1(\Theta)$, which would be missing from a dead-reckoning model.

$w_1(\Theta)$ is continuous, but very sparse. The representation of the distribution of object perturbations can be done without recourse to finite elements if a graphical modelling system of the constructive solid geometry (CSG) type is available (e.g see AMBLER & POPPLESTONE, 1975). A member of the set of signals is then the CSG representation of the unperturbed scene differenced with that of the perturbed scene. Such a volume can be stored quite economically by identifying that set of CSG primitive 'instances' that comprise the rigid object perturbed, storing a cartesian 6-tuple to represent the magnitude and direction of composite centroid, linked with the record of those instances. The sparseness of the p.d.f arises from the likely movement constraints upon most of the objects in a real assembly cell. We can see that it is at least theoretically possible to know the number of degrees of freedom the scene possesses at i in T, although the process is daunting.

$F_1(V|S(\Theta))$ is the stumbling block. $S(\Theta)$ implies that we know the forward transformation of the perturbation into the signal. Although this is knowable, it is a sizeable problem. A systematic elicitation of $F_1(V|S(\Theta))$ requires simulation, as it is not practical to examine the anticipated range of $S(\Theta)$ by physical experiment.

3.3. Pre-learning the noise

It is now apparent, in the light of our original desire for a cheap and relatively simple sensor, that an unsupported detector structure is not amenable, due to the complexity of the knowledge representation process. The required on-line and off-line computational effort could conceivably dwarf the resources required by the entire supervisory system into which the sensor is integrated. The question we must ask is;

"Are we using all the knowledge available to us ? "

We have so far ignored the fact that ,if we 'teach'[8] the measurand in the unperturbed state, we have two measurands for any state i in T, namely $V_i(H_0)$ and $V_i(H_1);H_0=N;H_1=S+N$. We merely have to decide if the two measurands differ. The problem seems solved immediately !, although on closer inspection we find that this is not entirely true. An assumption that was hitherto implicit must now be stated ;

The mean signal to noise ratio is constant for any S,N and is at an 'acceptable' level.

Although the behaviour of $N=N_i;i=1...T$ has been ignored, -it is a small set compared with the signal space- without a model for $F_1(V|S(\Theta))$ we do not know how well-conditioned the signal is over Θ. This amounts to saying that a monotonic list of Θ, ranked by magnitude of perturbation, does not need to result in a smooth sequence for $S(\Theta)$[9] More plainly, we don't know when the signal-to-noise ratio will be adequate to indicate a practical value for $V_i(H_1)$ that is measurably different from the $V_i(H_0)$ over all the desired range of V. This can be expressed as a requirement for every object in the scene to project a strong enough object wave to a receiver to allow detection. Note that this is a much relaxed occlusion criterion compared to an imaging system. Consider the rudimentary example of a cube, illuminated, diffusely, by a carrier property. Reception must always result in sensory occlusion[10] of part the object. For the reception requirement of this sensor system, it suffices that physical continuity exists within the cube. If it is displaced, all facets will project new 'object waves' as tokens of the displacement and no sensor occlusion occurs.

It is intuitively appealing to consider that the signal statistics will be of the same form as the noise, being the transform of a simple displacement from one 'noisy' state to another. Such a result is indeed valid (PARZEN 1962) if the transformation is quasi-linear. An analytic treatment is not given here.

3.4. Multiple views

A more direct solution to alleviating an unacceptable signal-to-noise ratio at a fixed measurement site is simply to have more than one site[11]. With spatial separations much greater than the wavelength of the carrier we have many independent signal sources, each asserting either H0 or H1. We can then turn our attention to making the decision process an integrable part of the intelligent supervisory function by finding a proven means of combining such redundancy.

3.5. Integrating probabilistic sensor data into a framework of boolean facts

The method chosen is essentially that of GARVEY, et al. 1981, after SHAFER 1976. Following the formalism presented; essentially the terminology of Dempster's rule for combination of evidence; we must satisfy two criteria. Firstly, the belief functions[12] must be independent for each knowledge source. Adequate spatial separation ensures this condition is met. Secondly, all events in the frame of discernment[13] must be exclusive. This readily follows, as H0=~H1. This allows us to create an overall evidential interval for the hypotheses H0 and H1 by combining the evidential intervals for each receiver. It also allows any uncertainty in the measurand, corresponding to quantizing-noise or other processing artefacts, to easily be accounted for when computing the overall support for either proposition. Yet further, it also accounts for a non-degenerate distribution of $N=N_i;i=1\ldots T$ for repeated measurement at a given i, i.e $\langle N\rangle\diamond$ the null vector; corresponding to artefacts in the off-line measurement also. Probability distribution functions (p.d.f's) are established incrementally throughout the lifetime of use of the sensor on a specific cell. The initial p.d.f's for the hypothesis pair, H0 H1 , for every state in T are degenerate ; $P(V(r_j))=1;V(r_j)=a;=0;V(r_j)\diamond a$ where a is the taught value, and uniform $;\langle V(r_j))\rangle=a;$ respectively. The latter expresses the shallowness of our expectations of likely constraint upon the workcell components. The uncertainty in the measurement of $V(r_j)$ is the interval over which the p.d.f is integrated to find the total probability mass. The two masses under each interval are normalised such that the sum is unity, followed by a post-normalisation to account for ΔV. This gives the belief function. We can conceive of three domains for system performance ;

a) 'well-resolved'; distinct p.d.f for H0, tending to $\delta(\langle V(r_j)\rangle)$, broad p.d.f for H1.
b) 'just-resolved'; both p.d.f's of similar character and mean.
c) 'unresolved'; both p.d.f's identical.

In case a) the system is well defined and the belief function commits it's belief to the expected hypothesis for each measurement totally. Cases b) and worse, c) will commit similar amounts of belief to each hypothesis in each measurement.

We reason that device performance is inadequate at that view angle if the sensor repeatedly asserts H0 in this fashion for many different $V(r_j)$ (we know it is), by possessing an evidential interval which is at variance to the pooled evidence from the other sensor sites. Such permanent sensory occlusion can easily be detected from an inspection of the signature sequence for the task being monitored.

4. Conclusions

A specification for a cheap, simple, probabilistic, remote-sensing device for monitoring task state errors in robotic assembly workcells has been presented, together with a discussion of the evolution of an adequate strategy for the methodology-of-use and evaluation-of-performance for candidate designs. The knowledge elicitation task for an unsupported binary detector is shown to be impractically complex. With the simple comparison against off-line data, the problem becomes tenable.

5. Acknowledgements

Grateful thanks are due to colleagues Dave Barnes, Nigel Hardy, Mark Lee, Vivien Robinson, Jem Rowland and Andy Woodward who have, at various times, focussed their attention upon this work. This work is in part funded by the ACME[14] Directorate of the United Kingdom Science and Engineering Research Council as Special Grant GR/D 37852, "A software development package for intelligent supervisory systems" in collaboration with Thorn-EMI Central Research Laboratories[15].

6. References

Albus J.S., "Brains, Behavior and Robotics", Chapter 5, BYTE Publications, Peterborough, N.H. 1981.

Brooks R.A., "Symbolic error analysis and Robot Planning", Int. J. Robotics Research, vol.1, no.4, pp29-68, 1982.

Barnes D.P., Lee M.H. and Hardy N.W., "A control and monitoring system for multiple-sensor Industrial Robots", Proc. 3rd Int. Conf. on Robot Vision and Sensory Controls, pp471-479, Cambridge, MA, November 1983.

Lee M.H and Orgill C.H, "Machine constraint taxonomy and the free-transfer problem", TR-RRG-68/86 ,Computer Science Dept., UCW Aberystwyth, 1986

Hallam J.C.T, Kwa J.B.H and Howe J.A.M, "Rule-based surface classification using specular sonar reflections", DAI-288, Dept. of Artificial Intelligence, University of Edinburgh, Scotland.

Middleton D., "Introduction to Statistical Communication Theory", chapters 18-20, McGraw-Hill, 1960 .

Requicha, A. A. G.,"Representation of Tolerances in Solid Modelling: Issues and Alternative Approaches", in "Solid Modelling by Computers. From Theory to Applications", Pickett M.S. & Boyse J.W. (ed.), pp3-22, Plenum Press, 1984

Ambler A.P, & Popplestone R.J, "Inferring the positions of bodies from specified spatial relationships", Artificial Intelligence, vol.6, pp157-174, 1975.

Parzen E. "Stochastic Processes", Theorem 1A, pp16-17, Holden-Day, 1962

Garvey, T.D, Lowrance, J.D, and Fischler, M.A., "An inference technique for integrating knowledge from disparate sources", Proc. 7th IJCAI, pp 319-325, Vancouver, 1981.

Shafer, G., "A mathematical theory of evidence", Princeton University Press, Princeton, New Jersey, 1976.

7. Notes

[1] i.e, those non-deterministic processes which cause computed tolerance bounds to be unexpectedly exceeded.

[2] either by virtue of the processing overhead degrading the task cycle time, or by the hostile nature of the assembly environment, etc.

[3] The volume of the workcell is divided into a mesh of cubical elements; either containing part of an object or nothing. In the limit, as the cube volume tends to zero, we have the continuous workcell state.

[4] A number of methods of implementation are possible and one, using ultrasound, is currently under development at UCW.

[5] see MIDDLETON 1960.

[6] the probability of saying something has moved when it hasn't.

[7] the notation is after MIDDLETON 1960.

[8] an activity synoptic with traditional arm location teaching, hence with little production penalty.

[9] A complete treatment of the 'ill-posedness' of deterministic inverse scattering problems, of which this system is an instance, is presented in ROSS et. al. 1980

[10] By 'sensory occlusion' we mean that a constraint upon the observability of a measurand has been identified by ; a) domain of sensing and b) a bound on the parameters of the constraint.

[11] If the receiver is simple enough there is little penalty for <u>not</u> implementing an information-redundant system.

[12] functions that distribute probability mass amongst a number of hypotheses, here H0 and H1 .

[13] the set of hypotheses

[14] Applications of Computing in Manufacturing Engineering.

[15] Thorn-EMI Central Research Laboratories, Dawley Road, Hayes, Middlesex, UB3 1HH.

Development of End–Effector Sensors for Industrial Robots

T.N. Moore, Ph.D., P.Eng., J. Jeswiet, Ph.D., P.Eng., R.L. Ten Grotenhuis, and W. Nshama

Department of Mechanical Engineering, Queen's University, Kingston, ON, Canada

1 INTRODUCTION

The introduction of the present generation of robots into manufacturing processes has certainly resulted in impressive productivity gains in many industrial sectors. However, it has become apparent that significant improvements must be made in the versatility of such industrial robots. There are two problems in robot control and sensing. The first problem: The capability of control equipment has outstripped the ability of engineers to apply it. The second problem: The capability and sophistication of control equipment has far outstripped the reliability and accuracy of the measurements they rely on. More accurate, reliable and innovative sensing systems will simplify control problems and make robots more versatile.

A variety of sensors which can be used in robots exist at present. Position sensors are available to measure joint position. Sensors for measuring force and torque by means of strain gauges have existed for some time. A wrist sensor, with three axes of force sensing and three axes of torque sensing, is available for use in assembly operations. Proximity sensors using ultrasonics or lasers are also being tested and made available to the robot industry. Computer vision has also made great strides and very sophisticated systems will be available for the next generation of robots. Although many sensing devices are now available there is still a need to expand the different types available.

Two additional types of sensors are described in this paper, one a tactile sensor for simulating human touch and the other a force/temperature sensor for metallurgical environments.

2 SENSOR DEVELOPMENT

At Queen's Mechanical Engineering, work has been undertaken to develop two different types of sensors which will increase the versatility of robots in manufacturing processes. One sensor may be placed in the tactile sensor category and the other is a combination force/temperature sensor. These sensors will be available for the next generation of robots which require the integration of enhanced proximity, range, tactile, force, temperature and vision sensors into their control systems.

In the research program at Queen´s the sensors are used with a 115 Kg capacity, cylindrical coordinate, hydraulic robot shown in reference (1).

3 CARBON-FIBRE-BASED TACTILE SENSOR

Tactile sensors have the potential to greatly enhance the capabilities of robot systems since they would allow the robot to "sense" the surface features of an object, its orientation, and the force being applied to it. Such attributes would greatly improve a robot´s ability to assemble intricate and delicate apparatus.

This section of the paper will describe the preliminary work undertaken in the development of a carbon-fibre-based tactile sensor. To begin, the design requirements for such a sensor are discussed, the properties of carbon fibres are reviewed and, finally, the construction of the carbon-fibre-based tactile sensor is detailed together with a summary of preliminary test results.

3.1 The Harmon Ideal

Harmon (2) has compiled the results of an extensive survey regarding various aspects of tactile sensors. In the survey, he gathered the opinions of those in the manufacturing and robotics industries as well as researchers. The analysis of the responses yielded a detailed description of what was thought to be typical of a practical sensor. Harmon also classified various categories of end use and described the characteristics of sensors suitable for the associated tasks. From this an outline of the desired tactile sensor characteristics was created.

Many of the desired tactile sensor characteristics were selected as design targets for the subject tactile sensor. In particular, minimum hysteresis was desired (certainly no greater than the quantization error of the analog-to-digital conversion stage), a minimum dynamic range of at least 1000:1, a threshold sensitivity of $2gm/cm^2$, a bandwidth of 100Hz or better, and a spatial resolution of 1 to 2mm.

Other design criteria not directly related to the mechanical specifications include wear characterstics, environmental stability, repeatability, drift, and need for calibration. In the manufacturing environment, these characteristics often define the utility of the sensor. This is especially true for tactile sensors as they are often operated in harsh conditions.

Numerous materials have been utilized to construct tactile sensors inlcuding conductive elastomers such as anisotropically conductive silicone rubber (ACS) and polyvinylidene fluoride (PVF2) (3,4,5-7) in various configurations.

These materials have shown promise; however, they do have inherent drawbacks. Nearly all have a restricted dynamic range and are plagued by lack of repeatability while others such as ACS demonstrate appeciable hysteresis. Other problems have included lack of robustness and degradation with time and heat.

Other technologies that could serve in the tactile domain are capacitive and magnetic type sensors (8,9), the Induced Vibratiion Tactile Sensor (10), the thermal-based tactile sensor (11) and the optical profile sensor (12). The relative strengths and weaknesses of these sensors are discussed in Reference (13).

3.2 Carbon Fibre

Another potential material for tactile sensor construction is pyrolyzed aramid fibre, better known as carbon fibre. Larcombe appears to be the first to have used carbon fibre as a sensing medium (14).

It can be shown that many of the characteristics of carbon fibre as a sensory medium are directly related to the fibre geometry as opposed to chemical or electrical effects. The fibres have a diameter of 7 to 30 microns, high strength and a high modulus of elasticity. Carbon fibre also conducts electricity. When pressure is applied to a group of fibres, they begin to elastically deform, the contact area between the fibres increases and the inter fibre resistance decreases. As the pressure increases, the resistance drops off in a non-linear fashion.

The high strength of these fibres results in a very wide dynamic range: Larcombe reports 4 to 5 decades. Fibre that is useful for tactile sensors is sold in either a ribbon form of 2-5 mm wide by 1 mm thick, or in a woven cloth, 50 mm wide by 1 mm thick, much like fibre glass mats. The ribbon form, called a "tow", usually has six thousand to ten thousand individual fibres. These fibres are subject to noise under low load conditions. The phenomena is a result of seismic and microphonic disturbances of the fibre junctions. As the number of fibres increases the noise drops off due to statistical averaging, the overall resistance increases as the junctions act as resistors in series.

3.3 Prototype Sensor

On the basis of these results, it appeared that carbon fibre was a good candidate for further sensor development. It was therefore decided to design and build a carbon-fibre-based tactile sensor.

A relatively simple prototype sensor was conceived to permit exploration of the effects of matrix thickness, orientation and geometry. In addition, it was desired to investigate various hardware aspects such as construction techniques, analog-to-digital conversion, filtering, multiplexing and layout of the element array. A schematic of the sensor is shown in Figure 1.

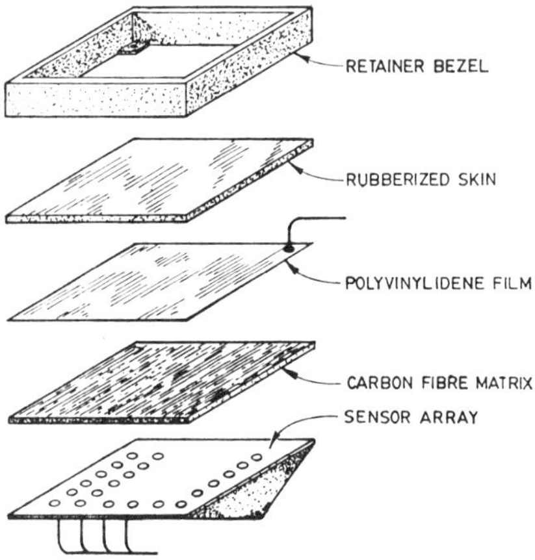

Fig 1 Exploded view of the carbon-fibre -based tactile sensor.

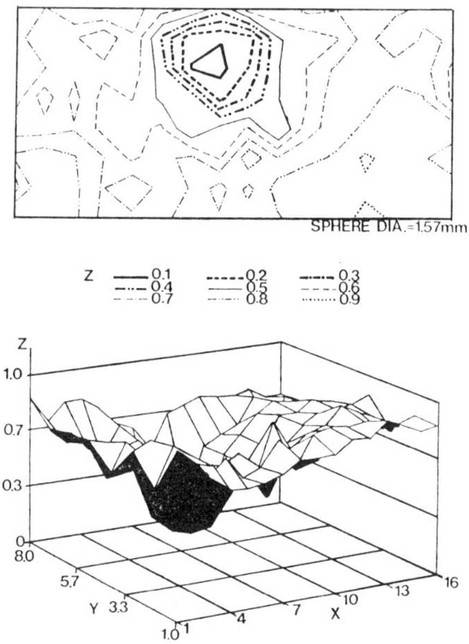

Fig 2 Graphical presentation of the sensor output.

The sensor was constructed from printed circuit board and solid strand 18 gauge insulated copper wire. A pattern was drilled into the PC board of 16 columns by 8 rows giving an array of 128 elements. The matrix was based on 1.524 mm centres. The resultant dimensions of the array are 23.88 mm x 11.68 mm. The wire was cut to short lengths, stripped of some insultation and inserted in the PC board. After this operation was complete, the wires were bonded by using low viscosity epoxy and a vacuum technique to remove air bubbles. The epoxy lends mechanical strength to the board to prevent damage under load. The resultant block was ground, filed and sanded to expose the circular cross section of the wire inserts at the board surface. At the back of the board, instrumentation wire was soldered into place.

At this time, a section of surgical rubber glove is used as a protective skin above the electrode. The entire assembly is held in place by a retaining rim which screws to the back.

The output from the sensor is sent via ribbon cable to an A/D converter board mounted in a 8086-based microcomputer. The A/D converter has a 12 bit resolution and a throughput to memory rate of 35kHz. This permits a reading of all 128 sensor outputs in approximately 3.7 ms or at the rate of 270 Hz.

At the present time the information from the sensor is read to memory and subsequently simple processing of the stored information is undertaken. This is not a "real-time" process, although work is progressing to speed-up processing and image display. Typical graphic images are shown in Figure 2 of a 1.57 mm diameter steel ball pressed onto the sensor.

The z-coordinate in Figure 2 is a normalized parameter representing the relative magnitude of the pressure applied at a given location on the sensor. In this instance the lower the value of Z the higher the applied pressure. The top illustration represents the sensor output data as a contour plot while the lower illustration presents the same information as a surface plot. The data shown has not been smoothed or "thresholded", although work is underway to determine if this would be advantageous and, if so, what form such manipulation should take.

Future work will include study of deflection versus load characteristics, shear sensitivity, temperature effects, frequency bandwidth, hysteresis and spatial resolution. In addition, other researchers will be investigating the potential for automated object recognition using the sensor's output data.

4 FORCE/TEMPERATURE SENSING GRIPPER

Gripper force/temperature sensing devices which may be used in heavy industrial, metallurgical environments are also being developed. A review of the literature shows that most activity in robot gripper sensors is in the area of tactile sensors and their application to assembly environments, indeed one well funded researcher has the objective to develop a robot which can be used in a kitchen environment, using the latest tactile and vision systems (some might consider this a harsh environment). However, there appears to be a dearth of information on sensors which can be used in heavy industrial/metallurgical environments, even though robots have been used successfully in foundry environments to grind flashing of castings. To increase the use of robots in the heavier industrial environments more rugged sensors are required.

The idea for a force/temperature sensor was recently described at the 6th International Conference on Robot Vision and Sensory Controls (15). It was proposed to use the principle espoused by Kannel and Dow (16). In rolling lubrication studies they used titanium, a metal that is temperature sensitive and relatively pressure insensitive, and manganin, a metal that is pressure sensitive while being relatively temperature insensitive. For the robot sensor it was first thought a grid of each metal, titanium and manganin, could be vapour deposited upon an insulating layer that had been vapour deposited upon a metal substrate. After a series of vapour deposition trials it was found that the insulating layers, suggested by Kannel and Dow (16), were inadequate because they were either too brittle or the resistance between metal substrate and vapour deposited metal was too low. As a result a ceramic, alumina, has been substituted for the metal substrate. The grid shape has also been simplified for the first part of the sensor development. The advantage of the ceramic substrate is it has high hardness, has a high temperature resistance and vapour deposited metals will bond to it. However the disadvantages are the ceramic is very difficult to cut, shape and polish, it does not conduct giving a very poor heat sink, and it is brittle.

4.1 Sensor Details

The vapour deposition process was not as simple as was originally envisioned and it took over twenty iterations, at eight hours per iteration, to achieve a satisfactory vapour deposited strip, each of titanium and manganin. The final experimental sensor is illustrated in Figure 3. The surface layer which protects and insulates the metal grid is also alumina. Once the sensor system has been fine tuned a more complicated grid will be used. For the vapour deposition process to be successful it was found the ceramic surface had to be highly polished and even more important the cleanliness of the ceramic surface had to be perfect.

A cement was used in the sensor to 1) attach the metal and ceramic substrates to each other, 2) keep the platinum lead wire in position in the ceramic substrate, 3) keep the porcelain insulation tubes in position. Finding a cement which was compatible with the vapour deposition process was also time consuming and included materials such as muffler cement in the search. The cement chosen appears to be working well so far.

A problem also arose in getting the vapour deposited material to adhere to the copper lead wires. A platinum paste was used with some initial success, but after a bit of use the copper and platinum became detached. This was probably due to oxidization of the copper at the 900 degree celcius firing temperature which was used. Nichrome and chromel wire were also tried but in all cases there was a high contact resistance at the joint. A combination which was finally found to give good adherance with low negligible contact resistance was platinum wire leads with gold as the connecting metal between the wire lead and the vapour deposited strip.

4.2 Gripper/Sensor System

The kinematic gripper illustrated in reference (15) has been built and the force/temperature sensor has been mounted on it and is presently undergoing tests. Figure 4 shows the current version of the gripper-sensor assembly.

The gripper will be operated by two pneumatically operated cylinders which are supplied by a pressure control valve which, in turn, is controlled by a Zenith computer. The computer will also be connected to the force/temperature sensor - the sensory feedback. The pressure control valve has its own feedback via the RS-232

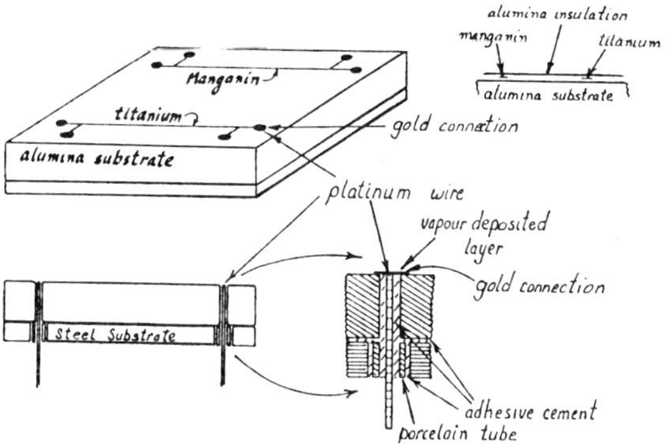

Fig 3 Details of the Force/Temperature sensor.

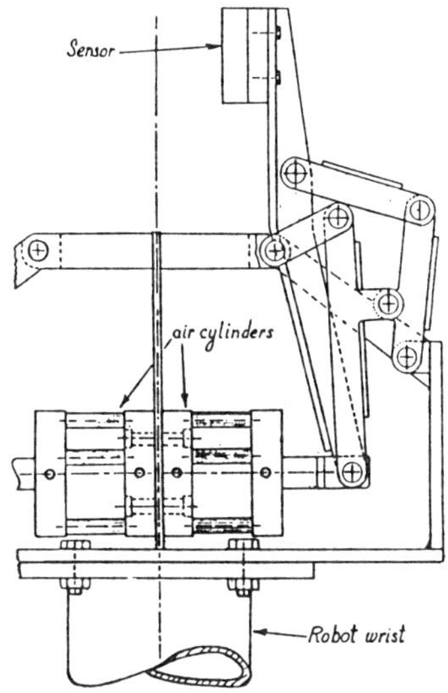

Fig 4 Details of the kinematic gripper-sensor.

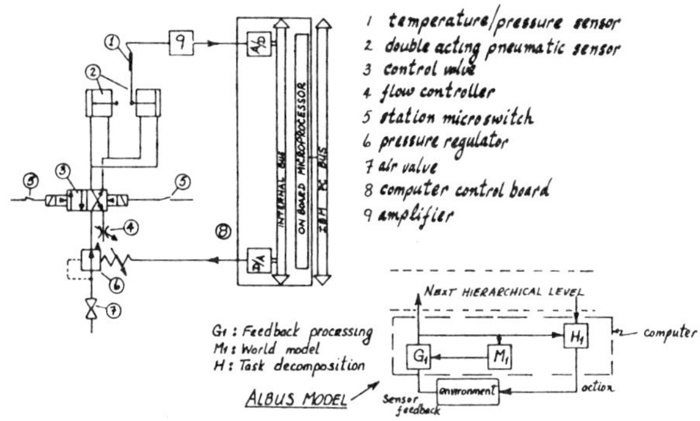

Fig 5 Details of the sensor-gripper feedback system. Also shown is the Albus computer-controller model.

link and this can be used as a check on the sensor which is being developed. The system is illustrated in Figure 5. It may be seen that this system fits into the computational hierarchy suggested by Albus (17) as shown in the same figure. This system, including the sensor, is being installed at the time of writing and the results will be available at the conference.

5 CONCLUSION

It may be seen that once development of these sensors is complete, two valuable feedback tools will be available for industry, one for fabrication and the other for the metallurgical industry. These sensors will contribute to increasing robot versatility and the necessity for increased productivity and quality control.

REFERENCES

(1) J. Jeswiet and T.N. Moore. Meeting the Robotics Challenge. 5th Canadian CAD/CAM and Robotics Conference, 1986, Toronto. SME Technical Paper No. MS86-727.

(2) Harmon, L.D., Automated Tatile Sensing, Intl. Jour. of Robotics Res., 1982, 1, No. 2, 3-32.

(3) Hillis, W.D., A High Resolution Imaging Touch Sensor, Intl. Jour. of Robotics Res., 1982, 1, No. 2, 33-44.

(4) Robertson, B.E. and Walkden, A.J., Tactile Sensor System For Robotics, Proc. 3rd Intl. Conf. on Robot Vision and Sensory Controls, Cambridge, Mass., 1983, 327-332.

(5) Sessler, G.M., Piezoelectricity in Polyvinylidene Fluoride, Jour. Acoustical Soc. of Am., 1981, 70, No. 6,

(6) Brown, T.D. and Maratori, D.R., Miniature Piezoresistive Transducers For Transient Soft Body Contact Stress Problems, Experimental Mechanics, 1979, , 214-219.

(7) Purbrick, J.A., A force transducer employing conductive silicone rubber, Proc. 1st Intl. Robot Vision and Sensory Control Conf., 1981, 73-77.

(8) Boie, R.A., Capacitive Impedence Readout Tactile Image Sensor, IEEE Intl. Conf. on Robotics, Atlanta, 1984, 370-378.

(9) Hackwood, S., Beni, G. and Nelson, T.J., Torque Sensitive Tactile Array For Robotics, Proc. 3rd Intl. Conf. on Robot Vision and Sensory Controls, Cambridge, Mass., 1983, 363-369.

(10) Nevill, Jr. G., Doty, D. and Schildwachter, E., Elternative Skin Geometries and Materials For Induced vibration Touch Sensors, Sensors '85, Detroit, Mich. Soc. Manufacturing Eng., 1985.

(11) Russell, R.A., A Thermal Sensor Array to Provide Tactile Feedback For Robots, Intl. Jour. of robotics Res., 1985, 4, No. 3, 35-39.

(12) Beni, G., Hackwood, S. and Rin, L., Dynamic Sensing For Robots - An Analysis and Implementation, Proc. 3rd Intl. Conf. on Robot Vision and Sensory Controls, Cambridge, Mass., 1983, 249-255.

(13) Ten Grotenhuis, R.L. and Moore, T.N., A Carbon-Fibre-Based Tactile Sensor, Proc. Machine Monitoring Sensors (For Untended Manufacturing), Chicago, 1986, SME Tech. Paper No. MS86-139.

(14) Larcombe, M.H.E., Carbon Fibre Tactile Sensors, Proc. 1st Intl. Conf. on Robot Vision and Sensory Control, Stratford-Upon-Avon, U.K., 1981, 273-276.

(15) J. Jeswiet and W. Nshama. A Robot Force/Temperature Sensing Gripper for Heavy Industry. 6th International Conference on Robotic Vision and Sensors, IFS(conferences), Paris 1986, pp 152-156.

(16) Kannel and Dow. The Evolution of Surface Pressure and Temperature Measurement Techniques for Use in the Study of Lubrication in Metal Rolling. ASME Paper No. 74-Lubs-7.

(17) J. S. Albus, C. R. McLean, A. J. Barbera and M. L. Fitzgerald. Hierarchical Control for Robots in an Automated Factory. Robots 7, 1983, pp 13.29 to 13.43.

A Management Perception—Need for Quality Control

Dr. S.P. Puri* and Suran N. Dwivedi†

*Chairman, Division of Business and Social Sciences, Barber-Scotia College, Concord, NC 28025, USA
†Associate Professor, College of Engineering, University of North Carolina at Charlotte, Charlotte, NC 28223, USA

Quality Control is very significant to the manufacturing concern because it relates to customer satisfaction.

The quality of the product is the key to the success and increased profitability for any business organization. Additionally the customers may also influence people's purchases of the product by telling of the problem with the product.

Quality Control is conducted by machines, humans, or both. The inspection in quality control by the human being is not 100 percent effective. The human error is caused by monotony, fatigue and other job-related factors.

As technology advances manufacturing firms turn more and more to machine inspection for quality control. The computers make highly accurate inspection possible in automated manufacturing process. For example, machines that produce wire, are often attached to gauges that measure the thickness of the wire, as it comes out of the machine. When the wire is too thick or too thin, this information is electronically relayed to the computer, which makes adjustments in the machine through electronically controlled devices.

Certain types of inspection are not possible except by human visual inspection. Time, cost, human fatigue and other factors often result in human inspection being performed on a sample basis.

All areas of production are important for quality control; and U.S. industry now is working hard to meet the new world standards set by the Japanese. America's leadership in quality has been going down for years, reports one writer:

"There's no question that the Japanese have set new world standards," says Robert E. Cole, who has worked in Japanese factories and now directs the University of Michigan's Center for Japanese Studies. "Their best factories are better than our best factories."

There has been a lot of duscussion and reference to quality control circles and zero defects. Although the concept of zd was initiated in the U.S. about a quarter of a century ago, most western countries have installed zd programs and are working well. The quality control slogan has captured the favorites of all progressive manufacturers. The Germans and the Japanese have recognized not only the value of quality control but also the value of team work and close relationship between function and organizational levels. There is growing evidence of the need, importance and the opportunity for ongoing worker-management cooperation to achieve the best results. According to Professor Robert Cole at the University of Michigan most Japanese consumer electronics firms have established active quality control programs in Japan in 1981. He contacted five Japanese owned television plants in U.S. to find out whether they are using such circles. Somewhat unexpectedly, he found that none of them have initiated QC circles for various reasons. This suggests lack of progress in the U.S. operation.

However there is plenty of evidence which support the benefits of such programs. The Ford Motor Company, which involves their employees in similar programs, found that workers role expectations and relationships with the management have changed dramatically in several plants. There was also significant change in Ford management's attitude toward workers, and their appreciation of the contribution they can make.

As we review that the Germans and Japanese manufacturing firms we find that they have taken a strong position for quality control, the U.S. has no choice but to follow them if they have to be competitive. There are no short cuts along the road. The sacrifice and the investment in quality control has to be made to achieve a substainable competitive advantage, once gained, will remain unassailable, even for a short time, is totally incorrect.

Many U.S. companies such as Zenith Radio Corporation take pride in quality control and use the slogan "The quality goes in before the name goes on." The QC is the attempt to assure both management and the customer that its product have been designed, produced, and marketed to meet high standards. The U.S. companies' efforts for the quality control information from the engineering production, and marketing point of view needs more elaboration. Certain standards must be maintained in production process. For example, a General Motors assembly line that produces Buick automobiles must use uniform parts that meet certain specific quality standards. Likewise the assembly parts must fit and work together correctly.

Dr. J.M. Juran who is internationally known as an expert on quality control has written several articles on Japanese and western quality control methods. He has compared the quality of the products and has identified many differences while evaluating the specific merchandise.

Although he has commended the western innovation particularly in the production of TV such as remote controls, random access, automatic tuning, smaller bulk, etc--along with degree of just gadgetry, but at the same time he has indicated that the western countries have two to four times the failure rate of the Japanese sets.

Another important person who is also well known in quality control circle is Dr. W. E. Deming. His speciality--statistical quality control--can be indispensable in modern manufacturing. Deming also places the blame in the US quality control on the management and not the workers.

According to him the management is more financially oriented and robs the hourly worker of his right to indulge in his quality of work. The Texas Instrument Company has started incentive of pay increase and bonus to their executives by the quality of the products their operations turn out. The Chrysler Corporation has also placed more emphasis on quality control. "All the plant managers who's operation miss a prescribed quality target for a month, must answer to a committee of executives. All these programs are an indication of the sense of urgency and need for QC." This growing emphasis on quality strategy results from mounting evidence that quality has replaced price as the key to improving market share and profit margins. In a recent study of some two-thousand business units by Strategic Planning Institute of Cambridge, Mass., suggests that improving product quality is an effctive way to gain market share. They also found out that this result five times greater profit margin. Mr. John Welch Jr., Chairman of General Electric Company made a statement: "Quality is our best insurance of customer allegiance, our strongest defense against foreign competition, and the only path to sustained growth and learning."

No program can be effective unless there is a complete commitment for the quality control program by the top management. However the main responsibility for meeting the quality targets will be vested with the line managers.

However there are few other things which need to be stated regarding the responsibility of quality with the quality control department. It is obvious that the US companies want to improve the quality to match with the Japanese but they do not place responsibility in the hands of production department. The present system of placing the responsibility of quality with the quality control department need to be discontinued. They must follow the Japanese method by placing the quality control in the hand of production department. The Americans are very fond of reorganizing and responsibility. In the event they decide to reassign a responsibility, they change several standards, measures, and controls to assure the preformance of change of responsibility. Our studies indicate that it is not necessary to tell the manager with the responsibility of performance. All that is necessary is to provide incentives and the measures of performance which indicate the rewards for achieving certain levels of performance. Simply stated, it means that the reassignment of responsibility for quality to the product department will improve the quality of the product.

In reality most Japanese have learned about the quality control from western quality control techniques and concepts. Most Japanese have invariably refer to the works of two Americans, Joseph M. Juran and W. Edwards Deming. These persons have influenced the quality control concept throughout the world. They have been consultants to various countries, they have also visited Japan as lecturers several times. Most Japanese think Deming as the father of Japanese quality control. In fact a prize has been named after him which is given to the Japanese manufacturers exhibiting the most impressive quality performance. This award is nationally televised.

The Germans who thought that their position was impregnable in their methods for high quality products, found that the Japanese imports were running about 10 percent of their market in 1980's and were threatening to inundate them. Most Americans had also a very poor view of the Japanese product but both these countries found a stunning reversal of their thinking of the Japanese product. It may however be pointed out that it took Japanese manufacturers twenty-five years of hard dedicated work to achieve this position. The high quality is not something that results from a few management decisions; it requires the creation of complex, all· incompassing, interacting, "Systems". The development of such systems require the commitment of top management. The Japanese just do not inspect the quality of the product but they build in it. The Japanese management is very concerned and they always thinking of quality in planning of the product design stage. Interminable discussion between engineering, production, quality, assurance and marketing take place before and design is finalized. They believe that once production begins the standards must be maintained. They make sure that thinking quality in training workers so that they become capable of delivery high quality. The workers make sure the parts are defect free. They work meticulously, knowing that any defect will be spotted and they will be responsible. Besides they also take a great pride in high quality product which is constantly reinforced by management attitudes and actions.

Although knowing fully well that the Japanese use Quality Control Circles widely, but however, Japanese subsidiaries in western countries do not use quality control as much as in Japan. This is not on the priority list. Quality Control Circles may have only a modest effect however, the workers can solve at least 15% of the problems and the rest must be handled by the management of the engineering staff. Most of the Japanese clan have an impression accord of their contribution to quality improvement, method improvement, morale-enhancing improvement and to worker motivation. There is no question that Quality Control circles during the 1960's era have begun to make helpful contribution year after year. The Quality Control Circles also has a tendency to bring people with common interest together which leads to natural formation of people with common interest. These informal groups of workers can relate to common problems which can minimize hinderance in the production. These valuable sources of employees' involvement suggest that the Quality Control will be very beneficial. These may not amount to significant improvements but they are responsible for bringing the last defects out of a production system. Years of training and plant-wide effort in accordance with the total Quality Control concepts has contributed to Japan's cycle of rapid and continual quality improvement. In fact Quality Control Circles formed on such a premise could be very beneficial in the real Quality Control Program.

Quality Circles are not the same as work simplification, nor are they identical with either zero defeats or employee suggestion programs, but the resemblances are close. The Quality Control Circle is basically a formally organized group of workers and their procedures are usually structured although details may vary depending upon the objectives that are being emphasized. Methods and quality objectives are generally important in Japanese Quality Control Circles.

However, the Japanese were aware after the second World War II that their goods and merchandise are considered of low quality. They determined to make special efforts to improve their image of qualtiy products. They immediately started to rebuild and recover the sufferings caused by the U.S. through plant shutdowns and losses of the whole industries resulting in social and economic upheavals. This was also a serious blow to their prestige which was the main cause of making changes and becoming compatible in the world market. The Japanese labor force was more interested in building Japan economically. The government was also very supportive of these efforts. As a result the Japan's wage advantage and government's supportive action produced very favorable economic results. Their success sent shock waves to industries in the western world, particulary United States

The big automakers such as GM and Ford started looking into the causes for Japan's success They hired consultants to crisscross the country of Japan and do marketing survey so that they can find reasons or to catch up with the Japanese management development program on Japanese techniques in every city.

The Japanese also got started on their Quality Control Crusade by using the knowledge from the american experts such as Deming and Juran, and a host of total quality control concepts and techniques. In fact the Japanese Quality Control System to seize this opportunity to start their own. There has been other factors relating to the institutions and the structural differences between Japan and United States, relating to personal savings, private investment, technical literacy, antitrust laws, government industry cooperation, and employment mobility. There are also other factors which favor the tax benefits to the Japanese industrialist as compared to the United States. The Japanese tradition of life time employment nurtures dedicated employees.

Another way of dealing with the quality control, is to discuss quality problems. In U.S. companies the production worker does not talk with quality control personnel; as a result, potential problems are not brought to the surface. As a result defects are shunted off to be reworked, and the pressure to meet the deadlines make it psychologically difficult for quality inspectors to delay the process because of minor quality problems. This would not be permitted in Japan. Feedback from production workers, quality assurance workers, salesmen, vendors, and customers are encouraged. The field service personnel often report to the manufacturing manager rather than to the sales manager, as they do in most United States companies. Cross-functional training are common in Japan for the attainment of quality control.

The Japanese make intensive effort to screen incoming parts and material. In fact they inspect 100 percent and feedback the results to the suppliers. This puts a tremendous pressure on the suppliers to improve the quality. The Japanese not only put pressure on the supplier they also work with them to try to ascertain why problems have arisen in the organization and help them develop approaches to solve them by conducting seminars and workshops. This system reinforce their skills and expectation.

The American concept of acceptable quality is not adopted by the Japanese, or by the Germans, the Japanese believe that no defect is acceptable. In one of the famous speeches of Mr. De Gaulle he referred that Japanese make a virtue out of necessity. A senior Japanese official once remarked that one percent defect rate may sound okay in U.S. but it is not acceptable in Japan. It simply means that if one sells ten-thousand units of a product, one hundred of them will be defective. One group of senior executives of a large American company, who studied the relationship between productivity and quality in plants around the world has observed that the defect-free operations promote product improvement. They have also found that a two percent reduction in defects is usually accompanied by a ten percent increase in productivity. This simply sugget that fewer defects mean more output without increase in cost. This also indicates the removal of defective items can eliminate many problems and will cost less to manage. Simply stated it will decrease waste and inventory. The morale of the employees go up and everybody feels comfortable and proud of the products.

Conclusion:

A. The top management should become serious in QC.

B. There must be a total commitment and of massive of training programs for the first line and middle management. They must understand the new techniques which can help them to find out the problems which causes defects in the products. This will help them to eliminate the problems.

C. There should be more coordination and cooperation in manufacturing of more reliable products.

D. Incentives should be provided for promotion, pay increases, bonus etc. for improving quality control and achieving the goals and objectives.

Selected References

Anon. 1982. Quality: a special report. Reprint from Business Week, (Nov. 1, 1982 issue).

Anon. 1981. Japan's strategy for the '80s. Reprint from Business Week (Dec. 14, 1982 issue).

Cole, R.F. 1980. Learning from the Japanese: Prospects and Pitfalls. Management Review, (Sep):22-28.

Deming, W.E. 1981. What Top Management Must Do. Business Week, (Jul. 20):19-21.

Hayes, R.H. and Wheelwright, S.C. 1984. Restoring Our Competitive Edge: Competing Through Manufacturing. New York: John Wiley & Sons.

Juran, J.M. 1979. Quality for Better Product Assurance and Reliability. Reprint from Quality, 027F (Jan.-Feb. issues).

Leonard, F.S. and Sasser, W.E. 1982. The incline of quality. Harvard Business Review, 60,5:163-171.

Schonberger, R.J. 1982. Japanese Manufacturing Techniques. Nine Hidden Lessons of Simplicity. New York: Free Press.

Digital Image System Software for Automated Part Identification

R. Radharamanan and Hong Shen

Department of Mechanical Engineering, San Diego State University, San Diego, CA 92182, USA

ABSTRACT:
In this paper a software model has been developed on image processing for a digitial system which contains a video camera, digitizer, and an IBM XT personal computer. The software includes noise reduction and signal enhancement to reduce the environmental influence of the vision area, then edge detection and part identification to extract and characterize the image. Through this modelling system, the original binary 2D projection image data coming from the camera through digitizer can be processed for providing output of objects' size, shape, position, and orientation which can be used to make the decision about the part being inspected and commands the robot, which is connected to the system, to pick up the object at the predetermined location. This developed computer model will increase the flexibility of the stiff, ponderous robotic arms and makes them suitable for industrial operations.

KEY WORDS: Image processing, vision system, edge detection, part identification, computer model, robot.

1. INTRODUCTION

In recent years, low cost machine vision system is widely applied as an improvement to robots so that they can be used more in manufacturing environment. The technical problems that directly affect the robot performance include: accuracy, dynamic performance, sensors/vision, control systems, world models, robot software, system integration, and mobility [10]. As part of the solution to the above problems, one of the research projects in our robotics laboratory deals with the vision controlled robot which is used for automated part recognition. An industrial robot is interfaced with an IBM personal computer and the conveyor system with automatic stop device, which in turn connected to video camera, digitizer, and interface through the computer. The object moving on the conveyor will pass under the video camera before stopping at the stop device. The image sensors of the camera, scanning the object, transform the light they sense into computer readable signals. The digitizer and the interface attached to the camera aid in image processing (Fig. 1).

To develop a software model for the low cost machine vision system which is expected to detect, locate, and identify objects in any position or orientation within the viewing area, two important criteria should be met: one is to save the computer memory, and another is to accelerate the response time. Several techniques have been developed for these purposes on different systems. It has been noticed that most of them use edge detection since this algorithm organizes significant local intensity changes into more global abstractions. Yet the problems encountered with local edge operators are widely known [2]. Even in the binary image, the small spatial extent of the operator relative to the events they are designed to detect is still a big problem. Therefore, algorithms of searching confined area where object exists are necessary. Only after the subregions are found, the successory process, like noise reduction, boundary extration etc., will be performed effectively.

In this paper, the problem cited above is solved by dividing the viewing area into subregions, then tracking the boundary into the region until a closed contour is finished. A software model has been developed which includes noise reduction and signal enhancement to reduce the environmental influence at the vision area using binary algebra operators. Since the data of the boundary is stored in chain structure, the noise reduction operation is processed only on the boundary. This will significantly reduce the computation time [11], provided that the noise-to-signal ratio is small. For part identification, several geometric properties are calculated and compared with standard image data; generally, these properties are: centroid, orientation angle, area, and perimeter [3]. In the program developed, all these properties are easily obtained through boundary data. By simulating the modelling system, the original binary 2D projection image data coming from the camera through digitizer can be processed for providing output of objects' size, shape, position, and orientation, which can be used to make the decision about the part being inspected and commands the robots to pick up the object by then waiting at the stop device and place it at the predetermined location.

2. THE SOFTWARE

2.1 General System Software:

The software package allows the user to execute machine vision system. The sensor based camera head is connected via differential interface to the controller board, which employs a bit slice processor, plus a microcontroller to implement the host system interface. The IBM XT compatible PC moves the image data to system memory by Direct Memory Access (DMA) transfer. Software routines implement a high level interface to all camera functions. After the coordinates of the objects are stored into memory, the succeeding phases including subregion dividing, object detection, edge detection, object location, and object identification are processed. The assumptions about the objects in the viewing area are:

1. all the objects are significant to a certain degree of magnitude;
2. the objects contain no holes;
3. the objects are static when photographing.

2.2 Subregion Dividing:

Given a viewing area {block (1, 1) - (2^k, 2^k), two pointers are tracking the middle column and middle row of the area (Fig. 2). Assume P(x, y) represent the intensity value for pixel (x, y). That is,

$$P(x, y) = \begin{cases} 1 & \text{if dark} \\ 0 & \text{if light} \end{cases}$$

If two consecutive values of P(x, y) = 1 are not found by the pointers, then the region being inspected is divided into two subregions and respectively the above procedure is repeated; otherwise, it is assumed that there is an object present in this block, then go to subroutine "object detection".

2.3 Object Detection:

Within a block ((startcol, startrow) - (endcol, endrow)), two pointers are used for tracking the object. Column pointer goes from left to right on certain row; row pointer goes from top to bottom on certain column. The column pointer works based on the following rules:

1. When tracking on ith row, if two consecutive values of P(x, y) = 0 are found then the pointer at (j+2, i) goes down to the (i+1)th row, backup two columns where the pointer starts at (j, i+1).
2. If the pointer goes down from (i-1)th row to ith row and stays at (j, i), then it goes to the left first, until it finds two consecutive values of P(x, y) = 1. Then it will go back to (j, i) and continuously searches at the right side.
3. When the pointer passes the boundary of the object, the program stores the addresses (j, i) if P(j, i) = 1.

Figure 3(a) shows how the column pointer works. The dots mean dark points, circles represent light points, while the crosses indicate the locations where the column pointer has passed. The row pointer works very similar to the column pointer. These two pointers together detect the whole contour of the object and save the data into memory [Fig. 3(b)].

2.4 The Structure of the Stored Boundary Data:

As shown in Fig. 3(b), the objects detected are stored into storage with only a single layer boundary. Array col(i), row (i) - i =1, 2, ..., n is used, where i indicates t he sequence order. Col (i), row(i) represent the column and row of ith data. To identify different objects, the numbers i = 1, 2, ..., n are divided into latest(j), and last(j) where j = 1, 2, ..., k (latest (j) is the starting data of jth object; last(j) is the ending data of jth object). To separate data in different rows, the number i = latest(j), ..., last(j) are divided into t parts, where $t = l_j, l_{j+1}, ..., ll_j$ and i = sub(t) is the starting data on t-th row. Whenever a new data is detected, it is stored by using a insertion sort, with the row number the first key, the column number the second key. If the same data have been saved previously, then the new data is cancelled to avoid duplication. Finally, the total data after tracing object should look like as shown in Fig. 4.

2. 5. Noise Reduction:

Noise in the viewing area produces effects such as irregular boundaries, small holes, missing corners and isolated points. To modify an intensity of a pixel (x, y), centered among 8-neighbors (Fig. 5), two boolean equations are evaluated at (x, y). They obey the following rules:

1. If majority of (x, y)'s neighbors are 1, then P(x, y) is 1;
2. If majority of (x, y)'s neighbors are 0, then P(x, y) is 0.

In the program developed:

$$p(x, y) = p(x, y) + p(x, y-1).\{p(x-1, y) + p(x+1, y)\} + p(x-1, y).p(x+1, y).\{p(x, y-1) + p(x, y+1)\} \quad (1)$$

$$p(x, y) = p(x, y).[p(x, y-1).p(x, y+1).\{p(x-1, y) + p(x+1, y)\} + p(x-1, y).p(x+1, y).\{p(x, y-1) + p(x, y+1)\}] \quad (2)$$

If these formulae are applied only at the detected boundary, it will save a lot more time than applying them in the whole viewing area. Since it is assumed that the objects are significant, this procedure is good for further image extraction.

2.6. Object Identification:

Several significant properties are used to identify the object. First the number of array which is used to store boundary intensity expresses the perimeter of the object. This is one of the most important features to distingush different objects. After that, the position and orientation of the object should be calculated in order that the robot can pick it up accurately (Fig. 6). The position is defined by the centroid ($\bar{x}$, $\bar{y}$). In the program developed, $\bar{x}$ and $\bar{y}$ are calculated by:

$$\bar{x} = \{\sum_{i=min.col}^{i=max.col} (max.\ row - min.\ row + 1)\ x_i\}/(\text{number of array}) \quad (3)$$

$$\bar{y} = \{\sum_{i=min.row}^{i=max.row}(max.col - min.col + 1)y_i\}/(\text{number of array}) \quad (4)$$

In order to find the orientation, the maximum principal axis of the object's projection area is calculated. The inertia along the vision coordinates are given below:

$$I_{xx} = \{\sum_{i=min.col}^{i=max.col}(max.row - min.row + 1)x_i^2\}/(\text{number of array}) - (\text{number of array})\bar{x}^2 \quad (5)$$

$$I_{yy} = \{\sum_{i=min.row}^{i=max.row}(max.col - min.col + 1)y_i^2\}/(\text{number of array}) - (\text{number of array})\bar{y}^2 \quad (6)$$

$$I_{xy} = \{\sum_{i=min.row}^{i=max.row} y_i \ .\ \sum_{j=min.col}^{j=max.col} x_j\}/(\text{number of array}) - (\text{number of array})\bar{x}\bar{y} \quad (7)$$

The angle counterclockwise from the row axis to the principal axis is :

$$\theta = (1/2)\tan^{-1}\{(2I_{xy})/(I_{yy} - I_{xx})\} \quad (8)$$

The flow chart of the computer program developed is shown in Fig. 7.

3. RESULTS AND DISCUSSIONS

The program developed has been tested by creating a graphic picture on the screen and using this picture to simulate a vision photograph. The pictures of original image and processed boundary image with or without noise are shown in Fig. 8. Noise points are generated by a random function Y = RND * (X), where X is any point within the field area and Y is the noise point. The calculated geometric properties are also shown in the figure. Table 1 lists the error of each calculated geometries for different noise ratio. The noise ratio is defined by (number of noise points)/(area of viewing). From the table it is seen that the program can not identify the object if the noise ratio is greater than 10%, but it works very well if the noise ratio is about 5%. Fig. 9 shows the graphs under three different noise levels. As the noise points increase, the boundary detecting method used in the program will gradually loose its accuracy. In practice, generally the noise level is below 5%. If there is a very rough background, it should be substituted before picturing the image. As for the geometric property, under 5% noise ratio, the centroid has a good accuracy. The relative errors of the perimeter and the area indicate that when two objects are compared, the standard image parts' characters should have greater difference than those errors due to noises. Further, it is tested by simulation that parts with different shape, size, and orientation can be predicted by this program accurately if there is no noise. It is also important to provide appropriate lighting facility to avoid rough background.

4. CONCLUSIONS

The program developed can identify any parts with different shape, size, and orientation on the clear background. It can identify parts with more than 10% difference in shape, size, and orientation, and less than 5% noise ratio on the background. The running speed and the storage is adequate for the IBM-XT computer and applicable for robot's real time working tasks. The method developed can be modified further in the directions:

1. increase the running speed by improving the data structure of the storage;
2. increase the flexibility by adding more functions which will enable communication between operator and machine;
3. work on real robot integrated environment.

5. ACKNOWLEDGEMENT

The authors are thankful to San Diego State University Foundation for the financial support provided through Grants-in-Aid for Research Program (1987-88) for developing the software discussed in this paper.

6. REFERENCES

1. Alliney, S., and Morandi, C., "Digital Image Registration Using Projections," IEEE Transactions on Pattern Analysis and Machine Intelligence, Vol. PAMI-8. No.2, pp.222-233, March 1986.

2. Burns, J.B., Hanson, A.R., and Riseman E.M., "Extracting Straight Lines," IEEE Transactions on Pattern Analysis and Machine Intelligence, Vol. PAMI-8, No.4, pp.425-455, July 1986.

3. Castleman, K.R., "Digital Image Processing," Prentice-Hall, Inc., Englewood Cliffs, New Jersey, 1984.

4. Forrest, A.K., "Robot Vision," Physics Technology, 17, pp.5-9, 1986.

5. Fu, K.S., Gonzalez, R.C., and Lee, C.S.G., "Robotics : Control, Sensing, Vision, and Intelligence," McGraw-Hill, 1987.

6. Goldstein, L.J., "Advanced Basic and Beyond for the IBM PC," Robert J. Brady Co., 1984.

7. Lunscher, W.H.H.J., and Beddoes, M.P., "Optimal Edge Detector Design I : Parameter Selection and Noise Effects," IEEE Transactions on Pattern Analysis and Machine Intelligence, Vol.PAMI-8, No.2, pp. 164-177, March 1986.

8. Lunscher, W.H.H.J., and Beddoes, M.P., "Optimal Edge Detector Design II : Coefficient Quantization," IEEE Transactions on Pattern Analysis and Machine Intelligence, Vol.PAMI-8, No.2, pp178-187, March 1986.

9. Ronse, C., and Devijver, P.A., "Connected Components in Binary Images : The Detection Problem," John Wiley & Sons Inc., 1984.

10. Roski, S., " Data Structure on the IBM PC," Brady Communication Company, Inc., 1985.

11. Unger, S.H., " Pattern Detection and Recognition," Proc. IRE, Vol.47, No.10, pp.1737-1752, 1959.

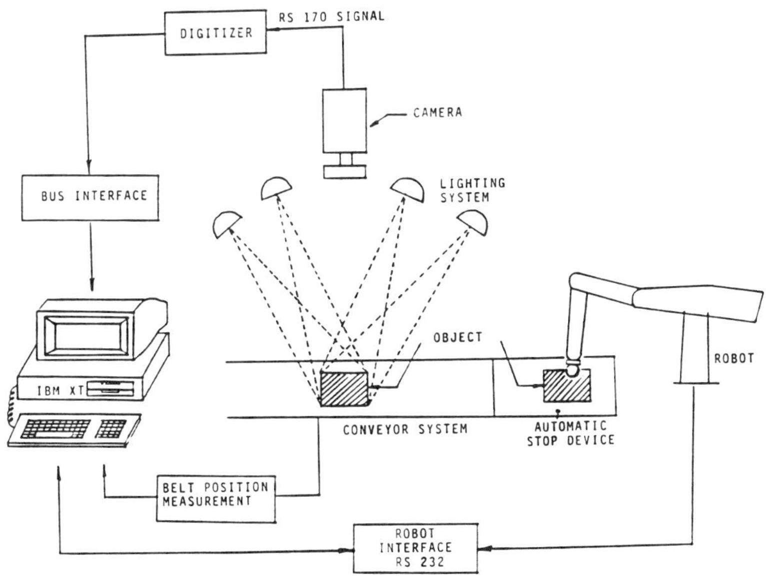

FIG. 1 VISION CONTROLLED ROBOT FOR AUTOMATED PART RECOGNITION

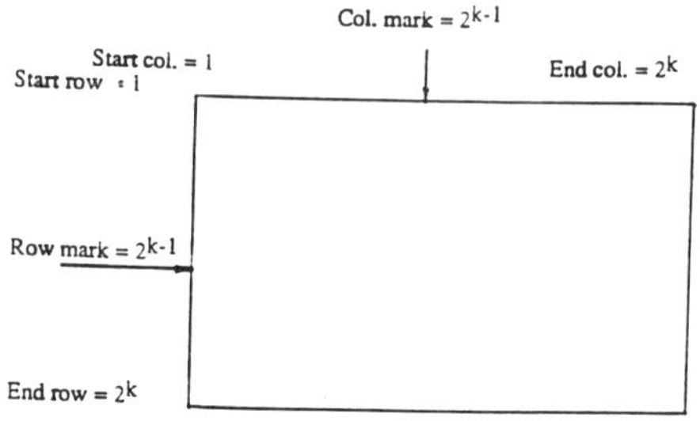

Fig. 2 Pointers searching object in subregion

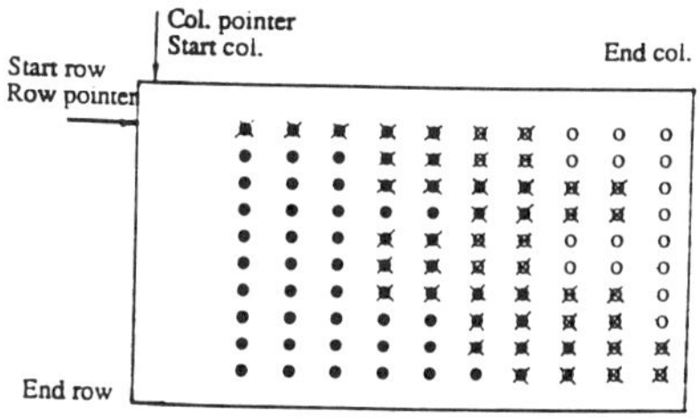

Fig. 3(a) Working algorithm of column pointer to detect edge of the object

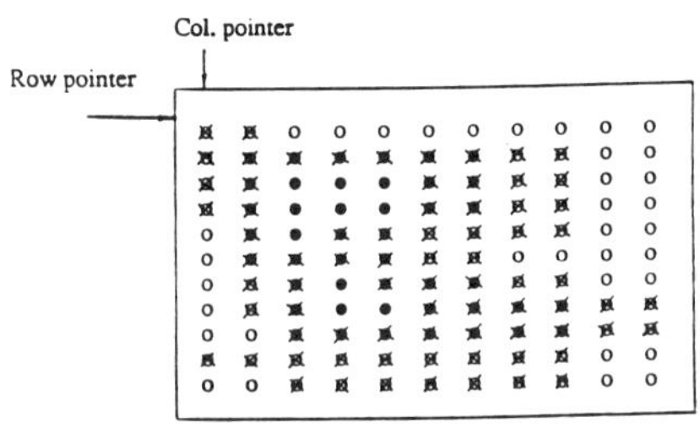

Fig. 3(b) Column pointer and row pointer detect the whole edge of the object

	t	Mini. col(t)	Max. col(t)
Mini. row	l_j	i=latest$_j$, latest$_j$+1, ...	
	l_j+1	i=Sub(t), Sub(t)+1, ...	
	....	....	
Max row	ll_j	i=Sub(t), Sub(t)+1, ...	i=last$_j$

Fig. 4 Data structure of the detected object's edge

x - 1, y - 1	x, y - 1	x + 1, y - 1
x - 1, y	x, y	x + 1, y
x - 1, y + 1	x, y + 1	x + 1, y +1

Fig. 5 8-neighbor's pixel (x, y)

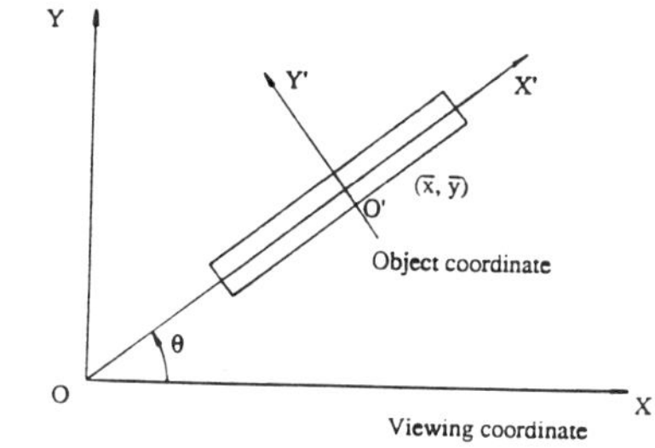

Fig. 6 Position and orientation of the object relative to viewing coordinate

Table 1. Error of each calculated geometries for different noise ratio

Noise level	50 points	75 points	100 points
Ratio	4.90%	7.30%	9.80%
$\Delta\bar{x}$	3.63%	15.06%	
$\Delta\bar{y}$	4.43%	7.68%	
$\Delta\theta$	18.90%	28.30%	
Δs	7.10%	16.70%	
ΔA	11.50%	47.50%	

Noise ratio = (number of noise points)/(area of viewing)

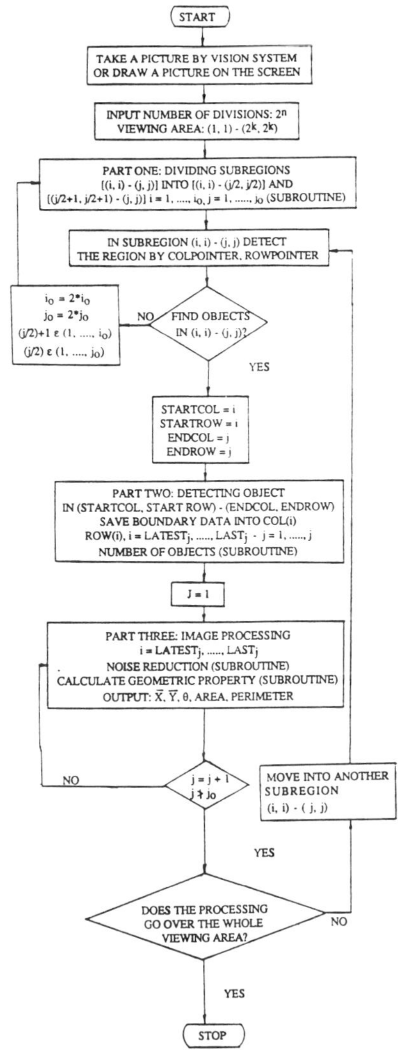

Fig. 7 Flow chart of the computer program

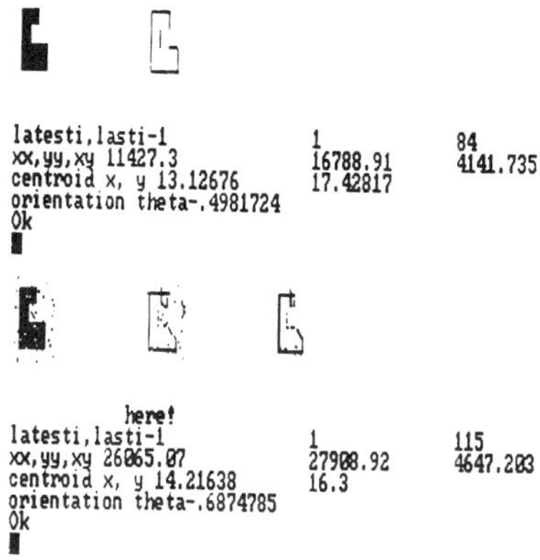

Fig. 8 The pictures of original image and processed boundary image with and without noise

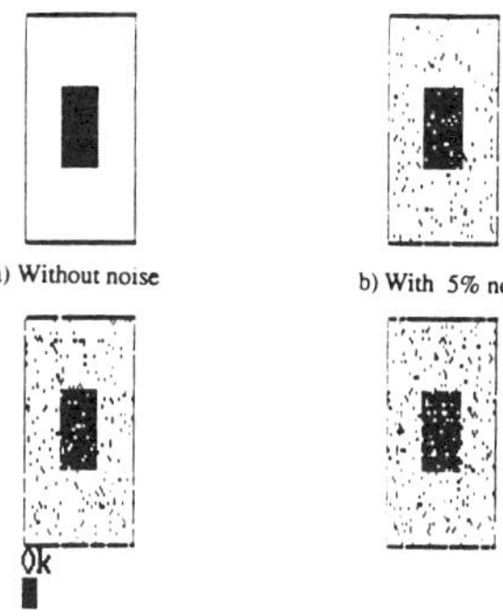

a) Without noise
b) With 5% noise
c) With 7% noise
d) With 10% noise

Fig. 9 Graphs under different noise levels

Modern Inspection Techniques for Computer Aided Quality Control

C. P. Sarathy

Professor of Mechanical Engineering, College of Engineering, Anna University, Madras 600 025 India

INTRODUCTION

Control of quality of a finished engineering product has been performed traditionally by manual inspection methods and statistical sampling procedures. Manual inspection methods are generally monotonous and time consuming apart from the fact that the product to be inspected is to be removed to a distant inspection area, which often constitutes a bottleneck in the manufacturing schedule. On the other hand, the use of statistical sampling procedures involve the risk of some defective parts slipping through the inherent drawback in the above mentioned traditional methods of quality control is that the defects in the product are detected after they are made and hence they must be scrapped or reworked at a cost which is often greater than their original manufacturing cost.

There are many socio-economic and technical factors at work in the modernization of quality control methods. The economic factors include the high cost of inspection procedures and the desire to eliminate time-consuming inspection, while the social factors include ever increasing demand by customers for near perfection in the quality of manufactured products, resulting in a number of expensive product-liability legal cases and Government regulations, all making the product cost high. The technical factors consist of several important advances which have been made in inspection automation. Principal among these advantages have been the tremendous growth in the application of microprocessors driving the quality control function towards what we are calling Computer Aided Quality Control (CAQC), Computer Aided Inspection (CAI) and Computer Aided Testing (CAT) are the subsets of CAQC.

COMPUTER AIDED QUALITY CONTROL

The principle objectives of Computer Aided Quality Control (CAQC) are :

1. To improve product quality
2. To increase productivity
3. To reduce production delay
4. To reduce the product cost

The strategy for achieving the objectives mentioned above is primarily to automate the inspection process through the application of computers combined with advanced sensor techniques. The main feature of this methodology is to undertake inspection on a 100 percent basis rather than sampling. Generally such type of inspection is made on line almost as part of the production processes.

The present discussion includes a few of the latest developments in both contact and non-contact inspection techniques, CAD/CAM data base can be used as a quality control function. Computer Aided Inspection (CAI) and Computer Aided Testing (CAT) are merely extensions of basic principles of quality control, but performed automatically using latest computer and sensor technology.

The important features of Computer Aided Quality Control are :

- inspection and testing will be accomplished on a 100 percent basis rather than by a sampling procedure.

- inspection and testing will be integrated into the manufacturing process.

- non-contact sensors used can complete the inspection in a small fraction of a second.

- on-line non-contact sensors together with the computerized feed back control system has the capability for making adjustments to the process variables in NC machines.

- on-line quality control with 100 percent inspection associated there is a considerable increase in the productivity ; reduction in production time and product cost.

- manifestation of automation in CAQC makes its place, manual inspection activity will be reduced and the QC personnel will have to become more computer-wise and technologically sophisticated to operate the more complex inspection and testing machinery and the automated methods.

Quality Control is achieved through inspection and measurements of the component. These activities is carried out with the help of gauges, measuring devices and testing apparatus manually. Computer Aided Inspection (CAI) and Computer Aided Testing (CAT) are merely the counterparts of manual inspection and measurements. For the above purpose suitable sensors are to be employed. These sensors will sense the dimensions and make the inspection and then provide the information to the computerized feed back control systems and these systems will be capable of making adjustments to the process variables based on analysis of the data collected by the sensors. The data analysis would include statistical trend analysis. Data can also be plotted on a control chart, which then can be used to manipulate the machine. There are different inspection techniques to be associated with Computer Aided Quality Control. In the first type there will be actual contact of the sensors with the components. These usually involve the use of coordinate measuring machines (CMM) and most of these machines today are controlled by computers. A CMM generally consists of a table holding the component in a fixed position and a movable head which holds a sensing probe. The probe can be moved in all the three directions. During operation the probe is brought into contact of the component in all the three dimensions and their lengths measured. In general an accuracy upto 0.005 mm can be achieved. In modern CMM, all the operations are computer controlled. These include the capability for automatic work part alignment on the machine table. Although the inspection time is reduced by about 10 percent in CMM compared to traditional manual inspection methods, there is nevertheless wasted time associated with transporting the parts to a distant place where a CMM is located. This method of inspection becomes less attractive that there is some wear and tear of the probe and work piece resulting in wrong inspection results. The second method (viz) non-contact inspection, is attractive alternative to contact inspection methods.

NON-CONTACT INSPECTION METHODS

Non-contact inspection methods have several advantages as

mentioned below :

- o eliminates the repositioning of the work piece
- o faster in operation
- o mechanical wear and tear is eliminated
- o eliminates the possibility of damage to the work piece due to non-contact probes

The non-contact inspection is generally divided into two groups viz. optical and non-optical categories. The optical systems are always the dominant type of non-contact method. These work on the principles of microelectronic technology and microprocessors are used for processing the sensing signals. The improved performance and reduced cost of the system makes the optical inspection method more and more economically feasible. Some of the important optical systems are described. The optical systems use some form of light sensor element or photosensitive material such as photocells, photodiodes or photographic sensitive paper.

OPTICAL INSPECTION METHODS

The important optical inspection methods include machine vision scanning laser beam devices and photogrammetry. The brief description of these are given below :

o Machine Vision : This is also called as microprocessor based television system, which basically consists of a television camera and a digital computer with an interface preprocessor. The unit digitizes the picture obtained by the camera and analyzes the image by comparing it with the stored data in the memory. The main limitation of this type of system is its limited capability to accomplish accurate measurements in terms of the image resolution. A second limitation with the system is its capacity to sense contrasting light ness and darkness in terms of black and white. This imposes requirements on lighting used for illuminating the object. Third and an important limitation is its capability to recognize and discriminate the features of different objects in terms of stored data.

o Scanning Laser Beam Devices : In a scanning device, laser is used as a coherent light beam which can be projected great distances without much diffusion. The system as such relies on the measurement of time than light. A simple schematic diagram of arrangement for scanning laser beam device is shown in sketch (1). It consists of a laser light source and a rotating mirror, which deflects the beam in order to sweep across the object under investigation. The light sensing unit is placed at the focus of the lens system. The time lapse corresponding to the interruption of the light beam is measured to determine the required dimension of the component with the knowledge of the beam sweeping velocity.

A microprocessor can be used to directly furnish the dimension by converting the lapse time into a linear measurement. In an automated production, such an inspection system can be utilized for automatic rejection of dimension defective part.

o Photogrammetry : In the photogrammetry a three dimensional data of a component is evolved from a pair of photographs obtained at different angles. In an inspection process, the two photographs are read by a device called monocomparator and coordinates and position are established. The data are then analyzed using a computer to extricate the desired information. A general arrangement of photogrammetry is shown in sketch (2). An inherent draw back of this system of inspection is the need for photographs, which is no doubt inconvenient and time consuming procedure in the inspection. An improvement of the photogrammetry being thought about presently is to obtain the images from cameras in a stereoscopic configuration to send visual data directly to a computer for analysis.

NON-OPTICAL INSPECTION METHODS

In the non-optical inspection methods, a non-optical sensors are employed for the purpose of measurements. The general types of non-optical systems are: electrical field techniques, radiation techniques and ultrasonic techniques.

In the electric field technique, the component to be inspected is placed in the electrical field and its effect is observed and analyzed. Location, part dimension, part thickness and other characteristics can be measured with reference to the probe, which is generating the electric field. Eddy current systems can be used to detect cracks, voids and other flaws in the material. There are three types of electrical fields (viz) reluctance, capacitance and inductance. The reluctance type of transducers indicates the presence and distance from the probe of a ferromagnetic substance ; the obvious limitation in this system is that the component under inspection should be electromagnetic. In a capacitance type of inspection probe the measurement is based on the variable capacitance from part or probe. The capacitance is inversely proportional to the distance between the probe face and the component. The capacitance transducer can be used to detect a variety of materials and the material should be an electrical conductor. In inductance transducers, an alternating magnetic field is subjected on the component, when Eddy currents are generated in the component. The magnetic field produced by these Eddy currents affects the impedance of the coil which determines certain characteristics about the component.

X-ray radiation can be used in the inspection processes in metal working industries. The amount of radiation absorbed by metal can be used to predict its thickness and certain other characteristics. A typical application of x-ray techniques is in a rolling mill to measure the thickness of the plates going through the rolls so that proper adjustments can be made in the rollers.

Ultrasonic technique in inspection work involves use of high frequency sound waves. A typical application of ultrasonics is in the non-destructive testing of materials. The method involves in producing a high frequency sound wave and directing it on the component. The reflected sound pattern is analyzed in a computer for making measurements of the component.

CONCLUSION

Computer Aided Testing and Quality Control will be very helpful for automation. At the lowest level, the computer would be used simply to monitor the test and analyze the results. At a much higher level of automation are computer integrated test cells with a series of testing stations connected by a material handling system. CAQC is also helpful for monitoring flexible manufacturing systems and for effecting changes in the product design.

REFERENCES

1 Mikell P. Groover and Emory W.Zimmers Jr.'Computer Aided Design and Manufacturing', Prentice Hall of India, 1986.

2 Monks J.G. 'Operations Management', McGraw Hill Book Company, Newyork, 1977.

3 Hughes, T. 'Winning with Better Quality: Bulls Eye for Digital Inspectors', Production Engineering, July 1982.

4 Chin R.T. 'Automated Visual Inspection Techniques and Application : A Bibliography', Pattern Recognition, 1982.

*

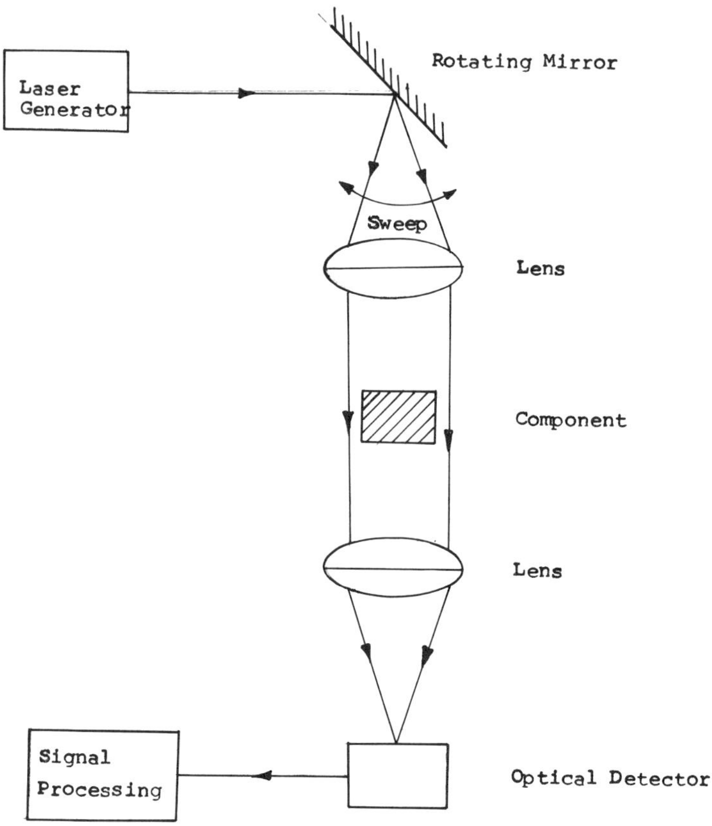

Sketch (1) Schematic Diagram Showing the Scanning Laser beam inspection System

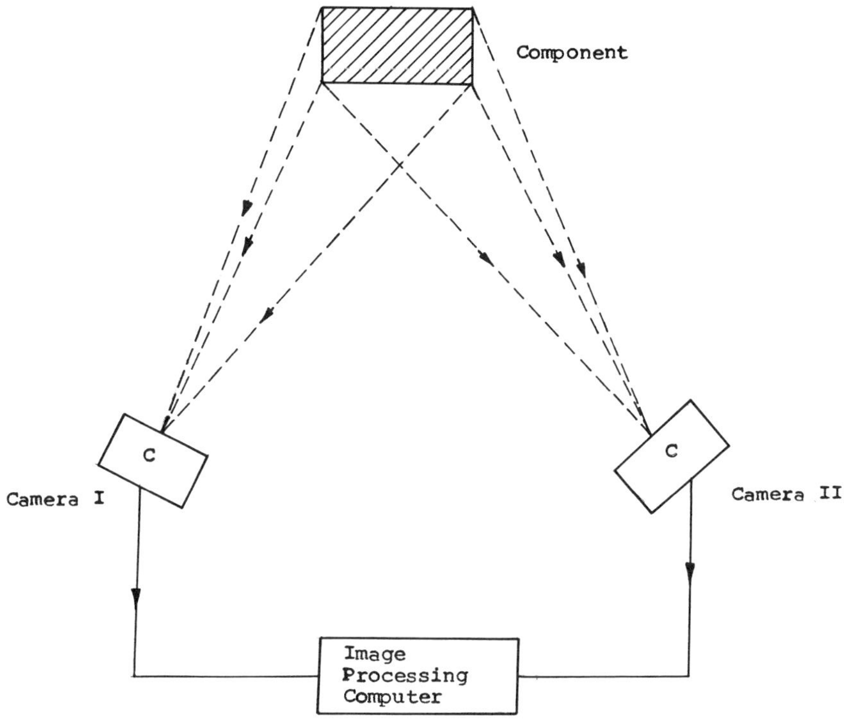

Sketch (2) Photogrammetry Inspection System

AI AND EXPERT SYSTEMS

The Use of Expert Systems in CIM

Andrew Young

Director, Northern Telecom Santa Clara, CA 95054, USA

ABSTRACT

This paper will review the applicability of expert systems to Computer Integrated Manufacturing (CIM) and explore a new type of CIM system in which the use of expert systems might be the only way these systems can be put into commerical operation.

There are many CIM system elements available today such as Computer Aided Design (CAD), Computer Aided Manufacturing (CAM), Production Activity Control (PAC), Manufacturing Resource Planning (MRP II), etc. Each of these systems can benefit from the use of Expert Systems to help in decision making and rules application. In some commercial packages this is being done today.

There is a new class of CIM system - Computer Aided Administration (CAA) - which is just beginning to be understood and defined. Expert Systems seem to be particularily well suited to supporting some information control and distribution rules typical of an administration environment which classical programing techniques cannot easily cope with.

This paper is based on work done at Northern Telecom on CIM administration systems.

INTRODUCTION

CIM and expert systems are both new technologies based on the use of computers to improve efficiency and productivity in areas where, previously, humans had worked without computer assistance. In the case of CIM, computer automation is being applied to the administration of the paper based information which flows between all aspects of the manufacturing process. With expert systems, automation is being offered in those areas where 'expert' human based reasoning can be captured in the form of rules and coded into an application.

Northern Telecom, like a lot of American companies, has been in the process of developing and implementing a CIM system over the past few years. The CIM development plan called for a combination of opportunistic bottom-up implementation of 'islands of automation' with simultaneous top-down development of the data control and delivery infrastructure. The 'islands' included CAD/CAM, MRP II, production control and others, while the infrastructure involved databases, communication networks and administration systems.

The administration systems have turned out to be the most complex of all the CIM components. This is not because the others are trival but because administration systems automate labor intensive, cross functional, paper based, control and decision making activities linking all the other CIM elements together. Since paper based administration systems follow procedures and rules, it was felt that an expert systems approach might make the automation task easier.

CIM has many definitions. In this paper CIM will be limited to the control and distribution of manufacturing information where manufacturing includes the product design and engineering process as well as the production process. The term, expert system, also has many definitions depending on the chosen concept borrowed from the psychology of learning and reasoning or the newer field of artificial intelligence. The simplest form of expert system, though, is based on defining the rules for task completion and this will be the form refered to in this paper.

In general, discussing CIM applications with other companies and vendors is easier than discussing expert system applications. This is not so much due to the complexity of the subject as to the secretiveness of the companies doing expert system related industrial research. Most companies find that expert systems, if completely and properly implemented, contain proprietary information about product design and/or production processes significant to their competitive capability, This being the case, the discussion of where expert systems are being used in CIM is limited to those applications known to the author.

Definitions

As stated above, there are many definitions of CIM and expert systems. In order to set the framework for the use of these terms in this paper, the following definitions will apply:

- Computer Integrated Manufacturing is a business philosophy which integrates all aspects of the manufacturing process from product design and introduction to manufacturing through volume production and repair. This integration is achieved by the control and distribution of information from the begining to the end of the process. In order to support this end-to-end information delivery system, most, if not all, activities which comprise the manufacturing process must be computer based. This means that all information - product design, manufacturing process and administrative data - must be computer based.

- expert system, as used in this paper, is a very simple concept - it is the rules executor for an application package. Historically, an expert system has been available on a seperate computer for consultation but the newer systems are being made available as linkable modules to an applications systems. In the case of an administration system, the bulk of the activity is not rules execution so the expert system must act in a supporting role helping the main program when required to do so.

- the term, Computer Aided Administration (CAA), has been used in this paper to designate those CIM tools which manage, control, and distribute and information about the information used in CIM applications. The control and distribution of product and manufacturing data is managed through the use of databases and comunication networks and in CIM, which automates and integrates data flow throughout the manufacturing process, managing this administration data must also be automated. The concept becomes clearer if the phrase 'paperless factory' is used which suggests that all data, even administrative data, is computer based.

USE OF EXPERT SYSTEMS IN CIM

Expert systems are being used in many areas of CIM. T diagram shows a typical CIM environment where data can flow from product design and engineering through the data administration system to use on the shop floor during product production. Each of these areas will be briefly described and the role or potential role of expert systems will be described. The major thrust of this paper, however, is to review the of the administration system, so this will be covered separately in the next section.

Product Design

The major activity in the design area is the creating and documenting of information describing the product are the relationship of its individual parts. For companies such as Northern, involved in electronic part assembly, this information describes integrated circuits, printed circuit boards, mechanical assemblies, components, cabling, etc.

There are a wide range of commercial and in-house Computer Aided Design systems to help product developers do a better job producing more complete and consistent information. None of the major commercial CAD suppliers, though, are offering expert systems with their packages although a number have stated that they are doing research in this area.

There are, however, numerous in-house activities in this area. A number of large companies are testing expert systems for VLSI cell connectivity, electrical and mechanical design rules verification, component selection, and even product design where the product is a variant of a group of a well defined product family.

Design Engineering

In design engineering, a product can be conceptualized and/or tested. This might involve simulating the functionality of an integrated circuit or a complete electronic system. Mechanical designs can be stressed using finite element mesh analysis and the actual fit of mechanical components can be viewed using kinematics.

Again, there are a large number of commercial systems on the market and, again, none of the larger commercial suppliers have introduced expert systems into this area.

Expert systems can be of major benefit here but the rules will be more difficult to introduce than with other areas of CIM. This is due to the nature of the task which is to try and find problems in the product before it reaches the field. There will need to be very good feedback loops so that the found problems can be captured and the rules entered into the expert system

Manufacturing Engineering

Manufacturing engineering involves all the activites required to transform design information for use on the production floor i.e production of numerical control tapes. There are also a significant number of value-added activities such as generation of full fault coverage test programs and the design of the manufacturing process required to produce a specific product.

There are large numbers of commercial systems in this area and quite a few of them are now beginning to offer expert systems with their products

Of all the CIM areas, manufacturing engineering is the most advanced in the use of expert systems. Applications range through simulation of manufacturing facilities, the generation of process plans, and interactive programming for robots on the shop floor allowing them to respond quickly to changes in product mix or resource availability.

Production, Planning and Control

These areas of manufacturing activity are very closely linked and were among the first CIM areas to have data automation. The automation of administration information is generally as limited here as elsewhere in CIM applications. The major activities are the planning of production, responding to changes in the production schedule due to order and/or production floor changes.

There are a large number of commercial suppliers in this area and some have introduced expert systems into their offerings. The majority of these systems are for dynamic shop floor scheduling, control of flexible manufacturing system cells, robot control to respond to changes, product test systems, automated data collection and the planing activities related to the production of complex many part systems. And there is, of course, the systems which are used to comfigure finished products before they are shipped.

As in the design and manufacturing rngineering areas, there are numerous in-house developments which companies are experimenting with in order to improve their productivity. These systems tend to be in the same areas as the commerical systems.

CIM FUNCTIONAL ARCHITECTURE

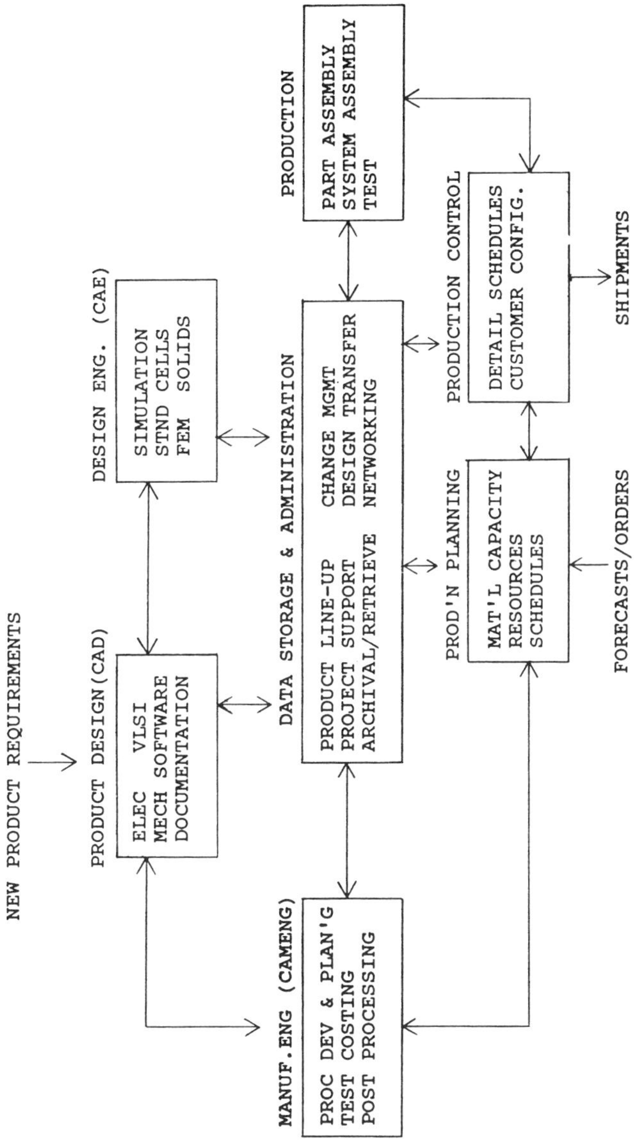

COMPUTER AIDED ADMINISTRATION

In the paper world, there are two levels or layers of manufacturing information which need to be automated if CIM is to be 'paperless'. The first level is the basic product and manufacturing information produced by the CAD systems and used by the shop floor systems. The second level or administration layer contains information about the first level of data such as attributes i.e. ownership, release date, document type and information on relationships of the data elements to each other or how these data elements are managed i.e. product line up, version and vintage control, approval routing for changes. The information on this second level allows the the organization to manage the product and the related control processes. It should be noted that the two levels of information follow different paths through the organization and, in general, the data is treated differently.

Most CIM islands of automation provide control for the product or manufacturing information created within that system. In a number of systens, some second level information control is provided, especially as this is the layer where information security is managed. In general, though, data which relates to product control processes or administration procedures is not part of the system and data which deals with relationships between information elements of a cross application nature are not available. It is in this administration environment that expert systems will prove to be very useful.

Administration Environment

The administration environment found today in most companies provides the control needed to manage a complex product delivery process which is constantly being subjected to changes. These changes occur in the product itself as new features are added or in the manufacturing process as new equipment or processes are introduced onto the production floor. In addition, it is often necessary to keep track of what version of the product is being shipped, where it is going, and responding to requests for upgrades and additions to the installed products. All of these activities, and more, require administration information which is related to the actual product information but is in addition to it.

In most companies, this administration data is maintained on paper and is kept current through labor-intensive updating procedures. In a lot of cases, this information is not synchronized with the product data to which it is related and can contain errors not found in the original product data generated on a computer system. One of the factors which encouraged companies to move into CAD and CAM systems was the opportunity to improve product data quality and reduce the opportunity for introducing human errors. There is the same need and the same opportunity for administration data.

An Example

At Northern Telecom, we sometimes say our business is change. We are constantly adding new features to our products and we are always updating the basic hardware and software to improve performance and meet changing customer needs. The decision to design and introduce these changes is managed by a complex administration procedure to ensure that the changes can be made, that they will be beneficial to our clients, and that they will not cause production problems. The information used in this process includes product and manufacturing information as well as a large amount of administration data about the product. To approve the change, this information is routed through various departments in the organization and the approval path will vary depending on the type of change, its significance, and other factors.

Due to the progress of the CIM development plan, most of the product data is produced by a CAD system and used on the shop floor by CAM systems. The first level of information - product and manufacturing data - is therefore computer based but the administration data is managed on paper meaning that any system developed to handle these changes would be very inefficient.

The problem was analysed and a systems concept developed to support computer access to the existing product and manufacturing data as well providing for automation of some of the administration data. With the defined systems concept, it was possible to manage the approval routing mechanism for the changes but the variability of these routes based on product type and change significance was very difficult to implement using classical programming languages. The system is now being built using an expert system.

Role of the Expert System

The expert system is being used to hold a set of rules which define how a change is to be routed and under what circumstances this routing will change. The routing is an administrative procedure. The route can not only change based on the type of change, but also due to who is available to sign it, what other products might be affected by the change, and differences in design and manufacturing processes which might have occured since the product was introduced. Basically, the routing can be considered dynamic and subject to change at any time and in any way. An expert system is ideally suited to this kind of application because the rules (in this case the routes) can be changes independent of the application (in this case, change approval).

This example represents a complex administration system but it is felt that this is probably typical of most administration systems if they support a CIM environment. Expert systems, therefore, will become a major part of this area of CIM data integration.

CONCLUSIONS

There is no doubt that expert systems can be used in CIM to great advantage although most of these systems are still in the reseach stage. When implementing CIM, all levels of data must be automated to provide complete and effective control. The administration data which supports the product and manufacturing data must also be automated. Because of the nature of the administration activity, systems for managing administration data will benefit from the use of expert systems.

Expert Systems Application for Optimal Production Scheduling

Roderick E. Allen[*], and Hamid R. Parsaei[†]

[*]Camden Wire Company, Camden, NY 13401, USA
[†]Department of Industrial Engineering, University of Louisville, Louisville, KY 40292, USA

ABSTRACT

Many small to mid-sized corporate entities in the American manufacturing industry arena are currently involved in the evaluation and/or implementation of strategic projects, initially promoting enhanced productivity and flexibility in production operations, and consequently yielding remuneration in the form of rejuvenated competitive strength and corporate marketshares. Promising strategic alternatives, with few exceptions, venture in some degree towards the incorporation of automating technology in production resources, ranging from limited "isles of automation" - to group technology concepts and flexible manufacturing systems (FMS) - and in some cases to factory-wide computer-integrated manufacturing systems (CIMS/CIFMS).

A barrier which commonly emerges in the evaluation/justification of such ventures, in addition to employee/union/management interaction and support factors, is comprised of overburdening investment requirements and affiliated risk factors. Any economic evaluation, despite weighting assigned to qualitative criteria, will most assuredly hinge the success of proposed ventures upon the estimated ability of production/project management to effectively expedite the initial implementation and successive optimal utilization of automation transformations proposed for the production systems, thus minimizing immanent financial risks and burdening, particularly if a gradual or incremental approach to assimilation is employed.

As with any significant modification/alteration of a stable manufacturing system, the adaptive ability of conscripted personnel challenged by the inherent adversities may be portrayed as a "learning curve". The objective of production management, in relation to such ventures, is to maximize the slopes of learning curves representative of planning, scheduling, and control efforts via development and institution of effective procedures/methodologies which realistically optimize machine/resource utilization and generate critical cash flows within a minimal time frame. From this perspective, the need becomes readily apparent for an efficient means with which to interpret and utilize the inordinate magnitude and complexity of anticipated shop floor data/permutations in eliciting sufficient decision-support information as mandated by decision-making performance levels critical to project success.

This paper will present an approach to production scheduling and shop floor control, specifically targeted for application in such hostile environments, which explores and examines the feasibility of artificial intelligence (AI) technology/expert systems (ES) applications in seeking solutions to the addressed problem.

Assistance rendered via the proposed ES application will presumably provide production management with the "tools" to not only ensure timely attainment of critical performance levels/cash flows, but ultimately the uniform and "holistic" decision-making required to maintain these "optimal" performance levels/cash flows as well.

INTRODUCTION

Current trends now evident in both the marketplace, and to some extent in more immediate consumer industries, will presumably result in the eventual widespread assimilation of flexible/automating technology throughout American manufacturing industry, culminating in the CIM-based "factories of the future". In order to retain/regain competitive strength and healthy marketshares in the shadow of market pressures to enhance productivity, product quality, and flexibility, and to minimize lead times associated with ever-increasing demands for part design/product line changes, many manufacturing firms will be driven to abandon the once dominant "transfer line" production systems in favor of FMS/CIMS technologies more capable and better adapted to meeting the challenges presented by current and future manufacturing scenerios (refs. 1-2).

While such advanced manufacturing technologies offer great promise in attaining goals established by strategic planning for many firms, the proported benefits wielded by proposed incorporations in existing production operations are commonly over-shadowed by the immanently adverse consequences and risk factors often affiliated with such implementations (ref. 3).

VENTURE RISKS

Whether proposed ventures involve the installation of simple, isolated flexible/automated work stations ("isles of automation") on the shop floor, or implementations of entire FMS/CIMS production environments, provisions must be made for the effective management of hostile production scheduling/control environments inherent with these ventures in attaining the performance levels/cash flow generation requirements established by proposal justification. From the view of production management/control personnel, regardless of the degree of advanced manufacturing technology involved in the strategic ventures proposed for a specific production system, the projected manufacturing environments fabricated by such ventures and their effects upon scheduling/control efforts may be adequately represented (in scale) for analysis/evaluation at the level of common fundamental components--namely, the flexible workcell.

At the work cell level appropriate scheduling/control schemes may be devised which may very closely approximate optimal utilization of production resources within the bounds of the cited unit, however mathematical programming/deterministic solutions have yet to be identified for these (and similar) situations. When considering the production system in its entirety, scheduling/control problems become insurmountable for the human expert due to the shear magnitude and complexity of associated "search spaces" involved. Assistance becomes mandatory for production management, charged with incorporating the combinatorially complex decision-support information within decision-making routines utilized in the performance of daily production scheduling/control tasks. Without adequate and reliable assistance critical cash flow estimates, upon which venture justifications are founded, may not be realized, and in light of initial overwhelming capital investment requirements such additional financial burdening could prove disastrous for the firm (refs. 4-6).

SCHEDULING/CONTROL

Upon introduction of flexible/automating transformations to a particular production system traditional optimization techniques/methodologies previously employed are subsequently rendered ineffective in managing the intractable job shop/intermittent characteristics of the infused production environment. The exponentional magnitude and complex interrelational nature of "search spaces" inherent with the pursued structural and behavioral characteristics of production operations eludes the simple, general application of algorithmic "tools" for scheduling/control assistance. Optimal solutions may only be approximated for a particular desired performance objective (due date performance, machine utilization, etc.) via heuristic pruning of the seemingly unbounded, dynamically complex search trees associated with resource assignments (loading) and sequencing tasks involved in the scheduling of production operations. Yet, even this commonly employed "rule of thumb" approach becomes overwhelming complicated and grossly inefficient when various expediting procedures (lot-splitting and overlapped scheduling) are instituted, multiple resource (machine) assignments become feasible, and when underlying decision-support information is probablistic/incomplete in nature (refs. 3-4).

ES ALTERNATIVE

Within constraints imposed by available technologies and required decision-making time frames, the only available feasible alternative to the impractical/impossible exhaustive search ("generate-and-test") approach to providing production scheduling/control assistance appears to reside in the realm of artificial intelligence (AI), more specifically that branch involving expert systems (ES) programming applications. While conventional computer technology applications are best suited for problem areas receptive to mathematical programming solutions, and would require extensive software development/revision allocations in such dynamic application environments, emerging computer technology in the form of ES programs appears most promising and of immense value in addressing problems encountered with the evolution of manufacturing industry toward the CIM "factories of the future" (ref. 5).

In an expert system, the expertise of human problem-solvers, including all relevant factual, intuitive, and procedural knowledge, is coded into a knowledge base (KB) for the particular application domain. This knowledge base, which is comprised of separate constituents for representation of both declarative and procedural knowledge, is distinctly segregated from the inference engine (control strategy), which governs the selection and application of appropriate rules as required for specific problem-solving tasks. An important feature of this programming structure/architecture, in addition to the power inherent with the knowledge represented, is the significant advantage provided for proposed dynamic application environments, as frequent modifications/revisions of both declarative and procedural components of the KB may be readily accomplished without regard for sequential or procedural considerations.

Recent developments in the areas of ES "shells" and ES "building tools," many of which possess an intact "general purpose" inference engine and furthermore provides assistance in the coding of the KB, now bring the promise of ES technology assistance out of the research laboratories and onto the shop floors of manufacturing industry. This emergence has been further enhanced by efforts to minimize AI literacy and "knowledge engineering" requirements, and to avoid unnecessary hardware expenditures, as evidence by several available "tools" which facilitate the "user-friendly," inexpensive development of functional ES on PC hardware by relatively untrained personnel.

In consideration of these developments in ES technology, and both current and projected manufacturing/market environments, the high-priority consideration of such applications in manufacturing becomes justifiable, if not essential. As is often stated, the key to CIM and to the livelihood/prosperity of American manufacturing industry may be found in emerging ES technology (refs. 7-10).

PROTOTYPING

As stated previously, production scheduling/control parameters may be adequately represented in scale at the workcell level in the addressed production systems. Decision-support information and associated decision-making "knowledge" utilized in corresponding scheduling/control tasks at this level may be presumed to be generally representative of the entire production system, in respect to ES development, due to the uniformity of desired scheduling performance objectives commonly evidenced throughout the components comprising such systems. Therefore, based upon this supposition, the development of ES prototypes for evaluation/analysis of their feasibility in specific application environments may be considerably simplified by initially limiting the scope of the problem domain to the boundaries of the workcell. This approach, providing for the acquisition of experience and the identification of suitable knowledge representation techniques at a relatively simple level, is made plausible by the structure and nature of ES, which allows for subsequent expansion of the prototype system by simply appending to the declarative and procedural components of the knowledge base.

While selection of an appropriate ES building "tool"/package, based upon the AI programming and "knowledge engineering" skills of available personnel as well as other specific application constraints, should be a prime consideration in prototyping ventures, the value of group technology (GT) classification and coding of parts in avoiding replication of duplicate or near-duplicate part designs/processing parameters becomes quite obvious, if not essential. Furthermore, if such information regarding part design/part processing may be retrieved from existing databases/files, whether GT classified or not, the capability to query these external files should also be a priority factor in the planning phase (refs. 9-11).

Upon commitment to prototype development and selection of the appropriate ES "tool," the "knowledge engineering" phase in the construction of the KB now demands the attention and cooperation of all involved. Input from human expert(s), in regard to rules, heuristic techniques, and information/data required in the performance of scheduling/control tasks, is extracted and coded within the corresponding components of the knowledge base, via knowledge engineering inquisitions. The ultimate system performance and value is entirely dependent upon this phase, and as knowledge engineering skills are a rare commodity, this often is a matter best approached through consultation.

CONCLUSION

Regardless of the methods employed in prototype development, the eventual testing and validation of the prototype system will yield a valuable tool to assist production management in attaining realistically "optimal" and uniform performance levels/cash flows from the production systems with which they are charged. Computer speed, accuracy, thoroughness, and uniformity, in combination with human expertise, obviously are qualities needed in today's manufacturing environments.

The capability of ES to be easily revised to reflect modifications/alterations of the production system, to provide multiple, ranked scheduling alternatives, to provide explanations as to how such solutions were obtained, and finally to provide "real time" monitoring/control of production operations, certainly represent the type of assistance mandated by hostile production environments. Their ability to "intelligently" accommodate large volumes of complex data and apply inexact reasoning techniques in specific problem-solving domains provides the help that is needed in manufacturing industry, as growth in automation toward the CIM systems of tomorrow continues.

REFERENCES

1 M.P. Groover, M. Weiss, R.N. Nagel, and N.G. Odery, Industrial Robotics, McGraw-Hill Book Company, New York, 1986.

2 M.P. Groover, and E.W. Zimmers, Jr., Computer-Aided Design and Manufacturing, Prentice-Hall, Inc., New Jersey, 1984.
3 W.F. Fabrycky, P.M. Ghare, and P.E. Torgersen, Applied Operations Research and Management Science, Prentice-Hall, Inc., New Jersey, 1984.
4 W.J. Stevenson, Production/Operations Management, 2nd Edition, Irwin, Illinois, 1986.
5 S.M. Alexander, Knowledge-Based Expert Systems: Their Application to Production Management, Production and Inventory Management, Fourth Quarter, 1985, pp. 109-113.
6 E.L. Fisher, Expert Systems Can Lay Groundwork for Intelligent CIM Decision Making, Industrial Engineering, March 1985, pp. 78-83.
7 S.R. LeClair, W.G. Sullivan, Justification of Advanced Manufacturing Technology Using Expert Systems, AIIE Conference Proceedings, 1985, pp. 362-368.
8 M.W. Parks, Expert Systems: Fill in the Missing Link in Paperless Aircraft Assembly, Industrial Engineering, January, 1987, pp. 37-45.
9 N.J. Nilsson, Principles of Artificial Intelligence, Tioga Publishing Co., California, 1980.
10 F. Hayes-Roth, D.A. Waterman, and D.B. Lenat, An Overview of Expert Systems, Addison-Wesley Publishing Co., Massachusetts, 1983.
11 Mr. Emrich, "Artificial Intelligence: Update 1986," Manufacturing Systems, February, 1986, pp. 22-30.

Inaccuracy Mapping of 2-D Surfaces for Robot Off-Line Programming

Ed Turner and W. Edward Red

System Automation Laboratories, Brigham Young University, Provo, UT 84604, USA

ABSTRACT

While the repeatability of a given robot can be determined and bounded, the accuracy of the same robot will vary with spatial position and orientation and is difficult to measure. Not knowing the accuracy of a given robot greatly limits the flexibility of using robotic workcells and limits the use of off-line programming in planning workcell tasks. By using sensor systems to capture robot to part relative geometry information it is possible to map the inaccuracy of a robot in a localized area. Using such error mapping techniques eliminates the need to define the accuracy of the robot in global terms.

INTRODUCTION

There currently are three basic techniques which can be used to program robots. These are the teach pendant or on-line mode, sensor driven systems, and off-line programming. However, with each method of programmming comes certain limitations which can greatly affect the flexibility of the robot workcell.

Robot repeatability and robot inaccuracy are two positioning measures critical to understanding the limitations of each programming method. Repeatability of a robot is defined as the ability of the robot to return to a given point from another given point within a specified tolerance. The inaccuracy of the robot is its ability to go to a commanded point in global space and be within allowable tolerances.

The on-line teach pendant mode utilizes the robot repeatability to overcome the robot inaccuracy problem but is severely limited in its flexibility and requires expensive tooling and a considerable investment of personnel time. Sensor based systems use feedback to overcome repeatability and accuracy limitations but are complicated, require significant computational power which tends to slow the process, and are not reliable. Finally, the current off-line systems use global techniques and are limited by the robot inaccuracy.Because it is difficult to define a robots base frame and because of robot arm structural errors, off-line programming of robot tasks which demand accuracy are not generally applied.

In the typical manufacturing process, parts are made and assembled relative to each other and not in relation to some abstract global reference frame. This raises the question of why should robots be accurate in relation to some global coordinate frame when the task is to assemble parts relative to one another.

One approach used in achieving greater robot accuracy has been to define the kinematic parameters of the robot and then somehow compensate in software for any kinematic errors [4]. Mooring [2] has shown that joint axis misalignment can greatly affect robot positioning accuracy and has developed methods to identify axis position and orientation errors [3]. His simulation methods have not fully accounted for measurement error and his measurements are conducted in global coordinates. Mooring briefly mentions the use of a relative coordinate system but it is located at the base of the robot. Whitney [5] presents an improved and involved calibration method but still does not overcome measurement errors to achieve acceptable accuracy data. Ahmad [6] defines a mathematical model which will allow the robot controller to compensate for second order nonlinear kinematic effects, but again in absolute space and still does not account for any manipulator dynamic effects in predicting positioning accuracy. These techniques all attempt to define the robot inaccuracy using global methods.

This paper will demonstrate methods which have been developed for overcoming the robot manipulator inaccuracy problem using relative positioning in the defined workspace instead of global positioning.

INACCURACY MAPPING OF PARTS

A successful off-line programming system will use methodologies which are based on relative information, relying on the fact that most robots are repeatable within acceptable tolerances, and on the fact that parts on which the robot operates have been previously manufactured to some satisfactory tolerance. By guaranteeing that the part to be worked on will reach the robot workcell within some predetermined tolerance, it is possible to use off-line programming to command the robot to move to a reference calibration feature on the part, thereby capturing the rigid body inaccuracy information. With the establishment of a reference feature the robot can now be commanded to move to other predetermined features on the part. Knowing the relative geometry between the part features, procedures are implemented to define the inaccuracy of the robot relative to the part on which it operates. Correction routines are now implemented which correct for the inaccuracy of the robot in localized regions, thus allowing the practical application of off-line programming methods.

The methods developed in the Systems Automation Laboratories at Brigham Young University use the robot and attached sensor devices to map the inaccuracy of the robot within the part area on which the robot operates. The typical sensoring devices used would include a vision system and a depth sensor or touch probe.

The mapping of the localized inaccuracy of the robot would not be necessary for each part which enters the workspace if the part is positioned repeatably within a predetermined tolerance. This allows the sensoring equipment to be turned off during assembly, thus speeding the assembly process by compensating for robot inaccuracy using software modification of the off-line programs.

CALIBRATION METHODS

When attached to the robot distal link the vision system camera terminal control frame (TCF) must be determined. Methods were developed [7] which can determine the camera TCF using simple features such as solid circles that do not have to be globally positioned.

Once the TCF is determined inverse kinematics routines can be implemented to correctly position the camera frame near the part feature. The Automatix AV-3 vision system was calibrated with the aid a calibration plate machined to a tolerance of ±0.0005 inches. The dimensions of the calibration plate features are shown in Figure 1.

The relative pixel size of the y-pixel to the x-pixel was determined by locating two dots on the calibration plate within the vision system camera frame. Their respective locations in pixel coordinates were then determined by statistically averaging over a number of tests to obtain the ratio of the y-pixel distance to the x-pixel (0.9941393).

The distance between pixels in real world coordinates also was required. By using the same two dots and by knowing the physical distance separating the dots, the statistical average obtained from the tests was 23.20101 pixels/mm.

The origin of the coordinate frame of vision system was located at an x-pixel of 192.0 and a y-pixel of 127.5. Because the vision system was used a coordinate frame was determined which related the feature captured by the vision system to the TCF frame [7] of the camera which was attached to the robot distal link. This transformation involved getting the calibration plate coordinate frame and the vision system coordinate frame in the coordinate frame of the camera TCF, equation (1).

$$T_V^f = T_V^c * T_C^f \tag{1}$$

where:

T_v^f = transformation from the calibration feature frame to the vision system frame

T_v^c = transformation from the camera TCF frame to the vision system coordinate frame

T_c^f = transformation from the calibration feature frame to the camera TCF frame

INACCURACY MEASUREMENT

The relative positioning method has great potential because an understanding of robot inaccuracy is only needed in the localized part area on which the robot operates. When the robot is making sweeping moves between workstations, the inaccuracy of the robot is not critical

However, because great accuracy is required in actual assembly, such as in a pick or place move, the inaccuracy of the robot needs to be defined within this small area. The procedures which follow were successfully demonstrated on a SEIKO RT-3000 series robot which is repeatable to ±0.001 inches.

By using the calibration plate of Figure 1, the robot inaccuracy propogation over a localized area was determined by first moving the camera over the reference calibration feature. From this feature the necessary information was captured which allowed for a part rigid body error correction for all subsequent moves of the robot. The rigid body error correction applies the same translation and orientation correction to all the features used in inaccuracy mapping of the robot relative to the part.

Once the rigid body error transformation was determined the robot was commanded in the vision system software to move to the next feature and capture the inaccuracy data for that feature. By accessing a data base [7] all eight features were mapped by the vision system. Table 1 lists the test data obtained in determining the robot inaccuracy. As seen in Figure 2 the robot error increases proportionally (near linear) with distance from the reference calibration feature.

INACCURACY CORRECTION

Since the localized position inaccuracy is a near linear function of distance, the following linear equations were developed to compensate for robot inaccuracy:

$$Ex = A_1x + B_1y \tag{2}$$
$$Ey = A_1x + B_2y \tag{3}$$

To solve for the constants it is necessary to capture the inaccuracy information from two independent points in the workspace area. This was accomplished by first moving the camera to the reference calibration feature to determine the rigid body error correction and then the camera was moved to two other features, designated i and j, from which all the necessary information was obtained to solve equations (2) and (3). The error equations at object i are:

$$Ex_i = A_1x_i + B_1y_i \tag{4}$$
$$Ey_i = A_2x_i + B_2y_i \tag{5}$$

where x_i and y_i are the Cartesian moves of the robot and Ex_i and Ey_i are the position errors captured by the vision system at object i. Similarly, the same information was captured at object j. Now sufficient information is available to solve for the constants A_1, B_1, A_2 and B_2 using Kramers rule. For the experiment summarized in this paper, features 2 and 6 represent, respectively, objects i and j.

Next, the database information was modified to correct for predicted positional errors in location of each feature as the robot was commanded to move to all the features. Table 2 lists the error after the robot has been corrected for inaccuracy. Comparing Tables 1 and 2, it can be seen that the maximum robot inaccuracy of 0.0172" was improved to 0.0025" at a distance of 14.142" away from the reference feature. This is near the camera resolution and not realistically attainable. A statistical analysis of tolerance stack-up would imply that these techniques could be applied to processes not requiring greater than 0.005" of positional accuracy.

Robot orientation compensation presents more difficulty. As seen in Figure 3 or Table 1, the orientation inaccuracy of the robot is not a function of distance from the reference calibration feature. The orientation can be attributed to sources such as gear backlash, joint compliance, vision system algorithm limitations, etc. The test results show in Table 2 show the orientation inaccuracy after a linear equation for orientation correction similar to (4) and (5) was applied. Note that orientation inaccuracy was corrected at the reference features 2 and 6 but not at the other features.

Table 3 contains test results from an experiment in which the linear error correction is again applied for position inaccuracy correction whereas the feature orientation inaccuracy is corrected by the initial rigid body error only. As might be expected, the maximum orientation error of 0.822 degrees is similar to that found in Table 1.

SUMMARY

As can be seen from the results the robot positional inaccuracy can be mapped over localized areas and corrected using simple linear equations which allow for very precise location in positional accuracy. This approach for error correction greatly enhances the ability to use off-line programming techniques in the use of robots in a manufacturing environment.

REFERENCES

[1] Turner, Ed, "A Vertical Assembly Methodology That Accounts for Robot Inaccuracy," Masters Thesis, Brigham Young University, Provo, Utah, August, 1987.

[2] Mooring, B. W., "The Effect of Joint Axis Misalignment on Robot Positioning Accuracy," Proceedings of the ASME Conference on Computers in Engineering, pp. 151-155, Chicago, Illinois, 1983.

[3] Mooring, B. W., "An Improved Method for Identifying the Kinematic Parameters in a Six Axis Robot," Texas A&M University.

[4] Hayati, S., "Robot Arm Geometric Link Parameter Estimation," Proceedings of the 22nd IEEE Conference on Decision and Control, pp. 1477-1483, December 1983.

[5] Whitney, D. E., C. A. Lozinski, and J. M. Rourke, "Industrial Robot Calibration Method and Results," Proceedings of the ASME on Computers in Engineering Conference, pp. 92-100, Las Vegas, Nevada, August, 1984.

[6] Ahmad, Shaheen, "Second Order Nonlinear Kinematic Effects and Their Compensation," IEEE, pp. 307-314, July, 1985.

[7] Wang, Xuguang, "Off-Line Integration Techniques for Robot Path Planning," Proceedings of the Second International Conference on Robotics and Factories of the Future, San Diego, California, July, 1987.

Table 1: Robot Inaccuracy

index	d(in)	Δx(in)	Δy(in)	e(in)	Δθ(°)	type
1	5.000	-0.00100	-0.00806	0.00866	-0.18189	feature
2	10.000	-0.00465	-0.01223	0.01308	-0.59117	feature
3	11.180	0.00211	-0.01189	0.01208	0.05415	feature
4	14.142	0.00873	-0.01482	0.01720	0.08954	feature
5	11.180	0.01187	-0.00961	0.01527	-0.17764	feature
6	10.000	0.01287	-0.00330	0.01329	-0.64040	feature
7	5.000	0.00546	-0.00181	0.00575	-0.72860	feature
8	7.071	0.00467	-0.01026	0.01127	-0.21365	feature

Table 2: Error correction with orientation correction

index	d(in)	Δx(in)	Δy(in)	e(in)	Δθ(°)	type
1	5.000	0.00008	0.00126	0.00126	0.15870	feature
2	10.000	-0.00060	-0.00076	0.00097	0.03391	feature
3	11.180	0.00088	-0.00153	0.00177	0.98821	feature
4	14.142	0.00175	0.00186	0.00255	1.35796	feature
5	11.180	0.00183	-0.00109	0.00213	0.65912	feature
6	10.000	-0.00070	-0.00099	0.00121	-0.05611	feature
7	5.000	-0.00018	0.00070	0.00072	-0.30861	feature
8	7.071	0.00086	-0.00079	0.00117	0.46107	feature

Table 3: Error correction with rigid body orientation correction

index	d(in)	Δx(in)	Δy(in)	e(in)	Δθ(°)	type
1	5.000	0.00080	0.00047	0.00093	-0.12178	feature
2	10.000	-0.00099	0.00039	0.00106	-0.61422	feature
3	11.180	-0.00039	0.00000	0.00039	0.00642	feature
4	14.142	0.00026	-0.00023	0.00035	0.05244	feature
5	11.180	0.00109	-0.00008	0.00109	-0.33948	feature
6	10.000	-0.00010	0.00089	0.00090	-0.82204	feature
7	5.000	0.00028	0.00052	0.00059	-0.60533	feature
8	7.071	0.00063	-0.00052	0.00082	-0.05080	feature

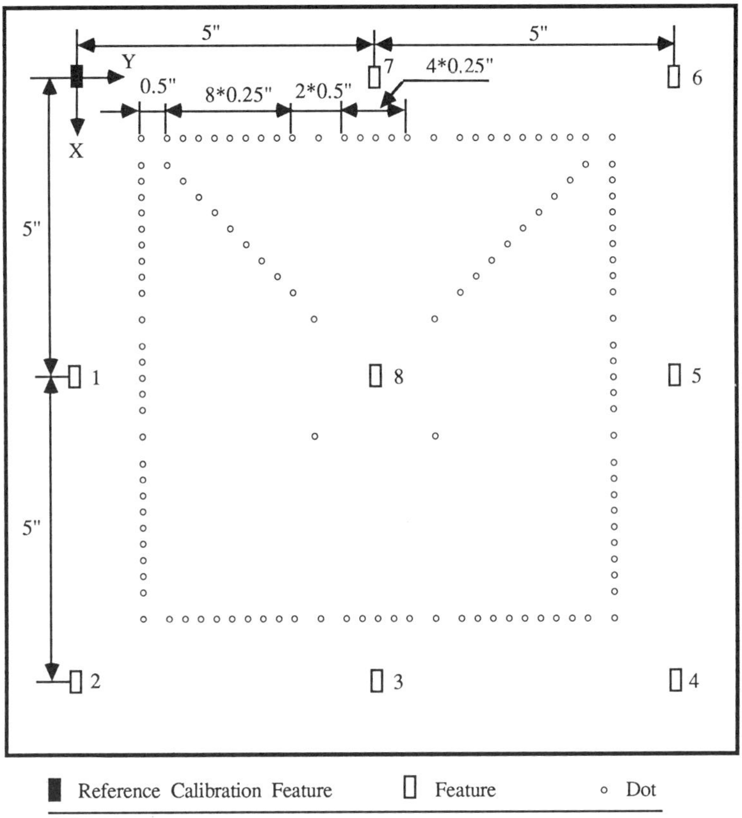

Figure 1: Calibration Plate

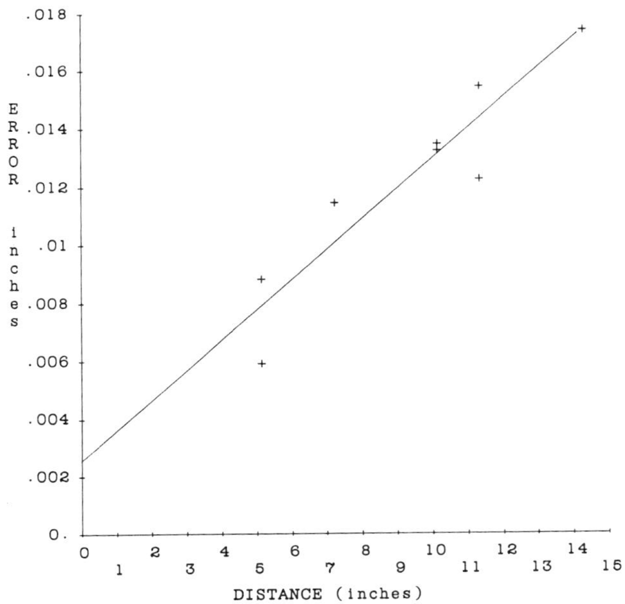

Figure 2 : Position Inaccuracy as a Function of Distance

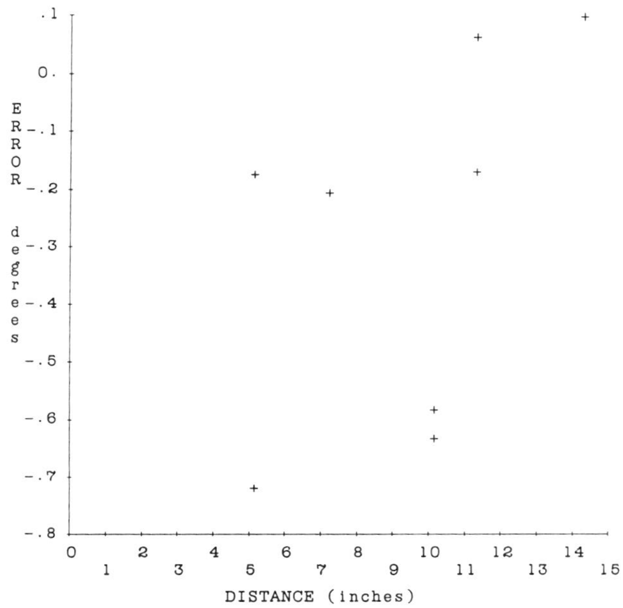

Figure 3 : Orientation Inaccuracy as a Function of Distance

Diagnosis of Robotic Workcells by Behavioral Models

Peiya Liu, Ming-Yee Chiu, Cheoung N. Lee, and Steven J. Clark

Siemens Research and Technology Laboratories, Princeton, NJ 08540, USA

Abstract

Much current research in AI approaches to diagnosis [DeKleer 83] focuses on purely electronic or mechanical systems. These techniques may not apply to systems in which hardware and software interact. In many computer control systems such as real-time systems, process control, and robotic manufacturing cells, functional behavior of hardware is mainly determined by the application-specific software. In addition, the hardware behavior may make changes in the workcell environment, which may be detected by sensors, and in turn modify the control flow of software. In this paper, a principle of modelling hardware behavior in such environments is described. An effective procedure is presented to construct causal chains from the behavioral model for diagnostic reasoning about hardware faults. Furthermore, a comparison of the causal chain and traditional watchdog timer techniques is made.

1. Introduction

In general a robotic workcell may consist of a variety of simple to fairly complex components ranging from on/off mechanical switches or actuators to intricate tooling or gripper devices. The functional behavior of hardware in a robotic workcell is influenced by how it interacts with software. This software may consist of multiple control programs which are called processes. The control programs will run in parallel and may actuate mechanical devices or wait for signals from sensors. The communication and synchronization among programs are either directly through software flags or indirectly through hardware actuators and sensors. The troubleshooting of a robotic system is a process to identify which physical components in the workcell are faulty. The troubleshooting process uses information from sensors, software flags, actuating signals, and the point at which each robot control program is stuck. The workcell is stuck either because all programs are waiting for signals or because of a crash.

A system called SDE (Siemens Diagnostic Engine), based on a principle of modelling workcell behavior, has been developed at Siemens for robotic workcell diagnosis. It has shown positive results and we intend to apply this technique to different domains of interaction environments.

2. Modelling Mechanical Behavior in Robotic Workcells

Obtaining a satisfactory diagnostic capability requires accurate description of mechanical device behavior, especially in a hardware-software interaction environment. In SDE, the behavior of mechanical devices, ranging from simple actuators to robots themselves, is represented by physical processes. The behavior of control programs is described by computer processes. A physical or computer process is further modeled by an finite state machine which consists of a set of state transition descriptions. Each state transition description is in the form: *tn x s0* --> *s1* . Given a transition tn, the process will change its state from *s0* to *s1*.

A list of signals may be associated with states, parts, or tools for indicating status. These associations are modeled by a tuple <S, Signal_List>, where S stands for a device state. The types of workcell signals used in this paper are (1) F/T for force/torque sensor signal, (2) s10xx for proximity sensor signals, (3) s20xx for software flags, and (4) sxx for actuation signals. An association <S, Signal_List> means that Signal_List is used to detect S. If S is a device state as shown in <SbeltFlip=on_the_left, {s1019_off}>, then the association means whenever SbeltFlip is on the left, s1019 is off. For notation simplicity, sensor signals may directly represent states of a device. For example, <s1019_off> is same as the previous device state association. Symbol "-" means "don't care state", and an empty set {} means either "does not use tool", or "does not carry parts".

A knowledge base of these state-based descriptions will be used for diagnosis of a robotic workcell. This state-based approach follows previous research [Chiu 85] to further clarify the sensor and state relationships. This information is crucial to elicit causal relations from the model for robotic workcell diagnosis.

To illustrate the principle of state-based descriptions, a workcell that assembles laundry pump motors is used as an example. In the laundry pump motor assembly cell, several types of peripheral and different kinds of robot actions are setup for assembling the motor.

Physical Processes for Peripheral Behavior

SbeltFlip shown in Figure 2-1 is a device which takes a subassembly, assembled upside-down, and turns it right-side-up so the robot can pick it up and assemble it to the rest of motor. From the description of SbeltFlip below, we see that if signal s20 turns on, then SbeltFlip moves to the left and picks up the subassembly {sbelt spring bush}. At the same time, proximity sensor s1019 will turn on since it is located to the left. If signal s20 turns off, SbeltFlip carries the subassembly to the right hand side, and sensor s1020 turns on because of the subassembly brought to the right. At the same time, s1019 will turn off. Note that the functions of s1019 and s1020 are different. Signal s1019 is to detect the device state while signal s1020 is to detect the parts present at that state.

The spring separator (SpringSep) is a device which presents springs (coming to it continuously from a bowl feeder), to the robot just one at a time. When s17 turns off, it allows the line of springs to move into a "holding area". When s17 turns on, it allows the foremost spring to continue the presentation point while holding back the rest of the *springs*.

SbeltFlip:
```
s20_on x  (-  ) -->(s1019_on)
s20_off x (s1019_on) -->
          (s1019_off s1020_on)
```

SpringSep:
```
s17_on  x (ready)-->(go)
s17_off x (go) -->(ready)
```

Physical Processes for Robot Actions

Robot actions are modeled as a set of physical processes. Strip models the process of the robot placing a subassembly, consisting of the sbelt, spring, and bush that it is holding in a special gripper, into a fixture for subsequent assembly. SpringPick models the robot picking up the spring in a special gripper. The force/torque sensor (F/T) is used to monitor the operation and can signal that no part was present, a crash, or the assembly operation is normal.

Strip:
```
StripCmd x (start)   -->(finish)
```

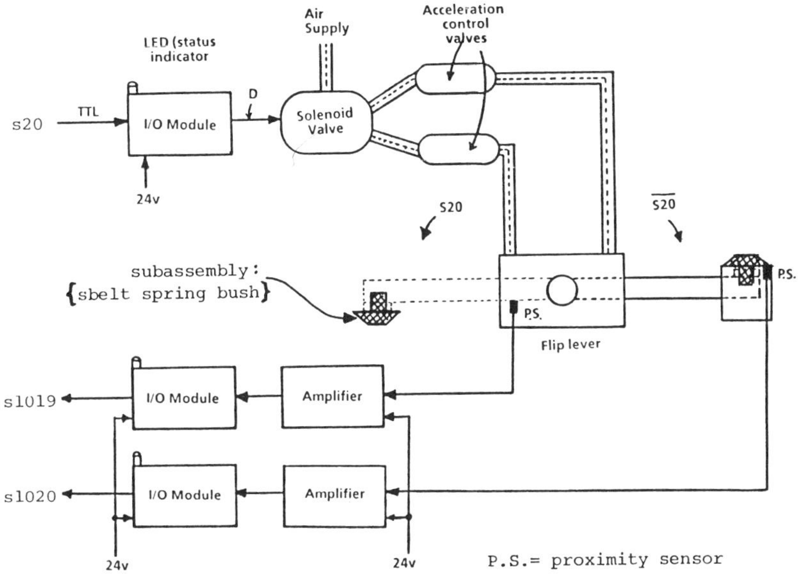

Figure 2-1: SbeltFlip

SpringPick:
```
SpringPickCmd x (start) --> (<finish {F/T}>)
```

ToolPick:
```
ToolPickCmd x (start) -->(<finish {F/T}>)
```

Computer Processes for Control Programs

A control program is in a *synchronization state* if the program waits for software flags or sensor signals. The control programs are modeled by computer processes using the same state transition form as *precondition* x *(CP=S) --> (CP=Q)*, where S and Q are two consecutive states of computer process CP. Usually, software flags or actuation signals are attached to a computer process state to indicate actions the computer process makes during the transition. An example is the association <CP=R, {s2012_on s19_on}> in the transition: s2019_on x (CP=Q) --> (<CP=R, {s2012_on s19_on}>), This means Q is a synchronization state for computer process CP since CP waits for software flag s2019 to turn on at that state. At state R, CP sets s2012 on and actuates certain devices by setting s19 on. The symbol *True* represents no precondition. Some computer processes are shown below.

Pspring(A peripheral control program for handling part *spring*):
```
s2019_on x (Pss=40 ) -->(<pss=10,{s17_on s2018_on s2019_off}>)
True     x (pss=10 ) -->(<pss=20, {s18_on s17_off}>)
s1017_on x (pss=20 ) -->(<pss=30, {s18_off}>)
s1018_on x (pss=30 ) -->(<pss=40, {s2018_on}> )
```

Amotor(The main robot control program for assembling motor):

```
True x (ms=59) -->(<ms=60, {StripCmd s2021_on}>)
s2021_on x (ms=60) -->(ms=61)
```

3. Construction of Causal Models from Behavioral Models

The idea of diagnosis based on causal models is not new. However, construction of causal models from behavioral model for robotic workcell diagnosis is new. Current construction of causality structure in AI can be categoried as (1) from mathematical model [Kuipers 85], (2) by experience [Reiger 77], [Koton 85] (3) from hardware structure and function [Davis 82, Davis 83, Davis 84, Hamscher 83]. But, in robotic workcells, the causality mainly comes from the actuation behavior. The order of process actions and sequences of parts assembly are the causal relations. The behavioral information about interactions between computer and physical processes is crucial for workcell diagnosis. It is useful to abstract shallow causal relations from the deeper workcell behavior model. Once a workcell is stuck, these causal relations will provide efficient and useful diagnosis paths.

SDE uses multiple finite state machines to describe functional behavior of hardware and software interaction in the robotic workcells. Shallow causal chains are automatically constructed from such behavior descriptions. Thus, each shallow chain has an inherent reason rooted in the behavior descriptions. Many expert systems use empirical rules. If these empirical rules are generated from the deeper first principle model, it has the potential to know where the expertise is in the expert system and when the expert system might fail in the applications. This meta-knowledge of expert systems is much more easily provided by this approach.

Generally speaking, the basic reason for computer processes being stuck is that all the processes are waiting for unsatisfying input signals. The diagnosis of workcells is a process to identify which faulty physical processes caused all actuation sequences to be stuck. The *actuation causal relationship* of a program synchronization point is a directed acyclic AND/OR graph rooted at that point as in Figure 3-1. Physical process PP2 needs to be actuated by both two signals - one is set up by computer process CP1 at state S1, and another is set by CP2 at S2. Physical process PP5 needs to be actuated by either PP4 or PP3. Computer process CP at S is waiting for a precondition which needs be set up by CP3 at S3 and PP1, or only by PP5 at certain state. Any solution graph [Nilsson 80] in the AND/OR graph will be an *actuation causal chain* shown in Figure 3-2.

CP1, CP2, and CP3 are called *actuating processes of CP at S*. If process CP is stuck at S, then CP is called a *waiting process* and the actuation causal chain is called a *stuck chain of CP=S*. A stuck chain of CP=S is called an *actuated stuck chain of CP=S* if all actuating processes of CP at S have activated the physical processes when CP is stuck at S.

4. Diagnostic Reasoning about Robotic Cells

4.1. Diagnosis from Actuation Causal Chains

The principle for diagnosis from actuation causal chains is to check whether or not stuck chains are actuated. Since actuating processes may be stuck before actuating a causal chain, and stuck chains are not actuated, there may be no faulty physical process in the chain. However, if actuated, there must be faulty physical processes in the chain. It is assumed that there are no logic bugs in the control programs.

To apply this principle to diagnosis, we need to identify which stuck causal chains are actuated. Note that stuck chains can be found by noting if the states of stuck computer processes are synchronization states. There

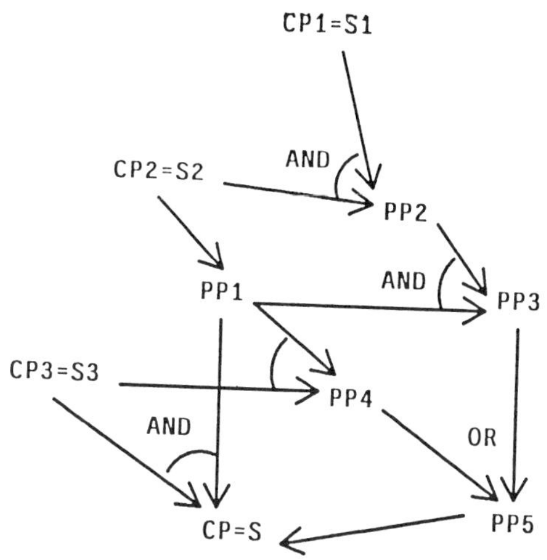

Figure 3-1: An Actuation Relationship at Synchronization State S

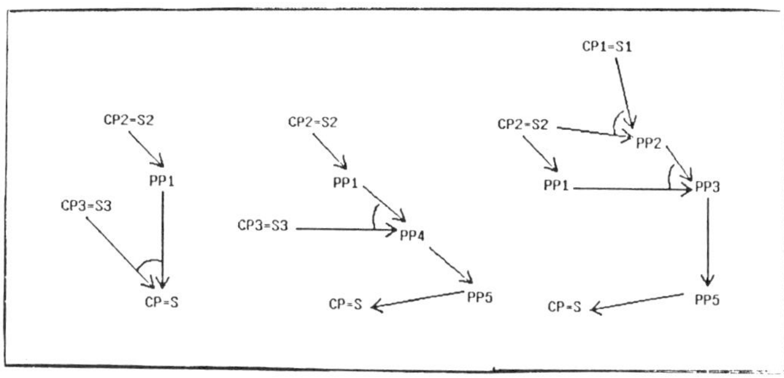

Figure 3-2: Actuation Causal Chains

may be multiple stuck chains when a workcell is stuck. In practice, the following are ways to check if a stuck causal chain is actuated:

1. If only one stuck chain in the workcell, then the causal chain must be actuated.

2. If all actuating processes in a stuck chain are the same as the waiting process, then the stuck chain must be actuated.

3. Recording actuation signals in run time and getting the signal information to check if stuck chains are actuated when the workcell is stuck. The values of signals recording are meaningful only if all actuation causal chains are constructed as critical regions. In other words, after actuating processes actuate causal chains, they cannot actuate the same causal chains again until the process at the synchronization state is receiving signals. In robotic workcells, the principle of critical regions are often followed in order to get a clearer and simpler manufacturing order for parts and to avoid time-dependent process risks.

Very often, robot control programs and mechanical devices functions are setup by a synchronization scheme which can check if chains are actuated easily. For instance, a robot control program P1 at synchronization state S1 waits for a sensor signal, and this signal indicates the state of parts present, or the state of a mechanical device which is actuated by the same program P1 at synchronization state S2.

4.2. State-based Reasoning

Our robot diagnostic reasoning is based on finite state machine modelling of robotic workcell behavior. A transformation or elicitation is used to obtain causal chains from state-based modelling of workcell behavior. These causal chains are implemented in the form of condition-conclusion rules, such as *CP=S and sxx=on, ... --> suspect components*. If the control programs are stuck at certain synchronization states waiting for signals indicating state of mechanical devices, only rules related to actuation causal chains will be fired. Otherwise, the control programs are waiting for signals indicating part or tool status and both types of causal chain related rules may be fired.

These condition-conclusion pairs look like rules acquired from experienced experts or through first-hand experience in the field. More importantly, each rule has an inherent reason for the relation between conditions and conclusions rooted in the deeper level model. This novel approach to diagnosis provides smooth and consistent interaction between empirical rules and deeper model descriptions [Michie 82] [Koton 85] [Hart 82].

4.3. A Comparison of Watchdog Timer and Causal Chain Techniques

In robotic workcells, watchdog timers are often used to detect the faults. Typically, watchdog timers are used to set a time limit for every assembly step; the timer will be automatically reset after executing that step. The moment some assembly step takes longer than its time limit, program execution stops and an error is reported. Usually, the information about the steps which caused the timeout, and inputs of a step which are not satisfied can be displayed. The problem with this technique is the length of time for each step is not normally adjusted carefully. Accordingly, it is quite common to have the "wrong" watchdog timer expire first. Even if all the timers are set up correctly, the problem still exists. In the sequence control programs, watchdog timers can only be set up to measure the waiting time of the next transition input rather than the time of transition, which waits for its actuated physical actuation completion. This subtle situation is shown in Figure 4-1; in this case or many variations of this program structure, the timer TM1 can expire but the physical process PP1 is not the real culprit. From the information of TM1 or TM2, we cannot tell whether PP1 is actuated or not at that moment. If PP1 is not actuated, then it is not the culprit which caused TM1 timeout. In general, what the watchdog timers can provide are early warning signals that something is going wrong and workcells will hold, or crash if running is continued. The causal chain technique is a reasoning scheme to further localize real faulty processes from the global information about unsatisfying inputs, synchronization points, and actuation chains.

5. Related Work

Diagnosis in varied domains has attracted much attention [Genesereth 82] [McDermott 82] [Chandrasekaran 84] in the AI community. Particularly, there are works GDE [DeKleer 86] IN-ATE [Cantone 86] on general purpose diagnosis inference engines; our work naturally followed these approaches. In addition, we focused on a domain specific (robotic workcells) diagnostic engine. The goal of SDE is to formalize the diagnostic knowledge of robotic workcells and to develop a workcell diagnosis inference engine based on the formalization and characteristics.

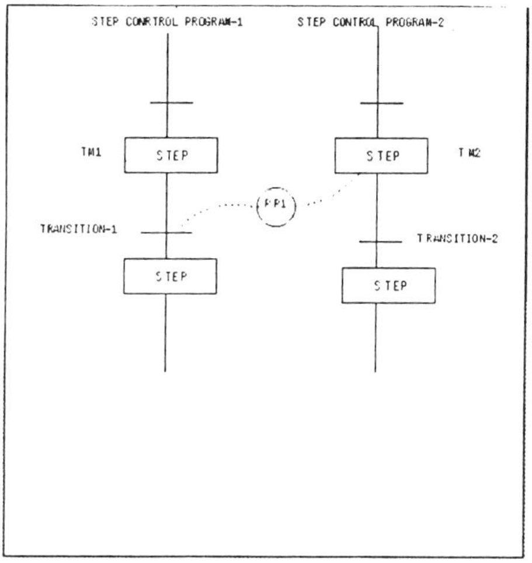

Figure 4-1: Unsolved Problems in Watchdog Timer Technique

In robotic workcells, the information about causal links derives from behavioral models. Usually actuation causal chains in robotic workcells have the following characteristics: (1) finite mechanical devices in a chain, (2) no loops (acyclic), and (3) sensors within a chain. Because of these facts, the diagnostic control used in GDE or IN-ATE is too general to be efficient because it cannot make use of sensor information. Furthermore, the minimal entropy and gamma miniaverage methods used in GDE and IN-ATE do not fit the robotic workcell environment for best test generation. These features in the robotic workcell diagnosis rarely exist in purely electronic or mechanical systems.

6. Conclusion and Future Work

A principle of modelling mechanical device behavior in hardware-software interaction environments has been described. Causality in workcells is identified for diagnosis. A prototype system of SDE is implemented in LOOPS on a Xerox Dandelion 1109 Lisp Machine. It has successfully been applied to a laundry pump motor assembly cell and can troubleshoot multiple faults. SDE is domain-specific in that it is a diagnosis development tool for robotic workcells. However, it is general-purpose in that it may be applied to any robotic workcell. We intend to apply the technique to different domains of interaction environment in the future.

7. Acknowledgments

We are grateful to Dr. Edward Devinney for his continued management support. Special thanks go to E. Niedermayr for indicating the importance of this research and the knowledge he provided during his tenure in our Princeton laboratories A. Marcantonio provided valuable workcell information. We also thank K. Winkelmann and J. Reimann for initial contributions to this diagnosis project.

References

[Cantone 86] Cantone, R. R., Lander, W. B., Marrone, M. P., and M. W. Gaynor.
IN-ATE: Fault Diagnosis as Expert System Guided Search.
In Bolc, L., and M. J. Coombs (editors), *Expert Computer System*. Springer-Verlag, New York, 1986.

[Chandrasekaran 84]
Chandrasekaran, B., and S. Mittal.
Deep versus Compiled Knowledge Approaches to Diagnostic Problem Solving.
International Journal of Man-Machine Studies, 19, 1984.

[Chiu 85] Chiu, M.Y., and E. Niedermayr.
Knowledge-based Diagnosis of Manufacturing Cells.
Siemens Forsch.-u. Entwickl.-ber. Bd., Spring-Verlag , 1985.

[Davis 82] Davis, R.
Diagnosis Based on Description of Structure and Function.
Proceedings of AAAI-82, Pittsburgh , 1982.

[Davis 83] Davis, R.
Diagnosis via Causal Reasoning: Paths of Interaction and the Locality Principle.
Proceedings of AAAI-83, Washington, D.C. , 1983.

[Davis 84] Davis, R.
Reasoning from Structure and Behavior.
International Journal of AI, 24, 1984.

[DeKleer 83] De Kleer, J.
AI Approaches to Troubleshooting.
Joint Services Workshop on AI in Maintenance, Boulder , 1983.

[DeKleer 86] De Kleer, J., and B. C. Williams.
Diagnosing Multiple Faults.
Proceedings of AAAI-86, Philadelphia , 1986.

[Genesereth 82] Genesereth, M. R.
Diagnosis Using Hierarchical Design Models.
Proceedings of AAAI-82, Pittsburgh , 1982.

[Hamscher 83] Hamscher, W.
Using Structural and Functional Information in Diagnostic Design.
Proceedings of AAAI-83, Washington, D.C. , 1983.

[Hart 82] Hart, P.E.
Direction for AI in the Eighties.
ACM SIGART Newsletter, 79, 1982.

[Koton 85] Koton, P.A.
Empirical and Model-based Reasoning in Expert Systems.
Proceedings of IJCAI-85, Los Angeles , 1985.

[Kuipers 85] Kuipers, B.
The Limits of Qualitative Simulation.
Proceedings of IJCAI-85, Los Angles , 1985.

[McDermott 82] McDermott, D., and R. Brooks.
ARBY: Diagnosis with Shallow Causal Model.
Proceedings of AAAI-82, Pittsburgh , 1982.

[Michie 82] Michie, D.
High-Road and Low-Road Programs.
AI Magazine, Winter 3(1), 1982.

[Nilsson 80] Nillsson, N.
Principles of Artificial Intelligence.
Tioga Publishing Company, 1980.

[Reiger 77] Reiger, C., and Grinberg, M.
The Declarative Representation and Procedural Simulation of Causality in Physical Mechanisms.
Proceedings of IJCAI-77, Cambridge , 1977.

A Spatial Modeling and Reasoning System for Robotics Research

Frank Pipitone

Navy Center for Applied Research in AI, US Naval Research Lab, Washington, DC 20375-5000, USA

Abstract

A new project is described whose goal is to build a software system which simulates, with appropriate simplifications, a complete robotic system. This system will serve, during its incremental development, as a research testbed for studying various pure and applied problems in perception, planning and learning in a three dimensional modeled world. Particular emphasis will be given to range-based vision, collision-free path planning, high-level task planning, and the application of machine learning techniques to these problems.

1. Introduction

This paper describes current plans for a new project at the US Naval Research Laboratory. The goal is to build a software system which can simulate, with appropriate simplifications, a "complete" robotic system. This system will serve, during its incremental development, as a research vehicle for studying various pure and applied problems.

A great deal of research has been performed in recent years on many specific areas of robotics, including high level control and planning, planning of grasps, collision-free path planning, and three dimensional visual and tactile perception. However, relatively little work has been done in integrating several of these capabilities into a system for research purposes. Two recent exceptions are the MIT Handey project [1] and the Carnegie-Mellon world-modeler's project [2]. The former includes physical components such as a manipulator with gripper and a laser rangefinder as well as software subsystems for grasp planning and execution, collision free path planning and execution, and three dimensional object recognition and localization. The latter is a software system involving the simulation of a set of objects governed by simple physics and sensory/effector interfaces to a controller. The controller is to be redesigned by various researchers to explore issues of learning and planning in the spatial world.

Our project is related to these two in that we are interested in exploring some of the perception, planning, and control problems addressed by them. However, in contrast to the Handey project, simulation will be used to avoid the need to support sensory and effector hardware. Range-based vision will be fully simulated, which the CMU project does not do.

We want the system to enable us to study several problems of key importance in advanced robotics. One is the use of range data for free-space characterization in mobile robot path planning and for the recognition and localization of known objects. Another is motion planning and control for mobile robot navigation and for manipulation. Also, the application of machine learning methods to the above problems will be pursued where appropriate. The simulation of interacting three-dimensional objects will require significant development effort, but this is not the main focus of our work.

2. Architecture

The major components will be a world modeling simulator, a robotic planning and control system (controller), and sensory and motor interfaces between them, as shown in Figure 1. These four modules could be used to describe almost any robotics project, and so we will describe each of them carefully. A human interface will provide the user with a high resolution color graphics display of the simulated world and allow him to request behavior traces, edit the world configuration, and send commands or knowledge to the controller. Such knowledge includes specific plans, planning heuristics, and other primarily procedural data.

The system will be implemented on a Sun 3 Workstation interfaced to a Silicon Graphics IRIS workstation. It will be programmed in LISP and C.

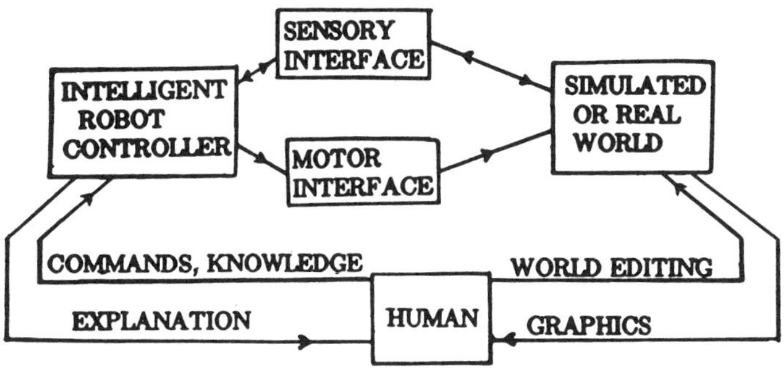

Figure 1. Overall System Architecture

3. The Simulator

Since we are not going to implement sensory and motor interfaces to the real world in this project, a simulator is necessary. Moreover, it will greatly facilitate experimentation, since it will be easy to modify and can be left running without fear of damage. This simulator will represent the world as a finite set of rigid objects, modeled as polyhedra, using octrees, Boolean combinations of linear inequalities graph-theoretic topological models, and compiled shape descriptors as needed for the various simulation tasks. The simulation will detect interferences and determine all contact points. These will be used in a simple physics simulator to estimate whether each object will be stable, sliding, rolling or in free fall. Approximate trajectories will then be followed until the contact points change so as to justify a new stability analysis. This physics simulator will be made as accurate as possible without severe speed compromise. Robot effectors will be modeled as other objects, except that they will be subject to position (and possibly force) control by the motor interface.

4. The Sensory and Motor Interfaces

The sensory interface will perform simulated rangefinding operations on the simulated world by raytracing. An octree representation of free space will be used for raytracing. The sensory interface will include simple touch sensing reflecting the collisions detected in the simulator between the effector and other objects. Other sensory modes, such as some simple form of photometric vision, may be included later if useful.

The motor interface will initially involve position control only, corresponding to changing the configuration parameters of the effector object and allowing the physics simulator to propagate the consequences. This seems sufficient to allow us to investigate many aspects of mobile robot planning, but in order to do manipulation planning, simulated force control will be implemented if the fidelity of the physics simulator is sufficient to warrant its use. In both motor and sensory operations, pseudo-random errors will be introduced to ensure that the controller performs robustly in the presence of uncertainty.

5. The Controller/Planner

The role of the controller is to plan and execute sensory and motor operations so as to achieve a dictated goal. This module will not be static. Its design will be the focus of the majority of our work as we use it to explore ways to achieve various navigation and manipulation capabilities in the simulated world. We will now describe some of the features we want to achieve in this module, first focusing on the vision subsystem and then on the planning subsystem.

Because spatial perception is of such crucial importance in advanced robotics systems, considerable effort will be given to a range-based vision subsystem of the controller/planner to allow experimentation with spatial planning methods in a testbed with rich perceptual feedback. One might argue that the vision problem needn't be addressed in this system because we could provide a "vision" function which simply finesses the three-dimensional recognition and localization problem by directly looking up object configuration coordinates in the simulator. We don't want to do this for two reasons. Firstly, we want to experiment with the intimate integration of vision operations with planning and control operations. For example, making one or two range observations in directions selected dynamically by the planner could efficiently resolve some specific ambiguity in the configuration of some object after performing a motor operation. Secondly, we want to do some vision work anyway. Specifically, we want to study the problem of recognizing and localizing objects using range data when the number of objects modeled is large. Much progress has been made recently in the case of a small number of objects [3,4].

The planning subsystem of the controller should incorporate a model of costs of planning, visual, and motor operations to minimize expected total task execution time. It will pursue goals in a hierarchical manner, taking advantage of problem decomposition where possible. It will use efficient algorithmic solutions for such subproblems as collision-free path-finding and perceptual tasks. Other subproblems will be solved by heuristically guided state-space search. As mentioned above, perception and action will be intertwined. That is, the planner will produce decision-tree-like plan fragments which alternate motor operations with confirming sensory operations. When a confirming sensory operation fails, the planner will generate another plan fragment. We expect to be able to make use of machine learning techniques to take advantage of the similarity between consecutive tasks found in most robotics applications. One potentially applicable learning approach is learning by analogy from previous experience [5], but it is premature for us to make claims regarding the most useful learning approaches for this domain.

6. Discussion of Goals

This work is motivated by a desire to address some of the problems which are the most significant obstacles to achieving the highly autonomous, adaptable and dextrous robotic systems of the future. This led to our focus on spatial perception and planning. However, these are subgoals of our primary goal of achieving efficient performance in an integrated robotic system. For example, we would like to be able to set up a configuration of objects in the modeled world, give the controller/planner a goal involving some modification of that configuration requiring many motions, and have the system use its computational resources efficiently to achieve the goal. This will require us to explicitly consider the tradeoffs among the many ways of combining perception and planning techniques to achieve high system performance. We expect that our work on optimizing this system level performance will be the most exciting and beneficial aspect of the project.

References

1. Lozano-Perez, T., Jones, J. L., Mazer, E., O'Donnell, P. A., Grimson, W. E. L., "Handey: A Robot System that Recognizes, Plans, and Manipulates", Proc. IEEE International Conference on Robotics and Automation, Raleigh, North Carolina, 1987.
2. Carbonell, J., Hood, G., "The World Modelers Project: Objectives and Simulator Architecture", in Machine Learning, a Guide to Current Research, ed. Mitchell, T. M., Carbonell, J. G., and Michalski, R. S., Kluwer, 1986.
3. Faugeras, O. D., Hebert, M., "The Representation, Recognition, and Locating of 3-D Objects", The International Journal of Robotics Research, Vol. 5, No. 3, Fall 1986.
4. Grimson, W. E. L., Lozano-Perez, T., "Model-Based Recognition and Localization from Sparse Range or Tactile Data", The International Journal of Robotics Research, Vol. 3, No. 3, Fall 1984.
5. Carbonell, J. G., "Derivational Analogy: A Theory of Reconstructive Problem Solving and Expertise Acquisition", in Machine Learning, Vol. II, ed. Michalski, R. S., Carbonell, J. G., and Mitchell, T. M., Morgan Kaufmann, 1986.

A Structural Framework for an Expert Authoring System

Cheickna Sylla,* and Ramaswamy Ramesh†

*College of Engineering, Drexel University, Philadelphia, PA 19104, USA
†School of Management, Buffalo, NY 14260, USA

ABSTRACT

We present a structural framework for an expert authoring system called EXAS. The purpose of this system is to improve the efficiency of producing and maintaining the technical information required to support equipment usage in a variety of applications. This automated system is designed to lower the cost of technical documentation supporting equipment in various installations. EXAS integrates two expert systems developed at the Department of the Navy. These two systems have been designed for editorial operations on written narrative material and automated page layout of procedure manuals, respectively. The integration of the two systems by EXAS provides the unique facility to produce graphically illustrated procedure manuals by an automatic process, starting from written narratives on product and procedure descriptions. We present the ideas underlying this integration, and discuss the salient features of the natural language processing and knowledge representation mechanisms in EXAS.

1. Introduction

In this paper, we present an approach to natural language processing within a framework of an expert system for authoring applications. The system under consideration is called EXAS, which performs editorial and layout functions in the authoring process. EXAS is an integration of two expert systems developed under the Naval Technical Information Presentation Program (NTIPP), which is an effort by the Navy to improve the efficiency of producing and maintaining information required to support equipment. The two systems that are integrated by EXAS are the Computerized Readability and Editing System (CRES, see Kincaid et al., 1981, Ramesh et al., 1985), and Job Performance Aids (JPA, see Sylla, 1983, Babu and Sylla, 1981). CRES is an expert editorial system that takes written narrative material as input, processes it, performs syntactical analysis, spelling corrections, lexical analysis, readability analysis and provides editorial suggestions on text simplification and improvement. CRES enables technical writers to adhere to writing rules contained in the military standards and guidelines developed at the department of the Navy. In some ways, CRES is similar to other text editing systems such as Bell Laboratories' Writer's Workbench (see MacDonald et al., 1982) and Westinghouse Electric Company's WRITEAIDS (see Kniffen, 1984). JPA and a related system called Paper Procedure Training Aids (PPTA, see Sylla, 1983), are two systems that perform automatic page layout operations in the preparation of technical manuals that are composed of graphically illustrated procedure steps.

The organization of this paper is as follows. Section 2 describes an overview of EXAS. Section 3 describes the natural language processing strategy adopted in EXAS. Section 4 provides the conclusions.

2. Operation of EXAS

The integrated expert authoring system begins processing from a narrative material written by an user. The material consists of two parts: a product description and a procedure description. EXAS is designed to process the input material written by an equipment designer and eventually produce an user manual consisting of graphically illustrated procedure tasks for the use of the equipment. The procedural steps employed in EXAS are as follows.

Step 1

EXAS will analyze, simplify and improve the text. This phase of EXAS is handled by the CRES module incorporated in it.

Step 2

If the text is for procedure training aids, EXAS will provide the following elements to the author upon request.

a. A breakdown of the text into logical sequence of task steps.
b. The logical illustration sequence of the steps obtained above with picture label relationships. This is done by the JPA module incorporated in EXAS.

The breakdown of the text into logical sequence of tasks steps requires a comprehension of the input text, and representing the knowledge contained in the text according to a set of semantic networks. For this purpose, EXAS employs a natural language processor and a semantic interpreter. The processor breaks a sentence down into its structural components, and the semantic interpreter extracts the knowledge from the sentences, assembles information, and builds semantic networks from which the procedural task steps are identified. In this integration, the entire procedure is broken down into a sequence of logical task steps and the components of the equipment that are needed to illustrate each task step. Using this network, JPA designs the procedure manual, and the layout of the illustration.

The purpose of this paper is to present the design and implementation of the natural language processor used in Step 2. We present the underlying concepts and details in the following section.

3. Natural Language Processing Strategy

We will describe some of the basic concepts of systematic grammar and analyze how it is applicable in our problem. Systematic grammar pays more attention to the way a language is organized into units, each of which has a special role in conveying meaning. In English, we can distinguish three basic ranks of units, the CLAUSE, the GROUP, and the WORD. In this analysis, the WORD is the basic building block. There are word classes like "adjective", "noun", "verb", and each word is an integral unit (that is, it is not chopped into hypothetical bits; valves = valve + s). Instead, we can view each word as exhibiting "features" (The word "valves" is the same basic item as "valve", but has the feature "plural"). Also, the words "took", "take", "taken", "taking", etc. are all the same basic word, but with different features such as "past participle", "infinitive", etc.

The next unit above the WORD is the GROUP of which there are 4 types (NOUN GROUP, VERB GROUP, PREPOSITION GROUP, ADJECTIVE GROUP). Each one has a particular function in conveying a meaning. NOUN GROUP (NG) describes objects, VERB GROUP (VG) carries complex messages about time and logical status of an event or relationship, PREPOSITION GROUP (PREPG) describes simple relations, while ADJECTIVE GROUP (ADJG) convey other kinds of relationships and descriptions of objects. Each GROUP can have "slots" for the words of which it is composed. (ex, NG has slots for "determiner", "numbers", "adjectives", "classifiers", and a "noun"). Each GROUP can also exhibit features, just as a WORD can.

Finally, the top of the rank is the CLAUSE. The clause is the most complex and diverse of the language, and is used to express relationship and events, involving time, place, manner, and many other aspects of meaning. Some of the CLAUSE features are shown in Figure 1.

We know that clauses are made up of groups, which are in turn made up of words. However, few sentences have this simple three-layer structure. Groups often contain other groups. Also clauses can be parts of other clauses and can be used as parts of groups ("the valve which has high pressure"). This phenomenon is called "rank shift", and is one of the basic principles of systematic grammar.

In the case of Verb group, after identifying the verb group, the voice of the verb group is determined to be either Active or Passive. Active places the semantic subject first, while Passive puts the semantic object first. If there is a PREPG beginning with "by", it is interpreted as the semantic subject, and the Clause has the feature Agent. The group structures are shown in Figure 2.

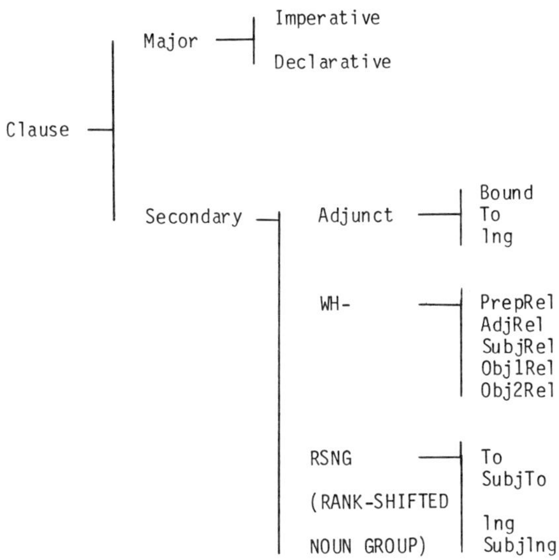

Ex)

PrepRel	:	the valve with which the engine blocks fluid
AdjRel	:	the method with which the text tells you a step
SubjRel	:	the valve which blocks the fluid
Obj1Rel	:	the process that I want
Obj2Rel	:	the process which the text gave me
To	:	The last step is to handle the valve
SubjTo	:	I wanted you to do the job
Ing	:	Building ship is hard work
SubjIng	:	They liked his doing the job

Figure 1. CLAUSE features

3.1 Parsing Strategy

In implementing a systematic grammar, we deal with a string of letters, and recognize patterns and features in it. We need interpretation rules which look at a pattern, identify its structure, and recognize its relevant features.

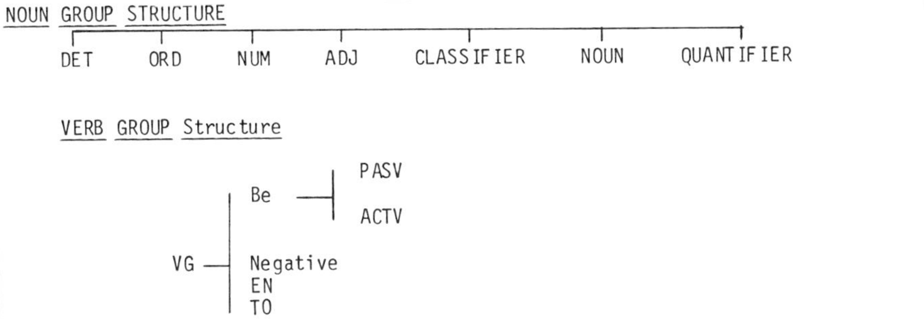

Figure 2. Group Structure

So, the parsing system should be an interpreter for a language for writing grammars. It is basically top-down, left-to-right parser. The parsing system uses special tools to handle complex constructions and irregular forms. For example, we can set up programs to define certain words like "and", and "or", as "demons", (see Winston, 1984) which cause an interrupt in the parsing process wherever they are encountered in the normal left-to-right parser order, in order to run a special algorithm for conjoined structures. In fact, the process can be interrupted at any point in the sentence, and other computations (either semantic or syntactic) can be performed before going on. These themselves may do more parsing, or they may change the course.

The ability to integrate semantics with syntax is particularly important in handling discourse, where the interpretation of a sentence containing "pronouns" may depend on the preceding discourse and knowledge of the subject matter.

This language process is not segmented into the operation of a parser, followed by the operation of a semantic interpreter. Rather, the process is unified, with the results of semantic interpretation being used to guide the parsing.

Therefore, in treating ambiguity, this process does not backup all the possible interpretations of a sentence, but looks for the first one and if that fails looks for the next one intelligently.

First of all, we can think of a grammar as a set of instructions for parsing a sentence in the language. A rule like: "NP -- Determiner Noun" can be interpreted as the instruction "If you want to find a NP, look for a Determiner followed by a Noun." Above grammar can be determined as shown in Figure 3.

As the procedure is followed, a parsing tree is built to keep track of the results. The basic function used in writing a grammar is Parse, which builds the structure of the parsing tree. Its arguments are syntactic features (including the name of a syntactic unit) specifying the type of node being searched for at that point in the parsing. It tries to add a constituent of the specified type to the parsing tree, using the words yet unparsed in a sentence, in left-to-right order. The different syntactic units, such as Clause and Noun Group each correspond to a grammar program for parsing them. If asked to parse one of these, Parse activates the program for that unit, giving it as input the part of the sentence yet to be parsed and the list of initial features. If the command is to parse something other than a unit, Parse interprets its arguments as a list of features which must be found in the dictionary definition of the next word in a sentence. If so, it attaches a node for that word to the parsing tree, and removes the word from the remainder of the sentence.

If not, it fails. If a Grammar program has been called and succeeds, the new node is attached to the parsing tree as a daughter of the node for the unit whose program initiated the Parse instruction. It it fails, the tree is left unchanged.

3.2 The Semantics

Given a syntactic parser with a grammar of English and a deductive system with a base of knowledge about particular subjects, the role of semantics is to fill the gap between them.

The semantic system must describe relationships at three different levels (word, group, clause). First it must define the meaning of word. We pointed out in parsing strategy that the full meaning of a word or concept cannot be defined in a simple dictionary entry, but it involves its relationship to an entire vocabulary and structure of concepts.

In the next level, groups of words also have meanings in the syntactic structures. We need an analysis of the ways in which English structures convey meanings.

Figure 3. Grammar Determination

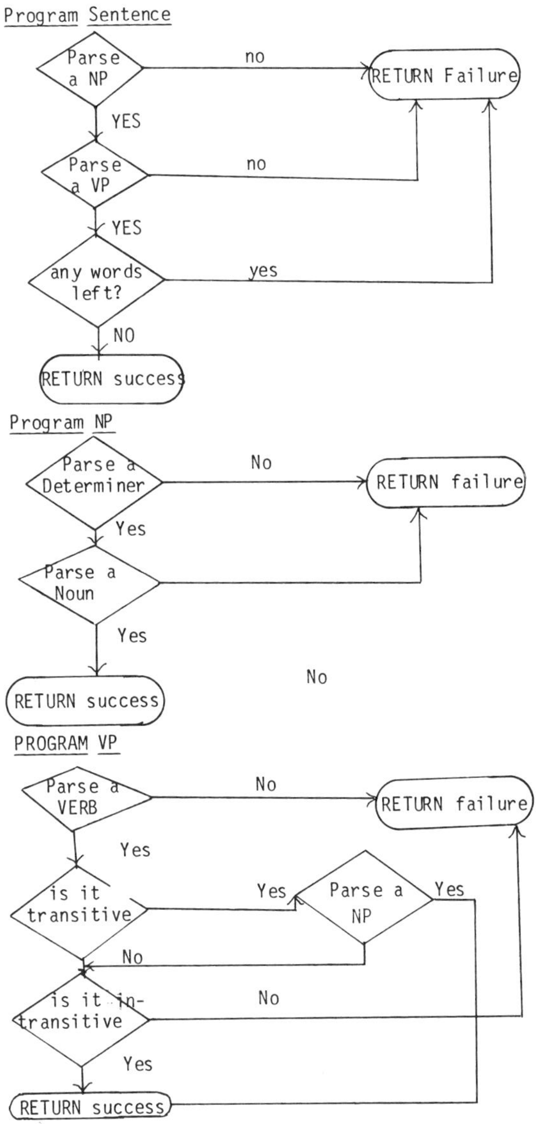

Finally, a sentence in natural language is never interpreted in isolation. A semantic system must describe how the meaning of a sentence depends on its context.

With these goals in mind, we will consider how to implement such a semantic system.

i) We use the Noun Group (NG) to describe objects. It contains a noun (which indicates the kind of objects), adjectives and classifiers (which describe further properties of the object) and a complex system of quantifiers and determiners (describe its logical status).

ii) We use the CLAUSE PREPOSITION GROUP (PREPG), and ADJECTIVE GROUP (ADJG) to describe relationships and events.

Since each of these semantic specialists can work separately, there is no need to wait for a complete parsing before beginning semantic analysis. (ex. NG specialist can be called as soon as NG has been parsed, to see whether it makes sense before the parser goes on).

Any semantic program will have full control over the deductive system and can even call the grammar to do a special parsing before going on with the semantic analysis. For this reason, it is very hard to classify the semantic analysis as "top-down" or "bottom-up".

Let us briefly look at the semantic structures used to describe the example sentences which we used in building a parsing tree.

A phrase such as "The emergency gear which you picked up, is supported by a big red gear release lever," would be built up from the descriptions of various objects. First, the NG parsing program finds the determiner "the" and noun "emergency gear."

At this point Semantic-NG is called and creates the object semantic structure. Notice that Semantic-NG is called before NG is finished, as soon as an initial semantic analysis is possible.

Next the NG program looks for a qualifier. However there are no qualifiers. Then the Clause program wants a VG, and from the parsing process, we already know that "is" is used as an auxiliary verb, and the main verb is "supported." Also we know that the sentence is passive, and the agent of the action is "gear release lever."

At this point Semantic-NG is called again to construct a semantic structure. After parsing the NG Qualifier "which you picked up," we know that the NG "the big red gear release lever" is the object to which the clause is being related. Semantic-clause knows this since the parser has noted the feature ObjRel. The definition of the verb "support" is called and a Relation semantic structure is built.

The semantic structure of the above two sentences is shown in Figure 4.

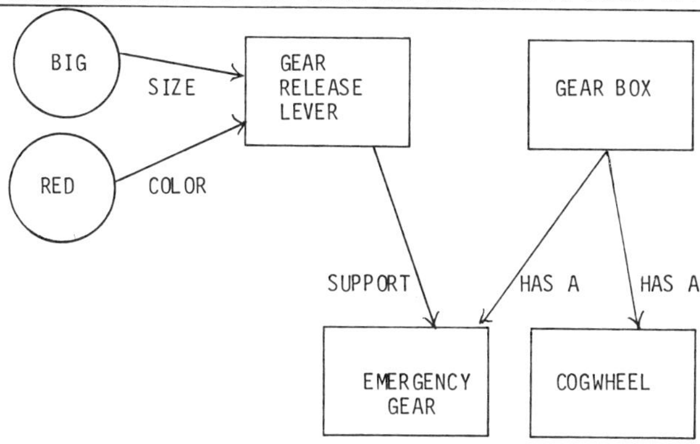

Figure 4. Semantic Structure

A semantic system must account for multiple meanings of word phrases, and sentences, and pick up each relevant meanings.

If a word has two meanings, then two semantic descriptions are built simultaneously and used to form two separate phrase interpretations. However, in this case there is a danger of a combinatorial explosion (see Winston, 1984). In such cases, the interpreter chooses the most appropriate meaning based on the semantic constructs built so far. Several strategies have been proposed in the literature for resolving such conflicts and is an area of considerable research interest. One of the solutions for this problem is to do the interpretation continuously.

If two different possible interpretations are encountered, humans immediately look in their memories to see which is meaningful in the current context. Since this system has a systematic grammar and the semantic specialists have full control, we can intermix the grammar, semantics and deduction easily, and therefore it is possible to do this kind of interpretation.

4. Conclusion

In this paper, we have described an overall framework of an expert authoring system and discussed the salient features of the natural language processor incorporated in it. EXAS integrates two expert systems developed at the Department of the Navy. These two systems have been designed for editorial operations on written narrative material, and automated page layout of procedure manuals, respectively. The integration of the two systems by EXAS provides the unique facility to produce graphically illustrated procedure manuals, starting from written narratives on product and procedure descriptions. This system, we hope, will have far-reaching effects in automated publishing in a wide variety of application areas. In particular, large-scale publishing houses such as the military will be the major beneficiaries of this automated process.

Bibliography

1. Babu, A. J. G. and Sylla, C., "COMALAY: Automated Layout of Naval Training Course Manual," Technical Report 81-B1, Department of Industrial Engineering, SUNY at Buffalo (1981).

2. Kincaid, J. P., Aagard, J. A., O'Hara, J. W. and Cottrell, L., "Computer Readability Editing System," IEEE Trans. on Prof. Comm., Vol. PC-24, No. 1, (1981) pp. 38-41.

3. Kniffen, J. D., "Computer-aided Editing - Present and Future," Proceedings of the 29th Annual Meeting, Technical Documentation Division, American Defense Preparedness Association, San Antonio, Texas (1984).

4. MacDonald, N. H., Frase, L. T., Gingrich, P. S. and Keenan, S. A., "The Writer's Workbench: Computer aids for text analysis," IEEE Trans. on Comm., Vol. 30 (1982).

5. Ramesh, R., Babu, A. J. G. and Kincaid, J. P., "Efficient Algorithms for Variable-depth Trie Index Optimization: Theory and Experimental Results," Working Paper No. 684, School of Management,SUNY at Buffalo (1986).

6. Sylla, C., "Computer Aided Layout of Job Performance and Procedure Training Aids," Ph.D. dissertation, Department of Industrial Engineering, SUNY at Buffalo (1983).

7. Terrel, W. R., "Computer Automated Page Layout (PLA) for Text-Graphic Materials," Users Guide (TAEG.T.R. 137), Orlando, Florida (1982).

8. Winston, P. H., Artificial Intelligence, Addison Wesley (1984).

Robot Selection Through Fuzzy Set Optimization

M.J. Wang, H.P. Singh, and W.V. Huang

Division of Industrial Engineering, Alfred University, Alfred, NY 14802, USA

ABSTRACT

Robots have made a significant impact in today's industry. Accompanied by the growing demand of industrial robots, the number of robot makers and suppliers is also increasing. However, the growing trend of robot applications causes more complexity in robot selection mechanism.

This paper seeks to introduce a robot selection method which is based largely on fuzzy set theory. The subjective factors of a robot are evaluated by fuzzy set membership function and the objective factors are evaluated by marginal value function. Both data then go through the fuzzy set multicriteria decision making algorithm and a final decision set is generated.

INTRODUCTION

Robots have played a significant role in today's manufacturing world. There is even a greater future ahead for robot as engineers begin to unwind the application spiral. Not only are robots used in manufacturing to help to reduce labor costs and to increase productivity and quality, but they are also used in situations which are risky and hazardous for human operator (e.g. welding, heat treatment, etc.).

The widespread applications and the increasing growth of the robot industry are now creating problems in a new direction. The selection mechanism for robots is becoming more and more complex as the number of robot vendors increases. Therefore, the prospective buyer of a robot commences a multicriteria decision making process with different robot attributes versus the alternative vendors. In general, the attributes of a robot can be classified into two major categories - objective and subjective factors. For objective factor, examples are cost, repeatability, load capacity, speed, gripping force, number of axes, etc. The subjective attributes include man/machine interface, vendors service contract, reputation, programming flexibility, robot language, etc. The usual robot selection process can be made either in a very simplistic manner which is rather subjective and random or by using a systematic selection method. However, not too many robot selection studies have been reported, and some of the methods fail to corporate the quantitative and non-quantitative factors into the multicriteria decision model.

This paper seeks to introduce a robot selection method which is rooted largely in the concept of fuzzy set theory. In this method, the subjective attributes of a robot are evaluated and expressed in terms of fuzzy set membership function and the objective attributes are evaluated through marginal value function applied by Jones et al. (1985). After the technical experts have gone through their evaluation on all the factors, the experts judgment data are processed following the fuzzy set multicriteria decision making algorithm developed by Hipel (1982). Then a final decision set will be generated.

FUZZY SET MULTICRITERIA DECISION MAKING ALGORITHM

The prescribed methodology can be administered through the following stepwise procedure:

1. Select the desired attributes for the robot to be purchased. Place the robot attributes into subjective and objective categories.
2. Identify the appropriate marginal value functions for each of the objective attribute (refer to Jones et al., 1985).
3. Tabulate the value of each objective attribute for each alternative.
4. Compute the value for each objective attribute of each alternative using the marginal value function.
5. Summarize each expert's judgment for all subjective attributes as a rating matrix.
6. Develop a modified pessimistic aggregated rating matrix from these rating matrices.
7. Combine values of the subjective and the objective attributes to form a new matrix C.
8. Develop a dominance matrix D of alternatives vs. alternatives for the combined matrix C.
9. Develop a modified aggregated weighting vector for all the

attributes involved. The transpose of this weighting vector is W^T.

10. Multiply W^T by C to obtain the final decision matrix.

METHOD - A HYPOTHETICAL EXAMPLE

Step 1: The important robot attributes were selected and categorized as in the following table:

Table 1: Important Attributes Being Selected for Evaluation

Objective	Subjective
Cost	Reputation
Repeatability	Vendor Service Contract
Load Capacity	Man/Machine Interface
Number of Axes	Programming Flexibility

Subsequently, five vendors were also identified.

Step 2: Considering the objective attributes, the marginal value functions were selected. The marginal value functions for the chosen objective attributes are shown in Table 2.

Table 2: Corresponding Marginal Value Functions for the Objective Attributes

Attributes	Marginal Value Function
Cost	$[(X_i^* - X_i)/(X_i^* - X_i^\circ)]^{1/2}$
Repeatability	$(X_i^* - X_i)/(X_i^* - X_i^\circ)$
Load Capacity	$(X_i - X_i^\circ)/(X_i^* - X_i^\circ)$
Number of Axes	$(X_i - X_i^\circ)/(X_i^* - X_i^\circ)$

where X_i^* denotes the highest attribute value

and X_i° denotes the lowest attribute value

Step 3: The objective attribute values for each robot vendor (alternative) identified for the decision model were then tabulated as in Table 3.

Table 3: The Objective Attribute Values for the Five Vendors.

Alternative	Cost (X $1000)	Repeatability	Load Capacity	Number of Axes
1	52	.012	150	2
2	86	.008	200	4
3	75	.016	160	4
4	92	.002	190	5
5	81	.010	220	3

Step 4: Using the marginal value functions of Table 2 and the objective values of Table 3, a scaled value for each attribute of each alternative were computed. The results are shown in Table 4.

Table 4: Scaled Value for the Objective Attribute of the Five Vendors

Alternative	Cost	Repeatability	Load Capacity	Number of Axes
1	1.0	.285	0	0
2	.921	.571	.714	.666
3	.758	0	.142	.666
4	0	1.0	.571	1.0
5	.524	.428	1.0	.333

Step 5: For subjective attributes, a questionnaire is evaluated by each of the K technical experts. The experts rate each attribute against each alternative from the scale of 0 to 1. The rating is considered as the fuzzy set membership function assigned by the expert. A rating matrix will then be developed from each expert judgment.

Step 6: An aggregated rating matrix is calculated to represent K technical experts opinion. Using the fuzzy set operation, a modified pessimistic aggregated matrix is computed from:

$$r_{ij} = \frac{1}{2}\left[(r_{ij}^{1} \cap r_{ij}^{2} \cap \ldots \cap r_{ij}^{K}) + \left(\frac{1}{K}\sum_{k=1}^{K} r_{ij}^{k}\right)\right] \text{ (Hipel, 1982)}$$

The results for this modified pessimistic aggregated matrix are shown in Table 5. The value in the modified pessimistic aggregated matrix is considered as the fuzzy set membership function for each subjective attribute against different alternatives.

Table 5: The Modified Pessimistic Aggregated Matrix

Subjective Attribute	Alternative 1	2	3	4	5
Reputation	.615	.565	.465	.450	.485
Service Contract	.505	.505	.620	.620	.350
Programming Flexibility	.345	.385	.530	.530	.175
Man/Machine Interface	.275	.310	.575	.515	.475

Step 7: A MXN matrix C of the combined attributes is established as shown in Table 6, where M is the objective and subjective attributes, and N is the number of vendors (alternatives). For this example, the C matrix is of the order (8X5) with each r_{ij} entry scaled from 0 to 1. Note Table 4 and Table 5 are combined to form Table 6.

Table 6: Combined matrix C of Scaled Values for Objective and Subjective Attributes

Attribute	Alternative 1	2	3	4	5
Cost	1.0	.921	.785	0	.524
Repeatability	.285	.571	0	1.0	.428
Load Capacity	0	.714	.142	.571	1.0
Number of Axes	0	.666	.666	1.0	.333
Reputation	.615	.565	.465	.45	.485
Service Contract	.505	.505	.75	.62	.35
Programming Flexibility	.345	.385	.51	.53	.175
Man/Machine Interface	.275	.31	.4	.575	.425

Step 8: A dominance matrix D is calculated for the 5 alternatives of the C matrix. This D matrix is used for comparison. An element d_{ij} of this matrix emphases the number of attributes for which the jth alternative dominates over the ith alternative. The dominance matrix derived for this example is shown in Table 7.

Table 7: The Dominance Matrix D of the C Matrix

	Alternatives					
Alternative	1	2	3	4	5	Total
1	-	5	5	6	4	20*
2	2	-	3	5	12	12
3	3	4	-	5	4	16
4	2	3	3	-	3	11
5	4	6	4	5	-	19
Total	11+	18	15	21	13	78

*The number of times the first alternative is dominated by the other alternatives.

+The number of times the first alternative dominates the other alternatives.

Note, the higher the column sums, the lower the row total and vice versa.

Subtract the row sums from the column sums. This gives:

A1	A2	A3	A4	A5
-9	6	-1	10	-6

The higher this difference, the better the alternative. Rank the alternative in descending order. That is:

A4 A2 A3 A5 A1

Step 9: It is also recommended that a questionnaire be designed such that a weight be assigned to each attribute. The weighting is scaled from 0 to 1. Similarly, a modified pessimistic aggregated weighting vector can be computed. The transpose of this vector can also be derived. For this example, W^T is:

$$W^T = (.106, .151, .121, .117, .135, .157, .126, .086)$$

Step 10: Finally, a weighted aggregated vector V will be calculated.

$$V = W^T \times C$$

This V will be a matrix of order 1 X N. For this example, the result gives

[.378, .579, .517, .610, .460]

A1 A2 A3 A4 A5

The ranking order from the decision matrix is

A4 A2 A3 A5 A1

The result suggested that alternative 4 will be the best choice for the company.

DISCUSSION AND CONCLUSION

According to Hipel (1982), the advantage of using this fuzzy set approach is that fuzzy set membership function can be used to quantify both subjective and objective factors. However, this method chose to use marginal value function for the objective factors and fuzzy set membership function for the subjective factors, and to combine both factors for further fuzzy set operations. Since the marginal value function can reduce the subjectivity being caused by using membership function for the objective factors, it seems that the validity of this fuzzy set multicriteria modeling method will be improved.

One approach that can help a decision maker to validate the final results is to perform a sensitivity analysis. The analysis can be performed by changing the input data or using different aggregation procedure to see whether the same results are obtained. If sensitivity analysis reveal conflicting results then further data collection and analysis need to be carried out to assure concrete conclusions to be determined (Hipel, 1982).

In order to find a robot that perfectly fits the needs of the company, a task analysis prior to the selection analysis is essential. Through detailed task analysis, the minimum level of functions and capabilities required by the robot can be specified. Subsequently, a few robot candidates which can meet the specific manufacturing needs will be determined. Then fuzzy set multicriteria decision-making analysis is performed to combine the experts opinion. Finally, a robot which can best serve the needs and the interests of the company will be selected.

REFERENCES

1. Jones, M.S., and Malmborg, C.J. and Agee, M.H. Decision Support System Used for Robot Selection. Industrial Engineering. Vol. 17, No. 9 September 1985, pp. 66-73.

2. Hipel, K.W. Fuzzy Set Methodologies in Multicriteria Modelling. In M.M. Gupta and E. Sanchez (eds.) Fuzzy Information and Decision Process. North-Holland Publishing Company, 1982.

Robot Programming Languages—A State of the Art Survey

K. Srihari and Dr. M.P. Deisenroth

Department of Industrial Engineering and Operations Research, Virginia Polytechnic Institute and State University, Blacksburg, VA 24061, USA

The brief history of textual robot programming language development through the last two decades is traced. Individual languages are then reviewed with respect to their inherent characteristics such as data types, control structures, use of sensors, interfacing capacity with external machines, etc. Reviewed languages range from WAVE released in the early 1970's to AML/X developed in 1986.

1.0 A Historical Introduction to Textual Robot Programming Languages

In 1973, WAVE was developed as an experimental language for robot programming in research applications at the Stanford Artificial Intelligence Laboratory [5,29]. AL was developed in 1974 [13]. Based on ALGOL, it illustrates the factors that have to be considered for flexible control of mechanical manipulation. The first two robot languages developed at the IBM Watson Research Center was Emily and ML [5]. They were used in several assembly applications. AUTOPASS, a high level programming system for computer controlled mechanical assembly, was also developed here. The Watson Research Center was also instrumental in the development of the AML family [18,40]. These languages are used to program the IBM range of robots such as the 7535, 7545, and the 7565.

A Robot Programming System was developed by SRI International which supported the RPL language [20,26] and is capable of controlling robots and machine vision systems among other industrial automation equipment. JARS, based on Pascal, was developed by academia to control robots engaged in solar cell assembly [10]. VAL was the first language to be designed and implemented in industrial applications and was designed for the Unimation Puma Robot in the latter part of 1978 [5,20]. The initial version, developed in 1975, was based on WAVE. In 1976 VAL was revised as VAL II [20,36]. In the latter part of 1986, VAL III was introduced.

The Manufacturing Control Language (MCL) was developed by McDonnell Douglas in 1981 [43]. Developed by Automatix Inc., RAIL can accommodate a machine vision interface [30,31]. RSS, developed in 1982, describes robot actions by declaring in software servo processes that control the robot's degree of freedom [14]. In the same year, General Electric introduced HELP, a Pascal based robot programming system that could control multiple arms in assembly tasks [20]. In 1983, American Robots introduced the AR-BASIC robot programming system to use the interpretive facility of Basic in robot programming [16,17].

In 1983 LMAC, a software system to control robots operating in a flexible manufacturing system, was developed [21]. HIROB is a high level language that has the syntax of English. This language, developed in an academic atmosphere, facilitates easy communication with the robot [4]. The MARTI programming system was introduced in 1984 [32]. This system consisted of a robot language independent, interactive, user friendly interface, and a robot dependent language generator. KAREL, a language for robot control was introduced in 1985 by GMF Robotics, [23]. In 1985, ASEA Robots of Sweden introduced ARLA, a textual robot programming language intended for use in a highly automated factory [7]. RoboTalk was developed and released in 1985 [34]. The Universal Robot Control System (UROCS) was developed in 1985 in an academic atmosphere [22]. KARMA, a knowledge base programming technique, uses an AI approach and was first presented in 1984 [24].

2.0 Robot Programming Languages - A Review

2.1 WAVE

Developed at the Stanford Artificial Intelligence Laboratory between 1970 and 1975, WAVE was one of the first robot programming languages to be developed [5,29]. It specifies robot motions, provides force and torque control and can interface with external vision systems. WAVE provides for symbolic description of robot manipulator actions with end effector positions described in terms of Cartesian coordinates. This system functions at two levels: the planning level and the execution level. Manipulator programs are assembled by this system into a trajectory file analogous to the object module. This is interpreted during the execution mode. The system can control two robot arms simultaneously, which simplified the task of programming assembly operations. WAVE does not provide for collision avoidance.

2.2 AL

This successor to WAVE was also designed at Stanford [13] and was developed for specific manipulator tasks such as assembly of an object from parts. AL has a file structure similar to ALGOL. Transportable in nature, it contains a source language similar to ALGOL, a translator to convert programs into executable code, and a runtime system to control manipulators and external devices. AL allows nesting of subroutines and parameter passing. A general purpose text macro facility is also available. AL can control multiple arms and execute several processes simultaneously. It does not provide for collision avoidance. Today, commercial arms are being controlled by the AL system. A new version of AL, written in OMSI Pascal, is available for industrial applications.

2.3 AUTOmated Parts ASSembly (AUTOPASS)

Developed by IBM, AUTOPASS is a task level language with structured programming constructs [26]. The user would generate an assembly plan, and code it into an AUTOPASS source program. A description of the initial state of the assembly world would be described through the geometric design processor. The AUTOPASS compiler processes the source code one statement at a time.

Embedded in a subset of PL/I, this language has access to many of the control and data types of that language. There are four types of AUTOPASS statements: state change statements, tool statements, fastener statements, and miscellaneous statements. The ambiguities caused by the high level of this language are addressed by the compiler which operates at two levels. At the upper level, the syntax is checked, operands are evaluated and general feasibility checks are performed. At the lower level, the compiler attempts to generate object code. This system permits multiple manipulators to be used, supports parallel processing, and could interface with complex sensory systems. However, it is still unimplemented and does not take into consideration collision avoidance or emergency decision making.

2.4 Robot Programming Language (RPL):

This language is supported by the Robot Programming System (RPS) developed by SRI International [19,20]. This system can control robots, vision systems, and several other types of external equipment used in an industrial atmosphere. Intended for application in small and medium sized automatic manufacturing systems, RPL was developed to provide an inexpensive operating system that could be used in robot control.

The power and complexity of RPL is somewhere between that of Basic and Fortran. A RPL program would consist of an optional common block, a mandatory named program block, and optional subroutine blocks. Since RPL has only one executable statement, it does resemble Lisp, though in a superfluous manner. The system provides debugging features such as comprehensive error messages, manual interrupts, trace facility, single stepping, breakpoints, and saving and reloading procedures. RPL has been used to control a Unimation Puma 550 arm and a Machine Intelligence vision system, a Unimate 2000A and 2000B hydraulic arms and a SRI vision system.

2.5 Manufacturing Control Language (MCL)

An extension of APT, MCL has the capacity to control machine tools, robots, sensing devices, parts feeders, and other devices such as robotic workcells [43]. It has the capacity to control any robot. MCL systems can take logical decisions based upon real time sensory data and it can interact with vision systems. It allows the programmer to use different frames of reference to define points. These definitions would be interpreted by lower level control systems with respect to cell hardware. The system operates in a hierarchical manner. At the highest level of the system, programmers create control programs independent of cell equipment. At the lower level, postprocessing of programs for specific cells is carried out. Each cell would contain a unique collection of robots, machine tools, sensory devices, and other manufacturing equipment.

2.6 KARMA

This language represents a novel approach to robot programming [24]. The user has to tell KARMA the tasks that the robot is required to perform. Using AI techniques, the system would search through its knowledge base, synthesize inferences made, clarify any ambiguous statements, determine the appropriate sequence of operations and teach the robot the requisite procedure. Although the scope of the problem, its goals and objectives have been defined, a great deal of research is required before this system could become entirely operational. This approach does have the potential to reduce the time and effort required to program a robot in an industrial environment. It is proposed to develop a graphical interface to this system to help the user interface with this system, and to reduce the time required to produce acceptable software.

2.7 ASEA Robot LAnguage (ARLA)

The ASEA off-line programming system consists of a high level language, ARLA and a personal computer [7]. The menu driven, user friendly man-machine interface enables the operator to easily learn and use the system. The system includes an editor, a compiler, a linker, and a file manager. In addition to programs, it is possible to edit process data and robot data off-line and transfer it to the robot controller. ARLA is a specialized high level language that contains motion control functions, I/O handling, arithmetics, and other logical functions.

2.8 AR-BASIC

Developed in 1983, this language is used by American Cimplex to control their Merlin series of robots [16,17]. This system consists of five components: the

BASIC language, position definition and control statements, Input/Output support for devices, text editing capacity, and memory and filing methodology. AR-BASIC has high level commands to specify robot manipulator arm position, to control the arm's motion, and to conduct I/0 operations. Motion control statements in AR-BASIC allow control of type of tool tip motion, speed, current frame of reference, and current tool tip definition.

This language provides the user with a full screen text editor that allows examination and modification of any segment of the software. The linking of the text editing and software execution environments enhances the utility of the software development environment. This interpretive system supports interactive program debugging. Debugging features provided include breakpoints, single stepping, partial program execution, and display of descriptive error messages. An on-line help facility is also provided. This language has the capacity to interact with external vision systems.

2.9 VAL & VAL II

This real time robot control system was first developed in 1975 [20,35,41]. It was initially based on WAVE. In 1976, VAL was rewritten, and in the latter part of 1978 it was offered commercially as an integral part of the PUMA robot control. Since it is an interpretive language, the editing, compiling, linking, and loading functions are not separate entities. Hence, the editor is also used in debugging, since any changes to the software can be made and tested quickly. VAL has no structural constructs, has limited subroutines capabilities, but does allow branching. Commands in VAL resemble the English language, and are indicative of the task to be performed. Language unity is not a strength of VAL.

VAL II contains extended features which include: network communication facilities, an increased interfacing capabilities with external sensors, enhanced math capabilities similar to that found in a high level language, a better operator interface, realtime path modification based on internal and external signals, and a capability to perform robot and process control activities simultaneously [6,36]. VAL II attempts to integrate flexible robot control with communication facilities, application support, and interface with sensors while simultaneously making the system easier to program and operate.

2.10 KAREL

Created and implemented at GMF Robotics in 1985 [23,42], KAREL's objectives were to develop a user programming system that would support software intended for special applications beyond the capability of the basic controller, and a language that would interface with robot controllers, machine vision systems, and cell controllers. KAREL permits both off-line and on-line program development. It is a high level programming language that offers certain features of Pascal along with some characteristics of earlier robot programming languages, and with certain unique features. It supports four basic scalar data types namely integer, real, string, and boolean. Structured data types are also supported. A variety of motion statements are provided by KAREL. It is possible to add special motion control and monitoring clauses to KAREL to temporarily alter a motion type or to specify conditions that need to be monitored throughout the motion. The READ and WRITE statements in this language provide for both formatted and unformatted I/O operations. Process control in a KAREL system is attained through digital and analog I/O. Up to 128 separate digital points can be supported in a normal KAREL system. Analog I/O is used for applications such as welding controllers, sensory inputs, and for general process control.

2.11 CIMPLER

This language is used for programming all GCA robots with a CIMROC controller [9]. It provides for relational, arithmetic, logical, and boolean operators. The data types supported are integer data, integer arrays, location data, and I/O data. It is a structured language and programs can be executed either sequentially or concurrently. Motion control statements allow the tool tip to be moved to a desired location relative to the current origin from the current location of the tip. It is also possible to move tool tip a specific incremental distance relative to the current tool tip location. The motion control statements provide for a specific path to be defined through a sequence of points. This language allows for the velocity and acceleration of manipulator motion to be controlled and varied. This system provides for communication with external controllers and computers. It can interact with external sensors and vision systems.

2.12 RAIL

Designed by Automatix Inc., this language was designed to control both vision and manufacturing equipment [20,30,31]. It was designed to control a robotic arc welding system, a materials handling system, and a parts inspection and visual robot guidance system. This language has the standard program structures similar to those found in PASCAL. Robot RAIL, an extension of RAIL, contains the special purpose features that allow RAIL to interface with external equipment such as sensors, conveyors, etc. Vision RAIL, yet another extension of RAIL, contains the features that help this system interface with vision systems. The features provided by this language include: arithemtic, logic, and comparative expressions, a function library containing functions such as sine, cosine, arc tangent, etc., and easy methods to access I/O lines connected to external equipment. RAIL is implemented as an interpreter and allows the user to define a function statement that would combine a group of commands into a single command.

2.13 The AML Family

This set of languages is a commercial offering of the IBM Corporation [8,40]. AML constants and variables have a specific data type that could be either scalar or aggregate. Scalar object types include integer, real, string, reference, identifier, and label types. An aggregate object could consist of an ordered set of scalar objects. AML commands can be classified as fundamental subroutines, calculational routines, robot and sensor I/O commands, system interface commands, and data processing commands. Subroutines could exist as user written code, or they may exist as primitive elements of the base system.

AML/E is only a small subset of AML and is used to control the IBM 7545 robot, whereas AML is used for the larger IBM 7565 robot [2,3,40]. AML has been interfaced with external vision systems and has also been coupled with NC machining systems. AML/X has been defined as an eXperimental eXtension of the AML language [1]. It is intended to be a general purpose programming language with its origin in robotics.

2.14 Lesser Known Languages

HIerarchical ROBot Command Language (HIROB) offers the user the facility of communicating with a robot in English, commanding the robot to perform a desired sequence of operations [4]. It requires the user to design command sequences in a hierarchical top-down fashion. After a sequence has been comprehensively defined, it must be saved in order that it could be used as a part of different command sequences. Using English removes the need for an expert programmer except at the lowest level. A user friendly editor has been designed to help the user create, develop, modify, and test HIROB commands.

The Horyu Off-line Programming System I (HOPS-I) was developed as part of an off-line, micro-computer based programming system for industrial robots. A flexible system that could be adapted to several robot types was developed [39]. This language runs on a CPM-86 operating system and is based on BASIC. This language has twenty instructions. The program developed by the user can be edited on a commercial editor that is compatible with CPM-86 system. HOPS-I can be adapted to control many different robot systems. A trace facility is provided to help in software debugging.

JARS was developed in order to provide a high level robot programming language to help carry out research in manipulation and control of robot arms and programming system [10]. Created in 1979, JARS is Pascal with additional features such as robot specific types, variables, and subroutines. JARS is used to control a modified 'Stanford Arm' and is also interfaced to a Puma 560.

Marti aims to provide an independent, interactive, user friendly operator interface and a language generator that is robot dependent [32]. This system has been writen in Pascal and provides a robot dependent interface and a high level user interface. This system presents the possibility of standardizing robot programming in a shop in which different types of robots are used. It is expected that this would have significant importance in the development of machine and robot control due to its modular and multi-level programming structure.

Developed at Purdue University [29], PAL has an ALGOL like structure and uses a Cartesian coordinate system. Since it uses coordinate transformations exclusively, PAL requires the user to be familiar with transformation determination. Since PAL allows only two character variables, it is not suitable for the novice user. It does not support subroutines nor does it allow comments to be imbedded within the program. It is an interpreted language. PAL does not interact with external devices such as sensors nor does not consider parallel execution of tasks.

The Rhino series of robots that Robo Talk was developed to control is aimed at the educational, experimental, and hobby segments of the robot market [34]. This language is organized into five segments: the editor, the interpreter, the direct execution mode, the filer, and the teach pendent support routines. The code for Robo Talk is written in the C language, with trace and debugging features to help a user. Debugging features identify the statement in which the error has occurred and the type of error. Since this language is interactive, it is useful to teach a novice the use of a robot, and for instructing the operator on the principles of programming.

UROCL was created as part of a flexible robot programming system [22] and is subdivided into two levels. Fundamental UROCL is the basis of this programming language. There are seven data types and the language provides for conditional branching and loops. Advanced UROCL provides for the interfacing of this language with robots, vision systems, and sensors, and allows parallel processing. A robot motion or a vision task could specify an option which would allow the system to continue processing the following program sequence without waiting for the completion of execution of the current task. This language includes various control statements for tasks such as gripper operation, etc.

Based on BASIC, YB-BASIC was developed for users of the YEWBOT distributed robot system for efficient control and operation [25]. This is a realtime operating system that can simultaneously handle multiple robots and programs. This language has more than 150 extended statements for sub-systems such as robot manipulator control, interaction with machine vision systems, and external sensor interfacing. Debugging features include breakpoints, trace facilities, and single step execution. A robot simulator is also provided. This language can detect interrupts issued by sub-systems in the user program. These interrupts are used by the error handling and recovery programs to maintain reliability and safety in the operation of automated assembly lines.

3.0 Conclusions

New languages to control robots are under development every day. Although teaching robots is still a very popular method of robot programming, the need to use robots in complex assembly and machine loading tasks makes it imperative that the role of textual robot programming will increase significantly over the years. It is obvious that the demand for more sophisticated robot applications would cause more sophisticated and better textual textual robot program languages to be developed. An English like, easy to understand language with structured programming constructs would be ideal. The facilities and features that this language would provide will be synchronous with the demand posed by future robot applications. Robots will have to interface with vision sensors and tactile sensing equipment. Offline robot programming would also mean that offline databases would have to be developed. Robots will be called upon to make intelligent decisions in uncertain situations. The use of natural language processing in robot control is an interesting area for future research.

BIBLIOGRAPHY

1. AML/X Users Manual, IBM Thomas J. Watson Research Center, Manufacturing Research Department, NY, 1986.

2. IBM 7565 Manufacturing System - A Manufacturing Language Reference and Screen Editor, IBM Corporation, FL, Second Edition, 1983.

3. IBM 7565 Manufacturing System - A Manufacturing Language Concepts and Users Guide, IBM Corporation, FL, First Edition, 1982.

4. Bock, P., "English as a Command Language for Robot Control (HIROBI)," Robotica, Vol. 1, No. 1, 1983, pp. 19-24.

5. Bonner, S. and Kang, G.S., "A Comparative Study of Robot Languages," IEEE Computer, Dec. 1983, pp. 82-86.

6. Braganca, C. A. J. and Sholl, P., "VAL-II - A Language for Hierarchical Control of a Robot Based Automated Factory," Robotica, Vol. 3, No. 4, 1985, pp. 265-272.

7. Brantmark, H. and Ramstrom, K. G., "ASEA Offline Programming System - A User Friendly Implement," Proceedings - 15th ISIR, Tokyo, Japan, 1985, pp. 741-749.

8. Buckley, S. J. and Collins, G. F., "A Structured-Programming Language," Handbook of Industrial Robotics, Edited by S. Y. Noff, John Wiley and Sons, pp. 381-403.

9. CIMPLER Programming Manual for the CIMROC Computer Integrated Manufacturing Robot Control, Version 611.0, GCA Corporation - Industrial Systems Group, Naperville, IL, 1984.

10. Craig, J. J., "JARS: JPL Autonomous Robot System," Robotics and Teleoperators Group, Jet Propulsion Laboratory, Pasadena, CA, 1980.

11. Derby, S., "Off-Line Programming of Two Industrial Robots," Proceedings - SME Robots VIII Conference, June 1984, pp. 20-65 -20-76.

12. Deisenroth, M. P., "A Survey of Robot Programming Languages," Proceedings - Fall Industrial Engineering Conference, December 1985, pp. 191-195.

13. Finkel, R., Taylor, R., Bolles, Paul, R. and Feldman, J., "An Overview of AL, A Programming System for Automation," Proceedings - Fourth International Joint Conference on Artificial Intelligence, June 1975, pp. 758-765.

14. Geschke, C. C., "A System for Programming and Controlling Sensor-Based Robot Manipulators," IEEE Transactions on Pattern Analysis and Machine Intelligence, Vol. 5, No. 1, pp. 1-7.

15. Gibson, W. E., "Programming and Maintenance of Industrial Robots," Proceedings - SME Robots IX Conference, June 1985, pp. 17-82 - 17-93.

16. Gilbert, A., Pelton, G., Wang, R. and Motiwalla, S., "AR-BASIC - An Advanced and User Friendly Programming System for Robots," Proceedings - SME Robots VIII, Vol. 2, Detroit, MI, 1984, pp. 20-47 -20-64.

17. Gilbert, A., Pelton, G., Wang, R. and Motiwalla, S., "Programming a Robot in Basic," Computers in Mechanical Engineering, Vol. 2, No. 6, 1984, pp. 28-33.

18. Grossman, D. D. and Short, W. M., "AML - Much More Than a Language," Proceedings - SME Robots IX, Vol. 2, Detroit, MI, 1985, pp. 18-14 - 18-22.

19. Gruver, W. A., Soroka, B. J., Craig, J. J. and Turner, T. L., "Evaluation of Commercially Available Robot Programming Languages," Proceedings - SME Robots VII, Vol. 2, Chicago, IL, 1983, pp. 12-42 - 12-57.

20. Gruver, W. A., Soroka, B. J., Craig, J. J. and Turner, T. L., "Industrial Robot Programming Languages: A Comparative Evaluation," IEEE Transactions on Systems, Man, and Cybernetics, Vol. SMC-14, No. 4, July/August 1984, pp. 565-570.

21. Haurat, A. and Thomas, M. C., "LMAC: A Language Generator System for the Command of Industrial Robots," Proceedings - SME Robots VII, Chicago, IL, 1983, pp. 12-69 - 12-78.

22. Hsiao-Fuh Tan and Fang-Yuan Chang, "A Flexible Robot Programming System - UROCS," Proceedings - 15th ISIR, Tokyo, Japan, 1985, pp. 725-732.

23. Introduction to the KAREL Programming Language, GMF Robotics Corporation, Research and Development, Troy, MI, 1985.

24. Kirshbrown, H. R. and Dorf, R. C., "KARMA - A Knowledge-Based Robot Manipulation System," Proceedings - SME Robots VIII Conference, Vol. 2, Detroit, MI, 1984, pp. 20-36 - 20-46.

25. Kuwuhara, J., Matsumoto, T. and Ono, Y., "Design Concept and Architecture of a Distributed Robot System Using a High-Level Robot Language," Proceedings - 15th ISIR, Tokyo, Japan, 1985, pp. 707-714.

26. Lieberman, L. I. and Wesley, M. A., "AUTOPASS: An Automatic Programming System for Computer Controlled Mechanical Assembly," IBM Journal of Research and Development, Vol. 21, No. 4, 1977, pp. 321-333.

27. Mujtaba, S. and Goldman, R., "AL User's Manual - Third Edition," Computer Science Department - Stanford University, Palo Alto, CA, 1981.

28. Park, W. M., " The SRI Robot Programming System (RPS)," Proceedings - SME Robots VII, Vol. 2, Chicago, IL, 1983, pp. 12-21 - 12-41.

29. Paul, R. P., "WAVE: A Model Based Language for Manipulator Control," The Industrial Robot, Vol. 4, 1977, pp. 10-17.

30. RAIL Software Reference Manual for AUTOVISION Systems, Automatix Inc., Billercia, MA, 1984.

31. RAIL Software Reference Manual for Robovision and Cybervision Systems, Automatix Inc., Billerica, MA, 1985.

32. Ranky, P. G., "Programming Industrial Robots in FMS (A Survey with Particular Reference to Off-Line, High Level Robot Program Generation using VAL, VAL-II, AML, and MARTI)," Robotica, Vol. 2, No. 2, 1984, pp. 87-92.

33. Rembold, U. and Blume, C., "Programming Languages and Systems for Assembly Robots" Computers in Mechanical Engineering, Vol. 3, No. 1, 1984, pp. 61-68.

34. Sandhu, H. S. and Schildt, H., "Robotalk: A New Language to Control the Rhino Robot," Robotics Age, Vol. 7, No. 9, 1985, pp. 15-19.

35. Shimano, B. E., "VAL: A Versatile Programming and Control System," Proceedings - COMPSAC 79, pp. 878-883, 1979.

36. Shimano, B. E., Geschke, C. C., Spalding, C. H. and Smith, P. G., "A Robot Programming System Incorporating Real-Time and Supervisory Control: VAL-II," Proceedings - SME Robots VIII, Vol. 2, Detroit, MI, 1984, pp. 20-103 - 20-119.

37. Shimano, B. E., Geschke, C. C., Spalding, C. H. and Smith, P. G., "VAL-II: A Robot Programming Language and Control System," Robotics Research, Edited by Paul, R. and Brady, M., MIT Press, 1984.

38. Soroka, B. I., "Why Can't Programming Languages Do?," Proceedings - SME Robots VII, Vol. 2, Chicago, IL, 1983, pp. 12-1 - 12-8.

39. Sugiyama, T., Yaguchi, S., Yoshimura, H. and Ikemoto, M., "Robot Controllers for Off-Line Programming System," Proceedings - 15th ISIR, Tokyo, Japan, 1985, pp. 715-722.

40. Taylor, R. H., Summers, P. D., and Meyer, J. M., "AML - A Manufacturing Language," International Journal Of Robotics Research, Vol. 1, No. 3, 1982, pp. 19-41.

41. Users Guide To VAL, Version 12.4.6, 398H, Unimation Inc., Danbury, CT, May 1982.

42. Ward, M. R. and Stoddard, K. A., "KAREL - Programming Language for the Factory Floor," Robotics Age, Vol. 7, No. 9, 1985, pp. 10-14.

43. Wood, B. O. and Fugelso, M. A., "MCL: The Manufacturing Control Language," Proceedings - SME Robots VII, 1983, pp. 12-84 - 12-96.

MOBILE ROBOTS/ROBOTIC DEVICES

Integration of Material Buffering Devices in an Automated Factory

Douglas Fishman, Harry Scott, and Robert Bunch

National Bureau of Standards, Gaithersburg, MD 20899, USA

INTRODUCTION

In a hierarchically structured automated factory, most components are part of a larger, higher-level module of the factory. For example, robots and machine tools may be considered part of machining workstations. Several machining workstations may then be considered to be components of a manufacturing cell. An automated shop could be composed of more than one manufacturing cell, and so on. However, some equipment components do not fit neatly into any one module of the factory. One such component is a material buffering device (MBD) that acts as an interface between a workstation and a facility-wide material transport system. This paper addresses integrating such devices into an automated factory.

An architecture for integrating material buffering devices into an automated factory has been implemented in the Automated Manufacturing Research Facility (AMRF) at the National Bureau of Standards [Albus, J.S. et al., 1984]. The AMRF currently consists of machining and materials handling workstations that are coordinated by a higher-level Cell Controller. The workstation controller for a machining workstation coordinates equipment-level components, such as NC machine tools, robots, and automated fixturing components, to perform typical machining workstation tasks. The workstation controller for a material handling workstation coordinates its equipment-level components, such as robot carts and a factory inventory storage device, to deliver materials to workstations.

In early operations of the AMRF, a robot cart controlled by the Material Handling Workstation (MHWS) delivered trays of materials to the "material transport side" of roller tables in both the Horizontal Workstation (HWS) and the Turning Workstation (TWS). The robots in those workstations accessed trays from the "machining workstation side" of the roller tables. A simple semaphore system was developed and used for the roller table interface. This interface had been sufficient for the case of the roller table delivery stations. However, a generic control strategy for integrating material buffering devices between workstations was developed that supports not only the roller table integration but also integraton of a complex automated storage and retrieval system.

Following is the description of an architecture for integrating material buffering devices in an automated factory. The implementation of this architecture in the Horizontal Workstation is given as an application. Included is a description of the data structures and the workstation commands that will be used in the integration of a Material Buffering Controller (MBC) into the AMRF.

ARCHITECTURE

A number of observations have influenced this architecture. MBDs have two supervisory controllers that coordinate their activities. This puts them in a class of equipment that has a different role from other types of devices already integrated on the shop floor. The MBDs require coordination with an equipment-level component of a machining workstation (a robot) to allow insertion and removal of objects. In this way, MBDs seem to be equipment-level components of workstations in the same way as robots, NCs, and fixture systems. Unlike robots, NCs and fixture systems, however, the MBDs also require coordination with activities outside of the workstation (cart arrivals).

This difference significantly affects how control of an MBD is carried out. While it is reasonable for a machining workstation controller (WSC) to coordinate robot actions with the movement of trays on the workstation side of the MBD, it is not reasonable for a machining workstation controller to coordinate the movement of trays in the MBD with the facility transport device (cart), since the cart is an equipment-level component of a different workstation (the Material Handling Workstation or MHWS).

One method of coordinating this activity is to coordinate it at the next higher level of control, that is, at the Cell Control level. Cell commands such as SHIP TRAY or RECEIVE TRAY would be given to a WSC. Since no WSC coordination would be required, the WSC would then simply pass the command to the MBD controller to effect tray movement on the material transport side of the MBD. Similarly, Cell commands such as PICKUP TRAY or DELIVER TRAY would pass through the Material Handling Workstation Controller (MHWSC) to cause the required cart actions. In this case, the status returned to the Cell by a workstation controller would be the status of an equipment-level component. The result is that the Cell controller would be coordinating equipment-level components of workstations rather than coordinating workstations.

In addition to the problem of a machining workstation controller's involvement in the coordination of a material handling workstation's task, the WSC handling of tray-oriented commands is inappropriate. Material handling workstations use trays of objects, so commands to a MHWSC are logically given in terms of trays. Machining workstation do not use trays of objects, they use the objects themselves. Therefore, it is more appropriate for a WSC to receive object-oriented commands.

The alternative approach described here provides independence between the activities of a facility-wide material transport system and the activities of a machining workstation. This approach will be called the "shared resource method".

The shared resource method is based on the observation that MBDs do not "belong" to any workstation. They are mechanical buffers that serve as mechanical interfaces between workstations. From the point of view of an MBD, there is no reason to favor grouping it in either the machining workstation or in the material handling workstation. An MBD has two mechanical interfaces, one on the machining workstation side and one on the material transport side. The function of the MBD and its controller is simply to service the material storage and retrieval needs of both the machining workstation and the transport system. Both mechanical interfaces are symmetric in function.

In this architecture an MBD and its controller, a Material Buffering Controller, are then considered shared resources by a machining and a material handling workstation. The machining workstation and the material handling workstation issue commands to and receive status information from the Material Buffering Controller. Each controller uses its own mailbox interface (Figure 1).

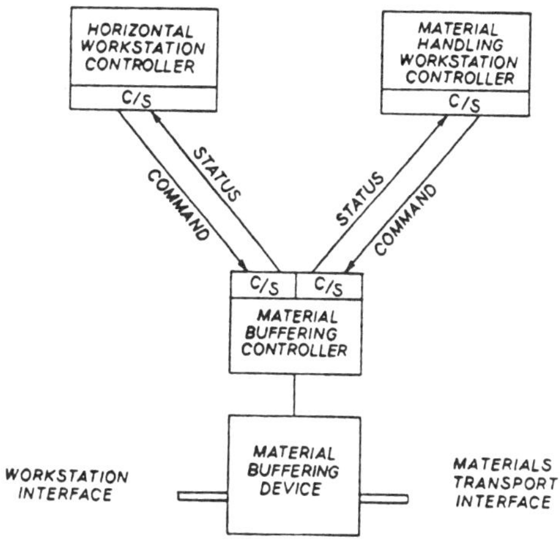

FIGURE 1 WORKSTATION INTERFACES TO A MATERIAL BUFFERING CONTROLLER

There are similarities between the way that the AMRF Data Management System manages data and the way that the Material Buffering Controller manages material. The AMRF Data Management System is a shared resource that stores and retrieves data for many AMRF systems during operation of the facility [Mitchell and Barkmeyer, 1984]. When systems update and access data in real-time, there exists an implicit guarantee that the Data Management System provides will be accurate. Similarly, the

Material Buffering Controller provides material storage and retrieval functions for two workstations. It stores and retrieves objects and updates information regarding the objects' storage status in real-time. Therefore, just as the AMRF Data Management System is an arbitrator for data access among the many AMRF systems, the MBC is an arbitrator for material access between workstations. The MBC resolves conflicts such as attempts by both workstations to remove the same object simultaneously. Figure 2 shows the MBD as a shared resource for two workstations.

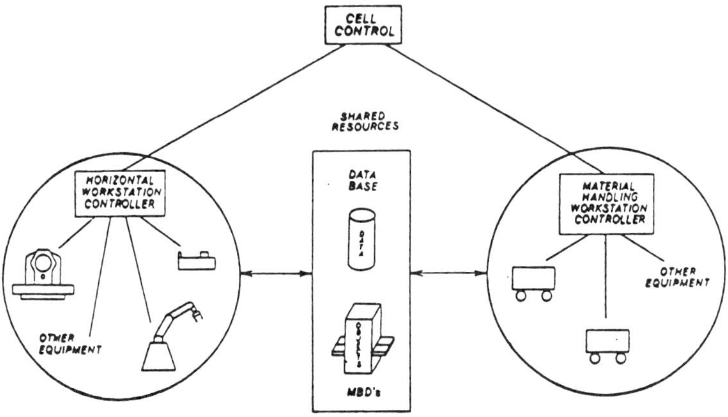

FIGURE 2 SHARED RESOURCES IN AN AUTOMATED FACTORY

ADVANTAGES

Two major forms of system independence have been gained through this architecture. First, there is an independence between the Material Buffering Controller's execution of machining and material handling workstation commands. The machining workstation is not affected by the way in which objects get into or out of the MBD on the material transport side. For example, none of the following material handling activities would change the machining workstation behavior:

- 1 tray of 12 objects is delivered by cart
- 12 trays of 1 object each are delivered by cart
- 2 trays of 6 objects are delivered by a person (as in a standalone workstation)
- A cart with a mobile robot removes or deposits 1 object from/into a tray
- A stack of trays is delivered as a unit into some kind of MBD
- The physical type of MBD is changed
- Trays are temporarily stored at one workstation's MBD although they are ultimately intended for another workstation

Because of this independence, trays can be delivered and picked up at the workstation while the workstation controller is executing cell commands. Any attempts to accesses an MBD simultaneously will be arbitrated by the Material

Buffering Controller using simple semaphore protection. Worst case delays in access to the MBD are purely MBD-dependent and are identifiable and predictable.

Second, there is an independence between the Material Buffering Controller's interfaces to the machining and material handling workstations. It is likely that a material handling workstation would handle all object movement about the factory in tray groupings. Thus, an interface between the MBD controller and the MHWSC could be implemented using commands such as:

PRESENT TRAY 14

RECEIVE TRAY 14

Individual workstations on the other hand, could greatly benefit by a different sort of interface to the MBD. For example, the implementation in the Horizontal Workstation allows object names to be used rather than tray indentifications. For example:

PRESENT TOOL TAP 502

The Material Buffering Controller (MBC), using data base information, decides what it must do to make that object available for workstation action and then does it. With the automated storage and retrieval system (ASRS) device that is currently in the Horizontal Workstation, this will result in a tray presentation. Notice that this command does not refer to a tray but instead to the object itself.

IMPLEMENTATION IN THE HORIZONTAL WORKSTATION

Figure 3 shows the integration of a material buffering system as a shared resource between the Horizontal and the Material Handling workstations.

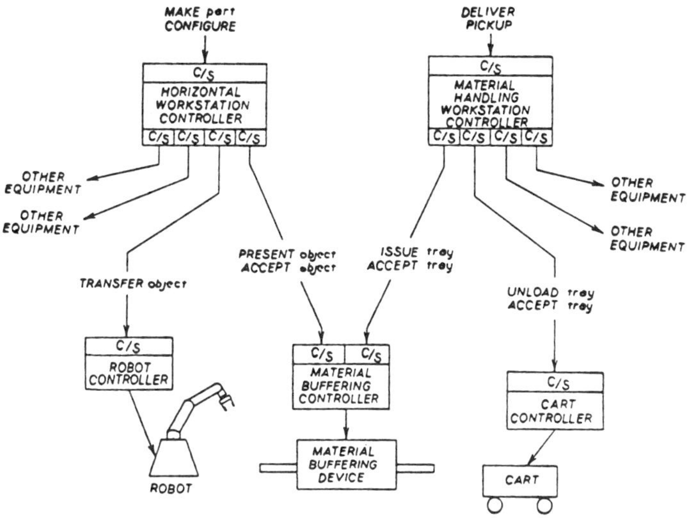

FIGURE 3 MBC INTEGRATED AS A SHARED RESOURCE IN THE HORIZONTAL WORKSTATION

In Figure 3, above, typical commands between control levels are also shown. The control architecture shown earlier in Figure 2 can be implemented in different ways to integrate various MBDs.

The Material Buffering Devices (MBDs) in the Horizontal Workstation are two Automated Storage and Retrieval Systems (ASRSs). Each MBD is controlled by a Material Buffering Device Controller (MBDC). The higher-level Material Buffering Controller (MBC) coordinates the operations of the Material Buffering Device Controllers to provide material storage and retrieval functions based on commands from the Horizontal and Material Handling workstation controllers. The control structure of the HWS buffering system is shown in Figure 4.

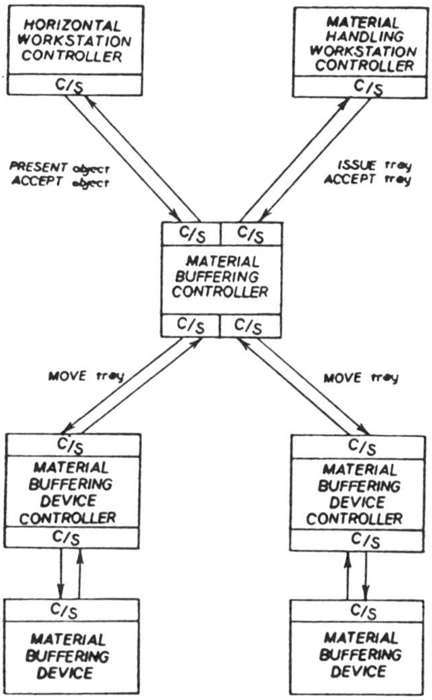

FIGURE 4 CONTROL STRUCTURE OF THE HWS MATERIAL BUFFERING SYSTEM

The ASR devices in the Horizontal Workstation (Figure 5) use elevator mechanisms to store trays on shelves. Each device has two delivery locations, one on the machining workstation side and the other on the material transport side. The delivery location on the material transport side is tailored specifically to meet the mechanical specifications of the robot cart; the delivery location on the machining workstation side provides unobstructed robot access. The two MBDs, placed next to each other in the workstation, are independent buffering devices and cannot exchange trays.

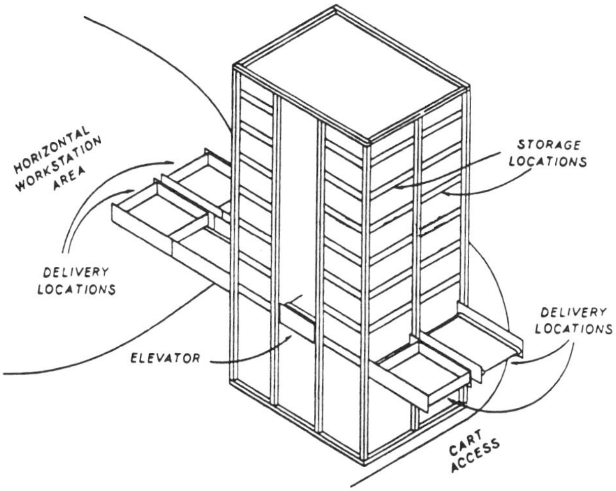

FIGURE 5 BUFFERING DEVICE USED IN THE HORIZONTAL WORKSTATION

The MBC responds to commands placed in either command interface, one for Horizontal Workstation Controller commands and the other for Material Handling Workstation Controller commands. The execution of one workstation's command may be delayed if the MBC is executing a command from the other workstation. The worst case delay is MBD-dependent and measurable.

DATA STRUCTURES

The data structures required to manage the materials stored in the MBDs are described below.

MBC Operating Status

The MBC Operating Status provides the current state of the storage and delivery locations in both MBDs. One occurance exists for each storage and delivery location available in the MBDs. The number of storage locations available is determined by the shelf spacing, and may be altered only by an off-line configuration.

The data structure consists of the following elements:

name	possible entries
LOCATION	SHELF 26 / DELIVERY 4
TRAY ID	T1114 / P78345
STORAGE TYPE	06 / TOOL / PART
ASSIGNED	YES / NO
OCCUPIED	YES / NO
LOCKED	YES / NO

LOCATION - Describes a position in an MBD that is logically ASSIGNED to a tray or one that is physically OCCUPIED by a tray.

TRAY ID - A unique serial number of a tray in a particular location. A TRAY ID element will have an entry when its corresponding LOCATION is ASSIGNED or OCCUPIED.

STORAGE TYPE - Refers to the tray size that a storage location can accommodate. STORAGE TYPE could be indicated directly in inches or implied by traits such as TOOL or PART.

ASSIGNED - Every tray in an MBD is ASSIGNED to a storage location. While a tray is at a workstation delivery location, its storage location is reserved by a tray assignment. The MBC will not store a tray at a location that is assigned to another tray.

OCCUPIED - Indicates whether a tray is physically present at a storage or delivery location.

LOCKED - While a workstation is accessing a tray at a delivery location, the tray is LOCKED at that delivery location. The lock prevents the tray from being transferred to another delivery location.

Source and Destination Specification Tables

The Horizontal Workstation Controller (HWSC) uses object-oriented commands to retrieve items from and to store items in the MBDs. For example,

PRESENT 10/32TAP
ACCEPT VALVE STEM

More than one tray can contain an object-type requested by the HWSC. Similarly, several trays could contain an acceptable location for storage of an object. The MBC uses the Source and Destination Specification Tables to find the appropriate tray to present when executing object-oriented commands.

When commanded by the HWSC to present an object to the workstation delivery location, the MBC accesses the Source Specification Table to determine the tray that it should present. When the tray is presented, the MBC passes a status parameter to the HWSC that indicates the location (sector) from which the object should be removed.

When commanded by the HWSC to present a location for an object, the MBC will access the Destination Specification Table to determine the tray that it should present. When the tray is presented, the MBC passes to the HWSC a status parameter containing the sector in which to place the object.

Each occurance of the Source Specification and Destination Specification Tables will have the following elements:

element name	possible values
OBJECT TYPE	BORING BAR / IGES BLANK
OBJECT SN	T0003.BBAR / P92547.IGES
TRAY ID	T1114 / P78345
SECTOR	1 / 2 / 6

Tray Table

In a future implementation of the MBC in the AMRF, the MBC will use information in Tray Tables to determine suitable storage locations for incoming trays. Either the HWSC or the RCS will access Tray Tables to determine locations within a tray in which to find or place an object.

When commanded by MHWSC to accept an incoming tray, the MBC will use the tray's size along with the storage sizes of all unassigned delivery locations (from the MBC Operating Status) to determine a suitable storage location for the tray.

When storing an object in a tray, either the HWSC or the RCS will determine the tray and sector number in which to place the object from the Destination Specification Table. The number of sectors into which a tray is divided affects the locations of the individual sectors. Therefore, either the HWSC or the RCS will obtain the total number of sectors from the Tray Table to determine the exact location of the individual sector.

Each Tray Table will have the following elements:

name	possible entries
TRAY ID	T3478 / P7822
TRAY TYPE	TOOL / PART
SIZE	16 / 08
#SECTORS	04 / 12

COMMANDS

This section consists of a brief description of the commands that can be issued by the Material Handling Workstation and Horizontal Workstation controllers to be executed by the Material Buffer Controller. The parameters of the command strings that are enclosed in brackets are optional.

Also note:
The MBC always locks a delivery location after presenting a tray to it. This lock means that the workstation on the other side of the MBD will not have access to the tray until the workstation that requested the tray "releases" the delivery location. When trays are delivered to the MBDs by the MHWS, they are not locked and are available to the HWSC until they are requested by the MHWSC.

The commands from the MHWSC to the MBC are:

CLEAR FOR	object-type, serial #, [delivery location]
ACCEPT	object-type, serial #, delivery location
CANCEL-ACCEPT	delivery location
PRESENT	object-type, serial #
RELEASE	delivery location
ISSUE	object-type, serial #, delivery location

Commands from the HWSC to the MBC are:

PRESENT	object-type, object, [serial #]
ACCEPT	object-type, object, serial #
RELEASE	delivery location

Example Parameter Values:

object-types:	TOOL, PART, TRAY, END-EFFECTOR
objects:	1/2DRILL, BRACKET-BLANK, PART-GRIP
serial numbers:	T711 (tool), P309 (part), TT48 (tool tray)
delivery locations:	DELIVERY 1, DELIVERY 2

The commands and their purposes are listed below:

Material Handling Workstation Commands

CLEAR FOR To clear a delivery location for an incoming tray.

ACCEPT To activate the MBD conveyor at the specified delivery location so that a tray can be transferred onto the MBD.

CANCEL-ACCEPT To turn off the conveyor at the specified MBD delivery location. A CANCEL-ACCEPT is used by the MHWSC if the cart malfunctions, preventing a tray issue to the cart.

PRESENT To present a tray at a delivery location for cart pickup.

RELEASE To unlock a delivery location, making the tray available to the HWSC. The MHWSC uses a RELEASE when it decides not to pickup the tray that it previously commanded the MBC to present.

ISSUE To transfer a tray from an MBD to a cart. Before the MHWSC gives this command to the MBC, it should command the cart to activate its conveyor to receive a tray.

Horizontal Workstation Commands

PRESENT To move the tray containing the specified object to a delivery location for robot access.

ACCEPT To move a tray containing a location for the specified object to a delivery location.

RELEASE To unlock the HWS delivery location, making the tray available to the MHWSC. The HWSC uses this command to the MBC to release a delivery location after accessing the trav following a previously issued PRESENT or ACCEPT command.

CONCLUSION

We have presented an architecture for integrating material buffering devices in an automated factory. The architecture is designed to support those material buffering devices which serve as interfaces between a workstation and a facility-wide material transport system.

A Material Buffering Controller has been implemented in the AMRF using the generic control architecture for integrating MBDs as shared resources. The data structures and the command and status interfaces that are required for the integration have also been defined.

The architecture supports integration of material buffering devices with widely differing physical characteristics. In the case of integrating a simple passive buffer device such as a roller table, an MBC would not have to perform complex buffer management functions. Therefore, the limited data required for its tasks could be managed by the Material Handling Workstation Controller. The MBC would then only be responsible for arbitrating access to the roller tables between workstations.

Perhaps most importantly, the shared resource method of integrating a Material Buffering Controller into an automated factory provides independence between the activities of a factory-wide material transport system and the activites of a machining workstation.

REFERENCES

Albus, J.S., Barbera, A.J., Fitzgerald, M.L., Kent, E., McLean, C., McCain, H., Bloom, H., Haynes, L., Furlani, C., Barkmeyer, E., Mitchell, M., Scott, H., Bloomquist, D., Kilmer, R., A Control System for An Automated Manufacturing Research Facility, Robots 8 Exposition and Conference, Detroit, Michigan, June 4-7, 1984.

Barbera, A. J., Fitzgerald, M.L., Albus, J.S., "Concepts for Real-Time Sensory-Interactive Control System Architecture," Proceedings of the 14th Southeastern Symposium On Systems Theories, April, 1982

Mitchell, M., Barkmeyer, E., "Data Distribution in the NBS Automated Manufacturing Research Facility," Proceedings of the National Symposium on Advances in Distributed Data Base Management for CAD/CAM, NASA Publication 2301, April 1984.

Scott, H., Strouse, K., "Workstation Control in a Computer Integrated Manufacturing System," Proceedings of AUTOFACT 6 Conference, Anaheim, California, October 1-4, 1984.

Robotics System—Elephant Trunk Type Elastic Manipulator Combined with a Quadruped Walking Machine

A. Morecki, K. Jaworek, W. Pogorzelski, T. Zielinska, J. Fraczek, and G. Malczyk

Team for Robotics and Biomechanics of Engineering, Warsaw University of Technology, A1.Niepodleglości 222, r.212, 00-663 Warsaw, Poland

Introduction

The last years have brought several publications in the field of combined system, namely manipulator located on a mobile robot. This paper describes the design of mechanical and control systems, which will be able to perform different manipulation and locomotion activities. A method of analysis of a manipulator with flexible links, employ finite element method, is proposed. The legs of the walking machine imitate the limbs of mammals (rabbits, dogs, horses), so forelegs are different from hind legs. The mathematical model of such a walking machine includes the model of the mechanical part of the system (this part executes the movement) as well as the control part of the system. For the execution part a concept of indirect control is introduced. In this paper same problems concerning the following of a given trajectory and the so called balancing procedure are presented.

1. The model of an elastic manipulator of an elephant trunk type.

The design of the arm is based on a study of anatomy and physiology of an elephant trunk and qualitative model of its muscular structure treated as an actuating system. There exist several propositions of constructing the arm. Those propositions differ in construction details as well as in methods of control. The authors are not able to decide, which of the propositions will give the best results from the point of view of motion abilities, simplicity of manufacture and minimalization of costs. Moreover the purpose of the arm should be taken into consideration. This arm has to be fixed to a walking machine. The authors imagine, that the arm may be used as an auxiliary executive organ or a flexible ontrigger of an optical sensor (scene recognition). The structure consists of,so as the model,of three segments.Each of them consists of the elements of the same type.The elements are composed of a helical spring fixed between two dished disks.The transmission of motion takes place when the spring bends.Such a construction has large motion abilities and mantains a continous structure of motion.Power is transmitted from the actuators fixed to the body by means of the cords to the elements of the arm. The cords are located at 120° intervals of the circumference of an element. Fig.1. shows the proposition of the structure of the arm of a manipulator. The arm consists of 7 elements divided into 3 segments marked I,II and III. The segments contain 1, 2 and 4 elements successively. The power is transmitted by means of 9 cords. Further investigations of this matter will take into consideration the following :

- the selection of construction and method of control,
- the formulation of a mathematical model of the arm of a manipulator,
- the assessment of dynamical parameters of the model by computer simulation,
- the construction of a technical model,
- the examination of the technical model,
- the comparison of results obtained from computer simulation and examination ; modification of the model,
- the construction of an elastic manipulator,
- the investigation of the device fixed to a walking machine.

For the sake of analysis of dynamics the elephant trunk type robot, a computer program, based on finite element method combined with the analysis of the trunk's global movements, was implemented. The algorithm is based upon works /2/ and /3/. Each segment of the robot is modelled by finite element method. The interactions between the segments are modelled as holonomical constraints. The segments of the same shape have the same grid of finite elements. Great number of degrees of freedom makes necessary the transformation of coordinates expressed in ortho - cartesian coordinate frame into model coordinates / the so called method of model superposition/. Equations for the manipulator are formulated by Lagrange's method, and the constraints equations are taken into account by exploiting Lagrange's multipliers. The obtained equations are integrated by β-Newmark's method. The transformation matrix between orthocartesian coordinates and model coordinates is evaluated classicaly utilizing author's formulas. The algorithm takes into consideration in--material damping as well as the construction damping /friction/,which usually is greater than the first one. The obtained program is within the computational power of an IBM PC/XT microcomputer.

2. The control of motion machine MK-4.

Fig.2. shows the side view of the machine with an elastic manipulator mounted on it. Each leg has two degrees of freedom and its motion is caused by two reciprocating hydraulic actuators. Work concerning the control of the machine is being led in three directions :

- the analysis of relationship, the fulfilment of which is the condition for execution of the given strategy of motion (elaboration of mathematical model and computer simulation of motion),
- the elaboration of a method of controlling the leg on the executive level(digital control system-extremal controllers),
- the discussion of principles of operation of higher levels of machine's control(specification of the concept of machine's intelligence).

The machines mathematical model was described in works /4 /and /5 /. The model consists of the control and mechanical part. The mechanical part executes the walk(it is its goal) and the control part organized the operation of the mechanical part, so that the walking will be possible.

More precisely :

- the control part (eg.the computer) evaluates the variables called the theoretical states of legs.

The theoretical states are the equivalents of real states ,which are defined as the position of leg ends expressed in appropriate coordinate frames. The functioning of the control part is presented as an execution of a general formula :

$$p^{*j+1} = F(p^{*j},j) \ , \ j = 0,1,2,... \qquad (1)$$

where : p^{*j} - discrete theoretical state of the walking machine,defined as a set of theoretical states of its leg ends(this state is described by the leg ends' coordinates conculated by the control part)

The form of the function F (the method of generating states) depends on constraints such as eg.: constraints of motion space, the type of gait(the sequence of leg transfers), the trajectory of body's motion, the forseen configuration of the terrain, a method of stabilization of machine's position eg. conservation of static stability margin /5/.The simulation took into consideration only statically stable motion based on quadruped crawl /5/. Changeing slant of the terrain and constrained leg's motion space were considered. Under favourable conditions the machine moves by quadruped crawl. The principle of generating theoretical states (1) is determined then by incremental relationship /5/. However external influence, such as change of slant of the terrain or small obstructions, can create a situation, in which the device cannot execute the planned gait. This situation takes place when static stability margin decreases below the minimal value. The machine executes then a sequence of motions called balancing. The goal of balancing is to obtain a position of legs coinciding with a position characteristic of the considered gait. The gait parameters during balancing (such as the lenght of step) are different from those during normal gait and are chosen so that the static stability margin is not surpassed. The balancing process is planned in the control part. A limited number of leg transfers and shifts of the body in the direction of motion or opposite to it are considered. The possibility of execution and the range of each motion is limited by the necessity of maitaining an adequate static stability margin. Fig.3. shows the three of motion considered by the control part. To the general relationship (1) corresponds a description of the balancing process given as an incremental function :

$$p^{*j+1} = p^{*j} + \{ R(!), Q(!),\ W(!),\ h,\ W'(!) \} \qquad (2)$$

where : ! - set of gait parameter values (balancing has to convey the legs to the position characteristic to the gait with given parameters),

h - the hight of leg's lift,

Coefficients R(!),Q(!), W(!), W'(!) determine the changes in values of coordinates of leg's ends. The method of balancing has been proposed as an alternative to trial and error methods of planning motion. Simulation did not reveal any flaws in the method.

3. Digital control system of the MK-4 walking machine.

Measuring and control system of the executive level were elaborated for the MK-4 walking machine. Fig.4.shows the block diagram of the system.

The control part was subdivided into automatic blocks :

- the control microcomputer with interfaces,
- ultrasonic transmitter /receiver block (T/R),
- leg's electronics (LE),
- driving unit with sensors.

The structure of the digital system is based upon the chosen control algorithm. Paper /4/ assumed that the machine's executive level is controlled by extremal controllers. Each leg possesses its controller. The extremal control algorithm is stored in the memory of the microcomputer in the form of adequate programs. Each of the extremal controllers is connected to the higher level of the control system - called trajectory level /4/. Two constant parameters a_{10} and a_{20} , that are transmitted from the trajectory level, are necessary for proper functioning of each controller. The pair (a_{10}, a_{20}) is necessary to execute the motion task assigned to the executive level (leg and hydraulic drive). The executive level control algorithm requires continous cooperation with the ultrasonic transmitter /receiver (T/R) block interfaced to the control processor T/R block supplies the parameter pair (a_1, a_2), which indirectly describes the method of executing the motion task by the machine's leg. Each of the machine's legs is equipped with three hydraulic actuators. One of them shifts the leg sideways and the other two cause the leg's motion in the sagittal plane. The extremal controller's principle of work is such, that after chooseing the leg's motion in the sagittal plane, the controller, taking as the basis the pair of parameters (a_{10}, a_{20}) and current parameters (a_1, a_2), controls each of the hydraulic alternately. Each cycle of controller's work is checked by the control processor. Failure states cause either the examination of the state of the environment of the leg or signalize the type of the failure. The extremal controller algorithm was tested by simulating the executive level /4/ with positive results.

4. Further development of the control system.

Further developments aim at describing structural and functional properties of a multilevel hierarchic control system capable of controlling the walking machine. This system should have the properties of : adaptation, learning and intelligence. Personality and multilevel structires of personality are considered /6/. The structure is subdivided into the following levels : execution (the lowest level), of individual features, independence (adaptation) and intelligence (fig.5.). The system recognised by the machine as its environment and its organism is called the world of the machine. Personality has two aspects:structural (personality as a system of different elements) and functional (personality as a characterization:of the course of a process,methods of assuming an attitude towards itself and the external world,attitude and its realization). The machine in its functioning should exhibit a certain degree of intelligence.Work in the field of machine intelligence is in initial phase /7/. Our approach creates the possibility of dealing with walking machine's intelligence in a way adapted from psychology. Machine's intelligence consists in understanding the practical situation it encounters and finding purposeful reactions to those situations. Intelligence becomes a functional part of personality. By intelligence we understand one of the functions of a homeostat. This homeostat will have several levels of personality.

5. Conclusion.

The goal of the conducted research is the design of a walking machine with an elastic manipulator mounted on it. Fig.2. shows such a machine. The parts of this system are currently being designed.

REFERENCES

/1/ Morecki A.,Malczyk G.: Mathematical Model of A Flexible Manipulator of the Elephant's-trunk-type. Preprints of Ro.Man.Sy'86.Six CISM-IFToMM Symposium on Theory and Practice of Robots and Manipulators, Sept. 9-13,86

/2/ Fraczek J.:"The analysis of dynamics of a robot with elastic links". 1st National Conference on Robotics. Scientific Papers of the Institute of Technical Cybernetics.Na 66,vol 2. Wroclaw 1985

/3/ Haug,E.J.,C.Wu,S.Kim : "Dynamics of Flexible Machines:A Variational Approach".IUTAMM/IFToMM Symposium Dynamics of Multibody Systems. Udine,Italy,Sept.,16-20,1985.

/4/ Jaworek K.,Morecki A.,Pogorzelski W.,Zielinska T."Some problems of the design and control of a quadruped walking machine MK-4" VII-th IFToMM World Congress on Theory of Machines and Mechanisms.Sevilla, Spain,Sept.1987/in printing/

/5/ Zielinska T.:Modelling of gait of the quadruped walking machine /Ph.D Dissertation/,Warsaw University of Technology,Febr.1987 /Supervisor Morecki A./

/6/ Pogorzelski W."MK-4 walking machine as system with intelligence" XI National Conference on Theory of Machines and Mechanisms",Zakopane 1987 /in printing/

/7/ Aleksander I.:Designing Intelligent Systems.Kogan Page.London 1984/ /1986.

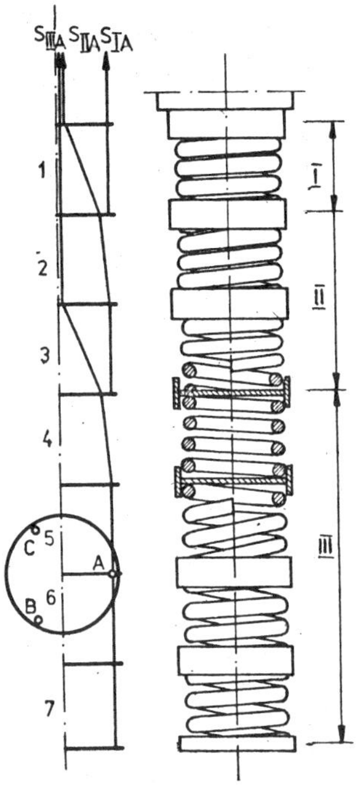

Fig.1. A proposition of elastic manipulator arm and the structure of actuating cords.

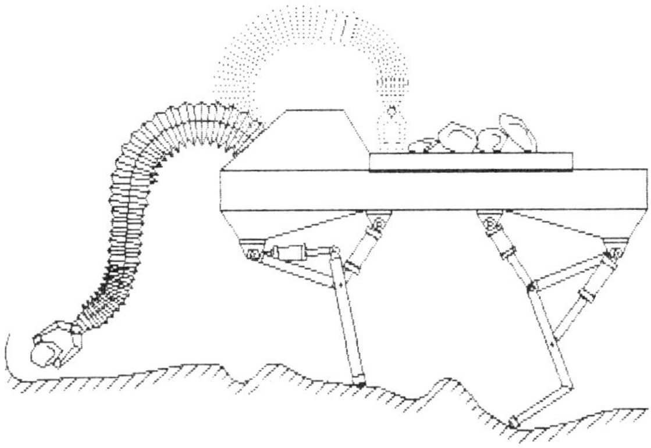

Fig.2. The side view of the machine.

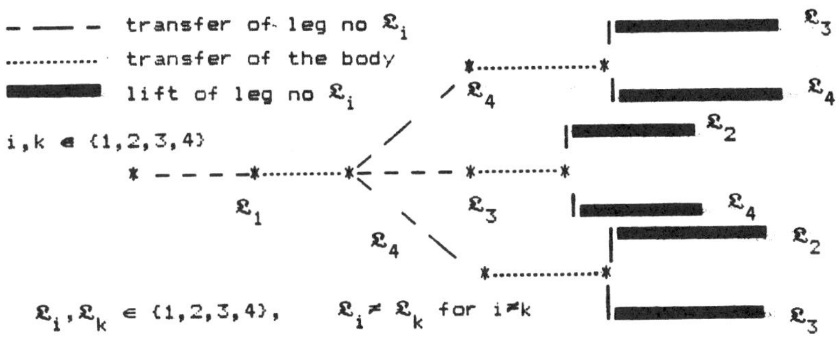

Fig.3. The tree of possible motions during balancing.

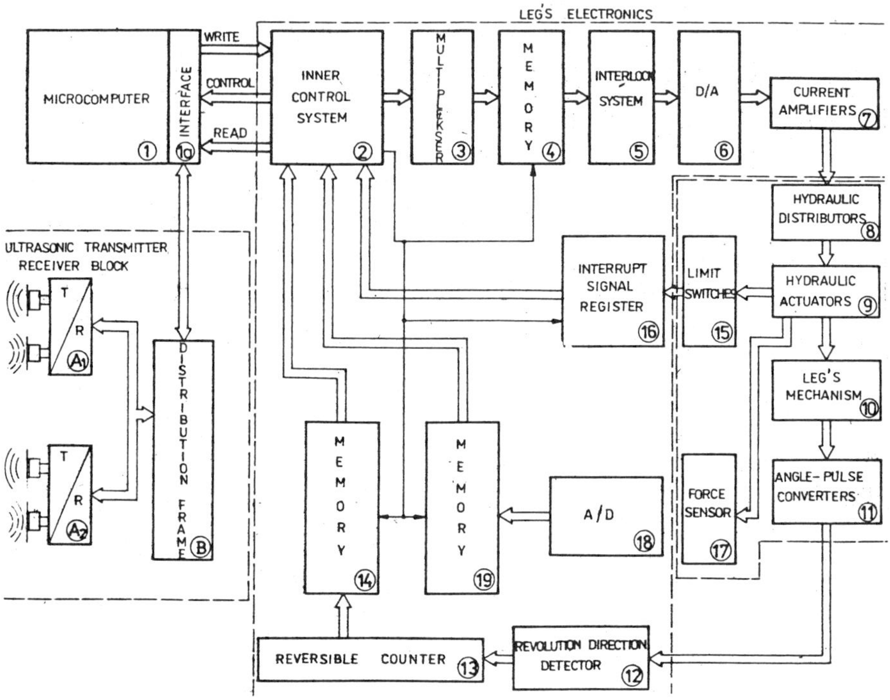

Fig.4. Block diagram of measuring and control system of the execution level (of the machine MK-4).

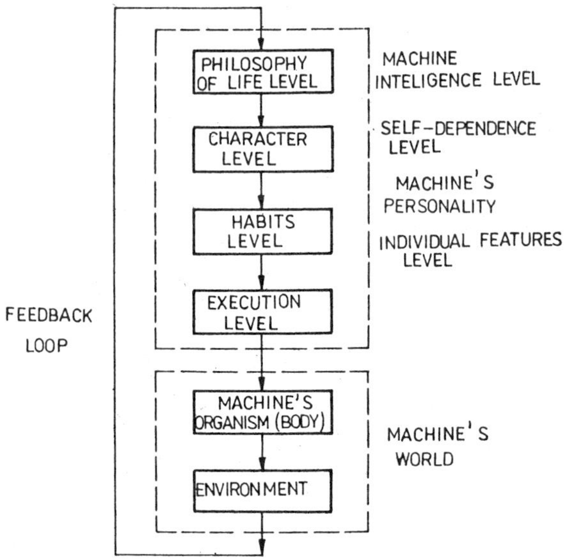

Fig.5. The structure of walking machine's personality.

Bipedal Balancing: A New Approach

John B. Shaw, Ming Hsu, Michael Mulder, Neldon Wagner, and Normal L. Kerth

University of Portland Applied Research Center, Portland, OR 97203, USA

Introduction

Progress in the development of a walking robotic biped depends on an understanding of the process of achieving and maintaining balance. Based upon the inverted pendulum model, Raibert has built a hopping monoped able to remain upright [1]. More recently, Furusho and Masubuchi built a bipedal walking machine [2] able to demonstrate a true walking gait. However, as with Raibert's hopping machine, the motion of the machine is constrained to the sagittal plane. No one has yet demonstrated a biped able to maintain balance in three dimensions.

This paper presents a new approach to balance which conceptually is readily applicable to providing balance in both the sagittal and frontal planes. Further, it suggests that the mechanisms needed to maintain balance may in theory be separated from the means of locomotion. In contrast to previous work, this approach does not require movement of the legs, but rather achieves balance by the motion of internal parts as, for example, the actions of a high wire walker rapidly moving his arms to maintain balance. With full control over balance, the legs may be used for motion, with little or no concern for balance.

Principles of Balance

Consider first a system with internal wheels. In the absence of gravity, the angular rotation of a wheel will cause the rest of the system to have an angular rotation in the opposite sense to conserve angular momentum. Thus, every part of the system will move, except for points on the rotation axis which must pass through the center of mass. In order to keep any other point from moving, an external force will have to be applied (such as when this point touches the ground). But such an external force, by Newton's law for multiple particle systems, acts on the system as though it were applied at the center of mass. Thus, a force applied where the system touches the ground will accelerate the center of mass. In the presence of gravity, this force may be used to counter gravity and achieve balance. This effect also follows from an internal linear motion of some part.

* This project partially supported by NSF under grant number DMC-8516873

An Example

A simple example of this technique utilizes a wheel connected to one end of a support bar, the other end of which acts as a foot (figure 1). The foot is free to pivot about an axis parallel to the axis of the wheel. A sensor at the foot measures the angle of the support leg relative to the ground, while another sensor measures the angular velocity of the wheel.

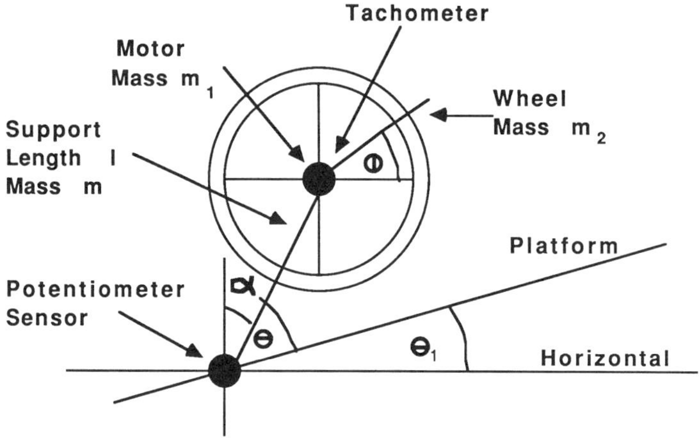

Figure 1. Equilibrium research test bed.

In operation, the motor accelerates the wheel causing a change in the system angular momentum. In the absence of external torques the total angular momentum of the system, that of the wheel and the support, must remain constant. Acceleration of the support structure opposite to the wheel acceleration accomplishes this. Considering now the external gravitation torque, the wheel acceleration can be used to cancel or overcome the support acceleration due to the gravitational torque.

Mathematically, the system may be described and analyzed by use of its Lagrangian in conjunction with multivariate control theory. The Lagrangian for this system is given by

$$L = T - V = (I\dot{\phi}^2/2 + I_1\dot{\theta}^2/2) - [glM\sin(\theta + \theta_1)]$$

where I is the moment of inertia of the system about the base, I_1 the moment of inertia of the wheel, and M is given by $M=m/2 + m_1 + m_2$. The Lagrangian equations of motion are then

$$I\ddot{\phi} = \tau$$
$$I_1\ddot{\theta} + glM\cos(\theta + \theta_1) = -\tau$$

where τ, the motor torque for a permanent magnet d. c. motor, is given by

$$\tau = aV - b\dot{\phi}$$

in terms of the applied voltage, V, and the instantaneous motor speed. A linear control equation specifies V as a function of the angles,

$$V = c(\theta - \theta_0) + d\dot{\theta} + e\dot{\phi}$$

where θ_0 is an arbitrary angle sensor calibration constant.

These equations may be converted to a set of first order, linear, differential equations linearized about $\theta = \pi/2 - \theta_1$, i. e. a vertical position, in the following form:

$$\begin{aligned} I\dot{u} &= (ae - b)u + adv + ac\theta - ac\theta_0 \\ I_1\dot{v} &= -(ae - b)u - adv - (ac-glM)\theta + [ac\theta_0 - glM(\pi/2-\theta_1)] \\ \dot{\theta} &= v \end{aligned}$$

where $\dot{\phi}$ has been replaced by u.

These equations may be solved by matrix techniques involving a solution for the eigenvalues. Control theory states that a stable solution exists if the real parts of all eigenvalues are less than zero. This is guaranteed if the Routh-Hurwitz criteria are met, which for this system require:

$$e > b/a \qquad d > \frac{(ae-b)(I_1)}{Ia} \qquad c > \frac{(ae-b)(glM)}{a[adI-(ae-b)I_1]} + \frac{glM}{aI_1}$$

These conditions are easily met experimentally. The system then approaches a final equilibrium position given by:

$$\phi_{t=\infty} = \frac{ac(\pi/2-\theta_1-\theta_0)}{(b-ae)} \qquad \theta_{t=\infty} = \pi/2-\theta_1 \qquad \text{or} \qquad \theta+\theta_1 = \pi/2$$

The last equation states that the final angle of the system will be 90 degrees, i.e. vertical, regardless of the slope of the base or the sensor calibration point. The first equation indicates that when equilibrium is achieved, there will be a residual wheel angular velocity.

Experimentally, our test bed verifies these results. The most interesting result, confirmed by experiment, is that having once obtained the vertical equilibrium position, the platform may be tipped (no sensor reads this angle) and the system will again seek and attain a true vertical equilibrium. Similarly, if the angle sensor calibration potentiometer is abruptly changed, the system tips and then reestablishes vertical equilibrium. Thus the system, if employed in a moving biped, is capable of finding the vertical, and positioning the center of mass there under conditions of changing terrain, aging sensors, and variations in loading or mass distribution.

A slight modification to the above theory, the inclusion in the voltage equation of a term proportional to $\varnothing$, has also been examined. Stability is again achievable, but now the wheel will be at rest when the system attains a balanced state, no matter what the angle of the terrain (platform).

Another Viewpoint

An alternative approach to the previous example starts with the two Lagrangian equations of motion, but adds an equation specifying the movement of the center of mass, rather than the voltage directly. Consider the following set of equations:

$$I\ddot{\theta} + glM\cos(\theta + \theta_1) = -I\ddot{\varnothing}$$

$$\ddot{\alpha} = -a\alpha - b\dot{\alpha} \quad \Rightarrow \quad -\ddot{\theta} = -a[\pi/2 - (\theta + \theta_1)] + b\dot{\theta}$$

$$\ddot{\varnothing} = -c\varnothing - d\dot{\varnothing}$$

The first follows by eliminating the applied torque, τ, from the two Lagrangian equations. The second requires the center of mass to tend toward the vertical position, and the third requires the wheel velocity to tend toward zero. These three equations overspecify the problem since there are only two variables. The latter two may be combined into

$$e\ddot{\varnothing} - \ddot{\theta} = -ce\varnothing - de\dot{\varnothing} - a[\pi/2 - (\theta - \theta_1)] + b\dot{\theta}$$

With appropriate choice of constants, this two equation system is stable. Using sensors to monitor θ and $\varnothing$, their derivatives may be obtained numerically or electronicly and the equations then determine $\ddot{\phi}$, from which the required motor torque and voltage follow. This approach develops the equations in a more intuitive manner which readily generalizes to more complex systems.

Conclusion

This paper has described an approach to balance based upon the motion of internal parts of a system rather than movements of the support base. Application of control theory adequately characterizes the system, and a working model confirms the results.

References

1. Raibert, M. H., et al, Dynamically Stable Legged Locomotion. Progress Report: October 1982-October 1983, Leg Laboratory, The Robotics Institution and Department of Computer Science, Carnegie-Mellon University, Pittsburgh, Pennsylvania, December 1983.

2. Furusho, J., and Masabuchi, M., "Control of a Dynamical Biped Locomotion System for Steady Walking", Journal of Dynamic Systems, Measurement, and Control, vol. 108, 1986, pp. 111-118.

On the Conversion of an Industrial Robot to a Mobile Robot

Kostia Mandel*, Prof. Neil A. Duffie†, and Prof. Bahram Ravani†

*AT&T Bell Laboratories, Columbus, OH 43213, USA
†Mechanical Engineering Dept. University of Wisconsin–Madison, Madison, WIS 53706, USA

1. BACKGROUND

Some of the factors that restrict the utilization of FMSs to their true potential are related to the limitations of the available material handling systems for FMSs. Two systems have been used extensively in FMSs, AGVs (Automated Guided Vehicles) or industrial robots, each having severe limitations. The AGV, while capable of traveling long distances and figure out the best route, has poor docking (arriving to a station) accuracy, and is dependent on external devices for loading and unloading. The Industrial Robot, while being flexible, accurate and dexterous, is operating in a very limited working space. There is a great need for truly flexible, variable mission, industrial material handling system.

It only seems natural, and logical, that such a system should combine the advantages of the AGV and the industrial robot, while eliminating the disadvantages of each one. An industrial robot mounted on an AGV should be able to perform almost all the stationary robot's tasks like loading / unloading tool magazines, pallets, etc. In addition, such a mobile robot will be able to perform scores of additional tasks - move materials and tools between stations, serve the stations upon arrival, and much more. The commercially available industrial mobile robots have very restricted motion - they are moving on rails. Although this setup has some advantages over a stationary robot, it does not lend itself easily to modifications, and the robot is restricted only to linear motion on the rails. Hence, it is not the truly flexible material handling system that is sought after.

This paper presents a study about one of the aspects of constructing an industrial mobile robot - improving its accuracy.

2. INDUSTRIAL MOBILE ROBOT OVERVIEW

A mobile robot (consisting of an industrial stationary robot and an AGV) should have an accuracy of at least ± 1.0 mm in order to perform a major portion of the tasks a stationary robot performs today (like machine tending). However a docking (arriving at a station) accuracy of ± 25 mm is considered to be a good one for an AGV. Since robot tasks are executed with respect to its base (the AGV in the case of the mobile robot), the accuracy of the mobile robot becomes ± 25 mm which is completely unexceptable. The accuracy of the mobile robot has to be impro ved by more then an order of magnitude, if it is to be considered as a substitute for a stationary robot in some cases.

Two approaches can be taken to improve the mobile robot's poor accuracy. The first approach assumes that there is one Robot Task per World Task, while the second approach does not make that assumption.

2.1 WORLD TASK, ROBOT TASK AND ROBOT PROGRAM

World Task is the specification of what the user wants the robot to do. It is the set of operations a robot is required to perform, specified in a reference frame fixed to the "world", and is independent of the robot kind and location.

Robot Task is a set of end effector locations, duties, and velocities that the robot has to go through in order to execute a certain world task. Robot task is defined in terms of the robot frame (usually attached to the robot's base) for a specific kind of robot at a specific location.

Robot Program is a synchronized set of commands issued to the joint actuators. Every command in the robot program is specified for a specific joint with respect to its current position.

Among these three terms there are two sets of relationship: between Robot Task and Robot Program, and between World Task and Robot Task. The relationship between Robot Task and Robot Pogram is defined by the robot's Kinematic Equations: the inverse solution of the Kinematic Equations transforms any Robot Task to the corresponding Robot Program. The relationship between a World Task and a Robot Task is defined by the offset between the reference (world) frame in which the World Task is specified, and the robot frame (usually attached to the robot's base) in which the Robot Task is specified. For stationary robot this relationship is simple - the two frames are either identical, or the offset between them is known as the robot location is fixed in the reference frame. For a mobile robot the relationship between a World Task and a Robot Task is not fixed due to the docking errors of the AGV. Each time the mobile robot will arrive to a certain station, its location will be perturbed about its theoretically designated location.

At this point, the two approaches for improving the mobile's robot accuracy mentioned earlier can be restated. In the first approach, the problem of the docking inaccuracy of the mobile robot can be overcome by reducing the mobile robot's problem to a stationary one. An external system (either mechanical fixture that will use force, or a 3-D servo positioning device that will use fine controlled motion) will position the mobile robot at the designated location in every docking. According to the second approach, the robot task should be modified according to the docking error. The offset between the designated location of the robot, for which the robot task was specified, and its actual location will be evaluated in each docking. The offset (expressed as a transformation matrix) will then be used to modify the original robot task such that it will perform the desired world task from the new location, using the modified robot task.

In this paper a solution that uses the second approach to solve the docking problem of a mobile robot is presented. This approach was chosen over the first one due to its flexibility, expendability, and low implementation cost.

3. IMPROVING THE ACCURACY OF A MOBILE ROBOT

The offset between the actual location of a mobile robot (frame S in figure 1) and the designated robot location (frame Q in figure 1) can not be measured as the Q frame is imaginary when the robot is at the S frame (and vise versa). Thus the transformation between the Q frame and the S frame can not be derived directly. Each frame however, can be related to a fixed reference frame attached to the station being served (frame F in figure 1). The offset between the Q and the F frames

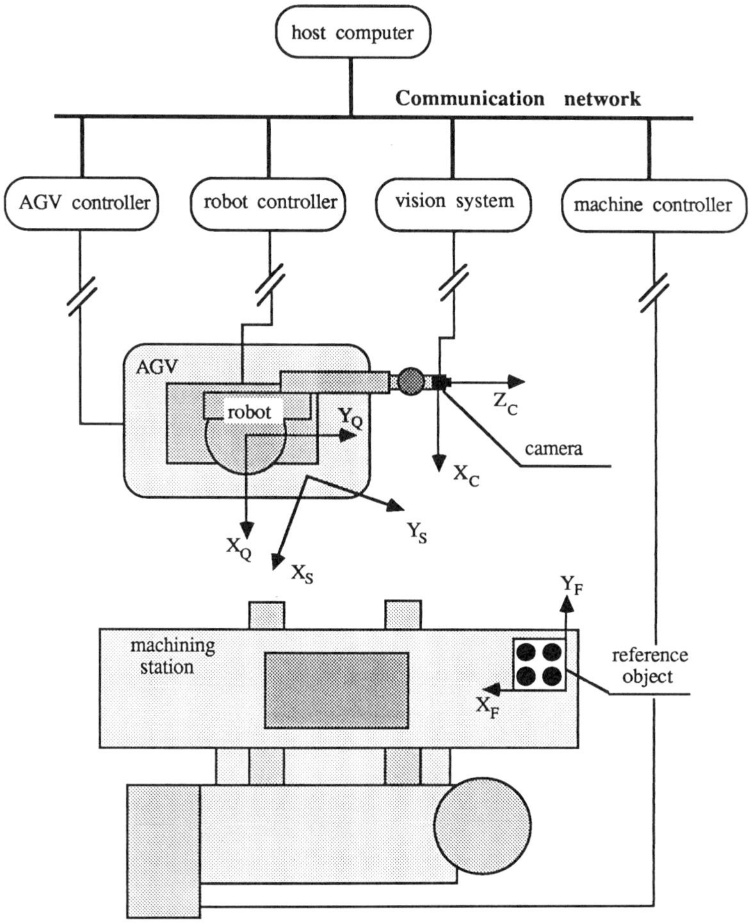

Figure 1 The experimental system with the physical locations of the frames used

can be determined either from physical measurement as in the case of teach by example, or be predefined as in the case of off-line programming. Then the original robot task can be transformed from the Q to the F frame (be specified in the F frame). In subsequent dockings, when the mobile robot will be in location S, the offset between the fixed frame F and the actual frame S will be sensed and the appropriate transformation will be derived. That transformation will be used to transform the robot task, then specified in the F frame, to the S frame wherc the robot will actually be. This process is illustrated in Fig. 2.

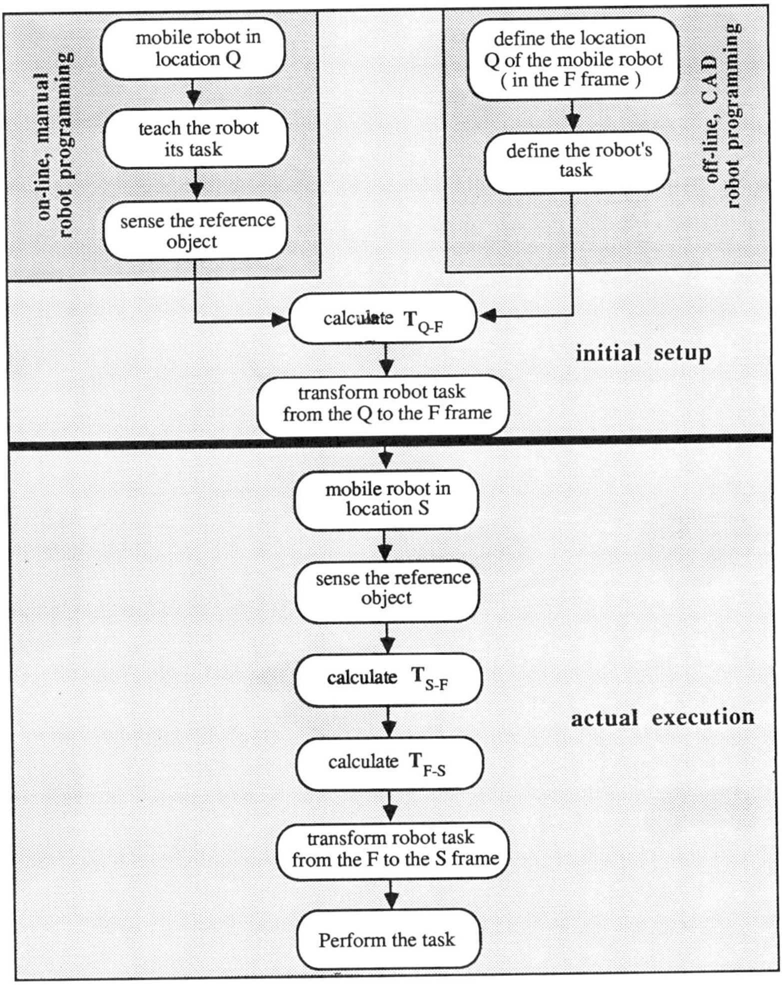

Figure 2 The process of transforming a Robot Task from the designated to the actual robot location

The relationship between the coordinates of points in any two frames (say Q and F) can be expressed in the form of a homogeneous transformation matrix

$$\underbrace{\begin{bmatrix} X_1 & X_2 & \cdots & X_n \\ Y_1 & Y_2 & \cdots & Y_n \\ Z_1 & Z_2 & \cdots & Z_n \\ 1 & 1 & \cdots & 1 \end{bmatrix}}_{F} = \underbrace{\left[\begin{array}{ccc|c} & \mathbf{R} & & \bar{L} \\ \hline 0 & 0 & 0 & 1 \end{array}\right]}_{Q\text{-}F} * \underbrace{\begin{bmatrix} X_1 & X_2 & \cdots & X_n \\ Y_1 & Y_2 & \cdots & Y_n \\ Z_1 & Z_2 & \cdots & Z_n \\ 1 & 1 & \cdots & 1 \end{bmatrix}}_{Q} \qquad (1)$$

Two methods can be employed to calculate the theoretical transformation matrix between two frames from theoretical position, in both frames, of a reference object [1,2]. In the first method the transformation matrix is decomposed into three homogeneous rotation matrices and a homogeneous translation matrix. A sequence of pure rotations is assumed and geometrical relationships are used to find the rotation angles corresponding to the assumed rotation sequence. The rotation angles are substituted back into the rotation matrices that when multiplied in sequence define the rotation portion of the transformation matrix (**R** in Eq. 1). The translation vector is then calculated by solving the three linear equations that relate the coordinates of one point in both frames. In the second approach the twelve unknown entries of the transformation matrix are calculated directly, from the known position of four reference points, by solving the twelve linear equations Eq. 1 is composed of ($n = 4$). Because real-time solution is desired, the direct evaluation approach has been chosen for the calculation of the theoretical transformation matrix. It requires the solution of only linear equations while the assumed rotation sequence approach requires the solution of six non-linear equations. The derivation of the theoretical transformation matrix, based on the theoretical locations of reference points in both the mobile frame and the fixed frame, is detailed in [3]. Two sensors were used to measure the locations of the reference points: trigger probe and vision system. Each sensor has different characteristics thus requiring a different data processing algorithm for calculating the transformation matrix.

In theory the available data is absolutely accurate all the time, in reality it is not. Errors in measured position data have significant impact on the calculation of the transformation matrix. Two problems have to be addressed when derivation of transformation matrix from actually measured data is attempted:

- How to make sure that the calculated transformation matrix is a rigid body, homogeneous transformation matrix and not simply a coefficient matrix that solves the linear set of equations in Eq. 1.
- How to make sure that the calculated transformation matrix is the most stable one and will be less effected by perturbation in the measured position data than any other transformation matrix.

The derivation of transformation matrix from actually measured data is discussed in the next section.

3. PERTURBATION ANALYSIS

The inaccuracies in the calculation of the transformation matrix are from two sources: hardware and software. The hardware related error sources include among others the finite resolution of the sensor, the accuracy of the robot, and the accuracy of the measuring devices used. The software related errors include the roundoff errors in the calculations of the reference points and the accuracy of the matrix inversion algorithm. Since the implementation of the proposed method was limited to existing hardware, whose modifying was not the scope of this project, it was decided to enhance the accuracy of the algorithms by addressing the software issues.

In [3], both algorithms (for the vision system and for the touch trigger probe) determine the transformation matrix (after all the coordinates of the reference points were calculated, with their inaccuracies) by matrix inversion. In Eq. 1, when the coordinates of exactly the same reference points are given in two frames - $\mathbf{P}_Q$ and $\mathbf{P}_F$, using matrix inversion will result in calculating the rigid body transformation matrix between the two frames - $\mathbf{T}_{Q\text{-}F}$. This transformation matrix is the only solution to the linear set of equations in Eq. 1. If however, the coordinates of the reference points are perturbed by more than the accuracy of the matrix inversion algorithm employed, a rigid body is no longer described from two frames. The calculated matrix will not be a rigid body transformation matrix but rather some coefficient matrix, with no geometrical meaning, that will solve the linear set of equations in Eq. 1.

It becomes obvious that rather than a deterministic tool, a statistical tool for calculating the transformation matrix is needed. The statistical method should be a stable one and able to minimize the influence of errors in the data on the calculated matrix. This means that based on perturbed data, the calculated matrix should be a rigid body transformation matrix, and that variations in the data will change very little, if at all, the calculated rigid body transformation matrix - minimize δ in Eq 2.

$$\| \mathbf{T}_{Q\text{-}F} * \mathbf{P}_Q - \mathbf{P}_F \| \leq \delta \qquad (2)$$

The most familiar statistical approach to the problem of determining the coefficients approximating a solution based on data is the Least Squares approach [4]. The LS drawback is that the calculated $\mathbf{T}_{Q\text{-}F}$ does not necessarily minimize the error δ. A different method for solving LS problems is the Singular Value Decomposition [5]. The SVD is considered to be [6] "...the most reliable method for computing the coefficients of general LS problem...and the most effective in dealing with errors in the data, roundoff errors, and linear dependence.".

Briefly, using the SVD technique an m x n matrix A can be decomposed into three special matrices

$$\mathbf{A}_{m \times n} = \mathbf{U}_{m \times m} * \Sigma_{m \times n} * \mathbf{V}^T{}_{n \times n} \qquad (3)$$

Those three special matrices allows the direct calculate the Pseudoinverse of the matrix **A** - $\mathbf{A}^P$.

$$\mathbf{A}^P = \mathbf{V} * \Sigma^P * \mathbf{U}^T \qquad (4)$$

The pseudoinverse is related to the LS problem in Eq. 2 by the fact that the $\mathbf{T}_{Q\text{-}F}$ that minimizes δ is defined as

$$T_{Q\text{-}F} = P_F * (P_Q)^P \tag{5}$$

Because of its advantages, the Singular Value Decomposition technique was chosen as the statistical method to solve for the transformation matrix between two frames based on position data, in both frames, of fixed reference points on a rigid body. A detailed discussion about SVD and its applications can be found in [4,5,6].

4. EXPERIMENTAL RESULTS

The simulation of the touch trigger probe algorithm showed that the accuracy of the 3-D position data (0.12 μm [7]) of the reference points, combined with the algorithm, were accurate enough to determine the transformation matrix by matrix inversion directly. On the other hand, from simulation of the vision system algorithm it was obvious that statistical method was needed as the only results that were reached were coefficient matrices with no geometrical meaning. Based on those two results, the vision system for data collection and the appropriate algorithm were implemented, and the implementation details can be found in [3].

A sample result from one of the different conditions under which the transformation matrix was calculated, is shown in Table 1. In this setup, Z_C and Z_F were pointing at the same direction (see Fig. 1) and the C frame was rotated by -30^0 around the Z_C axis. The distance between the C frame and the F frame was 800 mm while the X and Y offsets between the two frames were not measured as it was determined that the error there is insignificant with respect to the error in the Z offset.

5. SUMMARY

The problem of compensating for the poor docking accuracy of an AGV used in an industrial mobile robot was discussed. A software solution to that problem was presented in which the basic idea is to modify the Robot Task according to the docking error. Two algorithms, for two data collection devices were developed: one that employs touch trigger probe mounted as a robot end effector, and one that uses vision system for that purpose. The need for a statistical method that should be able to calculate the transformation matrix from actually measured (perturbed) data was demonstrated and such a method was presented. Finally, results from an experiments that confirm the validity of the proposed method were presented.

6. BIBLIOGRAPHY

[1] Rogers, D. F., and Adams, J. A., "Mathematical Elements for Computer Graphics", 1978.

[2] Kane, T. R., Likins, P. W., and Levinson, D. A., "Space Craft Dynamics", McGraw-Hill, New york, 1983.

<table>
<tr><td colspan="2">Rotation by -30^0 around Z_C Z_m = 800 mm</td></tr>
<tr><td>Theoretical matrix T</td><td>calculated matrix $T + \delta T$</td></tr>
<tr><td>$$\begin{bmatrix} 0.866 & 0.5 & 0.0 & ? \\ -0.5 & 0.866 & 0.0 & ? \\ 0.0 & 0.0 & 1.0 & 800.0 \\ 0.0 & 0.0 & 0.0 & 1.0 \end{bmatrix}$$</td><td>$$\begin{bmatrix} 0.868 & 0.482 & 0.0 & -138.675 \\ -0.495 & 0.862 & 0.0 & -36.461 \\ 0.019 & -0.087 & 1.0 & 805.433 \\ 0.0 & 0.0 & 0.0 & 1.0 \end{bmatrix}$$</td></tr>
<tr><td>Parameters</td><td>Error matrix δT</td></tr>
<tr><td>Z1 = 796.37 $\theta 1 = -0.23^0$
Z2 = 794.43 $\theta 2 = 0.46^0$
Z3 = 804.09 $\theta 3 = -1.18^0$
$\theta 4 = -0.33^0$
average angular error = -0.32^0</td><td>$$\begin{bmatrix} 0.002 & 0.018 & 0.0 & ? \\ 0.005 & 0.004 & 0.0 & ? \\ 0.019 & 0.087 & 0.0 & 5.433 \\ 0.0 & 0.0 & 0.0 & 0.0 \end{bmatrix}$$</td></tr>
</table>

Table 1 Results of the transformation matrix calculation from actually measured data

[3] Mandel, K., "On-Line Correction of Mobile Robot Docking Errors", M.S. Thesis, University of Wisconsin - Madison, August 1985.

[4] Lawson, C. L., and Hanson, R. J., "Solving Least Squares Problems", Prantice -Hall, New Jersy, 1974.

[5] Golub, G. H., and Reinsch, C., "Singular Value Decomposition and Least Squares Solutions", Numerical Matematics, Vol. 14, 403-420, 1970.

[6] Foesyth, G. E., and Moler, C. B., "Computer Methods for Mathematical Computations", Prantice-Hall, New Jersy, 1977.

[7] Duffie, N. A., and Halldin, G. W., "The use of tactile sensing and computer graphics in the automation of mold production for plastics processing", Progress report #1 to NSF, December 1983.

Simulation of Steady Human Walking and its Application to Artificial Locomotive System

T. Yamashita*, and Y. Tagawa†

*Department of Control Engineering, Kyushu Institute of Technology, Tobata. Kitakyushu 804, Japan
†Department of Mechanical Engineering, Kurume Institute of Technology, Kamitsu, Kurume 830-91, Japan

Abstract

The idea of the zero moment point (ZMP) that has been often used in both the simulation of human walking and the control of a bipedal robot is extended. Our model which consists of eight elements can simulate both the normal and the abnormal human walking. The model has a zero moment joint (ZMJ), which cannot generate moment during the stance phase, in one leg. Assuming that the motions of the lower torso and the lower extremities are the same pattern as in the normal, various simulations are carried out. Main results are the following: the motion of a point of application of floor reaction in the leg with the ZMJ is restrained, and becomes greater as the ZMJ locates higher position in the leg; the lost function at the ZMJ is mainly compensated by the normal joints located in the lower position than the ZMJ. The simulation results by our model will be useful to a synthesis of artificial leg, an assessment of the abnormal walking, and a design of the bipedal walking machine.

Introduction

Locomotive robot is one of active research projects in recent years. Some successful achievements have been reported in static or jumping locomotion, but a dynamic locomotion resembling to human walking has not been realized yet. Human walking can be achieved by a coordinating control of many degrees of freedom, but its control mechanism has not been understood deeply enough to be applied in a design of a robot. Simulations of human walking characteristics may provide valuable information for the control mechanism and the design of locomotive robots.

The characteristics of normal human walking have been also actively simulated in recent years by using a computer[1,2,3]. The analysis of abnormal walking, however, is very difficult because an effecfive simulation model has not been developed yet. Development of a simulation method which can be applied to various walking conditions is a challenging subject in the field.

This paper presents a model which can simulate both normal and abnormal human walking. ZMP method[4], in which the fact that moments about a point of application of a floor reaction are zero is utilized, is effective for the simulation of the human walking. This idea is expanded further in our model. Assuming that a ZMP is located at a ideal joint in one leg of the model, the joint can transmit force to adjacent bodies connected but cannot generate the moments during the stance phase to control the motions of the bodies. The joint having a ZMP is called zero moment joint (ZMJ). In this paper human walking having a ZMJ is called abnormal walking. The moments at the remaining active joints compensate the lost function at the ZMJ and exhibit different characteristics, compared to the normal walking, in both legs. Motions of the lower torso and the lower extremities in the model are assumed to be the same pattern as in the normal walking. The influence of the unbalanced moments between two legs results in a different motion of the upper torso from the normal.

Mathematical model

Our dynamic model consists of 8 massive elements, which represent the upper and lower torsos, the thighs, the shanks, and the feet. Adjacent elements are coupled by ideal joints. The body, which are supported by either one or two legs at the same time depending on the supporting condition, are allowed to move in three-dimensional space. The motions of each element are controlled by the moments M_{iX}, M_{iY}, and M_{iZ} applied at the i-th joint. The ZMJ in the abnormal case is defined as one of the joints in the right leg, that is, ankle, knee, and hip joints.

The equations of motion for each element of the model are derived from the Newton's mechanics: equilibrium equations of force and moment. These equations of motion can be transformed to a form suitable for a specific purpose. For example, equation of motion for the upper torso, which consists of the inertia forces and moments of the each element, is obtained. Eq.(1) shows the equation of motion for the upper torso in either the normal or the abnormal walking. The subscript j in the equation shows the location of the ZMJ: the value of j, 1, 2, and 3, corresponds to the ankle, knee, and hip in the right leg, respectively.

$$\underline{M}^5 = \{ \rho_{j3} R_l(t) \sum_{i=1}^{3} \underline{\xi}_i - R_l(t) (\sum_{i=1}^{3} \rho_{j(i-1)} \underline{\xi}_i + \rho_{j3} \Delta \underline{H}_r) - \sum_{i=1}^{4} \rho_{j(i-1)} \underline{\xi}_i$$

$$
\begin{aligned}
&- \rho_{j3}\underline{\varepsilon}_5 \} \times (m_5\underline{g} + \underline{F}^5) - \sum_{i=1}^{4} \{ (\sum_{k=0}^{i-1} \rho_{j(k-1)}\underline{\xi}_k + \rho_{j(i-1)}\underline{\varepsilon}_i) \\
&\times (m_i\underline{g} + \underline{F}^i) \} + (\sum_{i=1}^{3} \rho_{j(i-1)}\underline{\xi}_i + \rho_{j3}\Delta\underline{H}_{\gamma}) \times \{ R_{\gamma}(t) \sum_{i=1}^{3} (m_i\underline{g} \\
&+ \underline{F}^i)_{l} - R_{l}(t) \sum_{i=1}^{4} (m_i\underline{g} + \underline{F}^i) \} + \rho_{j3} \sum_{i=1}^{3} \{ - \underline{M}^i_{l} + (R(t) \sum_{k=1}^{3} \underline{\xi}_k \\
&+ \sum_{k=i}^{3} \underline{\xi}_k - \underline{\varepsilon}_i)_{l} \times (m_i\underline{g} + \underline{F}^i_{l}) + \underline{\xi}_{il} \times R_{l}(t) \sum_{k=1}^{4} (m_k\underline{g} \\
&+ \underline{F}^k) \} - \sum_{i=1}^{4} \rho_{j(i-1)}\underline{M}^i
\end{aligned}
\qquad (')
$$

where $\rho_{ji} = \begin{cases} 0 & \text{in } j > i \\ 1 & \text{in } j \leqq i, \end{cases}$

$\underline{F}_i$:force vector acting on i-th element at i-th joint,
$\underline{F}^i$:inertia force vector of center of mass of i-th element,
$\underline{M}_i$:moment vector acting on i-th element at i-th joint,
$\underline{M}^i$:inertia moment vector about center of mass of i-th element,
$\underline{g}$:vector of gravitational acceleration,
m_i :mass of i-th element,
$\underline{\varepsilon}_i$:position vector from (i-1)-th joint to center of mass of i-th element,
$\underline{\xi}_i$:position vector from (i-1)-th joint to i-th joint,
$\Delta\underline{H}$:position vector from right hip joint to left hip joint,
R(t) :supporting function,
subscripts:
γ , l :indicate right and left legs, respectivly.

The ZMJ at the j-th joint of the right leg cannot generate active moment, then equilibrium equation of moment at the j-th joint is equal to zero: the equation containing terms of elements above the j-th joint becomes differential equation for the upper torso; the equation containing terms of elements below the j-th joint becomes restrained condition of motion for the point of the application of the floor reaction in the leg with the ZMJ. Eq.(2) shows the restrained condition.

$$
\begin{aligned}
&- \sigma_{j0}R_{\gamma}(t)\underline{\xi}_1 \times \underline{R} - \sigma_{j0} (\underline{\xi}_1 - \underline{\varepsilon}_1) \times (m_1\underline{g} + \underline{F}^1) - \sum_{i=2}^{3} \sigma_{j(i-1)}R_{\gamma}(t)\underline{\xi}_i \\
&\times \underline{R} - (\sum_{i=1}^{2} \sigma_{ji}\underline{\xi}_{i+1}) \times (m_1\underline{g} + \underline{F}^1) - \sum_{i=2}^{3} \{ (\sum_{k=i}^{3} \sigma_{j(k-1)}\underline{\xi}_k) \\
&\times (m_i\underline{g} + \underline{F}^i)\} + \sum_{i=2}^{4} \{ \sigma_{j(i-1)}\underline{\varepsilon}_i \times (m_i\underline{g} + \underline{F}^i) \} \\
&+ \sum_{i=1}^{3} \sigma_{j(i-1)}\underline{M}^i = \underline{0}
\end{aligned}
\qquad (2)
$$

where $\sigma_{ji} = \begin{cases} 1 & \text{in} \quad j > i \\ 0 & \text{in} \quad j \leq i. \end{cases}$

Flow of simulation for the steady human walking in both normal and abnormal cases is shown in Fig.1. Because the model has so many degrees of freedom, the number of unknown variables in the equations exceeds the number of the equations. To make the equations closed system, reliable kinematic data obtained experimentally in the normal walking are given into the model: the motions of the lower torso and of the point of application of the floor reaction, and the foot angle during the stance phase are predetermined. The motions of the remaining elements are determined as solutions of algebratic or differential equations.

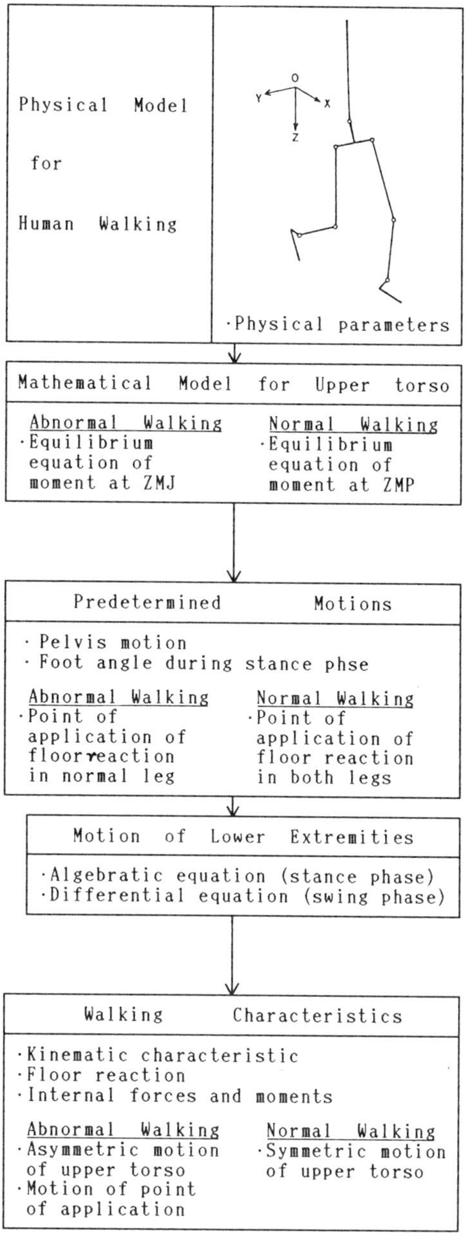

Fig.1 Flow of simulation

Simulation results

The rotational motion of the upper torso and the walking characteristics are simulated. Because the simulated results in the normal walking show good agreements with the experimental results [5,6], results showing differences between the normal and the abnormal walking are summarized in this paper: the floor reactions; the moments at the ankle, knee, and hip joints in both legs; the motions of the point of the application of the floor reaction.

The characteristics are expressed in dimensionless forms derived from the mathematical model: the forces are divided by weight of the model; the moments are divided by multiplication of the model weight and the leg length; the motion of the point of the application is divided by the leg length. The time is normalized by one walking cycle time. Walking parameters, that is, step length S_X, walking cycle time T, and double support period T_d, correspond to ordinary walking of normal person.

floor reaction:

The floor reactions to both feet in the X and Y coordinates are shown in Fig.2. The ZMJ at the hip has the greatest influence on the X component in the right leg, R_{Xr}. The influence of the ZMJ on R_{Yr} shows a consistent pattern, that is, the magnitude of the floor reaction becomes greater as the location of the ZMJ becomes higher position in the leg. The ZMJ gives less influence on the components of the left leg.

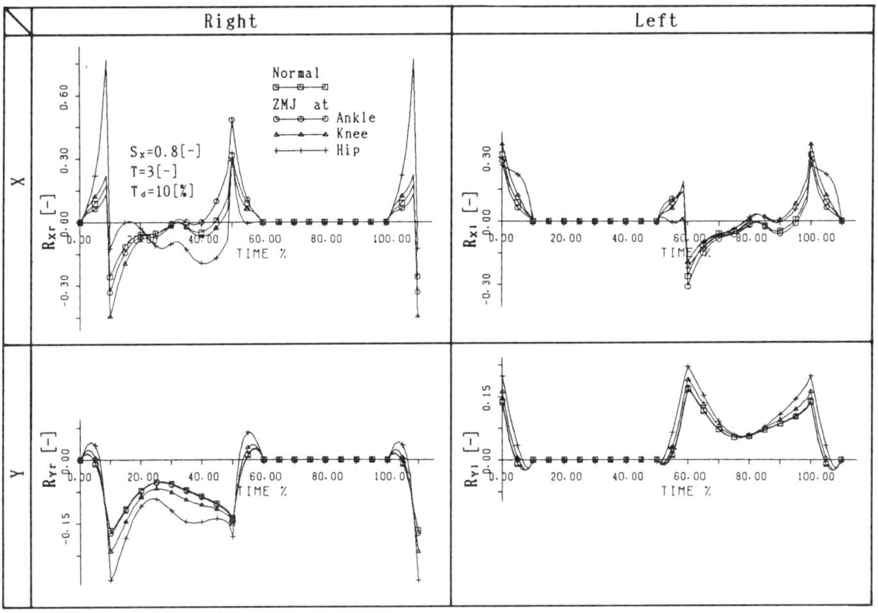

Fig.2 Floor reactions

moment:

The moments at the ankle, knee, and hip joints in both legs are shown in Figs.3 and 4. Fig.3 shows the moments about the X axis (M_{Xi}.). The characteristics of the normal walking show the same pattern in each joint of two legs, respectively. As shown in the moments at joints in the right leg, the joints that are located below the ZMJ show a different pattern: a reversed moment is generated. The moments about the Y axis (M_{Yi}.) shown in Fig.4 exhibit the different pattern at each joint. The moments seem to be coordinating each other in a complexed fashion. The ZMJ gives a greater influence on the normal joints located below the ZMJ. The influence of the ZMJ on the normal joints in the left leg is very small.

The lost function at the ZMJ is mainly compensated by the normal joints located in the lower position than the ZMJ. The ZMJ at the hip has the greatest influence in the compensatory action at the active joints.

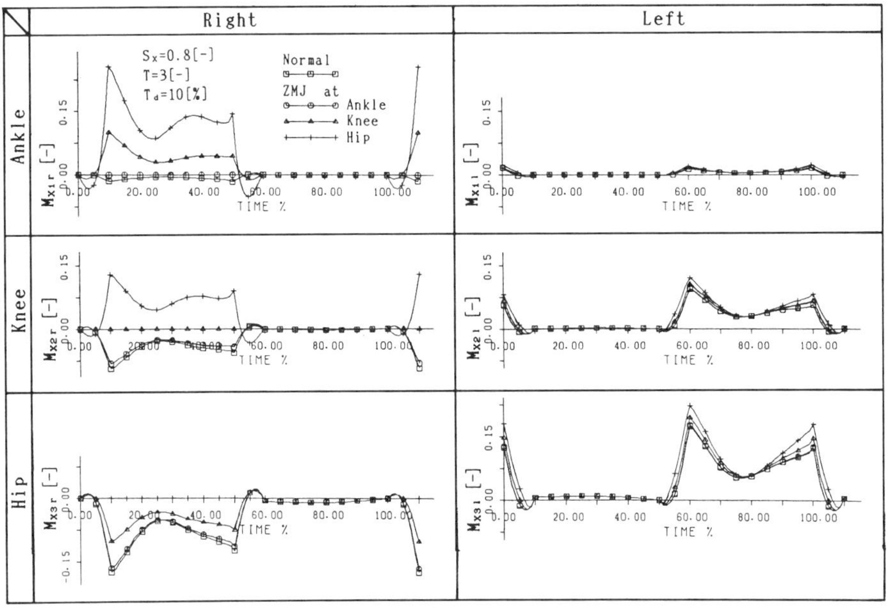

Fig.3 Moment at each joint in both legs - moments about X axis

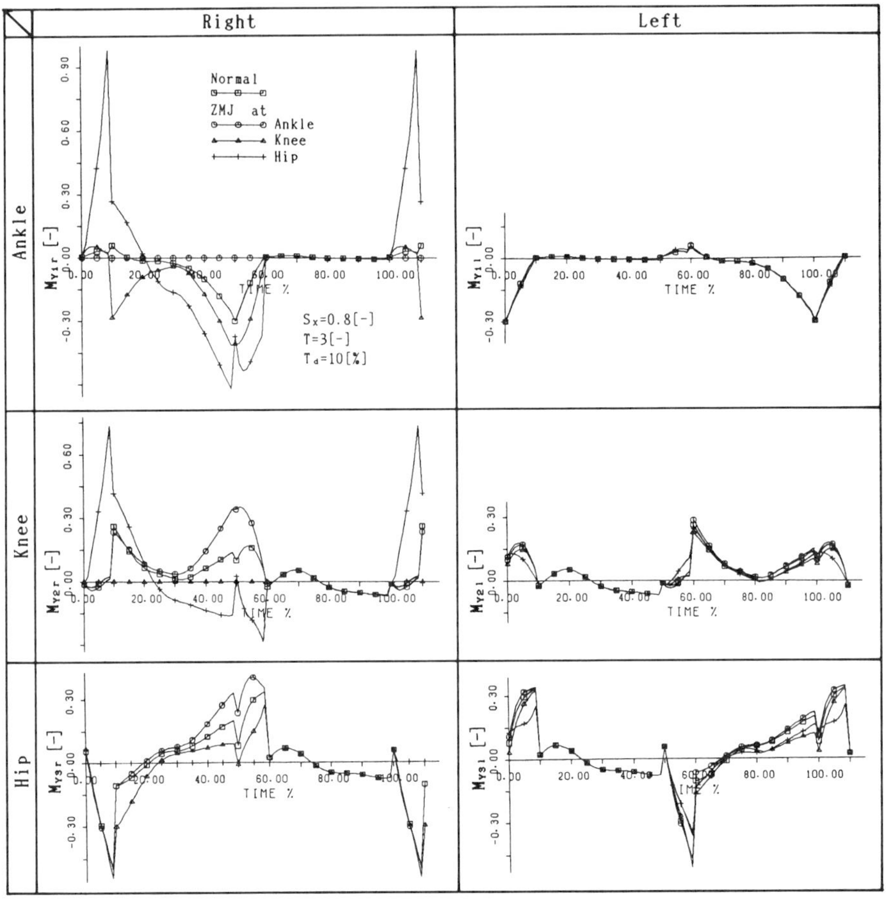

Fig.4 Moment at each joint in both legs - moments about Y axis

motion of point of application:

The motions of the point of application of the floor reaction are different according to the position of the ZMJ (see Fig.5(a)~(d)). Symbols in each figure are marked at every 5% of the walking cycle time. As the ZMJ locates higher position in the right leg, the point of the application moves

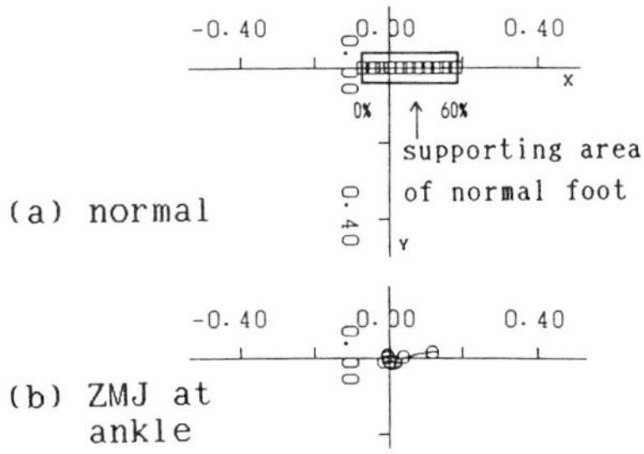

Fig.5 Motions of point of application

in a wider area. In other word a wide supporting area is needed to walk.

(c) ZMJ at knee

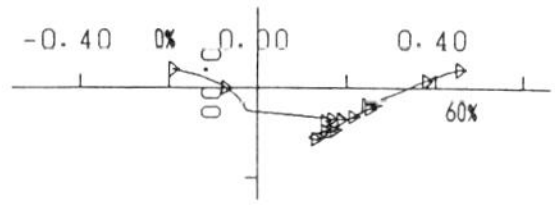

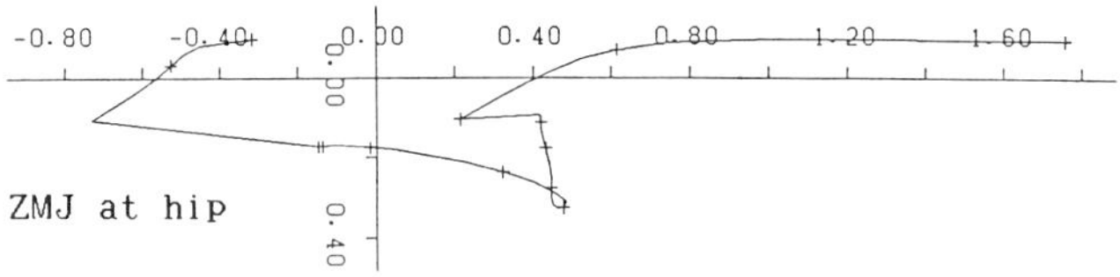

(d) ZMJ at hip

Fig.5 Motions of point of application (continued)

Conclusions

The results of simulation of the human walking by the model having a ZMJ in one leg are presented. Assuming that the motions of the lower extremities and the lower torso are the same pattern as in the normal, simulations are carried out. The following results were obtained:

(1)The motion of the point of the application in the foot of the leg having the ZMJ is determined by the equilibrium of the moment, and becomes greater as the ZMJ locates higher posiotion in the leg. Therefore a wider supporting area than normal foot is needed to walk.

(2)The ZMJ at the hip gives the greatest influence on the compensatory action at the active joints.

(3)The lost function at the ZMJ is mainly compensated by the normal joints located in the lower position than the ZMJ.

References

[1] C.K.Chow and D.H.Jacobson, "Studies of Human Locomotion via Optimal Programming", Mathematical Biosciences,Vol.10,239-306,1971.

[2] T.Yamashita, K.Kabashima, and Y.Sakatani, "Control of Macro-Model to Simulate Human Level Walking", Preprints of Second CISM-IFToMM Symp. on Theory and Practice of Robots and Manipulators, 183-192, Polish Scientific Publishers,1976.

[3] Y.Tagawa and T.Yamashita, "Characteristics of a Multi-Body Model for Human Level Walking", Biomechanics, 8-A, 373-379, Human Kinetics, 1983.

[4] M.Vukobratovic and D.Stokic, "Dynamic Control of Unstable Locomotion Robots", Mathematical Biosciences, Vol.24, 129-157, 1975.

[5] Y.Tagawa and T.Yamashita, "Simulation of Human Steady Walking on Level Ground", Trans. of the Society of Instrument and Control Engineers, Vol.20, No.2, 167-174, 1984 (in Japanese).

[6] Y.Tagawa and T.Yamashita, "Analysis of Main Characteristics for Human Walking by Multi-Body Model", Trans. of the Society of Instrument and Control Engineers, Vol.20, No.6, 567-569, 1984 (in Japanese).

A Navigation Control System for an Autonomous Mobile Robot

Mark B. Kadonoff and R. Warren George

Denning Mobile Robotics, Inc., Woburn, MA 01801, USA

I. Introduction

The Denning Sentry I is an autonomous mobile robot configured for use as an electronic security system. A mobile security robot can provide comprehensive volume protection of an office building, warehouse, or factory at a lower installation cost than for a fixed intrusion detection system. The security robot's sensor inputs can be correlated to provide a higher probability of detection than uncorrelated sensors generally employed by fixed systems. The security robot's patrol routes can be modified as customers' facilities change, without requiring extensive rewiring. Finally, additional sensors can be easily added for detecting fire, smoke, or toxic gases.

The capabilities of a mobile security robot are useful in many other applications. Mobile robots are being developed for parts handling in warehouses and flexible manufacturing systems, for inspection and exploration in hazardous environments, for military reconnaissance and bomb disposal, and as aides to the physically handicapped.

A mobile security robot employs many of the functions of an *autonomous mobile chassis*. Autonomy is defined as independence from any external electrical or computational resource. We consider the attainment of autonomous mobility as an important milestone in development of a security robot. Once the robot is capable of moving along the patrol route ordinarily taken by a security guard, its mobile chassis can be easily applied to other applications such as parts-handling or inspection.

An autonomous mobile chassis has three navigational capabilities.

- It can move to any specified location within a known environment.
- It can avoid unexpected obstacles in its path.
- It can improve its knowledge of the environment through exploration.

In this paper, we will discuss the approach we have taken towards achieving these capabilities. Section II describes the Denning Sentry I. Section III examines the navigation executive and world model which permit navigation within a known environment. Section IV describes the navigation techniques used for local obstacle avoidance. Section V outlines some mechanisms which will enable the robot to extend its knowledge of the environment.

II. Physical Aspects

II.A Introduction

We briefly describe the physical aspects of the Denning Sentry in order to give the reader an understanding of the software requirements for an autonomous mobile chassis.

The Denning Sentry I is pictured in figure 1. In addition to the robot, a Communications Station provides an operator interface, and connections to an external alarm and security communications system. A Charge Station provides automatic recharging capability, and storage when not on patrol.

Figure 1. The Denning Sentry I

II.B Mechanical Systems

The Sentry I's mechanical systems consist of a locomotion system and chassis. The locomotion system consists of three parallel wheels, which drive and steer simultaneously. This kind of transmission, called a *synchro-drive*, enables omni-directional motion while requiring only two degrees of freedom.

The chassis consists of a body and a head section. The body stays stationary as the wheels turn, and acts as an orientation reference over short distances. The head is attached to the steering train, and always faces the direction of travel. Slip-rings provide power and communications to the head, and allow for continuous rotation.

II.C Electrical Systems

In the Denning Sentry I, both drive and steering motors are DC permanent magnet servomotors driven by PWM (pulse width modulation) amplifiers. Two PID (proportional-integral-derivative) motion controllers are used for position and velocity control. An emergency stop system disables motor power upon contact with bumper switches mounted on the fiberglass skin.

Electrical power is generated by three 80 amp-hour 12 volt, gel cell, batteries. High efficiency DC/DC switching converters provide appropriate voltages and currents to the sensors and computers. A power distribution system allows for power-up sequencing and automatic fault detection.

II.D Computer Systems

The Sentry uses a distributed computer architecture, with multiple processors communicating over a common serial bus. Each processor has an *expert module*, which is responsible for performing a particular sensing, control, or computation task. Each processor resides on an STD bus, with local memory. Currently, four Hitachi 64180's handle the ultrasonic and optical sensing, position estimation, intrusion detection, and internal status monitoring tasks. Two Motorola 68701's are dedicated PID motion controllers. A Motorola 68010 with a real-time multi-tasking operating system (OS-9) runs the navigation software and communications interface. A 3.5" floppy drive is used for loading application software and storing patrol logs.

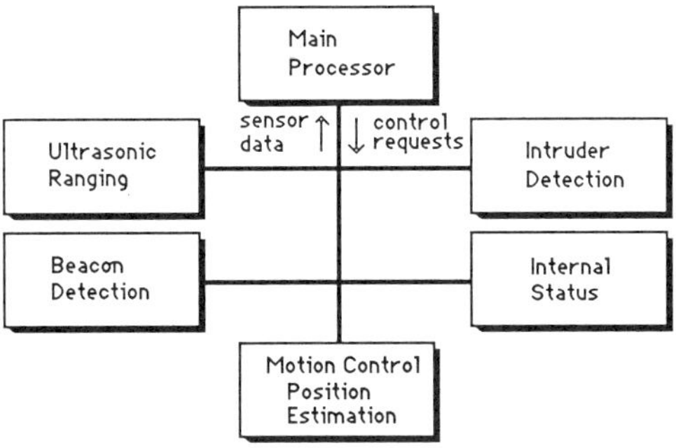

Figure 2. Computer Architecture

II.E Sensors

The sensors on the Denning Sentry are divided into three groups:

- navigation sensors
- internal status sensors
- intrusion detection sensors.

Navigation sensors include ultrasonic sensors for local collision avoidance, an optical beacon-tracking system for navigating along a path, and an odometry system for computing dead-reckoned position.

Internal status sensors include sensors for measuring chassis pitch and roll, battery voltages, motor currents, and ambient and internal temperatures.

Intrusion detection sensors are used for the security application, and include microwave doppler-effect motion sensors, passive thermal infrared sensors, fire, heat, and toxic gas sensors, and audio/video feedback for the operator at the communication station.

III. Navigation Executive and World Model

III.A Introduction

The goal of our navigation system is to move the Sentry I to a named place in the known environment. The robot can respond to the following commands:

- goto (goal-list)	visit each of the named places in the goal list, then stop
- patrol (goal-list)	visit each of the named places in the goal list, then repeat
- random (goal-list)	continously visit the named places in a random order
- return (base-node)	return to the charge station to recharge

Several conditions can occur which will cause the robot to stop and report its status to the Communications Station. These are:

- an obstacle is blocking the path and cannot be avoided
- a beacon which was expected is not in view, and the path is labeled 'dangerous'
- a possible intruder has been sighted

III.B Internal Representation

The navigation executive is constructed as three nested hierarchical procedures. These are called the:

- planner
- global navigator
- local navigator

Each procedure constructs the spatial representation required for the succeeding level. Figure 3 is an example of this internal representation depicted at each level of the navigation system.

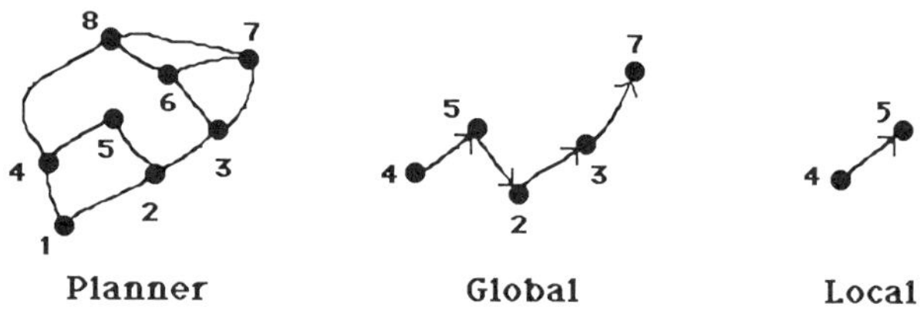

Figure 3. Internal Representation

In the Planner, the external environment is represented as a graph structure, called the Map. Nodes of this graph structure are reference locations, and arcs are navigable paths between them. The planner obtains a goal node from the user, and using the Map, together with knowledge of its current position, constructs a sub-graph detailing the shortest distance route to the goal.

In the Global Navigation, the sub-graph consists of a consecutive list of paths leading to the goal. Global navigation receives this Pathlist from the planner, and traverses the list from start to finish. At each node, global navigation changes headings, recalibrates estimated position, and runs stationary intrusion detection routines.

In the Local Navigation, each path element is a vector in two-dimensional space, whose coordinate system is defined by the robot's current position. The local navigator moves the robot to the specified point, avoiding obstacles along the way.

Figure 4 shows data structures for each level in our example.

```
map:
  node1:
    path 1→2        pathlist:
    path 1→4          path4→5         path4→5:
  node2:              path5→2           distance   100 ft
    path2→1           path2→3           angle      45°
    path2→3           path3→7           velocity   1 ft/sec
    path2→5                             width      8 ft
     •
     •
     •

    Map               Pathlist            Path
```

Figure 4. Map Data Structures

In the next section, we describe navigation techniques that can be run by the Local Navigation procedure.

IV. Navigation Techniques

IV.A Introduction

Robot control systems require both high-level planning and low-level collision avoidance. Ideally, a continuum of navigation methods provide higher and higher levels of competence [Brooks 85]. More capable methods are usually more expensive, so a demand exists for fast simple navigation algorithms. Multiple control techniques running simultaneously can produce a synergy of behavior, if arbitrated properly. Therefore, our goal is to produce a number of these techniques, each capable of exhibiting some well-defined behavior. Each technique can be debugged and tested independently of any other, thus allowing improvements without obsoleting or breaking previously working code.

IV.B Local Navigation Techniques

Experimentation in our lab and in customer sites has yielded several behaviors which are useful in a wide set of circumstances.

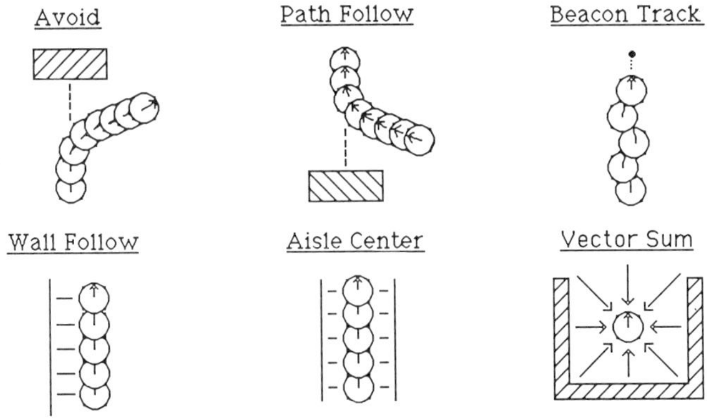

Figure 5. Local Navigation Techniques

Each technique can be viewed as a control system **G**, operating on sensor data **S**, and producing **Ø** and **V**, the desired turn angle and drive velocity.

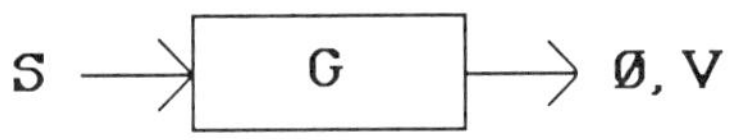

Figure 6. Navigation Control System

We would like to build parallel networks of these control systems. Each plant $\mathbf{G_i}$ operates on sensor data **S**, and produces $\mathbf{Ø_i}$ and $\mathbf{V_i}$. As shown in Figure 7, some arbitration mechanism is then required.

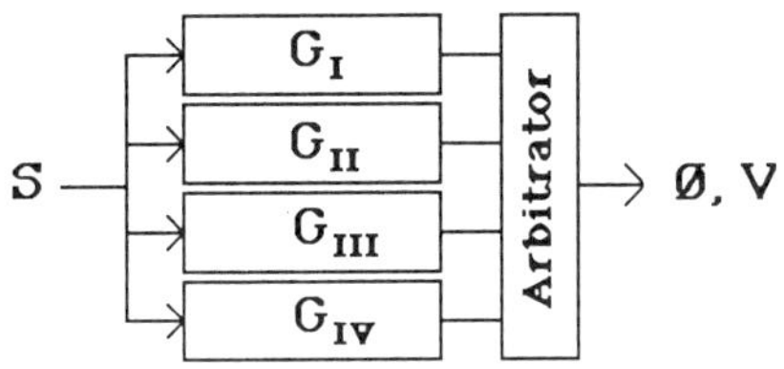

Figure 7. Paralleled Navigation Techniques

IV.C Arbitration

The arbitrator must be aware of each method's inadequacies, and ignore its outputs at those times. A systematic approach is needed for encoding this information. This is where we feel some form of knowledge-based system is required. We define the boundary conditions outside which the outputs of each technique should be ignored. The arbitrator uses these limits, and the current outputs, to decide who should be in control at any moment.

As an example, if we are trying to arbitrate among the avoider, path follower, and beacon tracker, we could specify the following prioritized set of rules:

1. if the avoider's output $\mathbf{Ø_t \neq 0}$, use avoider
2. if the distance from path **abs (Y)** > $\mathbf{Y_{min}}$, use path follower
3. if the distance from path **abs (Y)** < $\mathbf{Y_{min}}$, use beacon tracker

In this case, the robot will always avoid obstacles in its path, maneuver back on course when the avoider forces it more than $\mathbf{Y_{min}}$ off of the path, and beacon track otherwise.

V. Extensible World Models

Notice that our world model as described in Section III said nothing about the robot's surroundings, only navigable paths through it. Geometric representations can be located at nodes to store knowledge of local surface features. This would allow the robot to incrementally extend its knowledge of the environment. We are attempting this with two different approaches, probability grids [Moravec 85], and line segment-based models [Crowley 85]. The initial application for these models is to increase the locations where the robot can update its estimated position. Both of these models are capable of matching new sensor interpretations with stored models, thereby reducing positional uncertainty. Position recalibration is performed at nodes by the global navigator. As shown in Figure 8, each model is located over a node, and a matching process produces position offsets from the center. The robot then moves to eliminate this error. We make the assumption that nodes are uncluttered enough to allow repositioning.

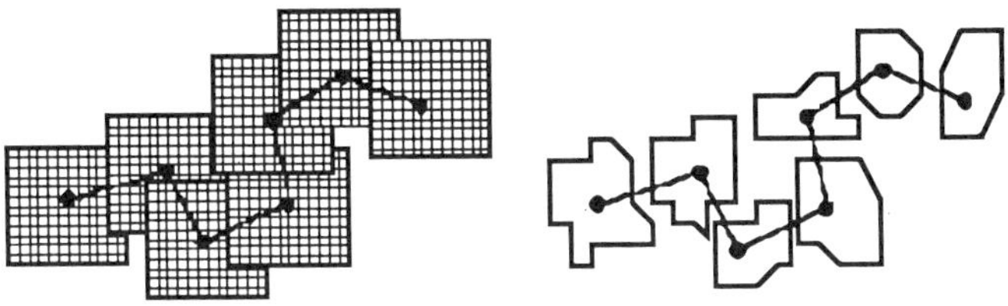

Grid-Based Model **Segment-Based Model**

Figure 8. Map Structure With Volume Models

Both approaches have been demonstrated using real sonar data, but neither has yet been implemented as a commercial system. The grid-based approach is expensive, both for processing and map storage, but can build accurate models with sufficient sensor data. Line segment models work well in relatively regular environments, but break down when trying to represent complicated scenes.

VI. Conclusions

We have attempted to present the key ideas behind the navigation software system for the Denning Sentry. The algorithms described here are operating successfully in a commercial product currently installed in several beta-test sites. By providing a beacon system for maintaining estimated position, and predefining navigable paths through the environment, the robot is able to patrol a building for 12 hours without stopping.

We consider the fundamental elements of the system to be:

- the navigation executive and world model
- the use of multiple navigation techniques operating in unison
- the arbitration mechanism for combining multiple techniques
- the approach towards integrating extensible world models

Our hope is that these elements, and others yet to be developed will lead to a generalized methodology for mobile robot control systems.

Bibliography

[Brooks 85] "A Robust Layered Control System For A Mobile Robot", Rodney A. Brooks, A.I. Memo 864, Massachusetts Institute of Technology, Spetember, 1985

[Crowley 85] " Navigation for an Intelligent Mobile Robot", James L. Crowley, *IEEE Journal of Robotics and Automation,* RA-1, March 1985, 31-41

[Moravec 85] "High Resolution Maps from Wide Angle Sonar", Hans P. Moravec and Alberto Elfes, *IEEE Conference on Robotics and Automation,* St. Louis, March 1985

ROBOT APPLICATIONS

Case Study of the Implementation of a Vision System in a Robotic Assembly Application

Thomas M. Kisko and Eginhard J. Muth

Department of Industrial and Systems Engineering, University of Florida, Gainesville, FL 32611, USA

Introduction

This paper describes the practical and theoretical aspects of interfacing a GE Optomation II vision system with an IBM 7565 robotic arm. We discuss our experiences in implementing such an interface, which was part of a project to demonstrate robotic assembly of personal computers. The objective was to interface the two systems such that:

- The vision system can find reference marks on incoming pallets.
- The robot can identify, in its frame of reference, where the pallet reference marks are located.
- The robot can move to specified points in the pallet frame of reference, given the location of the reference marks.

The implementation was carried out in the Industrial and Systems Engineering Department's Computer Integrated Manufacturing Laboratory, at the University of Florida. The equipment in the laboratory includes an Optomation II vision system, donated by the General Electric Company, and an IBM 7565 manufacturing system, donated by the IBM Corporation.

The Hardware Interface

The first step in interfacing two hardware systems is to determine the built-in interface capabilities of each system. The standard version of the GE vision system has two 8 bit parallel I/O ports (8 bits in and 8 bits out) and an RS232C serial port. Additional parallel and serial ports can be obtained as extra cost options.

The IBM Series I that controls the 7565 has four 16 bit parallel input ports, four 16 bit parallel output ports, and an IBM 3780 bisynchronous communications port. The 3780 port is designed to send/receive files from/to disk. AML (A Manufacturing

Language : the robotics language of the 7565) does not offer RS232C communication capabilities.

Given these constraints, we elected to interface the two machines through two 8 bit TTL compatible parallel ports. Because neither machine has extra bits dedicated to hand shaking, the ports were implemented as:

7 input data bits + 1 input control bit
7 output data bits + 1 output control bit

The ports of the two systems were connected directly through a 16 wire cable. To increase reliability in an industrial setting, the use of optical isolators is recommended.

It would have been nice to have a higher level communication interface between the two machines (e.g. MAP). This application is yet another example of the need for interface standards for manufacturing equipment.

Communication Protocol

Once the hardware was interfaced, we had to develop a protocol for communication. The low level 7-bit transfer protocol which evolved is described next.

On the GE vision system, the input control bit is named IBMFLAG and the output control bit is named GEFLAG. On the IBM 7565, the input control bit is named GEFLAG and the output control bit is named IBMFLAG.

The procedure to send 7 bits of data from the vision system to the robot is as follows:

Step	Vision system	Robot
Initial Conditions	GEFLAG=0	IBMFLAG=0
Step 1	Output 7 data bits and set GEFLAG=1	Wait for GEFLAG=1
Step 2	Wait for IBMFLAG=1	Get 7 data bits and set IBMFLAG=1
Step 3	Set GEFLAG=0	Wait for GEFLAG=0
Step 4	Wait for IBMFLAG=0	Set IBMFLAG=0

The procedure to send 7 bits of data from the robot to the vision system is similar:

Step	Vision system	Robot
Initial Conditions	GEFLAG=0	IBMFLAG=0
Step 1	Wait for IBMFLAG=1	Output 7 data bits and set IBMFLAG=1
Step 2	Input 7 data bits and set GEFLAG=1	Wait for GEFLAG=1
Step 3	Wait for IBMFLAG=0	Set IBMFLAG=0
Step 4	Set GEFLAG=0	Wait for GEFLAG=0

This protocol was written in AML on the 7565 and in VPL (GE's Visual Programming Language) on the vision system. Procedures named GET7BITS and PUT7BITS were implemented for both systems. INITCOM routines were also written to initialize the communication ports.

Because both AML and VPL are interpreters, the communication rate between the two systems is slow. In our application, the time required to request and obtain the coordinates of an object is about two seconds. During this period seventeen 7-bit transfers occur and the vision system "takes" and processes the picture.

A constraint of this protocol is that it is not full duplex. Communications cannot proceed in both directions at the same time. The protocol is further limited because the higher level communications modules on each machine must be synchronized with respect to which machine is to send or receive.

Neither of these constraints turns out to be a limiting factor in the higher level information exchange that was developed.

Information Exchange

After low level communication was established between the machines, a protocol had to be developed to let the robot instruct the vision system to find a reference mark and send back its coordinates.

As background information, the GE vision system has a windowing capability. The camera attached to the system has a field of view that can be, for computational purposes, restricted to a user specified window. The system also has the capability to identify a set of objects within a window and report the area and coordinates of each.

We took advantage of these features. The high level protocol that was developed passes a window definition (coordinates of upper left and lower right corners) and a search object size (measured in square pixels) from the robot to the vision system.

The vision system then returns the coordinates of the location of the object within the defined window that has a size closest to the specified size. The protocol is outlined below:

Step 1 Robot sends a 1 to start the point transfer protocol.
Step 2 Robot sends the window upper left X coordinate.
Step 3 Robot sends the window upper left Y coordinate.
Step 4 Robot sends the window lower right X coordinate.
Step 5 Robot sends the window lower right Y coordinate.
Step 6 Robot sends the specified object size.
Step 7 Vision system processes information.
Step 8 Vision system returns size (area) of the selected object.
Step 9 Vision system returns X coordinate of object.
Step 10 Vision system returns Y coordinate of object.

In step 7, the vision system processes the information using standard VPL statements. The process consists of setting a window, setting a threshold, taking a picture, and selecting from the set of objects in the window, the one that most closely matches the specified size.

To implement the above protocol, a method for transferring 14 bit integers was developed. The 14 bits are transferred by calling the appropriate 7-bit routines twice. The number to be sent is broken into a low 7-bit part and a high 7-bit part. After the two transfers the two parts are recombined. For example, in order to send the value 1000, the low part is 1000 AND 127 = 104 and the high part is the integer part of 1000 divided by 128 = 7. To recombine the parts, the high part is multiplied by 128 and the low part is added, e.g. 7 * 128 + 104 = 1000.

Vision System to Robot Transformation Fundamentals

The vision system and the robot have different coordinate systems. The vision coordinate system is based on the field of view of the vision system camera which is mapped to a pixel coordinate system that is 932 by 752 pixels. The location X=466, Y=376 is in the center of the field of view.

The robot's coordinate system is a three dimensional system with a horizontal XY plane and the Z axis pointing up. In our application, the vision system camera is mounted above the robot, looking down. In this configuration, the vision system sees images of the robot's XY plane. For initial discussion purposes, assume that items in the vision system's view are all at the same elevation.

In this configuration, the vision system's X and Y axes are not necessarily aligned with the robot's X and Y axes. There is normally a translational, a rotational, and a scaling difference between the two coordinate systems. Thus, the need exists for a methodology to transform coordinates in the vision XY frame to coordinates in the robot XY frame.

The next section describes the mathematics of performing coordinate transformations. The sections following it discuss the application of the transformation methodology.

Transformation Mathematics

Consider a two dimensional coordinate frame, r, and a second coordinate frame, v, that lies in the same plane as the r frame. We wish to find the XY coordinates of a point, $\mathbf{p}^r$, in the r frame, given the coordinates of the same point, $\mathbf{p}^v$, in the v frame. The equation

$$\mathbf{p}^r = \mathbf{o}^r + \mathbf{T}^r \mathbf{p}^v \tag{1}$$

describes the transformation from the v frame to the r frame. $\mathbf{o}^r$ is a two element offset vector that represents the location of the origin of the v frame as measured in the r frame. This vector provides the translation part of the transformation. $\mathbf{T}^r$ is a 2x2 matrix which performs the rotation and scaling between the two frames.

For the general case (i.e., translation, rotation and scaling), to determine the values of $\mathbf{o}^r$ and $\mathbf{T}^r$ (6 unknowns), we need to know the coordinates of three separate calibration points in both frames. Application of equation 1 to the three points produces six equations in six unknowns. Let **p1**, **p2**, and **p3** denote the three points. To solve for o^r and T^r, we begin by rearranging equation (1).

$$p^r = [o^r \; T^r] \begin{bmatrix} 1 \\ p^v \end{bmatrix} \tag{2}$$

Then for our 3 points, p1, p2, and p3:

$$[p1^r \; p2^r \; p3^r] = [o^r \; T^r] \begin{bmatrix} 1 & 1 & 1 \\ p1^v & p2^v & p3^v \end{bmatrix} \tag{3}$$

Solving for $[o^r \; T^r]$ we get:

$$[o^r \; T^r] = [p1^r \; p2^r \; p3^r] \begin{bmatrix} 1 & 1 & 1 \\ p1^v & p2^v & p3^v \end{bmatrix}^{-1} \tag{4}$$

Applying the Transformation Methodology

The first step in using the results of the previous section was to develop a method of determining both the vision and the robot coordinates of three calibration points. A small white dot (.25" diameter) was afixed to the gripper of the 7565. Then, by moving the gripper (and the dot) to three predetermined locations, and requesting the vision system to locate the dot, the necessary information is obtained.

However, there was one problem with this approach. Because of the physical limitations on the range of motion, the robot cannot move the calibration dot far enough down in the Z direction to be in the same XY plane as the reference marks on the incoming pallets.

Working in Alternative Planes

In order to get around this problem, the mathematical transformation methodology had to be extended so that it could be applicable to planes at elevations different than the feasible elevation range of the calibration dot. This was accomplished by taking a fourth calibration point whose elevation is different from that of the other three points, and then using triangulation to find the transformation that applies to the plane of the pallet mark. Figure 1 depicts the applicable geometry.

Referring to the figure, ZL (Z low) is the level of the pallet reference marks, ZM (Z mid) is the level of the first three calibration points, ZH (Z high) is the level of the fourth calibration point, and ZC (Z camera) is the elevation of the focal point of the camera.

The objective is to find the transformation for points in the ZL plane. Below is an outline of the procedure we use:

I. Find the vision transformation for the ZM plane.
   A. Move the robot calibration dot to the first calibration point at the ZM level, with coordinates XR1 and YR1.
   B. Have the vision system report the vision coordinates of the point, XV1 and YV1.
   C. Move the robot calibration dot to the second calibration point at the ZM level, with coordinates XR2 and YR2.
   D. Have the vision system report the vision coordinates of the point, XV2 and YV2.
   E. Move the robot calibration dot to the third calibration point at the ZM level, with coordinates XR3 and YR3.
   F. Have the vision system report the vision coordinates of the point, XV3 and YV3.

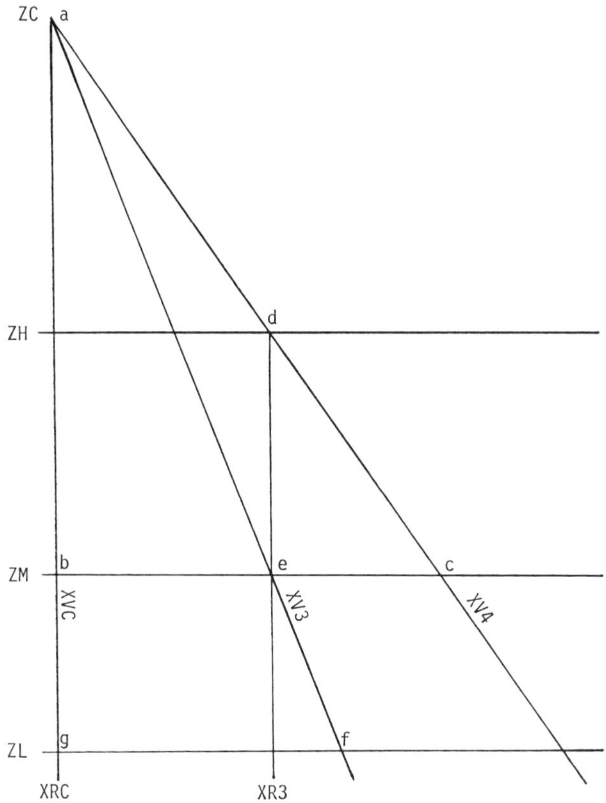

Figure 1: Triangulation Geometry

G. Use the values obtained with equation (4) to find the transformation at level ZM.

II. Use triangulation to compute the vision coordinates of the center point (that unique point in the field of view at which a vertical movement of a calibration dot does not change the dot's coordinates in the vision frame). The center point will be close to the optical axis of the camera.

A. Move the robot calibration dot vertically from the latest calibration point to the ZH level. Do not change the robot X and Y coordinates.

B. Have the vision system report the vision coordinates of the point, XV4 and YV4.

C. Note the similar triangles, abc and dec, in Figure 1. Using those triangles, solve for XVC and YVC.

$$XVC=XV4 - \frac{(XV4-XV3)(ZC-ZM)}{ZH-ZM} \tag{5}$$

$$YVC=YV4 - \frac{(YV4-YV3)(ZC-ZM)}{ZH-ZM} \tag{6}$$

III. Find the robot XY coordinates, XRC and YRC, of the center point. This can be done by applying the transformation of step I.

IV. Find the robot XY coordinates, in the ZL plane, of the three vision points used in calculating the ZM vision transformation. Use the similar triangles of Figure 1, abe and agf.

V. Find the vision transformation for the ZL plane, given the vision and robot coordinates, at the ZL level, of three separate points.

It is possible to apply this technique to several different planes of reference. This would be appropriate if vision reference marks were expected at different, but known, elevations.

The process of determining the vision transformation needs to be done only once, presuming that the camera and the robot base/frame of reference do not change. However, due to vibration or other industrial environmental factors, calibrations usually do not remain valid over extended periods of time. A prudent practice would be to make daily calibrations.

Using the Transformation

The purpose of a vision transformation is to allow the mapping of vision points into corresponding robot coordinates. This is done by applying equation (1).

In our application, the pallets move into their position in the robot's envelope within a certain tolerance. By using the vision system and the transformation, the

XY coordinates of the pallet reference marks can be determined, thus providing information on the exact location of the pallet.

We use two reference marks per pallet. One represents the origin of the pallet frame of reference, and the other is a point on the positive Y axis of the pallet frame of reference. Once the robot coordinates of the two reference points are determined, we apply a transformation between the robot frame and the pallet frame of reference. See [1] for more information on this transformation technique.

Conclusion

Our work was carried out with a specific vision system and robot. It is obviously difficult to give general rules for applying vision systems in a manufacturing setting. We do hope, however, that others who are faced with similar applications may benefit from what we have presented.

The vision system interface work, as part of a project to demonstrate robotic assembly of personal computers, was supported under a grant from the IBM Corporation. The GE Optomation II vision system was a gift from the General Electric Corporation. This support is gratefully acknowledged.

References

1. Kisko, T. M. and Muth, E. J., "Redesign of a Personal Computer for Automated Assembly," Proceedings of the IIE International Electronics Assembly Conference, Boston, Mass., September, 1986.
2. Kisko, T. M. and Muth, E. J., "Multiple-Stage Assembly of Personal Computers in Robotic Workcells with Vision Support," Proceedings of the IIIE International Conference on Robotics and Automation, San Francisco, CA, April, 1986.
3. Kisko, T. M., "General Purpose Coordinate Transformations in AML", Proceedings, CAD/CAM, Robotics and Automation International Conference, Tucson, 1985
4. Paul, R. P., Robot Manipulators: Mathematics. Programming, and Control, The MIT Press, Cambridge, Massachusetts, 1982.
5. "Optomation II VPL Language Reference Manual", General Electric Company, Syracuse, New York, 1981
6. "IBM 7565 Manufacturing System : A Manufacturing Language Concepts and User's Guide",IBM Corporation, Boca Raton, Florida, 1983
7. "IBM 7565 Manufacturing System : A Manufacturing Language Reference",IBM Corporation, Boca Raton, Florida, 1983

The Effect of Mechanical Vibrations on the Robotic Assembly Process

Kalyan Ghosh and Claude Lincourt

Université du Québec à Trois-Rivières, Trois-Rivières, PQ, Canada

Abstract

Insertion is the most common task during assembly in mechanical and electronics industries. Insertion with very small clearance is easier to accomplish in a reliable manner if remote center compliance devices are used. Incidence of jams can be further reduced by lowering the coefficient of friction at the contact points. This is possible by producing mechanical vibrations in the longitudinal direction, so that a lower kinetic friction coefficient comes into play. On the other hand, vibrations in the transverse direction create some relative movement between the peg and the hole, so that insertion can be achieved even if the initial placement is not very accurate.

INTRODUCTION

Assembly is one of the major cost items in the manufacture of finished goods. The assembly costs often account for more than 50% of the total manufacturing costs [1] and this figure is likely to increase unless appropriate types of automation could be used. The development of the assembly line by Henry Ford in the early part of this century helped to increase the productivity of assembly workers quite considerably by eliminating the inefficiencies and by proper planning of the work.

However, in the conventional assembly line, the individual assembly operations are usually done manually and hence the process becomes expensive as the labor cost increases. Since some years back, attempts have been made to introduce hard automation equipment to accomplish some assembly tasks. Manual workers were replaced by automatic workheads for some of the simpler tasks, but the operators have been retained for more complex tasks (requiring special skills and judgement) which would be uneconomical to mechanize.

These automated assembly machines have become quite popular for mass production. But a major problem with these systems is that they are dedicated machines and they lack the flexibility to be used economically for the manufacture of products in small lots. Robots provide the opportunity to introduce automation in assembly systems while assuring flexibility. Some pioneering work on APAS (Adaptable Programmable Assembly System) was done by the Westinghouse Electric Corporation in the late 1970s [2] and the feasibility of using automated programmable assembly system for batch assembly was examined. They investigated the assembly of fractional horsepower motors by an integrated system using robots, vision and tactile sensors, multipurpose end effectors, and programmable part presenters and fixtures. The whole system operated under a hierarchical control system using a supervisory minicomputer and local microcomputers. This project demonstrated that the concept was feasible, leading to a conclusion that the future automated systems will be a combination of hard automation and robots, and that inspection will probably have to be done concurrently with assembly operations to make the concept economically feasible.

TYPES OF TASKS CARRIED OUT DURING ASSEMBLY

It is important to examine the individual tasks that are carried out during the assembly of a product. Kondoleon [3] studied the assembly of a number of commonly-used products such as a refrigerator compressor, a toaster oven and a bicycle brake. Figure 1 [3] shows the typical tasks executed during the assembly of these products and Figure 2 [3] shows the direction of attachment of the parts, as well as the frequency of each type of assembly tasks. It is seen that, in the assembly of these mechanical devices, insertion-type tasks are by far the most common followed by assembly using screws. Similar conclusions were reached in examining the assembly of automotive engines [4], though use of press-fits are more frequent in that case. Electronics industries use printed circuit boards (PCB) and many of the components are installed by insertion of leads into holes in the PCB. The size of this industry is quite large; for example, the total value of the PCB assembly activities in the United States in 1983 was estimated to be $7 billion [5].

The above discussion indicates that insertion of the peg-in-the-hole

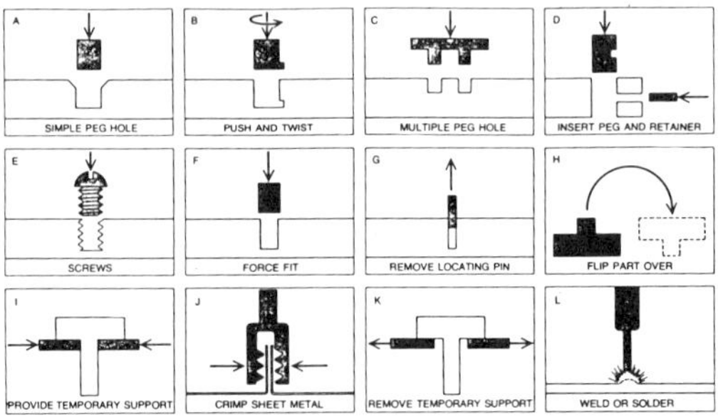

Figure 1. Typical Manufacturing tasks executed during the assembly of some commonly-used products

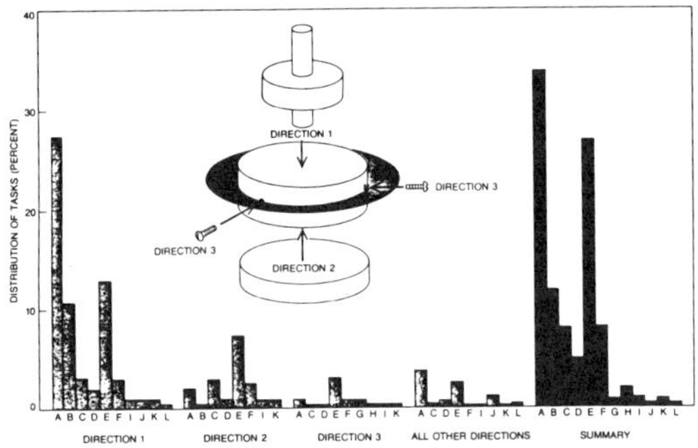

Figure 2. Direction of attachment of the parts, as well as the frequency of each type of assembly task

type is the most common assembly task in discrete product manufacturing and successful automation of these tasks is of considerable interest for the industry. We will examine in detail this type of assembly tasks in the following paragraphs.

ASSEMBLY CONDITIONS DURING INSERTION

The assembly of components by insertion is strictly a positioning problem. Perfect knowledge of the parts being assembled and a perfect positioning device would make the task of assembly a trivial matter. Unfortunately, industrially manufactured parts are subject to dimensional variations. In addition, the realization of a near-perfect positioning device conflicts with cost and flexibility considerations required for a general purpose assembly system.

Use of chamfer at the entry point of at least one of the components to be assembled contributes significantly to the success of the assembly process. Figure 3 [6] illustrates the four stages of the basic peg-in-the-hole assembly process. In this figure, (a) illustrates the two components approaching with some lateral and angular misalignment, (b) shows the action of the chamfer in correcting the lateral misalignment, (c) illustrates the 1-point contact, and (d) the 2-point contact. Figure 4 [6] shows the initial position of parts prior to assembly. In this figure, c = clearance ratio = $\frac{R - r}{R}$

ε_0 = lateral alignment error

and θ_0 = angular alignment error.

Figure 5 [6] is a plot of the angle of inclination of the peg as a function of the insertion depth ℓ and this diagram illustrates very clearly the helpful effect of the chamfer in achieving the insertion. Figure 6 [6] shows the applied forces and reactions in two-point and one-point contact.

Here, f_1 and f_2 = contact forces at the points of contact

μ = the coefficient of friction

F_x = lateral force at the tip of the peg

F_z = force of insertion

and M = moment at the tip of the peg.

Figure 7 [6] shows the condition that the forces and moments have to satisfy for the peg to slide into the hole [7,8]. The peg will enter into the hole without jamming, if the conditions of operation are represented by points situated inside the parallelogram. Jamming will occur for points falling outside this area. The size of the non-jamming region can be expanded by reducing the value of μ, the coefficient of friction. Figure 8 [6] shows a peg supported in a compliant manner with lateral springs and L_g is the distance between the tip of the peg and the center of compliance. The smaller the value of L_g, less is the possibility of jamming with two point contact. The Remote Center Compliance (RCC) device, developed by the Draper Laboratory [9,10], is designed to be fitted on the robot wrist and the gripper

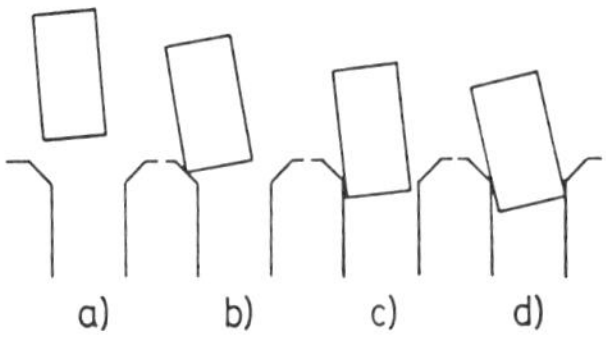

Figure 3. Four stages of the basic peg-in-the-hole assembly process

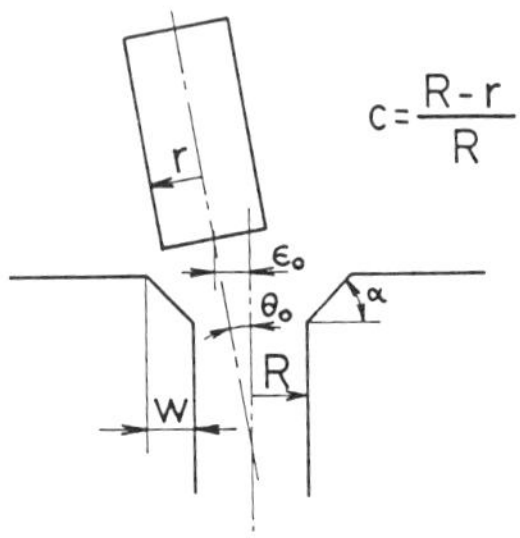

Figure 4. Initial position of the parts prior to assembly

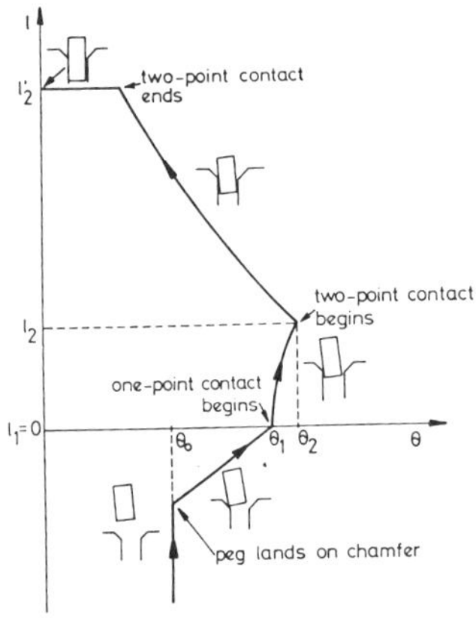

Figure 5. Variations of the angle of inclination of the peg as a function of the insertion depth

$$F_z = \mu(f_1 + f_2)$$
$$F_x = f_2 - f_1$$
$$M = f_1 l - \mu r (f_2 - f_1)$$

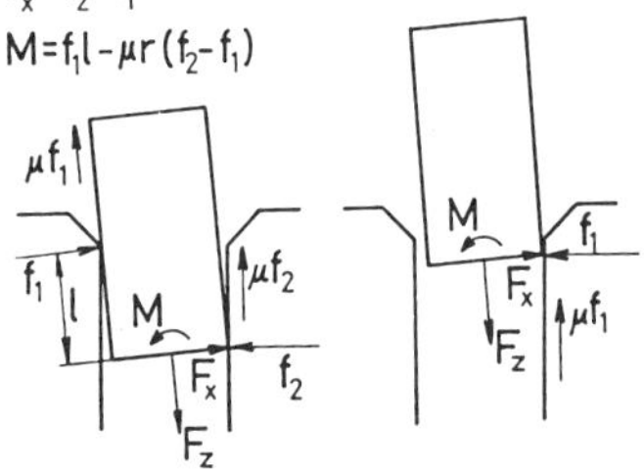

Figure 6. Applied forces and reactions in two-point and one-point contact

(which holds the part) is bolted on to the RCC. Because of its unique mechanical construction, this device provides a controlled amount of float to enable positional alignment during assembly. In particular, the RCC permits a certain amount of rotation of the peg about its center of compliance, which is situated near the tip of the peg. RCC devices are now commercially available from companies such as Barry Wright Corporation (Watertown, Massachusetts) and Lord Corporation (Erie, Pennsylvania).

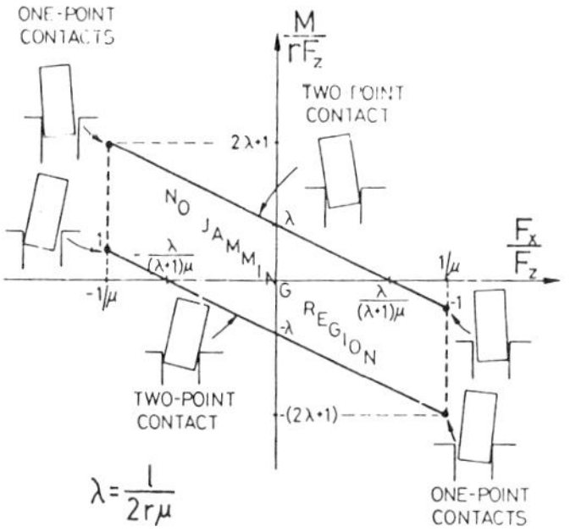

Figure 7. Conditions to be satisfied for successful assembly by insertion

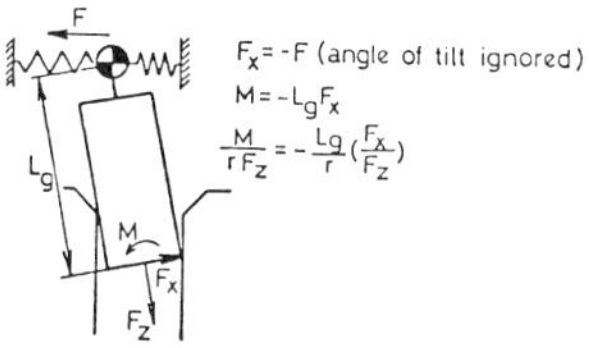

Figure 8. Forces and moments on a peg supported in a compliant manner

Apart form the use of passive means such as the RCC device, a method using active impedance may also be used. Sensors provide readings of forces and moments during the insertion operation, and these values could be used to correct the position of the peg during insertion using a certain algorithm. For the general case, this requires the ability to make small controlled movements in all six degrees of motion. To achieve the required resolution and simultaneously make high-speed corrective motions using active control, a convenient method is to attach to the assembly robot an auxiliary device capable of making small, high-speed, high-resolution motions. The active systems are more expensive and tend to be much slower than the passive systems, and we feel that the passive systems offer greater advantages for general assembly work. However, in using the passive systems, means should be developed to reduce the force necessary for insertion.

IMPORTANCE OF WORKSTATION RELIABILITY IN AUTOMATED ASSEMBLY SYSTEMS

As we have discussed in the preceding paragraphs, it is highly desirable to automate assembly systems in order to reduce the manufacturing cost of a product. However, the operational reliability of the individual workstations is an important consideration in achieving these cost benefits in practice. Jamming of parts during insertion is a common problem in automated assembly equipment and if a person has to be continuously present near the machine to free the jams, no particu-

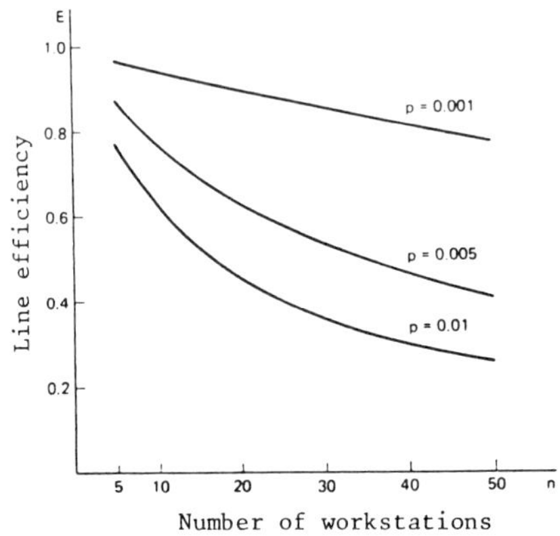

Figure 9. Relationship between line efficiency (E) and number of workstations (n) for various workstation breakdown rates (p)

lar advantage is gained by using the automated equipment. In addition, with the increase of the number of workstations in an automated assembly line, the line efficiency drops very rapidly as the individual workstation breakdown rates increase. From Figure 9 [11], we see that for an automated assembly line (without internal storage buffers) containing 10 workstations, the line efficiency drops from 92% to 62%, if the workstation breakdown rate increases from 0.001 to 0.01. It will be very hard to economically justify an assembly line which will remain inoperative 38% of the time. Ensuring high operational reliability of individual workstations is an absolute necessity to be able to use automated assembly lines in production. Hence it is important that insertion operations be realized without jamming.

USE OF VIBRATIONS TO FACILITATE ASSEMBLY BY INSERTION

In the preceding discussion, it was pointed out that the presence of chamfers at the entry surface and use of compliant component supports are very helpful in achieving assembly by insertion in a reliable manner. As seen in Figure 7, the incidence of jamming (and consequent insertion failure) can be reduced if the coefficient of friction can be decreased.

The phenomenon of friction between solids is usually explained using the adhesion theory and the real area of contact between the two surfaces is determined by the contacting asperities. The components to be assembled to produce a certain product could be of the same material or a different one. The coefficient of friction can be expressed by the following relationship [12,13]:

$$\mu = \frac{s}{p - \frac{2W_{ab}}{r}\cot\theta}$$

where s = bulk shear strength of the softer of the contacting materials

p = penetration hardness

W_{ab} = surface energy of adhesion

r = average junction radius

θ = surface roughness angle

If a and b are the two materials in contact, approximate values of W_{ab} are as follows [13]:

$W_{ab} \approx 2\gamma_a$, if a and b are identical

$\approx \frac{1}{2}(\gamma_a + \gamma_b)$, if a and b are incompatible.

This explains why the coefficient of friction is higher when the two sliding surfaces are of the same material. Also, for very smooth surfaces, the angle θ becomes very small and the friction coefficient increases. On the other hand, it starts to increase for very rough surfaces because of the need to lift one surface over the asperities on the other. Friction is lower in the intermediate range of roughness, from about 10 to about 100 micro-inches RMS. Machined components might have burrs on them and they could cause significant increase of the force of insertion during assembly. As regards the influence on the friction coefficient of relative motion between contacting surfaces, the kinetic friction coefficient decreases with the increase of the sliding velocity as indicated in Figure 10 [13]. For sliding velocities v higher than about 0.01 cm/sec, the kinetic friction coefficient μ_k is given by [13]:

$$\mu_k = Cv^{-1/10}$$

where C is a constant.

We have reviewed some important aspects of the theory of friction of solids in the previous paragraph. The materials used for constructing

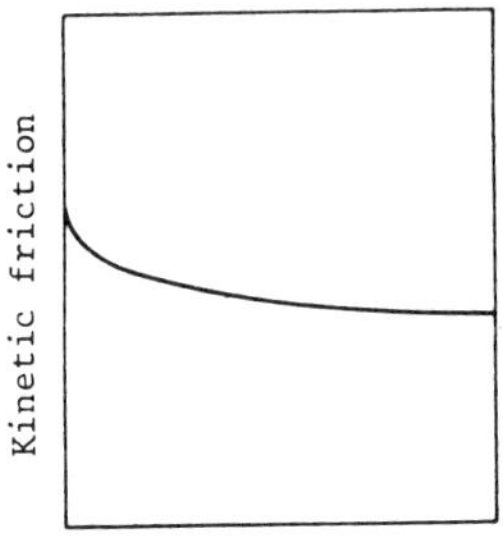

Figure 10. Relationship between the kinetic friction coefficient and the sliding velocity

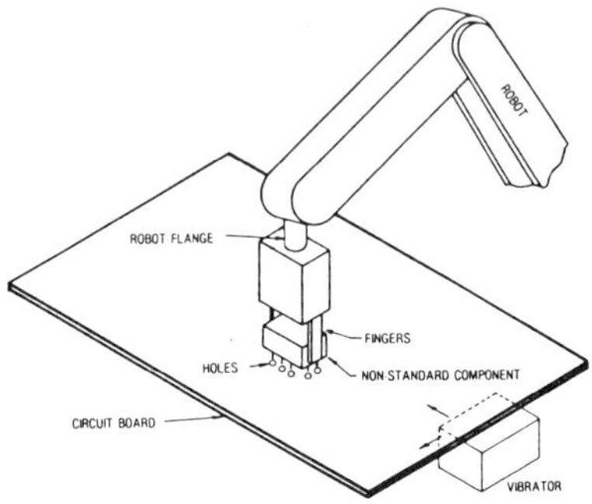

Figure 11. Basic hardware for the vibratory insertion Process. The board is vibrated while the robot places the component at the nominal insertion point. The component drops into the board when holes and pins align.

the two components to be assembled by insertion are specified in the design stage and no changes can be made during manufacturing. However, the friction coefficient can be reduced during assembly operation by setting one of the components in vibration. By choosing the characteristics of the vibratory system properly, sufficiently high values of the RMS velocities can be obtained to be able to significantly reduce the friction coefficient. This will result in reduced insertion force and more importantly, reduced probability of jamming during insertion. This idea has been used by the Canadian Atomic Energy Commission for inserting nuclear fuel pellets in containing tubes. The tube is subjected to mechanical vibrations at its natural frequency and in that condition, the fuel pellet easily enters into the tube, though the clearances are very small.

In another application [14], vibrations were used to improve the lead insertion reliability during the assembly of nonstandard electronic components on Printed Circuit Boards. There are often problems in inserting rigid-leaded cube components in holes with tight lead/hole clearances because of tolerance stack-up and imprecise positioning by the robot. In the vibratory insertion process, the PCB is vibrated in a tranverse direction as illustrated in Figure 11 [14], so as to create a relative motion between the part and the board. The robot picks up a component from the feeder and brings it to the nominal insertion location on the board (which is being vibrated). Eventually, the board holes "find" the component leads. When the holes and leads align, the component drops into the board. The use of vibrations greatly facilitated the insertion task and insertion reliabilities in excess of 99.8% were realized under normal operating conditions. From the time the component first contacts the board surface until the component inserts, the elapsed time was typically 0.25 seconds.

CONCLUDING REMARKS

The use of vibrations have a beneficial effect in realizing the assembly by insertion. Vibrations can be applied in a longitudinal or a transverse direction relative to the axis of insertion. Experimental work that we have conducted indicate that the coefficient of friction can be substantially reduced by imposing vibrations in the longitudinal direction. This will significantly reduce the incidence of jamming in industrial situations, where there are some dimensional inaccuracies and where the alignment is imperfect. Vibrations in the transverse direction will create a relative motion between the peg and the hole. This will allow alignment to take place and insertion will be achieved in a reliable manner. Compared to the active systems, other advantages of such passive systems using vibrations are simplicity, lower cost and greater speed of operation.

BIBLIOGRAPHY

1. Boothroyd, G. et al, "Automatic Assembly", Marcel Dekker, Inc., New York, N.Y., 1982, 378 pages.

2. Stauffer, R.N., "Westinghouse Advances the Art of Assembly", Robotics Today, February 1983, pp. 33-36.

3. Nevins, J.L. and D.E. Whitney, "Computer-controlled Assembly", Scientific American, February 1978, pp. 62-74.

4. Ghosh, K., A. Caracciolo and W. Cwycyshyn, "Robotic Applications in the Assembly of Automotive Engines", ROBOTS 11 Conference Proceedings, Robotics International of SME, Dearborn, Michigan, 1987, pp. 7-1 to 7-13.

5. Ghosh, K. and M. Bonenfant, "Robotic Assembly of Printed Circuit Boards", 4th Canadian CAD/CAM & Robotics Conference Proceedings, Canadian Institute of Metalworking, Ancaster, Ontario, 1985, pp. 7.1 to 7.6.

6. Brezina, M. and M. Khan, "Robotic Assembly of Compliantly Supported Rigid Parts", 4th Canadian CAD/CAM & Robotics Conference Proceedings, Canadian Institute of Metalworking, Ancaster, Ontario, 1985, pp. 7.7 to 7.14.

7. Simunovic, S.N., "Parts Mating Theory for Robot Assembly", SME Technical Paper MS79-259, Society of Manufacturing Engineers, Dearborn, Michigan, 1979, 11 pages.

8. Whitney, D.E., "Quasi-Static Assembly of Compliantly Supported Rigid Parts", Journal of Dynamic Systems, Measurement, and Control, Transactions of the ASME, Vol. 104, March 1982, pp. 65-77.

9. Watson, P.C., "Remote Center Compliance System", U.S. Patent No. 4 098 001, July 4, 1978.

10. Lane, J.D., "Applying the Remote Center Compliance Device in Assembly", Robotics Today, Spring 1980, pp. 38-42.

11. Groover, M.P., "Automation, Production Systems, and Computer-Aided Manufacturing", Prentice-Hall, Englewood Cliffs, N.J., 1980, 601 pages.

12. Ghosh, K., "Design and Development of a High-speed Strand Annealing Plant for Bright Processing of Austenitic Stainless Steel Wire", Doctor of Engineering Thesis, Dartmouth College, Hanover, N.H., 1970, 384 pages.

13. Rabinowicz,E., "Friction and Wear of Materials", John Wiley, New York, N.Y., 1965, 244 pages.

14. Hoffman, B.D. et al, "Vibratory Insertion Process: A New Approach to Non-Standard Component Insertion", ROBOTS 8 Conference Proceedings, Robotics International of SME, Dearborn, Michigan, 1984, pp. 8-1 to 8-10.

Integration of Vision, Transportation and Assembly Tasks in a Flexible Assembly Cell Using Robot Cooperation

M. Staroswiecki, M. Bayart, and D. Jolly

Centre d'automatique. Bât P2, L.A. C.N.R.S. n° 370, Université de Sciences et Techniques de Lille Flandres Artois, 59655 Villeneuve D'Ascq Cedex–France

A description of real time management system for a two robots assembly cell is presented. The architecture is hierarchical and distributed . We first present the supervisor system which determines tasks for the individual robots in order to realize a given assembly. The supervisor makes use of information being transmitted from a lower level by a vision system, a conveyor belt controller and the controllers of the robots. We then define the communication and synchronization between the different levels.

INTRODUCTION

A flexible assembly cell in which several robots with multiple abilities cooperate has been developped at Lille University. Parts to assemble arrive randomly on a conveyor belt and are recognized by a vision system. The aim of the control is to ensure cooperation between robots in order to minimize one or several objectives.

Such an application requires a hierarchical control to supervise the whole cell (3). The vision system, the conveyor belt and the robots being governed by local controllers. A communication system is necessary between those controllers and the central supervisor.

1. HARDWARE DESCRIPTION

The experimental flexible assembly cell contains the following elements (fig 1) :

- Two cooperating robots (SCEMI 4CO6)
- A vision system controling up to 8 cameras (VISIOMAT)
- A conveyor belt

Each of these systems is controled by it own local computer, the overall coordination is devoted to a main computer (HP 1000) which is on the other hand linked to the workshop control computer.

The main computer acts as the supervisor of the cell and also computes the trajectories of the robots. A serial link provides him with the vision informations and with the position informations given by the conveyor belt. Two high speed IEEE 488 links allow him to transmit the movement parameters to the axle controls of the robots, which are based on a MOTOROLA 68000 specialized systems.

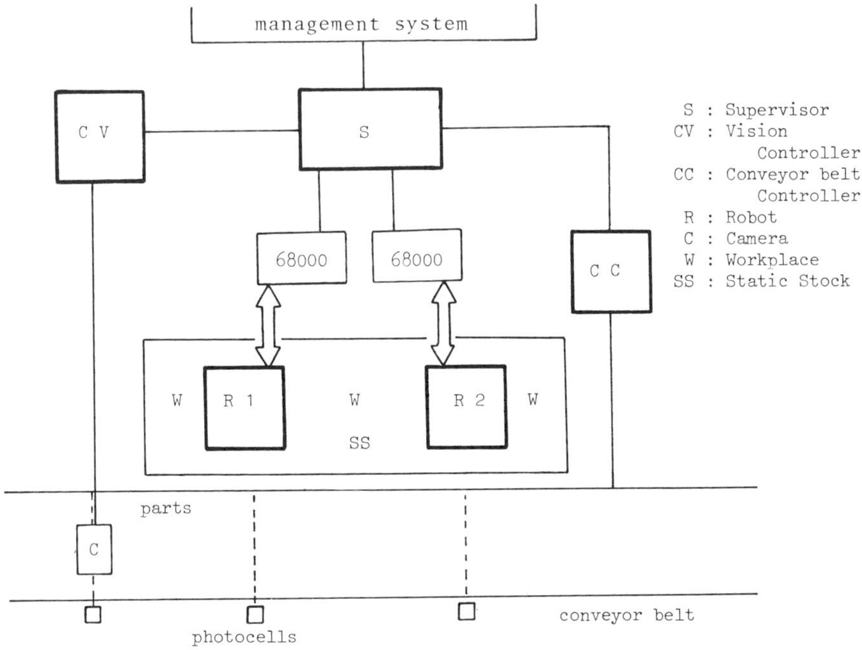

fig. 1

The disposition of the robots is such that it offers three workspaces in the cell, one being common. Their programmation is performed through a high level robotic language, LM,(4),(5) which has been developped at the IMAG in France.

The vision system is composed of a image digitalizer, a contour extraction and is based on INTEL's 8086 processor. It's programmation is performed through a high level language, LV (Vision Language).

The belt conveyor constitutes a dynamic stock containing the parts which enter the cell and are conveyed till the robots.

Such an handware structure allows to produce simultaneously one or several products. The camera which is placed at the input of the cell allows the identification of the entering parts. The dynamic stock, comprised between the camera and the first robot, constitutes a waiting queue of recognized parts, on which the decision procedures will apply. A static stock is also created, which allows to keep within the cell some parts for a future use.

2. THE SUPERVISION SYSTEM

The supervision system is composed of a decision system, an information system, and a communication system. It is organized in a modular way, the activation of each module being a function of the different states in which the cell can be. The state of the cell and its evolution are

2.1. The information system

This system receives, processes and memorizes the data which are necessary for the control of the assembly cell. It computes the state of the cell and of the environment for the decision system. More precisely its function is to update the states of the robots, of the assembly, of the dynamic and the static stocks.

a - State of the robots

The state of each robot is defined at time t by a variable BUSY (0 or 1) and by a list which gives its abilities.

The updating of this state is done when the robot begins a task (BUSY = 1) as well as when the task is finished (BUSY = 0). The updating of the abilities list is made through the control and diagnosis modules. The abilities list remains unchanged if the task has ended correctly but is reconfigured if the cell enters some degraded working mode.

b - State of the assembly

The assembly of a given product supposes that the cell "knows" the sequence of operations or tasks to perform. This sequence is described by means of a graph which gives the list of the operations to be performed and the synchronizations or ordering relations between them.

The state of the product under assembly is given by the concatenation of the states of the different tasks which constitute its description.

The state of a given task is modified by the following events :

- arrival of a new part in the cell
- correct execution of a task by one or both robots.

c - State of the environment

The environment is constituted by both the static and the dynamic stocks. The static stock is common to both robots and both stocks are represented by the list of the parts they contain, each part being described by an identificator and by some position and orientation parameters. These parameters are elaborated by the vision system.

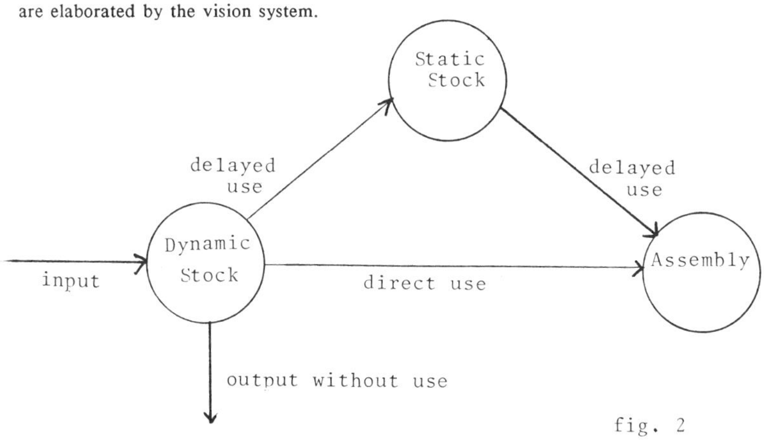

fig. 2

The state of the stocks is modified at each entering of part in the cell and also as a consequence of the use of a part already existing in the cell. Different uses can be defined, following the general scheme (fig 2).

2.2 THE DECISION SYSTEM

a - The task allocation

The information system gives as an output, the list of the excutable tasks of the assembly. The decision system will have to choose, among this list a couple of tasks to be allocated to both robots as soon as they become free. The choice will take into account some optimality criteria in order to minimize the idle time of the robots and by the way to maximize the production of the cell. Two dynamical allocation algorithms have been developped.

The first one tries to optimize the flow of operations through the graph describing the assembly. It is based upon a multicriteria optimization approach, and associates to each candidate task its execution time, the number of its successors, the fraction of them whose resources are present, etc ...(1),(6).

The second one is based on a closed loop waiting queue model, in which the optimization algorithm tends to maximize the length of the queue, i.e. the quantity of work waiting to be processed in the cell (7).

The allocation module is activated at each modification of the waiting queue, i.e. when a part enters the cell, when a robot ends correctly a task.

b - The diagnosis

At the end of each task execution the diagnosis module decides whether the task has ended correctly or not. This decision is based upon the informations provided by an execution spy, or control module. Following the output of this module, the cell goes on working or calls a human operator for help.

2.3. The communication system

The communication system allows the flexibility of the cell by means of the link with the workshop control computer. The latter computes the overall production plan of the workhop and sends to each of the cells the informations which allow the configuration of the supervisor according to the products to assemble : sequence of the operations, parameters of the low level tasks, identification of the resources associated to each task, number of products to assemble, etc ...

On the other hand, the communication with the human operator is etablished on request of the diagnosis module.

The result can then be a re-initialization of the cell (in the case of a fatal defect) or a continuing of the work in a degraded mode which is decided by a humam operator. The list of the abilities of the robots is then updated.

3. FUNCTIONAL ARCHITECTURE

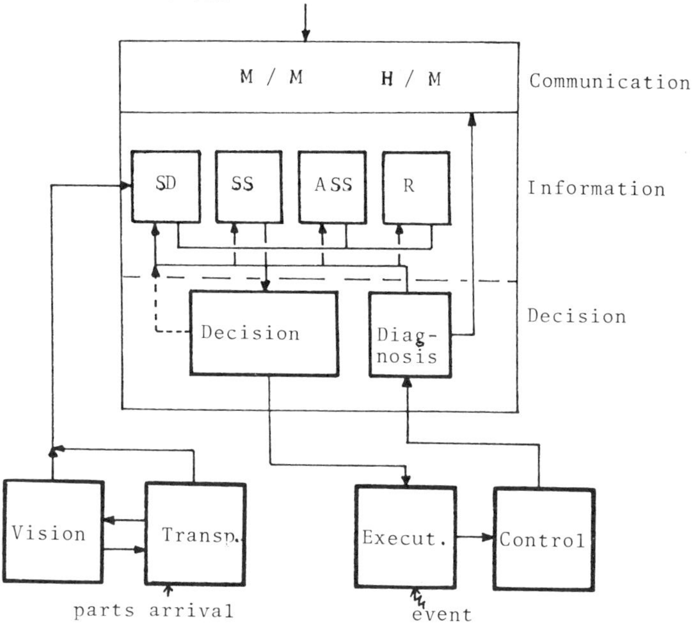

fig. 3

The fig. 3 defines the different functions which are performed in the cell, as well as their interconnexions, and the computer levels at which they are processed.

Let us give some details concerning these functions.

3.1. PARTS IDENTIFICATION

This function takes into account the arrival of parts within the cell and gives to the function UPDATE DYNAMIC STOCK the following informations : (reference of the part, initial position, orientation). This function is activated by a STOP of the transportation function caused by the detection of a part entering the cell. The event END OF IDENTIFICATION re-enables the transportation function and at the same time activates the function TRANSMISSION TOWARDS INFORMATION SYSTEM.

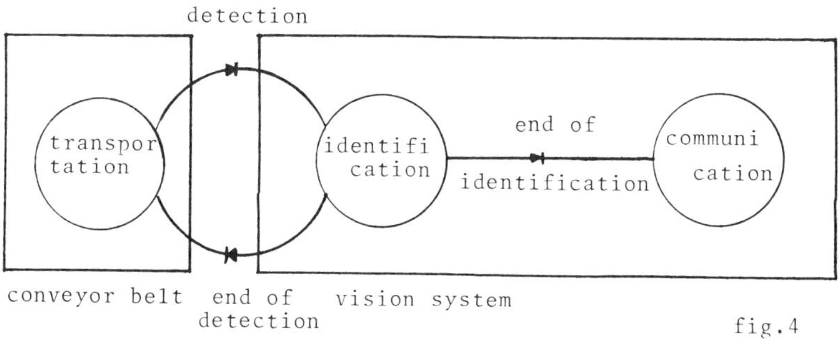

fig.4

The fig. 4 shows the decomposition of the identification function and its relation with the transportation.

3.2. TRANSPORTATION FUNCTION

Its aim is of course to move the parts from the input of the cell towards the robots but it has also to present to the information system the values of the actual position of the parts in the Dynamic Stock.

The events governing the MOVE state of conveyor belt are :

- the detection of an entering part at the input of the cell. This event activates the identification function.

- a part of the Dynamic Stock is in front of one of the robots and the OUTPUT WITHOUT USE DECISION has not been taken.

3.3. EXECUTION FUNCTION

The tasks which are allocated to the robots by the decision module are choosen by the allocation algorithm among the set of the executable tasks at the considered time. Such a task may concern one of the robot or both.

In the first case, a task is described as a sequence of elementary operations at the effector level, following the general scheme (fig. 5).

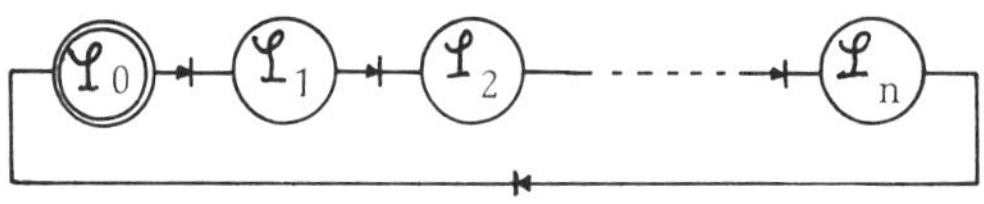

fig. 5

At each of these elementary tasks corresponds a parametrized procedure written in the LM language. The execution function fills up the parameters table, loads into a stack the code to be interpreted and runs the LM interpretor.

When each of the two robots are controled in that mode, the resulting operation is parallel processing without synchronization (this supposes nevertheless some kind of supervision in order to generate collision-free trajectories).

In the case when the realization of a task needs the participation of the two robots, the resulting parallel processing has to be synchronized. Two modes are then possible :

The discrete time synchronization is traduced by the implementation of a finite number of "rendez-vous" in the two sequences of elementary operations which are interpreted in parallel, and needs a supervision module whose aim is to synchronize the interpretation of the two stacks which contain the two LM codes.

The second mode leeds to a continuous time synchronization in which the referencials of the two robots are linked in permanence, from the beginning to the end of the task. The supervision module linked to this mode has to control the two trajectory generators as well as the set of the sensors linked to the two robots.

The fig. 6 shows the different sub-functions which intervene in the execution function.

Whatever the execution mode, the interpretation of the LM code continues as long as the control module (the spy) does not detect any anomaly. The END OF INTERPRETATION event activates then the diagnosis function. Following to the case, this latter enables the continuation of the assembly (by the confirmation of the updating of the state (cf 3.4.) or calls the human operator for an outer decision.

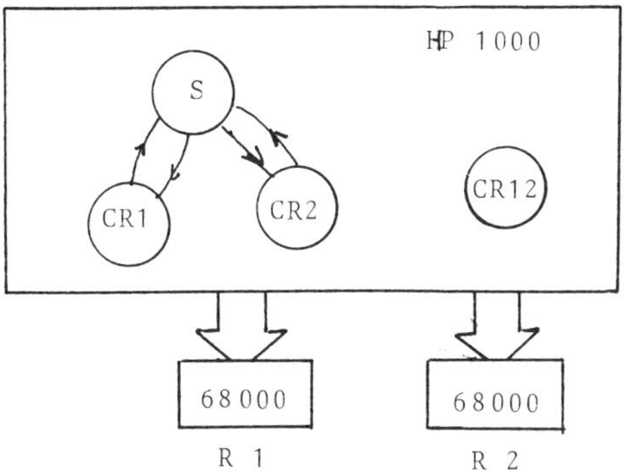

fig.6

S : Synchronisation

3.4. UPDATING OF THE STATES

This function provides the decision system with the information which are needed in order to optimize the tasks allocation. Taking into account the computation time needed for this allocation, the decision is made provisionally, using the estimated state of the cell as it will be when the decision will become applicable. This future state is estimated on the basis of the previous decisions (arrow--→ in fig 3). At the end of the execution of each allocated task, the estimated state is confirmed or not by the diagnosis function. In the case of confirmation, the following processes take place :

- the state of the cell is updated according to the previously estimated one.
- the provisionally allocated task is effectively transmitted for execution.
- the new future state is estimated in order to compute the next task is allocate.

4. CONCLUSION

The structure of an assembly cell in which several robots cooperate has been presented. The real time management system which controls the cell is organized in a modular way. The upper level consists of the supervisor system which includes the information, the decision and the communication sub-systems. The lower level comprises the controllers for the vision system, the conveyor and the robots. Presently, these modules are under implementation in the experimental cell, developped at Lille University.

References:

(1) Problèmes de décision dans une cellule de production flexible utilisant la coopération entre robots. M. DJEGHABA, Th de Doctorat en Automatique, U.S.T.L. Lille mars 1986

(2) A management system for a flexible assembly cell using robot cooperation. M. DJEGHABA, M. STAROSWIECKI, International Conference on Systems, Mans and Cybernetics. Nov. 1985, Broadway (Tucson) ARIZONA.

(3) Controlling multirobot application fron UNIX. V. DUPOURQUE, O. ISHACIAN, 16 Th International Symposium of Industrial Robots 30 sept- 2 oct. Bruxelles, Belgium.

(4) Réalisation d'un support expérimental de recherche pour le projet robotique PANDORE. Définition et implantation du langage LM. E. MAZER, Th 3ième cycle I.N.P.G. Janvier 1981

(5) Le langage LM. Manuel de référence. J. F. MIRIBEL - E. MAZER Editions Cépadues - 1984

(6) Tasks scheduling by multicriteria optimization in a flexible assembly cell using robot cooperation. M. STAROSWIECKI, M. DJEGHABA, M. BAYART. 15 Th International Symposium of Industrial Robots Sept 1985, TOKYO, JAPAN.

(7) Dynamic scheduling of a queuing system with application to the control of a robotized assembly. M. STAROSWIECKI, Y. YANG, M. BAYART. Seventh International Congress of Cybernetics and Systems, Sept 1987, Londres, England.

An RCS Application Example: Tool Changing on a Horizontal Machining Center

John C. Fiala and Albert J. Wavering

Robot Systems Division, National Bureau of Standards, Gaithersburg, MD, USA

Abstract

A large six-axis manipulator performs material handling tasks for a horizontal machining center in the National Bureau of Standards' Automated Manufacturing Research Facility. The manipulator is controlled by the Real-time Control System (RCS), a hierarchical control system developed by the Robot Systems Division at NBS. This manipulator is used to load and unload parts for the machine and, recently, to change tools on the machine. This paper describes the development of tool changing as a new RCS application, including the creation of new commands to be sent to RCS from a workstation controller and the techniques used to program manipulator movements. The integration of force sensing to monitor tool insertion and removal as well as the addition of a changeable end effector for handling tools are discussed.

I. Introduction

Development of the Horizontal Workstation (HWS) began in 1981 as the first of the Automated Manufacturing Research Facility's machining workstations [1,10]. Today, the HWS is a highly flexible subsystem that has the ability to change the tool set-up of its machining center. The HWS consists of a horizontal machining center (HMC), an industrial robot, programmable fixtures for holding workpieces [11], and a material buffering device (MBD) for interfacing the workstation to the material handling system [9]. All of the equipment is coordinated by the Horizontal Workstation Controller (HWSC) [7,8].

The industrial robot, a five-thousand pound, six-axis, hydraulic manipulator, is controlled by the Real-time Control System (RCS) [15]. A brief overview of RCS is given in Section II. In the HWS, RCS is mainly responsible for effecting the loading of part blanks from the

MBD to the fixtures, the refixturing of partially machined parts, and the unloading of finished parts into the MBD. It has been noted, however, that a limitation with workstations is that the tool magazine of the machining center can hold only enough tools to make a few parts [14]. This greatly reduces the workstation's flexibility. To overcome this limitation, the robot can be used to automatically reconfigure the workstation by changing the tools in the magazine.

This paper describes the development of tool changing as an RCS application. Section III discusses the hardware developed for tool loading and shows how it is easily integrated into the system. The addition of new commands to the controller is discussed in Section IV. Finally, Section V describes the control data generated for the new application.

II. The Real-time Control System

RCS is a Robot Systems Division product of the National Bureau of Standards [2,3]. The system can be described as a collection of control processes organized in a hierarchy. These control processes execute non-sequentially in a distributed computing environment. The processes communicate through fixed-field, common memory buffers.

A control process in RCS is a functional component that communicates with other processes in the system only through explicitly-defined interfaces. The principle control structure of RCS is shown in Figure 1. Task, Subtask, E-move, and Prim are the processes involved in decomposing commands from the workstation into simple actions. The gripper control processes interface the end effectors [6,12] to the controller. These end effectors are connected to and disconnected from the robot by the quick change device [5].

The control processes of RCS are distributed among various processors. Most of the system executes on a set of processors residing on a single bus as shown in Figure 2. However, the vision system and active pedestal are complex enough to require their own hardware systems. RCS sends commands to the pedestal via the factory network [13] and makes sensory processing requests of vision via the high-speed 589 DMA interface. The division of the computational work load between several processors ensures that every process can complete one cycle of execution in the time between updates to the robot controller.

The lowest level of RCS, the Robot Interface process, transmits position updates to the robot controller every 40 ms. The Prim(itive) control process computes the knot points of trajectories and sends them to the Robot Interface. Prim also processes the sensor inputs appropriate to its operation, e.g. joystick inputs. The E(lemental)-

move process coordinates its subprocesses, i.e. Prim, the grippers, and the quick change, and queries the vision system. The commands to E-move from Subtask take the form of generic manipulation tasks such as Move-to-Object, Locate-Object, and Grasp-object. Subtask generates these commands as a decomposition of more general commands from the Task level like Get-Object or Place-object. Subtask is also responsible for coordinating the movements of the active pedestal and the robot. The Task process receives its commands from the workstation controller via the factory network. These commands define the principle operations of the robot in the workstation, e.g. Transfer, Move, Refixture. The commands include parameters such as the type of object to be moved, and the source and destination locations for the object.

This control hierarchy was developed for moving in free space, that is, in regions where there is limited contact between the manipulator and its environment. Thus, almost all commands are decomposed, at the Prim level, into sequences of Goto-Point or Go-Thru-Point commands. This is sufficient for most operations of the robot, but for assembly operations such as tool changing, force sensing is needed.

III. Tool Changing Hardware

The feasibility of changing tools with the robot was addressed by Rippey and Vranish [4]. They designed hardware and tested it with a simple teach/playback controller. The main components developed were a gripper for grasping tools, a tray for transporting and holding tools, and a load/unload position on the horizontal machining center.

The tool gripper, shown in Figure 3, was designed to grasp the standard tool holders in which tools are mounted. The tool gripper fingers locate around the flange of the tool holder. This provides a very stable grasp and leaves a space between the fingers for the tool drum fingers to grip the holder during the exchange of the tool between HMC and the robot. As demonstrated in Figure 4, the tool trays hold the tools upright so that the robot can easily obtain this grip. (For more on the gripper and tool tray designs see [4,5,6].)

The load/unload station of the tool drum can be seen in Figure 5. The large, white plate is a low-friction Teflon* surface that is used to control the position of the tool gripper fingers. The fingers are compliantly attached to the gripper. By pressing the fingers against the strike plate, the robot is steadied and small errors in robot position will not affect the exchange, Figures 6 and 7. The tool pocket has a mechanism for actuating the tool drum fingers, and sensors for detecting the position of its fingers and presence of a tool. In addition, a chamfered tab is used to acquire a notch of a tool holder to secure its orientation.

To load a tool, the robot moves the tool straight into the tool pocket, the chamfered tab acquiring the inboard notch of the tool holder, until the gripper reaches the strike plate. Then the tool drum fingers grip the tool, the robot releases the tool and slides outward from the tool drum along the strike plate. When the robot's fingers are free of the tool flange, the robot moves off the strike plate to a safe position. To unload a tool, this process is reversed.

During testing of this mechanical system, Rippey and Vranish found that some form of sensing was needed to prevent disasters. For example, if the robot tries to insert a tool into a pocket while the tool drum fingers are closed, the robot will drive into the closed fingers and cause extensive damage. A force sensor used to prevent disasters must be capable of surviving substantial loads so that it is not accidentally destroyed by the hydraulic robot. The sensor should be able to detect several directions of force and torque as well. To provide this sensing, a JR^{3*} custom force/moment sensor was installed between the fingers and actuator of the tool gripper, Figure 3. The stainless steel, six-axis device has a 900-pound capacity in the axial direction, the direction in which the robot is most stiff.

Integrating the force sensor into RCS required getting conditioned sensor readings into the Prim control level. To do this, unused quick change signal lines were connected to the force sensor. The corresponding lines across the quick change interface were fed into strain gage conditioners. The conditioned signals were plugged into the A/D port of the Prim/E-move board, where they are read by routines in the Prim process.

Figure 1 shows that each gripper is controlled by a separate control process for that gripper. A new control process for the tool gripper was created and compiled onto the computer board with the other gripper control processes. The modularity of RCS allows this to be done simply, with minimal modification to existing control processes. An interface to the Tool Gripper process was created in E-move. E-move sends commands to the Tool Gripper and receives status back. RCS has provisions for quickly creating these communication buffers and incorporating them into a control process [2]. Gripper control signals were brought into the Tool Gripper process in the same way the force sensor was integrated into Prim.

IV. Control Process Programming

Four new Task-level commands were added to RCS to enable the robot to perform tool changing. If it were not desired to monitor forces during tool changing motions, the application could have been easily added using the existing Transfer, Move and Position commands. As noted previously, however, such force monitoring is required to prevent disasters when the robot moves in contact with fixed workstation equipment, and the new commands make this possible. Two of the new Task commands, Insert-tool and Release-tool, provide for loading a tool into the drum. The other two, Acquire-tool and Remove-tool, were created for unloading tools. Four corresponding commands were also added to the Subtask and E-move levels of RCS. Two new Prim level commands, Go-while-not-detected (Go-While) and Calibrate-sensor, provide the desired guarded move capability.

The Task-level Insert-tool command causes the robot to move the tool into the tool transfer location, monitoring the force during the final segment of the insertion. Release-tool directs the robot to open the gripper fingers, perform a force-guarded slide away from the tool along the strike plate, and depart from the strike plate.

Similarly, for tool unloading the command Acquire-tool directs the robot to make a force-guarded move to the strike plate followed by a guarded slide along the plate toward the tool. Acquire-tool also causes the gripper to close after the slide to the tool is complete. Remove-tool results in the withdrawal of the tool by the robot, again with forces being monitored to detect any jamming or improper contact.

The Horizontal Workstation Controller (HWSC) provides coordination between the robot tool changing motions and the opening and closing of the tool drum fingers. It accomplishes this by sequencing the issuance of commands to RCS and the High-Level Machine tool Controller (HLMC). The command and status protocol between the HWSC, RCS and the HLMC provides a natural means to effect the handshaking needed for a successful tool transfer.

Adding a command to a control process essentially means creating an additional branch possibility for its primary decision process. This branch represents an additional capability which executes if the new command is selected. Each command has a secondary decision process to determine what commands will be sent to subordinate processes based on sensory, status, and other conditions. This is how RCS performs task decomposition. Addition of new commands is facilitated by the use of modular, generic structures for decision processing routines. Adding a new command does not affect the execution of existing commands at all.

The E-move level performs most of the decomposition of the tool change commands. E-move breaks the commands down into elemental motion segments such as Delta-moves and Go-while moves. Delta-move and Go-while are Prim-level input commands which specify moves relative to a particular location and sensor-guarded moves, respectively. E-move also sends a Calibrate-sensor command to Prim prior to each guarded move to negate the effects of tool weight and normal contact forces from the force readings. In addition, the E-move level coordinates gripper actions to grasp and release the tool appropriately.

At the Prim level, the Go-while and Calibrate-sensor commands were added to provide the guarded move capability needed for tool changing. Although created for tool changing, they may also be useful for other operations. Go-while allows a sensor to be monitored during a move, and if the sensor exceeds a threshold value then the motion is halted and an error status message is sent. The Prim level responds with a "done" status only if the robot successfully moves to the goal position without generating a detected error. The sensor to be monitored during a Go-while move is specified by name as one of the parameters contained in the input command from E-move. Any sensor integrated into the Prim level may be monitored, and individual force/moment components from the force sensor may be selected by naming them as different "sensors". The other new Prim command, Calibrate-sensor, takes a baseline reading from the force sensor with the robot in position prior to a Go-while and uses this to calculate the threshold values for the guarded move.

The new tool change commands which have been added allow for force monitoring during critical motions. Unlike the free-space commands, however, the Task-level tool changing commands are not generic—they may not be executed with an arbitrary object at an arbitrary location. The commands do, however, represent a first step in the development of such a set of generic commands for assembly-type tasks where contact with fixed objects is anticipated. The next step will be to define appropriate data structures and modify the decision processing to allow the object and location of an Insert or Remove command to be specified by the workstation controller.

V. Data Programming

Whenever the robot is to manipulate a new object, data which describes how to handle the object must be entered into RCS. This section describes the types of data which were entered to program tool changing. RCS provides special data structures and forms for entering the required information. The available data structures and the details of RCS data entry have been described in a previous paper [2] and will not be repeated here.

The data for new RCS applications falls broadly into two categories: 1) locational information and 2) object-related information. The locational information entered for tool changing consists of:

- The position and orientation (pose) of the robot at the tool transfer location
- The robot pose for a location near the tool transfer position, but safely back and away from the machine
- Offsets used for the tool changing motions, which are performed relative to the tool transfer pose
- The locations of the tool holders in the tool trays, which define a one-dimensional array
- Intermediate trajectories to define collision-free paths between the tool trays and the transfer location, including velocity and acceleration parameters for each segment

Creating the object-related data for tool changing required the definition of:

- An object called TOOL
- The approach, grasp, and depart grip sizes for TOOL
- The association of the tool tray array with TOOL
- Position offsets to be used for different grips of TOOL at different locations
- Approach and depart trajectories for different grips of TOOL at different locations, including points moved through and velocity/acceleration parameters

Although the tool change commands described contain a substantial amount of information about how and where to perform the operation, they are still data-independent in that locations, offsets, and the like are referred to symbolically. The control procedures do not contain any numeric values to describe movements, locations, or trajectory parameters. Only one robot pose, the tool transfer position, had to be taught for tool changing. The "safe" location was defined by adding an offset to this pose. The locations of tools at the MBD are defined by the tool tray array, which specifies the positions relative to a taught pose common to all tray configurations. The use of relative moves makes it relatively easy to fine tune the tool change procedure by making small adjustments to the transfer pose.

VI. Conclusion

The RCS-controlled tool changing operation described provides the workstation with the tooling flexibility needed to automate production of a wide variety of parts. Force sensing and guarded-move capabilities were added to ensure the safe performance of this assembly-type task. A new gripper control process was also integrated into the system. The new commands and data which have been added provide easily-modified control of the robot tool-changing motions and a basis for implementing more generic RCS assembly commands.

VII. Acknowledgements

Development of the NBS Real-time Control System is partially supported by funding from the Navy Manufacturing Technology Program.

*Certain commercial equipment is identified in this paper to adequately describe the experimental facility. Such identification does not imply recommendation or endorsement by NBS.

References

[1] Simpson, J. A., Hocken, R. J., Albus, J. S., "The Automated Manufacturing Research Facility of the National Bureau of Standards," Journal of Manufacturing Systems, Vol. 1, No. 1.

[2] Barbera, A. J., et al., "RCS: The NBS Real-time Control System," Robots 8 Conf., Detroit, June, 1984.

[3] Fitzgerald, M. L., et al., "Real-time Control Systems for Robots," 1985 SPI Nat. Plastics Conf., Chicago, June, 1985.

[4] Rippey, W. G., Vranish, J. M., "Robotic Tool Changing in a Horizontal Workstation," Robots 9 Conf., Detroit, June, 1985.

[5] Vranish, J. M., "'Quick Change' System for Robots," Robots 8 Conf., Detroit, June, 1984.

[6] Bunch, R. W., Vranish, J. M., "'Split Rail' Parallel Gripper," Robots 9 Conf., Detroit, June, 1985.

[7] Scott, H., Strouse, K., "Workstation Control in a Computer Integrated Manufacturing System," Autofact 6 Conf., Anaheim, CA., Oct., 1984.

[8] Scott, H., Strouse, K., "Horizontal Workstation Controller Task Programming and Execution," AMRF Internal Report, Sept., 1985.

[9] Fishman, D., Scott, H., Bunch, B., "Integration of Material Buffering Devices in an Automated Factory," 2nd Int. Conf. Robotics and Factories of the Future, San Diego, CA., July, 1987.

[10] Albus, J. S., et al., "A Control System for an Automated Manufacturing Research Facility," Robots 8 Conf., Detroit, June 1984.

[11] Slocum, A. H., Peris, J., Donmez, A., "Development of a Flexible Automated Fixturing System," SME Conference on Flexible Manufacturing Systems, SME Technical Paper #MR 86-128, Feb., 1986.

[12] Myers, D. R., "Sensory-Interactive Control of an Instrumented Gripper," Internal Report, Robot Systems Division, 1984.

[13] Jones, A. T., McLean, C. R., "A Proposed Hierarchical Control Model for Automated Manufacturing Systems," Journal of Manufac-

turing Systems, Vol. 5, No. 1.

[14] Zygmont, J., "Flexible Manufacturing Systems: Curing the Cure-all," High Technology, Oct., 1986.

[15] McCain, H. G., "A Hierarchically Controlled, Sensory Interactive Robot in the Automated Manufacturing Research Facility," 1985 IEEE Int. Conf. on Robotics and Automation, St. Louis, March, 1985.

Fig. 1. Hierarchy of Control Levels.

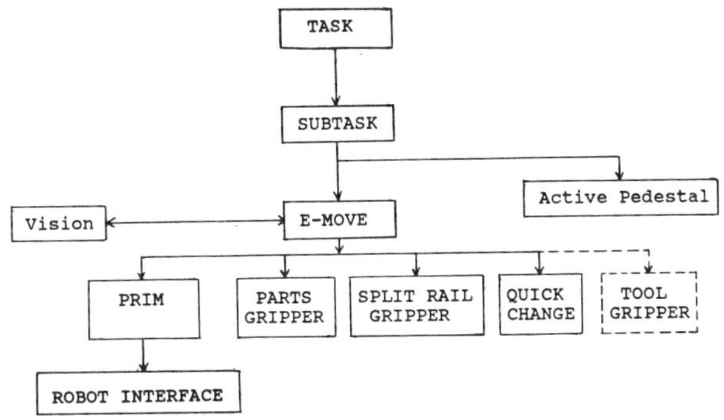

Fig. 2. RCS Hardware Configuration.

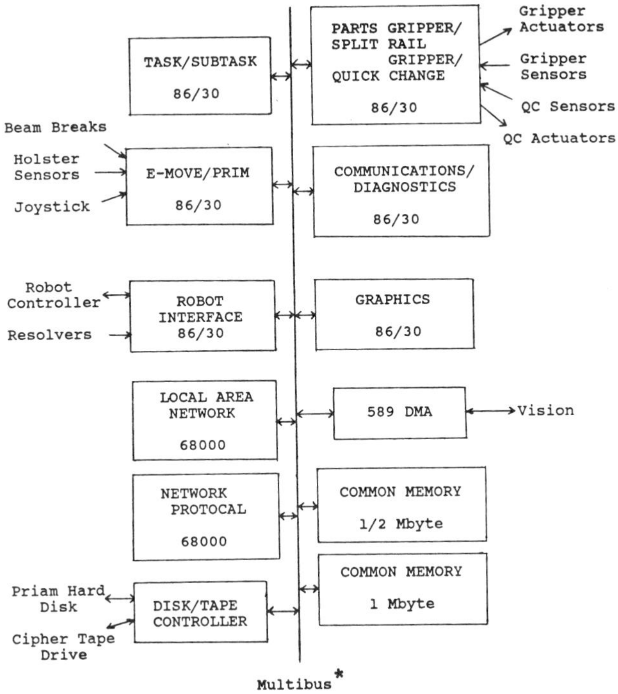

Fig. 3. Tool Gripper.

Fig. 4. Robot Approaching Tool in Tray.

Fig. 5. Tool Drum Load/ Unload Station.

Fig. 6. Robot Gripping Tool in Drum.

Fig. 7. Overall View of Tool Transfer.

The Low Volume Robot Assisted Assembly of Engine Components

R.J. Stamp*, S.G. Andrews*, J.J. Hill*, T.G. Mills†

*Department of Engineering, Bristol Polytechnic, Bristol, UK
†Lister–Peter Limited, Dursley, UK

INTRODUCTION

The established benefits of automated assembly include quality improvements, increased return on capital employed, and the gains resulting from a steady manufacturing rate. Unfortunately, it is generally only those producers of repetitive components in high volumes that can attain these benefits. Many manufacturing organizations, producing fewer components, would also like to achieve these benefits but are unable to justify the capital expenditure considered to be necessary for a fully automated assembly system.

The Engineering Department of Bristol Polytechnic, UK, has recognized this problem area and is researching techniques that will enable a reduction in the volume throughput necessary to justify assembly automation. All current work is based around flexible sensor driven robots, and includes the automation of the fabrication of sheet metal enclosures, techniques for the assembly of printed circuit boards and the assembly of Diesel engines. A prototype system for the low volume assembly of engines is discussed here.

THE LOW VOLUME ASSEMBLY OF ENGINES

These developments have taken place on the behalf of a major UK based manufacturer of Diesel engines, which produces approximately 40,000 cylinders per annum of the type under consideration. In this particular design of engine, the cylinder is produced remote from the crankcase, and each completed engine includes from one to four cylinder assemblies.

It is the intention to develop individual processes for all the aspects of engine assembly, which will finally be integrated into one assembly cell and used in the production environment. The robot used in the cell will be required to perform actual assembly operations, rather than the solely manipulative tasks of transporting components between holding areas and special purpose, hard automation, assembly processes. Each robot will be required to accomplish several different assembly functions with a range of different components, using interchangeable end effectors.

It is envisaged that the cell will be operated with a cycle time of several minutes and that it will mainly be concerned with the assembly of a single engine, rather than as is the case in high volume assembly operations, where several products are in the cell at the same time, each in a different stage of completion. It is proposed that parts will be delivered to the cell in batches and stored in pallets or gravity fed conveyors which hold sufficient capacity to provide for several hours untended operation. The robots will be required to select and inspect components prior to assembling them.

Research into robot assisted assembly techniques is currently proceeding on several main fronts. One of these, the insertion of a piston sub assembly comprising the connecting rod, gudgeon pin and retaining clips together with the piston, into the cylinder barrel, is considered here. Orientation of all components and compression of the piston rings prior to assembly is also required.

PROTOTYPE ASSEMBLY SYSTEM

Cylinder barrel

When used in the manufacturing environment the larger component parts of the Diesel engine, including the cast iron cylinder barrel will be stored in precision made pallets, designed to maintain the positional accuracy of individual components to within 1 mm. The prototype assembly system includes a small storage pallet of components which includes four barrels and four piston sub assemblies (plate 1).

Verification of component presence and precise location in the pallet is performed by a robot mounted tactile probe. The barrel is handled with pneumatic gripper capable of expanding within the cylinder bore to grip the barrel. A static tactile sensor is employed to ascertain the angular orientation of the asymmetric cylinder barrel, prior to the insertion of the barrel into a precision holdingfixture(plate 2).

Piston location

The production version of the assembly system is to include a procedure for the build up of a sub assembly incorporating the piston, complete with five piston rings, the engine connecting rod, the gudgeon pin and two pin retaining clips. Individual components will be supplied by gravity fed conveyors. The sub assembly orientation will be known accurately after build. At the prototype stage of development it is necessary to sense the true orientation of the piston assembly prior to assembly using a robot mounted tactile probe.

An offset combustion chamber in the crown of the aluminium piston, together with the location of the gudgeon pin, dictate the angular orientation of the piston. The tool used to manipulate the cylinder barrel is also used to hold the piston. The piston is secured by a pair of external jaws operating on the outer diameter of the crown. Use of combined tooling which manipulates both the major component parts of the assembly reduces the need for repeated tool changes and saves valuable assembly time for more productive tasks.

Orientation of the piston rings

The piston rings are sufficiently slack in the ring grooves to enable light mechanical forces to push them into the grove and around the piston. In the uncompressed state the gap between the two ends of the ring is 20 mm. It is necessary to locate and then to move the ring relative to the body of the piston so that the gaps on the adjacent rings are 180 degrees apart and are in line with the gudgeon pin.

Several techniques for the orientation of the piston rings have been investigated. A previous design required the robot to rotate the piston while keeping the ring in contact with a tactile sensor. This enabled location of the ring gap. After determination of the gap positions on all five rings, the robot control program optimised the total movement required to complete the orientation, each ring in turn being held stationary while the piston was rotated. The ring orientation system in current use is faster in operation than the above procedure and incorporates a spring loaded plunger which slides over the rotating piston ring and into the ring gap. It prevents rotation of the ring so that orientation may be accomplished. The orientation of each ring in turn takes considerably longer to accomplish than any other task in the assembly operation. The prototype system is being refined so as to check and orientate all five rings simultaneously, using these proven principals.

Piston ring compression

A pneumatically actuated piston ring compressor is used which clamps and compresses the five rings while supporting the piston directly above, and in line with, the cylinder barrel holding fixture. The piston sub assembly is orientated and positioned above the compressor which is then closed temporarily to ensure that the connecting rod will clear the cylinder bore. This is necessary because the connecting rod is free to swing. The piston assembly is lowered so that the connecting rod enters the bore and the compressor is returned to the open state. The piston is then lowered further until the piston rings are within the jaws of the compressor. The compressor is operated a second time to compress the rings. The robot releases the piston, which is held steady by the compression fixture, and positions itself so that the gripper tool may act on the crown of the piston and push the piston assembly vertically down through the compressor and into the cylinder barrel.

The complete assembly is then held by expanding the gripper onto the inside diameter of the barrel and the assembly is transported to the pallet which originally contained the cylinder barrel. The piston insertion sequence is illustrated in plates three, four and five.

REQUIREMENTS OF THE ASSEMBLY ROBOT

The assembly robot

For an industrial robot to be suitable for this assembly task a number of criteria must be satisfied. The robot should be capable of repeatedly positioning components accurately, for example, it is required that the piston must be located relative to the barrel within 0.25 mm. The robot must be also capable of carrying a varied payload up to approximately 15 kg for the gripper and piston/barrel sub-assembly. The robot should be able to produce the required component manipulations such as the orientation of the piston rings. Due to the philosophy behind this type of application, components are not continually brought in and out of the assembly cell with conveying systems. Hence, a large robot working volume is required to accommodate pallets of components in addition to sensing and assembly fixtures. The prototype assembly system features a six axis jointed arm configuration robot supplied by Reflex Automated Systems and Controls Ltd, UK. The robot provides two continuous wrist axes with an internal pneumatic coupling which allow complex manipulations to be made without pneumatic supply line or mechanical fouling. This robot offers an over the shoulder reach mode which provides a full hemispherical working envelope. The robot has been found to have a repeatability better than 0.1 mm and working speeds of up to 300 mm/S using straight line motion.

Sensing strategy

A number of sensing operations are required throughout the assembly task to determine component presence, position and orientation. The sensing strategy implemented in this application is based on the use of a number of simple analogue and digital sensors, which have no in built intelligence. These have been interfaced directly to the robot controller using the available input/output facilities. Various software routines within the robot control program are then used such that a given sensor may be used for several different operations within the assembly task.

This strategy offers the advantages of a simple and flexible hardware sensing scheme which allows for modifications to sensing operations by changes in the control software only. The sensors which have been included in the prototype system to date are:

1) A fixed tactile probe located within the workspace to which components may be presented in order to confirm component presence, establish gripped component orientation and locate component features such as the gudgeon pin hole in the piston.

2) A robot mounted tactile probe which is used to establish the position of pallets and components relative to the robot and locate component features such as the piston combustion chamber.

3) A specially designed microswitch for confirming piston ring presence and locating the gap in the piston rings.

These sensing units currently provides all of the orientation and positional information required for the assembly operations described here. Additional proximity sensors are currently being incorporated into the main gripping unit which will reduce some of the trips made to the fixed probe for simple operations such as confirmation of component presence. An infra-red digital signal transmission system is under development which will allow a number of proximity sensors to be fitted in to the gripper without the problems associated with the fouling of electrical connections.

Programming system requirements

The tactile probes used in this application provide a digital 'ON' signal when they are in contact with a component. Currently available robot control systems provide two possible methods of using this signal for sensing operations such as component location. The first, referred to as polling, involves the robot performing incremental moves and checking the state of the probe signal between each move. As a result of the need for the system to process the move and signal checking language statements between each move, this method tends to be slow and can result in discontinuous robot motion. However, it does allow a part to be located with the accuracy of the distance traveled during the incremental move.

The second method relies on the robot control system providing an interrupt facility. A digital 'ON' signal may be then used to stop robot motion and force a user defined subroutine within the robot control program to be executed. This interrupt facility efffectively allows robot motions of the form 'Move Until Sensor Triggered' to be programmed. A critical factor in the success of this method is the length of the delay between the controller receiving the 'ON' signal and bringing the robot to a halt. When locating a component this delay affects the velocity with which the part may approached and the accuracy with which the component is located.

The tactile probes used here, allow an overtravel of 10 mm. The maximum robot velocity which allows the the robot to come to a halt within this distance is approximately 100 mm/s. Slower approach velocities provide less overtravel. This study has shown that a compromise between speed and accuracy can be achieved when using tactile probes for part location by combining interrupts for rough location with polling for final accurate location. This method is used in the prototype system to locate the center of the piston in approximately 8 seconds with an accuracy of 0.1 mm.

Another programming facility which has been found to support the use of sensors is the ability to modify reference frames, robot positions and tool offsets based on sensory information obtained during execution of the task. Robot positions are stored as coordinate points and orientation angles relative to a set of coordinate axes known as a reference frame. In some robot systems this reference frame is fixed, and robot motions are relative to this fixed point. The robot used in the prototype system offers the facility of relocatable reference frames. This allows a given set of robot point definitions to refer to different positions in the robots workspace by moving the reference frame. In this application errors due to the manual positioning of pallets are compensated for by using the position of the pallet as a reference frame. Component positions on the pallet are predefined relative to the pallet reference frame. After the pallets are manually positioned a software routine is executed which instructs the robot to use the robot mounted tactile probe to locate a number of faces of the pallet. This location data is used by the routine to define a reference frame. Robot motions during subsequent pallet loading and unloading operations are then made relative to this frame allowing the components to be accurately located despite any error in the positioning of the pallet.

Tool offsets are modified during program execution to simplify robot manipulations and to compensate for positional errors in grasping components, for example, after grasping the piston assembly, its center point is found relative to the gripper using the fixed tactile probe. This center point is used to define a tool offset which effectively shifts the tool center point to the center of the piston. Wrist orientations such as those required to rotate the piston in contact with a probe for ring sensing, are then easily made despite the fact that the piston is not held concentric with the wrist.

This application has required an extensive amount of software development in order that the robot may perform the many subtasks within the application. The use of a textual programming language which supports subroutine structures has simplified the development task. A subroutine facility allows the software to be developed and tested in small manageable modules. The use of representative location names such as BARREL_PARK and ABOVE_PISTON_1 has also helped to produce readable and maintainable robot control programs.

CONCLUSION

The prototype system for orientation and insertion of the piston sub assembly into the cylinder barrel is operating successfully, and development is now concentrating on the modifications to the system necessary to render it suitable for industrial use. Research into other aspects of the engine assembly is continuing and the design of a system for the build up of the piston sub assembly is at an advanced stage.

The use of a robot with the appropriate mechanical, control and programming feacilities, to undertake assembly tasks has been shown to be feasible and to result in a reduction in the complexity of individual assembly stations, although with some slight reduction in speed. It is considered however that it is practical to use an industrial robot as the main component part of a flexible manufacturing system for the low volume automatic assembly of Diesel engines.

ACKNOWLEDGEMENTS

The authors wish to thank Lister-Peter Ltd, Dursley, Glos, UK, for their support of this work, and the students and technicians of Bristol Polytechnic Engineering Department who have contributed to the design and construction of this installation.

Plate 1 - The prototype system.

Plate 2 - Sensing component orientation.

Plate - 3

Plate - 4

Plate - 5

Illustration of piston insertion

3 - Location of piston over barrel in locating fixture.

4 - Compression of piston rings.

5 - Pushing of piston into barrel.

A Reconfigurable Multicomputer System for Robotics Applications

Subramaniam Ganesan

Department of Computer Science and Engineering, Oakland University, Rochester, MI 48063, USA

ABSTRACT

Architecture of a reconfigurable cache-based multicomputer system for intelligent machines/robotics applications is discussed. To overcome the memory access latency and conflicts, extensive cacheing of instructions and data is performed in each processor node. This architecture utilizes distributed common memory, multiported RAM, and private cache storage between the local distributed common memory and the local processor. A simple interface circuit which permits multiple microprocessors to share a common memory is described. Theoretical model for task execution time, for the multiprocessor system is presented.

I. Introduction

One approach to supercomputing is through parallelism at the processor level, i.e., connecting several microcomputers and concurrently running programs on all of them. A number of multiprocessing systems with different configurations are in existence [1-5]. Multiprocessor architectures such as Binary N-cube; Hyper-cuber, Array, Tree, Ring, Pyramid, Pipe-line and many others are reported in the literature [6-10]. Reconfiguring the multimicroprocessor interconnection to suit the problem to be solved, is highly desirable. A reconfigurable system is one which can change processor interconnection architecture (say from array to pipe line) to suit the application. The multiprocessor configuration is tailored to the performance and reliability requirement of a particular class of applications. Pipelined computations, pattern matching algorithms, fourier transform computations, matrix vector multiplication and others can be done efficiently with linear array architecture than with tree structure. While, some search algorithms, divide-and-conquer computations and many others are best executed by a binary tree configuration. Thus multicomputer systems that can logically provide array, tree, ring or other configurations, dynamically as required, are of practical importance. Such systems can be used to realize multiple, logically distinct architectures using the same physical interconnection. The advantage of such a design is that the logical configuration can be dynamically changed to best fit the application in hand. Certain fault-tolerant network architectures have been proposed recently in this framework of dynamically reconfigurable architectures [10]. Normally the reconfigurability is achieved by means of interconnection network or switching elements and global controllers [11-13]. This paper describes the architecture of a reconfigurable multimicroprocessor system for intelligent robotics applications.

II. A Reconfigurable Multiprocessor System

A reconfigurable system is one which can change processor interconnection architecture to suit the problem to be solved or to eliminate faulty processor. Normally the reconfigurability is achieved by means of interconnection network or switching elements and global controllers. A block diagram of the general-purpose building block (or node) of the proposed multiprocessor system is given in figure 1. The node architecture is given in figure 2. The reconfiguration of the system to various interconnection architecture is shown in figure 3. Only the links shown need be activated to obtain the different structures (ring, pyramid, etc.). The multiprocessor system consists of a multi I/O ported processor module (P-module) and a multiport memory module (M-module). Multiported distributed common memory with private cache between local bus and distributed common memory is used. The P-ports of the processor are connected to the M-port of its own M-module and the M-ports of other building blocks to provide shared memory access. The M-modules are the distributed common memory. The local distributed common memory is connected to the local Processor P-module through a cache-memory. The private memory and I/O devices dedicated to the processor within the P-module can not be shared. For each processor the distributed common memory is seen as one single linearly addressable structure. The addressing scheme is designed that any common memory cell has the same address for all the microprocessors. For each microprocessor, the lower two block addresses select its private memory and the local distributed common-memory. The memory access control circuit connected to each processor bus checks the memory address and appropriately grants access to its private memory, local distributed common memory or passes the request to the concerned multiported distributed common memory. Distributed common memory reduces the bus contention problem. Mutually exclusive access and priority in getting access to distributed common memory are provided by the arbiter circuit within the multi-ported M-module. The private cache consistency (coherence) problem is taken care by the processor [14].

The use of cache memory increases the processor speed and reduces the frequent contention for the common memory. Cache-multiprocessor systems do not in general, employ the single common cache organization or the cache buffered memory organization because of their drawbacks [14]. So, the organization with processor private caches is selected for the present emplementation. The private cache coherence problem is solved by (i) prohibiting interprocess cooperation, or (ii) Intercache signals conveying the change of cache contents that are used to mark copies of the same cache blocks in other caches as invalid and unusable because of inconsistencies.

Simple hardware circuits for multiport memory arbiter are discussed in [15,16]. Fig. 4 shows the block diagram of a multiported memory. Fig. 5 shows the interface circuit. When a processor requires an access to the multiported memory, it produces request pulse through the I/O port. The request lines enable the three state gates U1-C, and U1-d, whose output now turns low. This high to low transition sets U2-b or U3-a depending on the Q output of U2-a. Its low to high transition sets flip-flop U2-a. The Q outputs of flip-flops U2-b and U3-a in turn determine the state of the two grant lines, G1 and G2. Access to memory is allowed when a processor's grant line is high. A processor requesting access to common memory has to wait till its grant line goes high.

The AND gates U5 implement the serial daisy-chain priority scheme and ensure that processor 1 (local processor) has higher priority than processor 2 and so on. If request signals are sent simultaneously by 2 or more processors, the processor with highest priority gets the grant signal, although the flip-flops U2-a, U3-a,.. are all set. Once a processor gets access to the memory, this interface circuit ensures that the operation is executed without interruption. The processor with access to the memory, clears the semaphore flip-flops U2-a, U2-b, U3-a,.., at the end of its memory transaction.

III. Concurrent Processing Performance Evaluation

Performance depends on hardware, architecture and algorithm [17,18]. Hardware factors are:

1. Memory size.
2. Processor speed.
3. Communication speed.
4. Packet size.
5. Interrupts.

$$S = \text{Speed up} = \frac{\text{time on 1 node}}{\text{time on N nodes}}$$

$$\text{Efficiency} = E = \frac{S}{N} \quad \text{where } N = \text{number of nodes.}$$

E=1 means that using N processors in parallel increases computing speed by a factor of N. Speed up is limited by

1. The idle time due to imperfect load balancing.
2. Waiting time due to communication latencies in channels and message forwarding.
3. Processor time for processing and forwarding messages.

Let f = Overhead measure due to the above 3 factors then $S = \frac{N}{1+f}$

IV . Model For The Task Execution Time

Throughput of distributed signal processors is a direct function of the tasks cycle time. A simple model for the total task execution time is presented here. We formulate the problem with the following representation. The decision variable X_{ik} is definedas

X_{ik} = 1 if task i is assigned to Kth processor, P_k .
= 0 otherwise.

Also we define

t_{ik} = Execution time of task i on P_k
t_{cik} = Communication time for i_{th} task between P_k and common memory.
w_i = waiting time, if any, to get ith task allocated
t_{Hik} = Handshake sequence time for ith task for P_k

We like to minimize the total task's cycle time. This is equivalent to minimizing the maximum execution time of the microprocessors (P_1 to P_k).
This problem can be written as

$$T = \text{Min of } T_{max}$$

$$T_{max} = \max_K \sum_i \left\{X.t_{ik} + X_{ik}\, t_{cik} + X_{ik}\, t_{Hik} + X_{iK}W_i\right\} \qquad (1)$$

where $\sum_K X_{ik} = 1$, i.e., all tasks must be assigned once, and

If the execution time and communication time of a task i, is directly proportional to the number of elements in the task (Y_i), then we can substitute the following in (1)

$$t_{ik} = Y_i\, t_k \text{ and } t_{cik} = Y_i\, t_{ck}$$

where t_k is the execution time for a unit element in P_k, t_{ck} is the communication time of a unit element in the ith task.

V . Conclusion

In this paper, system architecture of a reconfigurable cache-based multiprocessor system with distributed common memory is discussed. A simple hardware circuit for multiport memory interface and a model for task execution time have also been presented. This system is suitable for intelligent robotics applications and many others.

Reference

[1] J. Tuazon, J. Peterson, M. Pniel and D. Liberman, "Caltech/JPL Mark II Hyper Cube Concurrent Processor," Proceeding of International Conference on Parallel Processing, Aug. 1985, pp 666-673.

[2] K. E. Batcher, "STARAN Parallel Processor System Hardware," Proc. AFIPS-NCC, Vol. 43, pp 405-410.

[3] S. H. Fuller and S. Harbison, "The C.mmp Multiprocessor," Technical Report, Carnegie Mellon University, 1978.

[4] F. Ozguner and M. L. Kao, "A reconfigurable multiprocessor architecture for reliable control of robotic systems," IEEE Int. Conf. Robotics and Automation, 1985, pp 802-806.

[5] S. H. Liebowitz and J. H. Carson, "Multiple Processor System For Real Time Applications," Prentice-Hall, Inc., 1985.

[6] J. C. Peterson, J. O. Tuazon, D. Lieberman and M. Pniel, "The Mark III Hypercube-ensemble Concurrent Computer," Proc. International Conference on Parallel Processing, August 1985, pp 71-73.

[7] C. Seitz, "The Cosmic Cube," Communications of ACM, January 1985, pp 22-33.

[8] S. B. Wu and M. T. Liu, "A Cluster Structure As An Interconnection Network For Large Multimicrocomputer Systems," IEEE Trans. Computer, Vol. C-30, No. 4, April 1981, pp 254-264.

[9] Creve Maples, "Pyramids, Crossbars and Thousands of Processors, "Proc. Int. Conf. Parallel Processing, August 1985, pp 681-688.

[10] D. K. Pradhan, "Fault-tolerant Computering," Prentice Hall, 1986.

[11] L. D. Coraor and P. T. Hulina, "A Reconfigurable Multiprocessor," Proc. Int. Conf. Parallel Processing, August 1985, pp 649-651.

[12] M. Lee, E. Fiene, C. Wu, G. Brown, and N. Bagherzahdeh, "Network Facility For a Reconfigurable Computer Architecture," Proc. 5th Int. Conf. Dist. Comp. Sys., Denver, May 1985, pp 264-271.

[13] W. Handler, E. Maehle, and K. Wirl, "DIRMU Multiprocessor Configurations," Proc. Int. Conf. Parallel Processing, August 1985, pp 652-656.

[14] S. Ganesan, S. M. Mahmud, S. R. Vishnubhotla, "Solution to Cache Coherency Problem in a Mulitprocessor System," Tech. Report TR 8701 CSE-1, Oakland University, Jan. 87.

[15] D. Sundararajan, M. O. Ahmad, and S. Ganesan, "Multi-processors," IEEE Tran. Circuits and Systems, June 1985, pp 620-622.

[16] Trey West, "Dual-ported RAM For 68000 Microprocessor," Motorola Application Note, AN-881, 1983.

[17] D. Vrsalovic, D. P. Siewiorek, Z. Z. Segall, E. F. Gehringer, "Performance Prediction For Multiprocessor System," Proc. Int. Conf. Parallel Processing, 1984, pp 139-146.

[18] A. Norton and G. F. Pfister, "A Methodology for Predicting Multiprocessor Performance," Proc. Int. Conf. Parallel Processing, 1985, pp 772-781.

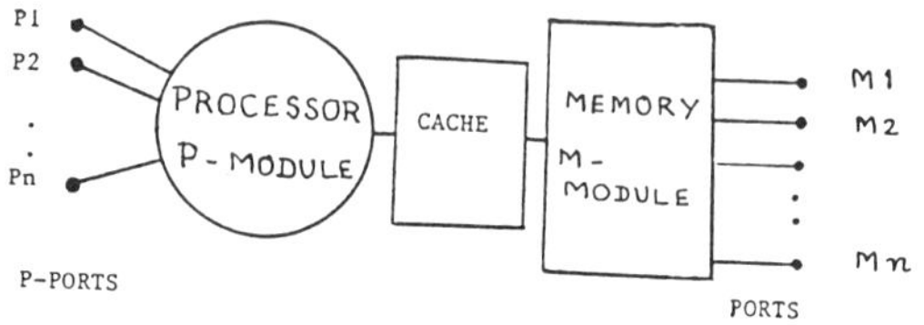

FIG.1 NODE BLOCK DIAGRAM

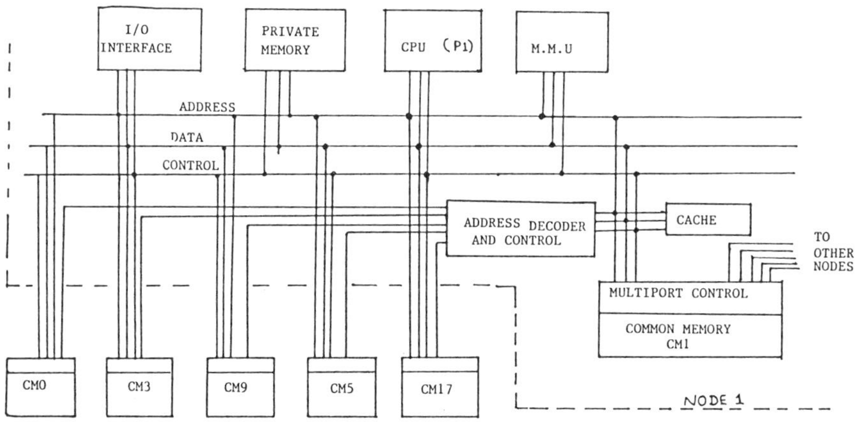

FIG. 2 NODE 1 ARCHITECTURE

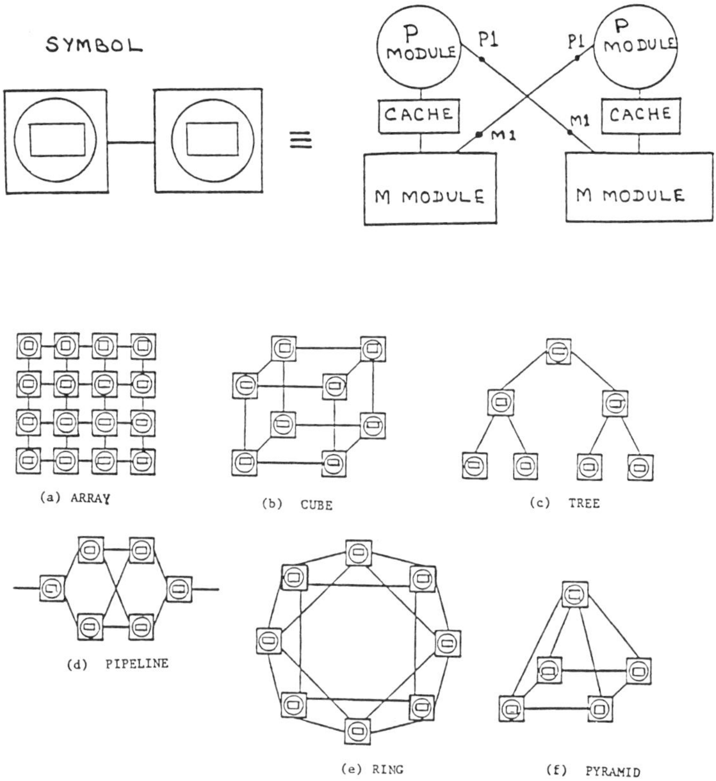

FIG.3 RECONFIGURATION OF INTERCONNECTION

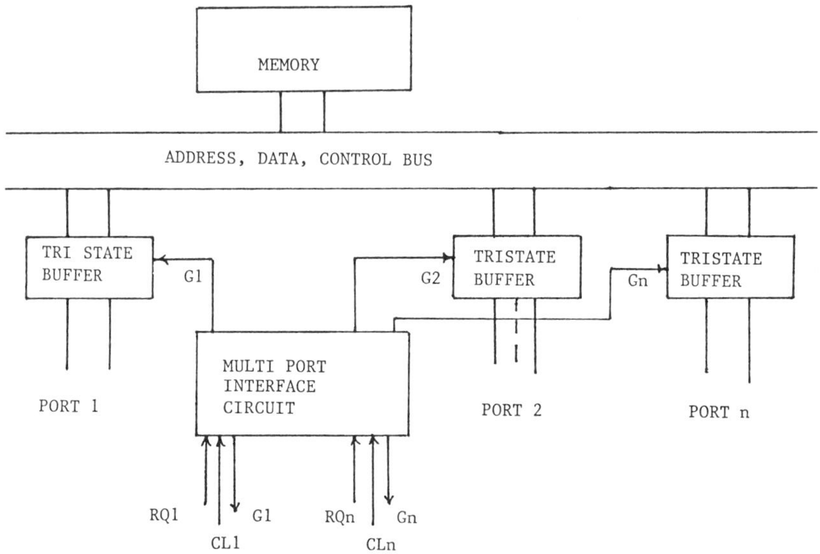

FIG. 4 BLOCK DIAGRAM OF MULTIPORTED MEMORY.

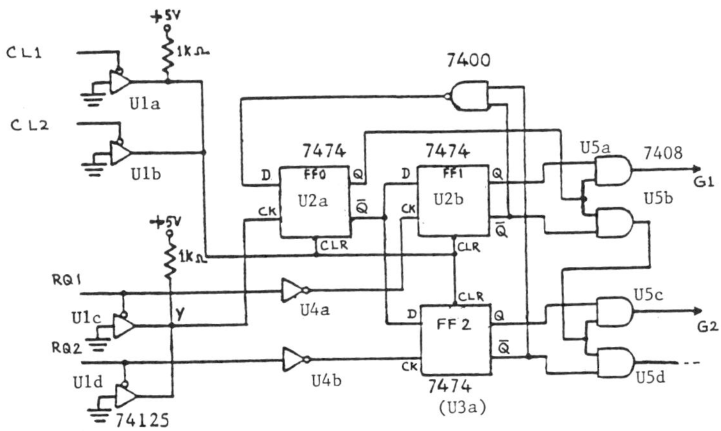

FIG. 5 MULTI-PORT INTERFACE CIRCUIT.

Real–Time Functional Programming

Peter D. Grogono

Department of Computer Science, Concordia University, Montréal, PQ, Canada H3G 1M8

1. Introduction

We can envision a real-time program as a system of concurrent processes. While concurrent processes suggest an efficient implementation, reasoning about them is difficult. An abstract representation based on a functional notation makes reasoning easier. The functional program that represents a particular real-time system can be transformed into an equivalent system of concurrent processes.

This research was suggested by previous work. Kluge and Schlutter have shown how λ-expressions can be converted to Petri nets [7]. Davis introduced the concept of a *continuous variable* [4]. A continuous variable conceptually changes continuously in time; an implementation is obliged only to ensure that its value is up-to-date whenever it is used. In a functional program, a continuous variable can be represented by a stream to which values are appended as they become available.

For the real-time application of music synthesis, Dannenberg proposed a functional language, Arctic [3]. Arctic programs include explicit references to timing constraints, whereas we have attempted to abstract timing constraints out of real-time programs. Each approach has merit in an appropriate context.

Abramsky and Sykes have designed and implemented an extended SECD machine [1] which responds to real-time events such as mouse movement. The SECD-M machine actually creates the streams defined in the source program. Our method differs in that we transform functional programs to processes without streams. The use of transformation to eliminate finite lists has been studied in depth by Wadler, who shows how finite lists can be eliminated from a well-defined class of functional programs [9].

2. Real-Time Programs

Fig. 1 shows a simple, real-time system represented as a flow graph. We use this system as an example throughout the paper. Each box represents a process. Inputs x and y are summed by process S, yielding a signal a. Signal z is filtered by process F, which selects values that exceed a constant value k. The process labelled W is a 'when-gate' which sends the sum $x+y$ to the output b whenever it receives a signal from F. In summary, the system produces the sum $x+y$ continuously at output a, and the same value at output b whenever $z > k$.

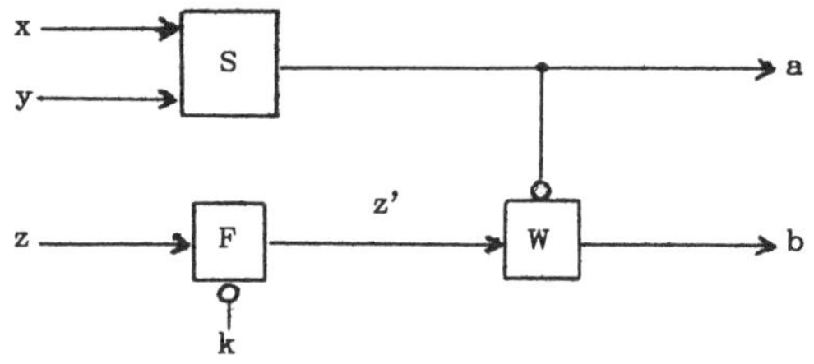

Fig. 1 A simple real-time system

The notation we use to describe the real-time program that implements this system is CSP [6] extended with ports [8]. The following code describes the action of the summing process *S* of Fig. 1.

```
class sum (X, Y → Z)
   do X?x → Y?y → Z!x+y
   [] Y?y → X?x → Z!x+y od
```

There may be several summing processes in a system. Each will be an instance of the class *sum* and will use the same code. The class *sum* has as its formal parameters two input ports, *X* and *Y*, and one output port, *Z*. Like all the processes that we consider, it is coded as a non-terminating loop bracketed by do and od. Alternative actions within a do body are separated by []. The code between do and [] in the example above may be paraphrased as 'if port *X* is ready, then receive datum *x* from it; then receive datum *y* from port *Y* and send the sum *x+y* to the output port *Z*'. The infix symbols '?' and '!' denote 'receive' and 'send' respectively. The code between [] and od handles the situation when port *Y* is ready before port *X*. The other process classes of the system can be defined in a similar way.

The following code describes the entire system by declaring instances of processes and connecting them by arguments.

```
S(x, y → a) : sum
F(z → z') : filter(< k)
W(a, z' → b) : when
```

Each line introduces an instance of a process class. The network connections are indicated by argument lists. For example, the process *S* of class *sum* sends a signal *a* that is received by process *W*. The process *F*, which filters values exceeding the constant *k*, has one input and one output port. The class *filter* requires an additional parameter which is a predicate. In this case, the predicate is $\lambda x.(x > k)$. We reverse the inequality, obtaining $\lambda x.(k < x)$; change to prefix notation, obtaining $\lambda x.(< k\ x)$; and write this function in the simpler form $(< k)$.

A system is called *data-driven* if the timing of events is determined by the arrival of data at its inputs, or *demand-driven* if the timing is determined by a request for output. The CSP model is neutral in this respect: two processes communicate when both are ready; a process cannot request another to communicate; and there are no implicit queues between processes. This form of communication is sometimes called a *rendezvous*.

3. Functional Programming

Functional programs are usually written as a collection of mutually recursive equations. The equations satisfy conventional algebraic laws. Consequently, it is relatively simple to transform functional programs and to reason about them [2]. Functional programs specify the temporal sequence of evaluation only in the weak sense that an expression must be evaluated before it is required for further computation. For real-time systems, functional programming provides a useful separation of concern: the equations specify the results of the computation, leaving the implementation free to determine the sequence of evaluation.

Components of a list are created when required and discarded when they are no longer available. We do not consider termination and therefore the lists are conceptually infinite. Lists processed in this way are called *streams*.

We write functional programs according to the schema

F x y ... = (a, b, ...).

F is a function with stream parameters *x, y, ...* . Its value is defined to be a tuple of stream expressions. The stream expressions *a, b, ...* consist in general of applications of functions which are either primitive or whose definitions appear later in the program. For example, the following program corresponds to the network of Fig. 1.

```
network x y z = let a = sum x y
                in (a, when a (filter (< k) z))
sum x y = cons(+ (car x) (car y))
              (sum (cdr x) (cdr y))
filter p x = if p (car x)
              then cons (car x) (filter p (cdr x))
              else filter p (cdr x)
```

In the first line, we use *let a = sum x y* so that we do not have to write *sum x y* twice in the second line. This corresponds to using the output of process *S* twice in Fig. 1. The functions *cons, car,* and *cdr* are borrowed from LISP. We do not include tests for the null stream because our streams are infinite. The unbounded recursion corresponds to the do ... od loops of the processes. The function *when* is discussed in Section 5.

4. Transformation

We demonstrate, using Fig. 1 as an example, that it is possible to transform the functional representation of a real-time system into the CSP representation. The CSP code can then be compiled using standard methods.

The first step in the transformation is to infer the types of parameters. In programs that can be translated to real-time systems, parameters are constants, scalars, or streams. Functional parameters are almost always constants that can be eliminated by symbolic unfolding. In the example, we have an application *filter (< k) z* and a definition for the second-order function *filter*. We introduce a new function, *filt-k*, obtained by replacing *p* by *(< k)* and *filter p (cdr x)* by *filt-k (cdr x)* in the definition of *filter*.

```
filt-k x =  if k < (car x)
                then cons (car x) (filt-k (cdr x))
                else filt-k (cdr x)
```

Different applications of a second-order function such as *filter* yield distinct first-order functions such as *filt-k*. We are simply reversing the abstraction process that led to *filter* in the first place.

The next step is to replace non-atomic arguments by variables with appropriate bindings. In the example, *when a (filt-k z)* is a non-atomic argument of the (implicit) tupling operator. We replace *(a, when a (filt-k z))* by

```
let b = when a (filt-k z) in (a, b) .
```

This expression still contains a non-atomic argument, namely *filt-k z*. A second transformation yields

```
let b =
    let z' = filt-k z in when a z'
in (a, b) .
```

Since *b* is not a free variable of *filt-k z*, we can further transform this expression to

```
let z' = filt-k z
    in let b = when a z'
        in (a, b) .
```

The function representing the network is now in the desired form: all parameters are streams, and there are no non-atomic arguments. We can show the result of the transformation process more clearly by omitting the keywords <u>let</u> and <u>in</u>.

```
a = sum x y
z' = filt-k z
b = when a z'
return (a, b)
```

Assuming that each function can be represented as a process, we can infer the network topology.

S(x, y → z) : sum
F(z → z') : filt-k
W(a, z' → b) : when

Having identified the processes, we must generate code for them. For the class *sum*, we have

sum x y = cons(+ (car x) (car y))
(sum (cdr x) (cdr y)) .

We identify occurrences of *(car p)*, where *p* is a parameter of the function, and replace them by local variables.

sum x y = <u>let</u> *cx = car x, cy = car y*
<u>in</u> *cons (+ cx cy) (sum (cdr x) (cdr y))*

The next step is to look for instances of 'tail recursion modulo cons' [9, 10]. There are several patterns that can occur. In this case the pattern is *cons H (F T)*, in which: *H* is an expression containing variables bound to the stream headers, obtained during the previous step; *F* is a recursive call of the function we are transforming; and *T* is an expression containing the stream tails (that is, expressions of the form *(cdr p)*, where *p* is a parameter of *F*). This pattern corresponds to the process

<u>do</u> *receive stream heads → send H* <u>od</u> .

In this case, the match yields the following process:

<u>do</u> *X?cx → Y?cy → Z!cx+cy* <u>od</u> .

In this code, we have introduced input ports *X* and *Y*, corresponding to the stream parameters *x* and *y*, and a single output port *Z*, corresponding to the value yielded by the function.

This process has an asymmetry which is not present in the functional definition: it always receives from *X* before receiving from *Y*. It is simple to recognize this situation and generate

<u>do</u> *X?cx → Y?cy → Z!cx+cy*
[] Y?cy → X?cx → Z!cx+cy <u>od</u> .

The transformation of *filt-k* proceeds in a similar way. After extracting stream headers, we obtain

filt-k x = <u>let</u> *cx = car x*
<u>in</u> <u>if</u> *cx > k*
<u>then</u> *cons cx (filt-k (cdr x))*
<u>else</u> *filt-k (cdr x)* .

The expression *cons cx (filt-k (cdr x))* matches the pattern we have already seen. The expression *filt-k (cdr x)* in the else branch of the condition represents 'no action' in the process because nothing is sent to the output stream. The code for the process is

```
do X?cx;
  if cx > k → Y!cx
  [] cx ≥ k → skip fi od .
```

5. Non-determinism

The word 'non-determinism' is used with several different meanings in Computer Science. We use it in the following sense: an expression *e* is *non-deterministic* if it does not necessarily yield the same value each time it is evaluated.

Non-determinism destroys *referential transparency*, the property that an expression can be replaced by another expression with the same value. It is a principle of functional programming that the expressions $e + e$ and *let* $x = e$ *in* $x + x$ are equivalent. If *e* is non-deterministic, this is not the case because *e* is evaluated twice in the first expression (possibly yielding different values) but only once in the second expression. We must avoid transforming an expression of the second kind into an expression of the first kind; for example, we must not unfold a function that has non-deterministic arguments.

The process *when* cannot be transformed by the method described in Section 4 because it is non-deterministic. In the example of Fig. 1, all we can say is that every component of stream *b* is also a component of stream *a* and that the order of these components in the two streams is the same. We have the choice of retaining *when* as a primitive operator which cannot be defined explicitly in the functional program, or of introducing a simpler primitive in terms of which *when* and other non-deterministic functions can be defined.

We introduce the primitive, non-deterministic operator [] in the functional language. Informally, *x* [] *y* evaluates to either *x* or *y*. A precise definition is: (*x* [] *y*) = *z* if and only if $(x = z \vee y = z)$. We use the symbol [] in both the functional and process notations. These uses are related but not identical.

There are simple implementations of [] that satisfy the definition. For example, an implementation could always return the left argument, or it could return the left and right arguments alternately. Our interpretation of [] is that it returns the first argument that becomes available in real time.

The definition of *when* using [] is

```
when x c =
      (let cc = car c in cons (car x) (when (cdr x) (cdr c)))
   [] (let cx = car x in when (cdr x) c) .
```

This function transforms, according to the rules described in Section 4, to

<u>do</u> *C?cc* → *X?cx* → *Z!cx*
[] *X?cx* → *skip* <u>od</u> .

We can use [] to define other non-deterministic functions, including the merging function called 'fair merge' by Abramsky [1] and 'interleave' by Henderson [5].

merge x y = (*cons* (*car x*) (*merge* (*cdr x*) *y*))
[] (*cons* (*car y*) (*merge x* (*cdr y*)))

The corresponding process is

<u>do</u> *X?cx* → *Z!cx*
[] *Y?cy* → *Z!cy* <u>od</u> .

It accepts a value from either input stream and passes it directly to the output stream. Merging the streams *1 2 3 4* ... and a b c d ... might yield the stream

1 2 a b c *3 4* d ... .

The output stream contains all the components of both input streams in their original order. *Merge* is non-deterministic because the way in which the values are interleaved may vary between runs.

6. Conclusions

Certain real-time systems can be coded as functional programs. These programs are easier to understand, reason about, and transform than corresponding programs written in a lower level, imperative language. They can, however, be mechanically transformed into CSP notation, in which form they can be efficiently compiled. Most real-time systems are non-deterministic. This can be accommodated by adding a non-deterministic choice operator to the functional language. The penalty is the loss of referential transparency, although most substitutions are still valid. Further work is required to characterize precisely the real-time systems that can be represented in this way.

Acknowledgements

I am indebted to V.S. Alagar, Geetha Ramanathan, and Paul Voda for helpful discussions about this work. The research was funded by a grant from the Natural Sciences and Engineering Research Council of Canada.

References

1. Abramsky, S. and R. Sykes. SECD-M: a virtual machine for applicative programming. In *Functional Programming Languages and Computer Architecture,* LNCS 201, Springer-Verlag 1985, 81-98.
2. Burstall, R.M. and J. Darlington. A transformation model for developing recursive programs. *J. ACM, 24,* 1, 44-67 (1977).
3. Dannenberg, R. B. Arctic: a functional language for real-time control. In *Symposium on LISP and Functional Programming,* ACM, 1984, 96-103.
4. Davis, E. A high-level real-time programming language. *Technical Report No. 145* and *Robotics Report No. 36,* New York University, 1984.
5. Henderson, P. Purely functional operating systems. In *Functional Programming and its Applications: an Advanced Course,* J. Darlington, P. Henderson, and D. Turner (eds), Cambridge University Press, 1982, 177-192.
6. Hoare, C.A.R. Communicating Sequential Processes. *Comm. ACM, 21,* 8, 666-677 (1978).
7. Kluge, W.E. and H. Schlutter. Petri net models for the evaluation of applicative programs based on λ-expressions. *IEEE Trans. Soft. Engrg., SE-9,* 4, 415-426 (1983).
8. Silberschatz, A. Port directed communication. *Comp. J., 24,* 1, 78-82 (1981).
9. Wadler, P.L. *Listlessness is Better than Laziness.* Ph.D. Thesis, Carnegie-Mellon University, 1984.
10. Warren, D.H.D. *An improved Prolog implementation which optimizes tail recursion.* DAI Research Report 141, University of Edinburgh, 1980.

A Computer Integration of Reconfigurable Fixtures and Drilling of Sheet Metal Parts

K. Youcef-Toumi*, W.S. Liu†, H. Asada‡

*Assistant Professor, Dept. of Mechanical Engineering, Massachusetts Institute of Technology, Cambridge, MA, USA
†Graduate Student, Currently with AT&T Bell Laboratories, Atlanta, GA, USA
‡Associate Professor, Kyoto University, Kyoto, Japan

Abstract

This paper presents an automated manufacturing system for the drilling of sheet metal parts. All stand-alone systems such as the robot, a set of reconfigurable fixtures, and the CAD/CAM workstation have been integrated into a Flexible Manufacturing System. This system analyzes and evaluates a given fixturing layout, and assembles the reconfigurable fixtures automatically using a robot manipulator. An optimum fixturing layout and assembly are achieved by examining the workpart from a stress-strain point of view. In addition, issues such as geometric constraints, yielding and buckling, database design, collision detection, fixturing sequential control, and the automatic assembly of fixture elements are considered. The computational and analytical concepts for the reconfigurable fixturing and drilling system are also presented in this paper.

1. Introduction

Workpiece fixturing is an important issue in flexible manufacturing systems. Although traditional jigs and fixtures are still used in highly automated manufacturing systems, they provide a major bottleneck for increasing the flexibility of the overall system. Conventional jigs and fixtures are typically single-purpose devices designed and manufactured for specific workpieces and machining operations. A significant amount of manual effort is required for setting up and changing the jigs and fixtures. This severely limits the overall flexibility of the manufacturing system and makes the process costly and time consuming particularly for low volume production applications. An example is the machining of stamped sheet metal parts for modern aircrafts. These curved and flexible sheet metal parts must be held rigidly in their true shape, holes must be drilled accurately at predetermined locations on the surface and the edge must be routed to shape. Since these parts have different shapes and sizes, a unique holding and drilling fixture is fabricated for each part. The part is clamped in this fixture manually, and a skilled operator, guided by the fixture, drills all holes and routs the edges. This procedure has a number of distinct drawbacks and would create a tremendous fixture storage problem, particularly in low volume, batch type of production.

The manufacturing of sheet-metal parts in particular is considered to be difficult to automate because of the complexities of the geometries involved (for example, size, surface contour, and outline etc.), the small lot size and the level of accuracy required. Past researches in alternative production methods for sheet metal parts were aimed at increasing the level of flexibility, effectiveness, and interactiveability of fixturing systems, through: (1) process automation, (2) fixturing automation, and (3) computer integration. Several Dedicated Automated Manufacturing Systems for sheet metal have been developed such as the Grumman Automation [15], the JOMAC project [16] and the General Dynamics Automation [12]. Fixturing automation is a fixturing unit which can be set up automatically and is adaptive to a multitude of different workpart geometries. Five major types of flexible fixtures have been developed in the past: matrix fixturing or array of discrete die elements [10] for metal forming applications of sheet metal parts, programmable clamps [4], flexible programmable vices [17] and modular fixturing systems, fluidized-bed vice [7, 8]. Most of the MFS [9, 5, 13] are designed to be set up by operators.

In addition to fixture characteristics, the workparts to fixture can generally be categorized into rigid bodies and flexible bodies. For rigid bodies fixturing, the fixture has to hold the workpiece against both gravity and machining forces. However, the fixturing of flexible bodies, such as sheet metal, not only requires the basic function of fixtures for rigid bodies, but also has to strengthen the part to avoid permanent deformations. Sheet metal parts usually exhibit low stiffness and consequently are easily distorted.

The objective of this paper is to present an automated manufacturing system for the drilling of sheet metal parts. All stand-alone systems such as the robot, a set of reconfigurable fixtures, and the CAD/CAM workstation have been integrated into a Flexible Manufacturing System. This system analyzes and evaluates a given fixturing layout from a stress-strain point of view and sets the reconfigurable fixtures automatically using a robot manipulator. The developed software consists of a database system and four major modules: a workpart geometry display, a routine for fixturing analysis, a fixture sequential control software, and a program to configure the fixtures automatically.

2. Reconfigurable Fixtures and The Drilling of Sheet Metal Parts

The hardware for the proposed reconfigurable fixture was designed and manufactured at M.I.T [1, 2, 18, 19, 6]. The design takes into account the automated assembly of the fixtures by a robot manipulator. The reconfigurable fixturing system consists of a fixture bed (baseplate), vertical supports (fixture elements) and two locating pins. The prototype of the fixturing system with a typical workpart is shown in Figure 2-1. The main idea of the robotic reconfigurable fixturing concept is to support the sheet metal from beneath with vertical support fixturing elements on a universal fixture base. The base consists of a flat plate with evenly spaced machined T-slots, as shown in Figure 2-1. Vertical supports are stored initially in a fixture magazine and then the robot grasps elements from the magazine and positions them onto the baseplate. Each slot of the baseplate has a tapered entrance to allow easy insertion of the fixturing elements. A typical element shown in Figure 2-2, is adjustable in x, y and z directions.

A typical sheet metal part, shown in Figure 2-3, has two tabs as a pair of reference points to locate the part which is a common practice in fuselage manufacturing. The sheet metal is placed and located accurately relative to the base by two pins, which constrain the workpart in x-y plane. Only one of the pins is fixed, and therefore its position is known. The second pin originally is located at its "home" position in the sliding track. It prevents the workpart from rotating about the first pin and is adjusted to the final location by the robot. During the assembly of the fixture, the vertical supports are inserted in the appropriate slot, slid along the desired position and oriented appropriatly. Therefore, the automatic assembly of these fixtures require the following parameters: slot number, sliding distance, orientation and the supporting shaft height. The assembly of the fixturing elements must take into account the collision and interference problem since the elements can share the same or adjoining slot. However, in dealing with flexible parts, simply constraining the six degrees of freedom is not sufficient to prevent undesirable deformations while loads are applied. Redundant supports are required so that the part will not distort from its original shape, even when no machine loads are present. This is especially important in a manufacturing environment requiring high accuracy. The design and assembly of the reconfigurable fixtures must address this inherent characteristic of sheet metals: the fixture must define the workpart shape completely. The possibility for the part to yield or buckle under the drilling loads must be checked.

The drilling process can be summarized as two distinct actions: one at the chisel and the other at the lips of the drill bit. The entire process can be modelled as a net thrust T and a torque M by assuming symmetry about the drill point. Several ways were developed to evaluate the thrust and the torque. The equations to determine the drilling thrust and torque along with the parameter definitions and values can be found in reference [14],

$$T = 2\ K_d\ F_f\ F_t\ B\ w\ +\ K_d\ d^2\ J\ W$$
$$M = K_d\ F_f\ F_m\ A\ W$$

3. Fixturing Analysis

3.1. Finite Element Procedure

Failure analysis constitutes a major element in the fixturing analysis. A major type of failure is the distortion of the part from its true shape after machining due to the inherent low stiffness of the sheet metal. This failure situation can be avoided by performing stress analyses of the system. Finite element analysis is used to check whether the workpart will sustain permanent deformation or buckle under the drilling loads, and to evaluate the fixturing layout considered.

In the model of Figure 2-3 the vertical supports are assumed to have point contact with the workpart. They only constrain the part in the Z direction. Two locating pins are considered as reaction forces to restrain the part in the X-Y plane. An eight-node double curved shell elements (S8R) , with four integration points, was used in modelling the sheet metal parts. The machining forces are modelled as forces and moments applied at the corresponding nodes. The vertical supports and the

Figure 2-1: Reconfigurable fixturing system for sheet metal parts

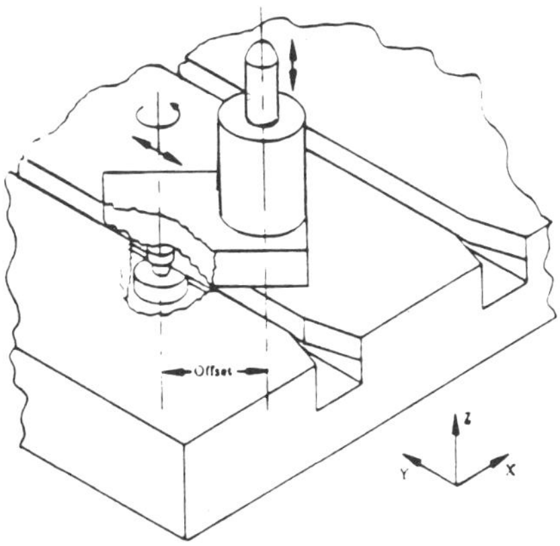

Figure 2-2: Modular and adaptable fixture for sheet metal parts

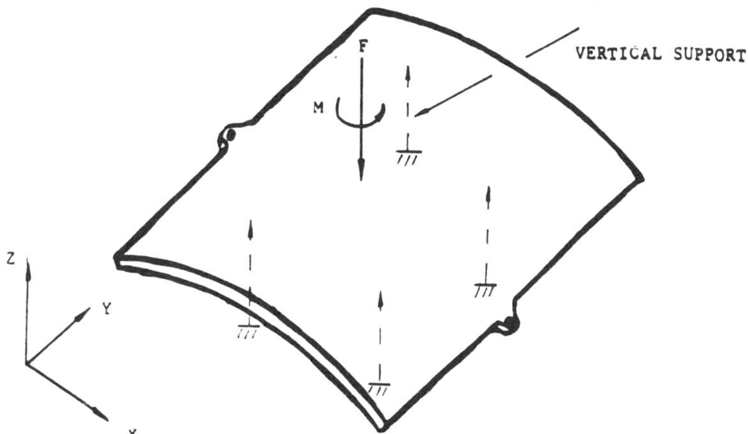

Figure 2-3: Physical model of sheet metal drilling

locating pins are introduced as part of the boundary conditions. The degree of freedom in the z-direction is constrained at the nodes where the fixtures are located.

3.2. Fixturing Layout Evaluation

Excessive stress level induced at any location within the sheet metal part will cause the part to deform plastically from its original shape. Three and four fixturing elements at different radii from the drilling point were considered. The stress distribution within the workpart for different layout configurations , were investigated. In addition, several plate curvatures were examined. Specifically, nine different radii of curvatures are investigated: ± 25 *in*, ± 50 *in*, ± 75 *in*, ± 100 *in*, and infinity. The fixture elements are located along the diagonal axes as shown in Figure 3-1.

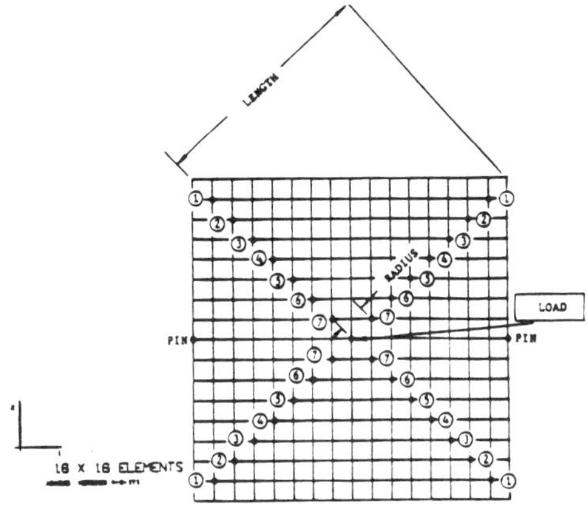

Figure 3-1: Fixturing layouts for four supports

The normal stresses in 11, 22 directions and shear stress as a function of fixture support location and the applied forces were analyzed. The simulations on yielding were performed for the following cases: flat plates, shells with positive radius of curvature (convex shells) and shells with negative radius of curvature (concave shells), [19].

Figures 3-2, 3-3 and 3-4 show the normal and shear stresses for the flat plate. The normal stresses in 11 and 22 direction at the loading point increase with the increasing radial distance of fixture elements from the center of the plate. These stresses are compressive when the fixtures are near the center of the plate because the applied forces are directly transferred to the fixtures resulting in small deflections. In this case the workpart fibers are compressed. However, the stresses are tensile as the fixture elements move away from the center of the plate. Due to the locating pins constraint, the plate stretches resulting in large deflections. Both normal stresses 11 and 22 at the fixtures increase slightly when fixtures are near the center of the plate, but stresses increase drastically as fixtures near the edges of the plate. The stresses are always tensile because material is pulled toward the loading point. The normal stresses at the pins are always compressive, and they decrease with the increasing radius of fixtures from the center of the plate. The shear stresses distribution at the point of pins and applied for are relatively constant and low, but the shear stresses at the fixtures increase as the radius of fixture from the center increase. After reviewing all the plots, the best fixturing layout is at a radius/length ratio of 0.55. Length and radius were defined previously in Figure 3-1. Several simulations have been conducted for convex and concave shells in reference [19]. The best fixturing locations for all positive-curvature shells are given by a radius/length ratio ranging from 0.45 to 0.65. As for the concave shells, any fixturing layout should be feasible as long as the fixture elements are within a 0.65 radius/length ratio. As for the three support fixturing system, the fixtures are placed 120 degrees apart and six fixturing layouts for each radius of curvature have been studied. All the collected data show that the stress profiles for all curvatures have more or less the same pattern as the four-point fixturing. However, the three-point support case has higher stress than the four-support fixturing as expected. The stresses at different locations, which are loading point, locating pins, and fixtures for different curves are given in reference [19].

The yielding analysis cannot ensure that the workpart will not deform from its true shape. Permanent deformation, which is still possible, can be caused by buckling where the workpart can suddenly snap through. The buckling phenomenon will happen when forces are applied to the convex faces. In order to analyze the buckling of convex shells, the critical load has to be determined. This can be done by controlling the displacement of the node at the center of the shell to move down a certain distance. The critical load is plotted against the radius measured from the center of the plate to the fixture locations. The performance for different radii of curvatures is illustrated in Figure 3-5. It can be seen that higher buckling force is found when fixtures are closer to the center. Thus it is easier to buckle when fixtures are further apart as expected. The smaller the radius of curvature, the larger the buckling force results.

4. Computer Integration

The hardware of the sheet metal machining cell consists of an IBM RS1 robot, a set of reconfigurable fixtures, and computer systems. The software package used is CIMSURF [3], a sculptured surface package developed for computer-aided design and manufacture of parts. This software package is used to define the surface of the sheet metal. From a wire-frame model, the user interactively inputs the locations of the holes to be drilled and the fixturing elements and the data files containing this information are generated automatically.

4.1. Database Structure

The package is made modular and programmed to achieve a high degree of flexibility. If different operations are required or different fixturing systems are used, the modifications in the program are minimal. The database architecture, as shown in Figure 4-1, is structurally divided into two major parts: (1) CAD database, and (2) fixture database. A CAD database contains the geometrical models of the workparts, information concerning the drilling holes, and the locations of the fixture elements. The fixture database consists of: (1) fixture element files, (2) fixture magazine file, and (3) fixture layout sequential file. The details of this database are discussed below.

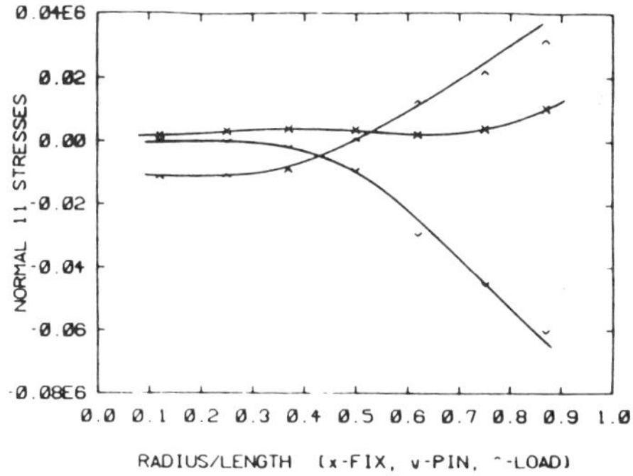

Figure 3-2: 11 normal stresses in flat plate

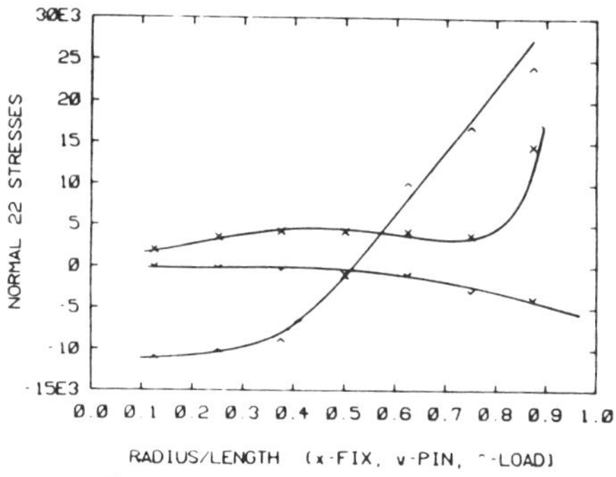

Figure 3-3: 22 normal stresses in flat plate

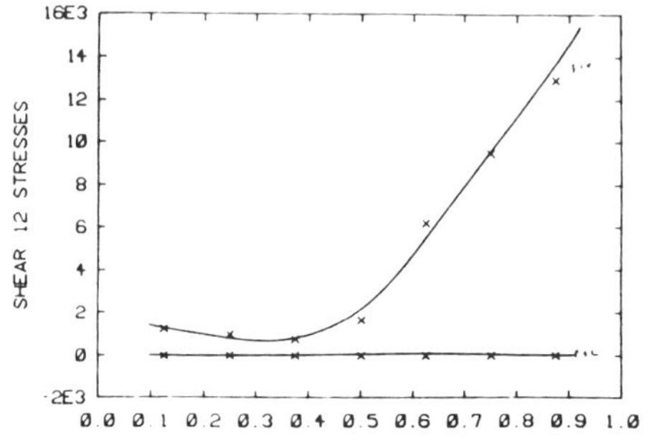

Figure 3-4: Shear stresses in flat plate

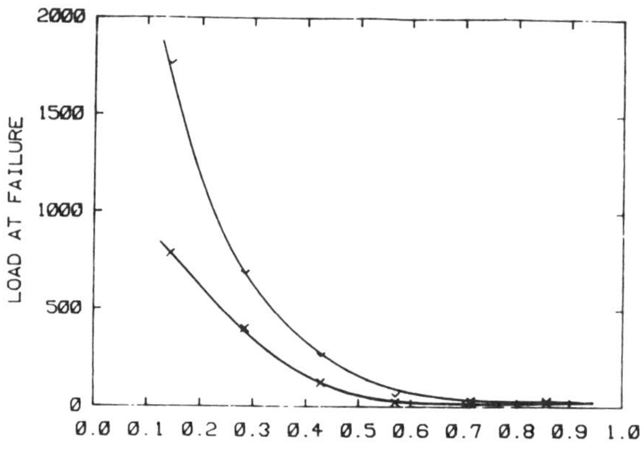

Figure 3-5: Buckling load for different radii of curvature

The fixture element files contain the elements specification file and the base-plate specification file. These files contain relevant physical dimensional data on different fixture element types and base-plate. The fixture magazine file stores information on the position and the availability of the fixture elements in the magazine. The fixture layout file contains a sequential list of assembly commands for a feasbile fixture layout in terms of the base-plate parameters.

4.2. Configuration of Fixtures

The purpose of the sequential control module is to check the feasibility of a fixture layout, to determine the baseplate parameters of the elements, to detect any collisions that may happen among elements during set-up of a fixture layout, and to arrange the sequential order of assembly. The flowchart for the software of this module is detailed in Figure 4-2. This package consists of five major subprograms: coordinate transformations, physical limitation checking, baseplate parameter generation, sequential ordering and collision detection. After these five computational operations are finished, a fixture layout sequential file is downloaded to the robot controller as an instruction for fixturing assembly. Collisions can be detected by processing the fixture sequential file line by line. Two pointers are used to indicate any two elements and the test is performed to all elements. The first level test checks whether the distance between the two pointed elements is within a possible collision region. The second level test basically checks whether the intersecting points of two pointed elements are within the two elements instantaneously.

The robot is instructed to assemble a fixture based on the orderly sequential file downloaded from the CAD/CAM computer. This software package, written in IBM AML [11] can be set up for any arbitrary fixturing layout. The fixture setup starts with positioning the vertical supports onto the fixture bed at the desired locations and then adjusting the locating pins. Programming the robot is done off-line.

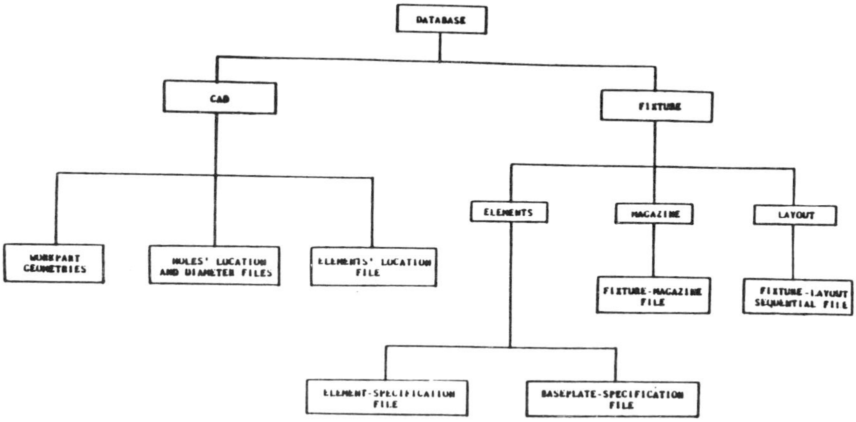

Figure 4-1: Database architecture

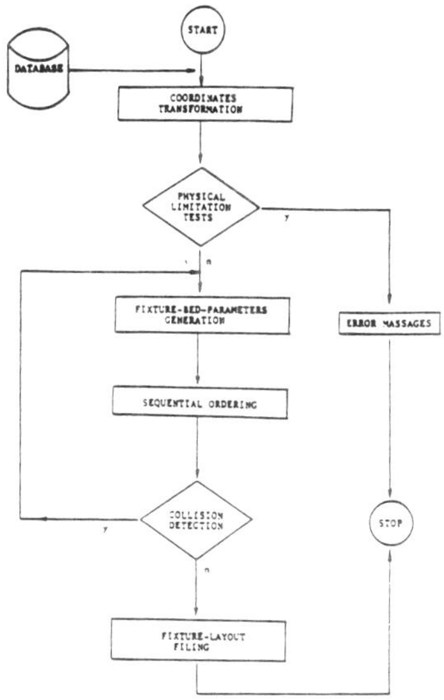

Figure 4-2: Fixturing sequential control flowchart

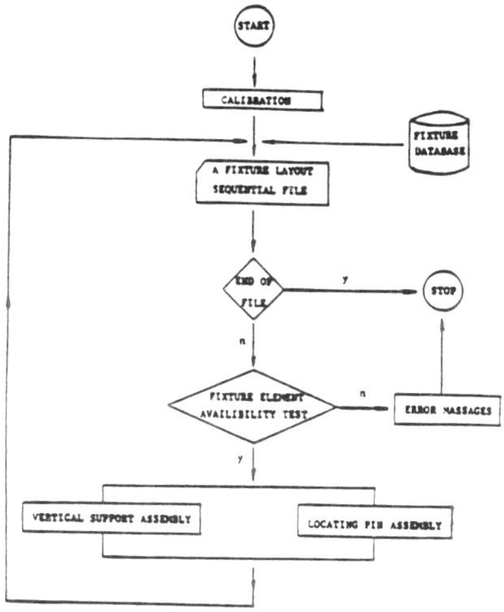

Figure 4-3: Fixture assembly programming flowchart

The programming technique in conjunction with the calibration procedure methods [6] compensate for changes in the working environment: (1) changes in position of the baseplate and the magazine; (2) fixtures wear; (3) changes in robot arm coordinates due to temperature deviation or some other effects.

5. Conclusion

A robotic sheet-metal fixturing system, which consists of an industrial robot, a CAD/CAM workstation and a reconfigurable fixturing system, has been integrated. The integrated software consists of four major modules: workpart graphic display, sheet-metal fixturing analysis, fixture sequential control and fixture assembly. A database for the entire package is developed.

Finite elements has been used to analyze the system consisting of the sheet metal parts, fixturing system and the drilling loads. The yielding and buckling of parts have been considered as major issues in selecting an optimum fixturing layouts for a given task. The criteria for designing an optimum fixturing layout was based on the stress distributions in the part. Different fixturing layouts for the supports, and parts with several radii of curvatures have been studied.

References

[1] Asada, H. and Fields, A.
Design of Flexible Fixtures Reconfigured By Robot Manipulators.
In *ASME Winter Annual Meeting Symposium on Robotics and Automation*. Miami Beach, Fl, 1985.

[2] Asada, H., Youcef-Toumi, K. and Fields, A.
Flexible Fixturing Using a Robot Manipulator: The design of Adaptable Fixtures for Sheet Metal Machining.
In *M.I.T Laboratory for Manufacturing and Productivity Report, LMP/RBT 85-02*, pages . , , 1985 .

[3] *Cimsurf User's Manual, CADLINC, Inc.*
Cadlinc , 1984.

[4] Cutkosky, M.R., Kurokawa, E., and Wright, P.K.
Programmable Conformable Clamps.
In *AUTOFACT 4 Conference and Exibition*. Philadelphia, PA, 1982.

[5] Drake, A.
Fixture Design: Working with Modules
1984.

[6] Fields, A. , Youcef-Toumi, K. and Asada, H.
Flexible Fixturing and Automatic Drilling of Sheet Metal Parts Using a Robot Manipulator.
In *In Proceedings of The 1986 Japan-U.S.A Symposium on Flexible Automation*, pages . Osaka, Japan , July , 1986.

[7] Gandhi, M. V. and Thompson, B. S. .
Phase-Change Fixturing for FMS.
In *Manufacturing Engineering 93* . December, 1984.

[8] Gandhi, M. V. and Thompson, B. S. .
Phase-Change Fixturing for Flexible Manufacturing System.
In *Manufacturing Engineering 4*. May, 1985.

[9] Gettelman, K.
Modular Fixturing Pushes Machine Uptime
1985.

[10] Hardt, D.E., Olsen, B.A., Allison, B.T., and Pasch, K.
Sheet Metal Forming with Discrete Die Surfaces.
In *Proceedings 9th North American Manufacturing Research.* , 1981.

[11] *A Manufacturing Language Concepts and User's Guide*
IBM Inc., 1982.

[12] Kusmierski, T.
Robot Development for Aerospace Batch Manufacturing.
In *Joint Automatic Controls Conference*. 1977.

[13] Lewis, G.
Modular Fixturing System.
In *Proceeding of the Second International Conference on Flexible Manufacturing Systems*, pages 451-461. London, England, , 1983.

[14] *Machinery's Handbook*
Industrial Press Inc., , 1984.

[15] Marx, W.
Robotic Drilling and Routing for Complex Aerospace Parts.
In *SME Conference on Applying Robotics in the Aerospace Industry*. St. Louis, MI, 1984.

[16] Molander.
Machining of Sheet Metal Plates in Aircraft Manufacturing - why use an industrial robot?
In *ROBOTS 8 Conference*. Detroit, MI, , 1984.

[17] Ogorek, M.
Workholding in the Flexible System.
Manufacturing Engineering , July, 1985.

[18] Youcef-Toumi, K., Asada, H., and Fields, A.
Flexible Fixturing Using a Robot Manipulator: Flexible Fixturing and Automated Drilling of Sheet Metal Parts Using a Robot Manipulator.
In *M.I.T Laboratory for Manufacturing and Productivity Report, LMP/RBT 86-01*, pages . , , 1986.

[19] Youcef-Toumi, K., Asada, H., and Liu, W.S.
Flexible Fixturing Using a Robot Manipulator: Computer Aided Analysis of Reconfigurable Sheet Metal Fixturing for Robotic Drilling.
In *M.I.T Laboratory for Manufacturing and Productivity Report, LMP/RBT 86-06*, pages . , , 1986.

Failsafe/Diagnosable Multiprocessor System for Robotics Applications

Subramaniam Ganesan, Sarma R. Vishnubhotla, and Syed M. Mahmud

Department of Computer Science and Engineering, Oakland University, Rochester, MI 48063, USA

Abstract

Multiprocessors are used in Robotic applications to obtain high speeds and error free computations. Fail-safe realization is one approach to highly reliable system design. In this paper we discuss a distributed technique for diagnosing a distributed-shared memory multiprocessor system for robotic applications, graph model and issues encountered in creating fault-tolerant system, i.e., self-checking computer design, use of natural redundancy available in a multiprocessor with minimum degration in performance, employing multilevel redundancy, and external fault recovery management with high reliability.

Introduction

Multimicroprocessors are used in Robotic applications to obtain high speed and error free computations. Achieving reliable operation becomes increasingly difficult with increased number of microprocessors in the system. Robots are used for a number of industrial tasks and the tasks are normally dynamically allotted to robots by means of communication signals. Since the action of robot is adaptive, i.e., sequence of movement of arms changes with the task and the objects to be moved, greater importance in the design of the different components of the robots like the sensors and controller, and the safety aspects are needed.

Fail-safe realization is one approach to highly reliable system design. In a fail-safe system, the output assumes one of predetermined safe values when a fault occurs. Concurrent fault detection is required in a multi-processor robot system to identify the transient and permanent faults. To recover from transient faults the following techniques may be used.

1. Single instruction roll-back.
2. Roll-back to programmer generated roll-back points where the various states are saved.
3. Restart the program from a known beginning point.

Each of the above steps takes more time to recover and may result in damages or safety hazards.

A number of papers have appeared on reliable/reconfigurable multiprocessor system management and analysis. [1-6].

The diagnosis technique very much depends on system architecture and Robot's sensors and tasks. The fault detection and location (or diagnosis) can be done by the following two techniques. 1. Centralized external processor behaves like a watch dog and monitors the other processors. 2. Distributed diagnosis, where each processor executes self diagnosis routines and also checks the neighboring processors.

A multiprocessor system consists of a large number of processing elements, connected to a number of memory modules through a crossbar switch, multiple interconnection network or multiple bus. The performance evaluation of the above cited multiprocessors have been reported extensively using analytic and simulation models [7]. Simultaneous consideration of both the performance and fault tolerant issues are important to evaluate a multiprocessor. The failure of a basic component should not bring the system down. The system should be able to detect any faulty module and should have the ability to reconfigure and operate at a degraded mode. Graceful degradation should be an inherent attribute of the system to provide fault tolerance.

Knowledge-based approach has been applied to the diagnosis of analog systems [5]. Utilization of knowledge-based testing and diagnosis of I/O ports, devices and sensor signals of a Robotic system is feasible. Initial model of the I/O ports, devices and signals are stored in shared memory. This model is automatically refined during successive fault-free test runs. This knowledge is represented in frames that are linked by cause-effect relation to build a diagnostic graph.

In the next section we discuss a graph model and a fault model for diagnosing a distributed shared memory multiprocessor system.

Graph Model for fault diagnosis

A system interconnection graph is an undirected graph showing the interconnection among processors of the system. (Fig. 1). A test graph shows the relations (or communication links between tested processors and testing processors. (Fig. 2). The arrow indicates the communication direction. When two way communication between every pair of processors exists, an undirected graph model is used. A testing

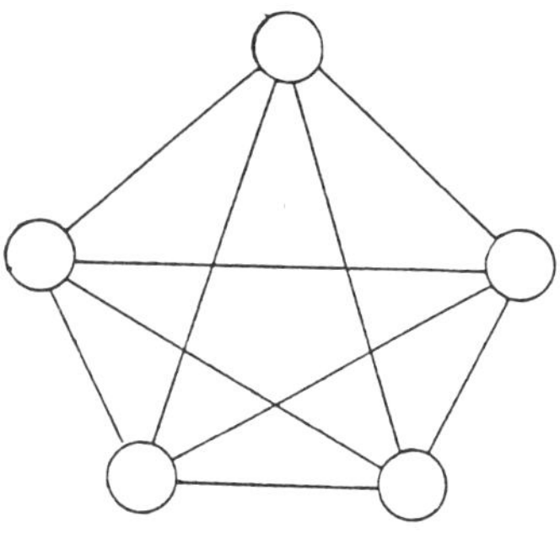

FIG.1 INTERCONNECTION GRAPH

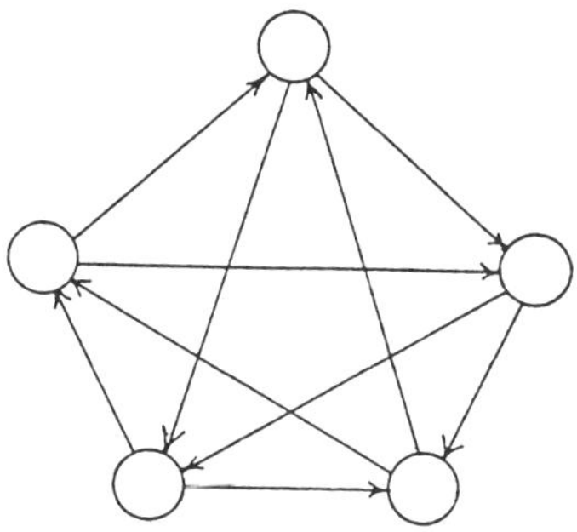

FIG.2 TEST GRAPH. EACH PROCESSOR TESTED BY 2 NEIGHBORS.

processor initiates self test routines in a tested processor and monitors the result. A tested processor receives a start signal, reports its status and reroutes any fault vectors from its testees to its testers.

Fault model similar to PMC [13] but with a few modifications is given below.

1. A faulty processor might modify fault vectors sent to it by its testees.
2. A processor may be tested by more than one neighboring processors, and results compared.
3. Only permanent faults are considered.
4. A fault free processor can become faulty during diagnosis.
5. Results of fault free processors always agree.
6. Results of fault free and a faulty processor always disagree.
7. Results of 2 faulty processors never agree.
8. Failed node is diagnosed immediately.
9. Link faults will not be treated as processor faults though it is difficult and time consuming to differentiate link faults and processor faults.

Table 1

Processor 1(P_1)	Processor 2(P_2)	Test Result comparison (P_1 tests P_2 and P_2 tests P_1)
fault free	fault free	1 (agree)
fault free	faulty	0 (disagree)
faulty	fault free	0 (disagree)
faulty	faulty	0 (disagree)

Table 1 shows the logic truth table for comparison test.

The multiprocessor network is modeled as a graph G(V,E) with nodes V representing processors and edges E representing the physical links. Depending on the desired behaviour of the multiprocessor several different fault tolerant criteria can be defined. The simplest criteria is the connectivity.

ARC connectivity between two nodes X and Y is the minimum number of arcs that must be removed from the graph to disconnect the nodes X and Y.

NODE connectivity is the minimum number of nodes when removed will make the graph disconnected.

The connectivity of the test graph of the fault tolerant multiprocessor should be equal to the minimum degree of its nodes. When the number of faults is equal to the connectivity of the test graph, then the diagnosis may be indecisive or not possible.

A graph G is K node connected if there exists K-node dijoint paths joining every pair of nodes of G. Shortest path between a pair of nodes is GEODESIC. Diameter of a graph is the length of the longest geodesic. The length of the shortest path between nodes x,y is denoted by d(x,y).

d(G) = diameter of graph G.

$= \max \{d(x,y)\}\ (x,y,)\ \gamma\ V(G)$

In a boolean n-cube multiprocessor, each node has n adjacent nodes and (2^n-n-1) nonadjacent nodes. Number of arcs = $n.2^n/2$. A pair of nodes is assigned fixed delay paths and alternate routes with different delays. Fig. 3a,b show the hypercube structure. Fig. 3c shows a hypercube with n=5 drawn to show the embedded ring and alternate routes between pairs of nodes.

As shown in Fig. 4, in a n dimensional cube the n-diagnostic distance of any pair of components is no greater than n+2 when $n \geq 3$. [10].

DIAGNOSABILITY: A system is said to be t-fault diagnosable if and only if each processor in the system can correctly identify all t faulty processors.
TRUSTABILITY OF DIAGNOSIS: It is the probability that the diagnosis system yields correct diagnostic result. A fail-safe diagnostic system is one which can correctly locate K faults and does not incorrectly locate if the number of faults is more than K. Then trustability of diagnosis = 1.
A non-fail-safe diagnostic system is one which can correctly locate upto K faults, but can not guarantee correctness of the result when number of faults is more than K. S connected K-graph is at least S-diagnosable. The diagnosability of a given K-graph can be computed by standard algorithms for connectivity analysis of graphs [12].

K DIMENSIONS	N NODES	n CHANNELS
0	1	0
1	2	1
2	4	4
3	8	12
4	16	32
5	32	80
6	64	192

$N = 2^K$; $n = K.N/2$

FIG.3a HYPERCUBE STRUCTURE.

TOPOLOGY

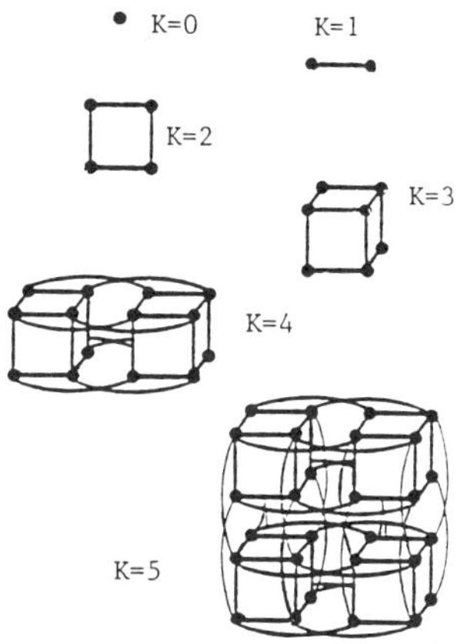

FIG. 3b.

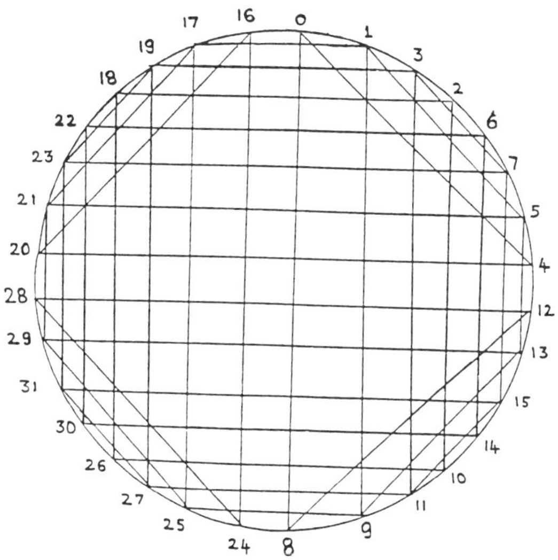

FIG.3c HYPERCUBE WITH K=5 DRAWN TO SHOW EMBEDDED RING.

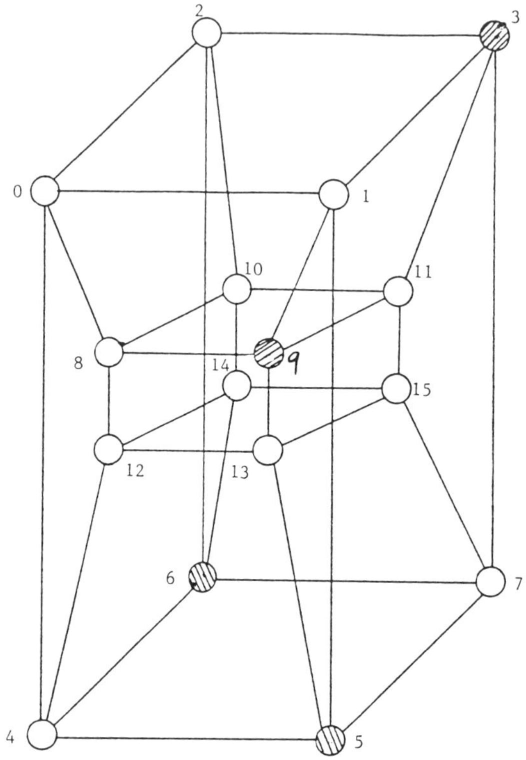

DIMENSION = 4

FAULTY PROCESSORS = 3,5,6,9.

Diagnostic Distance = D(1,7) = n+2 = 6.

FIG. 4 DIAGNOSTIC DISTANCE IN A 4 DIMENSIONAL CUBE.

<u>Surviving Graph</u>: Multiprocessor network is modeled as a graph G(V,E) with V nodes and E links. Let F be a set of fault nodes. Define G' as a graph of (V-F) nodes and links E' i.e., set of edges of G omitting those edges of G which have at least one end in F. G' is called a surviving graph (Fig. 5a,b). Routing P(x,y) is said to avoid F, if no internal node of P(x,y) is a member of F.

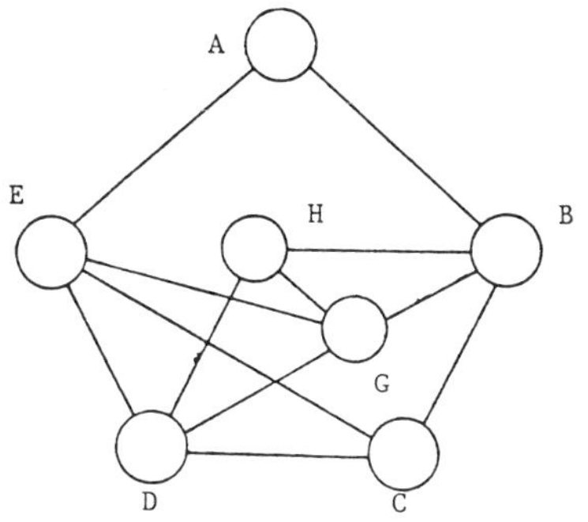

FIG.5a GRAPH G(V,E)

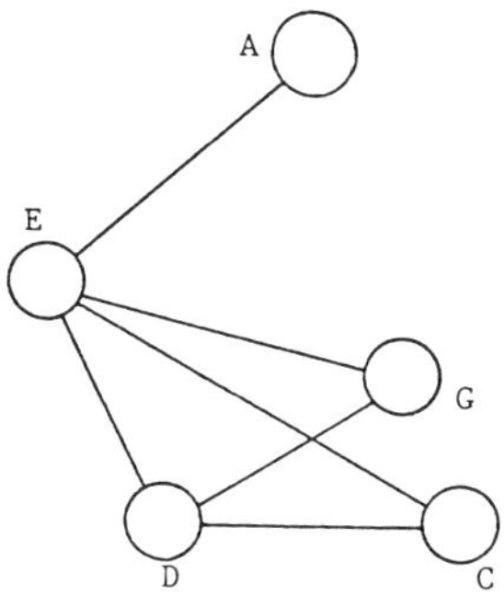

FIG. 5b SURVIVING GRAPH G' WHEN FAULTY NODES ARE H AND B.

Diameter of a surviving route graph was studied for some networks by A Broder and Dolev et al [8,9].

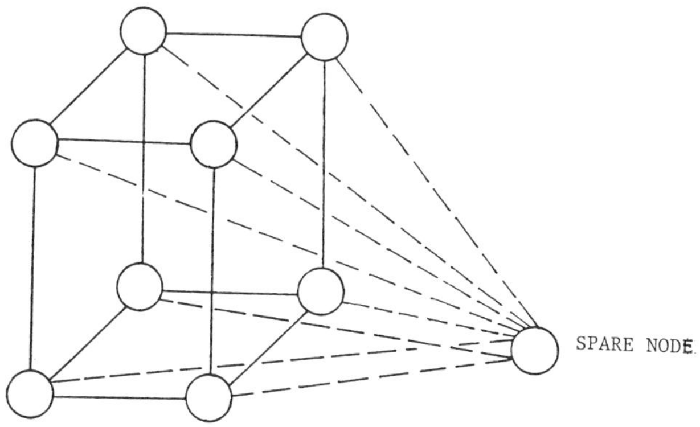

Fig. 6. Connection of spare processors to a 3-cube network.

Fig. 6 shows the connection of spare processor to a 3 dimension cube network [6]. Fault recovery is by reconfiguration of fault free nodes and spare nodes, to obtain the original interconnection graph. If there are no spare nodes, reconfigure the nodes to avoid the faulty nodes. This is called configurations with graceful degradation or fail-soft. Both types of reconfiguration is allowed only if the fault free nodes form a graph (called surviving graph) which is a subgraph of the original interconnection graph.

A practical self-checking multiprocessor for train control applications is described in [3]. A reconfigurable multiprocessor for reliable control of a six legged walking machine is described in [11].

Conclusion:

In this paper a few of the architectural, software and hardware schemes, problems and solutions encountered in creating a fault-tolerant multiprocessor system for robotics applications are presented. These include self-checking computer design, use of natural redundancy available in a multiprocessor with minimum degradation in performance, employing multilevel redundancy and external fault recovery management with high reliability. We also presented a graph model and fault model for a simple multiprocessor system.

REFERENCES

1. J. G. Kuhl and S. M. Reddy, "Distributed fault-tolerance for large multiprocessor systems," Proc. Symp. Comp. Arch. 1980, pp 23-30.

2. C. Y. Chiang and C. Wu, "Fail safe distributed fault diagnosis of multiprocessor systems," Int. Conf. Parallel Processing, 1986, pp 358-365.

3. W. G. Marshall and W. Forsythe, "Self checking multiprocessor module for train control applications," Electronic Letters, 1981.

4. J. Opatrney, N. Srinivasan, and V. S. Alagar, "Construction of large fault-tolerant communication network models," IEEE 1986 Fault-tolerant Computing Proceedings, pp 110-115.

5. J. Kriz, H. Sugaya, "Knowledge-based testing and diagnosis of analog circuit boards," Int. Symp. IEEE 1986 Fault-tolerant Computing Proceedings, pp 378-383.

6. D. A. Rennels, "On implementing fault-tolerance in binary Hypercubes," Int. Symp. IEEE 1986 Fault-tolerant Computing Proceedings, pp 238-243.

7. C. R. Das and L. N. Bhuyan, "Reliability Simulation of Multiprocessors Systems," 1985 Int. Conf. Parallel Processing Proceedings, pp 591-598.

8. A. Broder, M. Fisher, D. Dolev, B. Simons, "Efficient fault tolerant routings in networks," Proc. of 16th Annual ACM Symp. on Theory of Computing 1984, pp 536-541.

9. D. Dolev, J. Halpern, B. Simons, R. Shong, "A new look at fault tolerant Network Routing," Proc. 16th Ann. ACM Symp. Theory of Computer 1984, pp 526-535.

10. S. C. Chu and J. R. Armstrong, "Fault diagnosis by testable diagnosis array," IEEE Fault Tolerant Computing Conference 1986, pp 304-309.

11. F. Ozguner and M. L. Kao, "Are configurable multiprocessor architecture for reliable control of robotics systems," Int. Conf. Robotics and Automation, 1985, pp 802-806.

12. C. Yang and G. Masson, "An efficient algorithm for multiprocessor fault diagnosis using the comparison approach," IEEE Symp. Fault Tolerant Computing, 1986, pp 238-243.

13. F. P. Preparata, G. Metze and R. T. Chien, "On the connection assignment problem of diagnosable systems," IEEE Trans. Electron. Comput., Vol. EC-18, pp 848-854, Dec. 1967.

AUTOMATION AND INNOVATION IN MINING

A Concept of a Rock Imaging System for an Underground Computer-Controlled Drilling Machine

S.A. Johnson*, D.T. Borup†, and V.J. Hucka‡

*Adjunct Associate Professor of Electrical Engineering and Research Associate Professor of Bioengineering, University of Utah, Salt Lake City, UT, 84112-1183, USA
†Postdoctorate Fellow, Department of Bioengineering, University of Utah, Salt Lake City, UT, 84112-1183, USA
‡Professor of Mining Engineering, University of Utah, Salt Lake City, UT, 84112-1183, USA

ABSTRACT

An innovative and novel concept of an underground drilling machine based on the existing jumbo-type rock drilling machine is presented. The focus is on the computer control using feedback from drill sensors and nonintrusive imaging of the rock beyond the working face. The degree of computer control may be selected by an onboard operator in one configuration or by a remote operator in a second version.

The percussive and rotary action as well as the thrust, torque, and depth of penetration on several booms may be monitored and controlled simultaneously. The feasibility of providing nonintrustive imaging and remote sensing by acoustic and high frequency electromagnetic sources and receivers is discussed as well.

Ideas are explored as to how these sources and sensors may be incorporated into a special sensing system. The data from the sensing system is passed to a special computer optimized in features to apply imaging algorithms that detect and display graphical plots of images of the internal rock features, discontinuities and nonconformities. The remote sensing and computer control aspects will be synergistic with parallel development in standard mainstream robotic control and machine vision.

1. INTRODUCTION

Many attempts have been made to use sensing systems for detection of rock discontinuities or of cavities in the rock mass for mining purposes. A ground penetrating radar system (GPRS) has been tested by the U. S. Bureau of Mines in central Florida to detect subsurface cavities in the phosphate district [5].

The possibility of providing the drilling machine with nonintrusive imaging and remote sensing by acoustic, microwave and high frequency electromagnetic energy sources is discussed. The imaging algorithms are based on solving or "inverting" fundamental nonlinear integral equations that model the propagation and interaction of the incident energy fields in the rock mass. For some applications requiring only information on cracks, voids or interfaces, linear approximate solutions suffice. However, for determining the inhomogeneous rock properties, full nonlinear solution methods are required.

The basic equations modeling the propagation and interaction of these radiative energy fields in the general rock masses surrounding the tunnel are presented as well. Progress in solving these equations for rock properties and for image reconstruction is also discussed with numerical simulation examples. Initial imaging results showing inverse scattering images of acoustic bulk moduli, density, and absorption have been obtained [3]. Computing these images presently requires several minutes on a supercomputer. Electromagnetic dielectric constant and conductivity (using a scalar wave equation approximation) in two dimensions are also shown. Cross-sectional images of up to 64 by 64 pixels have been obtained.

2. HISTORY OF UNDERGROUND DRILLING

The purpose of blast drilling in mining is to disintegrate the rock by use of explosives. The invention of the first mechanical drill and the introduction of gunpowder dramatically improved mining productivity. In 1844 air-powered drill hammers were used in mines, and the consumed air had the advantage of contributing to the ventilation of the mine. Generally, the drilling methods can be classified according to the work of the bit as percussive drilling, rotary drilling (for soft rocks such as coal), and combination of rotary and percussive drilling.

The need for faster and more economical means to drill holes in hard rocks led to the development of drilling machines, particularly useful for driving tunnels. The first percussion drilling machine ever made was described by [9]. In 1849, a drilling machine was designed by J. J. Couch of Philadelphia. Compressed air was used to power a drilling machine to drill boreholes in a 12.8 km tunnel between Modane in France and Bardonecchia in Italy. Many other machines were designed and manufactured since the time of Couch's rock drilling machine.

The first generation of hydraulic rock drilling machines was unveiled in 1973 when two or more percussive rock hammers mounted on a mobile carriage created a so-called "jumbo drill." The efficiency of hydraulic jumbos, however, still largely depends upon the driller since it is difficult to simultaneously control all the integrated factors that affect the drilling (profile of the tunnel, cuts, location and alignment of drill holes and their spacing using electronic or mechanical methods).

3. COMPUTER-CONTROLLED JUMBO DRILLS

On the occasion of the 1986 Mining Show in Las Vegas, Nevada, two computer-controlled drill rigs were simultaneously introduced: The Datamatic rig, manufactured by the Finnish company Tamrock, and the Robot Boom drill rig, manufactured by Atlas Copco in Sweden. Both are fully programmable hydraulic percussive drilling machines [1,2]. Before drilling, a laser beam is used to position the boom (Fig. 1). The computer installed in the machine can then produce a drilling pattern relative to the correct coordinates furnished by the laser beam; the alignment is fast, because it does not depend on the position of the rig carrier, or its direction. The drilling pattern, showing drilled holes, the hole being drilled, and the next hole to be drilled is monitored by a computer program. The control system provides the driller with a magnified display of the cut and penetration rates.

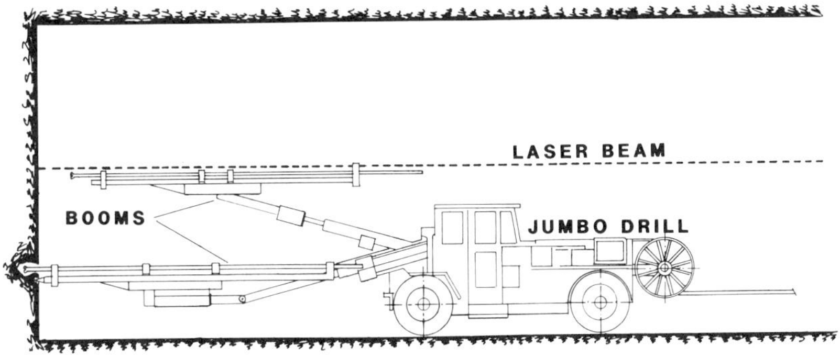

Figure 1. Jumbo drilling machine with laser alignment

In practical applications, the driller uses manual control for positioning the booms in the desired direction while the computer calculates and controls boom joint positions. The location, length, and direction of each hole and parallelism of individual drill holes are predetermined by use of the computer program. For this purpose, data on rock conditions must be known for programming the drilling cycle. Further, on the basis of the known rock conditions, the computer program can assess the drill bit wear and rig service requirements. Working conditions thus are improved by easier operation of the controls. The computer control also allows an easy excavation of curved walls and directionally curved tunnels. The control system is located in a sound-and-vibration-proofed cabin.

However, there is still a need to provide the information on the rock conditions which are indispensable for preparing the program. In changing strata, the computer input must be corrected, which takes time. This time can be drastically reduced if the drill rig could be equipped with a system which continuously detects the strata in front of the machine during the drilling and feed the data to the computer. Such information then could be delivered into a computing center where the modified program could be prepared and sent to the control system of the drill rig.

The results could also be recorded on a magnetic tape for further evaluation. The accuracy of drilled boreholes contributes to elimination of excessive overbreaks which results in less loading and transporting of waste. The precise length of boreholes permits an increase in the distances between the holes, thus reducing the number of holes to be drilled. The result is then the use of fewer explosives which improves the economy. In the following sections a concept of remote sensing is discussed which could be incorporated into such modern computerized drilling machines.

4. SENSING SYSTEMS FOR PRE-EXCAVATION SUBSURFACE OF TUNNELS

The feasibility of the ground penetrating radar was tested for use as an additional tool to detect underground cavities in dam sites of the Florida phosphate settling area. The Florida State Regulations require the identification of subsurface conditions. This is done by use of boreholes. The boreholes are drilled on 152-meter centers by use of 30.5 meters deep. A more accurate and economical method is investigated in order to reduce dam failures due to sinkhole development.

Ground penetrating systems have been tested in geological exploration. According to [5] in salt domes, where the radiofrequency attenuation is lower, discontinuities have been detected at about 975 meters.

The Bureau of Mines uses a microwave radar to investigate coal seam thickness, shape, and orientation [10]. Furthermore, the Bureau uses a short-pulse ground penetrating system capable of penetrating 33 m in a coal seam where voids were detectable at 16 m using a frequency of 10 to 60 MHz. In one test, an uncased borehole was detected through 8 m of coal. In these results, the conductivity of coal was high, resulting in poor propagation of electromagnetic (EM) waves [4].

5. NONINTRUSIVE IMAGING

In an ideal case of nonintrusive imaging, a set of sensors is placed on the working face of a tunnel in a way that they collect data to image a two-dimensional plane ABCD (Fig. 2). A specific but common case of a dry granite host rock is chosen as our model. As a secondary rock a vertical vein having thickness of one-fifth of the image plane depth, T, is considered. The spacing of sensors depends on the frequency used in the modeling. For 500 MHz frequency, the spacing would equal to about 0.08 to 0.15 m. Practically, however, it is not possible to install sensors on all four sides of the plane. A series of model tests have thus been executed with sensors located on three and four sides, e.g., on sides AB, BC, and CD of the plane ABCD for the three side case (Figs. 3 and 4).

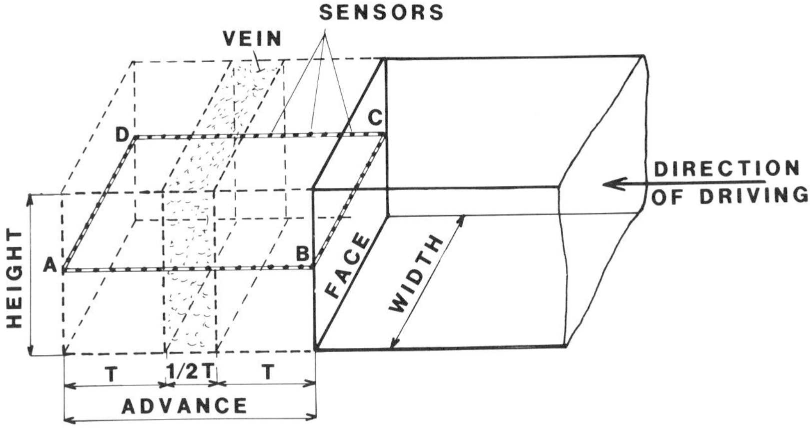

Figure 2 Location of sensors in the working face intersected by a vertical vein

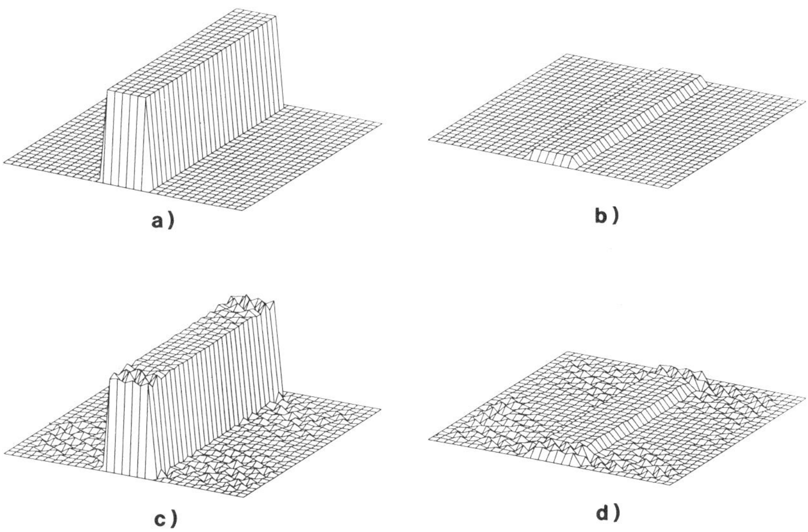

Figure 3 Full view test case (a: Re(γ_1) for the test object; b:-Im(γ_1) for the test object; c: reconstructed Re(γ_1) obtained by the full view imaging algorithm; d: reconstructed -Im(γ_1))

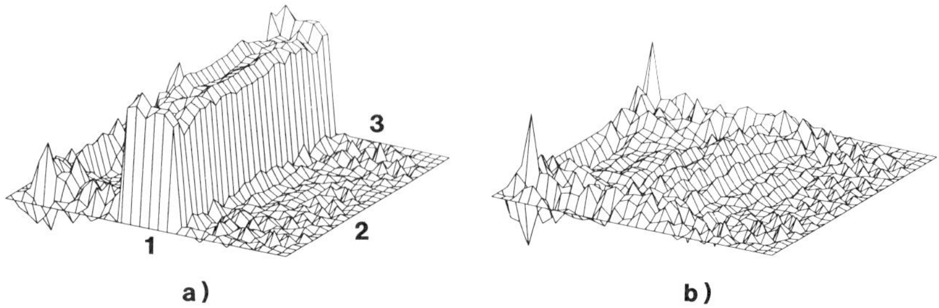

Figure 4 Three-side view test case (a: reconstructed Re(γ_1) obtained by the partial view imaging algorithm; b: reconstructed -Im(γ_1); the sides labeled 1-3 are the three sides along which transmitter and receivers are located)

5.1. Theory for Quantitative Imaging

The commonly used imaging methods based on radar or sonar principles produc qualitative but not quantitative images. For example, the strength of a rada return or echo is proportional to the product of many factors such as the range shape, size and conductivity of the scattering object. By quantitative imaging, we mean a computational method that transforms a set of measurements into a rectangular array of numbers whose values match one to one with some unique material property of a body to be imaged and whose coordinates (position) correspond to points on a plane through that body.

The production of such a quantitative image of one or more material properties requires the existence of **four** essential factors:

(1) an accurate model that relates the material properties of the object to be imaged to the measured data
(2) a set of measured data of sufficient extent and accuracy
(3) a method of inverting the model, either explicitly or iteratively, to obtain the material property from the data
(4) a way of introducing physically meaningful constraints on the possible solution images; especially when the quantity and quality of the data is lacking.

The nature of each of these factors and their interaction in rock imaging problems is described below.

The interaction of acoustic and electromagnetic energy with materials may each be modeled accurately with well-known wave equations. The electromagnetic wave equation model for rock with both dielectric and magnetic inhomogeneous regions is given by a pair of coupled integral equations [7]:

$$\mathbf{E}(\mathbf{x}) = \mathbf{E}^{(in)}(\mathbf{x}) + (k_0^2 + \nabla\nabla\cdot)\int\left(\frac{\varepsilon(\mathbf{y})}{\varepsilon_0} - 1\right) \mathbf{E}(\mathbf{y})\, G\,(\mathbf{x} - \mathbf{y})\, d^Q y$$
$$- i\omega\, \mathbf{curl} \int[\mu(\mathbf{y}) - \mu_0]\, H\,(\mathbf{y})\, G\,(\mathbf{x} - y)\, d^Q y, \tag{1}$$

$$\mathbf{H}(\mathbf{x}) = H^{(in)}(\mathbf{x}) + (k_0^2 + \nabla\nabla\cdot)\int\left[\frac{\mu(\mathbf{y})}{\mu_0} - 1\right] \mathbf{H}\,(\mathbf{y})\, G\,(\mathbf{x} - \mathbf{y})\, d^Q y$$
$$+ i\omega\, \mathbf{curl} \int [\varepsilon\,(\mathbf{y}) - \varepsilon_0]\, \mathbf{E}\,(\mathbf{y})\, G\,(\mathbf{x} - \mathbf{y})\, d^Q y. \tag{2}$$

Here $\mathbf{E}\,(\mathbf{x})$, $H\,(\mathbf{x})$, respectively, are the vector electric field and the vector magnetic intensity at point $\mathbf{x}$. $\mathbf{E}^{(in)}(\mathbf{x})$ and $H^{(in)}(\mathbf{x})$ are the incident electric and magnetic fields and are defined to be the fields present if $\varepsilon(\mathbf{y}) = \varepsilon_0$ and $\mu(\mathbf{y}) = \mu_0$, where $\varepsilon(\mathbf{y})$ and $\mu(\mathbf{y})$ are the respective scalar electric permativity and magnetic permeability at point $\mathbf{y}$. ω is the angular frequency. $G\,(\mathbf{x} - \mathbf{y})$ is the Green's function coupling $\mathbf{x}$ and $\mathbf{y}$. In 3-D, $Q = 3$ and $G(r) = [\exp(-ik_0 r)]/(4\,\pi\, r)$. The time dependence is given by $\mathbf{E}(t,\mathbf{x}) = \mathbf{E}(\mathbf{x})e^{i\omega t}$ and $\mathbf{H}(t,x) = He^{i\omega t}$. In 2-D, $Q = 2$ and $G(r) = H_0^{(2)}(r)/4i$.

The wave equation for elastic waves is given by acoustical mathematical models. Much of present acoustic imaging is based on modeling wave propagation by one or both of the following wave equations:

$$\nabla^2\mathbf{u}_l(\mathbf{x}) + k_l^2(\mathbf{x})\mathbf{u}_l(\mathbf{x}) = 0,\quad \nabla^2\mathbf{u}_s(\mathbf{x}) + k_s^2\,(\mathbf{x})\, u_s(\mathbf{x}) = 0\,. \tag{3}$$

Here $\mathbf{u}_l(\mathbf{x})$ is the longitudinal displacement and $\mathbf{u}_s(\mathbf{x})$ is the shear displacement at point $\mathbf{x}$. The longitudinal wave number is k_l and the shear wave number is k_s. The total displacement $\mathbf{u}$ is $\mathbf{u}(\mathbf{x}) = \mathbf{u}_l(\mathbf{x}) + \mathbf{u}_s(\mathbf{x})$. This model is incomplete because it does not predict coupling between longitudinal and shear waves (mode conversion) at interfaces or flaws. A more accurate model for wave propagation in solids is given by a coupled complex Navier-Stokes-like wave equation which is a function of the total displacement $\mathbf{u}$, the Lame' elastic constants μ and λ and the density ρ :

$$[\lambda(\mathbf{x}) + 2\mu(\mathbf{x})]\,\nabla\,[\nabla\cdot\mathbf{u}(\mathbf{x})] - \mu(\mathbf{x})\,\mathbf{curl}\,[\mathbf{curl}\,\mathbf{u}(\mathbf{x})] + \rho(\mathbf{x})\omega^2\mathbf{u}(\mathbf{x}) = 0\,. \tag{4}$$

Although this equation does describe the observed coupling between shear and longitudinal waves in inhomogeneous materials, it is **only** applicable to **isotropic** materials. We will not treat the interaction of electromagnetic or elastic waves with anisotropic materials in this paper.

For some rocks the magnetic dependence is negligible and we may set $\mu(\mathbf{y}) = \mu_0$ and the terms involving $\mathbf{H}(\mathbf{x})$ and $\mathbf{H}(\mathbf{y})$ may be dropped. Thus, equation (2) may be dropped in this case. Another simplification occurs if only the z-component of equation (1) is retained as in the so-called transverse magnetic or TM mode. The TM mode equation is also a scalar wave equation. Likewise, if the shear wave component is dropped in equation (4), further simplification occurs. Dropping the shear waves produces a scalar wave equation known as the acoustic wave equation [8]. The TM mode electromagnetic and the acoustic wave equations are isomorphic and may be represented by the same wave equation. This equation is

$$\psi^{(sc)}(\mathbf{x}) = \psi(\mathbf{x}) - \psi^{(in)}(\mathbf{x}) = k_0^2 \int \gamma_1(\mathbf{y})\, \psi(\mathbf{y})\, G(\mathbf{x} - \mathbf{y}) d^Q y$$
$$+ \int \gamma_2(\mathbf{y}) \nabla_y \psi(\mathbf{y}) \cdot \nabla_y G(\mathbf{x} - \mathbf{y}) d^Q y . \tag{5}$$

For the acoustic case, let $p(\mathbf{x})$ be the acoustic pressure. Then,

$$\psi(\mathbf{x}) = p(\mathbf{x}), \qquad \gamma_1(\mathbf{x}) = \gamma_\kappa(\mathbf{x}) = \frac{\kappa(\mathbf{x}) - \kappa_0}{\kappa_0} + i \frac{2}{\omega\kappa_0} [\kappa(\mathbf{x})/\rho(\mathbf{x})]^{1/2} \alpha(\mathbf{x}) , \tag{6,7}$$

$$\gamma_2(\mathbf{x}) = \gamma_\rho(\mathbf{x}) = \frac{\rho(\mathbf{x}) - \rho_0}{\rho(\mathbf{x})} . \tag{8}$$

Also note that for the acoustic case, for reasons based on tradition, the time dependence and Green's function are often complex conjugates of those used above for the electromagnetic case. For the TM electromagnetic case, let $E_z(\mathbf{x})$ be the z-component of electric field. Then,

$$\psi(\mathbf{x}) = E_z(\mathbf{x}), \qquad \gamma_1(\mathbf{x}) = \gamma_\varepsilon(\mathbf{x}) = \frac{\varepsilon(\mathbf{x}) - \varepsilon_0}{\varepsilon_0} - i\frac{\sigma}{\omega\varepsilon_0} , \tag{9,10}$$

$$\gamma_2(\mathbf{x}) = \gamma_\mu(\mathbf{x}) = \frac{\mu(\mathbf{x}) - \mu_0}{\mu(\mathbf{x})} . \tag{11}$$

Thus TM mode electric field E_z, permativity ε and magnetic permeability μ are isomorphic, respectively, to pressure p, compressibility k, and density ρ. We have demonstrated methods for inverting for both γ_1 and γ_2. However, for the goals of this introductory paper, we will retain only γ_1, knowing that γ_2 could also be treated simultaneously.

We have previously shown that equation (5), for $\gamma_2 = 0$ can be converted into two simultaneous matrix equations[3,6].

$$R^{(sc)}_{\omega\phi m} = \psi^{(sc)}_{\omega\phi}(\mathbf{x}_m) - \sum_j D_{\omega mj} \psi_{\omega\phi}(\mathbf{y}_j)\, \gamma_{(1)\omega}(\mathbf{y}_j) , \tag{12}$$

$$R^{(fld)}_{\omega\phi n} \equiv \psi_{\omega\phi}(\mathbf{x}_n) - \sum_j C_{\omega nj}\, \psi_{\omega\phi}(\mathbf{y}_j)\, \gamma_{(1)\omega}(\mathbf{y}_j) - \psi^{(in)}(\mathbf{x}_n) . \tag{13}$$

Equation (12) related the data to the field $\psi_{\omega\phi}$ and to γ_1 while equation (13) relates the field $\psi_{\omega\phi}$ at any point $\mathbf{x}_n$ to all points $\mathbf{y}_j$. The indices ω, ϕ, m, j, and n in equations (12,13) refer, respectively, to frequency, source position, detector position, pixel position, and pixel position. The use of these indices is selfconsistent and their bounds are given by $\omega = 1,\ldots,\Omega$, $\phi = 1,\ldots,\Phi$, $m = 1,\ldots,M$, $j = 1,\ldots,J$, $n = 1,\ldots,N$. These equations constitute the discrete model that we seek to invert for $\gamma_{1(\omega)}(\mathbf{y}_j)$ at each grid point $\mathbf{y}_j$. We have described ways of inverting this model in previous papers, and the details will not be given here. The general idea is to find the column vector $\mathbf{v} = [\gamma_{(1)\omega}(y_j),\ \psi_{\omega\phi}(y_j)]^T$ that minimizes an objective function, or functional $F(\mathbf{v})$, defined as one-half of the modulus squared of the residual column vector $\mathbf{R}$:

$$F = (1/2)\ \|\mathbf{R}\|^2\ , \qquad \mathbf{R} = [R^{(sc)}_{\omega\phi m},\ R^{(fld)}_{\omega\phi n}]^T\ . \tag{14,15}$$

Thus $\mathbf{v}$ is a column vector of $\Omega N + \Omega\Phi N$ components and $\mathbf{R}$ is a column vector of $\Omega\Phi M + \Omega\Phi N$ components. In most cases $\gamma_{(1)\omega}(\mathbf{y}_j)$ is independent of frequency, thus $\mathbf{v}$ has $N + \Omega\Phi N$ components and a necessary condition for a unique solution $\gamma_{(1)}$ is $\Omega\Phi M \geq N$.

The minimization process is carried out by the iterative steps

$$\mathbf{v}^{(k+1)} = \mathbf{v}^{(k)} + \delta\mathbf{v}^{(k)}\ , \tag{16}$$

where $\delta\mathbf{v}^{(k)}$ is the solution of the linear equation

$$J^{(k)}\delta\mathbf{v}^{(k)} = -\ \mathbf{R}^{(k)}\ . \tag{17}$$

Here $J^{(k)}$ is the Jacobian matrix given by

$$J^{(k)} = (\nabla R)^{(k)} = \left.\frac{\partial\left(R^{(sc)}_{\omega\phi m},\ R^{(fld)}_{\omega\phi n}\right)}{\partial\left(\gamma_{(1)\omega}(y_j),\psi_{\omega\phi}(y_j)\right)}\right|_{v\,=\,v(k)}\ .$$

5.2. Computer Simulation Test Cases

Thus far the invese scattering, imaging algorithm described above has been applied to 2-D problems only. An example of a test object for the electromagnetic case is illustrated in Fig. 3. Because of the isomorphic relationships of equations (5-11), acoustic imaging is therefore treated implicitly. The test problem consists of an infinite background material with dielectric constant, $\varepsilon_{r1} = e_1/\varepsilon_0 = 10.0$ and conductivity, $\sigma_1 = 10^{-3}$ (ohm meters)$^{-1}$ and an inhomogeneous dike of higher dielectric constant $\varepsilon_{r2} = 11.0$ and conductivity $\sigma_2 = 3.88 \times 10^{-3}$ (ohm meters)$^{-1}$. These values are typical for rocks such as granite. Selection of the imaging frequency is dictated by two main considerations. The maximum grid spacing that can be tolerated at this contrast for an accurate reconstruction is approximately $\Delta = \lambda_1/4$ where λ_1 is the wavelength in the background material. Since the number of grid samples is constrained by the available computer memory, the optimum imaging frequency is given by $f = Nc_1/(4D)$ for N x N image grid where D is the depth of the image space and c_1 is the velocity of electromagnetic waves in the background medium. In addition, because the background medium is attenuative, the skin depth must be large enough at the selected frequency so that the depth of penetration is sufficient. For the background properties given above a frequency of 500 MHz was selected for which $\lambda_1 = 19$ cm and so $\Delta = \lambda_1/4 = 4.74$ cm. At present the largest grid that we have been able to use is 64 x 64, but for the problem in Figs. 3,4 a 32 x 32 grid was used. The depth of the image is $D = N\Delta = 1.5$ meters. The skin depth at this frequency is $\delta = 17$ meters and so penetration is not a problem. For these parameters

$$\gamma_1 = \frac{\varepsilon_{r2} - \frac{i\sigma_2}{\omega\varepsilon_0}}{\varepsilon_{r1} - \frac{i\sigma_1}{\omega\varepsilon_0}} - 1 = 0.1 - i0.01$$

inside the dike and zero outside. The top two plots (a) and (b) in Fig. 3 show the real and imaginary parts of the test object.

For the first reconstruction, the image space was surrounded with 32 cylindrical wave sources evenly spaced along all four sides of the grid boundary. Receivers were located at all 124 boundary points. The real and imaginary parts of the resulting reconstruction using the previously described algorithm are shown in (c) and (d) of Fig. 3. The maximum point error in the real part of the reconstructed image is 8%. The imaginary part, which is ten times smaller than the real part, has a maximum point error of 60%. The largest errors occur near the boundaries of the grid. The computer time needed to generate this image was approximately three minutes on a Cray-XMP supercomputer.

For the actual mining problem, it is clearly not practical to position transmitters and receivers all the way around the image space. Fig. 4 shows the reconstruction obtained for transmitters and receivers located on three sides only. This could correspond to the use of transmitters and receivers loaded into two boreholes on the edge of the tunnel in conjunction with those on the tunnel face. For the case in Fig. 4, 33 transmitters and 94 receivers were evenly spaced on sides 1-3 only. As in the previous case, only one imaging frequency was used. The resulting reconstructed real and imaginary parts of γ_1 are shown in Fig. 4. The maximum error in the real part is now 20% and the imaginary part is severely distorted. Notice in particular the accumulation of error near the side AD that contains no transmitters or receivers. Nonetheless, the reconstruction accurately resolved the front edge of the dike discontinuity which may be of greatest importance in the mining problem. The back edge of the dike is adequately resolved. The average real part of the dielectric constant is accurate when averaged over a region a few pixels in radius in the center and front portions of the reconstructed image, and this may have commercial predictive value.

CONCLUSION

A concept of a nonintrusive imaging system for a programmable drilling machine type jumbo is presented. The imaging system is meant to provide the computer-on-board with information on rock changes and its discontinuities which are fed into the computer for altering the computer program for the machine.

The comparison of the reconstructions of Figs. 3 and 4 illustrate the well-known fact that the limited view problem (the case where the image space cannot be surrounded with transmitters and receivers) is ill-posed when compared with the full-view case. This represents a major challenge for the mining problem since in this case it may be impractical to locate transmitters and receivers anywhere other than the tunnel face. Experience with 1-D imaging problems and a small number of 2-D problems indicates that this difficulty may be resolved by the use of multiple frequency imaging. Also, the application of constraints such as maximum and minimum allowable values of dielectric constant and conductivity may need to be incorporated to improve image quality.

The extension of this work to data collected on the tunnel face only or to data from the face and one or two boreholes by these additional tools is a subject of future research. We also note that the true imaging problem is three-dimensional and not two-dimensional. Thus, the planar geometry used in our calculations is not representative of the tunnel geometry. Indeed, material is present above, below, and to the side of the borehole transmitters and receivers. The Green's functions suggested herein sample these extended 3-D spaces. In principal the accuracy of solutions to 3-D problems can be significantly improved by use of improved directional sources and matched Green's functions. The quantitative imaging methods demonstrated here represent a large step forward in model accuracy over present seismic and electromagnetic methods which are now used commercially. Therefore, we believe their commercial applications will be proportionally greater. The progressive improvement of data collection geometries sources, receivers, and software beyond that demonstrated here should provide additional incentive for applications for such quantitative imaging methods.

ACKNOWLEDGEMENTS

The study was partly funded by the Mining and Mineral Resources Research Institute of the University of Utah for which the authors are deeply appreciative. We also appreciate partial support from U. S. Public Health Service grants 5 R01 CA29728 and 5 ROI HL 34995.

REFERENCES

1. Anon.: Atlas Copco's Bull., Robot Boomer 135, Las Vegas (Nevada) Mining Show, October 5-9, (1986)
2. Anon.: Tamrock's Bull, 3rd Quarter, (Datamatic: Tamrock's Computer-Controlled Drilling Rig) 8. (1981)
3. Berggren, M. J., Johnson, S. A., Curruth, B. L., Kim, W. W., Stenger, F., and Kuhn, P. K.: Ultrasound Inverse Scattering Solutions from Transmission and/or Reflection Data. Proc. Intern. Workshops on Physics and Engineering of Comp. Multidim. Imaging and Processing, SPIE, **671**, 114-121 (1986)
4. Coon, J. B., et al.: Experimental Uses of Short-Pulse Radar in Coal Seams, Geophysics, **46**, No. 8, 1163-1168 (1981)
5. Church, R. H., and Webb. W. E.: Evaluation of Ground Penetrating System for Detecting Sursurface Anomalies, U. S. Bureau of Mines, RI 9004, 21, (1986)
6. Johnson, S. A., Zhou, Y., Berggren, M. J., and Tracy, M. L.: Acoustic Inverse Scattering by Moment Methods and Backpropagation, Proc. Conf. on Inverse Scattering: Theory and Application, SIAM (Ed. Bednar, J. B., et al.) 144-155 (1983)
7. Jones, D. S.: Methods in Electromagnetic Wave Propagation. Clarendon Press, Oxford (1979)
8. Morse, P. M. and Ingard, K. U.: Theoretical Acoustics, McGraw-Hill Book Co., New York (1968)
9. Stack, B.: Handbook of Mining and Tunnelling Machinery, John Wiley, New York, 742 (1982)
10. Wait, J. R., Chang, D., Lytle, R. J., Ralston, M. A., Umashhankar, K. R., and Wittman, R. C.: Analytical Bases for Electromagnetic Sensing of Coal Properties, U. S. Dept. of Energy, DOE FF 8972-1, July, 34 (1978)

Application of Artificial Intelligence and Robotics to Mining

C. Raman Janakiraman*, and B. Manas Das†

*Advanced Computing Technology Centre, The University of Calgary, Calgary, Alberta, Canada
†Coal Research Laboratories, Canada Centre for Mineral and Energy Technology, Federal Department of Energy, Mines and Resources, Devon, Alberta, Canada

ABSTRACT

Artificial Intelligence (AI) and Robotics technology have recently taken on an international flavour and become a significant topic in engineering research. Robotics has a great potential for assisting in the complex task of extracting and processing minerals from the ground. Just as the advent of computer technology has made a significant impact on computer aided design and planning, the advent of robotics will make a significant impact on the mining industry.

This paper will provide an overview of the rapidly evolving fields of artificial intelligence and robotics. The development of new theories and techniques and the potential of both these fields in mine design and planning are examined. A selection of areas for engineering applications of interest to the Canadian mining community are described.

1. INTRODUCTION

In recent years, Canadian research and industrial organizations like the National Research Council (NRC) and Canada Centre for Mineral and Energy Technology (CANMET), a Federal Government Division (Bartel, 1983) have investigated how Artificial Intelligence (AI) techniques and robotics technology can be applied to problems in mining. The application of high technology is one of the most effective methods for improving productivity and ensuring safe working conditions in a hazardous occupation such as mining. The idea of totally automating the extraction process, although not new, was not believed to be practical. Now, the fully automated and unmanned mine is probably only a decade away. Automation of coal mines began many years ago with the introduction of mechanical haulage devices, conveyors, and shearers on the coalface. In the mid -1950s the drive to mechanize underground operations, particularly those associated with the coalface, was accelerated, and by the mid -1970s the process was virtually completed. In parallel with this mechanization, attempts were made to automate coal transport systems and, to a lesser degree, monitor coalface operations. While some success was achieved with the former, little progress was achieved with the attempts to monitor coalface operations. These early failures can be attributed to (1) the methods in which monitored information was presented, (2) technical limitations of the equipment then available and (3) lack of communication technology to collect, transmit, process and display data.

The focus of this paper is on the applications of robotics to mining. The evolution of automated systems into robotic systems has taken place in two stages, as shown in Figure 1, as far as industrial applications are concerned:

-The first stage consisted of robotic systems distinguished by their versatility and flexibility; a large number of repetitive tasks were programmed into the robots. Most of the robots in use today fall into this category.
-The second stage of development began recently with the interactive robot, which interacts with its physical environment. This robot can be thought of as an "AI robot" in the sense that the interactions are due to the intelligence built into it.

In this paper, we refer to robotics in both senses and include processes associated with the robots. There are three main aspects of mining in which robots can be used: exploration, production and aiding humans with difficult or dangerous tasks. Using robots during exploration operations allows work to be carried out in hostile environments - for example, under water, in an underground mine, in radioactive areas, or in high temperature/humidity environments.

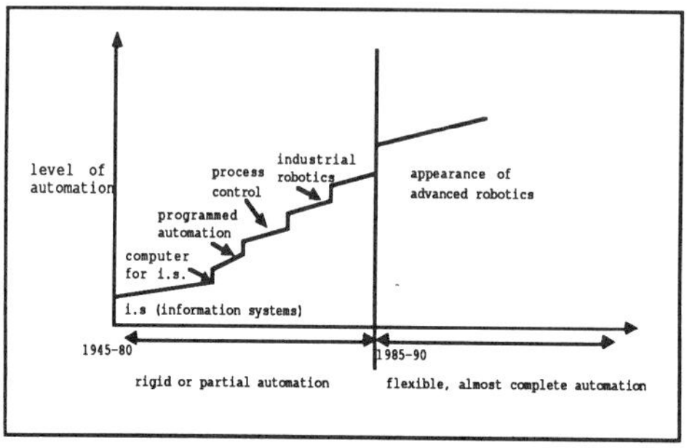

FIGURE 1. Graph showing the development of robotics technology

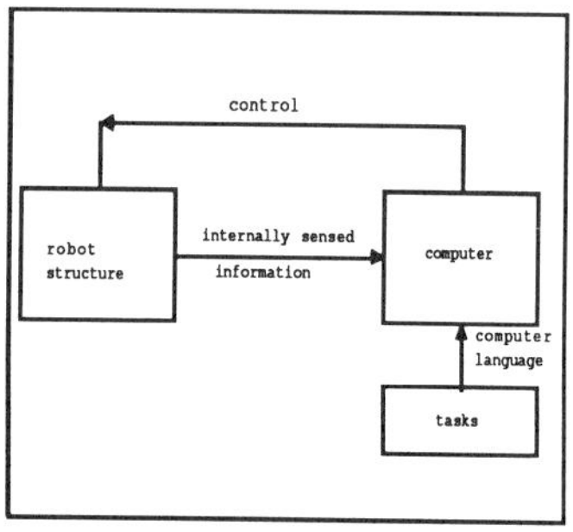

FIGURE 2. Schematic representation of of a typical industrial robot

This paper describes the mining applications of robotics, and then explores the possible use of teleoperation in mining, partly for exploitation, but mainly for the maintenance of automated systems in mines. The problems of determining how much automation is really needed for the integration of advanced technology to mining are then discussed, and various approaches are described.

2. ROBOTICS DEVELOPMENT AND INDUSTRIAL APPLICATIONS

Coiffet (1983) mentions that robotics developed as a result of the desire to create a device that can replace man in everything he cannot do or does not wish to do himself, but still needs or wants to do, without presenting any threat to his authority. See the schematic of typical industrial robot in Figure 2.

It has been emphasized in many publications and by different robot manufacturers that in the long run robots can do certain mining jobs faster than human operators, they are more consistent and more reliable than humans, and that they can work in hazardous environments. Moreover, they cost less than employing human workers for repetitive jobs - obviously there are tasks which only humans can solve. As pointed out by Strong (1980), unfortunately this requires a lot of training and hard work.

The environment in which a robot is placed and where it operates defines the surroundings for a fixed or variable position. In the environment, the robot encounters obstacles that are to be avoided and objects of interest upon which it must act.Thus, interaction is established between the robot and the environment. For example, if a robot is to operate a drilling machine which can bore through rock at angles of over 60 degrees to the vertical, by gripping the sides of its hole, it must drill out a new space and then move into it. In this way it can drill around corners, move in a curve, drill sideways or even uphill, allowing access to locations where it is not practical to drill straight down. The action of the robot being evaluated for rock drilling should monitor the drilling and rock conditions in an effort to forestall drill failure due to overload. A versatile 2-boom drill rig introduced last year can operate in a mine heading completely unattended.

Real robots possess, in varying degrees, two characteristic properties: versatility and adaptability to their environment.

1.Versatility: the structural/mechanical potential for performing varied tasks and/or performing the same task in different ways. All existing robots possess, to varying extent, this quality.

2.Adaptability: the ability to perform a task in a changing environment. Some of the roles for industrial robots described in this paper require that the robot be aware of the mining environment. Robot sensors developed so far are adequate for interpreting the environment in manufacturing industries, but considerable research and development work is required before they can be applied in mining.

The mining environment is very different from the environment in a factory, in which industrial robots currently function. Two characteristics of mining environments which impose demands on the designer are dust and the confined space in the mine headings where work is carried out . For this reason, as shown in Figure 3, it is necessary for the robots used in a mining environment to have more adaptability and "intelligence" than robots in other scenarios.

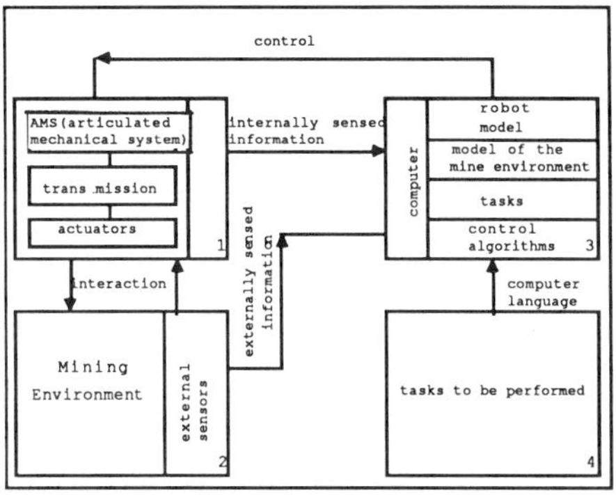

FIGURE 3. Schematic representation of robot function

A typical mining robot, therefore, consists of a robot structure with a computer (AI or knowledge-based system) built into it. In practice such mining robots can be used if the tasks to be performed are initially repetitive. When automatic environmental analysis is fully developed, a reliable robot that can react to its environment is possible. Mining robots equipped with limited performance visual systems or stress and strain measurement systems at affordable cost are now being developed, as described by Chironis (1986). These robots will be able to sense changes in their environments, and compensate for these changes in their performance of tasks.

3. MINING APPLICATIONS

Work has recently started in Canada (Pathak,1983) to evaluate and identify potential areas of application of AI and robotics in serving the industry needs. Research is focussed on Canadian hard rock mining, machinery manufacturing and services provided.

Robots can be employed either autonomously or in a teleoperation system, which involves a human directing the robot by remote control, shown in Figure 4.

Autonomous robots in a hostile environment could perform pre-programmed tasks, such as face monitoring, ground control and support, haulage control, and environmental control. At present, autonomous robots are employed to do only simple tasks, for they do not have AI built in to allow them to think and independently interpret their environment. Therefore, in current practice, teleoperation systems are used most often in exploration work. The potential role for AI and robotics systems is included in Table 1.

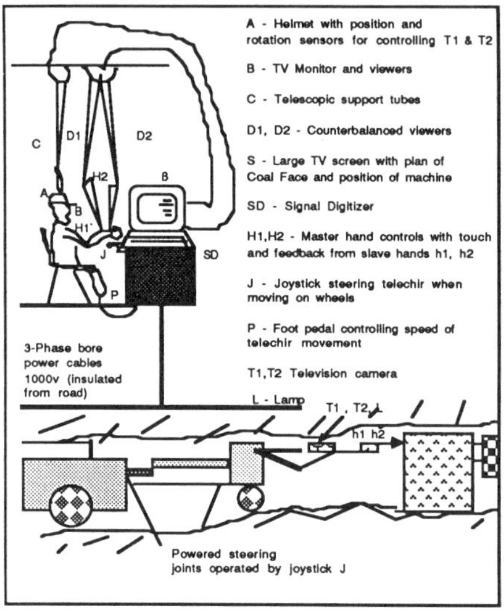

FIGURE 4. M.W.Thuring's concept of mining-teleoperation (From Thuring (1983) Robots and telechirs, Ellis Harwood)

Table 1. Potential role for AI and robotics systems in mining

Automated haulage systems;
Methods for the quick detection of methane;
Machines for detection of the divide between coal and rock;
Robots for manless mining, including robots for use in mine zones with high hazards (e.g., in zones with high methane emission, faults, or rock falls);
Automatic systems for the extraction of minerals at great depth;
Communication devices for underground mining and hauling operations (e.g., cellular phones, fibre optics, etc.);
Videosensors for environmental feedback; and
Undersea mining.

Without aiming to provide a full list of applications in this rapidly changing field, the major application areas for industrial robots in mining according to Bartel (1983) are contained in Table 2: It is important to remember that robotics assembly, or other system design processes for the mining industry, must start with the analysis and very often with the redesign of the parts to be assembled. Design for mining and for processing is of crucial importance. In this area CAD/CAM systems need to work with AI (expert systems) and should become increasingly intelligent.The best solution would be to take into account functional constraints and maintenance restrictions imposed by the mining environment. All activities at a mining site are to some degree interdependent. It is therefore essential that account is taken of the possible consequences of applying remote control and monitoring to a particular mining activity on other activities contained within the overall system. For example, if the effect is to cause an improvement in the efficiency and hence throughput of the haulage system beyond that with which the face operations can deliver, then it is probable that some further action will be required in the design of AI systems for face and haulage control.

Table 2. Application areas for industrial robots in mining

* Pick-and-place type operations, in which case point to point controlled devices are usually sufficient: machine loading and unloading, simple palletizing and warehouse service.
* Machining robots (which are the largest in terms of population) for spot welding of mining equipment.
* Machining robots for drilling and face monitoring operations, which require a rigid body and arm structure with adequate positioning repeatability. They often have to change tools and/or hands to accomodate different jobs.
* Cleaning and deburring applications with special purpose tools. Such robot applications often need adaptive force feedback sensing as well as automated robot tool changing.
* Roof bolting and screening machine.
* Automated material handling and inspection which is probably the most interesting and fastest growing area of applications. This is due to the potential of the rapidly growing market of the manufacturing industry and its complexity in programming, tooling, sensory feedback processing, interfacing with other devices and communicating with remote computers.

4. PROBLEMS IN THE INTRODUCTION OF ROBOTICS TO MINING

There are some problems, however, in introducing AI and robotics technology to the area of mining. Applications that affect mining are not readily available at this time, and technology would have to be adapted from other areas. Only recently have commercial industrial robots and tools become available. Also expertise is in short supply, compared with the number of potential applications to mining.

Pathak (1985) discusses the numerous factors unique to mining which impede the rapid introduction of new technology like robotics to mining. Hostile underground conditions prevent the easy introduction of robotics technology. Other factors that prevent rapid introduction include (1) the low level of awareness of potential applications, (2) the inefficiency of present methods based on conventional technology, (3) the fact that the introduction of high technology into mining is capital intensive, and (4) the dearth of technological talent in the industry.

There are numerous factors that are crucial to the success of an AI and Robotics proposal for the remote control and/or monitoring of mining processes. These include:

1.The level of awareness of the new technology and the perceived need by the user.
2.The introduction of any new technology must make a significant contribution to the improvedefficiency of the mining processes.
3.Lack of understanding of the capabilities and limitations of the evolving technologies.
4.The benefits to be gained from introducing these techniques do not always manifest themselves in terms of immediate financial return.

Present day robots do not possess a high degree of flexibility. The interactive and changing nature of mining operations demands careful consideration to possible future use. Memory capacity, storage devices (e.g., diskettes, disks, laser disks, etc.), software availability, knowledge-based shells, communication links (Local Area Networks, Network Access Devices), and engineering work stations all have to be evaluated in the light of the requirements of mining applications.

In addition, the advent of affordable computing has made it necessary to pay careful attention to the proper role of humans as well as computers. Several factors require careful consideration at this stage. As the industry structure changes human workers will move to new areas to assume increased responsibilities. This in turn will require computer aided training in robotics technology.

5. BENEFITS OF AI AND ROBOTICS TO MINING

The benefits that can be derived from the implementation of AI and robotics in mining will, of course, depend upon the particular mine environment in which the system is used. The following general benefits in relation to a hypothetical scenario based on AI applications are expected: (a) Continuous monitoring of the mine environment; (b) Immediate warnings of hazardous conditions and preventive action automatically enforced; (c) All necessary information is readily accessible for quick action; (d) Automatic shutoff of valuable equipment by sensors to prevent serious accidents and (e) Centralized information access (as shown in Figure 5) for monitoring and control.

The benefits of employing these techniques to monitor underground mines can be itemized as follows: They can be used to keep up with production changes and control machinery; Information can be displayed on computer-based control systems, making communication and remote control in the mining environment possible; Computerized systems can monitor and control conveyor belts, pumps, electrical switchgear, elevators, fans, bunkers and car loading; Flexible and adaptable equipment can be created for coal extraction without human presence underground. In addition, they can be used to deal with hazardous situations. They can monitor continuously for methane gas, carbon monoxide, and other harmful gases, air velocity, air pressure, humidity, dust and smoke. Remote controlled application is possible, with the worker at a safe distance from the face for tunneling in weak and fractured rock. Small mines can benefit by fewer workers in hazardous operations by the use of robots (or at least teleoperators).

6. FUTURE APPLICATIONS IN MINING

In light of these constraints, the current Canadian program is proceeding along the following avenues:

1.ongoing assessment of industry needs and the potential for using AI and robotics techniques,
2.the development of specific equipment in areas where there is a perceived need, and
3.greater emphasis on routine environmental monitoring functions, which could be addressed by robotics.

The future thrusts in the area of robotization include:

1.Adopting and improving technology that is being developed in other areas of engineering,
2.Accelerating development of technology necessary for robotization particularly connected with AI research, microelectronics and automation,
3.Producing devices with new properties (e.g., adaptable robots associated with CAD and CAM systems).

The goal of the government and industry in Canada, as discussed in a paper by Pathak (1985), is to address these problems by identifying AI technology pertinent to the mining environment, and supporting the research and development work in progress at various centres. The ultimate goal is to reduce potential delays in adopting the latest available technology and encourage the smooth transfer from laboratory to the industry.

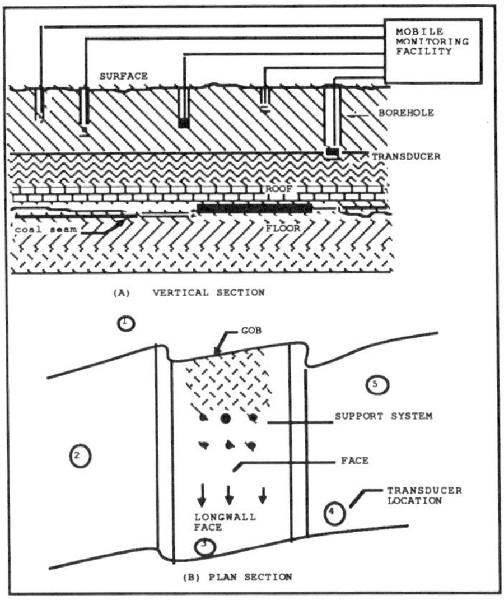

FIGURE 5. Simplified vertical and plan sections of a longitudinal coal mine and a typical surface geophone array for ground monitoring

The National Research Council of Canada is acting as coordinator (Burtnyk,1986) for Canadian participation in international programs in advanced robotics, with the focus of interest on potential mining applications. It is now clearly evident that if the aim is automation, the provisions and installation of expensive AI systems to control and monitor mining functions must be carefully evaluated. During the planning stage a full examination must be made of the system which is to be controlled and monitored, and subsequent action then taken to bring the standard of the system to a level which will accept automation. In Canada, according to Kasvand (1986), this examination is carried out by senior scientists and committee members of the appropriate disciplines in government, universities, and industry. Engineers from other groups are consulted with a view to integrate international cooperation in research and development applied to automated mining, if required. Action to implement proposals from the International Centre for Research Applied to Mining and Tunneling is carefully evaluated to determine its suitability to Canadian operations.

7. CONCLUSIONS

We have examined the availability of AI and Robotics from a problem solving point of view, and have seen that it can be adopted for mining and processing industries provided certain design criteria to satisfy mining environment are met. Since current robotics research is far from satisfactory, much more work remains to be done in Canada and elsewhere before fully automated mining is feasible.

We have provided a brief review of the many robotics applications planned with some comment on the experience gained with those systems. The industry, country-wide, is experiencing an extremely tough challenge. The need is crucial for improved productivity which will enable Canada to compete with other countries producing energy minerals.

The potential for the application and use of AI and robotics technologies is tremendous. Improvements in productivity are really only part of the story. Other important benefits that will occur are the following:

* cost savings - Reduction not only because of productivity improvements, but also because of designs optimized for mining, control of environment and the use of efficient face monitoring methods;
* product quality - Increased quality control due to reduction in human error and increased reliance on standardized methods and processes;
* elimination of hazardous, unpleasant jobs - Jobs in hostile environments will be automated and replaced by robots;
* increased ability to compete in world markets - Because of all of the above facts, minerals will be produced at a lower cost with higher quality and reliability.

The technology of robotics in mining is now emerging. Experience in the U.K., U.S.A., Sweden, Finland and Canada points to the fact that robotics is more than just a concept. The automated mine will come. The challenge in the 80's is adoption of developments before further implementation takes place. The future of the mineral industry in Canada will depend on how we address the challenge.

REFERENCES

Bartel, E.W., Industrial robotics potential applications to mining, Mining Technology and Policy Issues, American Mining Congress 1983 Mining Convention, September 2-14,1983, San Francisco, Cal.,16pp.
Burtnyk, N.,Scrimgeour, J.E., Udd, and J.,Pathak, Recent Developments in remote mining systems, 88th Annual General Meeting of the CIM, May 11-15, 1986, Montreal.
Chironis, N.P., Bureau continues to pour out useful medley of machines and ideas, Coal Age: 91(2); 60-68, Feb.1986, 4.
Coiffet, P., and M., Chirouze, An Introduction to Robot Technology, Hermes Publishing (France) 1982.
Kasvand, T., The state of robotics and its future in mining, 86th Annual General Meeting of the CIM, April 15-19, 1984, Ottawa.
Pathak, J., Hostile underground conditions prevent the consideration of mining robotics technology, Northern Miner, v.71,n.30,p.b2, Oct. 7, 1985.
Pathak, J., Robotics in mining - a Canadian view, Canada Centre for Mineral and Energy Technology, CANMET, Mining Research Laboratory, Report MRP/MRL 85-65, 12pp., May 1985.
Pathak, J., Application of advanced technologies in underground hardrock mining, Proc. of a Seminar, Robotics and Automation in the Mineral Industry, Nov.1,1983, Ottawa, Canada.
Robertson, G.M., The identification of potential applications for robotics and remote control systems in Canadian mining, Canada Centre for Mineral and Energy Technology, CANMET, Contract Report, July 1982.
Strong, D.R., Robotics and Mining, 82nd AGM, Canadian Institute of Mining and Metallurgy, Toronto, Canada April 20-23, 1980.

Remote Control of Drill and Blast Excavation and Demolition

Robert G. Lundquist*, Carl R. Peterson†, and Allan T. Fisk‡

*The Ohio State University Columbus, OH 43210, USA
†Massachusetts Institute of Technology Cambridge, MA 02139, USA
‡Foster-Miller, Inc. Waltham, MA 02254, USA

ABSTRACT

The development of a remotely-controlled drill and blast excavation system is desirable both because it is potentially more economical than mechanical excavation in the hardest rocks and because it will improve safety. A serious impediment to the development of such a system has been the problem of relocating and reentering previously drilled blastholes to place the explosives. The Integrated-Drill-Load-Shoot system overcomes this problem by loading the explosive through the drill steel as it is removed from the hole. This simplifies the design of the system by permitting the use of existing hardware with little modification.

The system originally arose from the development of a continuous drill and blast tunneling machine. Other applications may be found in such diverse areas as a longwall mining system for hard rock, demolition of structures under hazardous conditions, and even extraterrestrial mining.

INTRODUCTION

The increased use of mechanical excavators for harder rock in recent years is due, in part, to the cyclic nature of conventional drill and blast excavation. Mechanical excavation is not inherently cheaper or safer but is used because it offers faster completion times. This higher advance rate is due to the fact that many operations, i.e. excavation, muck removal, and lining, can be carried out simultaneously. In drill and blast excavation, drilling, charging, wiring, blasting, ventilation, mucking, and roof support are done sequentially. Also, the drilling and mucking equipment each fill the space at the end of the heading so that no other operations can occur and all equipment must be removed to protect it from fly rock during blasting.

There is also greater awareness of the high-risk zone near the face in both mining and tunneling. There have been increasing efforts to protect personnel from this zone either by protective systems such as cabs and canopies or by removing them and using remote control operation of the face equipment.

Previous work by Rapidex, now a division of Foster-Miller, Inc., has concentrated on changes in the approach to drilling and blasting to reduce or eliminate problems due to cyclicity. These changes require remote control of all face operations. The Rapidex concept uses a spiral drill round (Figure 1) which directs blast energy and fly rock laterally toward the walls of the tunnel rather than axially toward the equipment. The Rapidex machine successfully demonstrated the concept using pumpable liquid explosives and an inductively initiated delay blasting cap (Figure 2) pumped into the hole along with the explosives. Problems were experienced with the reliable delivery of measured quantities of these explosives. More serious problems were encountered in remotely relocating holes previously drilled and inserting the loading tube into them. This problem was not as tractable as it first appeared. Part was due to inadequate contrast of the remote viewing system. Another less troublesome problem was excessive flexibility of the explosive hose. These, of course, were fixable. Still another problem occurred when a hole was damaged by the shooting of another hole between the time the first hole was drilled an attempt to load it. This can be prevented only by shooting each hole as it is drilled, which limits the ability to drill and load simultaneously and thus slows the process. The worst problem occurred in holes near the bottom of the face. Imperfect mucking, drill cuttings from holes higher in the face, and water from both drilling and natural sources all collect at these bottom holes. Additionally these holes slant downward away from the face and fill up with sand and water. In conventional drilling and blasting the holes must be uncovered with a shovel and blown out with an air pipe before loading. The conclusion seems to be that is is impossible to reliably relocate and reenter a previously drilled blasthole. A system to overcome this problem is essential to the development of remote control drill and blast excavation and thus to the development of continuous drill and blast systems.

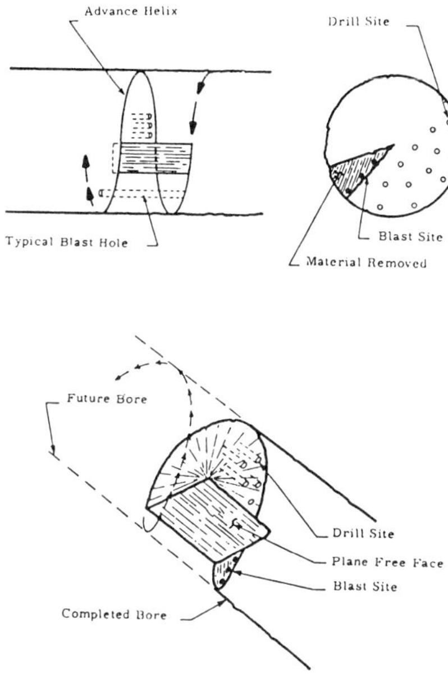

Figure 1. Basic Spiral Round

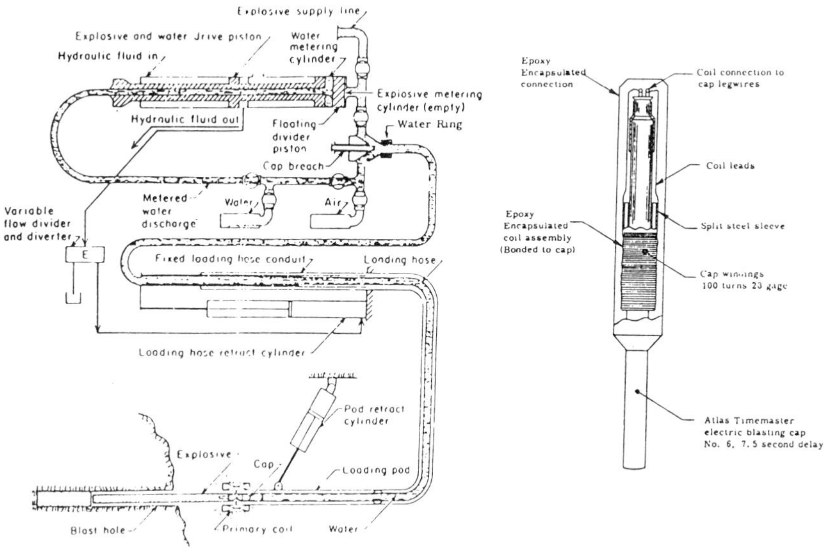

Figure 2. Liquid Explosive System and Inductive Blasting Cap

THE INTEGRATED-DRILL-LOAD-SHOOT SYSTEM

Given the impossibility of sufficiently reliable means to reenter the blasthole and given the fact that a drill steel is hollow and can be used to conduct explosives as well as the cuttings removal fluid, the solution to remotely controlled drilling and blasting becomes obvious. Combining the previously developed concept of a pre-initiated cap with the use of easier-to-meter ANFO explosives, both placed in the borehole through the drill steel as it is removed from the hole after drilling, creates the much simpler Integrated-Drill-Load-Shoot (IDLS) system (Figure 3).

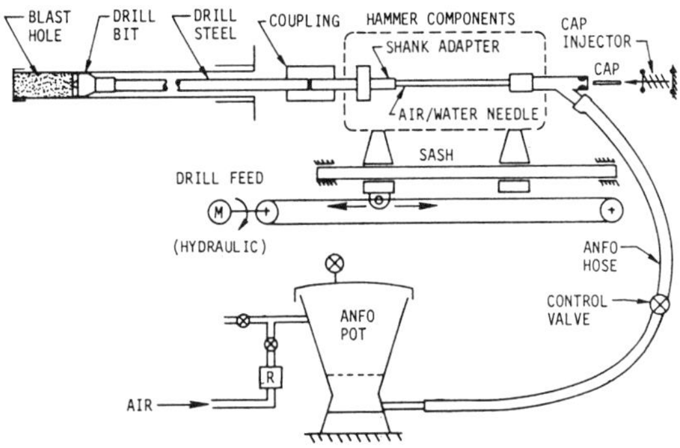

Figure 3. IDLS Schematic

At its simplest, IDLS consists of only three major components plus connecting hoses and valves. Two of these, the percussive drill and the ANFO feeder, are relatively minor modifications of existing hardware. The third, a cap feeder, is unique but quite simple. Special caps (Figure 4) are required but are simpler than the previous inductive caps and can be made by a cap manufacturer with only a small change in the production process.

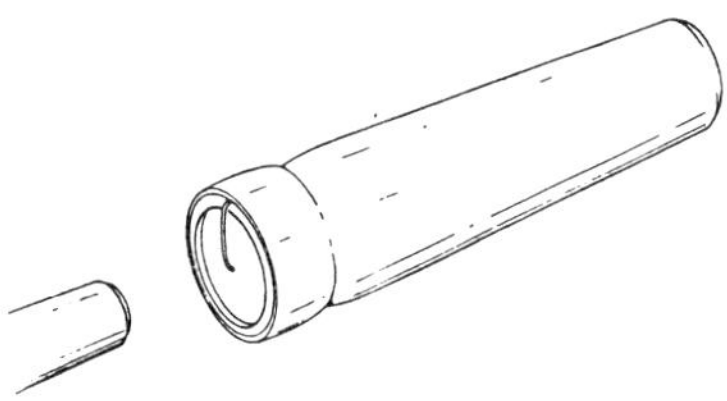

Figure 4. Special IDLS Blasting Cap

The hole is drilled as usual with either air or water flushing. It is then blown clean and dry with air. After a few inches of retraction the air can be switched to ANFO feed. The retraction rate and the ANFO feed rate must be matched to provide proper standoff between the nozzle (drill bit) and the emplaced ANFO column. A few inches before the end of the explosive column and initiated, long-delay, high-strength cap is fed into the airstream; imbedded in the remaining ANFO. Retraction is completed, the drill is moved aside, and after the preset delay time the hole is shot.

Recent tests by Foster-Miller, Inc. have demonstrated the feasibility of the IDLS concept. These tests have shown that blown ANFO can be delivered at reasonable rates through the drill, drill steel, and bit and placed in the bottom of the hole, that the results explosive column is consistently uniform, of a suitable density and particle size, and without voids of gaps, and that a cap of the size needed can also be delivered through the drill, steel, and bit and embedded in the ANFO column. Split aluminum tubes ten feet long were used as simulated boreholes so that ANFO quality and cap placement could be examined.

Control strategies for relating ANFO flow rates to retraction rates were also examined. Despite previous speculation that some complex strategy, such as a constant standoff between the nozzle and the compacted explosive column might be required, a simple approach which coupled a constant retraction rate with a regulated airflow to the ANFO feeder yielding an approximately constant explosive flow rate proved adequate. In this approach the average standoff distance is set by the delay between the start of retraction and start of ANFO feeding. Another even simpler approach in which the drill bit was "Parked" a few inches from the collar of the hole before starting explosive flow also was successful in providing acceptable ANFO placement and is presently the preferred method due both to its simplicity and the added safety in avoiding possible hangups during retraction. Further work on control of explosive placement is planned during 1987.

Also scheduled to begin soon is the construction of a full scale test IDLS excavation system which will conduct further proof of concept testing. Since remotely controlled drilling, ANFO and blasting cap placement, and shooting single holes without a blast shielding bulkhead have all been tested previously only the integrated testing of the entire drill and blast module remains.

APPLICATIONS

The initial impetus to the development of the IDLS system came from the need for improved methods for hard rock tunneling, both for civil construction such as railroad, highway, subway, water supply, sewer, and hydropower, and for the closely related mine development application. A conceptual drawing of a system for tunneling is shown in Figure 5.

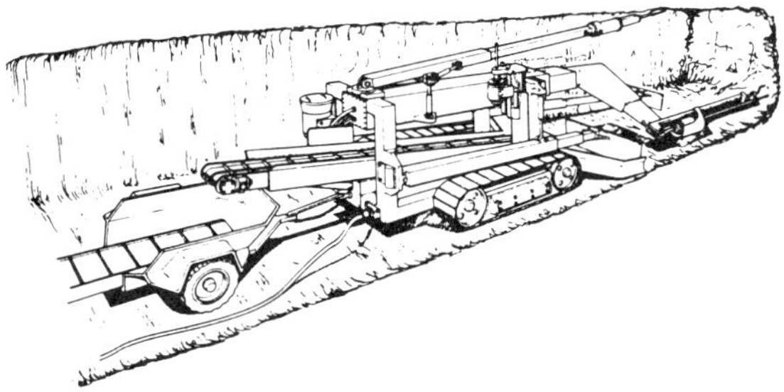

Figure 5. Conceptual IDLS Tunneler

it was originally felt that the use of the spiral round coupled with remote control drill, load and shoot, and much components using television feedback for most control functions could provide improvements in advance rate, economics, and safety in such applications. As previously described, problems with the remote explosive loading hardware diverted research attention away from the development of an integrated tunneling machine. This has resulted in the present focus on the IDLS excavation module which can now be viewed as an independent entity having a applications beyond the original tunneling machine.

The first of these applications is longwall mining. Mechanized longwall has had an important impact on both productivity and safety in underground coal mining. These same benefits cannot be directly applied to the mining of metal and nonmetallic bedded deposits because the mechanical shearers used for coal cannot economically cut hard rock. Longwall has been used with manual drill and blast breakage in hard rock but loses the benefits of mechanization. An IDLS module used in Place of the shearer in a longwall for hard rock (Figure 6), could provide increased safety, productivity, and advance rate even in the hardest rock. A system for continuous drill and blast mining of vein deposits has also been proposed (Figure 7). Other mining applications which have been examined include shaft sinking and raise boring.

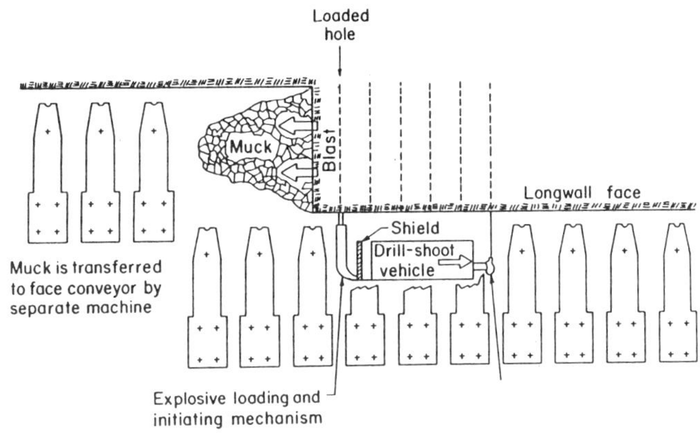

Figure 6. IDLS Hardrock Longwall

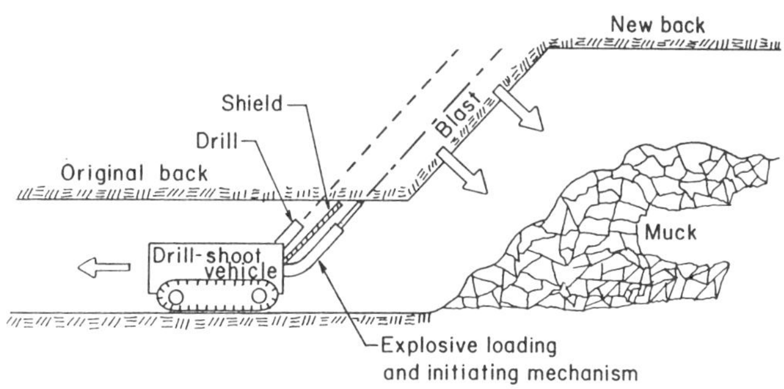

Figure 7. IDLS Shrinkage Stopping

A crawler-mounted IDLS module can provide remotely controlled demolition and excavation services under a variety of hazardous conditions. Some examples will illustrate this concept. The 1986 Mexico City earthquake left many buildings so badly damaged that no repair was feasible. Their very existence in an unstable condition presented a grave public safety hazard which required their immediate demolition. Brave volunteers were asked to enter these building to drill holes and place charges. IDLS could accomplish this remotely. A damaged nuclear reactor presents a similar hazard where remote drill and blast could break reinforced concrete without exposing personnel to radiation. Several recent volcanic eruptions have involved explosive excavation to build or remove lava dams to direct lava streams away from habitations. With careful selection of high temperature explosives and detonators and some care for temperature sensitive parts of the drill and crawler, a remote system could accomplish this task. Sites contaminated by hazardous materials are often excavated by remotely controlled machinery such as trenchers and front loaders. IDLS could add the capability to excavate consolidated materials by these methods as well.

The farthest reaching application of IDLS may be literally out of this world. A project presently underway at Ohio State University is examining possible excavation methods for mining on the moon and asteroids. IDLS is one of several candidates. It offers the advantage of high efficiency in a situation where time may be very limited. Major problems include cuttings removal from the hole and explosive selection for both transport safety and ease of delivery to the hole.

CONCLUSION

Drill and blast excavation still offers many advantages over mechanical methods. The Integrated system permits the automation of drill and blast so that these advantages may be realized in the mining, excavation, and demolition operations of the future.

CAD/CAM AND ROBOTICS EDUCATION/TRAINING

Survey of Robotics in Industrial Engineering Education

Herman R. Leep, P.E. and Deborah L. Anderson

Department of Industrial Engineering, University of Louisville, Louisville, KY 40292, USA

INTRODUCTION

Due to the steady increase in the utilization of industrial robots in the 1980's, additional qualified industrial engineers will be needed to install the robots, and to interface them with other automated equipment. In response to the increase in the utilization of robots, a growing number of universities with industrial engineering (IE) departments are revising their programs to include courses, instructional laboratories, and research facilities in robotics.

The major problems most universities have faced while introducing robotics into the curricula include recruiting new professors due to a shortage of instructors with a background in robotics, retraining current faculty members in robotics, finding good textbooks dealing specifically with robotics, and cooperating with industry [1]. Interaction with industry involves donation of equipment, participation in research projects, and development of cooperative education programs. One source of limited funding for capital equipment and faculty development has been the Manufacturing Engineering Education Foundation of the Society of Manufacturing Engineers. Also, the government has been providing millions of dollars to advance robotics education and research [2].

BRIEF REVIEW OF LITERATURE

To meet the demands being placed on universities by industry, many universities are introducing robotics courses, and are developing instructional laboratories and research centers to provide "hands on" experience to the students. In order to obtain a general overview, the robotic activities of the industrial engineering departments were examined by reviewing the 1985 edition of the American Society for Engineering Education (ASEE) newsletter from the IE Division [3]. Of the 62 IE departments throughout the country which responded, almost half (29) briefly described their activities in the robotics area. At least 19 of these universities mentioned their robotics laboratories. Fifteen schools indicated they had obtained one or more robots within the last 12 months (either through donations or acquisitions). Eight departments described new robotic courses which were being offered or developed at the undergraduate and/or graduate levels. Five departments discussed their involvement in robotics research projects.

Several universities have introduced robotics courses into their industrial engineering curricula. Two examples follow. At The Pennsylvania State University, the Industrial and Management Systems Engineering Department offers two graduate-level courses in robotics [4]. The course dealing with concepts and analysis covers end effectors, vision systems, stability and control, off-line programming, and simulation of robotic systems. The Industrial Engineering Department at The Cleveland State University has introduced a Manufacturing Option into their master's program [5]. One of the fourth-year courses includes the basics of robotics and the writing of control programs.

In a 1984 study by Thompson, 16 IE departments were surveyed concerning their activities in computer-aided manufacturing. Twelve of these universities were using robotic equipment in their laboratories. The instructional labs normally had Mircrobot and/or Rhino educational robots which were relatively inexpensive and required only a limited amount of space [6]. Most research facilities had a few full-sized robots. The major manufacturers of the robots were Cincinnati Milacron, IBM, and Unimation (usually PUMA), but there were a few robots from General Electric, Fanuc, Cybotech, ASEA, DeVilbiss, Prab, and Seiko. Major research projects involved robotic control using sensors and the development of manufacturing cells.

The most outstanding robotics research project reported in the 1985 issue of the American Society for Engineering Evaluation (ASEE) Newsletter from the IE Division was the one at New Mexico State University [7]. The three-year grant involves the development of knowledge-based robots with multiple sensory inputs, and substantial interaction with industry. Five faculty members and five students are supported by the project.

Four laboratories are available at The Pennsylvania State University to complement their robotics curriculum [8]. The equipment includes numerical control (NC) and computer numerical control (CNC) machine tools integrated with material handling and assembly robots. A welding robot and work positioner are also available.

The robotics laboratory at GMI Engineering and Management Institute houses ten full-sized industrial robots from seven different major vendors. Students are involved in designing and developing robotic systems, as well as, programming the wide variety of robots in this lab [9]. GMI also offers workshops for managers, engineers, and technicians [10].

The Center for Robotics and Automation at the University of Arkansas uses its robots for teaching, research, and service [11]. Three large-scale robots and four table-top robots are included in the Center.

Filley visited ten laboratories and study centers of accredited IE programs in every part of the United States, and reported his observations in 1984 [12]. As one example, the Center for Robotics and Integrated Manufacturing (CRIM) at the

University of Michigan combines the research and instructional activities of six engineering departments. The Center was established in 1981 and funding for the first two years was $1.9 million.

Another example is the Computer-Integrated Design Manufacturing and Automation Center (CIDMAC) started at Purdue University in 1982 [13]. The IE component of this Center is the Computer-Integrated Manufacturing (CIM) Laboratory which consists of two work cells. These work cells include a horizontal machining center, robots, a conveyor, and other material handling devices. They are used for experimental verification of concepts in automatic part production, handling, and product assembly.

A third example is the $16 million Engineering Excellence Research Center opened in 1983 at the Arizona State University [14]. The top floor of this six-story facility houses the Center for Automated Engineering and Robotics, which is operated by the IE Department and includes four laboratories.

Oregon State University was able to avoid the high investment cost of equipment, in terms of both capital and space, by developing a joint venture with several companies in the Pacific Northwest [15]. At these companies, students in the five-year program have the opportunity to work with equipment, including industrial robots, by participating in the Manufacturing Engineering Co-op Program (MECOP).

MAJOR OBJECTIVES OF STUDY

The overall goal of this study was to examine the current state of robotics curricula and research in industrial engineering throughout the country. Using research methodology, a representative sample of universities was surveyed to determine the extent to which robotics education is offered, the types of laboratory facilities which are available, and the general directions schools are taking in robotics education and research.

The specific objectives of the study were as follows:

1. Determine the degree to which robotics education is being offered in industrial engineering programs throughout the U.S.
2. Determine the degree to which robotics education is being offered in different regions of the country.
3. Identify methods for introducing robotics education.
4. Identify specific problems encountered during the introduction of robotics into industrial engineering programs.
5. Identify particular successes experienced as a result of introducing robotics into industrial engineering education.
6. Determine future plans and trends concerning robotics in industrial engineering education.

RESEARCH METHODOLOGY

The target of the survey was the 82 universities with an accredited degree in industrial engineering. This population was stratified by dividing the United States into four regions with 20 or 21 universities in each region for geographical distribution. The regions selected were North/East, North/Central, South, and West. (A region name with a slash means states were combined from two different areas to form a region.) The final distribution was obtained by balancing the number of schools in each region while maintaining coherent regions.

A random sample was drawn from each region using a table of random numbers to determine which schools to call. A standard sampling procedure was followed until ten schools were selected from each region.

A structured questionnaire was developed based on the literature review and accepted research methodology procedures. To establish the validity of the questionnaire, it was pretested to determine if the questions were providing the info selected from the literature review to pretest the preliminary questionnaire. The following schools were selected: University of Cincinnati, Kansas State University, University of Louisville, University of Massachusetts/Amherst, The University of Michigan-Dearborn, and University of Southern California.

The procedure for interviewing the participants started with preliminary telephone calls to determine if the university met the requirement of offering at least one robotics course. If so, the appropriate contact person was identified. A copy of the questionnaire and a cover letter were mailed to the contact person approximately ten days prior to the interview to allow that person sufficient time to consider the questions, and to locate the necessary information for the study.

SUMMARY OF RESULTS

The national survey provided insight into the current status of robotics education and research in industrial engineering programs. A summary of the results which were significantly different between the universities is presented in Table I. From the population of 40 industrial engineering departments, 24 offered at least one robotics course (minimum criterion for participation in the survey). The South Region had the largest percentage of participants (8 of 10), followed by the West Region (6 of 10), the North/East Region (5 of 10), and the North/Central Region (5 of 10). The regional concentration of programs may be related to the manufacturing industry activity in the corresponding region.

The majority of the participants represented public institutions. The only private institutions were Lehigh University, Rochester Institute of Technology, and University of Miami. Classification was based on The College Handbook, 1984-85 [16].

Most of the institutions (21 of 24) offered their robotics education solely through courses included in the curriculum. Both the University of Rhode Island

TABLE I - SUMMARY OF RESULTS

REG	UNIVERSITY	ORIENTATION OF ROBOTICS PROGRAM	NO. OF ROBOTICS COURSES	SEPARATE ROBOTICS RES LAB	MAJOR ROBOTICS PROG PROBLEMS
NORTH/EAST	LEHIGH	AR	4	YES	FU SP
	NC STATE	CO AR BR	2	NO	FU SP D1
	RHODE ISLAND	AR BR	3	YES	FA D1
	ROCHESTER	CO AR BR	2	NO	FU FA SP
	WEST VA	EDUC	2	NO	FU FA SP
NORTH/CEN	CINCINNATI	CO AR BR	4	NO	FU SP D1
	CLEVELAND	AR	2	YES	FU
	LOUISVILLE	CM AR	2	NO	FU FA SP
	PURDUE	CO AR	1	YES	FU
	WI-MILWAUKEE	CM AR	1	NO	FA
SOUTH	ARKANSAS	CO AR BR	2	NO	FU FA
	AUBURN	CO AR	2	YES	FU FA
	LAMAR	AR	2	NO	FU FA
	MIAMI	AR	1	NO	FU SP D1
	SOUTH FL	CO AR BR	4	YES	FU FA SP
	TX A & M	CO AR	1	YES	FU FA SP
	TX TECH	CO AR	3	NO	FA
	TX AT EL PASO	AR	2	NO	FU SP
WEST	ARIZONA	EDUC	1	NO	NONE
	MT STATE	CO AR	1	NO	FU D1
	NEBRASKA	AR BR	2	YES	FU FA SP D1
	ND STATE	CO AR	3	NO	FU SP
	OK STATE	CO AR BR	4	YES	FU FA D1
	OR STATE	EDUC	2	NO	FU SP D1

EDUC: EDUCATION ONLY
CO: CO-OP OPTIONAL
CM: CO-OP MANDATORY
AR: APPLIED RESEARCH
BR: BASIC RESEARCH

FU: LACK OF FUNDING
FA: RECRUITING OF FACULTY
SP: SPACE FOR EQUIPMENT
DI: INTERACTION W/DISCIPLINES
NONE: NO MAJOR PROBLEMS

and the University of Cincinnati offered robotics education as an option. Texas Tech University had a minor in robotics, and Montana State University was the only institution developing a program to award a degree in robotics. Montana State was also the only school which was currently not offering a graduate degree with a robotics course.

One question was specifically related to the objectives of the robotics program. Only three schools had programs mainly for education--without co-op or an emphasis on research. Fourteen institutions included either optional or mandatory cooperative education in their program. Of the 21 schools active in research, 13 classified their research as applied and the balance considered their research to be both basic and applied.

All but three universities taught their robotics courses to both undergraduate and graduate students. Montana State University offered their course at the undergraduate level only, while the University of Louisville and the Texas A&M University offered their courses at the graduate level only.

Almost half the schools (11 of 24) offered two courses. The maximum number of robotics courses available at a given school was four, and this number of courses was available at four schools.

Only three of the universities did not have a robotics instructional laboratory. One of the three schools without an instructional lab was planning one. The typical instructional laboratory had approximately 600 to 700 ft^2 (56 to 65 m^2) and housed equipment worth approximately $50,000 to 100,000. The following schools had laboratory equipment which exceeded $300,000 in value: University of Arkansas, University of Louisville, and North Carolina State University at Raleigh. The equipment included a mix of both educational and industrial robots.

In addition to an instructional laboratory, seven universities had a separate research laboratory for robotics. Also, the University of Rhode Island and the University of Nebraska-Lincoln used their laboratory for research only. Most of the research labs were larger than 1000 ft^2 (93 m^2), however two of the labs had less than 600 ft^2 (56 m^2) of floor space. The equipment in the majority of the facilities ranged in cost from $100,000 to 300,000, but exceeded $300,000 at the University of Rhode Island. Full-sized industrial robots were used by faculty members and graduate students to perform research in areas such as robotic control, robotic assembly using vision sensors, design and testing of robotic manipulators and end effectors, and artificial intelligence.

The majority of the schools (21 of 24) were including robots in networks with machine tools, material handling equipment, computers, and other devices. Fourteen schools also had plans to develop a fully-integrated manufacturing cell.

Major problems encountered in the 24 programs included lack of funding (at 20 schools), recruitment of qualified faculty with a background in robotics (at 13 schools), space for equipment (at 13 schools), and interaction with other disciplines (at 8 schools). External funding is essential for the acquisition of equipment and the establishment or expansion of facilities to house equipment requiring a considerable amount of floor space and high-bay areas. Some funding from state governments plus grants and research projects from companies such as IBM, GE, DEC, Timken, and Bethlehem Steel have made it possible to enhance both the research and educational activities. All but one university indicated that their robotics program depended on external funding.

The problem of recruiting faculty members with experience in the area of robotics could be traced to the increasing demand in industry for talented students graduating with a master's degree which is related to robotics. Current faculty members holding a Ph.D. are devoting more of their time to the development of laboratories, and the funding and documentation of research projects.

Several positive results were obtained by introducing robotics into the programs. The following examples can be cited: (1) student interest created larger-than-expected classes, (2) interaction with other departments increased, (3) interaction with private industry increased, and (4) highly-qualified graduates with a background in robotics are providing a valuable resource to the manufacturing sector.

An attempt was made to forecast the trends of robotics in industrial engineering education for the next few years. The following predictions were based on interviewing representatives from 24 randomly selected universities from all parts of the country: (1) robotics courses will be updated and added in the areas of introduction, applications, programming, control, design, and artificial intelligence, (2) existing laboratories for robotics instruction and research will be expanded and new labs will be built, (3) the quantity and quality of robotics research will increase, and (4) networking will be used to fully integrate the equipment in the robotics and manufacturing systems labs.

CONCLUSIONS

The following conclusions about the present status and the future trends of robotics in industrial engineering education were drawn from the results and are based on the objectives of this study:

1. Approximately 60 percent of the accredited industrial engineering programs in the U.S. offer at least one robotics course.
2. The largest percentage of robotics programs in industrial engineering was found in the South Region (at 80 percent of the schools), followed by the West Region (60 percent), the North/East Region (50 percent), and the North/Central Region (50 percent).
3. Most institutions offer their robotics programs solely through courses included in the curriculum. A small number of universities offer an option or minor in robotics, and only one of 24 institutions surveyed was developing a program to award a degree in robotics.
4. The major problems encountered during the introduction of robotics into industrial engineering programs were lack of funding (at 83 percent of the schools), recruitment of faculty with robotics experience (54 percent), space for equipment (54 percent), and interaction with other disciplines (33 percent).
5. The most encouraging successes resulting from the introduction of robotics were related to the stimulation of student interest, increased interaction with other departments, increased interaction with private industry, and the education of highly-qualified graduates in the area of robotics.
6. Robotics programs will continue to be enhanced through the courses which are offered, the facilities which are made available, and the research which is conducted.

REFERENCES

1. Soroka, B. I., "Robotics Education and National Needs," Paper No. ER83-11 presented at SME AUTOFACT 5 Education Forum, Detroit, MI, November 1983.
2. Stauffer, R. N., "The Government and Robotics Education," Robotics Today, Vol. 7, No. 3, June 1985, p. 28.
3. Lohmann, J. R. (Ed.), "Industrial Engineering Division Newsletter," American Society for Engineering Education, The University of Michigan, Ann Arbor, MI, April 1985.
4. "Graduate Study in Industrial Engineering," The Pennsylvania State University, University Park, PA, 1985.
5. Thompson, V. L., A Survey of Computer-Aided Manufacturing in Engineering Education, M. Eng. Thesis, Department of Industrial Engineering, University of Louisville, Louisville, KY, 1984, pp. 57-58.
6. Ibid., pp. 92, 96.
7. Lohmann, p. 30.
8. "Graduate Study in Industrial Engineering."
9. Thompson, p. 19.
10. "The Hands-On Approach to Integrating Robots in Manufacturing," Robotics International of SME, Dearborn, MI, February 1985.
11. Thompson, p. 55.
12. Filley, R. D., "Laboratories and Study Centers: New Directions for IE Education," Ind Eng, Vol. 16, No. 1, January 1984, pp. 52-72.
13. Ibid.
14. Ibid.
15. Leep, H. R. and Wilhelm, M. R., "CAM on Campus: How Good Is It?" CAD/CAM Technology, Vol. 2, No. 1, Spring 1983, pp. 11-16.
16. Lovell, S. (Ed.), The College Handbook, 1984-85, 22nd ed., College Entrance Examination Board, New York, NY, 1984.

Impact of Robotics and CAD/CAM on an Industrial Technology Curriculum

Abhay V. Trivedi

Assistant Professor, Industrial Technology, Southern Illinois University, Carbondale, IL 62901 USA

Introduction

Robotics and CAD/CAM are probably two technologies today that have made a tremendous impact in the manufacturing field. This impact has resulted in an increase in the job opportunities for people with expertise in the Robotics and the CAD/CAM area. The demand will continue to rise as the technology develops further and is implemented on a much wider scale than its current state. But tomorrow's experts have to be educated today. A subject common to debate is which of the traditional engineering or technology fields is to provide experts in the Robotics and CAD/CAM fields. There is no one answer. This paper looks at Robotics and CAD/CAM from an Industrial Technologist's point of view.

Diversification in Industrial Technology

The field of Industrial Technology, like Industrial Engineering, tries to address several areas of manufacturing. Traditionally, these areas have been machining, motion and time study, plant layout and design, production planning and control. An indepth look at several of these areas results in a clear understanding of the intricacies of the basic manufacturing process. The student of Industrial Technology learns about productivity and work improvement, better machine utilization, effective plant layouts, machining, and safety as part of his/her fundamental understanding of the manufacturing process. Perhaps this understanding of the fundamentals of the manufacturing process leads the Industrial Technologist to see Robotics and CAD/CAM not as stand-alone technologies, but as an integration of old and new ideas to better enhance the manufacturing goal--"increased productivity." As the need for more sophistication arises in the manufacturing industry, technology would be able to provide experts in the field of CAD/CAM, Robotics, and Artificial Intelligence. The Industrial Technology curriculum has grown rapidly in the past few years to provide upgraded knowledge to the students in manufacturing and to further provide additional knowledge to lead to tomorrow's breakthroughs.

Understanding CAD/CAM and Robotics as Tools of Technology

Recently, "automation" has come under heavy scrutiny, both from the industry's point of view and from an economic point of view. Suddenly, there is a feeling amongst experts that robots are not doing their jobs as expected. "Automation" is perhaps trying to achieve too much too soon. Instances are also quoted where plants relying less heavily on advanced technology tools have proven to be more productive than high technology plants implementing the "state of the art" CAD/CAM and Robotics technologies. The question then arises: Why are we unable to benefit greatly from implementation of such advanced technology as exists today? The answer is simple. Fault lies with lack of understanding the technology rather than the technology itself. Several factors could be attributed to the failure of correct implementation of high technology, amongst them, the inability to differentiate between manufacturing philosophy and manufacturing tools, the lack of integration between management and technology personnel, the lack of planning on the part of management, the inability to define specific goals and objectives, the open-ended time frome, and the overzealousness of research personnel are the key ones.

Perhaps all these problems undermine the strength and potential of such tools as CAD/CAM and Robots. The increased application of CAD/CAM technology has brought the design and the Manufacturing Engineer closer together. The result obviously is better designs leading to better quality products. CAD/CAM has further carved the path for the Robot technology to be an integral part of the manufacturing system rather than a stand-alone tool. Amongst the relatively new technology is CAD debugging of robots, Intelligent Assembly by robots using Machine Vision, hardware breakthroughs leading to better gripper design and lighter robot bodies. These have all begun to explore additional applications of Robots and CAD/CAM in the manufacturing industry. But this is just the beginning; the technology is spreading to other industries besides the manufacturing industry. CAD, along with Vision Systems and lasers, are fast becoming an important tool in the medical field. We have barely scratched the surface!

Industrial Technology - An Integration Approach

Over the years, technology has taught us that "implementation without planning" is disaster. Availability of computers has provided the planner of today with tools to organize his/her thoughts in a very formal manner. In the manufacturing industry alone, tools are now available to formalize every manufacturing process. Computer programs provide for generation of Quality Control chars, generation of route sheets, analysis of machine utilization, and worker utilization. Also, there is ample availability of MRP and MRPII software to organize the entire inventory and to run the production schedule effectively. An analysis of all these activities leads us into the acceptance or rejection of the CAD/CAM and Robotics technology.

A poor quality product suggests the need for improvements in design. Further, high machine and work force utilization, resulting in negative productivity, leads to the fact of poor process planning. Implementation of CAD/CAM for increased productivity would therefore be most appropriate as it can revolutionize the design process, could generate better process plans, and would also provide for additional tools in the form of finite element analysis, and stress-strain analysis, to put the design under stricter tests to bring about a quality product.

The impelmentation of robots in industry should rely heavily on a firm determination of need. Development of dynamic simulation models with animation capabilities would provide for a realistic look at the robot's capabilities. A model of such magnitude would often times be useful in changing the layout of the Robot workstation, thereby providing additional flexibility to the Robot. Often times, poor layout design and lack of proper feeding of the parts into the Robot workstation results in a low Robot utilization resulting in a lower Robot productivity.

Further, "automation" in parts have to be integrated effectively. An effective strategy would be to be able to integrate the CAD/CAM and Robotic technology by having a centralized computer data base. A central computer with application programs ranging from Quality Control to MRPII would provide as valuable inputs into the successful implementation of the CAD/CAM and Robotic technologies.

All problems cannot be solved by just pointing to the different variables involved in the manufacturing industry. However, if each variable is defined clearly, it becomes much easier to identify the cause of failure, and often, determining the real cause of failure in a manufacturing industry is the most difficult activity. Hopefully, Industrial Technology would be able to provide many solutions to ease the understanding of the fast moving, high technology, manufacturing industry.

Summary

As technology advances, the need for creating experts in the field increases. Several factors would have to be considered to create these experts of tomorrow in the field of technology. An important factor would be "integration" of different tools used in industry today to avoid "islands of automation." Industrial Technology provides for several tools to solve tomorrow's problems. It provides the student with a fundamental understanding of the Manufacturing System and incorporates CAD/CAM and Robotics right at the basis of the System. The impact of CAD/CAM and Robotics on Industrial Technology has been tremendous. It has changed in many ways the basic infrastructure of Industrial Technology. CAD/CAM and Robotics have resulted in enhancing the quality of the Industrial Technology program.

References

Allan, J. J. (Ed.), A Survey of Industrial Robots, Productivity International, Inc., Dallas, Texas, 1980.

Engelberger, J. F., Robotics in Practice, AMACOM (American Management Association), New York, 1980.

Flora, Philip C. (Ed.), International Robotics Industry Directory. Fourth Edition, Technical Data Base Corporation, 1984.

Nagel, R., Vanderburg, G., Albus, J., and Lowerfeld, E., "Experiments in Part Acquisition Using Robot Vision," Robotics Today, Winter, 1980-81, pp/ 30-35.

Oldroyd, L. A., "MCL: An APT Approach to Robotic Manufacturing," paper presented at SHARE 56, Houston, Texas, March, 1981.

Schreiber, R. R., "Robotics Research: The Next Five Years and Beyond," Robotics Today, December 1984, pp. 54-55.

Simpson, J. A., et-al, "The Automated Manufacturing Research Facility at NBS," Journal of Manufacturing Systems, 1982, Vol. 1, No. 1, pp. 17-31.

Tanner, W. R., (Ed.), Industrial Robots, Vol. 1: Fundamentals, SME, Dearborn, Michigan, 1979.

Undergraduate Robotics Education on a Shoestring

Wayne W. Walter

Mechanical Engineering Department, Rochester Institute of Technology, One Lomb Memorial Drive, Rochester, NY 14623, USA

Abstract

The paper traces the development of the robotics program in the Mechanical Engineering Department at RIT with absolute minimal financial support from the departmental budget. The paper discusses how student robotics projects at local industries have provided real-world problems for students to work on, as well as exposure to state-of-the-art hardware and software that RIT cannot afford. The paper also discusses some of the problems that have been worked out including initiating and maintaining academic/industry relationships, getting students access to equipment to the industrial site, and the question of students' patent rights. Some typical student projects are discussed.

Introduction

Initiating the formation of a robotics program for undergraduate education is a formidable task. With typical prices of industrial robots between $40K and $100K, such purchases on a typical university budget are out of the question. Although many universities look to industry for donations, the present economy makes this difficult. This paper traces the development of the robotics program in the Mechanical Engineering Department at RIT with virtually no departmental financial support. Our program began in 1983 with a student project at a local industry. At the present time, we have an upper division robotics course in our program, a robotics laboratory with five industrial robots and three educational robots, and relationships with a number of local industries that provide real-world industrial projects for our students to work on. These three areas are discussed below.

Initial Development Efforts

Our program got its start when one of our students, having just returned from ten weeks of cooperative employment at IBM Boca Raton, approached me with a request to do an independent study in robotics. At that time, I had no experience or hardware in the robotics area. I subsequently was able to place him with a project at a major automaker in Rochester. As a result of his success in establishing a robotic workcell around an IBM 7535 robot, I became well acquainted with the engineers at the site. This led to more student projects. It also led to summer employment for me, and a subsequent leave of absence for the fall quarter in order to finish my project involving the evaluation of a vision-guided robot bin-picking system obtained on consignment. I have subsequently been involved in many projects there one day per week during the school year, and essentially full-time during the summers. At about the same time, I became involved in the activities of the executive committee of the Rochester Chapter 290 of Robotics International. These interactions provided the initial expertise and industrial contacts to get started developing a robotics course.

The Course

Although initially offered on an independent study basis and then on an elective basis, the course is now a four credit Group I course in our program. This essentially means that it is one of a group of six courses from which students are required to choose three. It also means that fifty percent of the course must be comprised of a design project. As will be discussed in more detail in a later portion of the paper, most of these projects have been concerned with the design of tooling and fixtures for robot-based workcells. The balance of the course is made up of lectures on a selected text augmented by video tapes, films, guest speakers, and field trips. While forty percent of the student's grade comes from exams based on the fundamental principles from the lecture portion of the course, the greater majority, in fact, sixty percent, is based on the project grade. Although most of those involved have been mechanical engineering students, there have been industrial engineering and electrical engineering students enrolled. In some cases, these students have worked with mechanical engineering students on joint projects, providing a technical interaction not often found on campus.

The Laboratory

As a result of the relationship I established with the automaker, we presently have four Unimation PUMA five axis industrial robots and an IBM 7535 industrial robot in our lab. These are presently on indefinite long term loan to us from the automaker. The automaker initially sought to donate these robots, but it soon became apparent that the required paperwork on his part was prohibitive. Although loans greatly accelerated delivery of the robots, this arrangement is not without some unique problems regarding maintenance of the units. Since the University does not technically own the robots, it is sometimes difficult to obtain funds from our departmental budget for replacing boards in the controller, replacing potentiometers, and various other items affecting performance of the robots. I am fortunate, however, to have a technician available who is adept at repairing robots and other electromechanical devices, and a department head who provides much encouragement.

The lab also contains three educational robots including two HERO robots, one purchased and one donated, and a donated RBX robot. Although the students who assembled the HERO robots from kits spent approximately one hundred hours in the process, they felt that experience was a good one. In addition to the hands-on abilities it fostered, they also picked up some valuable microprocessor fundamentals. Subsequent to their assembly, however, these educational robots have not been well received by students using them for projects in the Robotics course. The HEROs require assembler language which is not particularly user friendly. They are quite slow so that simulations of machine loading and assembly tasks are difficult to do meaningfully. It is difficult to design, build, and test tooling for these robots, and it is difficult to utilize industrial type sensor input via input/output ports. Although the RBX robot is programmable in BASIC, it unfortunately suffers from many of the same shortcomings.

The robotics laboratory is presently used for programming assignments to provide hands-on experience with the equipment and for in-house projects involving the design, build, and test of tooling and fixtures for robots. In-house projects have provided the manpower to begin to build workcells around the five robots in the lab, although there remains considerable work yet to be done.

Student Projects

Although some students prefer to work on in-house projects in the robotics lab at RIT, the great majority prefer to work on real-world problems at local industries in Rochester even though this requires some travel away from campus each week. These problems are normally of great interest to the industries involved, and this provides much motivation for students. While much of their work has involved the design of tooling, grippers, and fixtures for robots in workcells, students have been involved in a variety of projects involving machine vision systems, material handling systems, sealant dispensing systems, communications, and sensors. A very brief description of some recent projects is given in Table 1.

Projects are initially arranged by phone through the contacts I have developed at local industries in Rochester. We talk about the possibilities, and I choose those which promise to be most suitable for a one-quarter, ten-week period. Students are assigned after matching the projects to the students' interests. They will typically work in small teams of two or three people. My experience has been that larger groups do not seem to be as successful. At this point, I will visit the site with the students and introduce them to an engineer who will be their key industrial contact and project coordinator.

The project coordinator is a key element in a successful project. He/she is usually the engineer responsible for the project at the site, and so is knowledgeable about its entire scope. The project coordinator will facilitate the students' entry to the plant, will help students locate needed resources and equipment, and together with the students' faculty advisor, provides the students with guidance and constructive criticism regarding their project. Their faculty advisor will help them to identify the specific tasks to be completed and who will be responsible for these tasks, and formulate a timeline to complete those tasks by the end of the ten week period. The project coordinator helps the students to establish the project priorities and design objectives. As a result, students get an appreciation for the concepts of design for manufacture and assembly, and design for production and factory-floor environments.

Students meet with their project coordinator and faculty advisor on a need basis, although it is expected that this will occur at least once per week. In many cases, students will spend up to eight hours per week at the site, depending on the nature of the project and they will typically expend about 100 hours of effort by the time their project is complete.

Safety of our students is a very important concern, and this is impressed upon them at the start of a project. Our students are covered under the Institute's insurance policy while at the site, and this is provided in writing to the company upon request.

Although most industrial projects have been of a nonproprietary nature, there has been one instance in which students were asked to sign a nondisclosure agreement. Students are fully informed of the ethical and legal requirements of signing such an agreement.

The question of patent rights is often an issue of concern to students working on an industrial project. Patent policy in this situation has only recently become clear. Although students do their projects almost exclusively at the industrial site, they are unfunded by the industry other than for relevant equipment purchases. In the past, most industries involved have taken the position that the students should not be prevented from pursuing a patent for their work. Most, however, have wanted a no-fee license to use the idea if it was developed on their site with their support. After some research, the University's lawyer has recently made it clear that the idea belongs jointly to the students involved and the faculty advisor. They have exclusive patent rights to the project, and may market the idea if they so desire. To date, one patent has been sought for a quick-change end-effector developed by a graduate student. In this case, the student declined the opportunity to apply for exclusive patent rights due to the substantial expense involved. Instead, the industry applied for the patent and named the student as a co-inventor. In another more recent case, a patent is being considered by students who developed a novel end-effector for handling sheet metal parts. See Figures 1 and 2, and the description of Project #3 and #4 in Table 1. It appears that they will also decline exclusive rights to the idea because of the high cost, and be content to be named on the patent applied for by the industry involved. It is important that the patent issue be discussed at the beginning of a project so that the students involved understand the policy under which they will be working.

Students are required to document their progress on their project by keeping a day-to-day log of their activities, and the time spent on those activities. Weekly progress reports are prepared by the group leader for the faculty advisor. Each project team submits a set of design drawings to their project coordinator at the end of the design phase. Although some students do their drawings on conventional drafting stations, many design teams construct their drawings on the industries' CAD terminals. Recently, many students have used RIT's new Intergraph CAD terminals for their drawings. The Mechanical Engineering Department has held three non-credit user seminars for students during the Fall, Winter and Spring quarters. Many of the students in the Robotics course have used these seminars to become proficient with the Intergraph software. As a result, they have been able to layout robotic workcells with ease.

If appropriate, after their drawings are approved, students either build the hardware in the departmental machine shop on campus or monitor the build-phase of their project at the industrial site. In the later case, they consult with toolmakers at the site who construct prototypes from the students' drawings. Once the hardware is built, students often have the opportunity to set-up and de-bug their devices by mounting them on an appropriate robot. During their interaction with the skilled tradesman, students get exposed to many of the concerns of union members regarding such topics as the impact of robotics and automation on the workplace, worker productivity, and remaining competitive in the present world market.

Although most of the students involved to date have been mechanical engineering students, we are starting to get more industrial engineering and electrical engineering students involved with mechanical engineering students on joint projects. This allows a technical interaction which does not often happen on campus. In addition, we are beginning to involve students in the Robotics course with students in the Advanced Mechanical Systems Design course on joint projects. A recent project involved both groups on the design of a three workstation robotic workcell to assemble an automotive headlight assembly. The Advanced Design students were responsible for the design of some of the more complicated tooling within one workstation, and the robotics students were responsible for design of the rest of the cell. An industrial engineering student provided a parts flow simulation through the cell, and helped a mechanical engineering student with the ergonomics necessary to do the overall three dimensional layout on the Intergraph. This layout is shown in Figure 3. In this project, the company involved was a special machines fabricator who provided the students with the functional specifications that the fabricator had just received from his customer. Although the students had ten weeks to complete their design, the company completed their design in four weeks and was concurrently building the workcell while the students were finishing their design. At the completion of their project, students were given the opportunity to visit the site to see the completed hardware just prior to shipment. They got a chance to discuss their design with the engineers at the site, and to find out why the fabricator had done some things differently.

At the conclusion of the ten week quarter, students provide the faculty advisor with a final written report, and make a formal presentation to the class and invited guests from the industries involved. Guests often include representatives from both the management and engineering staff of cooperating companies. Students are now becoming quite proficient in video taping their results using a new video system recently purchased by the Mechanical Engineering Department, and most groups show a tape as part of their final presentation. By the time these students take the Robotics course in their fifth year, all have already been out to industry on cooperative work assignments. In addition to earning needed funds to pay tuition, our students return to campus with "street-smarts" that make industrial student projects of the type discussed here easy for them to handle.

Conclusions

The program has worked out well for both the students and the companies involved. The students work with high-tech equipment which the university cannot afford on real-world problems of great interest to industry. In addition, while working in small design teams, students have a chance to interact with engineers and toolmakers at the industrial site. Students learn how to work effectively with both groups and to develop those interpersonal communication skills so important to a successful engineer. The companies get the students' time and ideas for free, and probably of more importance to them, get a chance to look these students over for possible permanent employment without any obligation on their part. Some students have obtained job offers as a result of the interactions provided by the projects. To date, the program has been in operation for three years, and many students have participated with local industries in this collaborative project effort. All projects have so far been well received by our students. Interactions of this type between industry and universities can be so beneficial to all concerned.

References

1. Wayne W. Walter, "Initiating University Robotics Education Through Industrial Student Projects", Robotics and Factories of the Future, Springer-Verlag, 1984, p677.

Table 1. Brief Description of Recent Student Projects in our Robotics Course.

Project No.	Description
1.	Overall workcell layout for the robotic assembly of an automotive headlight assembly; use of the Intergraph CAD System for 3D drawings; ergonomics required for design; computer simulation of parts flow and cycle time; control algorithm, sensors, and safety system.
2.	Workcell design for the robotic application of RTV sealant on the headlight assembly of Project #1; design of sealant dispensing system including pumps, hoses, nozzles, etc.
3.	End-effector design, build, and test for sheet metal applications; end-effector to maintain a grip on the part as it is bent in a press; uses a lockable/unlockable wrist. (See Figure 1).
4.	Second generation design, build, and test of an improved end-effector for the application of Project #3. (See Figure 2).
5.	Design and laboratory simulation of a single robot workcell to assemble a three-way power seat motor using an ADEPT robot; was to replace a robotic assembly line presently assembling the motor using 13 robots; layout drawings, using the Intergraph System.
6.	Design of sensor-based tooling and fixtures to allow software compensation for small misalignments of a robotically unloaded pallet; develop algorithm to recompute part location in the matrix as the pallet translates and rotates.
7.	Design, build, and test of tooling for the robotic assembly of printed circuit boards; tooling tested on an Intellidex robot in a development lab.
8.	Design, build, and test of tooling and a rotary index table for an IBM 7535 robot; table controlled via the robot I/O port.
9.	Design-for-assembly analysis of the IBM Proprinter; use of PC-based design-for-assembly software.
10.	Design of a mechanical packaging system for a low-cost load cell-based force sensing wrist.
11.	Design of a robotic workcell for the automation of chemical laboratory tests.

12. Software development to allow a disabled person to robotically feed themselves; develop required sensor system with controls.
13. Design, build, and test quick-change tooling for a robot performing chemical laboratory tests; allows interchanging fingers to handle test tubes, beakers, etc.
14. Develop the communication software to interface an ADEPT robot with two linear array cameras.

Figure 1. First generation end-effector for sheet metal applications. Shows the extension and rotation of the device needed to maintain a grip on a part as it is bent in a press.

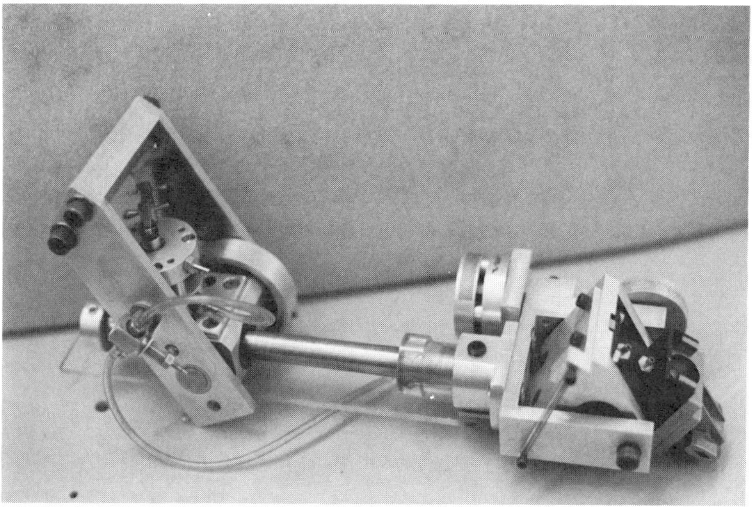

Figure 2. Second generation of the end-effector in Figure 1.

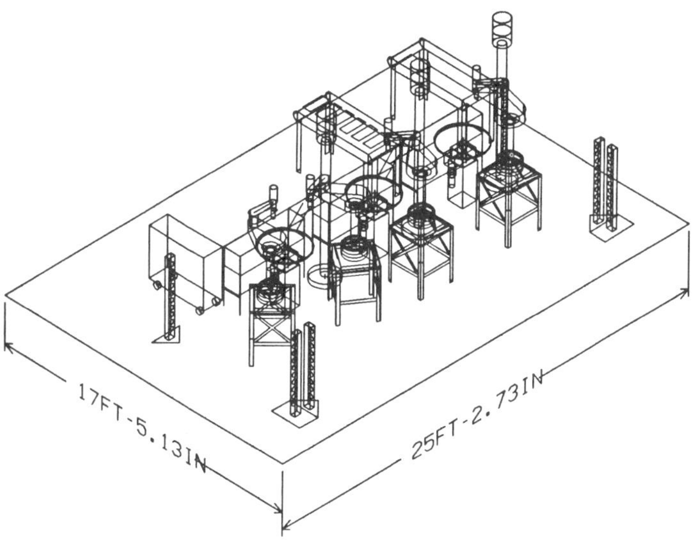

Figure 3. Robotic workcell done on an Intergraph workstation.

Development of a Robotics Course for Engineering Students

Jack Jeswiet[*], and Thomas N. Moore[†]

[*]Associate Professor, Mechanical Engineering, Queen's University, Kingston, ON Canada
[†]Assistant Professor, Mechanical Engineering, Queen's University, Kingston, ON Canada

INTRODUCTION

This paper describes the introduction of an industrial robot into the Mechanical Engineering program at Queen's University. Until the summer of 1985, the Mechanical Engineering Department had a Rhino Robot for demonstration purposes. The Department also conducted a special one week course on a yearly basis at another institution to give students exposure to a variety of industrial robots. This course is described in the 1985 ASEE conference proceedings [1]. With a variety of projects geared to studies of robot applications in the local industries, described in Proceedings of ASME/CIME [2] and the 1985 Proceedings of the ICPR [3], it was shown that Queen's Mechanical Engineering Students were getting a fairly good exposure to Robotics. However, with a very generous donation by General Motors of a 1980 hydraulic PRAB Robot (Model FA), it was found that some substantial changes occurred in the robotics area within the Department. The robot, which is nine feet high, has a seventy-two inch working radius and a payload capacity of 1125 N., required a large amount of space within the department. This space was designated after several meetings concerning rearrangement of already very crowded laboratory space. It also precipitated a series of laboratory space problems which have yet to be resolved. Within a period of two months after installation of the robot, there were three graduate student theses projects as well a variety of undergraduate Mechanical Engineering projects, dealing with various aspects of the robot. The introduction of this major piece of equipment into the department has gone relatively smoothly and has also precipitated the development of a new Robotics course to be offered in early 1987.

This paper describes the impact of the new Robotics course and equipment on the Mechanical Engineering program and discusses the various projects, both graduate and undergraduate that have been started.

FORMER ROBOTICS PROGRAM

In a previous paper [1], the short course in Robotics offered by Queen's University using the facilities of Centennial College was described. The course was introduced as a non-credit part of the program and was forty hours in duration. The equipment made available to us included a Cincinnatti Millacron T3, two Unimate Puma's, a mechanical/hydraulic PRAB 4300, a pneumatic ASEA MHU and several Rhino teaching robots. This was a wealth of equipment, considering that our department had just one Rhino teaching robot, which was put to good use in the short time available. The exposure to the equipment at Centennial gave our students a basic foundation from which to work in the Robotics field. Also, as described in other papers [2,3], experience with robots was gained in other ways, such as studies of robot applications, without the availability of actual on site hardware. Because of this growing interest in Robotics within the Department, there was a natural desire to expand. Indeed one graduate student had already started work on a special gripper sensor design [4]. There was also a desire to teach robotics within the department of Mechanical Engineering as a credit course.

ROBOTICS PROGRAM

It may be seen that the ground work for Robotics in Mechanical Engineering had been laid in the previous two years. As mentioned previously, a PRAB FA, 1125 N capacity, hydraulic robot was donated to the department by General Motors of Canada (early 1980's version). This is a relatively up-to-date hydraulic Robot which is used by a number of automobile manufacturers in the automated assembly of their cars and vans.

With the introduction of the Prab robot into the Department's laboratories in September '85, came an increase in the number of graduate students working in this area. It should be noted that the work being done on Robot sensors is an extension of research work that had been going on for some time in other areas such as Rolling Mills [5]. Within a few months, three graduate students were working in the laboratory on the theses topics of 1) a carbon fibre based tactile sensor [4,6]; 2) automated diagnostic systems for robots [9,8] and 3) a Robot gripper/sensor for measuring pressure/temperature at the gripper interface [9]. As can be seen, this work has already led to a number of publications.

In a very short space of time, it became apparent that other researchers in the faculty were also becoming interested in Robotics and that a variety of equipment was now available on campus. The PRAB and Rhino were available in Mechanical Engineering and a Unimate Puma was available, with a vision system, in Electrical Engineering. Courses for credit are also being introduced so that a student may map out a Robotics program of study if so desired. Consultation has occurred between the Departments regarding the contents of the courses so that they will complement, rather than conflict with, each other.

LABORATORY EXPOSURE

So far the final year graduating students have been exposed to the robot through a laboratory called the Design of an Experiment. In this laboratory, the students are given an experimental task which must be completed within four weeks. They are allotted one three hour time slot per week where they have access to all the technical equipment the department has to offer. At other times they are on their own.

In assigning tasks to be conducted upon the robot and its components, two references were kept in mind. The publication by Lau and Hocken at NBS [10] and Dubovski and Leu at Cornell [11]. Lau and Hocken describe the criteria by which one should judge the acceptability of a robot. These are:

1) Quasi-static repeatability
2) Quasi-static accuracy
3) Kinematic repeatability
4) Kinematic accuracy
5) Dynamic stability

Dubovski and Leu describe the major factors affecting the repeatability of a robot to be:

1) the link structure
2) the joint bearings and similar elements
3) the driving and transmission elements

Keeping the foregoing in mind, an experimental program was started. Six laboratories with five students per laboratory group were carried out.

The assigned tasks were as follows:

<u>Task 1)</u> Design and conduct an experiment which will determine the amount of backlash that occurs in the Prab robot and the possible effect this could have upon robot repeatability. See Figure 1 - gear 1

<u>Task 2)</u> The position of the robot arms is determined and controlled in part by a set of potentiometers mounted upon the Robot which feed back to the controller. Design and conduct an experiment to determine the accuracy of these devices and their effect on repeatability. See Figure 1 - potentiometer 2 .

<u>Task 3)</u> A set of moog valves, hydraulic control devices, are mounted on the robot. Design and conduct an experiment to determine the characteristics of these valves. See Figure 1 - moog valve 3 .

<u>Task 4)</u> The various link elements of the robot are driven by hydraulic motors. Very little information is available for these devices. Determine their characteristics. See Figure 1 - hydraulic motor 4 .

Figure 1. A photograph of the robot indicating points considered in student laboratories.

Task 5) The PRAB robot has a 1125N capacity. Carry out an analysis and conduct strain measurements which will determine what the probable design safety factors are on the robot arm. Also determine if the deflection caused by a 900N force on the robot arm, at full extension, will effect the repeatability. See Figure 1 - Robot arm 5 .

Task 6) Design and conduct tests which will determine the actual static and kinematic repeatability of the robot.

Figure 1 illustrates the different parts of the robot that each of the assigned tasks considered.

In each case, the students must submit a proposal after six days. These are evaluated and the students are advised of the suitability of their proposal the next day.

As one would expect, not all these tasks were successfully completed. For instance, Task 4, the reason for a lack of complete success was the failure of the test rig for the hydraulic motor prior to completion of alltests. However, students were able to obtain manufacturers performance curves and these varified the preliminary results they had been able to obtain.

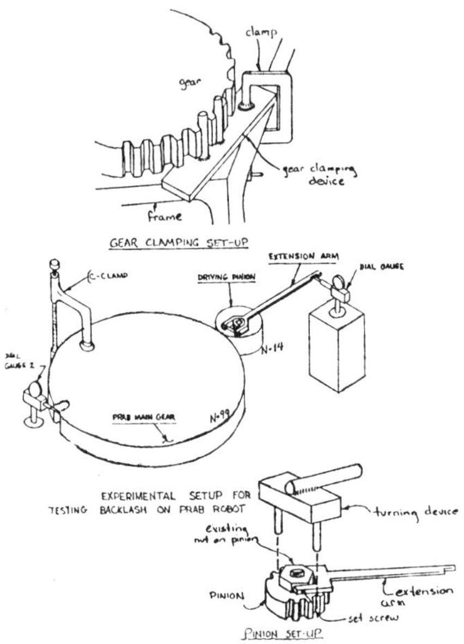

Figure 2. An illustration of the specially devised test for backlash.

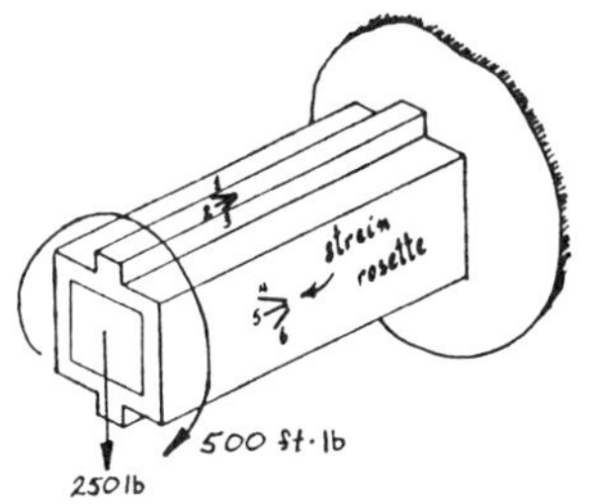

AVERAGED STRAIN READINGS ($\mu\epsilon$)

gauge	centre load 50lbs	100lbs	150lbs	left offset 50lbs	100lbs	150lbs	right offset 50lbs	100lbs	150lbs
1	4	9.5	14.5	4	7.5	11	6	12	18.5
2	13	28	42	14	27.5	41.5	14	26	42
3	3	10.5	15	7.5	14.5	21	4	7	11
4	1	2.5	4.5	-4	-9	-14	7	14.5	22
5	0	1.5	2.5	0	1	2	1	1.5	2
6	0	-1.5	-2	5	10.5	16	-6	-13	-19.5

Figure 4. Schematic details of the strain gauging with some results.

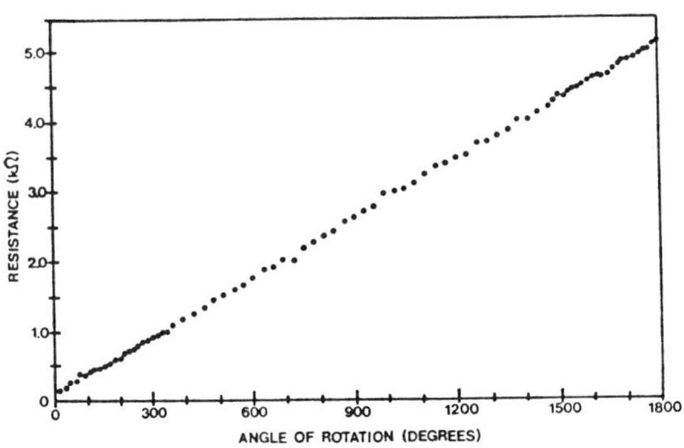

Figure 3. An example of a student graph showing potentiometer results.

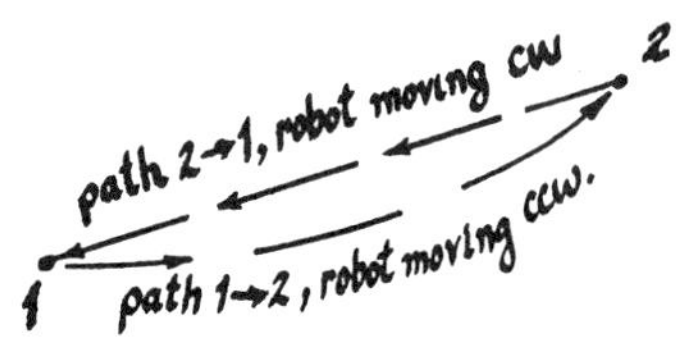

Figure 5. A schematic of the point to point motion tests conducted in task 6.

Task 1 addressed a problem that often occurs in machine design, namely, gear backlash. Figure 2 illustrates schematically the method of conducting the test for backlash which was found to be 0.112 mm. Shadow-graph tracings of wax profiles of the teeth showed them to be involute teeth, as was expected.

The results of Task 2 showed that, to the surprise of the students, the potentiometers were both very accurate and very linear devices, except at either end of this travel. Figure 3 shows an example of these results.

Task 5 required the laying of strain gauges and an experimental stress analysis on the part of the students. Its results are shown in Figure 4.

Tasks 1 to 5 all considered individual elements of the robot. Task 6 considered the action of all the elements working together. Due to thenature of this problem, "off the shelf" testing methods were not available,so the students spent a considerable amount of time thinking about how to solve the problem. Their solution was very simple, yet effective.

For the static test they simply used theodolites to measure how much the centre of a grid, mounted on the robot, was off from its previous position. In all cases, the static repeatability was well within the specifications of 1.25 mm´s quoted by the manufacturer. In the case of the kinematic repeatability, a camera was placed on the floor beneath the robot in a darkened room. Two point sources of light were placed on the arm 5 , shown in Figure 1, and they traced the path of the arm on the film in the camera which was held "open" for a series of sweeps. Kinematic repeability appeared to be within acceptable limits for this simple test but more tests need to be done. It was learned that although the robot is a point to point device, it would trace a different path depending upon the direction it was moving. Figure 5 shows an illustration of this motion schematically.

At the end of the allotted time period of four weeks the students are required to give a verbal presentation and submit a written report. Some of their results have been shown in Figures 2 to 4.

Other students did their Fourth Year Projects on subjects related to the robot. For instance, two students were set to the task of commissioning the robot properly and checking for any unforeseen problems which were not described in the operating manual. They were also given the task of designing and building a simple gripper, shown in Figure 6.

As part of the graduate research program two other special devices are being fabricated. One is a robot "wrist" which will be mounted with the gripper. This load cell senses forces along three axes and torques about the same three axes; it is shown in figure 7. Illustrated in figure 8 is the model of a kinematic parallel motion gripper which is presently being fabricated in our machine shop. This is the gripper mentioned by Jeswiet and Nshama [9] and will be the only one of its type actually in operation when it is completed.

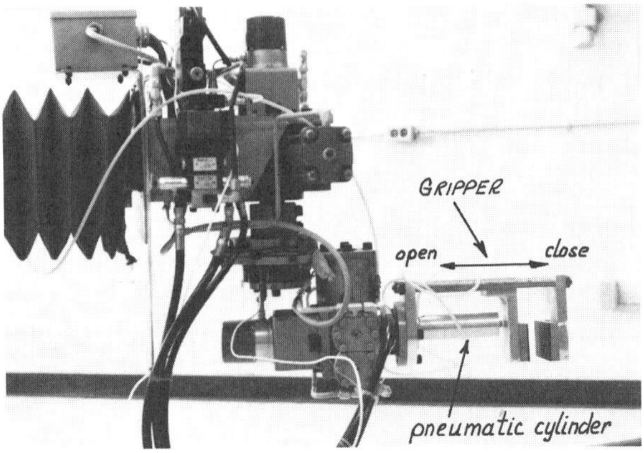

Figure 6. A simple open/close gripper

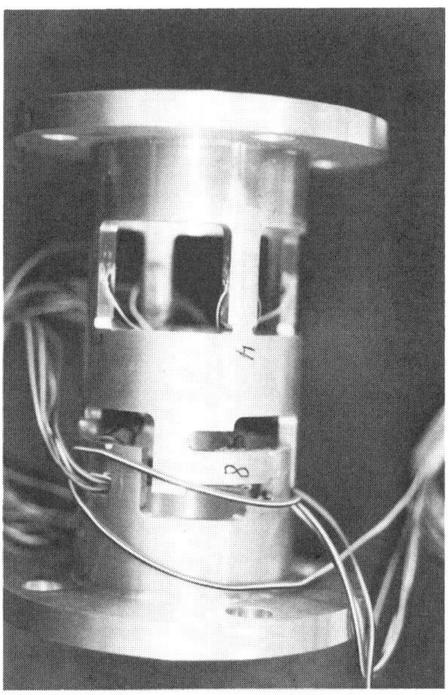

Figure 7. A photograph of the robot wrist

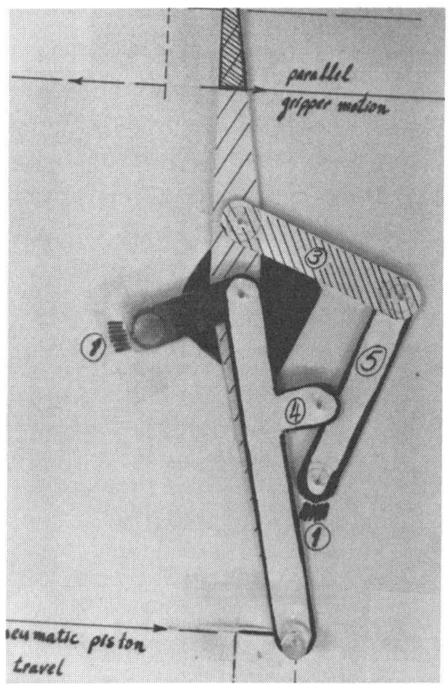

Figure 8. A cardboard model of the kinematic parallel motion gripper

Robotics Courses

New robotics courses have also been introduced. These are being undertaken with interdepartmental cooperation. The courses are:

- Introduction to Robotics and Robot Control - Electrical 448: Mathematical modelling, trajectory generation, position control, vision systems, robot programming languages.
- Industrial Robots, Mechanical 440/840: A double numbered undergraduate/postgraduate course): Manipulators, hydraulic and electric actuators, interfacing, designing for robot assembly, maintenance and diagnostics, economic justification and case studies.

These courses when joined with a normal Mechanical Engineering degree program, should provide the student with a good background in robotics if she/he so desires.

SUMMARY

The introduction of an industrial robot in the Departmental laboratories has provided students with an opportunity for "hands-on" experience in Robotics, while at the same time reinforcing the importance of a good understanding of fundamental Engineering principles, and problem solving techniques. At the same time, this new acquisition, and the interest it has developed, provided the impetus to undertake the development of an introductory course in Robotus and provide the basis for an active graduate research program in various aspects of industrial robotics.

REFERENCES

1. J. Jeswiet, A cooperative venture in robot education, ASEE Conf. Proc., Vol. II, Atlanta, 718 - 720 (1985).
2. J. Jeswiet, Robots in education, Proc. ASME Intl. Computers in Eng. Conf., Vol. II, Boston, 445 - 449 (1985).
3. R. Givens and J. Jeswiet, A study of small foundry automation, Proc. 8th Intl. conf. on Production Res., Stuttgart, 935 (1985).
4. R.L. Ten Grotenhuis and T.N. Moore, Development of a carbon fibre based tactile sensor, Sensors ´85, Detroit, SME paper MS85-994 10 p. (1985).
5. D. Britten and J. Jeswiet, A sensor for measuring normal forces with through and transverse friction forces in the roll gap, NAMRC XIV Proc., SME Trans., 355-358 (1986).
6. R.L. Ten Grotenhuis and T.N. Moore, A carbon fibre based tactile sensor, CASA/SME Mach. Monitoring Sensors Seminar, SME paper MS86-139, 15 p. (1986).
7. T.N. Moore, A health monitoring system for robots, Proc. of IASTED Intl. Sym., robotics and Automation, Santa Barbara, 30-34 (1985).
8. T.N. Moore, Automated Diagnostics for robot-based assembly systems, to be published in Proc. Sym. on Sensors and controls for Manufacturing, 1986 ASME. Winter Annual Mtg., Anaheim, Calif. (1986).
9. J. Jeswiet and W. Nshama, A robot gripper/sensor for measuring forces and temperatures at the gripper interface. 6th International Conference on Robotic Vision and Sensors, Paris (1986).
10. K. Lau and R.J. Hocken, A survey of current robot metrology methods, National Bureau of Standards, Gaithersburg. Maryland, Private Communication.
11. V. Dubovski and M.C. Leu, Effect of manipulator flexibility on static deflection, Proc. 4th Canadian CAD/CAM Robotics Conf., Toronto, 14.5-14.7 (1985).

SAFETY, HUMAN RESOURCES, SOCIAL ECONOMIC IMPLICATIONS

The Human Resources Policy Implications of Advanced Manufacturing Systems

Daniel Levi[*], Charles Slem[*] and Andrew Young[†]

[*]California Polytechnic State University, San Luis Obispo, CA, USA
[†]Northern Telecom, Santa Clara, CA, USA

The adoption and implementation of advanced manufacturing technology (AMT) will have a variety of impacts on human resources management and policy (Manufacturing Studies Board, 1986). This paper reviews the impact of technological change on eight human resources issues and recommends policy options which can be used to promote the acceptance and full use of AMT. The policy options are designed to encourage discussion on the ways to improve the acceptance and successful implementation of AMT and to strengthen human resources management in technologically oriented corporations.

The selection and development of these human resources issues comes from our work during 1985 and 1986. The research underlying this policy paper includes interviews with professionals and managers at a telecommunications manufacturing facility, analysis of Human Resources records at the facility, review and analysis of existing research on the human resources effects of AMT and workshops with Human Resources professionals.

There are a wide variety of human impacts of technological change which range from individual responses to changes in one's job to corporate-wide changes in Human Resources policy. We have structured these widespread effects into eight separate, but interconnected, issues which management should address in order to promote technological change. Rather than providing a single policy alternative, each issue contains a number of alternatives which can be used to facilitate technological change.

1) EMPLOYEE INVOLVEMENT IN TECHNOLOGICAL CHANGE ACTIVITIES

Using employee involvement teams to develop, adapt and implement AMT not only improves the quality of changes, but also leads to increased commitment and acceptance of change (Walton, 1985). Our analysis of employee involvement activities has identified three problem areas which interfere with effective employee involvement: communication and management skills of team participants, knowledge of how AMT change programs are connected, and coordination among related change programs.

The communication and management skills of team members can be improved in two ways. Training programs can be used to improve these skills. Participants could be sent to these training programs individually before they begin work on AMT teams, or the programs could be performed on the AMT change teams as a group. Second, a Human

Resources member could be assigned to each AMT team to act as a communications facilitator.

Increasing employee knowledge of how technological change programs are connected is difficult because of the complexity of the connections. Reliance on a general education program alone will be insufficient, but a comprehensive program may be too long and difficult for most people. The managers responsible for AMT should design specific presentations to AMT change groups on the groups' interconnections and have a representative on AMT change groups to answer questions about the effects of proposals on other groups' activities. An effective education program will improve the quality of employee involvement and enhance the integrating potential of AMT.

Improving coordination among AMT change programs will be difficult because the programs operate under different time frames and include different departmental representatives. Coordination can be improved by having an AMT change agent serve as a member of each team and encouraging overlapping group members among the AMT teams. Departments may consider assigning people as their AMT representatives for all technological change activities which affect them. These people can help to minimize coordination problems by being able to recognize how potential conflicts would affect their department.

While most of the problems with employee involvement are due to lower level employees' lack of knowledge or skills, upper management experiences the same types of problems. Upper management inadvertently creates scheduling and resource allocation problems for employee involvement groups because of their lack of knowledge about AMT program activities. The AMT change agents need to help this level to understand and coordinate AMT activities.

2) JOB SECURITY AND CAREER CONTINUITY

There is a paradox in many companies concerning job security issues. In practice, managers try to protect their employees from layoffs and they are humanely concerned about reducing the threat of job insecurity. But, most companies do not have a meaningful set of policies which attempt to deal with the problem of job insecurity and technological change. What is needed is a recognition by top management that minimizing job insecurity during technological change is in the best interests of the corporation as well as the workforce (Bolt, 1983).

This will require a change in perception by top management. Human resources need to be viewed as corporate assets which the company invests in and develops just like the new technology, rather than like commodities which can be bought and sold in the marketplace. In order to insure this investment, top management needs to take concrete actions to promote employee commitment to the organization.

Job insecurity needs to be recognized as a problem which usually reduces (not increases) productivity and quality control, and increases turnover including turnover of very good employees. As AMT develops and people are trained and upskilled to fill new positions, hiring new employees from outside will become more difficult and more expensive to train (regardless of their existing skills). Job insecurity also reduces the cooperation among workers which is needed to enhance the implementaion of AMT and increases resistance to technological change.

There are a variety of ways to deal with job security issues (Greenhalgh, 1983). Managers need to prepare their employees well in advance for the transition by openly informing them of planned changes. When changes occur, employees should be protected from layoffs while being given the opportunity to be trained for new positions or be transferred to another area. Attrition and the use of temporary workers can allow most reductions to occur without layoffs. This approach to dealing with change needs to be developed into a formal policy which guarantees that changes in technology will not lead to layoffs. While this requires aggressive planning and phased transitions into new technological systems, a no layoff policy should not be expensive in the short run, and will provide the benefits of a more capable and committed workforce in the long run.

Among the professional staff, job insecurity issues will be more difficult to solve. Because of the use of traditional Return-On-Investment justification methods, the introduction of new technology often creates the need to reduce professional staff to justify the investment (Kaplan, 1986). Managers and professionals will be asked to design themselves out of a job. Without a commitment from the organization about reassignment or career continuity, the future technological changes will be limited by employees trying to protect their positions. In addition, they will probably be able to protect their positions and prevent the organization from discovering what they are doing by controlling the flow of information.

Although a no layoff policy due to technological change will help to deal with this problem among professionals, this is only the first step. The central issue will be how to effectively manage displaced professional workers. In the short run, displaced professionals can perform other activities within their departments. However, a longer term perspective requires the development of a career management program which would include developing skills and expertise through training programs, job rotation among departments, and a proactive internal job bidding system which deals with career pathing for professionals and managers.

3) JOB DESIGN

Job design issues on the factory floor tend to operate fairly well because supervisors have experience with these issues. The main exceptions to this relate the use of new types of technology such as robots when job design is performed by engineers with limited experience supervising employees (Ostberg and Enqvist, 1984). If existing supervisors are given the authority and time to plan for new jobs, they are often capable of handling these issues. One problem related to job design is analyzing and scheduling training needs. It may be useful to include a Human Resources representative early in the process of designing jobs to help define and coordinate training programs.

It is more difficult to analyze job design issues for professional and managerial staff. It is important to recognize the need to make system changes easy for the users to operate and adaptable to their specific department's needs. The use of these two criteria (simplicity of use and local adaptability) show the value of using employee involvement to design AMT changes. Job design issues among the professional staff often do not seem to be a problem because professionals usually are not displaced or shifted to substantially new types of positions. Since technological changes are often justified by reducing personnel, this may be a more important problem in the future. Again, it may be useful to include a Human Resources representative early in this process.

4) TRAINING AND PROFESSIONAL DEVELOPMENT

There are three types of training and professional development issues which need to be addressed: specific job training, overall AMT system education, and communication and management skills.

For most new jobs created by AMT, existing employees will be selected as the best candidates. An evaluation process will need to be developed to help decide which employees will be trained for new positions (Downs, 1985). Besides developing the specific training programs, the trainers will need to address issues such as anxiety about learning how to operate new equipment (e.g. computer phobia) and language problems. One of the main problems with training programs are a lack of time and resources to adequately plan and implement the programs. Human Resources training representatives need to be included early in the change process to insure adequate employee evaluations, job design, and training program development and execution.

Educating people about the overall AMT system is necessary to reduce job insecurity, increase employee commitment, and help people to coordinate change programs. For most employees, a general education program using employee meetings and company publications should be sufficient. These general education programs will not be sufficient for members of AMT teams who need to know about how their activities will be connected to other changes. This issue may require policies which encourage overlapping members on change groups, change agents making regular presentations to groups about interconnections, cross department transfers of professionals to encourage an integrative perspective, and people designated as AMT representatives in relevant departments. If there develops a group of people throughout the organization who are involved in AMT activities in different programs, it may be useful to have regular meetings of them to deal with interconnection questions and issues.

Teamwork is essential for the planning and implementation of AMT. There is widespread recognition of this need for teamwork and some of the problems people have working in teams. Corporations sponsor a number of programs to improve teamwork (Zenger-Miller, 1985). However, training programs typically do not have the commited support of top management. Top managers often do not take these courses or practice open communication with their subordinates. Resources for these programs are one of the first activities to be cut back during a slow down. It is ironic that during slow periods when professionals and factory floor employees have the time for these programs, management removes the resources. Until management commits itself to the value of these programs, their impact will remain minimal and underestimated.

5) ORGANIZATIONAL COMMUNICATION

Although various programs may help to open up communication, the primary emphasis needs to be focused at the top of the organization. Top management needs to recognize the need for more open communication and act on it. These actions include more meetings with subordinates, increased use of internal publications, taking communications skills training with subordinates, and providing more resources to support communications training programs.

In addition, all managers need to increase their sensitivity to changes in employees' perceptions of their jobs. One approach to doing this is through improved organizational communication. Regular job surveys, informal communication checks, and communication skills training are necessary, but insufficient (Rowe and Baker, 1984). Other programs which encourage employee feedback, such as quality circles, are also helpful. However, top management needs to demonstrate the value and importance of open communication through their actions by being more open with their managers and professionals and the company as a whole.

6) SCHEDULING PROBLEMS

AMT will create a more sensitive working environment where problems will need to be handled more quickly. It will also allow better planning of change because of increased information and better analytical tools. The benefits of AMT may be lost to an organization if its staff is dealing with artificial emergencies, if training is ignored due to lack of time, or if information provided by the system is not used to form realistic schedules.

Developing unrealistic schedules is a widespread practice in many companies. The organizational culture promotes this practice with beliefs like "activity is good, and too much planning is bad", "people need lots of pressure to function well", and "if you give designers too much time, the product will have too many frills" (Davis, 1984). Often schedules are so typically unrealistic that most managers do not believe they have to be met. However, these unrealistics schedules force managers to speed up the development and implementation process in order to create the appearance of trying to meet the schedule.

The effects of unrealistic schedules will limit the development and the successful operation of AMT. Unrealistic schedules encourage a crisis management approach to problems which searches for the quickest, rather than the highest quality, solution. New products are designed to meet minimum criteria and then corrected later through changes on the factory floor. Process changes are introduced with limited employee training in order to be closer to schedule. Problems with training and job design are corrected after the new system has been implemented. The "gains" from unrealistic schedules get paid for later in the process by someone else or through increased changes in the product during the manufacturing stage.

Optimally, companies should attempt to develop and use more realistic schedules. AMT can help to do this by providing better planning tools and improved feedback on progress. An alternative approach is to try to deal with human resources issues such as training and job design earlier in the planning process, so that they are not ignored in the rush at the end. Emphasizing employee oriented quality control programs may also help. These quality programs help to identify and solve some of the technical and human resources problems created by unrealistic schedules (Lawler and Mohrman, 1985).

7) CAREER ADVANCEMENT

Technological changes will have a major impact on the careers of employees. In order to operate an AMT environment, employees will need to have new skills, greater commitment to the organization, and an increased sense of personal responsibility at work (Davis, 1983). However, the training and professional development activities necessary to create an AMT employee may cause some problems. Employees will have more marketable skills if they decide to leave. Career planning will be more difficult because of the higher rates of organizational change. Both of these factors will increase the importance of promoting commitment to the organization.

The impact of AMT on people's careers will increase career advancement problems. It will also make hiring from outside a more difficult and expensive alternative because of the uniqueness of the manufacturing environment. There is not a pool of AMT experienced professionals to hire from outside, so experienced employees will be in greater demand. Ignoring career advancement issues will become increasingly expensive (Edwards, 1986).

There are a variety of approaches to dealing with these problems. Developing a stronger career planning program, improving the internal job bidding system, encouraging job rotation across departments, and giving preferences to hiring from within could help to deal with these problems. However, the most important issue is the need to make changes in the development, recruitment, and selection of people to top levels of management. Executive development programs should be used to prepare people from within the organization to assume those positions and the selection process needs to be viewed as open and fair by the employees.

8) ROLE OF THE HUMAN RESOURCES DEPARTMENT

One of the reoccuring recommendations in this paper is to have a representative from the Human Resources Department on AMT development teams to deal with employee involvement, job security, job design, and training issues. An alternative approach is to have a human resources trained person who works directly as one of the AMT change agents. This person would serve as a liason between AMT activities and the Human Resources Department, but is directly responsible to manufacturing. This dual role can also be handled by having human resources professionals assigned to both departments.

These two approaches reflect a decision about the role of human resources. Technological change will have a significant effect on people and human resources practices. Should the Human Resources Department be an integral part of the technological change process, or should it respond to human resources issues as they are created by technological change? Should the human impact of technological change be handled by the Human Resources professionals, or should all managers and supervisors assume the roles of dealing with human resources issues? What are the best ways to get human resources issues handled early in the planning process in order to give them adequate consideration?

As AMT progresses, this issue should become less important. Over time, the process of making AMT successful should increase the human resources sensitivity and skills of most managers. The corporation will reward those managers who deal successfully with human resources issues, and they will serve as role models for the other managers. During this transition period, the central issue may be how can the Human Resources Department best facilitate and encourage the development of human resources oriented managers.

REFERENCES

Bolt, J. (1983) Job security: Its time has come. Harvard Business Review. 83, 115 - 123.

Davis, L. (1983) Workers and technology: The necessary joint basis for organizational effectiveness. National Productivity Review. Winter 1983, 7 - 14.

Davis, S. (1984) Managing Corporate Culture. Cambridge, MA: Ballinger Publishing Co.

Downs, S. (1985) Retraining for new skills. Ergonomics. 28, 1205 - 1211.

Edwards, C. (1986) Aggressive recruitment: The lessions of high-tech hiring. Personnel Journal. 65(1), 40 - 48.

Greenhalgh, L. (1983) Managing the job insecurity crisis. Human Resource Management. 22(4), 431 - 444.

Kaplan, R. (1986) Must CIM be justified by faith alone? Harvard Business Review. 86, 87 - 95.

Lawler, E. and S. Mohrman (1985) Quality circles after the fad. Harvard Business Review. 85, 65 - 71.

Manufacturing Studies Board (1986) Human Resource Practices for Implementing Advanced Manufacturing Technology. Washington, D.C.: National Academy Press.

Ostberg, O. and J. Enqvist (1984) Robotics in the workplace: Robotic factors, human factors and humane factors. In H. Hendrick and O. Brown (Eds.) Human Factors in Organizational Design and Management. North-Holland: Elsevier Science Publishers.

Rowe, M. and M. Baker (1984) Are you hearing enough employee concerns? Harvard Business Review. 84, 127 - 135.

Walton, R. (1985) From control to commitment in the workplace. Harvard Business Review. 85, 77 - 84.

Zenger-Miller (1985) Working: A development program for individual contributors. Cupertino, CA: Zenger-Miller, Inc.

Displaced and Excess Workers: Rebalancing the Workforce in the Factories of the Future

Charles M. Slem*, Daniel J. Levi*, and Andrew Young†

*California Polytechnic State University, San Luis Obispo, CA, USA
†Northern Telecom, Santa Clara, CA, USA

As corporations begin to create their "factories of the future" by implementing advanced manufacturing technologies (AMT) and complex integrating systems like computer integrated manufacturing (CIM), one of the major expectations is that there will be a sudden reduction in the number of individuals in the workforce. In addition to highlighting the myth of massive displacement, this paper presents a rebalancing strategy which may help clarify the issues and identify a range of tactics to effectively manage the organization during its transition.

The source of the myth about the necessity for an automatic reduction-in-force (RIF) lies in the outdated use of traditional return-on-investment (ROI) rationale for the justification of the large capital expenses associated with AMT. A RIF characterized by a wave of layoffs often becomes the primary indicator of success because it is one of the few factors which fits into the structure of the ROI calculations. As Robert Kaplan of Harvard (1986) and the National Research Council (1985) have recently pointed out, other factors which are not part of ROI computations represent more dynamic and important reasons for implementing CIM even though the "overhead" does not appear to be reduced. They include benefits such as improved quality, shortened design to production cycle, increased product flexibility, and increased competitiveness (e.g. increased share of the market).

Although new models are being developed to deal with this ROI justification problem (e.g. Kaplan, 1986), the obsession with reduction-in-force as an automatic consequence of CIM continues to have damaging effects on both the workforce and the successful implementation of the new technologies. And, for the first time in recent history, the reduction-in-force implications of technological change have risen to the supervisory, managerial, and professional-technical levels. What may not be clearly understood is that at these higher organizational levels, negative perceptions of the impact of CIM can mobilize powerful organizational forces to resist or undermine the successful implementation of advanced manufacturing technologies.

Does new technology automatically reduce the workforce? The history of the effects of automation and new technologies on the workforce is mixed, but in most cases new technology has increased the numbers of workers in the corporation or industry, especially if the new technology is an integral part of an aggressive corporate strategic plan. In some specific instances automation has eliminated large numbers of the workforce. For example, advanced technologies applied to agriculture led to the dramatic reduction of laborers.

Burnham (1982) points out that the excess workers created a workforce which in turn became critical for the success of the rise of modern manufacturing.

In other instances where it was assumed that automation would eliminate people, the number of workers actually increased. For example, it was predicted that the introduction of computers into bookkeeping and accounting would greatly reduce the need for accountants. Automation did reduce the need for people who mechanically collected financial data. The new technology replaced a large portion of that routine data collection function. However, the increased availability of financial information produced a greater demand for workers who could analyze financial data. As a result, the demand for accountants has risen as the result of the introduction of the new technology.

For most modern corporations using advanced manufacturing technologies as part of an active strategic plan, the primary effect of the new technologies is to cause shifts in the distributions of the various types of workers that make up the workforce rather than the reduction of the workforce. As new technology increases the productivity in one area, it tends to create greater demand for other kinds of skills in other areas (Hugh-Jones, 1969). Rebalancing the workforce becomes a more accurate description of the way in which successful corporations react to the impact new technologies.

DOWNSIZING AS AN ADAPTIVE CORPORATE STRATEGY

While reduction-in-force is not an automatic consequence of advanced manufacturing technologies, downsizing as a method of rebalancing the workforce to promote the development of a lean, highly productive, innovative, and flexible organization is an adaptive strategy in the current international market. Unfortunately, most senior managers consider the terms "reduction-in-force" and "layoffs" as synonymous. USING LAYOFFS AS A DOWNSIZING TECHNIQUE IS UNACCEPTABLY COSTLY AND A SIGN OF POOR HUMAN RESOURCE MANAGEMENT. As a part of corporate human resource management strategy, layoffs are only appropriate when an unanticipated and rapid change (crisis) in market conditions places the corporation's survival in jeopardy. This has occurred recently in the electronics industry as a result of rapid shifts in the market. The implementation of advanced manufacturing technologies and CIM, however, do not represent that kind of rapid change. Technological change is rarely so abrupt and usually appears in time lagged phases covering several years.

The unanticipated negative effects of using layoffs as a way to downsize in a non-crisis situation offset short term monetary savings and usually erode long term corporate strategies. In fact, layoffs are likely to be seen as early indications of organizational decline and deterioration. But as part of a long range and well managed corporate strategy, downsizing can be accomplished with no layoffs of permanent employees. The lead time from the planning for complex technologies to their actual implementation provides sufficient time to use at least 5 other techniques to downsize, including attrition, transfer, voluntary retirement, retraining and career counseling. When human resource rebalancing mechanisms are initiated in concert with a corporate decision to adopt a new technology, it is unlikely that layoffs are necessary to achieve the corporation's goals.

The Paradoxical Effects of Layoffs on the Workforce

Although downsizing may be a very adaptive aspect of corporate strategy, layoffs represent the most costly and ineffective means to achieve downsizing. By the end of the first year, apparent savings in the salaries of terminated employees are offset by unforeseen administrative costs (Greenhalgh, 1978). Although these unforeseen costs are significant, the psychological effects of layoffs on the remaining workforce can result in even greater costs to the corporation.

If a corporation has a history of using layoffs of permanent employees as a preferred mechanism to rebalance the organization, a specific type of psychological threat called "job insecurity" is experienced by its workforce during times of transition. Briefly, job insecurity (Greenhalgh and Rosenblatt, 1984) occurs when employees perceive an involuntary loss of either their job or some important feature of their current job (e.g. income, status, self-esteem). Employees experiencing job insecurity do not continue to perform their tasks in the same manner.

A review of the available research on the relationship between job insecurity and typical workforce reactions indicate what might be considered by some senior managers to be paradoxical effects. While one might expect increased employee effort in performance to prevent being laid off, the research has not supported that belief. The results revealed mixed employee reactions with decreased performance being the most consistent consequence of job insecurity. From the individual employee's perspective, poor business conditions or the introduction of new technology, coupled with rumors of a potential reduction-in-force using layoffs, produces increased levels of anxiety and/or depression. High levels of anxiety interfere with performance. People become so preoccupied with what might happen to them that they are not able to concentrate on the task at hand. In some cases productivity may remain high, but quality suffers and the rework rate soars because of distracting preoccupations. Depression, or anticipatory grieving over the loss of their job (whether they actually get laid off or not), can slow work performance. Work performance may also be slowed in order to "stretch out" the work at hand. Heroic efforts on the part of the supervisors to smooth out the transition are usually only partially successful. Their efforts may mask the negative effects which now might only appear in increased overhead or indirect services needed during the transition.

Many managers believe layoffs are good because they get rid of the poor performers. However, feelings of job insecurity in the remaining workforce also lead to the unwanted loss of valuable employees with scarce technical skills who would not have been laid off in the downsizing effort. In general, layoffs increase the remaining individuals' propensity to leave the corporation. But only those who have reasonable alternatives available in the job market will put those intentions into action. The people with viable alternatives tend to be high performers with scarce skills. With the increased voluntary turnover of high performers, the remaining workforce becomes less capable because poorer performers or people with lower level skills and no obvious alternatives are the most likely to stay with the corporation. They can only hope that they will not be affected by layoffs. In a tight economy, the negative effects of job insecurity are usually postponed until the economy improves and offers reasonable alternatives to the high performers. As a result, the loss of valuable employees is often puzzling because senior management does not see the causative link due to the passage of time.

Greenhalgh (1978) estimated the cost of attrition versus layoffs in an actual case of downsizing and found attrition to be financially superior to layoffs. Although salary savings were significant when the workers were abruptly terminated, layoffs still proved to be a more costly strategy. This was true even during the first year when salary savings had their biggest impact. In addition to hidden administrative costs such as severance pay and increased unemployment insurance, the psychological effects of job insecurity lowered productivity and caused higher rates of voluntary turnover of valuable employees. These employees had to be replaced. Not only was there a disruption in the performance of work groups who lost high performers due to voluntary turnover, the subsequent selection and training of the replacement employees were also important costs of layoffs to the corporation. In their study of internal job changers versus new hires, Feldman and Brett (1983) found that new hires worked fewer hours, required more informal training and social support and had greater difficulty in delegating responsibilities to subordinates than internal job changers. Shunk and Young (1986) and Berger (1985) have also identified substantial costs incurred in hiring and training new engineers and computer programmers. These costs are typically overlooked by management when considering downsizing strategies.

Resistance to change is another costly consequence of job insecurity. This resistance is even more frustrating to human resource managers because the change may have been designed to minimize the loss of employees by achieving greater organizational competitiveness. (For a more complete description of the relationship between threat and resistance to technological change, see Slem, Levi and Young, 1986.)

A MODEL FOR REBALANCING A WORKFORCE

Rebalancing, including downsizing, can be accomplished during a period of technological change without layoffs and their negative effects. If a proactive human resource perspective is an integral component of corporate strategy, rebalancing is an ongoing process which responds to forecasts about a number of long range changes that will affect its workforce. The may include not only technological change but changes in the marketplace, the introduction of new products, and projected personnel requirements. Any of these may initiate a rebalancing effort. The example outlined here will focus on new technology as a specific trigger for rebalancing.

I. Corporate Commitment to the New Technology.

A. Human resource considerations in selecting the new technology.

Consideration of the human resource implications of the shift to new technology should be included in corporate level strategic discussions. Strategic planners should be asking the following questions:

* Which technology is most consistent with corporate goals, culture and strengths of the current workforce?

* How might the technology be modified in order to take advantage of current workforce strengths?

* What categories of new skills are going to be required?

* What categories of current skills are no longer going to be necessary?

B. Selection of technology initiates active planning for workforce rebalancing.

* When the responsibility for the new technology begins to appear in the management objectives of plant managers (e.g. approximately 2 years before implementation), steps to begin human resource rebalancing should appear in the management objectives of those responsible for managing the change at the plant as well as corporate level.

II. Technological Planning Phase

A. The people responsible for human resource management should work with the people responsible for the development of the new technology. Input from employee involvement activities should be used to help shape the specific design, configuration and implementation of the new technology.

* Human resource issues are considered in the specific design of the technological system.

* As the technology is more precisely defined, more information becomes available for developing a rebalancing program.

B. Analysis of new skills.

* What skills will be needed in the new technical/manufacturing environment?

* What kinds of a skill gaps will the technological change create?

* Which positions/skills most closely approximate the new skills?

* How can new skills be obtained?

  - via internal selection (e.g. training).
  - via external selection (e.g. university graduates, etc.).

C. Identification of the likely skills and positions which will be displaced.

D. Initiate Attrition Program.

* Positions which will become excess in the future (18 months) will not be filled, or filled with temporary contract persons.

* In instances where it is important to fill an open but excess position with a permanent employee, selection should include consideration of the individual's transferability to a new position in the future.

E. Cross-Functional Training Increased.

* Current cross-functional training programs are adjusted to train potentially displaced persons for positions which will still be required after the technological change.

* As positions in cross-functional areas become available through attrition, replacements should come primarily from positions which will likely become excess.

F. New skills training programs planned and/or initiated

* Depending upon the amount of training required and the available information concerning the specific skill required to operate the new technology, a specific skills training program may be initiated. This planning may involve creating new training programs internally, or working with an educational institution (community college or university) to tailor a specific program to the new technology.

* If the shift in technology creates major skill level gaps (e.g. analog engineer to digital engineer), begin retraining programs (internal and external) which clearly identify selection criteria and training needs.

* If the shift represents major skill level gaps and the individual is not interested in retraining, make career counseling (including outplacement services) available.

III. Post Planning/Pre-Implementation Phase.

A. Assessment of Status.

* As the final plans for the new technology become known, an assessment of the consequences of current rebalancing strategy is made.

B. If there will still be excess or displaced persons, initiate following programs:

* Voluntary transfers to other plants which have need for their skill sets.

* Incentives for early retirement.

IV. Implementation Phase.

A. As the new technology is being implemented, unanticipated opportunities or issues will be identified by the implementation teams.

B. If an excess workforce remains,

* Voluntary temporary reassignment until attrition opens a more appropriate position. In some cases this could mean that work which has been traditionally contracted out (e.g.

maintenance or grounds) might be temporarily done by excess employees.

* Involuntary temporary or permanent reassignment within the corporation.

* Outplacement services for those who decline reassignment.

V. Reassessment Phase.

A. Adjusting to the response of the marketplace.

* If the effect of the new technology increases the share of the market as well as productivity, previously reassigned employees may decide to return to their original positions as increased demand allows.

* Temporary contract employees may become potential candidates for permanent positions.

B. Assessment of rebalancing procedures.

* Analysis of the organization's ability to rebalance the workforce without organizational disruption:

  - identify sources of disruption, and
  - develop plans to minimize disruptions during next rebalancing cycle.

* Examine the perceived equity in the rebalancing procedures by the workforce.

Although this model of rebalancing may require extensive modification for specific situations, it offers an integrated process initiated by the initial corporate commitment to adopt the new technology. In addition, it highlights the major elements of rebalancing and demonstrates the wide range of alternatives to layoffs which are available to management during transitions driven by technological change.

More importantly, the model demonstrates that career continuity as an integral element of a corporate human resource management strategy is not in conflict with the actual effects of technological change on the workforce. Rebalancing, rather than reducing the workforce, is a more accurate description of the major challenge to corporate management. Combined with the critical dimension of selection and integrated into a comprehensive human resource management program (see Levi, Slem and Young, 1987), rebalancing insures continuity and coordination of human resources during periods of change. When these human resource considerations are part of corporate strategy, the corporation is able to rapidly take full advantage of the potential of new technologies and to maintain or increase its competitive advantage in the marketplace.

References

Berger, John. (1985) Human resource tools for quality improvement. Quality Progress. 18, 21-28.

Bolt, James. (1983) Job security: Its time has come. Harvard Business Review. 83 (November), 113-130.

Burnham, D. C. (1982) Productivity: An Overview. In Salvendy, Gavriel. (Editor) Handbook of Industrial Engineering. NY: John Wiley & Sons.

Feldman, Daniel and Jeanne Brett. (1983) Coping with new jobs: A comparative study of new hires and job changers. Academy of Management Journal. 26, 258-272.

Greenhalgh, Leonard. (1978) A cost-benefit balance sheet for evaluating layoffs as a policy strategy. Unpublished paper, Amos Tuck School, Dartmouth College, Hanover, New Hampshire.

Greenhalgh, Leonard and Zehava Rosenblatt. (1984) Job insecurity: Toward conceptual clarity. Academy of Management Review. 9, 438-448.

Hugh-Jones, E. M. (1969) Economics and Technological Change. NY: Augustus M. Kelley Publishers.

Kaplan, Robert. (1986). Must CIM be justified by faith alone? Harvard Business Review. 86 (March-April), 87-95.

Levi, Daniel, Charles Slem and Andrew Young. (1987) The Human Resources Policy Implications of Advanced Manufacturing Systems. Second International Conference on Robotics and Factories of the Future. San Diego, California.

National Research Council. (1985) Computer-integrated manufacturing: Barriers and Opportunities. National Productivity Review. Spring, 170-179.

Shunk, Daniel and Hewitt Young. (1986) Industry and the university: A partnership in education and training of factory-of-the-future employees. Paper presented at the 1986 World Congress on the Human Aspects of Automation, Ultratech, Long Beach, California.

Slem, Charles, Daniel Levi and Andrew Young. (1986) The effects of technological change on the psycho-social characteristics of organizations. in Brown, O. and H. Hendrick, (Editors). Human Factors in Organizational Design and Management II. New York: North-Holland.

Human Health and Safety with the Robot

Malcolm C. Robbins

Department of Mechanical Engineering, San Diego State University, San Diego, CA 92181, USA

Abstract

The issue of human health and safety while reacting with robots is discussed. Several cases of accidents or death are cited with suggestions on how they could have been avoided. Sources of threat from robotics to human safety and health are discussed. Concepts for the design and installation of safe-guard devices are outlined.

Introduction

With the rapid trend towards use of robotics at the workplace, research and development is needed to prevent unnecessary injury or health problems to the worker who is operating or in proximity of robots. Already accidents have occurred with robotic usage which could have been prevented had suitable safety measures been taken, or regulations set and followed. In one case in the United States a jury awarded the family of a victim $10 million to compensate for the fatal injury of a factory worker.[1]

Health and safety standards associated with the design installation and operation of robots need to be identified and addressed before the major impact of robotic usage takes place. Here the rule of "an ounce of prevention is worth more than a pound of cure" must surely apply with just cause. It is easy to be overly optimistic about the success of robotics and fail to consider safety and health. Failure to do so will have potentially negative consequences on the decisions associated with the future of robots. Many projects are undertaken with optimism, resulting in the burdening of production facilities with costly but inefficient, dangerous, or unreliable systems. The resultant loss of jobs on the part of workers and failure of the system to improve productionprocess results in a no-win situation.

Already the subject of robots can be seen to arouse strong public sentiments associated with being replaced by a robot at the work place. If robots are to gain acceptance by the public without further negative sentiments, then the issue of safety needs addressing. The tremendous potential for robots presents temptations to ignore human safety hazards in the pursuit of profit. Robots are in the spotlight, but their future acceptance by the public will be largely dependent upon the ethics of the engineers and the regulations to govern health and safety associated with their design, installation and operation.

A survey of robot-related accidents based on a thirty month period between January 1976 to June 1978 was made in the steel industry in Sweden. [2] The survey found that the industry which used two hundred and seventy robots had fifteen accidents in the thirty month period. While the accidents were not described in detail, the survey analysis concluded that they were typical of accidents caused by robots. Swedish officials concluded that the accident rate could be considered as one accident per robot per year! Accidents were not described in detail, however the survey analysis concluded that the accidents were a result of two main reasons. Firstly, operators got carelessly too close to the robots while they were in cycle. Secondly, operators were incorrectly operating control buttons and were then struck by tools held by the robots. With simple perimeter guarding the first group of accidents could have been el iminated.

Another case which can be cited is that of a Japanese factory maintenance worker who was stabbed to death by a robot.[3] In this case, the robot control unit was designed without a mechanism to prevent unintentional operation. While selected for ;manual control, the worker accidentally brushed against the "on" switch, causing the robots claw-like appendage to pin him against a nearby automobile gear tooling machine. Here two simple steps of design would have prevented this death. (1) The control should have been designed differently, i. e., with a safety switch. (2) The emergency stop switch was, according to the report, not easily identifiable or conveniently located. When other workers rushed to help they were unable to stop the robots action.

Sources of Threat to Man from Robots

The safety and health hazards which robots pose to man are more complex than traditional manual or automated machinery. This increased complexity is due to several factors.

1. Robots have many degrees of freedom and are capable of a much wider range of movements.
2. Robots use both hardware and software, which forms a highly complex system giving rise to faults and hazards which are unpredictable.
3. The technology of robots is changing at an alarming rate and their range of applications is increasing at much higher speeds than traditional machinery ever did.

This results in a rapid rate of advancement in technology faster than present legislation governing health and safety is able to cope with, or is accustomed to.

4. The vast range of energy sources which robots are connected with. These can be categorised as follows:

   Kinetic Energy with its application to multidirectional moving robot parts such as arms, hands, spindles, and bodies which offer a potential for collision with an operator.

   Potential Energy is stored in like various grasp devices holding work which could fall should the robot release accidentally. Also there are pressure cylinders and piping which could bust or fly off striking the operator.

   Chemical and gas pose as a further safety and health threat due to multidirections a robot undertakes when handling such elements. Robots also present risk of fire and explosion when involved in such operations as spraying and welding.

   Thermal energy used for operations such as casing, forging, and heat treatment can also be a source of danger if the robot does not perform correctly.

   Electrical energy shock accidents are common to traditional machinery but are more liable with robotic use. Robots have many movable, multidirectional arms and hands, which hold or move the cables, welding wire, welding holders or other tools. Due to the constant hand and arm movements which these devices are attached to, rapid deterioration takes place, leading to frayed cables and broken devices. Not only is there a risk of operator shock, but also the possibility of power failure or short-circuit that could result in erroneous action of the robot.

   Ultra-violet, x-ray, infrared, and gamma rays are another energy source which robots are being introduced to, and pose as a major health and safety risk to man. Radioactive materials are another energy source which robots have been handling with the very idea of minimizing human exposure to radiation.

Engineering for Safe and Healthy Relations Between Man and Robot

As with traditional machinery, mechanical perimeter guards which prevent the operation from coming into accidental contact with the robot can be used. However, such mechanical guards do not give any protection for people working inside the fences. In the case of robots, the need for personnel to be inside the safe-guard perimeter should be considered. Certain operations, such as maintenance, programming or fault diagnosis may of necessity, need to be carried out while the robot is still active. Hence further safeguarding devices need to be incorporated if personnel are to work in safeton on robots which have to be manipulated under power in order to carry out their task.

Measures such as "hold to run controls", "deadmans buttons", and speed limitation switches will lead to a reduction of risks while the operatiors are working in close proximity of robots.

Safety Mats

Essentially two types of safety mat are available, both perform in the same manner, i. e., they stop machinery when someone steps on them. The method of operations are different. One is air operated and is simply a series of air filled tubes. Anyone stepping on the mat will squeeze the tube, causing like air pressure level to change and actuating a switch which will stop the machinery. The second type of safety mat activates a switching device when pressure is applied to any of the contacts within the mat construction. The use of these mats would allow for safety sectors within the boundary of the robot. With suitable programming the robot should not be able to move into the sector where the person is standing.

Passive infrared Sensor

A passive infrared sensor detects a sudden change in radiation in the infrared wave length area. Human infrared radiation is greater than the radiation from the ambient surroundings. This dedection device could be used in conjunction with the robot programming to prevent collision between operator and robot.

Lightbeam Curtains and Video cameras with image sensors are already widely used for human detection and could also be used in conjunction with the robot programming. Their best use would result if installed in such a way as to provide a detection zone directly around the robot hand or arm.

Distance and direction sensor

Personnel entering the perimeter of the robot area could carry a transmitter which emitted signals of short range. If the person became in danger of collision with the robot, emitted signals from the transmitter to the robot sensor unit would activate a stop.

Conclusion

It is clear to see that manufacturing technology throughout the world is typically changing with the advent of robotics. Great hopes are viewed with the expectations that robots will reduce tedious manual tasks, increase productivity, and undertake hazards tasks now performed by man. However interaction between robot and man is still needed for installation, programming and maintenance. It is at this interaction that potential safety and health problems for many will arise with the use of robotics. This paper calls for research, development, legislation and standards to prevent accidents and health problems to humans involved in working with robots.

References

1. "Jury awards faimly $10 million." Northwest Arkansas Times. Fayetteville, Ark. (Associated press article), August 14, 1983.
2. Steel industry, Sweden. (Surgimoto and Kawaguchi, 1983)
3. Tokyo (AP) 1981 Newspaper article. Outlined in Robotics for Engineers by Yovam Koren. McGraw-Hill 1985.

Author Index

AUTHOR INDEX